FORTSCHRITTE DER CHEMIE ORGANISCHER NATURSTOFFE

PROGRESS IN THE CHEMISTRY OF ORGANIC NATURAL PRODUCTS

PROGRÈS DANS LA CHIMIE DES SUBSTANCES ORGANIQUES NATURELLES

HERAUSGEGEBEN VON   EDITED BY   RÉDIGÉ PAR

L. ZECHMEISTER

CALIFORNIA INSTITUTE OF TECHNOLOGY, PASADENA

DREIZEHNTER BAND
THIRTEENTH VOLUME   TREIZIÈME VOLUME

VERFASSER   AUTHORS   AUTEURS

A. CHATTERJEE · A. R. H. COLE · W. GRASSMANN · T. NOZOE
S. C. PAKRASHI · R. J. PRICE · O. TH. SCHMIDT · CH. TAMM · G. WERNER
E. WÜNSCH

MIT 48 ABBILDUNGEN   WITH 48 ILLUSTRATIONS   AVEC 48 ILLUSTRATIONS

WIEN · SPRINGER-VERLAG · 1956

ISBN-13: 978-3-7091-8033-4   e-ISBN-13: 978-3-7091-8032-7
DOI: 10.1007/978-3-7091-8032-7

Inhaltsverzeichnis.
Contents. — Table des matières.

Natural Tropolones and Some Related Troponoids. By TETSUO NOZOE,

Recent Developments in the Chemistry and Pharmacology of Rauwolfia Alkaloids. By ASIMA CHATTERJEE and SATYESH C. PAKRASHI, University College of Science and Technology, University of Calcutta, India, and G. WERNER, Faculdade de Medicina de Ribereirão Preto, Universidade de São Paulo

Synthese von Peptiden. Von W. GRASSMANN und E. WÜNSCH, Max-
Planck-Institut für Eiweiß- und Lederforschung, Regensburg 444

Infrared Spectra of Natural Products.

By **A. R. H. COLE**, Nedlands, Australia.

With 23 Figures.

Contents.

I. Introduction.

Physical methods are being increasingly applied to the solution of structural problems in the field of organic chemistry (*25*) and the use of infrared spectroscopy is standard practice in most laboratories today. The early work of COBLENTZ (*34*) indicated that the absorption of infrared light was closely related to structural factors, but it is only in the past decade that instrumental developments have allowed full use to be made of these techniques. There is no doubt today that of absorption studies in all parts of the electromagnetic spectrum, the infrared measurements are generally the most useful for structural determinations. It is true that a more complete picture of molecular structure might be obtained through the use of X-ray- (*138*) or electron- (*26, 126, 161*) diffraction, but these methods mostly remain within the precincts of the chemical physics laboratory and have not been widely applied to the solution of everyday problems of organic chemistry. In the various fields of natural product chemistry, where compounds are often isolated in extremely small amounts, seldom crystalline in the first instance, infrared methods are particularly useful.

Fast recording infrared spectrometers became commercially available in about 1946, and much of the early work to be carried out was surveyed by RASMUSSEN (*134*) in this Series. The number of infrared papers in the literature today is far too great to attempt to cover in one article but it is hoped to give a broad picture of the field and to treat particular examples in some detail in order to draw attention to different applications.

In the early stages of post-war popularity of infrared spectroscopy, there was a tendency to regard the recording of such spectra, often with little systematic control of experimental conditions, as an end in itself. However, a firm basis for the application to structure determinations was laid by such authors as R. B. BARNES, R. C. GORE, H. M. RANDALL, R. S. RASMUSSEN, H. W. THOMPSON, G. B. B. M. SUTHERLAND, and V. Z. WILLIAMS; and recent years have seen a consolidation and critical appreciation of the field rather than the emergence of many new principles.

II. Methods.

1. General.

The interaction of chemical compounds with infrared light is related to the fact that all molecules are vibrating systems with a number of discrete vibrational frequencies. A polyatomic molecule will vibrate in a complicated manner but this complex motion can be resolved into a number of simpler components called *normal* or *fundamental vibrations*, and the number of these is determined by the number of atoms in the molecule. A non-linear molecule with n atoms will have $3\,n{-}6$ normal modes of vibration ($3\,n{-}5$ if linear) and the frequencies of some of these modes are given by the frequencies of the light absorbed. It is from the assignment of these absorption bands to specific vibrating entities in the molecule that structural information is obtained.

To interact with electromagnetic radiation, and thereby give rise to an absorption band, a vibration must involve a periodic change in the dipole moment of the molecule. Symmetrical molecules have some (symmetrical) vibrations which do not cause absorption and these are called "inactive" modes. Carbon dioxide is a simple example of such a compound. This has four normal modes illustrated by the following scheme:

(a) $\overset{\leftarrow}{O}{=}C{=}\vec{O}$ Symmetrical stretching

(b) $\overset{\leftarrow}{O}{=}\vec{C}{=}\overset{\leftarrow}{O}$ Unsymmetrical stretching

(c) (i) $\overset{\uparrow}{O}{=}\underset{\downarrow}{C}{=}\overset{\uparrow}{O}$

    (ii) $\underset{\oplus}{O}{=}\underset{\ominus}{C}{=}\underset{\oplus}{O}$ $\Bigg\}$ Bending

Only two bands, corresponding to (b) and (c) are found in the absorption spectrum, since (a) leads to no overall change in dipole moment. The bending mode (c) is described as "doubly degenerate", consisting as it does of two modes of equal frequency, at right angles to one another.

The masses of atoms and the forces holding them together are of such magnitude that the usual vibration frequencies fall in the range, $5 \times 10^{12}{-}1.2 \times 10^{14}$ cycles/sec. This corresponds to the absorption of light of wavelengths between 2.5 and 60 microns ($1\,\mu = 10^{-3}$ mm. $= = 10000$ Å). The position of an absorption band in the spectrum is not usually designated in terms of the absolute frequency in cycles/sec. but by reference either to the *wavelength* or the *wavenumber* of the light. A certain amount of controversy has arisen in the literature as to which of these units should be used. It is the author's opinion that the wavenumber is to be preferred since it is a frequency unit, more closely allied to the vibration frequency of the molecule under investigation than is the wavelength of the light. Furthermore, the wavenumber is much more readily comparable with Raman displacements (*85*) which are also related to molecular vibrations and hence to molecular structures. Usually, then, the terms "wavenumber" and "frequency" are taken as synonymous in the infrared field and the unit generally employed is the reciprocal centimetre (waves per cm. or cm.$^{-1}$). For converting from one system to the other there is the simple relation

$$\text{wavenumber (cm.}^{-1}) = 10^4/\text{wavelength } (\mu).$$

The main reasons why some spectroscopists have used wavelength rather than wavenumber are firstly that wavelength is generally used in ultraviolet work, and secondly that instrument manufacturers found it easier to construct spectrometers which recorded the spectrum on a linear wavelength scale because the dispersion of sodium chloride in the range 2.5–16 μ is much more nearly linear in wavelength than in wavenumber. In modern instruments, however, suitable cams are employed to scan the spectrum in such a way that the final record is linear in wavenumber. It is perhaps relevant to point out here that the dispersion of quartz in the ultraviolet region of the spectrum is roughly linear in wavenumber and if early workers had appreciated this fact together with the theoretical importance of frequency in relation to energy, infrared spectroscopists might have been spared a certain amount of confusion. However, in order to facilitate reading, all band positions in this paper are given in both frequency and wavelength.

The suggestion has been made (*189*) that, in order to give the unit of wavenumber a more euphonious title, and to pay a tribute to one of the pioneers of spectroscopy, it should be called the *kayser*, to be denoted by K. It is probable that this name will be universally adopted, but for the present, most journals still prefer cm.$^{-1}$

The nature of infrared work depends greatly upon the *size of the molecule* being investigated. For small molecules (*e. g.* those containing up to approximately twelve atoms), which are usually investigated in the gaseous state with spectrometers of high resolving power, the aims may be divided into three sections: (a) the assignment of the absorption frequencies to specific modes of vibration; (b) the explanation of the rotational fine structure of the individual absorption bands (leading to evaluation of the moments of inertia of the molecule, and hence to bond lengths and angles); and (c) the correlation of the vibration frequencies with a potential function which includes force constants related to the periodic changes in bond lengths and bond angles taking place during the vibration. If the last of these steps can be carried out it is possible to calculate the frequencies of inactive vibrations.

Normal vibrations are theoretically a function of the molecule as a whole and this is illustrated in the cases of those small molecules whose vibrations have been completely assigned (*84*). In large molecules the number of vibrations becomes too great to assign to particular modes, but many of them appear to be localised in specific substituent groups and are little altered by changes in other parts of the molecule. This is particularly true of vibrations involving (a) light or heavy atoms attached to the skeleton of the molecule or (b) bonds whose force constants are greatly different from those of neighboring bonds. Class (a) includes such bonds as OH, NH, CH and CCl, and class (b) the multiple bonds $C\equiv C$, $C=C$, $C=O$, $C\equiv N$ etc. This has allowed an empirical approach to the infrared spectra of complex molecules. From a study of the spectra of a number of similar compounds, the presence of a particular absorption band can be correlated with the presence of a chemical group, and this information then used to identify that group in new compounds.

Correlation charts of this nature have been published covering a wide variety of compounds (*6, 49, 162*). However, attempts to cover such a wide field have usually meant that the range of frequency assigned to a particular group is unnecessarily broad, and for structural work a few frequencies from compounds very similar to that being studied are much more useful than a large number of measurements on all different types of compounds which are grouped merely because they include a particular chemical entity. For example, in detailed work on ketosteroids (p. 35) much more can be gained if it is known that 11-ketones absorb between 1710–1718 cm.$^{-1}$ (5.85–5.82 μ) and 3-ketones between 1706–1710 cm.$^{-1}$ (5.86–5.85 μ) and so on, rather than knowing that aliphatic ketones absorb between 1660 and 1740 cm.$^{-1}$ (6.02–5.75 μ) (*49*).

2. Instruments.

Infrared spectrometers consist of a source of continuous radiation, cells to hold samples, a radiation dispersion system, a radiation detector with amplifying equipment and a recorder or output meter. Many of the components of present day spectrometers have been in use for many years and most of this discussion will be limited to fairly recent developments.

a) Radiation Sources.

The Globar and the Nernst Glower continue to be the two most popular sources and there seems to be little to choose between them.

The Globar is the more robust and has a fairly large heat capacity so that its temperature ($\sim$ 1400° K), and therefore its radiation output, is comparatively unaffected by transient draughts. It is turned on merely by switching on the electricity. The main disadvantage is that its mounts must be water-cooled, and it is perhaps undesirable to have water connected to such a complex and expensive piece of equipment as an infrared spectrometer, with consequent risk of flooding. The working temperature of the Nernst Glower ($\sim$ 2200° K) is higher than that of the Globar and this is an advantage particularly when working at short wavelengths. The Glower has a high negative coefficient of resistance and does not conduct electricity when cold. It must be heated to red heat either with gas or an auxiliary electrical heating coil when starting up, and the circuit must include barretter stabilizing lamps to control the current. If well shielded from draughts its emission is constant over long periods. Both types of source have a working life of 2–3 months.

In an attempt to obtain more energy, to allow the use of narrower slits with a consequent increase in resolving power, very high temperature sources such as a tungsten strip ($\sim$ 2900° K) (*160*) and the carbon arc ($\sim$ 3900° K) (*143, 144*) have been used by STRONG and his associates. The carbon arc appears too cumbersome for routine use but the tungsten strip source might well be adapted to commercial instruments.

b) Cells.

Absorption cells are available commercially but many spectroscopists prefer to make their own to suit their particular problems. Many aspects of laboratory technique including the construction of cells have been described in a valuable paper by LORD, McDONALD and MILLER (*116*). Most structural work has been carried out in the region of transparency of the sodium chloride prism ($2.5–16\,\mu$) and plates of this material are generally employed for cell windows. Because of the very long wavelengths of infrared light compared with visible and ultraviolet, optical tolerances are not at all strict and very simple polishing methods may be used. The choice of a cell depends on the state of the sample and this is discussed in more detail in Section 3 (p. 18).

c) Dispersing Systems.

Most infrared spectrometers in organic laboratories use prism dispersing systems. The prism materials and the spectral ranges in which they are generally used are as follows:

Material		Preferred Range
Lithium fluoride	2000–3800 cm.$^{-1}$ (5.0– 2.6 μ)	2000–3800 (5.0– 2.6 μ)
Calcium fluoride	1300–3800 cm.$^{-1}$ (7.7– 2.6 μ)	1300–2000 (7.7– 5.0 μ)
Sodium chloride	650–3800 cm.$^{-1}$ (15.4– 2.6 μ)	650–1300 (15.4– 7.7 μ)
Potassium bromide	400– 650 cm.$^{-1}$ (25.0–15.4 μ)	400– 650 (25.0–15.4 μ)
Caesium bromide	250– 650 cm.$^{-1}$ (40.0–15.4 μ)	250– 400 (40.0–25.0 μ)

Any of these prism materials is transparent at a higher frequency (shorter wavelength) than the preferred range indicated above but the resolving power of the spectrometer will not then be as great as when using a material mentioned earlier in the list.

Very few laboratories would be equipped with a complete set of prisms and there are a number of factors, both of convenience and economic, which determine the number of prisms which might be purchased. For instance, the lithium fluoride prism covers the range of only a few stretching vibrations (OH, NH, CH, C≡N) and most of these are well separated from other bands in the spectrum. Any close peaks of structural interest in the CH stretching region can usually be well enough resolved with a calcium fluoride prism which will also cover the regions of C=O, and C=C stretching and C—H bending, so that the latter material may be used for the complete range 1300–3800 cm.$^{-1}$ (7.7–2.6 μ). Again, the caesium bromide prism, although having slightly smaller resolving power than potassium bromide in the range 400–650 cm.$^{-1}$ (25.0–15.4 μ), is transparent to 250 cm.$^{-1}$ (40 μ) so that it might as well be used for the whole range 250–650 cm.$^{-1}$ (40.0–15.4 μ).

d) Detectors.

Most wide-range infrared spectrometers employ a fast vacuum thermocouple as detector and this converts the radient energy to electrical energy. The speed of response of the thermocouple is such that the light beam can be chopped at about 10–15 c. p. s. and tuned a. c. methods of amplification used. In this manner drifts in room temperature or spurious electrical signals are not recorded.

Another type of detector of high sensitivity is the GOLAY pneumatic detector (74). With this the infrared beam is absorbed by a receiver which makes up one end of a small gas chamber and the resulting gas expansion distorts a flexible mirror at the opposite end of the chamber. The flexible mirror is part of an auxiliary optical system involving a lamp, a line grid and a photo-cell. Variations in the radius of curvature of the mirror cause changes in the light passing through the grid and reaching the photo-cell so that the output of the photo-cell is proportional to the intensity of the infrared beam. This detector has a fast speed of response (time constant ∼ 0.003 sec.) as well as high sensitivity and is used with a. c. amplifiers.

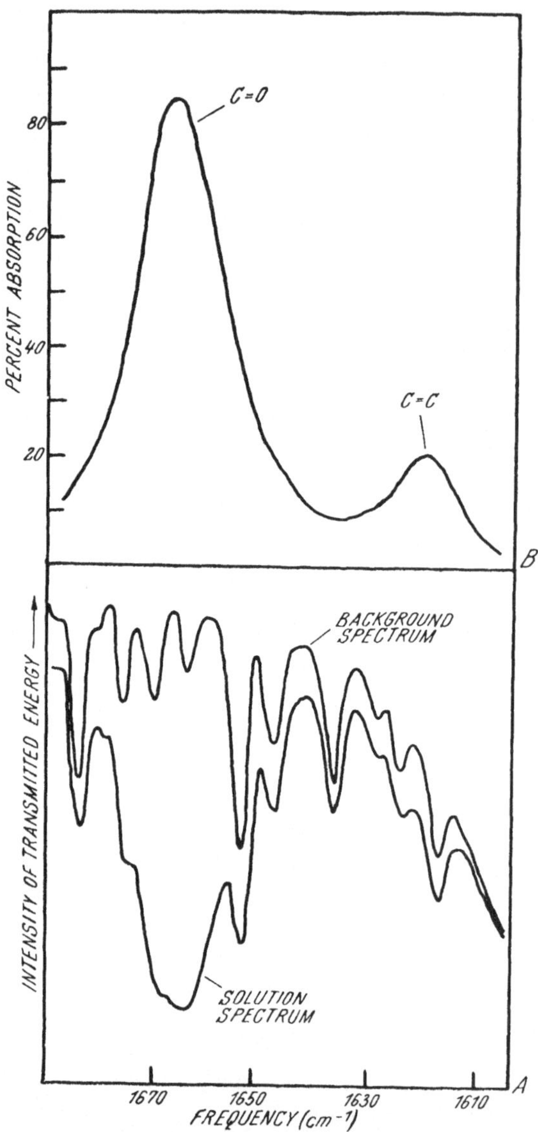

Fig. 1. Single beam spectrum in the region of atmospheric water vapour absorption: (A) instrumental record; (B) spectrum of solute plotted from A.

For the short wavelength region photoconductive cells of lead sulphide, selenide and telluride are extremely fast and sensitive. The last two, when cooled with dry ice or liquid air have sensitivities of the order of one hundred times that of the thermocouple. The high speeds of response allow chopping rates of about 500 c. p. s. and thus simplify amplification. However, they only cover the range from the visible to about 6 μ and are not very suitable for general organic work.

e) Single-Beam Spectrometers.

The single-beam spectrometer, as its name implies, has one beam of infrared radiation which passes through the sample, and the record

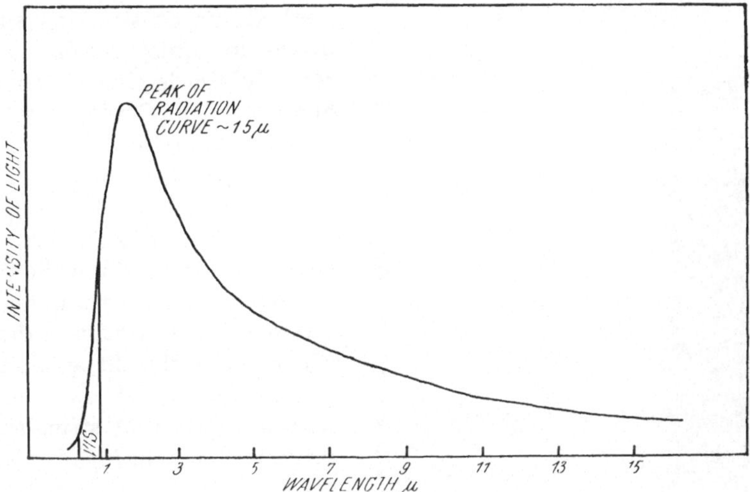

Fig. 2. Distribution of energy in the emission from a "black body" source.

obtained is that of the emission of the source with various absorption bands due to the sample, the solvent if one is used, and the atmosphere superimposed upon it. To obtain a true sample spectrum it is necessary to record also a "background curve" of solvent and atmospheric absorption and measure up the sample curve relative to this. Such a procedure is somewhat tedious and depends upon source emission and instrument characteristics remaining constant over a period of $^1/_2$–1 hr. The optical system of the single-beam type of instrument is well known (7, *134*) and need not be described here. Part of a spectrum in the region of water vapour background absorption is shown in *Figure 1*.

The energy from a "black body" source such as a Globar or Nernst Glower has the form of the curve in *Figure 2* with maximum emission near 1.5 μ. In the region beyond 2.5 μ, which is of greatest interest to chemists, the energy falls off approximately exponentially with wavelength. In order that a suitable amount of energy should reach the detector, the slits of the instrument must be opened, either continuously or in steps while the spectrum is being recorded. With single-beam instruments it must be possible to reproduce the rate of slit opening accurately at least twice while the background and sample spectra are recorded. This can be done in a number of ways. The Perkin-Elmer Corp. supplies a string slit drive (*187*) to suit the sodium chloride prism over the wavelength range 2.5–16 μ and methods of constructing such drives for other prisms have recently been described [Blount and Cole (*20*); Goulden (*78a*); Tolberg and Boyd (*167*)]. It is also possible to use small constant-speed motors to open the slit but these must be changed periodically while the spectrum is being scanned. Cannon (*29*) has made a drive which opens the slit at an exponential rate and can be used with any prism. A string-type drive is shown in *Figure 3* and spectra obtained with and without its use are shown in *Figure 4*.

The most significant improvement in recent years in single-beam infrared instrumentation has been the development of *multiple-pass monochromators* by Walsh (*80, 81, 171–175*).

In conventional Littrow-type monochromators, the light traverses the prism twice, but by arranging a pair of small mirrors at right angles near the slits, Walsh returns a part of the light to the prism up to three additional times, with a twofold increase in dispersion each time. By ingeniously placing the chopper between the right-angled mirrors only the light which undergoes the extra dispersion is chopped and although some "first-order" energy reaches the detector, the signal due to it is

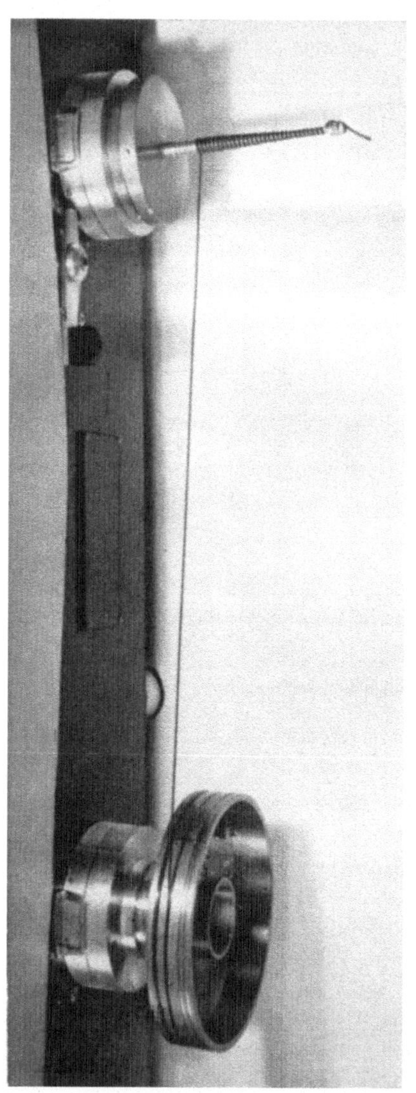

Fig. 3. Automatic slit drive, according to Blount and Cole. [From: J. Sci. Instr. 32, 471 (1955).]

not amplified and the recorder shows only the multiple-pass energy.
A further great advantage of this system is that, since only a very small
fraction of the light is chopped, false signals due to scattered light are
not recorded, thus eliminating one of the biggest errors in quantitative

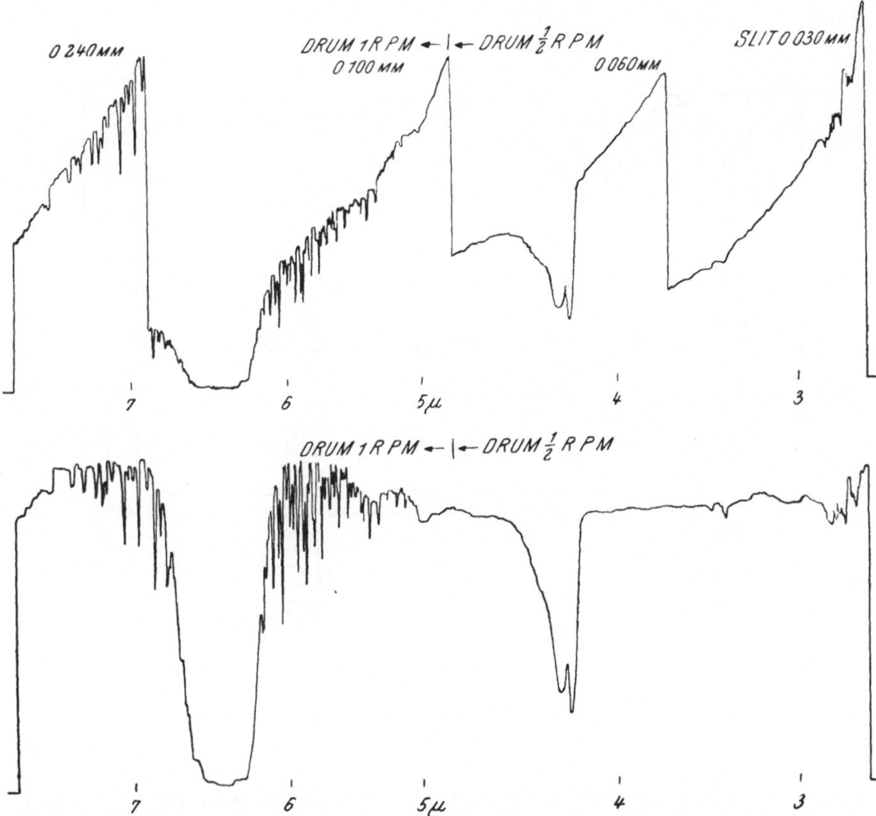

Fig. 4. Infrared spectra illustrating the use of automatic slit drive according to BLOUNT and COLE. 1.3 mm. cell containing CCl₄; CaF₂ prism. Upper curve, fixed slits; lower curve, slit drive in operation. [From: J. Sci. Instr. *32*, 471 (1955).]

infrared work at long wavelengths. The optical system of a double-pass
monochromator is illustrated in *Figure 5* (next page).

f) Double-Beam Spectrometers.

In order to obviate the necessity for replotting spectra and to eliminate
atmospheric and solvent absorption, double-beam instruments are
employed wherever possible today. Early models employed a "double-
beam in space" system [SUTHERLAND and THOMPSON (*158*)] where two

identical beams of light traversed the spectrometer one above the other, and impinged on two separate detectors. The sample was placed in one beam, a blank cell in the other, and when the amplified outputs of the detectors were compared potentiometrically, the record gave the percentage absorption of the sample. Since the beams were of equal length, atmospheric absorption was automatically eliminated.

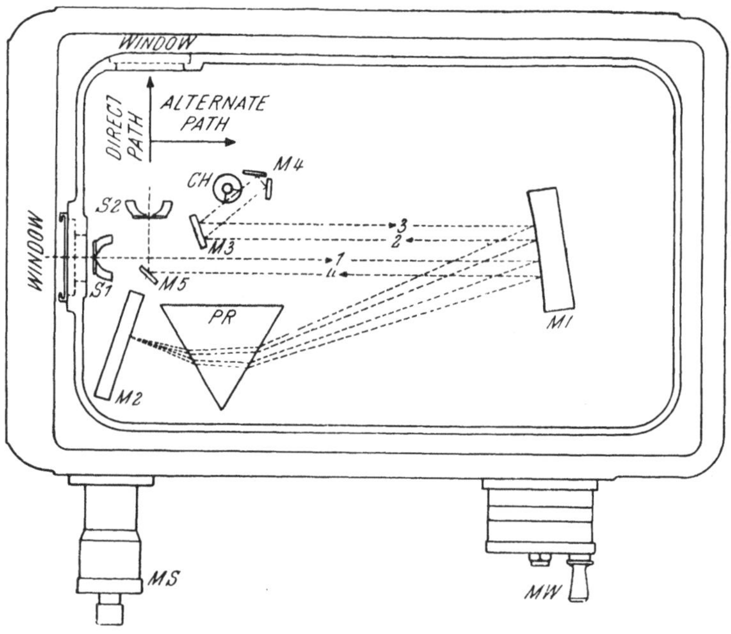

Fig. 5. Optical system of a double pass single beam monochromator (Perkin-Elmer Model 99). Walsh mirrors M_4, light chopper CH.

More modern instruments employ a "double-beam in time" method whereby one beam of light is sent alternately through sample and blank cells by a rotating or reciprocating mirror system (see *Figure 6*). The output from the single detector can either be separated into sample and blank components and the two compared potentiometrically as in the ratio-recorder type [AHLERS and FREEDMAN (*2*); HORNIG, HYDE and ADCOCK (*87*); SAVITSKY and HALFORD (*145*); WILD (*178*)] or the out-of-balance signal amplified and used to drive a servo comb-attenuator as in the optical-null instruments [BAIRD, O'BRYAN, OGDEN and LEE (*2a*); WHITE and LISTON (*176, 177*); WRIGHT and HERSCHER (*185*)]. In the latter type the out-of-balance signal only exists if the sample is absorbing relative to the blank cell. The comb attenuator is then driven into the blank beam until balance is restored, *i. e.* until the comb has blocked off an amount of light equal to that absorbed by the sample. The recorder pen is coupled mechanically to the comb and draws the spectrum as the latter moves into and out of the beam.

The problem of slit control is much simpler with double-beam spectrometers than with single-beam since it is now necessary only that sufficient energy should reach the detector to operate the attenuator and recorder system. Exact reproducibility from run to run is of minor importance. The ratio-recorder instruments have an advantage in this respect that a fraction of the energy of the blank beam may be taken off, amplified

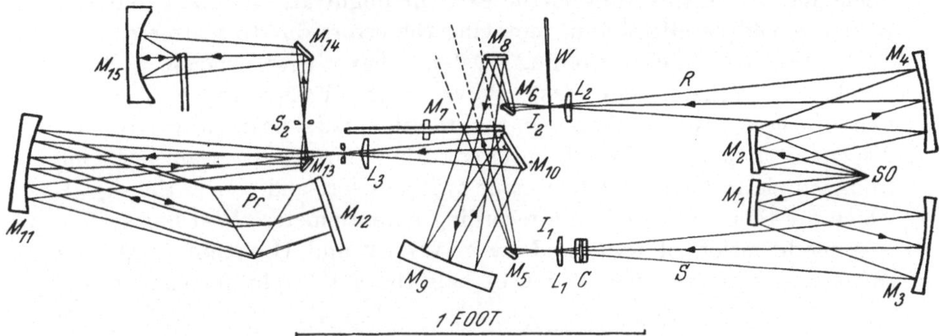

Fig. 6. Optical system of an optical null double beam spectrometer (Perkin-Elmer Model 21), according to WHITE and LISTON. The rotating half-mirror M_7 alternately sends the sample and blank beam through the monochromator. The comb attenuator W is coupled to the recorder pen. [From: J. Opt. Soc. America 40, 29 (1950).]

separately and used to drive a servo mechanism which will control the slits so that they admit a constant predetermined amount of energy [HORNIG, HYDE and ADCOCK (87); SAVITSKY and HALFORD (145)].

A warning might be given here that it is important to monitor double-beam instruments carefully when employing solvents with strong absorption bands of their own. In the region of such a band, neither beam will have enough energy to operate the recorder system and although the instrument may appear to be working properly, in reality the pen may be just wandering about on the recorder paper.

A fairly difficult problem in spectrometer design is the incorporation of WALSH's multiple-pass system into *double-beam* instruments. For the optical-null system which has no light chopper, two suggestions have been put forward to date. COLE (36) proposes that the paraboloid mirror be tilted slightly and the WALSH right-angled mirrors placed above the plane of the slit [ROCHESTER and MARTIN (139)] so that the first order radiation is separated in space from the double-pass beam. It is then no longer necessary to place the chopper between the WALSH mirrors in order to separate the two signals. This method has the advantage of simplicity and may be readily incorporated into any existing spectrometer, but suffers from the defect that it does not eliminate errors due to scattered light. These errors, however, are negligible in the wavelength range 2.5–$12\,\mu$ which is of considerable interest to chemists.

PLÍVA (*127*) has put forward a design for a double-beam, multiple pass spectrometer in which the sample and the balancing attenuator are placed in their respective beams after the first dispersion. This requires the construction of a complete optical system, and the placing of cells in the dispersed beam means that visible light is not available for alignment. Focussing of the spectrum on the exit slit might also be upset, but this system would be effective in removing the errors due to scattered light.

Multiple-pass ratio-recording systems have been suggested by a number of workers [ORR (*125*); PLÍVA (*127*); WALSH (*174*)]. The first two of these involve chopping the multiple-pass beam after its first dispersion at a rate which is different from the rate of beam swinging from sample to blank. Synchronous switches sort and rectify the signals which are then compared on a recording potentiometer. WALSH's system is a modification of that used by SAVITSKY and HALFORD (*145*) in a single-pass ratio recorder and uses half the slit height for each beam.

g) Diffraction Gratings.

As mentioned earlier, most infrared spectrometers used for organic structural work have employed prism dispersion systems. The echelette grating [WOOD (*180, 181*)] offers much higher resolving power but, mostly because of expense, has hitherto only found application in work on small gas molecules. Recently however, MERTON (*123*) has demonstrated a method whereby very good echelette gratings can be made quite cheaply in large numbers, and this has been developed by SAYCE and his collaborators (*51–53, 79*). It is now possible that they will be applied more widely in general infrared spectroscopy and experimental instruments have been described by a number of independent workers [COLE (*40*); GAUNT (*73*); GOULDEN (*78*); MENZIES (*122*)].

Gratings suffer from the inherent disadvantage that a fore-prism assembly or filter system is necessary for the elimination of overlapping spectral orders but these problems can be overcome [COLE (*40*); FRIEDMAN and GLOVER (*69*); GAUNT (*73*)].

h) Calibration.

The usual method of frequency calibration is by comparison with the spectra of gases which have been measured accurately on grating instruments. All the spectra necessary for prism calibration have been illustrated with a critical discussion in a valuable paper by CRAWFORD and his collaborators (*57*).

3. Sampling Techniques.

The infrared spectrum is dependent to some extent on the physical state of the compound. Complex organic materials can be studied in a number of different ways:

(a) Very thin layers of pure liquids.

(b) Solid films formed by cooling from the melt, provided, of course, that they can be melted without decomposition.

(c) Solid films formed by evaporation from solution.

(d) Pastes or mulls when mixed with paraffin oil.

The oil is used to reduce reflection and scattering of light by the crystals, and since the most suitable type of oil is the commercial product "Nujol" (medicinal paraffin oil), this method of preparing samples is usually referred to as a "Nujol Mull". The oil is transparent to infrared light except in the regions of C—H stretching and bending absorption. These parts of the spectrum, however, can be covered by using a fully fluorinated oil or hexachlorobutadiene as the mulling agent. Both these liquids, however, absorb strongly in spectral regions other than those of C—H absorption.

(e) Pressed plates, mixed with potassium bromide [SCHIEDT (*146*); SCHIEDT and REINWEIN (*147*); STIMSON and O'DONNELL (*156*)].

In this procedure a small amount of the sample is thoroughly mixed with pure potassium bromide crystals and the mixture subjected to a pressure of the order of 10 tons in an evacuated die. This results in a clear plate which is very convenient for infrared measurements. The potassium bromide is, of course, transparent over the entire range 4000–400 cm.$^{-1}$ (2.5–25 μ) so that it suffers none of the defects of Nujol. This technique has been used for quantitative analysis of alkaloids by BROWNING, WIBERLEY and NACHOD (*27*). FARMER (*60*), however, has warned that spectra measured on these discs might be vastly different from those obtained by using other techniques, especially when dealing with hydroxylic compounds.

(f) In solution in suitable solvents.

JONES (*95*) has pointed out possible inaccuracies in quantitative measurements carried out on mulls, due to variations in effective sample thickness, and these considerations would also apply to the pressed plates of potassium bromide if the preliminary mixing is not done carefully. More serious than this, however, is the fact that the spectra of solid samples are sometimes greatly perturbed by intermolecular hydrogen bonding and it is often difficult to correlate hydroxyl and carbonyl frequencies with structure because of this factor [COLE (*39*)]. Hence, if the sample will dissolve in a suitable solvent, it is usually much more satisfactory to measure spectra by using dilute solutions. Even with compounds which do not form hydrogen bonds, the frequency of a particular group is found to be much more constant from one compound to another when measured in solution than when solid samples are used. This situation arises because the absorption frequency is affected by the dielectric constant of the surrounding medium [BAYLISS, COLE and LITTLE (*15*); JOSIEN and FUSON (*113*); JOSIEN and LASCOMBE (*114*)] and it is obvious that if the frequency of a particular group is to be compared in a number of different compounds, it is preferable to study them all in one solvent than in the varying dielectric constants of their own crystals. This consideration applies also to liquids.

a) Solvents.

To be satisfactory as a solvent for spectroscopic work a liquid should have few absorption bands of its own. This condition is much harder

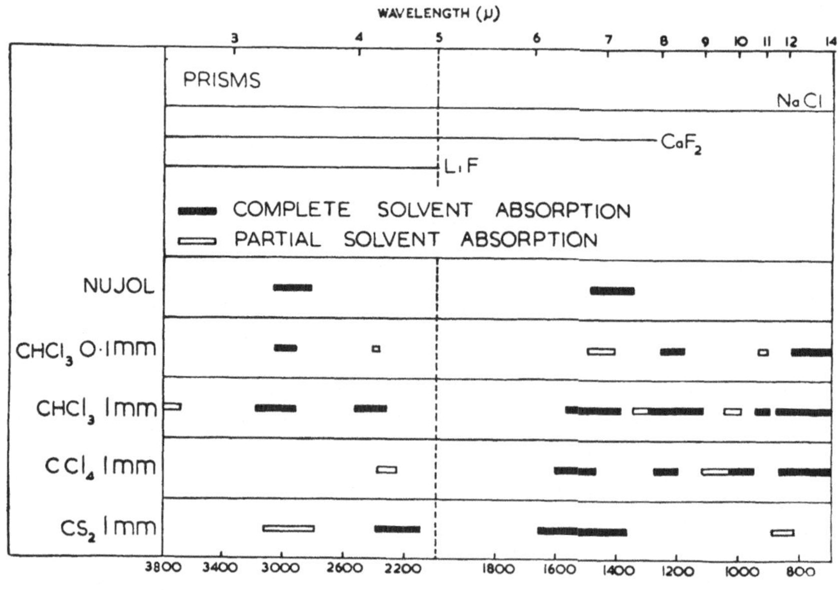

Fig. 7. Transmission ranges of prism materials, solvents and Nujol, according to COLE. [From: Rev. Pure Appl. Chem. (Australia) *4*, 111 (1954).]

to satisfy in the infrared region of the spectrum than in the quartz ultraviolet, since all organic compounds absorb infrared light strongly at one place or another.

It has already been mentioned that the number of absorption bands in the spectrum of a compound is proportional to the molecular size and inversely proportional to its symmetry. Therefore the most suitable solvents should have small symmetrical molecules. The only two liquids which fulfil this condition and combine satisfactory boiling points with reasonable solvent properties are carbon tetrachloride and carbon disulphide, while chloroform is the next best. The latter, although having a more complex spectrum of its own, is rather better than the others from the point of view of solubilities.

Figure 7 shows the most important regions of the spectrum in which each of these solvents is transparent, together with the ranges of the different prisms. It will be noticed that the most convenient arrangement is to use a solution in carbon tetrachloride with prisms of lithium and calcium fluoride for the range 3800–1300 cm.$^{-1}$ (2.6–7.7 μ), and a solution in carbon disulphide with the sodium chloride prism for the region 1300–650 cm.$^{-1}$ (7.7–15.4 μ). These solvents may be used in cells up to at least 1 cm. thick. Many compounds are sufficiently soluble in them to enable the spectrum to be measured in a 1 mm. cell. Compounds with more than two hydroxyl groups, those with a carboxyl group and some lactones are sometimes not sufficiently soluble in either of these solvents but they can usually be studied in chloroform.

Chloroform is usually supplied with a small amount of ethanol added as a stabilizer. Carbonyl and double bond absorption can be studied in the presence of the ethanol, but for work on hydroxyl absorption the stabilizer must be removed.

This is most readily done by pouring the mixture down a column of silica gel which has been prepared by heating at 140° in a vacuum for about one hour. Chloroform purified in this manner can be kept in dark bottles for a few days and any subsequent oxidation can be observed by the appearance of the phosgene carbonyl band at 1810 cm. $^{-1}$ (5.525 μ).

It is important to note that the frequencies of some bands may vary somewhat from one solvent to another and in precise work figures obtained from chloroform should not be compared directly with those obtained from carbon tetrachloride or carbon disulphide solutions. In the case of carbonyl absorption bands, the latter two solvents give approximately the same frequencies, while chloroform solutions give values from 5–20 cm.$^{-1}$ lower. If thin cells (0.05–0.2 mm.) and strong solutions are used, chloroform may be employed as solvent over most of the sodium chloride range.

The wider use of other solvents such as bromoform and methylene bromide has been advocated by TARPLEY and VITIELLO (*159, 159a*) and many other liquids by PRISTERA (*132*) but they do not appear to have been employed by many workers in the field.

b) *Size of Samples.*

It is usual to place the sample at or near a focus in the light beam so that very small amounts of material only are required. For most techniques, whether on solids or solutions, approximately 1–3 mg. is usually sufficient, although when dealing with pure liquids rather more than this amount is necessary to facilitate handling. A simple cell for solutions, which consists of a brass spacer and two thin sodium chloride plates is illustrated in *Figure 8*.

One of the best adhesives for sealing the windows on to the spacer is Seccotine which is soluble in water but not in any common infrared

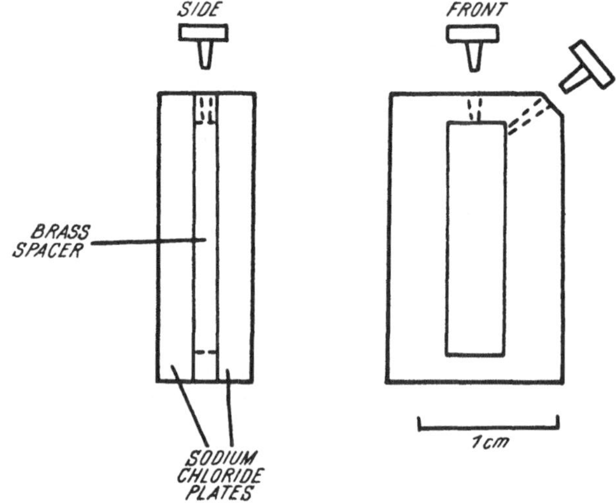

Fig. 8. Simple absorption cell for solutions.

solvent. If the plates are clamped lightly to the spacer and the Seccotine painted around the edge, capillary action allows a thin layer to run along the faces in contact, but not enough flows to contaminate the internal faces of the liquid compartment. This type of cell is readily filled and emptied with a hypodermic syringe, and this procedure facilitates the recovery of samples after spectral measurements. The small tapered stoppers, if lightly turned with pliers, effectively prevent evaporation of volatile solvents.

Cells of thickness between 1 and 10 mm. can be made in this manner but for thinner layers the spacer is too thin to drill at the edge. Lead foil is the commonest spacer material for thin cells and amalgamation with mercury forms a good seal if the cell windows are mounted under pressure. These cells are usually filled and emptied through holes in one window.

The study of *extremely small amounts of material* and of single crystals has been facilitated by the use of the reflecting microscope [BARER, COLE and THOMPSON (*3*); BURCH (*28*)]. Unlike refracting instruments,

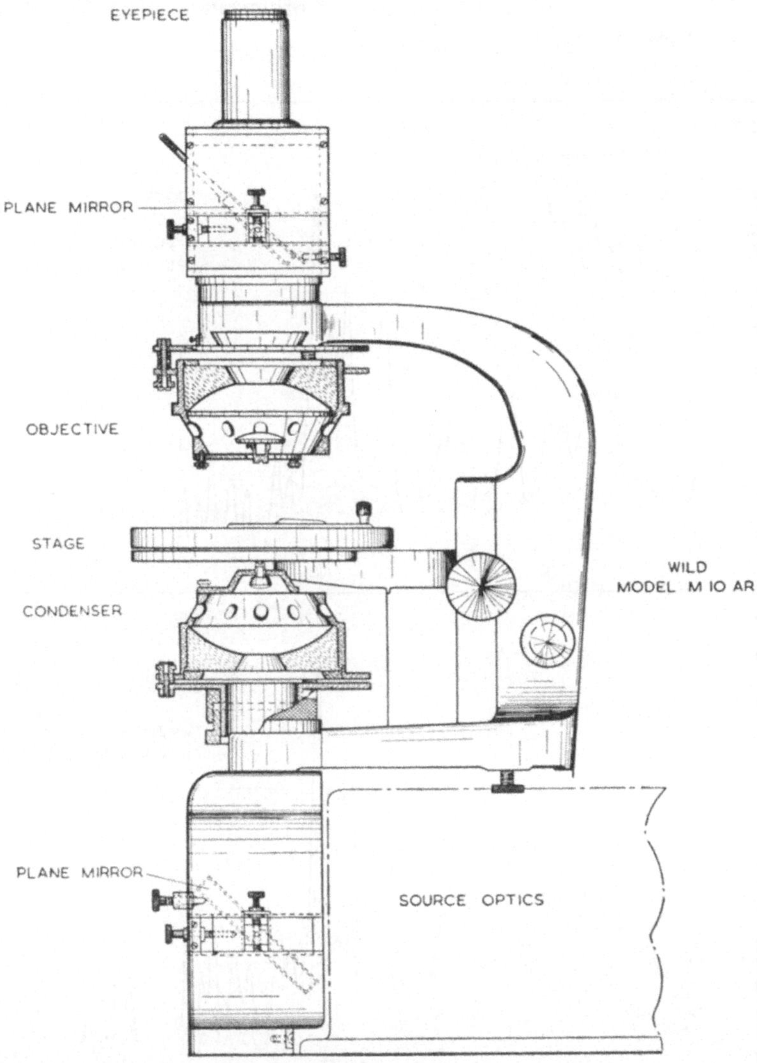

Fig. 9. Details of construction of a reflecting microscope for infrared spectroscopy, according to COLE and JONES. [From: J. Opt. Soc. America *42*, 348 (1952).]

the reflecting microscope is completely achromatic, since the focal lengths of the mirrors depend only on the geometry of the surfaces. They are therefore ideal for infrared work where the range of wavelengths is quite large. The sample can be focussed by using the visible light from the source and then is automatically in focus for any other wavelength. Such a microscope, constructed for infrared work is illustrated in *Figure 9*

[Cole and Jones (*44*)] and a number of others have been described in the literature (*21–23, 66, 75, 179, 188*).

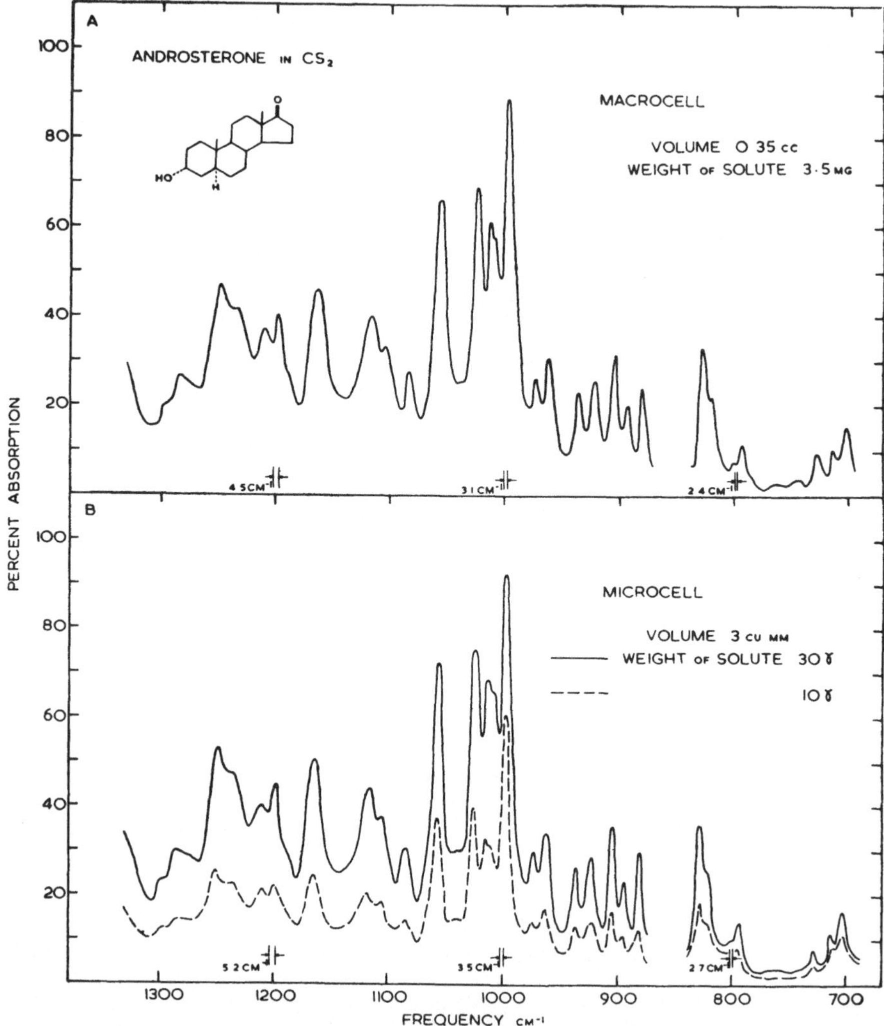

Fig. 10. Infrared spectrum of androsterone in carbon disulphide solution, according to Cole and Jones. (A) Conventional large cell. (B) Microcell with reflecting microscope. [From: J. Opt. Soc. America *42*, 348 (1952).]

In this manner it is relatively easy to measure an infrared spectrum on a sample of less than 10 micrograms. *Figure 10* compares the spectrum of the steroid androsterone obtained from samples of 10 and 30 micrograms

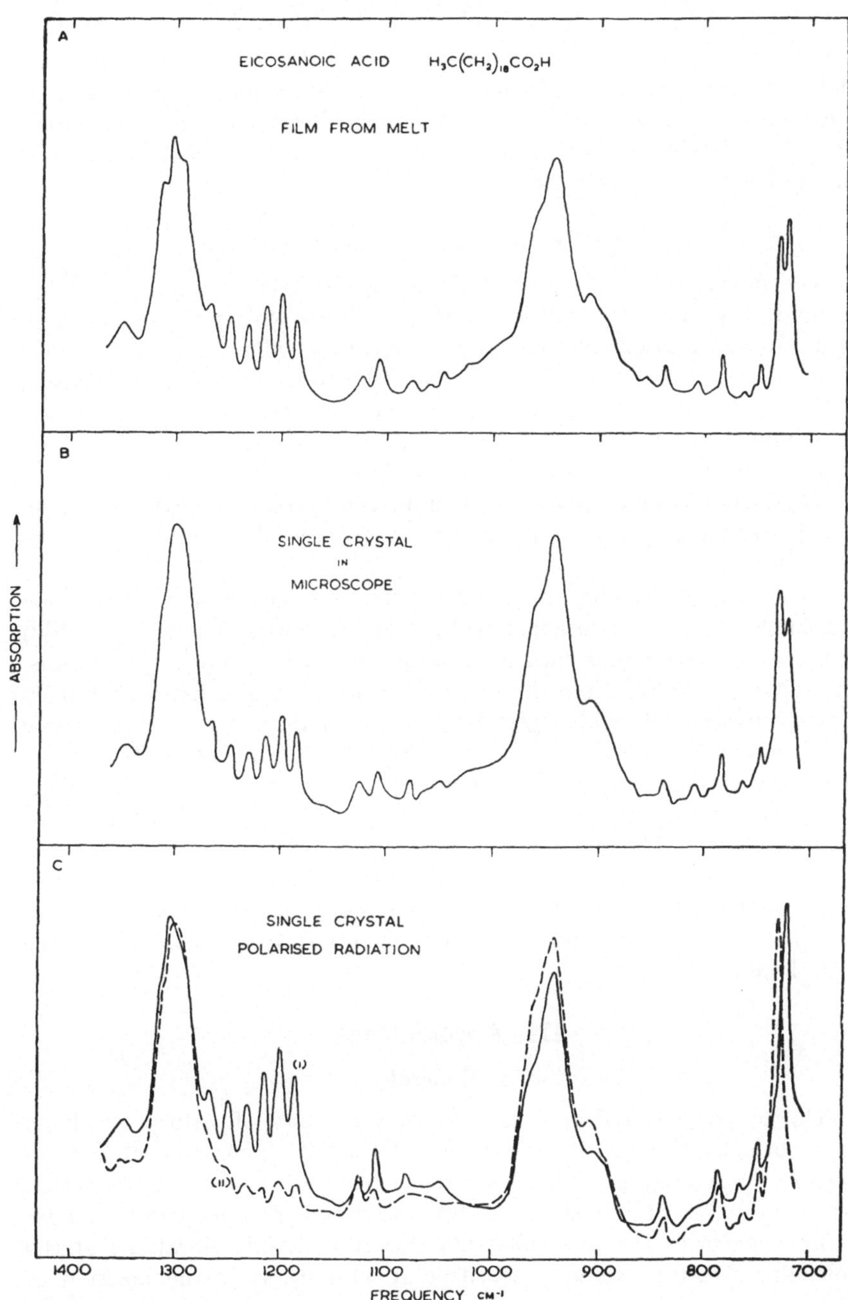

Fig. 11. Infrared spectrum of eicosanoic acid, according to COLE and JONES. (A) Solid film prepared by melting between sodium chloride plates. (B) Single crystal using reflecting microscope. (C) Same crystal as in (B) using polarized radiation. [From: J. Opt. Soc. America *42*, 348 (1952).]

in solution in a microcell (*44*) with the aid of a microscope with that measured by using a conventional large cell in the normal manner. Very little spectral resolution is lost, and of the 10-microgram sample (Figure 10 B) only about 2.5 micrograms was effectively in the light beam. The point has been reached where ability to handle small amounts of material is the limiting factor.

c) Use of Polarized Radiation.

The development of this technique has also greatly helped in studies of single crystals and oriented films and fibres with polarized infrared light. In an oriented solid only those vibrations which involve a change of dipole moment with a component parallel to the plane of polarization of the light can cause absorption. Hence, the measurement of spectra with light polarized respectively parallel and perpendicular to a given crystal or fibre axis can lead to information about the relative directions of certain bonds or certain vibrations in the system [ELLIOTT, AMBROSE and TEMPLE (*59*); JONES and SUTHERLAND (*92*); MANN and THOMPSON (*117, 118*)].

In *Figure 11* (**A**) and (B) the spectrum obtained with a microscope of a single crystal of eicosanoic acid is compared with that of a thin film of the same compound measured under normal operating conditions. The spectra obtained from the same crystal by using radiation polarized in two directions at right angles are shown in Figure 11 (C). Particularly to be noted is the dichroism of the regularly spaced group of bands between 1150 and 1350 cm.$^{-1}$ (8.70–7.40 μ) which are associated with vibrations of the methylene groups along the carbon chain [JONES, McKAY and SINCLAIR (*108*)]. The marked dichroism of the double band near 720 cm.$^{-1}$ (13.89 μ), due to the "rocking" vibration of the methylene groups, supports the hypothesis of SUTHERLAND (*157*) that they are related to the in-phase and out-of-phase rocking modes of two chains within the crystal unit cell.

III. Applications.

1. General.

The use of infrared spectroscopy in the study of natural products can be broadly divided into two main sections, viz. compound comparison and structural analysis. It has been mentioned that some of the absorption bands are due to functional groups and are very little affected by changes in other parts of the molecule. On the other hand, vibrations of the molecular skeleton are very sensitive to alterations in the position of substituent groups. With compounds consisting primarily of carbon, hydrogen, oxygen and nitrogen, the specific group vibrations are mostly found at the higher frequency end of the spectrum between 1350 and 3800 cm.$^{-1}$ (7.40–2.63 μ) while most skeletal frequencies are found below this range.

a) Compound Comparison.

The low frequency end of the spectrum ($<$ 1400 cm.$^{-1}$; $>$ 7 μ) has been aptly named the "fingerprint" region since no two different compounds have been found to have exactly identical patterns of absorption there. Optical isomers when studied *in solution* give identical spectra but in the solid state might exhibit differences due to crystal habit etc. This is particularly likely if one is comparing an optically active isomer with a racemic mixture. The final proof of the structures of many natural products has been obtained in this way from a comparison of the infrared spectra of the natural and synthetic material. Very good examples of this are given in the work of JOHNSON (*91*) and WOODWARD (*183*) and their collaborators. For estrone methyl ether (XIX, $R = CH_3$; p. 34) cf. *Figure 12*. The use of infrared spectra in these cases is often the only convenient means of establishing identity since natural *dextro-* or *laevo-* rotatory isomers and synthetic racemic compounds will most probably have different melting points.

Quite small differences in molecular constitution, produce marked changes in the spectra. In *Figure 13* are shown the spectra of four

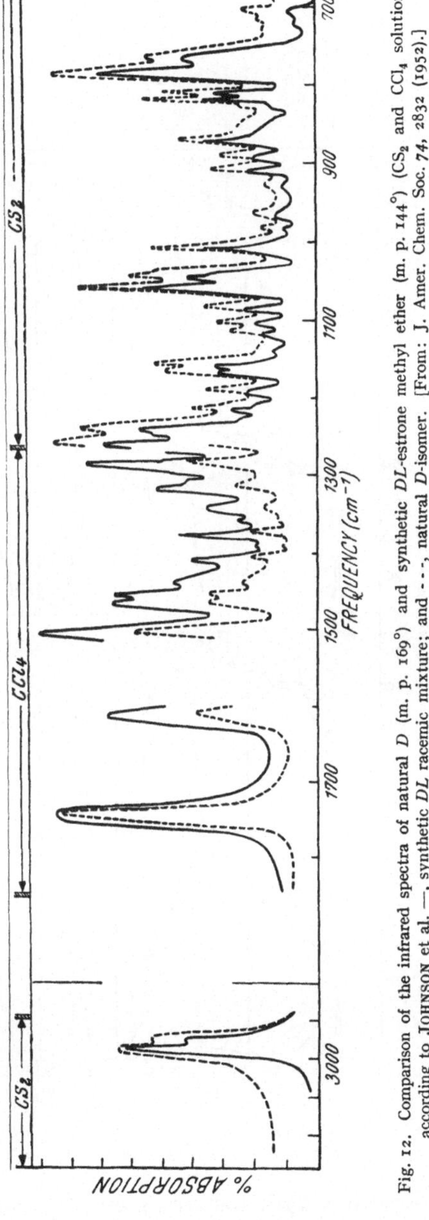

Fig. 12. Comparison of the infrared spectra of natural D (m. p. 169°) and synthetic *DL*-estrone methyl ether (m. p. 144°) (CS$_2$ and CCl$_4$ solutions), according to JOHNSON et al. —, synthetic *DL* racemic mixture; and - - -, natural *D*-isomer. [From: J. Amer. Chem. Soc. 74, 2832 (1952).]

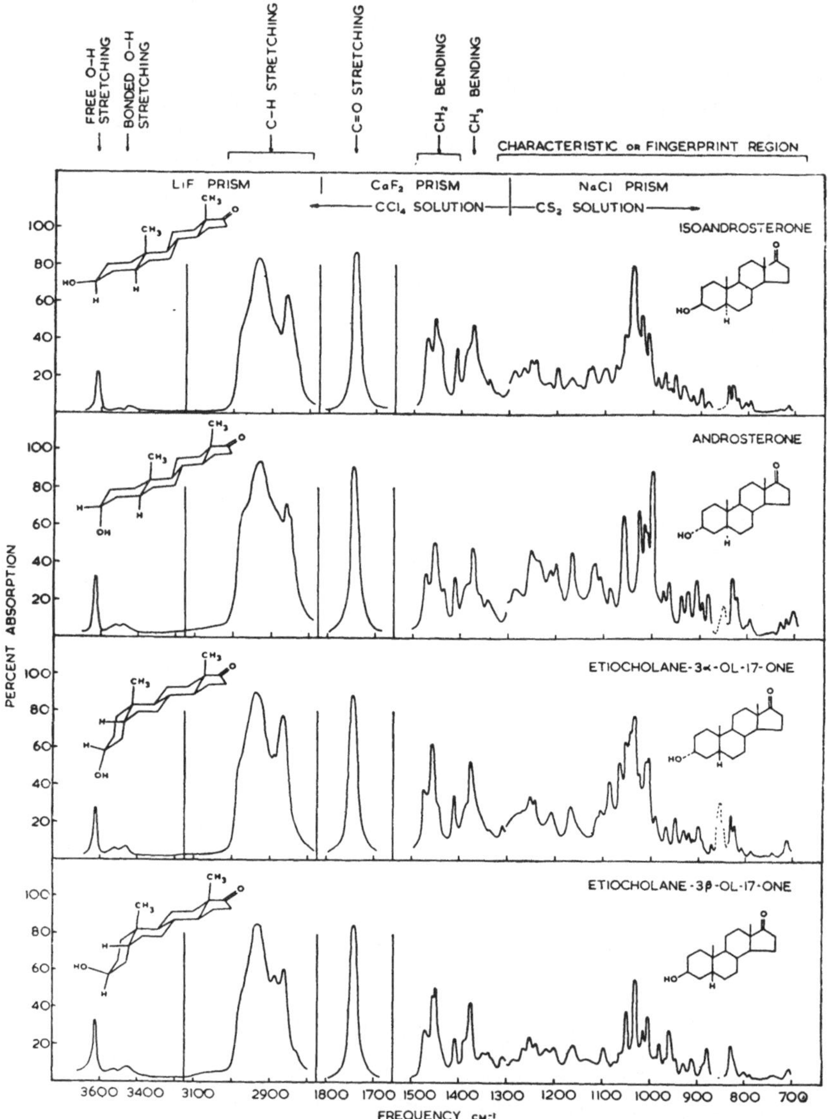

Fig. 13. Infrared spectra of four stereoisomeric 3-hydroxy-17-ketosteroids according to COLE. [From: Rev. Pure Appl. Chem. (Australia) *4*, 111 (1954).]

3-hydroxy-17-ketosteroids which differ from one another only in stereochemical configuration. The differences in the range 1400–700 cm.⁻¹
(7–14 μ) are unmistakable and it is quite clear from the complexity of
the curves, that if an unknown compound exhibited a spectrum identical

with one of these, then the probability that they are the same compound would be very high indeed.

Two factors combine to make the infrared spectrum so useful in fingerprint identification. Firstly, the very complex nature of the absorption pattern means that there are many more parameters to be satisfied — the frequency and intensity of many bands — than is the case with a melting point and mixed melting point. Secondly, it is not necessary to have the known substance on hand, since spectra can easily be compared with published curves. The number of collections of spectra is increasing and those of the American National Bureau of Standards, the American Petroleum Institute and the British National Depository are invaluable for this purpose. A very useful catalogue of steroid spectra has been prepared by DOBRINER, KATZENELLENBOGEN and JONES (54).

DOBRINER and his collaborators (56, 99, 115) have used this technique to great advantage in identifying steroid hormone metabolic products isolated from urine and separated by chromatography. By following a standardised procedure of chemical separation and chromatography they found that the steroids normally present in urine always came off the column in the same order and each could be identified in a few minutes by its spectrum.

b) Structural Analysis.

Useful though fingerprint identification is, it can only be applied to an unknown compound after chemical or other tests have indicated the broad class to which the substance belongs, and then only if the required known compound has been available before. With new compounds this procedure is not possible, and a great deal of natural product work is concerned with the determination of the molecular structure of new substances. The absorption bands due to vibrations of specific functional groups are very useful for indicating some of the major structural features of such compounds, or for following the course of reactions in synthetic and degradative work. In the spectra illustrated in *Figure 13*, the presence of the hydroxyl and carbonyl groups is shown quite clearly by the bands near 3630 and 1740 cm.$^{-1}$ (2.755 and 5.747 μ) respectively. Bands due to other groups are indicated at the top of the Figure.

Table 1 indicates the regions in which absorption bands related to most of the common chemical groups are found. Considerable discussion of these and other regions has been given by RASMUSSEN (134) and by BELLAMY (17), and we may assume that such correlations are well established. We shall be concerned here mainly with applications involving some of the small differences within these broad regions which are caused by slight changes in environment of the absorbing groups and which are particularly useful for structural work.

Table I. Positions of Characteristic Infrared Bands.

	Frequency	Wavelength
OH Stretching Vibrations		
Free OH	3590–3650 cm.$^{-1}$ (sharp)	2.786–2.740 μ
Intermolecular hydrogen bonds	3200–3550 cm.$^{-1}$ (broad)	3.125–2.820 μ
Intramolecular hydrogen bonds	3450–3600 cm.$^{-1}$ (sharp)	2.900–2.780 μ
Chelate compounds	2500–3200 cm.$^{-1}$ (very broad)	4.00 –3.125 μ
NH Stretching Vibrations		
Free NH	3200–3500 cm.$^{-1}$	3.12–2.85 μ
Hydrogen bonded NH ...	3070–3350 cm.$^{-1}$	3.25–2.98 μ
CH Stretching Vibrations		
$\equiv$C—H	3300–3340 cm.$^{-1}$	3.03 –3.00 μ
=C—H	3000–3100 cm.$^{-1}$	3.33 –3.23 μ
CH_3	2872 and 2962 cm.$^{-1}$ $\pm$ 10 cm.$^{-1}$	3.482 and 3.376 μ
CH_2	2853 and 2926 cm.$^{-1}$ $\pm$ 10 cm.$^{-1}$	3.505 and 3.418 μ
CH	2890 $\pm$ 10 cm.$^{-1}$	3.472–3.448 μ
SH Stretching Vibrations		
Free SH	2550–2600 cm.$^{-1}$	3.92–3.85 μ
Hydrogen bonded SH	very slightly lower	
C$\equiv$N Stretching Vibrations		
Non-conjugated	2240–2260 cm.$^{-1}$	4.46–4.43 μ
Conjugated	2215–2240 cm.$^{-1}$	4.51–4.46 μ
C$\equiv$C Stretching Vibrations		
C$\equiv$CH	2100–2140 cm.$^{-1}$	4.76–4.67 μ
C—C$\equiv$C—C	2190–2260 cm.$^{-1}$	4.57–4.43 μ
C—C$\equiv$C—C$\equiv$CH	2040, 2200 cm.$^{-1}$	4.90, 4.55 μ
C=O Stretching Vibrations		
Non-conjugated	1700–1800 cm.$^{-1}$	5.88–5.55 μ
Conjugated	1650–1750 cm.$^{-1}$	6.06–5.71 μ
Amides................	$\sim$ 1650 cm.$^{-1}$	$\sim$ 6.06 μ
C=C Stretching Vibrations		
Non-conjugated	1620–1680 cm.$^{-1}$	6.17–5.95 μ
Conjugated	1585–1625 cm.$^{-1}$	6.31–6.16 μ
CH Bending Vibrations		
CH_2	1405–1465 cm.$^{-1}$	7.12–6.83 μ
CH_3	$\left\{\begin{array}{l} 1355–1395 \text{ cm.}^{-1} \\ 1430–1470 \text{ cm.}^{-1} \end{array}\right.$	$\left\{\begin{array}{l} 7.38–7.17\,\mu \\ 6.99–6.80\,\mu \end{array}\right.$
C—O—C Vibrations in Esters		
Formates..............	$\sim$ 1175 cm.$^{-1}$	$\sim$ 8.51 μ
Acetates..............	$\sim$ 1240 cm.$^{-1}$	$\sim$ 8.07 μ
Benzoates	$\sim$ 1275 cm.$^{-1}$	$\sim$ 7.84 μ

	Frequency	Wavelength
C—OH Stretching Vibrations		
Secondary cyclic alcohols .	990–1060 cm.$^{-1}$	10.10–9.43 μ
CH Out-of-plane Bending Vibrations in Substituted Ethylenic Systems		
—CH=CH$_2$	{ 905–915 cm.$^{-1}$ { 985–995 cm.$^{-1}$	{ 11.05–10.93 μ { 10.15–10.05 μ
—CH=CH— *(trans)*	960–970 cm.$^{-1}$	10.42–10.31 μ
—CH=CH— *(cis)*	660–800 cm.$^{-1}$	15.15–12.50 μ
C\ C=CH$_2$ . . . . / C	885–895 cm.$^{-1}$	11.30–11.17 μ
C\ H C=C . . . . / \ C C	790–840 cm.$^{-1}$	12.66–11.90 μ
CH Out-of-plane Bending Vibrations in Substituted Benzene Systems.		
Five adjacent free hydrogen atoms	{ 730–770 cm.$^{-1}$ { 690–710 cm.$^{-1}$	{ 13.70–12.99 μ { 14.49–14.08 μ
Four adjacent free hydrogen atoms	735–770 cm.$^{-1}$	13.61–12.99 μ
Three adjacent free hydrogen atoms	750–810 cm.$^{-1}$	13.33–12.35 μ
Two adjacent free hydrogen atoms	800–860 cm.$^{-1}$	12.50–11.63 μ
One free hydrogen atom .	860–900 cm.$^{-1}$	11.63–11.11 μ
Carbon-Halogen Stretching Vibrations.		
C—F	1000–1400 cm.$^{-1}$	10.00–7.15 μ
C—Cl	600–800 cm.$^{-1}$	16.7–12.5 μ
C—Br	500–600 cm.$^{-1}$	20.0–16.7 μ
C—I	~ 500 cm.$^{-1}$	~ 20 μ

2. Steroids and Terpenoids.

The steroid hormones and vitamins [FIESER and FIESER (*62*)] play an important role in the control and development of animal life, and it is largely due to the interest of medical research laboratories that their chemistry has been investigated so thoroughly. Terpenoid compounds on the other hand [SIMONSEN (*150–152*)], although structurally very similar to the steroids, are obtained usually from plants, and at the moment most of them do not appear to possess the kind of physiological activity exhibited by many of the steroids. However, in recent years great interest has been taken in the triterpenoids (*5, 11, 18, 46, 68, 89, 93, 169, 182*) and it is possible that many of them will be useful in medicine, perhaps not directly, but as starting materials in the synthesis of physiologically active compounds.

A. Steroids.

The steroid and cyclic terpenoid substances are an ideal class of compounds for infrared study since they exist in very large numbers based on a relatively few types of molecular skeleton and differ mainly in the nature and position of a few fairly non-polar substituents. Moreover, the vast majority of these substances are either saturated or contain only a very few double bonds (usually isolated) so that polar effects are rarely transferred from one part of the molecule to another as is the case in long conjugated systems. Various aspects of the infrared spectra of steroids have been summarised by Jones and Dobriner (*99*), by Jones and Herling (*100*) and by Cole (*38*).

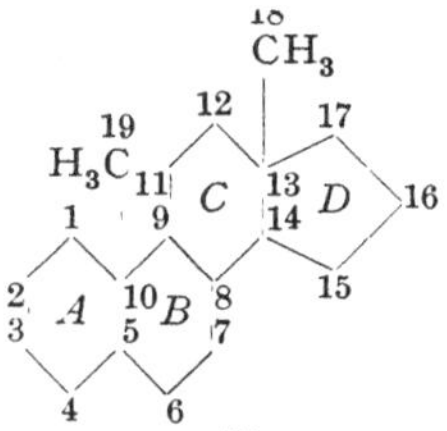

(I.) 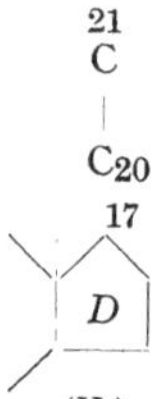   (II.) 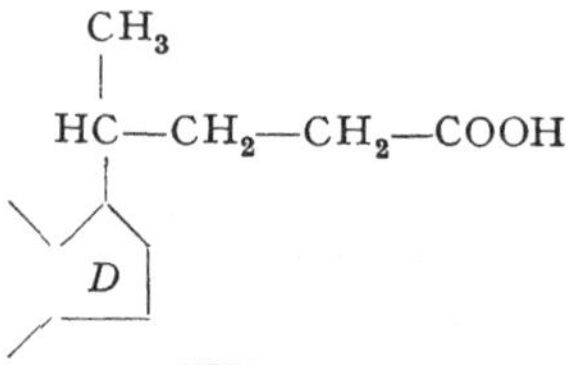   (III.)

As is well known, the basic steroid skeleton consists of the perhydro-*cyclo*penteno-phenanthrene ring system with angular methyl groups at $C_{(10)}$ and $C_{(13)}$ (I), and side-chains of varying complexities may be attached at $C_{(17)}$. The pregnane series, which includes the hormone progesterone, have the side-chain (II), the bile acids (III), cholesterol and its related compounds (IV), and ergosterol (V).

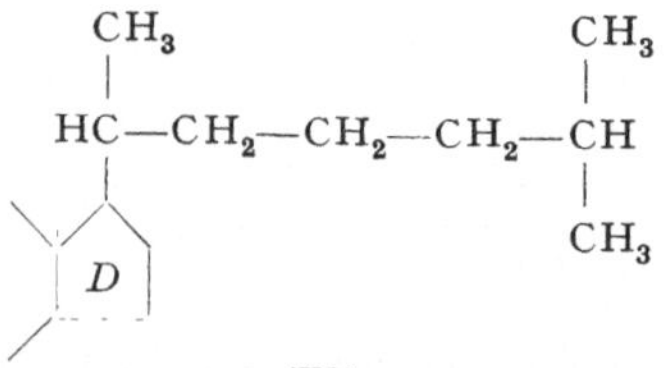

(IV.) 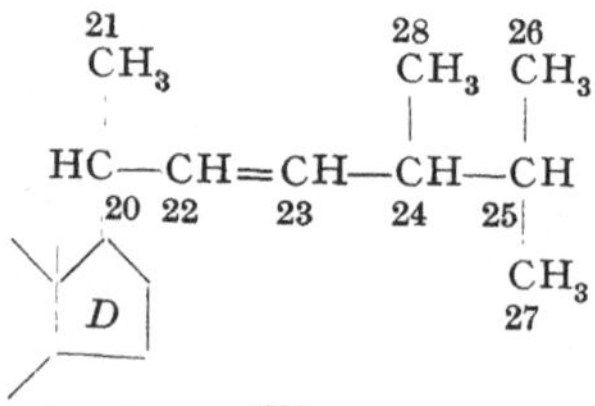   (V.)

Theoretically 64 stereoisomers of structure (I) could exist, when the possibility of *cis* and *trans* ring junctions is considered. In practice, however, most of the natural steroids have *trans* junctions between rings B and C, as well as between C and D, and it is only in the A/B ring junction that the various isomers differ. The usual method of indicating bonds above the plane of the ring system (β bonds) by full lines (VI) and those below (α bonds) [FIESER and FIESER (62)] by broken lines (VII)

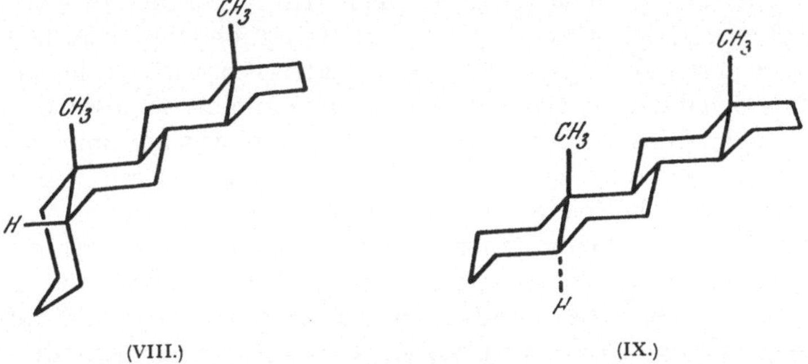

(VI.) *cis-A/B.*   (VII.) *trans-A/B.*

might profitably be replaced by side-view formulae (VIII, IX) which give a much better impression of structure. Much of the interesting evidence which has been presented in favour of the "chair" form of the *cyclo*hexane ring rather than the "boat" in this type of compound has been summarised by BARTON and ROSENFELDER (8, 9, 13).

(VIII.)   (IX.)

Further stereoisomerism is associated with the substituents, which may be connected to the ring carbon atom by either β or α bonds, although strictly they do not all project *above* or *below* the ring system. It is obvious from (X) that of the two substituent bonds from each carbon atom, one is approximately in the plane of the ring [equatorial (e)] and the other parallel to the axis XY and projects above or below the plane [axial (a)] [BARTON, HASSEL, PITZER and PRELOG (10)].

There are slight differences in chemical reactivity between axial and equatorial groups [BARTON (9)], the latter being thermodynamically more stable. This distinction is rather better than the "α" and "β" nomenclature. It should be remembered that at each carbon atom a

β bond projects above an α bond rather than saying that one is above
the ring system and the other below.

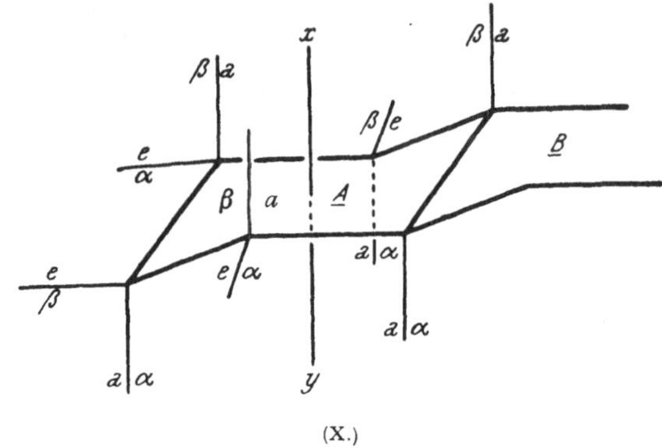

(X.)

Substituents. Most of the infrared work on steroids has been concerned
with the identification and location of hydroxyl groups, carbonyl groups
and ethylenic double bonds. Many of the hormones and vitamins have
a hydroxyl group at the 3-position and others can occur less frequently,
perhaps only in synthetic compounds, at positions 2, 5, 6, 7, 11, 12, 14,
17, 20, and 21. Since a molecule might easily contain two or three
hydroxyl groups, and at most of the above positions they may be either
axial or equatorial, it is readily seen that, even without introducing
other functional groups, there are a very large number of possible sterols.

Carbonyl groups can occur at any carbon atom which is not concerned
with a ring junction, and the most common positions are 3, 6, 7, 11,
12, 17, and 20. The bile acids contain the carboxyl group as shown in
formula (III), and many other carbonyl-containing systems may be
introduced in the course of chemical work. These include the methyl
ester, acetate, benzoate, formate and lactone functions, and infrared
measurements on these extra carbonyl groups are sometimes useful
in solving structural problems.

Unsaturation can occur in many positions around the ring system,
usually in the form of isolated double bonds but sometimes in conjugation
with other $C{=}C$ bonds or with carbonyl groups.

a) Hydroxyl Absorption. The band due to the hydroxyl stretching
vibration occurs near 3630 cm.$^{-1}$ (2.755 μ) in dilute solution (CCl$_4$) and
if high dispersion is used significant differences in frequency can be
found for primary (3640 cm.$^{-1}$; 2,747 μ), secondary (3629 cm.$^{-1}$; 2.756 μ)
and tertiary (3618 cm.$^{-1}$; 2.764 μ) types. Apart from this, infrared
measurements give little information about the position of the hydroxyl
groups in the ring system. More can be obtained by oxidising primary

and secondary groups to the corresponding carbonyl group and studying the spectrum of the product. Very little quantitative work has been carried out on hydroxyl groups, but provided the solutions are sufficiently dilute ($\sim$ 0.01 M) to prevent intermolecular hydrogen bonding, a measurement of apparent molar extinction coefficient* is usually enough to determine the number of hydroxyl groups in the molecule. Each

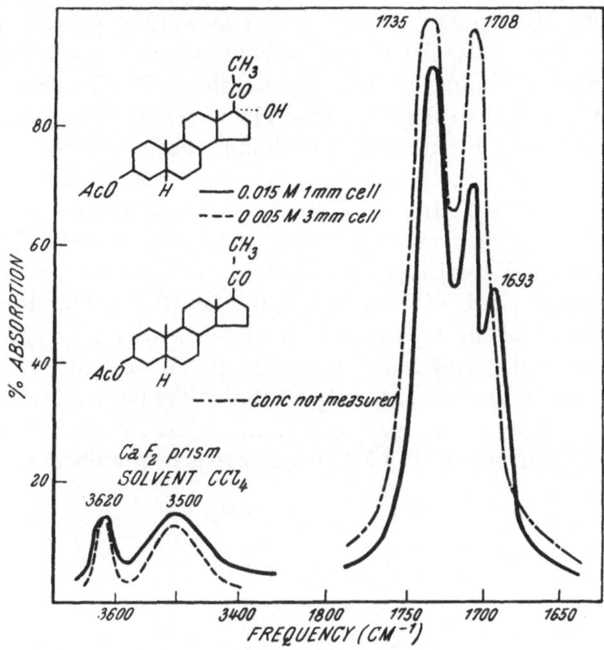

Fig. 14. Infrared spectra illustrating intramolecular hydrogen bonding, according to JONES et al. [From: J. Amer. Chem. Soc. 74, 2820 (1952).]

group contributes a value of ε of approximately 50 [COLE and THORNTON (47); SMITH and CREITZ (155)] and Beer's Law holds fairly well for this absorption.

Hydrogen bonding lowers the frequency of the hydroxyl stretching vibration and alters the form of the absorption band. In the solid state or in concentrated solutions, unless greatly hindered by neighboring groups, hydroxyl groups are always involved in hydrogen bonds, resulting in a broad band between 3200 and 3550 cm.$^{-1}$ (3.125–3.815 μ). In carboxylic acids which are usually dimerized, and chelate compounds

---

* $\varepsilon = (1/cl) \log_{10} (T_0/T)$, where c is the solute concentration in moles per litre, l is the cell thickness in cm., and T_0 and T are, respectively, the transmissions through the pure solvent and the solution.

such as enolic β-diketones, the OH absorption is broader still and lies across that of the CH groups between 2500 and 3200 cm.$^{-1}$ (4.00–3.125 μ).

Intramolecular hydrogen bonding is independent of the concentration of the solution and is very useful in studies of the stereochemical relationships between neighboring hydroxyl groups or between a hydroxyl and a carbonyl group. The absorption in this case is fairly sharp and not so greatly displaced from the free hydroxyl position as that described above. JONES, HUMPHRIES, HERLING and DOBRINER (*105*) have shown how studies of intramolecular hydrogen bonding may be used to identify a number of side-chains in the C_{21}-steroids. Besides lowering the frequency of the hydroxyl group this phenomenon lowers that of the other group involved in the hydrogen bond. For instance in *Figure 14* the spectrum of 3β-acetoxypregnan-20-one is compared with that of its 17α-hydroxy derivative. In the hydroxyl stretching region of the latter two bands occur, at 3620 cm.$^{-1}$ (2.762 μ) (free tertiary) and at 3500 cm.$^{-1}$ (2.857 μ) (associated hydroxyl group), and the relative intensities are practically independent of concentration. In the carbonyl region, in addition to the bands at 1735 and 1708 cm.$^{-1}$ (5.76 and 5.85 μ) expected for the acetate and 20-ketone (*Table 2*, p. 36) a band is found at 1693 cm.$^{-1}$ (5.91 μ), and the intensity of the 20-ketone band (1708 cm.$^{-1}$; 5.85 μ) is lower than in the spectrum of the reference compound. These extra bands are attributed to (XI A) which exists in equilibrium with (XI).

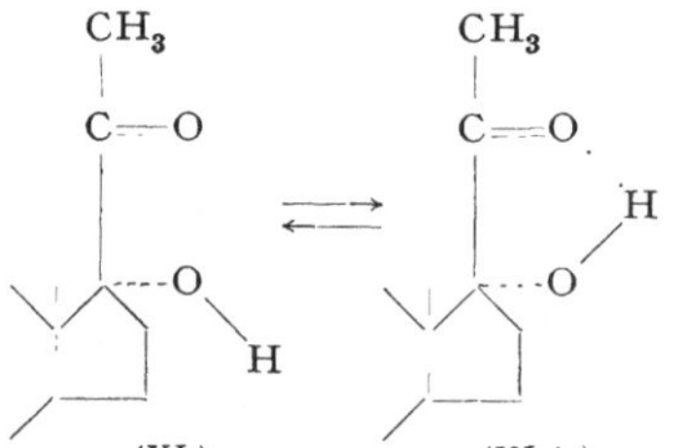

(XI.)   (XI A.)

Similar effects are observed in the spectra of the methyl ester of 17α-hydroxy-bisnorcholanic acid (XII, XII A) and of some compounds containing 12α-acetoxyl and 20β-hydroxyl groups (XIII, XIII A).

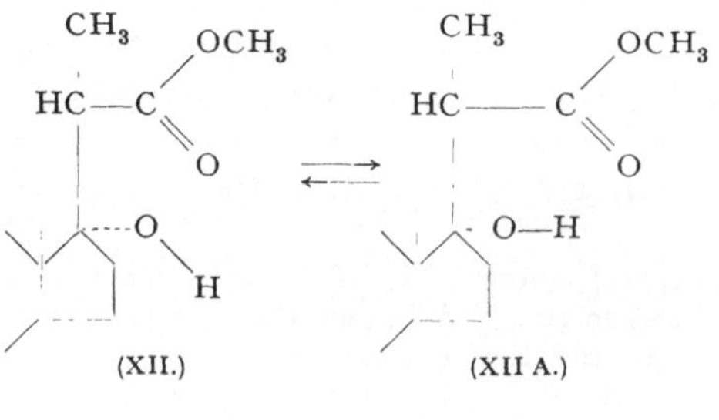

(XII.)   (XII A.)

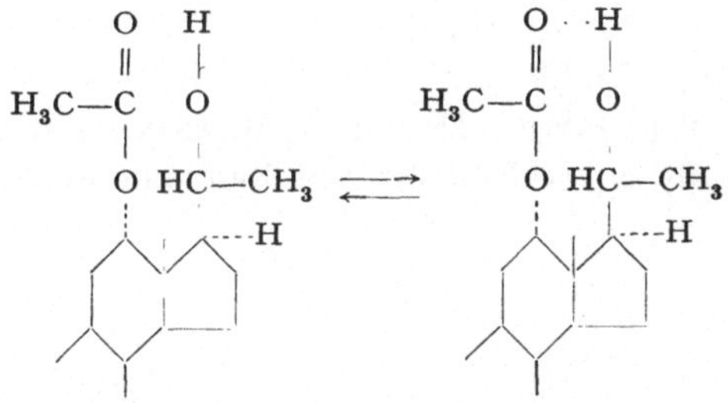

$$\text{(XIII.)} \rightleftharpoons \text{(XIII A.)}$$

An interesting application of the use of these effects in studying steroidal sapogenins has been described by ROTHMAN and WALL (*141*).

b) C—H Stretching Absorption. Most organic compounds have strong absorption in the range 2850–3000 cm.$^{-1}$ (3.50–3.33 μ) due to stretching vibrations of methyl and methylene groups. Although it is possible in certain cases to identify absorption peaks in this region with methyl and methylene groups separately (see Table 1, p. 26) [FOX and MARTIN (*63*); SHEPPARD and SIMPSON (*148, 149*)], considerable overlap of bands occurs and very little information of structural interest has been obtained here. FRANCIS (*64, 65*) has used intensity measurements to analyse this absorption but only in relatively simple molecules.

NOLIN and JONES (*124*) have been very successful in sorting out the C—H stretching absorption by using deuterium substitution in specific groups, but once again this has only been possible in simple compounds. Rather more information on methyl and methylene groups in complex molecules can be obtained from a study of their bending vibrations, and these are considered later (p. 42).

C—H groups attached to multiple bonds (acetylenic, olefinic or aromatic) absorb at a higher frequency and the absorption bands of these are very useful in structural work.

The absorption of (XIV) which is characteristic of most double bonds at ring junctions usually appears as a shoulder between 3010–3040 cm.$^{-1}$ (3.322–3.289 μ) on the high frequency side of the main C—H band, and might not be detected unless fairly good resolution is available [BLADON et al. (*19*); JOHNSON et al. (*90*); JONES et al. (*99, 106, 111*); HENBEST, MEAKINS and WOOD (*82*)]. When the double bond is in the five-membered *D*-ring as in the Δ^{14}-steroids (XV) the absorption of this C—H group is intensified and found at a slightly higher frequency (3055 cm.$^{-1}$; 3.273 μ).

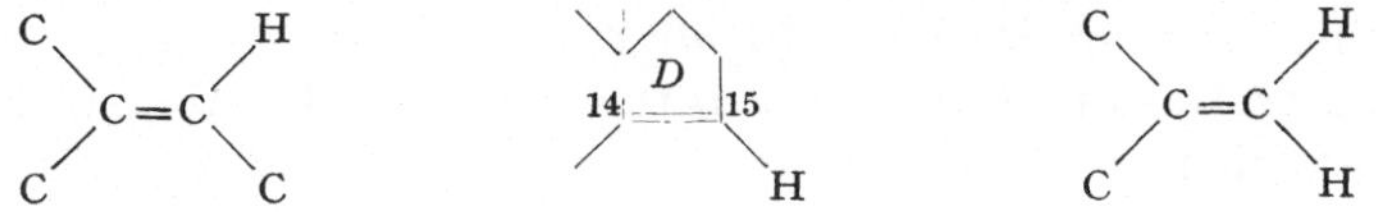

$$\text{(XIV.)} \qquad \text{(XV.)} \qquad \text{(XVI.)}$$

A type which is easily identified by its C—H stretching absorption is the 1,1-disubstituted double bond (XVI) which gives a well defined band close to 3070 cm.$^{-1}$ (3.257 μ). HENBEST, MEAKINS and WOOD (*82*) have made a special study of *cis*-disubstituted double bonds (XVII) in steroids

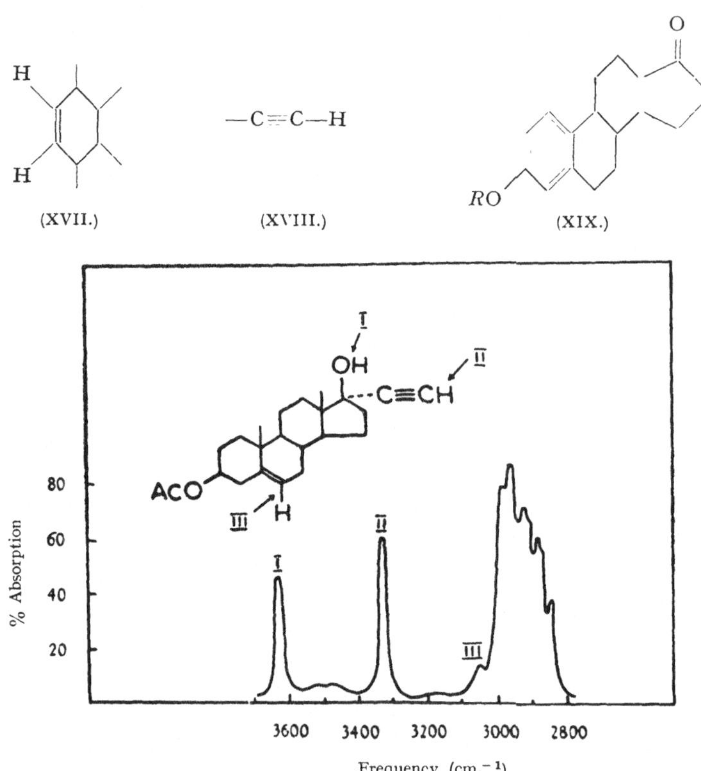

Fig. 15. Infrared spectrum in the region of O—H and C—H stretching vibrations (CCl$_4$ solution), according to COLE. [From: Rev. Pure Appl. Chem. (Australia) *4*, 111 (1954).]

and give the characteristic frequencies for Δ^1, Δ^2, Δ^3, Δ^6 and Δ^{11}-double bonds in the range 2995–3050 cm.$^{-1}$ (3.339–3.279 μ) (see Table 4, p. 41). The acetylenic C—H bond (XVIII) has a very high stretching frequency and absorbs at 3308–3310 cm.$^{-1}$ (3.023–3.021 μ) [*Figure 15;* JONES et al. (*106*)] while aromatic C—H bonds, which are found in estrogenic steroids (XIX) and benzoate esters, give two or three peaks between 3000–3100 cm.$^{-1}$ (3.33–3.23 μ). A number of these different types of C—H absorption are shown in *Figure 15*.

An absorption of a similar type is that due to a CH$_2$ group in a three-membered ring as found in the 3,5-*cyclo*steroids (i-steroids) (XX) [JOSIEN (*112*)]. This occurs near 3040–3055 cm.$^{-1}$ (3.289–3.273 μ) and

is very useful for distinguishing different degrees of substitution of
3-membered rings in large molecules [*Figure 16*; COLE (*37, 41, 42*)].

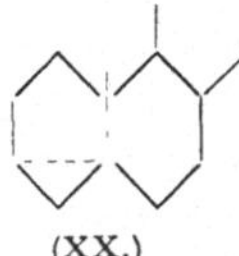

(XX.)

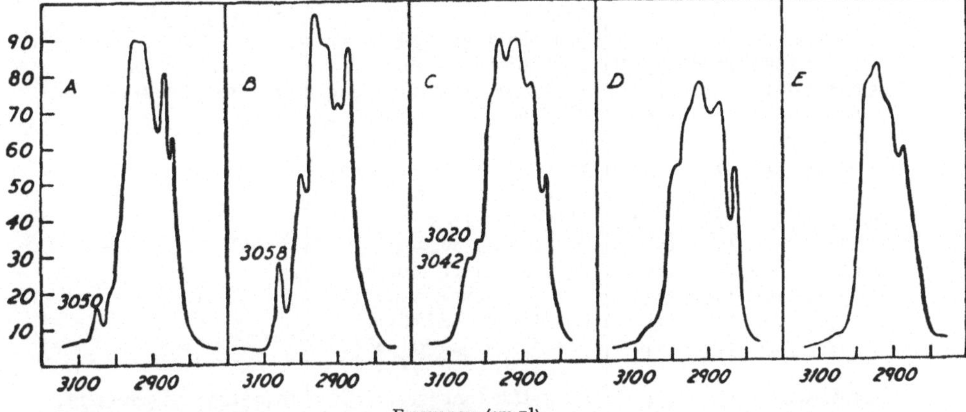

Frequency (cm.$^{-1}$)

Fig. 16. Infrared spectra in the C—H stretching region showing (in A, B, and C) the absorption near
3040—3055 cm.$^{-1}$ due to a CH_2 group in a three-membered ring, according to COLE: (A) 3,5-*cyclo*cholestane;
(B) thujane; (C) α-thujene; (D) car-3-ene; (E) lanosta-8,24-dienyl acetate. D has a fully substituted *cyclo*-
propane ring and E has no three-membered ring. [From: J. Chem. Soc. (London) *1954*, 3807.]

The C—D stretching absorption of synthetic deuterated steroids is
found in the range 2105–2275 cm.$^{-1}$ (4.75–4.40 μ) [DOBRINER et al. (*55*);
JONES, COLE and NOLIN (*98*); NOLIN and JONES (*124*)].

c) Carbonyl Absorption. By far the greater part of all infrared work
has been concerned with measurements in the region of carbonyl
stretching absorption between 1650 and 1800 cm.$^{-1}$ (6.0–5.5 μ) (*99, 102,
103, 105, 111*). This band is one of the most intense, and its exact position
has been correlated with a large number of different types of carbonyl
function. These frequencies are remarkably specific and values for
131 different types in steroids have been listed by JONES and HERLING
(*100*). Some of those encountered more frequently are given in *Table 2*.

Conjugation of a carbonyl group with ethylenic double bonds leads
to a lowering of frequency and an increase in intensity, both of the
carbonyl band and the double bond stretching band. With esters and
lactones, the presence of an unsaturated group on the opposite side of
the oxygen atom from the carbonyl group as in (XXI) causes an increase
in carbonyl frequency relative to the saturated compound. Thus, while

3*

Table 2. Carbonyl Absorption Positions in Steroids (*38, 100*).

CCl$_4$ and CS$_2$ solutions		Carbonyl Type	CHCl$_3$ solution	
μ	cm.$^{-1}$		cm.$^{-1}$	μ
5.615–5.627	1781–1777	γ-Lactone group in 20-spiro-(2′-oxa-3′-oxo-cyclopentano-) steroids	—	—
5.650	1770	Naphtholic 3-acetate	1762	5.675
5.659–5.669	1767–1764	Phenolic 3-acetate, propionate	1762–1753	5.675–5.705
5.669	1764	16,16-Dibromo-17-ketone..	—	—
5.688–5.695	1758–1756**	2,4-Dibromo-3-ketone.....	1752	5.708
5.688–5.701	1758–1754	*21-Acetoxy*-20-ketone*	1750–1745	5.714–5.731
5.688–5.701	1758–1754	$\Delta^{3,5}$-Diene-3-yl acetate ...	—	—
5.688–5.705	1758–1753	Δ^2-3-Enol acetate	—	—
5.688–5.721	1758–1748	Shoulder due to *monomeric* carboxylic acid*	—	—
5.695–5.718	1756–1749	$\Delta^{17(20)}$-20-Enol acetate....	—	—
5.698–5.708	1755–1752	Δ^{20}-20-Enol acetate	—	—
5.698–5.708	1755–1752	*12β-Acetoxy*-11-ketone* ...	—	—
5.701–5.708	1754–1752	Δ^{14}-17-Ketone	—	—
5.705–5.708	1753–1752	*11β-Acetoxy*-12-ketone* ...	—	—
5.705–5.721	1753–1748	*11α-Acetoxy*-12-ketone* ...	—	—
5.708	1752	*12α-Acetoxy*-11-ketone* ...	—	—
5.708–5.721	1752–1748	11,17-Diketone*	1742–1738	5.741–5.754
5.718	1749	$\Delta^{9(11)}$-11-Enol acetate	—	—
5.718	1749	16-Ketone..............	1742	5.741
5.718	1749	20,21-Diacetate	1739–1736	5.750–5.760
5.727	1746	Phenolic 3-benzoate, naphtholic 3-benzoate	1740	5.747
5.731–5.744	1745–1741	17-Ketone..............	1737–1733	5.757–5.770
5.734–5.741	1744–1742	δ-Lactone	—	—
5.741–5.760	1742–1736	*17β-Acetoxy*-20-ketone* ...	1733–1730	5.770–5.780
5.741–5.764	1742–1735	Methyl ester	1732–1728	5.774–5.787
5.750–5.764	1739–1735	Hexahydrobenzoate	1719	5.817
5.750–5.770	1739–1733	Acetate	1728–1719	5.787–5.817
5.757–5.764	1737–1735	2,2-Dibromo-3-ketone.....	1732–1725	5.774–5.797
5.760	1736	*17α-Acetoxy*-20-ketone* ...	—	—
5.760–5.800	1736–1724	21-Acetoxy-*20-ketone**	1727–1720	5.790–5.814
5.764	1735	11α-Bromo-12-ketone.....	—	—
5.764–5.770	1735–1733**	2-Bromo-3-ketone	1725–1722	5.797–5.807
5.764–5.770	1735–1733**	4-Bromo-3-ketone	1729	5.784
5.780–5.790	1730–1727	*12β-Acetoxy-11-ketone** ...	—	—
5.780–5.794	1730–1726	*11α-Acetoxy-12-ketone** ...	—	—
5.784–5.797	1729–1725	Formate	—	—

* These structures give rise to two carbonyl bands, both of which are mentioned in the Table. The group causing the absorption is italicized.

** The position of this band is dependent on the stereochemical relationship between the C=O and C-halogen bonds [JONES, RAMSAY, HERLING and DOBRINER (*109*)].

CCl₄ and CS₂ solutions		Carbonyl Type	CHCl₃ solution	
μ	cm.⁻¹		cm.⁻¹	μ
5.794	1726	11,12-Diketone	—	—
5.800	1724**	2-Iodo-3-ketone..........	—	—
5.800–5.824	1724–1717	Benzoate...............	1713–1710	5.838–5.848
5.804	1723	12-Ketoetiocholanic acid methyl ester	—	—
5.804–5.821	1723–1718	3,6-Diketone	—	—
5.807	1722	25-Ketone...............	1711	5.845
5.814	1720	12α-Acetoxy-*11-ketone** ...	—	—
5.814–5.848	1720–1710	11β-Acetoxy-*12-ketone** ...	—	—
5.817	1719	17α-Acetoxy-*20-ketone** ...	—	—
5.817–5.838	1719–1713	3-Ketone...............	1709–1700	5.834–5.882
5.817–5.838	1719–1713	*11,17*-Diketone*	1711–1707	5.845–5.858
5.838–5.848	1713–1710	7-Ketone...............	1707	5.858
5.821	1718	10-Aldehyde.............	—	—
5.824	1717	Δ¹⁶-*12,20*-Diketone*	—	—
5.824–5.828	1717–1716	24-Ketone...............	1709	5.851
5.828	1716	Δ¹⁵-17-Ketone	—	—
5.828	1716	17β-Acetoxy-*20-ketone** ...	1714	5.834
5.834–5.838	1714–1713	6-Ketone (*A/B trans* series)	—	—
5.834–5.851	1714–1709	12α-Bromo-11-ketone.....	—	—
5.841–5.862	1712–1706	12-Ketone...............	1705–1698	5.865–5.889
5.841	1712	4-Ketone...............	—	—
5.848–5.862	1710–1706	20-Ketone...............	1707–1698	5.858–5.889
5.848–5.869	1710–1704	11-Ketone...............	1705–1698	5.865–5.889
5.848–5.882	1710–1700	*Dimeric* carboxylic acid* .	—	—
5.855–5.862	1708–1706	6-Ketone (*A/B cis* series)	—	—
5.858–5.869	1707–1704	22-Ketone...............	1700	5.882
5.862	1706	11β-Bromo-12-ketone	—	—
5.865	1705	17α-Bromo-20-ketone.....	—	—
5.865–5.872	1705–1703	Δ⁸⁽¹⁴⁾-15-Ketone	—	—
5.875	1702**	2,4-Dibromo-Δ¹-3-ketone ..	1700	5.882
5.882	1700	Δ⁴-*3,6*-Diketone*.........	—	—
5.893	1697	2-Bromo-Δ¹-3-ketone	1685	5.935
5.893	1697**	2-Bromo-Δ⁴-3-ketone	1685–1680	5.935–5.952
5.900–5.921	1695–1689	3,5-*cyclo*-6-Ketone(*i*-steroid 6-ketone)	—	—
5.910	1692	Δ⁴-*3,6*-Diketone*.........	—	—
5.935	1685	16,17-Methylene-20-ketone	—	—
5.938–5.952	1684–1680	Δ¹-3-Ketone.............	1672–1670	5.981–5.988
5.938–5.952	1684–1680	Δ⁹⁽¹¹⁾-12-Ketone	1676–1671	5.967–5.984
5.945–5.974	1682–1674	Δ⁵-7-Ketone.............	1669–1666	5.992–6.602
5.949–5.963	1681–1677	Δ⁴-3-Ketone	1668–1660	5.995–6.024
5.959	1678	Δ¹⁶-*12,20*-Diketone	—	—
5.984–6.013	1671–1663	Δ¹·⁴-Diene-3-ketone.......	1666–1660	6.002–6.024
5.988–6.002	1670–1666	Δ¹⁶-20-Ketone	1662–1652	6.017–6.053
5.992–6.002	1669–1666	Δ⁴·⁶-Diene-3-ketone.......	—	—
5.999	1667	Δ⁸-7-Ketone	—	—
6.013	1663	Δ³·⁵-Diene-7-ketone.......	—	—
6.024	1660	Δ⁸-11-Ketone	—	—

saturated acetates absorb at 1733–1739 cm.$^{-1}$ (5.77–5.75 μ), the carbonyl band of enol acetates (**XXII**) is found near 1755 cm.$^{-1}$ (5.70 μ), and that of phenolic acetates near 1765 cm.$^{-1}$ (5.67 μ).

$$C\!=\!C\!-\!O\!-\!\underset{\underset{O}{\|}}{C}\!-\!C$$

(XXI.)   (XXII.)

The frequency ranges of the various carbonyl groups, although quite narrow (1–5 cm.$^{-1}$), overlap in many cases and it is sometimes necessary to examine bands in other parts of the spectrum to distinguish between ambiguous types. Five-membered ring ketones might be confused with methyl esters and acetates, but the ester groups have strong bands at lower frequencies [methyl ester 1135, 1155 cm.$^{-1}$ (8.81, 8.66 μ); acetate 1240 cm.$^{-1}$ (8.07 μ)] which enable them to be distinguished. Most of the six-membered ring carbonyls (at positions 3, 4, 6, 7, 11 and 12) and the 20-ketone group cannot be identified with certainty from their absorption in this region, but more information about some of them can be gained from the methyl and methylene bending region (p. 43). The Δ^{15}-17-ketone (1716 cm.$^{-1}$; 5.83 μ) can be distinguished from the six-membered ring carbonyls by its higher intensity (see *Table 3*) and by the very low C=C stretching frequency (1587 cm.$^{-1}$; 6.30 μ, *Table 4*, p. 41).

Table 3. Average Values for Integrated Absorption Intensities of Carbonyl Stretching Bands in Steroids (*110, 133*).

Carbonyl Type	Intensity (units)	Carbonyl Type	Intensity (units)
Saturated ketones:		Halogenated ketones:	
3-Ketone	2.55	2-Bromo-3-ketone	1.89
7-Ketone	2.16	2-Iodo-3-ketone	1.99
11-Ketone	2.21	4-Bromo-3-ketone	1.87
12-Ketone	2.27	2,2-Dibromo-3-ketone	1.80
16-Ketone	2.74	2,4-Dibromo-3-ketone	1.20
17-Ketone	2.69	Δ^4-2-Bromo-3-ketone	2.72
20-Ketone	1.79	Ester carbonyl:	
$\beta\gamma$-Unsaturated ketones:		Acyl acetate	3.24
Δ^5-3-Ketone	2.88	Phenolic acetate	2.75
Δ^{14}-17-Ketone	3.04	Methyl ester	3.13
$\alpha\beta$-Unsaturated ketones:		Benzoate	3.55 (*43*)
Δ^4-3-Ketone	3.65	Lactone carbonyl	
Δ^{15}-17-Ketone	3.67	δ-Lactone	4.1 (*43*)
Di-unsaturated ketones:			
$\Delta^{1,4}$-Diene-3-ketone	3.70		
$\Delta^{4,6}$-Diene-3-ketone	4.11		

Carbonyl Intensity measurements. RAMSAY and his collaborators (*110, 133*) have made a detailed study of the shapes and intensities of carbonyl absorption bands and their results are very useful in work on polycarbonyl compounds in which the various bands may not be resolved from one another. They have given values for the integrated intensity A, or the area under the absorption band, defined as:

$$A = (10^{-4}/cl) \int \log_e (I_0/I) d\nu$$

for a number of steroid types (*Table 3*) and among others have supplied all the necessary information for obtaining *true* absorption intensities

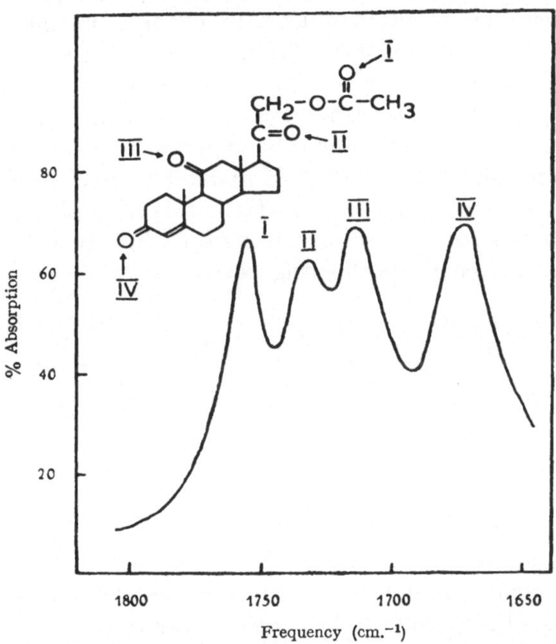

Fig. 17. Infrared spectrum of 11-dehydrocorticosterone acetate in the carbonyl stretching region (CS solution), according to JONES et al. [From: J. Amer. Chem. Soc. *71*, 241 (1949).]

corrected for limitations in instrumental resolving power. The band intensities are constant for each particular type and are additive in polycarbonyl compounds, provided the groups are far enough apart in the molecule so that their vibrations do not interact.

Three different situations may arise in polycarbonyl compounds:

a) **The groups absorb independently.** If the groups are separated by more than two carbon atoms they usually absorb at the same frequencies as in monocarbonyl compounds (*Table 2*, p. 36), and if the groups are of different types, the number of absorption bands tells the number of groups present. *Figures 17 and 20* (pp. 39, 46) give examples of this type.

b) The number of absorption bands is less than the number of carbonyl groups. Due to the natural width of carbonyl absorption bands the spectrometer will not resolve bands closer than about 10 cm.$^{-1}$. This is the case, for example, with 3,12-diketosteroids, but the additive intensity allows one to show that two groups are present.

c) The number of bands is as expected but their positions do not agree with monocarbonyl compounds. When groups with similar vibration frequencies are located close together in a molecule, interactions between them may lead to changes in their absorption frequencies [JONES and DOBRINER (99)]. Some of these changes are due to electronic interaction; e. g. the 21-acetoxy-20-ketone side-chain illustrated in *Figure 17* absorbs at 1758 and 1732 cm.$^{-1}$ (5.69 and 5.77 μ) whereas the groups separately might be expected to absorb at 1735 cm.$^{-1}$ (5.76 μ) (acetate) and 1707 cm.$^{-1}$ (5.86 μ) (20-ketone). Similar increases in frequency of 11,12-diketones (XXIII) and 11,17-diketones (XXIV) may be attributed to increase of ring strain when two doubly bonded groups are introduced in neighboring positions.

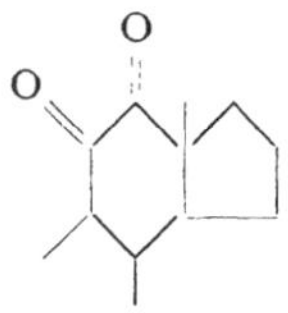

(XXIII.) 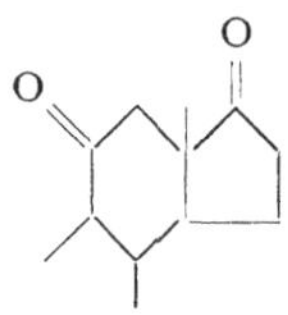   (XXIV.)

d) Ethylenic Double Bonds. Three regions of the infrared spectrum may be used to obtain information about ethylenic double bonds. The high C—H stretching frequency of ethylenic systems has been discussed earlier, and the other regions to be described are those of C=C stretching and C—H bending vibrations.

Since the bond has very little polarity, the stretching vibration of the C=C system does not interact strongly with infrared radiation and the resulting weak band also lies in a region of fairly intense atmospheric water vapour absorption (1580–1680 cm.$^{-1}$; 6.33–5.95 μ; Figure 1; p. 8). This means that when working with a single-beam instrument it is essential to dry the air in the spectrometer and also to record a background spectrum on each chart. Double beam spectrometers should be used wherever possible for this part of the spectrum.

In the steroid ring system there are numerous locations at which double bonds may be found and their stretching frequencies [BLADON et al. (*19*); HENBEST, MEAKINS and WOOD (*82*); JONES, HUMPHRIES,

Table 4. Characteristic Absorption Bands of Ethylenic Double Bonds in Steroids.

A. Isolated Double Bonds.

Position of C=C	C=C Stretching Vibrations ($CHCl_3$)		C=C—H Stretching Vibrations (CCl_4)		C—H Bending Vibrations (CS_2)		References
	μ	cm.$^{-1}$	μ	cm.$^{-1}$	μ	cm.$^{-1}$	
Δ^1	6.083	1644	3.339—3.310	2995—3021	14.29—13.26	700—754	(82)
Δ^2	6.050—6.035	1653—1657	3.296	3034	15.06—12.92	664—774	(82, 106)
Δ^3	6.072	1647	3.317	3015	14.90—12.94	671—773	(82)
Δ^4	6.035	1657	3.289	3040	12.35	810	(19)
Δ^5	5.999—5.981	1667—1672	3.300	3030	12.52—12.45	799—803	(19, 86, 106)
					12.32—12.29	812—814	
					12.05—11.90	830—840	
Δ^6	6.124—6.101	1633—1639	3.333—3.315	3000—3017	14.20—12.95	704—772	(82, 100)
Δ^7	6.010—6.002	1664—1666	3.319—3.289	3013—3040	12.09, 11.81	827, 847	(19, 106)
Δ^8	—	—	—	—	—	—	(19, 106)
$\Delta^{8(14)}$	—	—	—	—	—	—	(19, 106)
$\Delta^{9(11)}$	6.086—6.070	1643—1648	3.287	3042	12.09, 11.74	827, 852	(19, 100)
Δ^{11}	6.173—6.143	1620—1628	3.297—3.279	3033—3050	14.22—11.92	703—839	(82, 106)
Δ^{14}	6.075—6.070	1646—1648	3.273	3055	12.55, 12.39, 12.12	797, 807, 825	(19, 99)
Δ^{16}	6.169—6.135	1621—1630	—	—	—	—	(106)
Δ^{22}	6.010—6.002	1664—1666	—	—	10.31—10.27	970—974	(94, 100, 168)

B. Conjugated Double Bonds (106).

Position of C=C	C=C Stretching Vibrations ($CHCl_3$)	
	cm.$^{-1}$	μ
$\Delta^{3,5}$-Diene	1618, 1578	6.180, 6.337
$\Delta^{3,5}$-Diene-3-yl ester	1670—1671, 1639	5.988—5.984, 6.101
Δ^1-3-Ketone	1604—1609	6.234—6.215
Δ^4-3-Ketone	1615—1619	6.192—6.177
$\Delta^{9(11)}$-12-Ketone	1607	6.223
Δ^{15}-17-Ketone	1587	6.301
Δ^{16}-20-Ketone	1588—1592	6.297—6.281
$\Delta^{1,4}$-Diene-3-ketone	1621, 1603—1606	6.169, 6.238—6.227
$\Delta^{4,6}$-Diene-3-ketone	1616—1619, 1587	6.188—6.177, 6.301
$\Delta^{3,5}$-Diene-7-ketone	1627, 1598	6.146, 6.258

PACKARD and DOBRINER (106)] are given in *Table 4*. The fully substituted types (Δ^8, $\Delta^{8(14)}$), being symmetrical, give no detectable absorption. Conjugation with a carbonyl group lowers the frequency and increases the intensity of the C=C band. Carbon tetrachloride is suitable as solvent for compounds absorbing above 1630 cm.$^{-1}$ ($< 6.14\,\mu$) but its own absorption is troublesome below that frequency, since fairly thick cells must be used for double bond detection. Chloroform may be used for all double bond compounds, and C=C frequencies do not vary by more than about 2 cm.$^{-1}$ from one solvent to the other.

The degree of substitution of double bonds may be determined from the frequency of the bending vibration of the CH groups at the ends of the double bond [HENBEST, MEAKINS and WOOD (82); JONES, HUMPHRIES, PACKARD and DOBRINER (106); RASMUSSEN and BRATTAIN (136, 137); THOMPSON and TORKINGTON (163, 164)]. HIRSCHMANN and DAUS (50, 86) have made a special study of the Δ^5-trisubstituted types found in cholesterol compounds, while TURNBULL, WHIFFEN and WILSON (168), and JONES (94) have confirmed the *trans* nature of the Δ^{22} double bond in the side-chain of the ergosterol series (XXV).

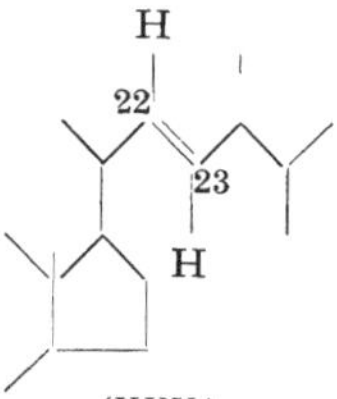

H
22
23
H

(XXV.)

As a general rule, increase of ring strain lowers the frequency of a cyclic double bond and increases that of an exocyclic C=C bond or a C—H group attached to a double bond [BLADON et al. (19)], and this is well illustrated by some of the values given in *Table 4*.

More information about double bonds can be obtained from the absorption of adjacent methylene groups which is described in the next Section, and this absorption will sometimes be helpful even in the case of the fully substituted types (Δ^8, $\Delta^{8(14)}$).

e) Methyl and Methylene Bending Vibrations. Much useful structural information can be obtained from the absorption bands in the range 1350–1470 cm.$^{-1}$ (7.41–6.80 μ) due to bending vibrations of methyl and methylene groups [RASMUSSEN (135)]. This region should be studied by using carbon tetrachloride solutions since carbon disulphide, chloroform and Nujol all absorb strongly here. Mulling agents such as perfluorokerosene or hexachlorobutadiene are also suitable if the sample is not soluble in carbon tetrachloride. The dispersive power of the calcium fluoride prism is particularly high in this part of the spectrum and a large amount of spectral detail can be resolved.

The CH$_2$ group has its bending frequency (XXVI) near 1450 cm.$^{-1}$ (6.90 μ) and the absorption due to this is well illustrated by the spectrum of *cyclo*hexane *(Figure 18 A)*. Methyl groups should have two bands, due respectively to symmetrical and antisymmetrical bending modes, but the latter ($\sim$ 1450 cm.$^{-1}$; 6.90 μ) is usually obscured by the CH$_2$

absorption. The symmetrical vibration (XXVII) leads to a band near 1375 cm.$^{-1}$ (7.27 μ) (e. g. methyl*cyclo*hexane, 1376 cm.$^{-1}$, Figure 18 B). These two types of absorption band are present in steroid spectra

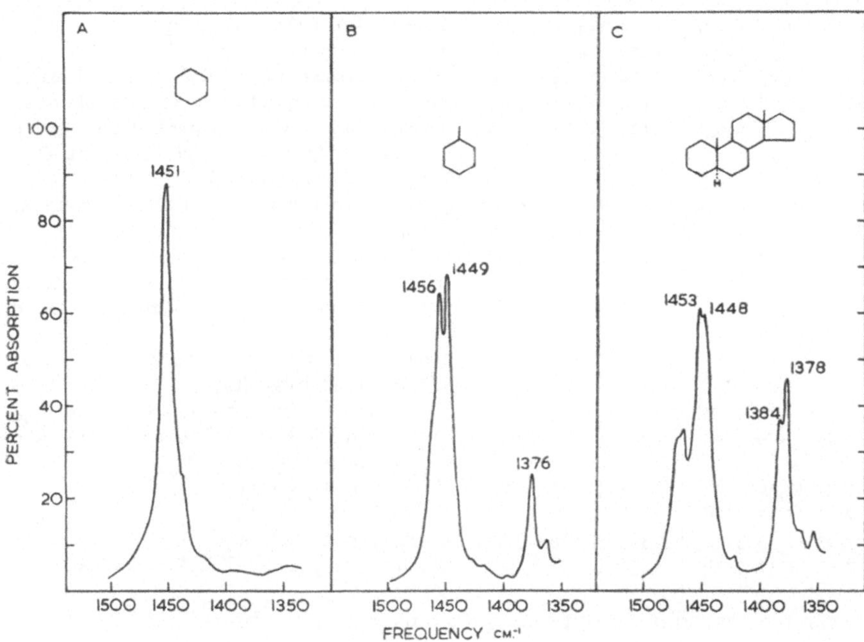

Fig. 18. Infrared spectra in the methyl-methylene bending region, according to JONES and COLE: (A) *cyclo*-hexane; (B) methyl*cyclo*hexane; (C) androstane. [From: J. Amer. Chem. Soc. **74**, 5648 (1952).]

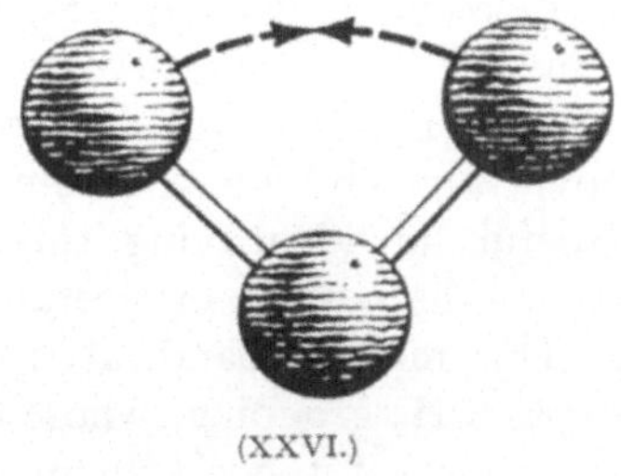

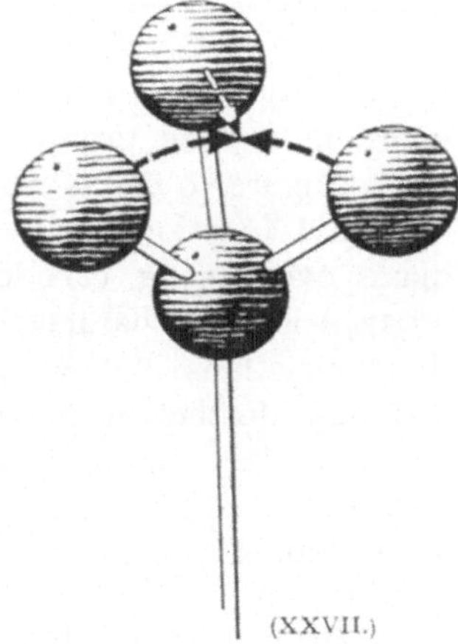

(Figure 18 C) and the study of a large number of compounds [JONES and COLE (97)] together with selective deuterium substitution [JONES, COLE and NOLIN (98)] has allowed bands to be assigned to individual CH$_2$ and CH$_3$ groups *(Table 5)*.

Some of these bands are useful in distinguishing between carbonyl groups which have very close carbonyl stretching frequencies. For

Table 5. Methyl and Methylene Bending Vibrations in Steroids (CCl_4 solution) (38, 97).

Wavelength (μ)*	Frequency (cm.$^{-1}$)*	Structure Assignment
6.81 (6.84–6.80)	1468 (1462–1470)	Side-chain methylene groups
6.90 (6.92–6.83)	1450 (1446–1464)	Ring system methylene groups
6.95 (6.97–6.94)	1438 (1435–1440)	Always present in spectra of methyl esters
6.95 (6.98–6.92)	1438 (1432–1445)	Methylene group adjacent to a double bond
6.97 (7.01–6.95)	1434 (1426–1438)	Ketosteroids with α-methylene group adjacent to a carbonyl group at $C_{(4)}$, $C_{(6)}$, $C_{(7)}$, $C_{(11)}$ or $C_{(12)}$
7.03 (7.07–7.01)	1422 (1415–1426)	3-Ketosteroid with α-methylene group at $C_{(2)}$ or $C_{(4)}$
7.10 (7.12–7.09)	1408 (1404–1411)	17-Ketosteroid with α-methylene at $C_{(16)}$, or quite generally, a methylene adjacent to a carbonyl group on a five-membered ring
7.22 (7.28–7.18)	1385 (1374–1392)	Angular methyl group between two six-membered rings ($C_{(10)}$)
7.25 (7.28–7.22)	1380 (1374–1386)	Side-chain methyl group at $C_{(21)}$ and $C_{(28)}$
7.26 (7.29–7.23)	1377 (1372–1383)	Angular methyl group between a five- and a six-membered ring ($C_{(18)}$)
7.27 (7.30–7.24)	1375 (1369–1382)	Acetate methyl group
7.31 (7.35–7.28)	1368 (1360–1374)	Gem. dimethyl group at end of side-chain
7.33 (7.35–7.30)	1365 (1360–1370)	Acetate methyl group
7.37 (7.37–7.36)	1357 (1356–1359)	Methyl group of —$COCH_3$ side-chain

example, carbonyl groups at positions 3 and 7 absorb near 1714 and 1710 cm.$^{-1}$ (5.83 and 5.85 μ) respectively, and these frequencies are too close for unequivocal identification. However, the adjacent methylene groups at $C_{(2)}$ and $C_{(4)}$ absorb near 1420 cm.$^{-1}$ (7.04 μ) (*Figure 19 A*) while that at $C_{(6)}$ absorbs near 1435 cm.$^{-1}$ (6.97 μ), and these bands serve to identify the carbonyls. The band near 1410 (7.09 μ), due to one or more methylene groups adjacent to a carbonyl on a five-membered ring (Figure 19 B), is particularly useful for identifying this type of carbonyl function in the presence of methyl esters or acetates which have overlapping carbonyl bands. The methyl band at 1357 cm.$^{-1}$ (7.37 μ) is very characteristic of the —$COCH_3$ grouping, whose carbonyl frequency falls in the same range as those of the six-membered ring ketones. Rather surprisingly, the methyl group of acetates gives *two* bands in this region, at 1375 and 1365 cm.$^{-1}$ (7.27 and 7.33 μ), but the cause of this doubling is not known. *Figure 20* shows the assignment of all bands in this region of the spectrum of a fairly complex steroid.

---

* The first figure is the average band position as determined from compounds in which there is no appreciable overlap with neighbouring absorption bands. The figures in parentheses indicate the extreme range found and include cases where the band may be reduced to an inflection by overlap.

f) Bands in the "Fingerprint" Region. With the exception of some fairly intense bands between 1200 and 1300 cm.$^{-1}$ (8.33–$7.70\,\mu$) related to acetates and other esters [JONES, HUMPHRIES, HERLING and DOBRINER (*104*); THOMPSON and TORKINGTON (*165*)], a band near 1000–1050 cm.$^{-1}$ (10.00–$9.50\,\mu$) due to the C—O stretching vibration

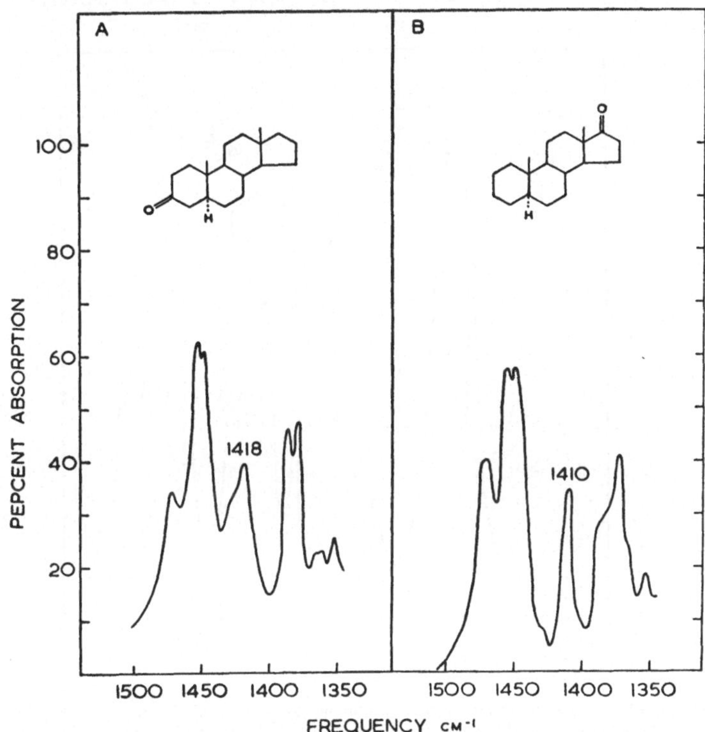

Fig. 19. Infrared spectra illustrating the effect of carbonyl groups on adjacent methylene groups, according to JONES and COLE: (A) androstan-3-one; (B) androstan-17-one. [From: J. Amer. Chem. Soc. *74*, 5648 (1952).]

of secondary alcohols [FURCHGOTT, ROSENKRANTZ and SHORR (*70*, *71*); ROSENKRANTZ and ZABLOW (*140a*)], and some bands between 1000 and 650 cm.$^{-1}$ (10.00–$15.4\,\mu$) due to out-of-plane C—H vibrations of ethylenic and aromatic systems, very little has been reported on the assignment of bands in the lower frequency region (1300–650 cm.$^{-1}$; 7.7–$15.4\,\mu$) in compounds composed of carbon, hydrogen and oxygen. Indeed, this may well be impossible in the case of molecules as complex as steroids. However, it has been shown that the intense peaks in this region are mostly due to parts of the molecules containing oxygenated groups.

Jones, Herling and Katzenellenbogen (*101*) have analysed the absorption patterns of a large number of ketosteroids and have listed the bands which appear to be characteristic of specific structures. No vibrational assignments were made, but it was pointed out that many of these bands will be useful in distinguishing between types of carbonyl functions which absorb at the same frequency in the carbonyl region.

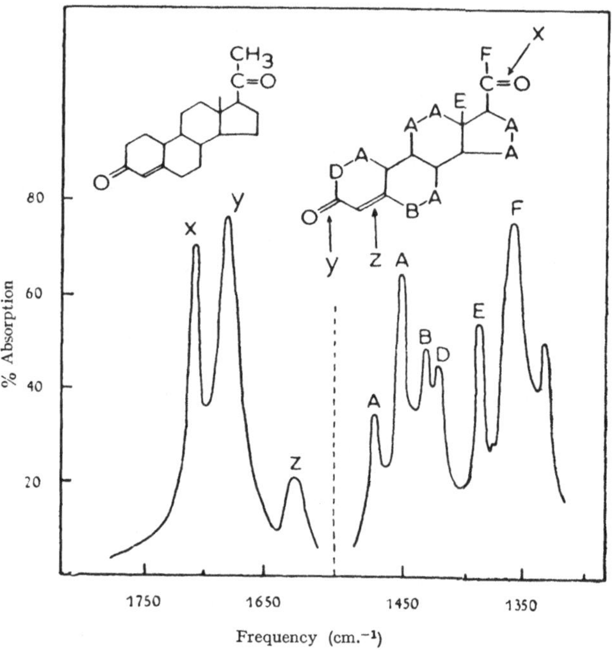

Frequency (cm.$^{-1}$)

Fig. 20. Infrared spectrum of Δ^4-19-norpregnene-3,20-dione in the carbonyl stretching and C—H bending regions (CCl$_4$ solution), according to Jones and Cole. (Letters on structural formula indicate the groups causing the particular absorption. [From: J. Amer. Chem. Soc. *74*, 5648 (1952).]

For example, the Δ^1-3-keto-allosteroids (XXVIII) and Δ^4-3-keto-steroids (XXIX) both have C=O stretching bands between 1684 and 1677 cm.$^{-1}$ (5.94 and 5.96 μ) but they can be readily identified by strong bands at 779–777 cm.$^{-1}$ (12.84–12.87 μ) and 867–862 cm.$^{-1}$ (11.53–11.60 μ) respectively. Similarly the $\Delta^{1,4}$-diene-3-ketosteroids (XXX; 888–887 cm.$^{-1}$; 11.26 μ) may be distinguished from the $\Delta^{4,6}$-diene-3-keto compounds (XXXI; 875–874 cm.$^{-1}$; 11.43 μ).

(XXVIII.)   (XXIX.)   (XXX.)   (XXXI.)

Similar analyses of this spectral region have been carried out for the steroidal sapogenins by EDDY and his collaborators (*58, 142, 170*) and by JONES, KATZENELLENBOGEN and DOBRINER (*107*). Frequencies

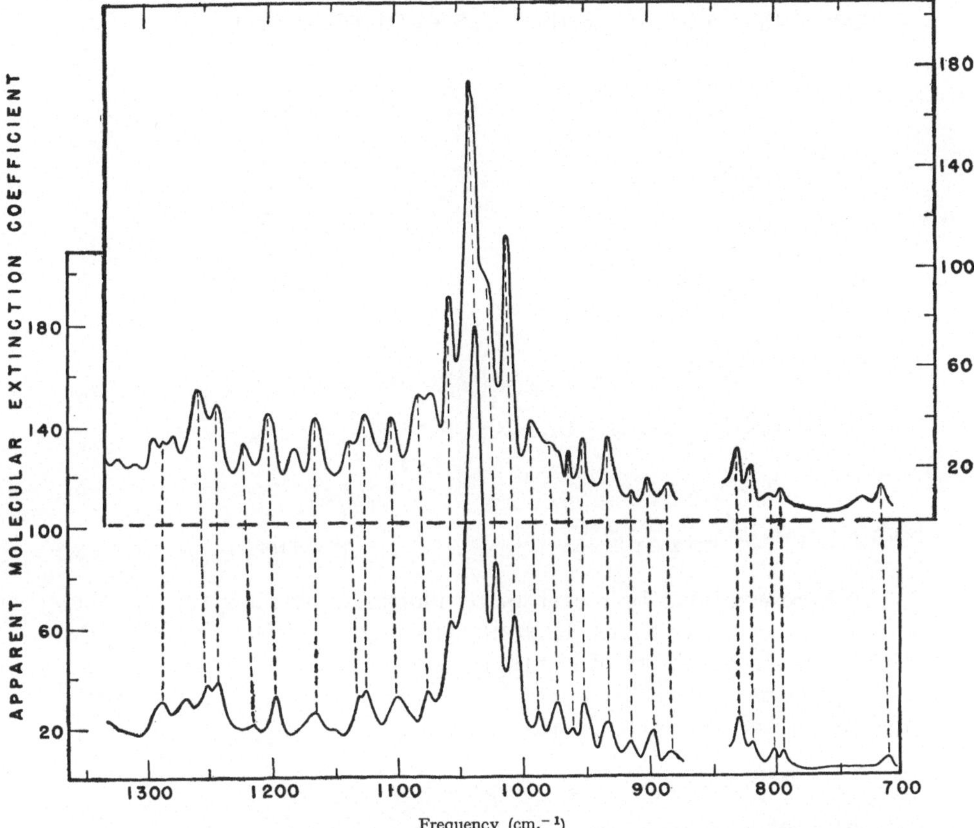

Fig. 21. Comparison of infrared spectrum of 3β-hydroxyandrostan-17-one (lower curve) with the curve obtained by adding the spectra of androstan-3β-ol and androstan-17-one and subtracting the spectrum of androstane (upper curve), according to COLE et al. [From: J. Amer. Chem. Soc. 74, 5571 (1952).]

are given for distinguishing between the normal sapogenins (XXXII) and the iso-sapogenins (XXXIII) and for identifying various substituent groups.

(XXXII.) Sapogenins.   (XXXIII.) Iso-sapogenins.

COLE, JONES and DOBRINER (*45*) have shown that the spectrum of a steroid containing two well separated oxygenated functions such as (XXXIV) can be approximated very well by the addition of the spectra of the two monosubstituted steroids (XXXV, XXXVI) and subtraction of that of the basic hydrocarbon (XXXVII) *(Figure 21)*.

O

(XXXIV.)   (XXXV.)

O

H   H
(XXXVI.)   (XXXVII.)

This procedure has been even better illustrated by the spectra of the steroidal sapogenins [JONES, KATZENELLENBOGEN and DOBRINER (*107*)]. This raises the possibility of identifying a new compound by comparison of its spectrum with a "synthetic" spectrum of this kind, but no such application has yet been described.

B. *Terpenoids.*

Although terpenoid compounds occur very widely and have been the subject of numerous chemical investigations, their spectra have not been studied as systematically as those of the steroids. This is perhaps because of the difficulty of obtaining a sufficiently large number of pure compounds based on the same molecular skeleton. Some acyclic terpenes were examined by THOMPSON and WHIFFEN (*166*) and re-examined by BARNARD, BATEMAN, HARDING, KOCH, SHEPPARD and SUTHERLAND (*4*), and by CARROL, MASON, THOMPSON and WOOD (*30*). A number of spectra of terpenes have also been presented in papers by PLÍVA and his collaborators (*128–131*), while the carotenoids have been studied by LUNDE and ZECHMEISTER (*116a*) (p. 58). The tetracyclic triterpenoids related to lanostadienol (lanosterol, XXXVIII) may be regarded as 4,4,14-trimethylsteroids and these

compounds offer a field for the extension of the steroid infrared work.
Similarly a large number of pentacyclic compounds related to α-amyrin

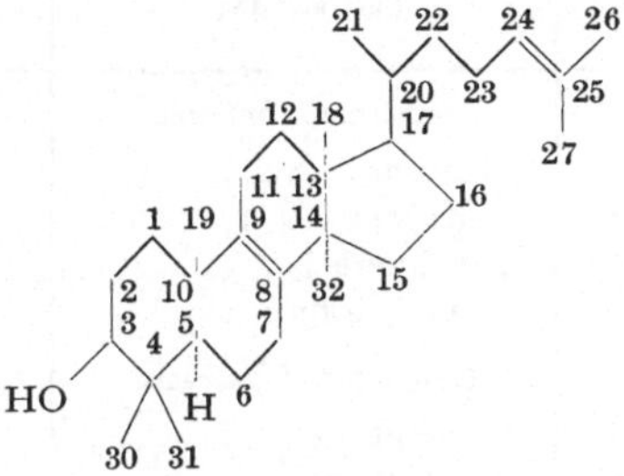

(XXXVIII.) Lanosterol.

(XXXIX or XL) [BEATON, SPRING, STEVENSON and STRACHAN (*16*);
MEISELS, RÜEGG, JEGER and RUZICKA (*121*); MEAKINS (*120*)], β-amyrin
(XLI) and lupeol (XLII) have now been made and infrared studies will
be useful in conjunction with chemical work in that field.

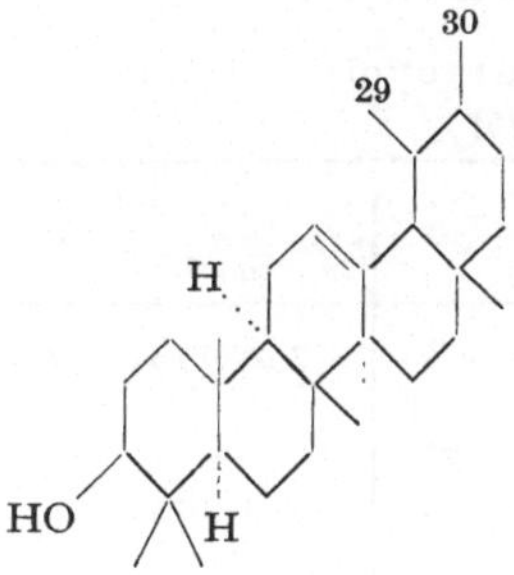

(XXXIX.)

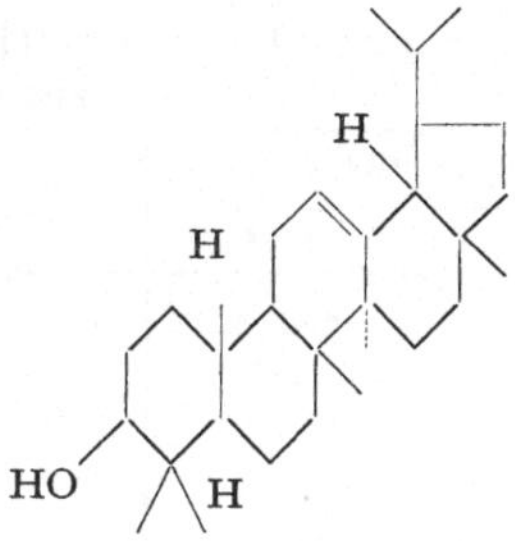

or

(XL.)

α-Amyrin.

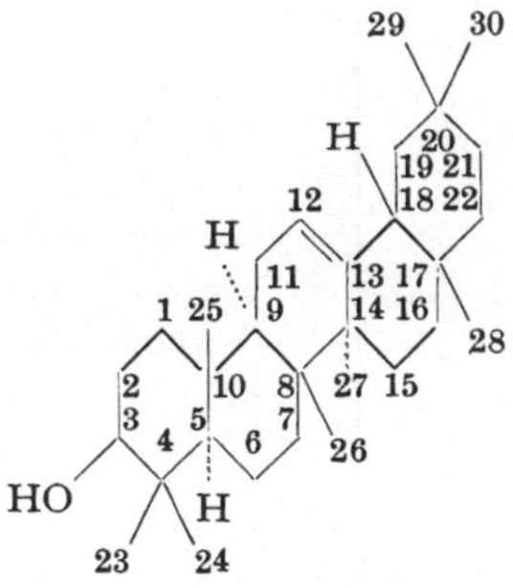

(XLI.) β-Amyrin.

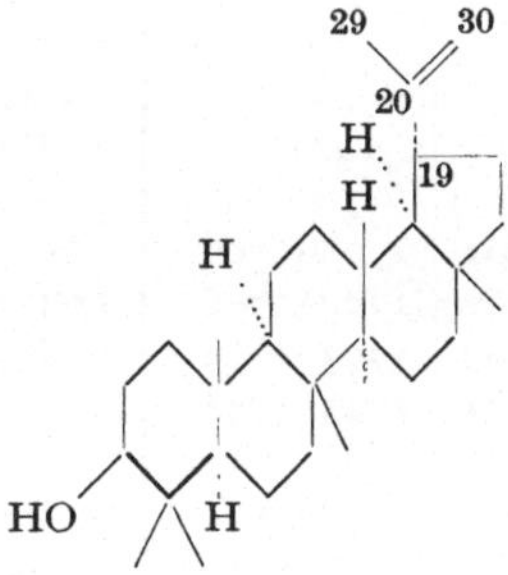

(XLII.) Lupeol.

The infrared spectra of the triterpenoids are very similar to those
of the steroids, with rather stronger absorption in the regions of methyl

Table 6. Carbonyl Band Positions for Tetracyclic Triterpenoids [Cole and Willix (*48*)].

CCl₄ solution		Carbonyl Type	CHCl₃ solution	
μ	cm.⁻¹		cm.⁻¹	μ
		Saturated ketones.		
5.855	1708	3-Ketone..............	1705	5.865
5.862–5.872	1706–1703	11-Ketone............	1703	5.872
5.855–5.865	1708–1705	7,11-Dione	1703	5.872
5.845	1711	3,7,11-Trione	1703	5.872
		Conjugated ketones.		
5.988–6.006	1670–1665	Δ^8-7-Ketone	1660–1653	6.024–6.050
5.995–5.999	1668–1667	Δ^5-7-Ketone	1663	6.013
6.035	1657	Δ^8-11-Ketone	1646	6.075
5.963–5.967	1677–1676	Δ^8-7,11-Dione.........	1673	5.977
5.977 6.046	1673 ⎱ 1654 ⎰	$\Delta^{5,8}$-Diene-7,11-dione ...	⎧ 1671 ⎩ 1648	5.984 6.068
5.747–5.757 5.931–5.938 6.039–6.061	1740–1737 ⎫ 1686–1684 ⎬ 1656–1650 ⎭	$\Delta^{5,8}$-Diene-7,11,12-trione	⎧ 1736–1732 ⎨ 1682–1681 ⎩ 1650–1646	5.768–5.774 5.945–5.949 6.061–6.075
6.072	1647	$\Delta^{5,8,11}$-Triene-7-ketone ..	—	—

Table 7. Carbonyl Band Positions for Pentacyclic Triterpenoids [Cole and Thornton (*46*)].

CCl₄ solution		Carbonyl Type	CHCl₃ solution	
μ	cm.⁻¹		cm.⁻¹	μ
5.634–5.643	1775–1772	γ-Lactone	1756–1751	5.695–5.711
5.747	1740	A,23,24-Trisnor-3-ketone (five-membered ring) .	—	—
5.757–5.774	1737–1732	Acetate	1725–1717	5.797–5.824
5.774	1732	Aldehyde	—	—
5.770–5.804	1733–1723	Methyl ester	1725–1714	5.797–5.834
5.794–5.804	1726–1723	Formate	1714	5.834
5.814–5.824	1720–1717	Benzoate.............	1708–1707	5.855–5.858
5.841	1712	19-Ketone (oleanane series)	—	—
5.841	1712	20-Ketone (30-norlupane series)	—	—
5.851–5.862	1709–1706	3-Ketone.............	1700–1698	5.882–5.889
5.858–5.862	1707–1706	11-Ketone............	1703	5.872
5.872–5.889	1703–1698	12-Ketone............	1696–1692	5.896–5.910
5.731–5.741 5.896–5.928	1745–1742 ⎱ 1696–1687 ⎰	Carboxylic acid........	1714	5.834
5.896	1696	Conjugated aldehyde ...	1690	5.917
5.851 5.910	1709 ⎱ 1692 ⎰	$\Delta^{13(18)}$-12,19-Diketone...	—	—
5.921	1689	$\Delta^{13(18)}$-19-Ketone (oleanane series)	1684	5.938
6.006	1665	Δ^{12}-11-Ketone	1652	6.053
6.013	1663	$\Delta^{12,18}$-Diene-11-ketone ..	1649	6.064

Table 8. Absorption of Methyl and Methylene Groups Adjacent to Carbonyl Groups in Tetracyclic Triterpenoids (CCl₄ solutions) [Cole and Willix (48)].

	cm.$^{-1}$	μ	Position of CH_2
3-Ketone	1427	7.01	2
11-Ketone	1433–1432	6.98	12
7,11-Dione	1438–1432	6.95–6.98	6, 12
Δ^8-7-Ketone	1420–1417	7.04–7.06	6
Δ^8-11-Ketone	1418	7.05	12
Δ^8-7,11-Dione	1429–1426	7.00–7.01	6, 12
$\Delta^{5,8}$-Diene-7,11-dione	1426	7.01	12
A-nor-3-Ketone (five-membered ring)	1407	7.11	2
20-Ketone	1357	7.37	(CH_3 absorption of —$COCH_3$)

Table 9. Absorption of Methyl and Methylene Groups Adjacent to Carbonyl Groups in Pentacyclic Triterpenoids (CCl₄ solutions) [Cole and Thornton (46)].

	cm.$^{-1}$	μ	Position of CH_2
11-Ketone	1430	6.99	12
3-Ketone	1429	7.00	2
12-Ketone	1420	7.04	11
A-Nor-3-ketone	1413	7.08	2
30-Nor-20-ketone	1354	7.39	(CH_3 absorption of —$COCH_3$)

Table 10. Absorption of Methylene Groups Adjacent to Ethylenic Double Bonds in Tetracyclic Triterpenoids (CCl₄ solutions) [Cole and Willix (48)].

Structure	cm.$^{-1}$	μ	Position of CH_2
Δ^8	1435–1434	6.97	7, 11
$\Delta^{7,9(11)}$	1432	6.98	6, 12
Δ^8-11-Ketone	1428	7.00	7
Δ^8-7-Ketone	1425	7.02	11
$\Delta^{2,5,8}$-Triene-7,11,12-trione	1412	7.08	2

stretching and bending vibrations. The increase in intensity of the CH_3 band at 2962 cm.$^{-1}$ (3.376 μ) makes small shoulders due to the tri-substituted double bond system (XIV) rather difficult to detect.

Carbonyl frequencies for a number of tetracyclic triterpenoids are given in *Table 6* and for pentacyclic compounds in *Table 7*. These are based on recent work in this laboratory and a more complete discussion will be published later. It should be noticed that the carbonyl frequencies are not all identical with those at equivalent positions on the steroid ring system, due most probably to the different mass distribution introduced by the extra methyl groups. Differences from steroid results are also apparent in the study of the "adjacent methylene" absorption (*Tables 8 and 9*, cf. Table 5, p. 44) in carbonyl and unsaturated compounds *(Table 10)*.

Quantitative measurements in the region of methylene bending absorption have been used to determine the number of CH_2 groups adjacent to a five-membered ring carbonyl group on compounds (XLIII) related to lanosterol [Barnes, Barton, Cole, Fawcett and Thomas (5)] and on (XLIV) derived from phyllocladene (diterpene) [Bottomley, Cole and White (24)].

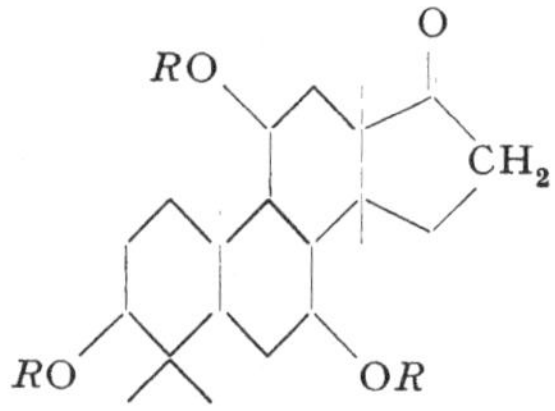

(XLIII.) R = Ac, Bz. 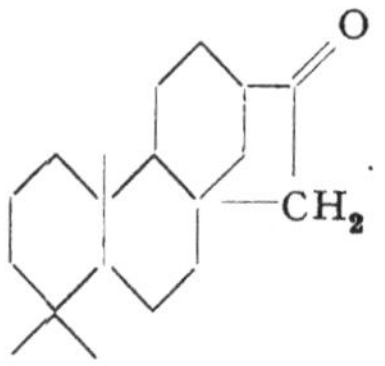 (XLIV.)

A very interesting application of quantitative measurements combined with selective *deuterium* substitution was described by Barton, Warnhoff and Page (12, 14). The tetracyclic triterpenoid *cyclo*artenol was shown by chemical tests to contain a *cyclo*propane ring close to $C_{(9)}$ and the final choice had to be made between five possible structures (XLV) –(XLIX). Of these, (XLVII)–(XLIX) were eliminated by Cole's

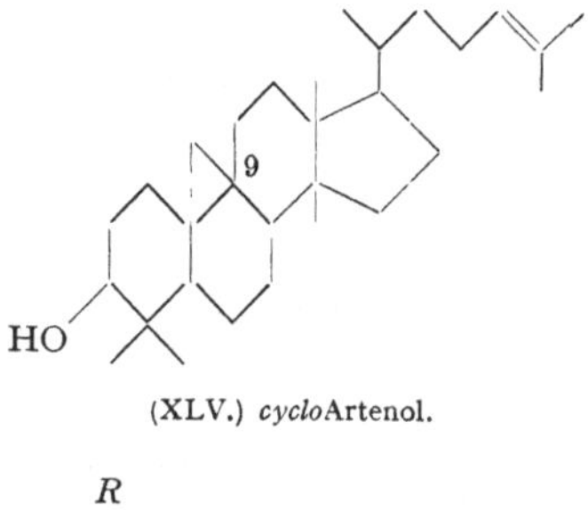

(XLV.) *cyclo*Artenol.

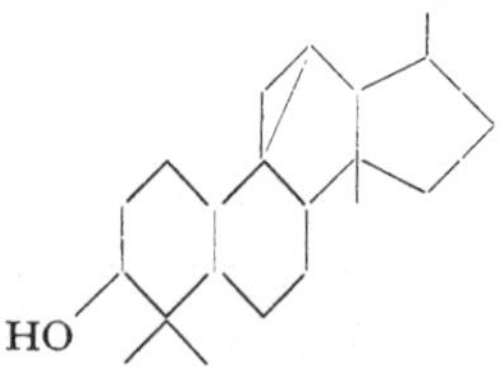

(XLVI.) 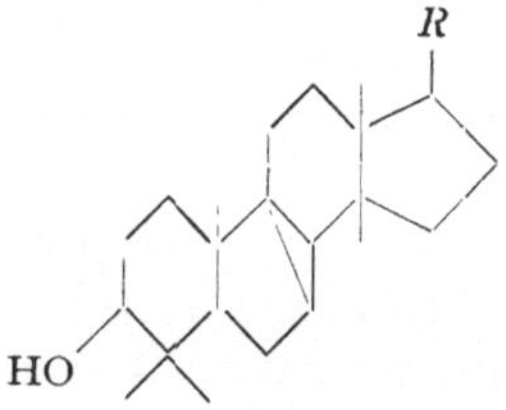 (XLVII.)

observation *(42)* of a band at 3042 cm.$^{-1}$ (3.287 μ) indicating that in *cyclo*artenol the *cyclo*propane ring contains an unsubstituted CH_2 group. Treatment of *cyclo*artenol with deuterium chloride opened the three-membered ring, and the reduced intensity of the methyl bending absorption at 1380 cm.$^{-1}$ (7.25 μ), compared with other triterpenoids, indicated that the deuterium had gone into a methyl rather than a methylene group. This meant that the product was (L) and not (LI), proving that *cyclo*artenol is represented by (XLV).

HENRY and SPRING *(83)*, at about the same time, reached this conclusion chemically by oxidizing the product to the conjugated ketone (LII) and showing by mass spectrographic analysis that the ketone still contained deuterium.

C. Studies of Stereochemistry.

In general, infrared measurements do not tell very much about the spacial arrangements of the skeletons and substituent groups in asymmetrical large molecules. In limited fields, however, some stereochemical information can be obtained.

a) Axial and Equatorial Hydroxyl Groups. FURCHGOTT, ROSEN-KRANTZ and SHORR *(70, 71)* noticed that the spectra of 3-hydroxysteroids varied systematically in the region 1000–1060 cm.$^{-1}$ (10.00–9.45 μ) depending on the orientation of the hydroxyl group and the nature of the *A/B* ring junction. These results were paralleled by the work of JONES, HUMPHRIES, HERLING and DOBRINER *(104)* on the absorption near 1240 cm.$^{-1}$ (8.07 μ) in the spectra of 3-acetoxysteroids. This band was found to be simple and symmetrical for some compounds and to consist of a number of peaks (complex) for others.

These effects were at first interpreted in terms of *cis* and *trans* relationships between the 3-hydroxy (or acetoxy) group and the 5-hydrogen

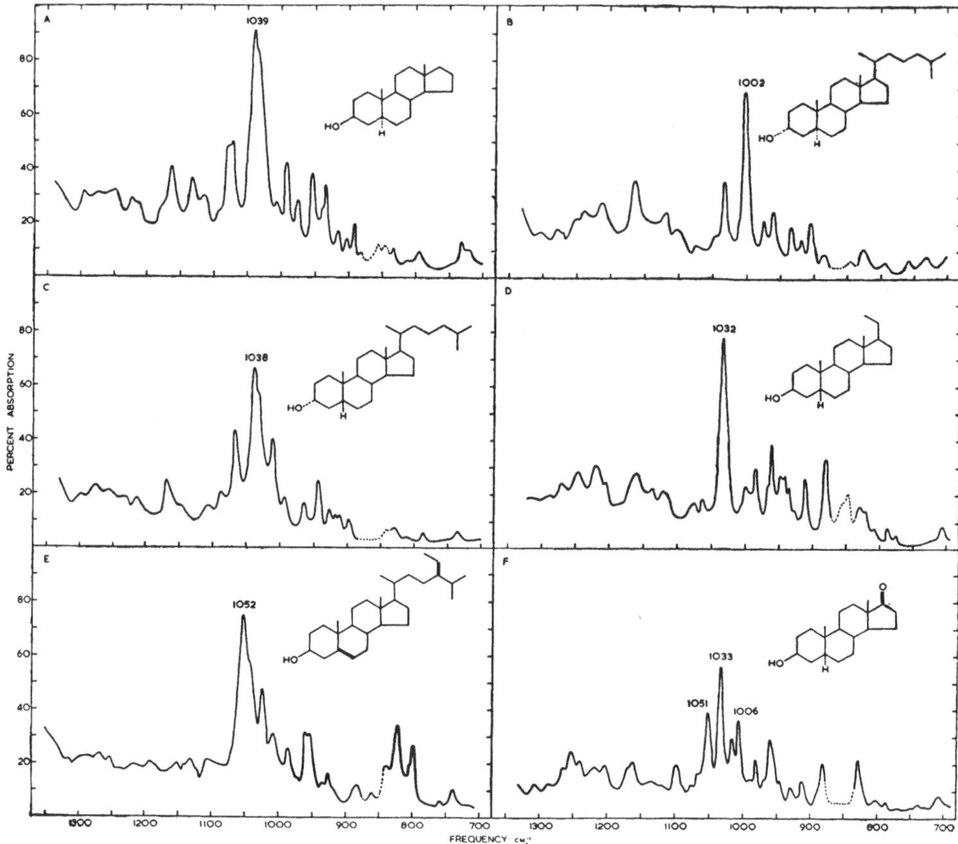

Fig. 22. Infrared spectra illustrating the intense band due to the C—O stretching vibration in stereoisomeric secondary hydroxyl groups, according to COLE et al.: (A) equatorial OH, *A/B trans;* (B) axial OH, *A/B trans;* (C) equatorial OH, *A/B cis;* (D) axial OH, *A/B cis;* (E) equatorial OH, Δ^5 C=C; (F) axial OH, *A/B cis,* in the presence of 17-keto group. [From: J. Amer. Chem. Soc. *74,* 5571 (1952).]

atom. Later results on 2- and 4-hydroxy compounds [FÜRST, KUHN, SCOTONI and GÜNTHARD (*72*)], some apparently anomalous results for lumistanol and a more complete study of the 3-hydroxysteroids [COLE, JONES and DOBRINER (*45*); HIRSCHMANN (*86*); ROSENKRANTZ and SKOGSTROM (*140*)] have made it apparent that these spectra are related to whether the substituent groups are axial or equatorial with respect to Ring *A*. Equatorial acetoxy groups give a simple peak near

1240 cm.$^{-1}$ (8.07 μ) and axial groups a set of two or three peaks between 1220 and 1265 cm.$^{-1}$ (8.20 and 7.91 μ), but no explanation has been given for the complex nature of the axial spectrum. Quantitative measurements led JONES et al. (*104*) to suggest that rotational isomerism might be responsible. It might be mentioned here that the band is always simple in the spectra of 3-axial-acetoxy tetracyclic triterpenoids where the presence of the gem-dimethyl group at C$_{(4)}$ must be taken into account [COLE (*43*)].

The relation between the frequency of the intense absorption band between 1000 and 1060 cm.$^{-1}$ (10.00 and 9.45 μ) in the spectra of the 3-hydroxysteroids *(Figure 22)* and the stereochemistry of the compounds can be explained more easily. This band is due to the stretching

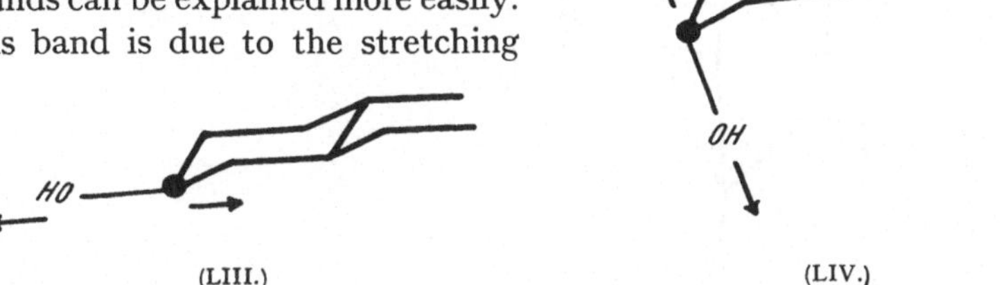

vibration of the C—OH bond and *Table 11* gives its position for a number of structures. In general the C—O frequency is higher for equatorial than for axial compounds and the reason for this can be seen from an examination of (LIII) and (LIV). During the vibration the C$_{(3)}$ atom moves towards the centre of the ring in the equatorial case and the resulting widening of the ring angles will produce a greater restoring

Table 11. Stereochemical Characterization of
3-Hydroxy-steroids (*45, 72, 104, 140a*).

A/B ring fusion	C—O bond configuration and conformation*	C—O frequency (cm.$^{-1}$) (CS$_2$ solution)	Wavelength μ	Acetate band type
trans	3β (e)	1037–1040	9.64–9.62	simple
trans	3α (a)	996–1002	10.04–9.98	complex
cis	3α (e)	1037–1044	9.64–9.58	simple
cis	3β (a)	1032–1036	9.69–9.65	complex
Δ^5 C=C	3β (e)	1050–1052	9.52–9.51	simple
Δ^5 C=C	3α (a)	1034**	9.67	complex
trans	2α (e)	1030–1035**	9.71–9.66	simple
trans	2β (a)	1010**	9.90	complex
trans	4α (e)	1040**	9.62	complex
trans	4β (a)	1000	10.00	complex

   * e = equatorial, a = axial.
  ** Nujol mull.

force (higher frequency) than in the axial compound where the motion involves a perpendicular bending of the ring which can be distributed right to the A/B junction. The L-shape of the A/B *cis* steroids (VIII, p. 29)

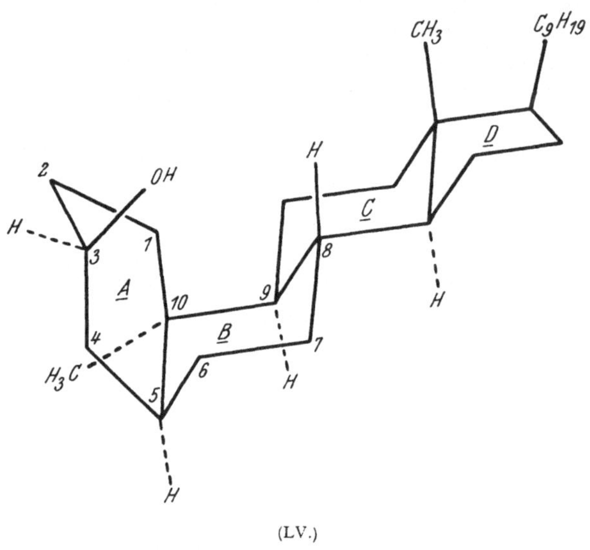

(LV.)

probably makes the ring system more rigid and is responsible for the smaller difference between the equatorial and axial frequencies compared with the A/B *trans* compounds.

Measurements in this region of the spectra of lumistan-3β-ol (1010 cm.$^{-1}$; 9.90 μ) and lumistan-3α-ol (epi-lumistanol, 1034 cm.$^{-1}$; 9.67 μ) have indicated that in the former the hydroxyl is axial, and when considered with the chemical knowledge of the molecule, this enabled the complete structure of lumistanol to be described [(LV); Cole (*35*)].

Changes in C—O frequency on epimerization of a hydroxyl group have also been used by Aebi, Barton and Lindsey (*1*) in the terpene field.

b) α-Bromo-ketones. Tables 2 and 3 (pp. 36, 38) include data on α-bromo-ketones from Jones, Ramsay, Herling and Dobriner (*109*) who have shown that the effect of an α-bromo-substituent on the carbonyl frequency and intensity is related to the stereochemistry of the C—Br bond. If the bromine substituent is equatorial (and therefore in approximately the same plane as the carbonyl bond) the carbonyl frequency is raised by about 17 cm.$^{-1}$, while if the substituent is axial there is no effect. Thus, the frequency of the 3-ketone group (1713–1719 cm.$^{-1}$; 5.84–5.82 μ) is raised to 1733 cm.$^{-1}$ (5.77 μ) in 2-bromo-

or 4-bromo-3-ketones and to 1756–1758 cm.$^{-1}$ (5.69 μ) in 2,4-dibromo-3-ketones. In 2,2-dibromo-3-ketones, however, where only one C—Br bond can be equatorial, the frequency is 1735 cm.$^{-1}$ (5.76 μ), which is not significantly different from the monobromo-compounds. Similar results have been found for 11- and 12-ketones. At the same time the intensity of the carbonyl absorption is reduced by the equatorial C—Br group but is virtually unaffected if it is axial. Some discussion has taken place as to whether the ring A in α-bromoketosteroids is more stable in the "chair" rather than the "boat" form [FIESER and DOMINGUEZ (61); JONES (96)].

3. Application of Infrared Spectroscopy to the Structure and Configuration of Long-Chain Polyenes.

a) Mycomycin.

The use of infrared methods in determining the molecular structure of the antibiotic mycomycin, $C_{12}H_9COOH$, by CELMER and SOLOMONS (31, 32) gives an excellent demonstration of the use of this tool in organic chemistry.

The presence of a monosubstituted acetylene (—C≡CH) is shown by the strong sharp C—H stretching band at 3280 cm.$^{-1}$ (3.049 μ) and the weak C≡C stretching band near 2040 cm.$^{-1}$ (4.90 μ). A prominent band near 2200 cm.$^{-1}$ (4.55 μ) is assigned to a disubstituted acetylenic bond and the ultraviolet spectrum indicates that the two triple bonds are in conjugation. Since one of the acetylenic bonds is terminal, the structure HC≡C—C≡C—C_3H_8COOH is thus established.

An allene grouping —CH=C=CH— is shown to be present by the band near 1930 cm.$^{-1}$ (5.18 μ) [WOTIZ and CELMER (184)]. The ultraviolet spectrum and the ready alkali-isomerization of mycomycin to isomycomycin which contains three conjugated acetylene linkages (31), suggests that the allenic group should be placed adjacent to the acetylenic bonds described above, i. e. H—C≡C—C≡C—CH=C=CH—C_5H_6—COOH.

A conjugated diene structure —CH=CH—CH=CH— with *cis-trans* stereoconfiguration is shown to be present by comparison with synthetic compounds [CELMER and SOLOMONS (33); JACKSON et al. (88)]. *Trans-trans* conjugated dienes have a single strong band near 988 cm.$^{-1}$ (10.12 μ) due to a C—H out-of-plane vibration and the *cis-trans* type are characterized by two bands of medium intensity at 948 and 982 cm.$^{-1}$ (10.55 and 10.18 μ).

The carbonyl frequency of the methyl ester of mycomycin is 1733 cm.$^{-1}$ (5.77 μ) (CCl$_4$ solution) indicating that it is not conjugated. Therefore the remaining CH_2 group must be interposed between the carboxyl group and the diene structure, so that mycomycin is represented by (LVI).

$$\overset{(5)}{H-C\equiv C-C\equiv C-CH=C=CH-CH=CH-\overset{(4)}{CH}=\overset{(3)}{CH}-\overset{(2)}{CH}=\overset{(1)}{CH-CH_2-COOH}} \cdot$$

(LVI.) Mycomycin.

The authors suggest that it has the 3-*trans*-5-*cis* configuration although the 3-*cis*-5-*trans* structure cannot be completely disregarded.

b) Carotenoids.

Cis-trans isomerism of conjugated systems has also been investigated by LUNDE and ZECHMEISTER (*116a, 116b, 186*) who studied a number of α,ω-diphenylpolyenes of the general formula

$$C_6H_5 \cdot (CH=CH)_n \cdot C_6H_5$$

and some carotenoid (polyisoprenoid) pigments, e. g. the β-carotenes (LVII). All the double bonds in the central chain of the diphenylpolyenes carry two hydrogens, and three infrared absorption bands have been related to the stereochemical configuration of these bonds: 1430–1410 cm.$^{-1}$ (7.0–7.1 μ, in-plane vibration of CH groups of a *cis* CC double bond); 779–772 cm.$^{-1}$ (12.84–12.95 μ, corresponding out-of-plane vibration); and 1000–945 (10.0–10.6 μ, out-of-plane vibration of a *trans* CH=CH grouping).

The different types of double bond in the carotenoids cannot be distinguished so readily since several of them carry a methyl group on one carbon (LVII). The authors mentioned have used the term "methylated double bond" for the $-(CH_3)C=CH-$ system and "unmethylated double bond" for $-CH=CH-$. Basically their problem in studying the carotenoids lay in being able to distinguish between methylated and unmethylated *cis* double bonds, and the spectra of four isomeric β-carotenes (*Figure 23*) may be taken as an example of their procedure.

All-*trans*-β-carotene (the natural product) has a strong band at 966 cm.$^{-1}$ (10.35 μ) due to the five *trans* $-CH=CH-$ groups, but has no absorption peak at 1379 cm.$^{-1}$ (7.25 μ) or near 779 cm.$^{-1}$ (12.84 μ), while central-mono-*cis*-β-carotene (structure established by synthesis by INHOFFEN, KARRER et al.) shows a strong band at 779 cm.$^{-1}$ (12.84 μ), slightly weaker (and split) absorption near 966 cm.$^{-1}$ (10.35 μ) and also has no peak at 1379 cm.$^{-1}$ (7.25 μ). The two other isomers, tentatively called neo-β-carotene U and neo-β-carotene B respectively show a strong single peak at 966 cm.$^{-1}$ (10.35 μ), no absorption near 779 cm.$^{-1}$ (12.84 μ) and both have a peak at 1379 cm.$^{-1}$ (7.25 μ) which is rather stronger in

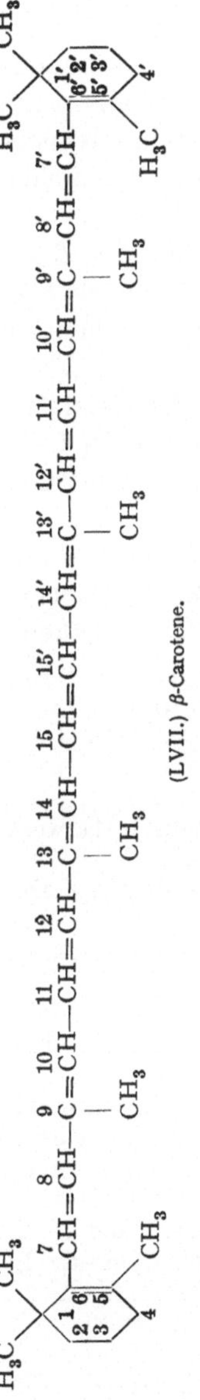

the B isomer. The splitting of the "*trans*" band (966 cm.⁻¹) in the spectrum of the central-mono-*cis* isomer appears to be related to the conjugation of a *cis* CH=CH group with a *trans* double bond.

Since the B and U isomers can only differ from the all-*trans*- and central-mono-*cis*-β-carotenes in *cis-trans* isomerism, the conclusion is drawn that the B and U isomers do not contain an unmethylated *cis* —CH=CH— group [absence of absorption near 779 cm.⁻¹ (12.84 μ)] and must therefore contain one or more methylated *cis* [—(CH₃)C=CH—] bonds. This is supported by the absorption band at 1379 cm.⁻¹ (7.25 μ) which is assigned to a bending vibration of a CH₃ group attached to a *cis* CC double bond, and further evidence for this is found in the spectra of a number of other similar sets of isomeric carotenoids.

With corroboration from visible and ultraviolet spectra neo-β-carotene U is designated as 9-mono-*cis*-β-carotene and the B isomer [with the higher intensity of absorption at 1379 cm.⁻¹ (7.25μ)] as 9,13′-di-*cis*-β-carotene.

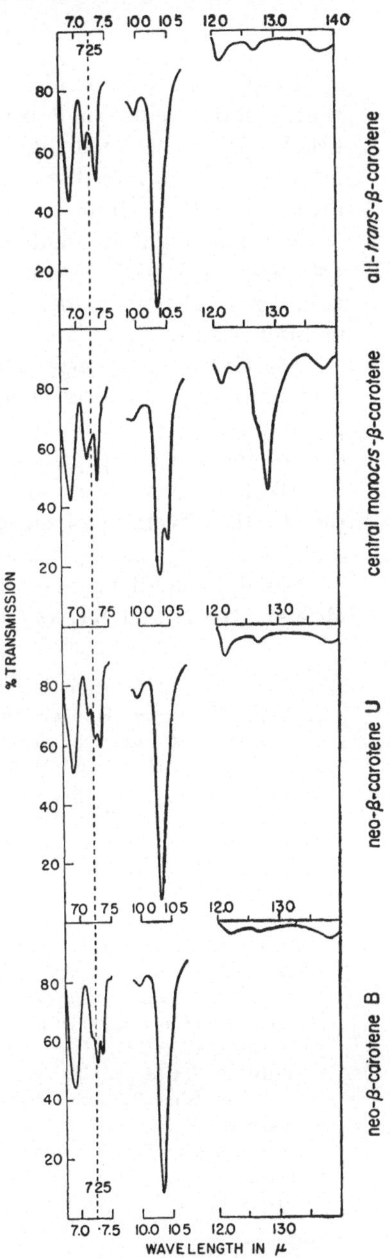

Fig. 23. Stereochemically important sections of the infrared spectra of *cis-trans* isomeric β-carotenes, according to LUNDE and ZECHMEISTER. [From: J. Amer. Chem. Soc. 77, 1647 (1955).]

IV. Conclusion.

Despite the volume of material presented in this report, it is obvious that certain fields have perforce been omitted or covered only very briefly. It has been felt that it was preferable to describe certain advances in detail, and to refer the reader to other papers for news of other fields, rather than to attempt to include too many diversified applications.

Mention might be made of the excellent book recently produced by BELLAMY (*17*) and the invaluable annual (now biennial) reviews by GORE (*76*). The latter, although brief in text are extremely comprehensive in bibliography.

The infrared spectra of hydrocarbons have been discussed by SHEPPARD and SIMPSON (*148, 149*), while GORE and WAIGHT (*77*) have given a good description of the application of polarized light to the study of proteins and polypeptides.

Alkaloids have been somewhat neglected in infrared work, perhaps due to the diversity of structures in that field, but MARION, RAMSAY and JONES (*119*) have published some useful details on a number of them.

Finally, some long-chain compounds have been studied by SINCLAIR, MCKAY, JONES and MYERS (*108, 153, 154*) and by FREEMAN (*67*).

References.

1. AEBI, A., D. H. R. BARTON and A. S. LINDSEY: The Relationship between Clovene and β-Caryophyllene Alcohol. Chem. and Ind. **1953**, 748.

2. AHLERS, N. H. E. and H. P. FREEDMAN: A Simple Ratio-recording Spectrometer. J. Sci. Instr. **32**, 61 (1955).

2a. BAIRD, W. S., H. M. O'BRYAN, G. OGDEN and D. LEE: An Automatic Recording Infra-Red Spectrophotometer. J. Opt. Soc. America **37**, 754 (1947).

3. BARER, R., A. R. H. COLE and H. W. THOMPSON: Infra-red Spectroscopy with the Reflecting Microscope in Physics, Chemistry and Biology. Nature (London) **163**, 198 (1949).

4. BARNARD, D., L. BATEMAN, A. J. HARDING, H. P. KOCH, N. SHEPPARD and G. B. B. M. SUTHERLAND: An Infra-red Spectroscopic Investigation of Double-bond Structure in Simple Acyclic Terpenes and Derivatives thereof. J. Chem. Soc. (London) **1950**, 915.

5. BARNES, C. S., D. H. R. BARTON, A. R. H. COLE, J. S. FAWCETT and B. R. THOMAS: Triterpenoids. IX. The Constitution of Lanostadienol (Lanosterol). J. Chem. Soc. (London) **1953**, 571.

6. BARNES, R. B., R. C. GORE, U. LIDDLE and V. Z. WILLIAMS: Infrared Spectroscopy. New York: Reinhold Publ. Corp. 1944.

7. BARNES, R. B., R. S. MCDONALD, V. Z. WILLIAMS and R. F. KINNAIRD: Small Prism Infra-red Spectrometry. J. Appl. Physics **16**, 77 (1945).

8. BARTON, D. H. R.: The Conformation of the Steroid Nucleus. Experientia **6**, 316 (1950).

9. — The Stereochemistry of *cyclo*Hexane Derivatives. J. Chem. Soc. (London) **1953**, 1027.

10. BARTON, D. H. R., O. HASSEL, K. S. PITZER and V. PRELOG: Nomenclature of *cyclo*Hexane Bonds. Nature (London) **172**, 1096 (1953).

11. BARTON, D. H. R. and P. DE MAYO: Triterpenoids. XIII. Phyllanthol, the First Hexacyclic Triterpenoid. J. Chem. Soc. (London) **1953**, 2178.

12. BARTON, D. H. R., J. E. PAGE and E. W. WARNHOFF: Triterpenoids. XVIII. The Constitutions of Phyllanthol and *cyclo*Artenol. J. Chem. Soc. (London) **1954,** 2715.

13. BARTON, D. H. R. and W. J. ROSENFELDER: The Stereochemistry of Steroids. IV. The Concept of Equatorial and Polar Bonds. J. Chem. Soc. (London) **1951**, 1048.

14. BARTON, D. H. R., E. W. WARNHOFF and J. E. PAGE: The Constitutions of Phyllanthol and *cyclo*Artenol. Chem. and Ind. **1954,** 220.

15. BAYLISS, N. S., A. R. H. COLE and L. H. LITTLE: Solvent Effects in Infrared Spectra: $C=O$, C—H, and C—C Vibrations. Austral. J. Chem. **8**, 26 (1955).

16. BEATON, J. M., F. S. SPRING, R. STEVENSON and W. S. STRACHAN: Triterpenoids. XXXIX. The Constitution and Stereochemistry of the Ursane Group of Triterpenoids. J. Chem. Soc. (London) **1955**, 2610.

17. BELLAMY, L. J.: The Infrared Spectra of Complex Molecules. London: Methuen. 1954.

18. BENTLEY, H. R., J. A. HENRY, D. S. IRVINE, D. MUKERJI and F. S. SPRING: *cyclo*Laudenol, a Triterpenoid Alcohol from Opium. J. Chem. Soc. (London) **1955,** 596.

19. BLADON, P., J. M. FABIAN, H. B. HENBEST, H. P. KOCH and G. W. WOOD: Studies in the Sterol Group. LII. Infrared Absorption of Nuclear Tri- and Tetra-substituted Ethylenic Centres. J. Chem. Soc. (London) **1951**, 2402.

20. BLOUNT, E. L. and A. R. H. COLE: Slit Drives for Single-beam Infra-red Spectrometers. J. Sci. Instr. **32**, 471 (1955).

21. BLOUT, E. R. and G. R. BIRD: Infra-red Microspectroscopy. II. J. Opt. Soc. America **41**, 547 (1951).

22. BLOUT, E. R., G. R. BIRD and D. S. GREY: Infra-red Microspectroscopy. J. Opt. Soc. America **39**, 1052 (1949).

23. — — — Infra-red Microspectroscopy. J. Opt. Soc. America **40**, 304 (1950).

24. BOTTOMLEY, W., A. R. H. COLE and D. E. WHITE: Infrared Spectra of Natural Products. IV. The Structure of Phyllocladene. J. Chem. Soc. (London) **1955**, 2624.

25. BRAUDE, E. A. and F. C. NACHOD: Determination of Organic Structures by Physical Methods. New York: Academic Press. 1955.

26. BROCKWAY, L.O.: Electron Diffraction by Gas Molecules. Rev. Mod. Physics **8**, 231 (1936).

27. BROWNING, R. S., S. E. WIBERLEY and F. C. NACHOD: Application of Infrared Spectrophotometry to Quantitative Analysis in the Solid Phase. Analyt. Chemistry **27**, 7 (1955).

28. BURCH, C. R.: Reflecting Microscopes. Proc. Phys. Soc. (London) **59**, 41 (1947).

29. CANNON, C. G.: A Universal Slit Drive Mechanism for Single Beam Infra-red Spectrometers. J. Sci. Instr. **29**, 50 (1952).

30. CARROL, M. F., R. G. MASON, H. W. THOMPSON and R. C. S. WOOD: Infra-red Examination of Terpenoid Substances. J. Chem. Soc. (London) **1950**, 3457.

31. CELMER, W. D. and I. A. SOLOMONS: Mycomycin. II. The Structure of Iso-mycomycin, an Alkali-Isomerization Product of Mycomycin. J. Amer. Chem. Soc. **74**, 3838 (1952).

32. — — Mycomycin. III. The Structure of Mycomycin, an Antibiotic Containing Allene, Diacetylene and *cis,trans*-Diene Groupings. J. Amer. Chem. Soc. **75**, 1372 (1953).

33. — — Mycomycin. IV. Stereoisomeric 3,5-Diene Fatty Acid Esters. J. Amer. Chem. Soc. **75**, 3430 (1953).

34. COBLENTZ, W. W.: Investigations of Infra-Red Spectra. Washington: Carnegie Institution. 1905.

35. COLE, A. R. H.: Infra-red Spectra of Natural Products. I. The Stereochemistry of Lumistanol. J. Chem. Soc. (London) **1952**, 4969.

36. — A Double-Beam Double-Pass Infrared Spectrometer. J. Opt. Soc. America **43**, 807 (1953).

37. — The Structure of *cyclo*Artenol. Chem. and Ind. **1953**, 946.

38. — Infra-red Spectra and the Structure of Steroids. Rev. Pure Appl. Chem. (Australia) **4**, 111 (1954).

39. — The Constitution of α- and β-Dipiperitone: Interpretation of Infra-red Spectra. Chem. and Ind. **1954**, 661.

40. — Performance of Merton-N. P. L. Gratings in a Commercial Infrared Spectrometer. J. Opt. Soc. America **44**, 741 (1954).

41. — Infra-red Spectra of Natural Products. II. Compounds Containing the *cyclo*Propane Ring. J. Chem. Soc. (London) **1954**, 3807.

42. — Infra-red Spectra of Natural Products. III. *cyclo*Artenol and Phyllanthol. J. Chem. Soc. (London) **1954**, 3810.

43. — unpublished results.

44. COLE, A. R. H. and R. N. JONES: A Reflecting Microscope for Infrared Spectrometry. J. Opt. Soc. America **42**, 348 (1952).

45. COLE, A. R. H., R. N. JONES and K. DOBRINER: A Relationship between the Stereochemical Configuration of 3-Hydroxysteroids and their Infrared Absorption Spectra. J. Amer. Chem. Soc. **74**, 5571 (1952).

46. COLE, A. R. H. and D. W. THORNTON: Infra-red Spectra of Natural Products. V. The Characterisation of Carbonyl Groups in Pentacyclic Triterpenoids. J. Chem. Soc. (London) (1956) (in press).

47. — — unpublished results.

48. COLE, A. R. H. and R. L. WILLIX: unpublished results.

49. COLTHUP, N. B.: Spectra-Structure Correlations in the Infra-red Region. J. Opt. Soc. America **40**, 397 (1950).

50. DAUS, M. A. and H. HIRSCHMANN: Δ^5-Pregnen-20-one. J. Amer. Chem. Soc. **75**, 3840 (1953).

51. DEW, G. D.: On the Preservation of Groove-form in Replicas of Diffraction Gratings. J. Sci. Instr. **29**, 277 (1952).

52. — On the Preparation of Plane Diffraction Grating Replicas from Helical Rulings. J. Sci. Instr. **30**, 229 (1953).

53. DEW, G. D. and L. A. SAYCE: On the Production of Diffraction Gratings. I. The Copying of Plane Gratings. Proc. Roy. Soc. (London), Ser. A **207**, 278 (1951).

54. DOBRINER, K., E. R. KATZENELLENBOGEN and R. N. JONES: Infrared Absorption Spectra of Steroids. New York: Interscience Publ. 1953.

55. DOBRINER, K., T. H. KRITCHEVSKY, D. K. FUKUSHIMA, S. LIEBERMAN, T. F. GALLAGHER, J. D. HARDY, R. N. JONES and G. CILENTO: Infrared Spectrometry in Metabolic Studies with Deuterium-labeled Steroids. Science (Washington) **109**, 260 (1949).

56. DOBRINER, K., S. LIEBERMAN, C. P. RHOADS, R. N. JONES, V. Z. WILLIAMS and R. B. BARNES: Studies in Steroid Metabolism. III. The Application of Infra-red Spectrometry to the Fractionation of Urinary Ketosteroids. J. Biol. Chem. **172**, 297 (1948).

57. DOWNIE, A. R., M. C. MAGOON, T. PURCELL and B. CRAWFORD, Jr.: The Calibration of Infrared Spectrometers. J. Opt. Soc. America **43**, 941 (1953).

58. EDDY, C. R., E. W. MONROE and M. SCOTT: Catalog of Infrared Absorption Spectra of Steroidal Sapogenin Acetates. Analyt. Chemistry **25**, 266 (1953).

59. ELLIOTT, A., E. J. AMBROSE and R. B. TEMPLE: Polarized Infra-Red Radiation as an Aid to the Structural Analysis of Long-Chain Polymers. I. J. Chem. Physics 16, 877 (1948).

60. FARMER, V. C.: The Pressed Disc Technique in Infra-red Spectroscopy. Chem. and Ind. 1955, 586.

61. FIESER, L. F. and X. A. DOMINGUEZ: Configuration of Steroid Bromoketones. II. 4β-Bromotestane-17β-ol-3-one Acetate and 2β-Bromocholestane-3-one. J. Amer. Chem. Soc. 75, 1704 (1953).

62. FIESER, L. F. and M. FIESER: Natural Products Related to Phenanthrene. New York: Reinhold Publ. Corp. 1949.

63. FOX, J. J. and A. E. MARTIN: Investigations of Infra-red Spectra. Determinations of C—H Frequencies ($\sim$ 3000 cm.$^{-1}$) in Paraffins and Olefins, with some Observations on "Polythenes". Proc. Roy. Soc. (London), Ser. A 175, 208 (1940).

64. FRANCIS, S. A.: Absolute Intensities of Characteristic Infrared Absorption Bands of Aliphatic Hydrocarbons. J. Chem. Physics 18, 861 (1950).

65. — Intensities of Some Characteristic Infrared Bands of Ketones and Esters. J. Chem. Physics 19, 942 (1951).

66. FRASER, R. D. B.: A o.8 N. A. Reflecting Microscope for Infrared Absorption Measurements. J. Opt. Soc. America 43, 929 (1953).

67. FREEMAN, N. K.: Infrared Spectra of Branched Long-chain Fatty Acids. J. Amer. Chem. Soc. 74, 2523 (1952).

68. FRIDRICHSONS, J. and A. McL. MATHIESON: Triterpenoids. The Crystal Structure of Lanostenyl Iodoacetate. J. Chem. Soc. (London) 1953, 2159.

69. FRIEDMAN, H. and C. P. GLOVER: The Optical Transmission of Additively Colored Alkali Halide Crystals in the Visible and Near Infra-Red. J. Opt. Soc. America 39, 795 (1949).

70. FURCHGOTT, R. F., H. ROSENKRANTZ and E. SHORR: Infra-red Absorption Spectra of Steroids. I. Androgens and Related Steroids. J. Biol. Chem. 163, 375 (1946).

71. — — — Infrared Absorption Spectra of Steroids. III. Progesterone and Pregnane Derivatives. J. Biol. Chem. 167, 627 (1947).

72. FÜRST, A., H. H. KUHN, R. SCOTONI, Jr. und Hs. H. GÜNTHARD: Infrarotspektren epimerer Alkohole und Acetoxy-Verbindungen. Helv. Chim. Acta 35, 951 (1952).

73. GAUNT, J.: A Simple Infra-red Grating Spectrometer for Use in Analysis. J. Sci. Instr. 31, 315 (1954).

74. GOLAY, M. J. E.: A Pneumatic Infra-red Detector. Rev. Sci. Instruments 18, 357 (1947).

75. GORE, R. C.: Infrared Spectrometry of Small Samples with the Reflecting Microscope. Science (Washington) 110, 710 (1949).

76. — Infrared Spectroscopy. Analyt. Chemistry 22, 7 (1950); 23, 7 (1951); 24, 8 (1952); 26, 11 (1954).

77. GORE, R. C. and E. S. WAIGHT: Infrared Light Absorption. In: E. A. BRAUDE and F. C. NACHOD, Determination of Organic Structures by Physical Methods, p. 195. New York: Academic Press. 1955.

78. GOULDEN, J. D. S.: The Use of Diffraction Gratings with an Infra-red Spectrometer. J. Sci. Instr. 29, 215 (1952).

78a. — Wavelength and Slit-drive Mechanisms for an Infrared Spectrometer. J. Sci. Instr. 30, 139 (1953).

79. HALL, R. G. N. and L. A. SAYCE: On the Production of Diffraction Gratings. II. The Generation of Helical Rulings and the Preparation of Plane Gratings therefrom. Proc. Roy. Soc. (London), Ser. A 215, 536 (1952).

80. HAM, N. S., A. WALSH and J. B. WILLIS: A Quadruple Monochromator. Nature (London) **169**, 977 (1952).

81. — — — Multiple Monochromators. III. A Quadruple Monochromator and Its Application to Infrared Spectroscopy. J. Opt. Soc. America **42**, 496 (1952).

82. HENBEST, H. B., G. D. MEAKINS and G. W. WOOD: Studies in the Steroid Group. LXVII. Infra-red Absorption of Disubstituted *cis*Ethylenic Centres. J. Chem. Soc. (London) **1954**, 800.

83. HENRY, J. A. and F. S. SPRING: The Structure of *cyclo*Artenol. Chem. and Ind. **1954**, 189.

84. HERZBERG, G.: Infrared and Raman Spectra of Polyatomic Molecules. New York: D. van Nostrand Co. 1945.

85. HIBBEN, J. H.: Raman Effect and its Chemical Applications. New York: Reinhold Publ. Corp. 1939.

86. HIRSCHMANN, H.: The Characterization of Δ^5-Unsaturated Steroids by Infrared Spectroscopy. J. Amer. Chem. Soc. **74**, 5357 (1952).

87. HORNIG, D. F., G. E. HYDE and W. A. ADCOCK: A Ratio-Recording Double-Beam Infra-Red Spectrophotometer with Automatic Slit-Control. J. Opt. Soc. America **40**, 497 (1950).

88. JACKSON, J. E., R. F. PASCHKE, W. TOLBERG, H. M. BOYD and D. H. WHEELER: Isomers of Linoleic Acid. Infrared and Ultraviolet Properties of Methyl Esters. J. Amer. Oil Chem. Soc. **29**, 229 (1952).

89. JEGER, O.: Über die Konstitution der Triterpene. Fortschr. Chem. organ. Naturstoffe **7**, 1 (1950).

90. JOHNSON, D. R., D. R. IDLER, V. W. MELOCHE and C. A. BAUMANN: Infrared Spectra of Some Cholestanol Derivatives. J. Amer. Chem. Soc. **75**, 52 (1953).

91. JOHNSON, W. S., D. K. BANERJEE, W. P. SCHNEIDER, C. D. GUTSCHE, W. E. SHELBERG and L. J. CHINN: The Total Synthesis of Estrone and Three Stereoisomers Including Lumiestrone. J. Amer. Chem. Soc. **74**, 2832 (1952).

92. JONES, A. V. and G. B. B. M. SUTHERLAND: Use of Polarized Radiation in Infra-red Analysis. Nature (London) **160**, 567 (1947).

93. JONES, E. R. H. and T. G. HALSALL: Tetracyclic Triterpenes. Fortschr. Chem. organ. Naturstoffe **12**, 44 (1955).

94. JONES, R. N.: The Stereochemical Configuration of the Δ^{22}-Ergostenyl Side Chain. J. Amer. Chem. Soc. **72**, 5322 (1950).

95. — The Absorption of Radiation by Inhomogeneously Dispersed Systems. J. Amer. Chem. Soc. **74**, 2681 (1952).

96. — The Steric Configuration of Brominated 3-Ketosteroids. J. Amer. Chem. Soc. **75**, 4839 (1953).

97. JONES, R. N. and A. R. H. COLE: The Characterization of Methyl and Methylene Groups in Steroids by Infrared Spectrometry. I. Correlations of Bending Frequencies with Molecular Structure. J. Amer. Chem. Soc. **74**, 5648 (1952).

98. JONES, R. N., A. R. H. COLE and B. NOLIN: The Characterization of Methyl and Methylene Groups in Steroids by Infrared Spectrometry. II. Methyl and Methylene Bending Frequencies in Steroids Labeled with Deuterium. J. Amer. Chem. Soc. **74**, 5662, 6321 (1952).

99. JONES, R. N. and K. DOBRINER: Infrared Spectrometry Applied to Steroid Structure and Metabolism. Vitamins and Horm. **7**, 293 (1949).

100. JONES, R. N. and F. HERLING: Characteristic Group Frequencies in the Infrared Spectra of Steroids. J. Organ. Chem. (USA) **19**, 1252 (1954).

101. JONES, R. N., F. HERLING and E. KATZENELLENBOGEN: The Infrared Spectra of Ketosteroids Below 1350 cm.$^{-1}$ J. Amer. Chem. Soc. **77**, 651 (1955).

102. JONES, R. N., P. HUMPHRIES and K. DOBRINER: Studies in Steroid Metabolism. VI. The Characterization of 11- and 12-Oxygenated Steroids by Infrared Spectrometry. J. Amer. Chem. Soc. **71**, 241 (1949).

103. — — — Studies in Steroid Metabolism. IX. Further Observations on the Infrared Absorption Spectra of Ketosteroids and Steroid Esters. J. Amer. Chem. Soc. **72**, 956 (1950).

104. JONES, R. N., P. HUMPHRIES, F. HERLING and K. DOBRINER: Studies in Steroid Metabolism. X. The Effects of Stereochemical Configuration at Positions 3 and 5 on the Infrared Spectra of 3-Acetoxy Steroids. J. Amer. Chem. Soc. **73**, 3215 (1951).

105. — — — — Studies in Steroid Metabolism. XII. The Determination of the Structure of the Side Chains of C-21 Steroids by Infrared Spectrometry. J. Amer. Chem. Soc. **74**, 2820 (1952); correction, p. 6319.

106. JONES, R. N., P. HUMPHRIES, E. PACKARD and K. DOBRINER: Studies in Steroid Metabolism. VIII. The Detection and Location of Ethylenic Double Bonds in Steroids by Infrared Spectrometry. J. Amer. Chem. Soc. **72**, 86 (1950).

107. JONES, R. N., E. KATZENELLENBOGEN and K. DOBRINER: The Infrared Absorption Spectra of the Steroid Sapogenins. J. Amer. Chem. Soc. **75**, 158 (1953).

108. JONES, R. N., A. F. McKAY and R. G. SINCLAIR: Band Progressions in the Infrared Spectra of Fatty Acids and Related Compounds. J. Amer. Chem. Soc. **74**, 2575 (1952).

109. JONES, R. N., D. A. RAMSAY, F. HERLING and K. DOBRINER: The Infrared Spectra of α-Brominated Ketosteroids. J. Amer. Chem. Soc. **74**, 2828 (1952).

110. JONES, R. N., D. A. RAMSAY, D. S. KEIR and K. DOBRINER: The Intensities of Carbonyl Bands in the Infrared Spectra of Steroids. J. Amer. Chem. Soc. **74**, 80 (1952).

111. JONES, R. N., V. Z. WILLIAMS, M. J. WHALEN and K. DOBRINER: Studies in Steroid Metabolism. IV. The Characterization of Carbonyl and Other Functional Groups in Steroids by Infrared Spectrometry. J. Amer. Chem. Soc. **70**, 2024 (1948).

112. JOSIEN, M.-L.: Étude infrarouge de quelques composés stéroïdes dans la zone de 3 μ. C. R. hebd. Séances Acad. Sci. **231**, 131 (1950).

113. JOSIEN, M.-L. and N. FUSON: Infrared Study of the Effect of Solvent upon ν (NH) in Pyrrole. J. Chem. Physics **22**, 1169 (1954); A New Analysis of the Effect of Polar Solvents Upon ν (NH) of Pyrrole. J. Chem. Physics **22**, 1264 (1954).

114. JOSIEN, M.-L. et J. LASCOMBE: Contribution à l'étude de l'influence des solvants sur les fréquences de valence mesurées par spectroscopie infrarouge. C. R. hebd. Séances Acad. Sci. **238**, 2414 (1954).

115. LIEBERMAN, S., L. B. HARITON, P. HUMPHRIES, C. P. RHOADS and K. DOBRINER: Studies in Steroid Metabolism. XIV. The Isolation from Human Urine of Δ^{9}-17-Ketosteroids. J. Biol. Chem. **196**, 793 (1952).

116. LORD, R. C., R. S. McDONALD and F. A. MILLER: Notes on the Practice of Infrared Spectroscopy. J. Opt. Soc. America **42**, 149 (1952).

116a. LUNDE, K. and L. ZECHMEISTER: A Study of the Infrared Spectra of Some Stereoisomeric Diphenylpolyenes. Acta Chem. Scand. **8**, 1421 (1954).

116b. — — Infrared Spectra and *cis-trans* Configurations of Some Carotenoid Pigments. J. Amer. Chem. Soc. **77**, 1647 (1955).

117. MANN, J. and H. W. THOMPSON: Absorption Spectra with Polarized Infra-Red Radiation. Nature (London) **160**, 17 (1947).

118. Mann, J. and H. W. Thompson: Infra-red Spectra and the Solid State: Measurements with Polarized Radiation. Proc. Roy. Soc. (London), Ser. A **192**, 489 (1948).

119. Marion, L., D. A. Ramsay and R. N. Jones: The Infrared Absorption Spectra of Alkaloids. J. Amer. Chem. Soc. **73**, 305 (1951); Cevine, a Correction. J. Amer. Chem. Soc. **74**, 270 (1952).

120. Meakins, G. D.: The Structure of α-Amyrin. Chem. and Ind. **1955**, 1353.

121. Meisels, A., R. Rüegg, O. Jeger und L. Ruzicka: Über die Konstitution des Ringes E und die Konfiguration des x-Amyrins. Helv. Chim. Acta **38**, 1298 (1955).

122. Menzies, A. C. G.: Recent Developments in Applied Optics. J. Sci. Instr. **30**, 441 (1953).

123. Merton, Sir Thomas: On the Reproduction and Ruling of Diffraction Gratings. Proc. Roy. Soc. (London), Ser. A **201**, 187 (1950).

124. Nolin, B. and R. N. Jones: The Infrared Absorption Spectra of Diethyl Ketone and its Deuterium Substitution Products. J. Amer. Chem. Soc. **75**, 5626 (1953).

125. Orr, S. F. D.: The Design of a Ratio-Recording Spectrophotometer with Double-Pass Monochromator. J. Opt. Soc. America **43**, 709 (1953).

126. Pirenne, M. H.: The Diffraction of X-Rays and Electrons by Free Molecules. Cambridge: University Press. 1946.

127. Plíva, J.: Design of Double-beam Spectrometers using Multiple Monochromators. J. Sci. Instr. **31**, 434 (1954).

128. Plíva, J. and V. Herout: On Terpenes. XVIII. Infrared Investigations of Terpenes. II. Collect. Czech. Chem. Communs. **15**, 160 (1950).

129. Plíva, J., V. Herout, B. Schneider and F. Šorm: On Terpenes. XLIII. Infrared Investigations of Terpenes. IV. Collect. Czech. Chem. Communs. **18**, 500 (1953).

130. Plíva, J., V. Herout and F. Šorm: On Terpenes. XXIV. Infrared Investigations of Terpenes. III. Collect. Czech. Chem. Communs. **16**, 158 (1951).

131. Plíva, J. and F. Šorm: On Terpenes. XI. Infrared Investigations of Terpenes. I. Collect. Czech. Chem. Communs. **14**, 274 (1949).

132. Pristera, F.: Solvents and Techniques in Infrared Spectroscopy. Applied Spectroscopy **6**, No. 3, 29 (1952).

133. Ramsay, D. A.: Intensities and Shapes of Infrared Absorption Bands of Substances in the Liquid Phase. J. Amer. Chem. Soc. **74**, 72 (1952).

134. Rasmussen, R. S.: Infrared Spectroscopy in Structure Determination and its Application to Penicillin. Fortschr. Chem. organ. Naturstoffe **5**, 331 (1948).

135. — Vibrational Frequency Assignments for Paraffin Hydrocarbons; Infra-Red Absorption Spectra of the Butanes and Pentanes. J. Chem. Physics **16**, 712 (1948).

136. Rasmussen, R. S. and R. R. Brattain: Infra-Red Absorption Spectra of the C_2 to C_4 Mono-Olefins and of 2-Methyl-2-Butene. J. Chem. Physics **15**, 120 (1947).

137. Rasmussen, R. S., R. R. Brattain and P. S. Zucco: Infra-Red Absorption Spectra of Some Octenes. J. Chem. Physics **15**, 135 (1947).

138. Robertson, J. M.: Organic Crystals and Molecules. Ithaca: Cornell University Press. 1953.

139. Rochester, J. C. O. and A. E. Martin: Improving the Performance of a Littrow-type Infra-Red Spectrometer. Nature (London) **168**, 785 (1951).

140. Rosenkrantz, H. and P. Skogstrom: Characteristic Infrared Absorption Bands of Steroids with Reduced Ring A. I. Tetrahydro Compounds. J. Amer. Chem. Soc. **77**, 2237 (1955).

140a. ROSENKRANTZ, H. and L. ZABLOW: The 9–10 μ Region of Infrared Absorption Spectra of Steroids in Relation to Chemical Structure. J. Amer. Chem. Soc. **75**, 903 (1953).

141. ROTHMAN, E. S. and M. E. WALL: Steroidal Sapogenins. XXII. Conversion of Hecogenin Acetate to a Hecololactone Derivative with the Adrenal Cortical Hormone Side Chain *via* Allopregnane-3β,17α,21-triol-12,20-dione Diacetate. J. Amer. Chem. Soc. **77**, 2229 (1955).

142. ROTHMAN, E. S., M. E. WALL and C. R. EDDY: Steroidal Sapogenins. III. Structure of Steroidal Saponins. J. Amer. Chem. Soc. **74**, 4013 (1952).

143. RUPERT, C. S.: A Water-Cooled Carbon Arc Source for Infrared Spectroscopy. J. Opt. Soc. America **42**, 684 (1952).

144. RUPERT, C. S. and J. STRONG: The Carbon Arc as an Infrared Source. J. Opt. Soc. America **40**, 455 (1950).

145. SAVITSKY, A. and R. S. HALFORD: A Ratio-Recording Double Beam Infra-Red Spectrophotometer Using Phase Discrimination and a Single Detector. Rev. Sci. Instruments **21**, 203 (1950).

146. SCHIEDT, U.: Zur Infrarot-Spektroskopie von Aminosäuren. II. Mitt. Eine verbesserte, zur infrarot-photometrischen Bestimmung von Aminosäuren und anderen polaren Verbindungen geeignete Präparationstechnik. Z. Naturforsch. **8 B**, 66 (1953).

147. SCHIEDT, U. und H. REINWEIN: Zur Infrarot-Spektroskopie von Aminosäuren. I. Mitt. Eine neue Präparationstechnik zur Infrarot-Spektroskopie von Aminosäuren und anderen polaren Verbindungen. Z. Naturforsch. **7 B**, 270 (1952).

148. SHEPPARD, N. and D. M. SIMPSON: The Infra-red and Raman Spectra of Hydrocarbons. I. Acetylenes and Olefins. Quart. Rev. Chem. Soc. (London) **6**, 1 (1952).

149. — — The Infra-red and Raman Spectra of Hydrocarbons. II. Paraffins. Quart. Rev. Chem. Soc. (London) **7**, 19 (1953).

150. SIMONSEN, Sir JOHN and D. H. R. BARTON: The Terpenes. Vol. III. The Sesquiterpenes, Diterpenes and their Derivatives. 2nd Ed. **Cambridge**: University Press. 1952.

151. SIMONSEN, Sir JOHN and L. N. OWEN: The Terpenes. Vol. I. The Simpler Acyclic and Monocyclic Terpenes and their Derivatives. 2nd Ed. Cambridge: University Press. 1947.

152. — — The Terpenes. Vol. II. The Dicyclic Terpenes and their Derivatives. 2nd Ed. Cambridge: University Press. 1949.

153. SINCLAIR, R. G., A. F. McKAY and R. N. JONES: Infrared Absorption Spectra of Saturated Fatty Acids and Esters. J. Amer. Chem. Soc. **74**, 2570 (1952).

154. SINCLAIR, R. G., A. F. McKAY, G. S. MYERS and R. N. JONES: Infrared Absorption Spectra of Unsaturated Fatty Acids and Esters. J. Amer. Chem. Soc. **74**, 2578 (1952).

155. SMITH, F. A. and E. C. CREITZ: Infrared Studies of Association in Eleven Alcohols. J. Res. Nat. Bur. Stand. (USA) **46**, 145 (1951).

156. STIMSON, M. M. and M. J. O'DONNELL: The Infrared and Ultraviolet Absorption Spectra of Cytosine and Isocytosine in the Solid State. J. Amer. Chem. Soc. **74**, 1805 (1952).

157. SUTHERLAND, G. B. B. M.: Some Problems in the Interpretation of the Infra-red Spectra of Large Molecules. Discuss. Faraday Soc. **9**, 274 (1950).

158. SUTHERLAND, G. B. B. M. and H. W. THOMPSON: Developments in the Technique of Infra-red Spectroscopy. Trans. Faraday Soc. **41**, 174 (1945).

159. Tarpley, W. and C. Vitiello: Infrared Spectra of Steroids. New Solvents for Steroids Difficultly Soluble in Carbon Disulphide. Analyt. Chemistry **24**, 315 (1952).

159a. — — Infrared Spectra of Steroids. II. The Influence of the Solvent on Carbonyl Frequency Correlations. Applied Spectroscopy **9**, 69 (1955).

160. Taylor, J. H., C. S. Rupert and J. Strong: An Incandescent Tungsten Source for Infrared Spectroscopy. J. Opt. Soc. America **41**, 626 (1951).

161. Thompson, G. P. and W. Cochrane: Theory and Practice of Electron Diffraction. London: Macmillan. 1939.

162. Thompson, H. W.: The Correlation of Vibrational Absorption Spectra with Molecular Structure. J. Chem. Soc. (London) **1948**, 328.

163. Thompson, H. W. and P. Torkington: The Infra-red Spectra of Polymers and Related Monomers. I. Proc. Roy. Soc. (London), Ser. A **184**, 3 (1945).

164. — — The Infra-red Spectra of Compounds of High Molecular Weight. Trans. Faraday Soc. **41**, 246 (1945).

165. — — The Vibrational Spectra of Esters and Ketones. J. Chem. Soc. (London) **1945**, 640.

166. Thompson, H. W. and D. H. Whiffen: The Infra-red Examination of Some Terpenoid Substances. J. Chem. Soc. (London) **1948**, 1412.

167. Tolberg, W. E. and H. M. Boyd: New Slit Drive for Beckman IR-2 Infrared Spectrophotometer. Analyt. Chemistry **24**, 1836 (1952).

168. Turnbull, J. H., D. H. Whiffen and W. Wilson: Stereochemistry of the Side-chain Double Bond in Calciferol and Related Steroids. Chem. and Ind. **1950**, 626.

169. Voser, W., M. V. Mijović, H. Heusser, O. Jeger und L. Ruzicka: Über die Konstitution des Lanostadienols (Lanosterins) und seine Zugehörigkeit zu den Steroiden. Helv. Chim. Acta **35**, 2414 (1952).

170. Wall, M. E., C. R. Eddy, M. L. McClennan and M. E. Klumpp: Detection and Estimation of Steroidal Sapogenins in Plant Tissue. Analyt. Chemistry **24**, 1337 (1952).

171. Walsh, A.: Design of Multiple-Monochromators. Nature (London) **167**, 810 (1951).

172. — Multiple Monochromators. I. Design of Multiple Monochromators. J. Opt. Soc. America **42**, 94 (1952).

173. — Multiple Monochromators. II. Application of a Double Monochromator to Infrared Spectroscopy. J. Opt. Soc. America **42**, 96 (1952).

174. — Design of Double-Beam Multiple Monochromators. J. Opt. Soc. America **43**, 215 (1953).

175. Walsh, A. and J. B. Willis: Multiple Monochromators. IV. A Triple Monochromator and Its Application to Near Infrared, Visible and Ultraviolet Spectroscopy. J. Opt. Soc. America **43**, 989 (1953).

176. White, J. U. and M. D. Liston: Construction of a Double Beam Recording Infra-Red Spectrophotometer. J. Opt. Soc. America **40**, 29 (1950).

177. — — Performance of a Double-Beam Recording Infra-Red Spectrophotometer. J. Opt. Soc. America **40**, 93 (1950).

178. Wild, R. F.: Automatic Ratio Recorder as Applied to Infra-Red Spectroscopy. Rev. Sci. Instruments **18**, 436 (1947).

179. Wood, D. L.: An Infra-Red Microspectrometer for Biological Research. Rev. Sci. Instruments **21**, 764 (1950).

180. Wood, R. W.: Diffraction Gratings with Controlled Groove Form and Abnormal Distribution of Intensity. Philos. Mag. [6] **23**, 310 (1912).

181. — Improved Diffraction Gratings and Replicas. J. Opt. Soc. America **34**, 509 (1944).

182. WOODWARD, R. B., A. A. PATCHETT, D. H. R. BARTON, D. A. J. IVES and R. B. KELLY: The Synthesis of Lanostenol. J. Amer. Chem. Soc. **76**, 2852 (1954).

183. WOODWARD, R. B., F. SONDHEIMER, D. TAUB, K. HEUSLER and W. M. McLAMORE: The Total Synthesis of Steroids. J. Amer. Chem. Soc. **74**, 4223 (1952).

184. WOTIZ, J. H. and W. D. CELMER: The Infrared Spectra of Some Allenic Compounds. J. Amer. Chem. Soc. **74**, 1860 (1952).

185. WRIGHT, N. and L. W. HERSCHER: A Double-Beam Percent Transmission Recording Infrared Spectrophotometer. J. Opt. Soc. America **37**, 211 (1947).

186. ZECHMEISTER, L.: Some Stereochemical Aspects of Polyenes. Experientia **10**, 1 (1954).

187. Perkin-Elmer Instruction Manual, Model 12C, p. 12 (1949).

188. Perkin-Elmer Instrument News **4**, No..4, 1 (1953).

189. Transactions of the Joint Commission for Spectroscopy; Minutes of the Rome Meeting. J. Opt. Soc. America **43**, 410 (1953).

(Received, December 12, 1955.)

Gallotannine und Ellagen-gerbstoffe.

Von **O. Th. Schmidt**, Heidelberg.

Mit 5 Abbildungen.

Inhaltsübersicht.

I. Einleitung.

Wenn über Fortschritte der Chemie der Gallotannine berichtet werden soll, so muß ein Ausgangspunkt gewählt werden, von dem aus die neu gewonnenen Erkenntnisse geschildert werden. Das ist in unserem Falle nicht schwer. Emil Fischer hatte für das chinesische Gallotannin (aus den Blattgallen von *Rhus semialata*) und das türkische Gallotannin (aus den Zweiggallen von *Quercus infectoria*) das bekannte Bauprinzip der polygalloylierten Glucosen herausgearbeitet. Aber weder ihm noch Karrer (*43, 44*) ist es möglich gewesen, aus dem Gemisch von Substanzen, die sie untersuchten, einzelne Individuen zu isolieren, deren exakte Konstitution man hätte aufklären und angeben können. Auch mit den modernen Methoden der Gegenstromverteilung und Verteilungschromatographie ist eine Auftrennung dieser Gemische offenbar bis heute noch nicht gelungen. Ähnlich liegen die Verhältnisse beim Sumach-Gerbstoff (aus *Rhus coriaria*), von dem Münz (*58*) gezeigt hat, daß der Hauptteil der Substanz aus einer Pentagalloyl-glucose besteht. Erwähnt man noch das Glucogallin und Tetrarin, die Gilson (*30, 30a*) aus chinesischem Rhabarber isoliert hat, und das Acertannin, ein Digalloyl-anhydrohexit aus koreanischem *Acer ginnale*, von Perkin und Uyeda (*65*), so hat man die im Jahre 1929 bekannten Forschungsergebnisse zusammengestellt, mit Ausnahme des Hamameli-tannins und der Chebulinsäure, die aber im nachstehenden Bericht noch ausführlicher behandelt werden.

Somit kann also das Jahr 1929 als Ausgangspunkt für die vorliegende Berichterstattung gewählt werden.

II. Gallotannine: Ellagsäure-freie Gerbstoffe.

1. Hamameli-tannin.

Der von Grüttner (*31*) 1898 entdeckte, schön kristallisierte Gerbstoff aus der Rinde von *Hamamelis virginica*, das Hamameli-tannin ist nach Untersuchungen von Freudenberg (*21, 28, 24*) eine Digalloylhexose, deren beide Gallussäuren getrennt an zwei Hydroxylgruppen des Zuckers gebunden sind, während die Carbonylgruppe an der Esterbindung wahrscheinlich keinen Anteil hat. Der Zucker ist eine Aldohexose mit verzweigter Kohlenstoffkette, wobei die Aldehydgruppe an der Verzweigung beteiligt ist, was aus dem Fehlen der Osazonbildung hervorgeht (*24*).

$$\begin{array}{c} CHO \\ | \\ HOCH_2-C-OH \\ | \\ HCOH \\ | \\ HCOH \\ | \\ CH_2OH \end{array}$$

(I.) Hamamelose.

Hamamelose ist α-Oxymethyl-*D*-ribose (I), wie O. Th. Schmidt (*74, 100, 101, 84*) festgestellt hat.

Die Konstitution des Hamameli-tannins selbst ist nicht mit Sicherheit bekannt. Es steht aber fest, daß die Aldehydgruppe des Zuckers nicht an der Bindung der Gallussäure beteiligt ist. Es ließ sich zunächst zeigen, daß das Hamameli-tannin leicht und ohne Verlust von Gallussäure zu Hamameli-tannin-methylacetal glykosidiert werden kann. In diesem Methylacetal liegt sehr wahrscheinlich ein furanoider Ring vor, denn nach der alkalischen Abspaltung der beiden Gallussäuren wird das Methyl-hamamelosid erhalten, dessen Säureempfindlichkeit auf ein Furanosid schließen läßt (74). Somit gewinnt Formel (II) für Hamameli-tannin die größte Wahrscheinlichkeit, und man könnte sich vorstellen, daß der Gerbstoff in der Pflanze aus zwei Molekülen 3-Galloyl-glycerinaldehyd bzw. aus 1 Mol. 3-Galloyl-glycerinaldehyd und 1 Mol. 3-Galloyl-dioxy-aceton entsteht.

$$
\begin{array}{l}
\text{HO} \\
\text{HO} \!-\!\!\!\bigcirc\!\!\!-\!\!C\!-\!O\!-\!CH_2\!-\!\overset{\displaystyle CHOH}{\underset{\displaystyle HCOH}{\overset{|}{\underset{|}{C}}}OH} \\
\text{HO} \qquad\;\; \overset{\|}{O} \\
\qquad\qquad\qquad\qquad HCO\!-\!\!\!-\!\!\!-\! \\
\qquad\qquad\qquad\qquad | \\
\qquad\qquad\qquad CH_2O\!-\!C\!-\!\!\!\bigcirc\!\!\!-\!\!OH \\
\qquad\qquad\qquad\qquad\;\; \overset{\|}{O}
\end{array}
$$

(II.) Hamameli-tannin (wahrscheinliche Formel).

2. Chebulinsäure.

Die Chebulinsäure wurde 1884 von Fridolin (29) entdeckt. Sie wird gewonnen aus getrockneten Früchten von *Terminalia chebula*, die als „Myrobalanen" im Handel sind, und ist der erste kristallisierte Vertreter der Gallotannin-Klasse gewesen. Als „Eutannin" ist die Chebulin-säure einige Zeit lang fabrikmäßig hergestellt worden. Die Untersuchung der Chebulinsäure durch Adolphi (1), Thoms (105), Richter (73), Fischer und Freudenberg (20), Fischer und Bergmann (18), Freudenberg (22), Freudenberg und Fick (25) sowie Freudenberg und Frank (26) hatten bis 1929 zu folgenden Vorstellungen geführt: 1 Molekül Glucose trägt 2 Moleküle Gallussäure und 1 Molekül einer sehr sauerstoffreichen „Spalt-säure", $C_{14}H_{14}O_{11}$ (22, 25, 73), die ihrerseits mit einem weiteren Molekül Gallussäure verbunden ist. Die Gesamtformel sollte $C_{41}H_{34}O_{27}$ sein.

Bei der partiellen Hydrolyse, die durch 24stündiges Kochen mit Wasser durchgeführt wird, entsteht 1 Mol. einer schön kristallisierten Digalloyl-glucose (22, 25), 1 Mol. Gallussäure und 1 Mol. „Spaltsäure". Die Haftstelle der beiden Gallussäuren in der Digalloyl-glucose ist unbekannt; in Betracht kommen vielleicht die Zuckerhydroxyle 3 und 6. Die Konstitution der „Spaltsäure" ist ebenfalls unbekannt, ebenso ihre Verknüpfung mit dem Zucker. Unbekannt ist schließlich der Ort der Bindung der dritten Gallussäure.

a) Chebulsäure („Spaltsäure" $C_{14}H_{12}O_{11}$).

Die „Spaltsäure" ist auch ein Bestandteil der Chebulagsäure (vgl. S. 97). Die Aufklärung ihrer Konstitution war vordringlich. Nach erfolgter Konstitutionsermittlung wurde der unspezifische Name „Spaltsäure" verlassen und dafür „Chebulsäure" vorgeschlagen (91), der im folgenden ausschließlich verwendet werden soll.

Chebulsäure kristallisiert nicht, liefert aber ein kristallisiertes Thallium(I)-salz [FREUDENBERG (22)] und ein kristallisiertes Brucinsalz [FREUDENBERG und FICK (25)], die beide zweibasische Salze sind. Anderseits verbraucht Chebulsäure 3 äquivalente Lauge (25). Die blaue Eisenchlorid-Reaktion weist auf das Vorliegen von drei vicinalen phenolischen Hydroxylgruppen hin. Damit stimmt überein, daß die trockene Destillation bei 320—340° und 12 Torr eine geringe Menge Pyrogallol ergab (25).

Chebulsäure ist optisch aktiv ($[\alpha]_D = +25{,}6°$, in Wasser, $c = 2$). Mit Diazomethan wird die (amorphe) Hexamethyl-chebulsäure erhalten, deren Hydrolyse zur (ebenfalls amorphen) Trimethyl-chebulsäure führt [O. TH. SCHMIDT, HEINTZELER und MAYER (85)]. Trimethyl-chebulsäure wird als dreibasische Säure titriert; sie liefert (amorphe) dreibasische Salze mit Barium oder Calcium (85) und (ebenfalls amorphe) Tri-p-bromphenacylester und Tri-p-phenyl-phenacylester [O. TH. SCHMIDT und MAYER (92)]. Durch Kochen mit äthanolischer Chlorwasserstoffsäure läßt sich Chebulsäure in den kristallisierten Triäthylester überführen, dessen Triacetat ebenfalls kristallisiert (91). Die weitere Konstitutionsaufklärung der Chebulsäure ist von O. TH. SCHMIDT und MAYER (92) durchgeführt worden. Bei der Einwirkung von methanolischem Ammoniak auf Hexamethyl-chebulsäure entsteht das kristallisierte Trimethyl-chebulsäure-triamid, dessen Formel $C_{17}H_{21}O_8N_3$ Veranlassung gegeben hat, die Bruttoformel für Chebulsäure, die früher mit $C_{14}H_{14}O_{11}$ angegeben war, in $C_{14}H_{12}O_{11}$ umzuwandeln.

Wie erwähnt, verbraucht Trimethyl-chebulsäure 3 äquivalente Natronlauge. Titriert man die Verbindung mit $n/10$-NaOH direkt oder nach dem Überalkalisieren, kalt oder heiß, so findet man tatsächlich diesen Verbrauch. Erhitzt man aber die Säure mit überschüssiger 2 n-NaOH auf dem Wasserbad, so entspricht der Alkaliverbrauch bei der Rücktitration mit Säure 4 Carboxylgruppen. Aus der heißen alkalischen Lösung läßt sich das (amorphe) Tetranatriumsalz $C_{17}H_{16}O_{12}Na_4$ mit Methanol ausfällen. Erhitzt man jedoch die neutralisierte Lösung auf dem Wasserbad, so wird sie wieder alkalisch unter allmählicher Freigabe eines Äquivalents NaOH. Dieses Verhalten läßt auf das Vorliegen eines außergewöhnlich festen Lactons schließen, wie es vom Phthalid (IV) bekannt ist [Guyot (33)]: Das Natriumsalz des o-Carboxy-benzylalkohols (III) spaltet beim Erhitzen der wäßrigen Lösung NaOH ab und geht in (IV) über:

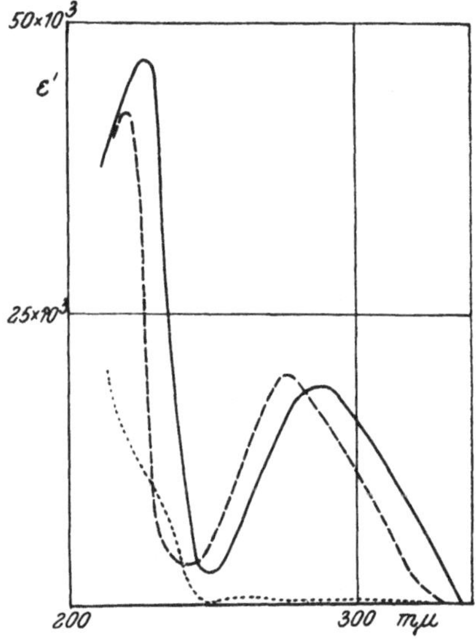

(III.) O-Carboxy-benzylalcohol (Na-Salz).   (IV.) Phthalid.

Analog darf man in der Chebulsäure eine vierte Carboxylgruppe am aromatischen Ring annehmen, die mit einer Hydroxylgruppe der Seitenkette zu einem sehr festen Lacton geschlossen ist. Diese Annahme wird bestärkt durch den Vergleich der Absorptionsspektren von Chebulsäure einerseits mit Gallussäure-methylester und anderseits mit Pyrogalloltrimethyläther (*Abb. 1*).

Nach diesen Feststellungen dürften 6 C-Atome der Chebulsäure, $C_{14}H_{12}O_{11}$, als aromatischer Kern, 3 C-Atome in 3 Carboxylgruppen und ein weiteres

(V.) Chebulsaure (Teilformel).

Abb. 1. Molekulare Absorptionsspektren in Methanol: - - - - - - Gallussaure-methylester; - - — „Spaltsaure"; und Pyrogallol-trimethylather*. [Aus: Liebigs Ann. Chem. *571*, 1 (1951).]

in einer Lactongruppe vorliegen. Von den 12 O-Atomen sind 3 in Pyrogallolanordnung am aromatischen Ring gebunden, 6 sind in den 3 freien Carboxylgruppen enthalten und die beiden restlichen in der Lactongruppe. Das führt zur Teilformel (V), wobei die Stellung der Carboxylgruppe am aromatischen Ring zunächst willkürlich wie in der Gallussäure angenommen worden ist. Die Seitenkette, $C_4H_5(COOH)_3$, ist gesättigt. Dadurch wird die frühere Formel für Chebulsäure (mit 14 H-Atomen) ausgeschlossen und die Formel $C_{14}H_{12}O_{11}$ bestätigt. Hier-

* Die Absorptionskurve des Pyrogallols (Landolt-Börnstein, Tabellen, 5. Aufl., Berlin 1935, 3. Erg., Teil II, S. 1616) gleicht sehr weitgehend der des Trimethyläthers.

mit ändert sich auch die Formel für Chebulinsäure und muß $C_{41}H_{32}O_{27}$ lauten, was auch mit den Analysen übereinstimmt.

Die Bindung zwischen dem aromatischen Kern und der C_7-Seitenkette ist nicht hydrolysierbar: Trimethyl-chebulsäure kann nach dem Erhitzen mit konz. Salzsäure oder Behandeln mit 60%iger Schwefelsäure oder 50%iger Kalilauge unverändert wiedergewonnen werden.

Der *Abbau* der Chebulsäure oder Trimethyl-chebulsäure mit Permanganat in stark schwefelsaurer Lösung bei 0° führt unter Zerstörung des aromatischen Ringes zu einer kristallisierten *Verbindung* $C_8H_8O_8$, welche die 7 C-Atome der Seitenkette enthält, vermehrt um ein C-Atom aus dem Benzolring.

Die Verbindung $C_8H_8O_8$ (Schmelzp. 207°, $[\alpha]_D = +105°$, in Wasser, $c = 3$) ist eine Tricarbonsäure. Aber auch sie enthält eine vierte Carboxylgruppe, die mit einer Hydroxylgruppe zum Lacton geschlossen ist. Auch dieses Lacton ist recht stabil. Man muß die Säure $C_8H_8O_8$ mit $n/10$-, oder besser mit n-NaOH einige Zeit auf dem Wasserbad erhitzen, um bei der Rücktitration mit Säure einen Laugenverbrauch zu erhalten, der vier Carboxylgruppen entspricht. Doch ist diese Tetranatriumsalz-lösung (im Gegensatz zur Lösung des Tetranatriumsalzes der Trimethyl-chebulsäure) beständig. Man erhält aus ihr das kristallisierte Salz $C_8H_6O_9Na_4$ ($[\alpha]_D = -5°$, in Wasser, $c = 1$). Es ist das Tetranatriumsalz einer Butanol-tetracarbonsäure $C_8H_{10}O_9$, die in freier Form zu isolieren nicht gelingt, weil sie sofort nach Freisetzen den Lactonring schließt.

Die Säure $C_8H_8O_8$ ergibt mit Diazomethan einen schön kristallisierten Trimethylester $C_{11}H_{14}O_8$ (Schmp. 81—82°, $[\alpha] = +117,5°$, in Methanol, $c = 0,9$), dessen Umsetzung mit methanolischem Ammoniak in der Hauptsache ein Tetraamid, $C_8H_{14}O_5N_4$, und aus dessen Mutterlauge ein Monolacton-triamid, $C_8H_{11}O_5N_3$, liefert (beide kristallisiert).

Da die Säure $C_8H_8O_8$ weder beim energischen Behandeln mit Mineralsäuren noch beim Erhitzen mit Methylanilin auf 180° Kohlendioxyd verliert, können alle Strukturformeln ausgeschlossen werden, die zwei Carboxylgruppen an einem C-Atom gebunden enthalten („Malonsäurestellung"). Es bleiben dann noch die Konstitutionen (VI), (VII) und (VIII), zwischen denen zu entscheiden ist.

```
(1)   O—CH—COOH          CH2—COOH          ———————O—CH—COOH
      |    |             |                        |
(2)   |   CH—COOH        O—C—COOH          HOOC—C—CH2—COOH
      |    |             |    |                   |
(3)   C—CH              |   CH—COOH              CH2
      ‖    |             |    |                   |
      O    |             C—CH2             —————————C=O
(4)       CH2—COOH       |
                         O

      (VI.)              (VII.)                  (VIII.)
```

Wird die Säure $C_8H_8O_8$ (oder ihr Triester) mit konz. Kalilauge unter Bedingungen behandelt, die einer Kalischmelze nahekommen, so werden neben Essigsäure 70% eines Mols Oxalsäure und 65% eines Mols Bernsteinsäure erhalten. Aus der (verzweigten) Formel (VIII) lassen sich diese Spaltstücke nicht erklären. Unter Zugrundelegung der Formel (VI) kann man sich diesen Abbau so vorstellen, daß nach der Lacton-öffnung zuerst 1 Mol. Wasser abgespalten wird (IX), dann Hydrolyse zu Essigsäure und Oxalbernsteinsäure (X) eintritt und diese schließlich durch „Säurespaltung" zu Oxalsäure und Bernsteinsäure zerlegt wird.

$$
\begin{array}{ccccc}
HC-COOH & & H_3C-COOH & & \\
| & & & & \\
C-COOH & \xrightarrow{\text{Hydrolyse}} & O=C-COOH & \xrightarrow{\text{„Säurespaltung"}} & HOOC-COOH \\
| & & | & & \\
HC-COOH & & HC-COOH & & H_2C-COOH \\
| & & | & & | \\
H_2C-COOH & & H_2C-COOH & & H_2C-COOH \\
\text{(IX.)} & & \text{(X.) Oxalbernsteinsaure.} & &
\end{array}
$$

Denselben Mechanismus mit den gleichen Zwischenstufen könnte man auch auf Formel (VII) anwenden. Daß Zitronensäure (XI) bei dieser Behandlung 1 Mol. Oxalsäure und 2 Mole Essigsäure ergibt, ist bekannt. Aber auch Iso-zitronensäure (XII) verhält sich ebenso, während Tricarballylsäure (XIII) nicht angegriffen wird.

$$
\begin{array}{ccc}
COOH & COOH & COOH \\
| & | & | \\
HO-CH & CH & CH_2 \\
| & \| & | \\
HC-COOH \longrightarrow & C-COOH \longleftarrow & HO-C-COOH \\
| & | & | \\
CH & CH_2 & CH_2 \\
| & | & | \\
COOH & COOH & COOH \\
\text{(XII.) Iso-zitronensaure.} & & \text{(XI.) Zitronensaure.}
\end{array}
$$

$$
\begin{array}{ccc}
 & COOH & COOH \\
 & | & | \\
 & CH_3 & CH_2 \\
 & | & | \\
HOOC-COOH \longleftarrow & O=C-COOH & CH-COOH \\
| & | & | \\
CH_3 & CH_2 & CH_2 \\
| & | & | \\
COOH & COOH & COOH \\
 & & \text{(XIII.) Tricarballylsaure.}
\end{array}
$$

Da die Abbausäure $C_8H_8O_8$ optisch aktiv ist, konnte eine Entscheidung zwischen den Formeln (VI) (drei asymmetrische C-Atome) und (VII) (zwei asymmetrische C-Atome) durch Synthese und Vergleich der physikalischen Eigenschaften langwierig sein. So führte zunächst der Vergleich im chemischen Verhalten zu einer Entscheidung, und zwar zugunsten der Formel (VI). Verbindung (VII) wurde synthetisiert [MAYER, WICK und O. TH. SCHMIDT (54)] und das erhaltene kristallisierte Gemisch der beiden Racemate in seinem Verhalten gegen alkalische Permanganatlösung mit der Abbausäure aus Chebulsäure verglichen. Die Abbausäure liefert bei dieser Reaktion 2,58 Mole Oxalsäure, die Säure (VII) aber nur 0,32 Mole. Die Reaktion dürfte bei der Abbausäure folgenden Verlauf nehmen (*Formelübersicht 1*), wobei über die Zwischenstufen (XIV) bis

$$
\begin{array}{lll}
\text{(1) } O\!-\!CH\!-\!COOH & O\!=\!C\!-\!COOH & HO\!-\!C\!-\!COOH \\
\text{(2) } \quad CH\!-\!COOH & CH\!-\!COOH & C\!-\!COOH \\
\text{(3) } C\!-\!CH & CH\!-\!COOH & CH\!-\!COOH \\
\text{(4) } O \quad CH_2\!-\!COOH & CH_2\!-\!COOH & CH_2\!-\!COOH \\
\qquad\text{(VI.)} & \text{(XIV.)} & \text{(XV.)}
\end{array}
$$

n-Butanol(1)-tetracarbonsaure(1,2,3,4)-lacton.

$$ HOOC\!-\!COOH $$

$$
\begin{array}{lll}
HOOC\!-\!COOH & HO\!-\!C\!-\!COOH & O\!=\!C\!-\!COOH \\
O\!=\!C\!-\!COOH & C\!-\!COOH & CH\!-\!COOH \\
CH_2\!-\!COOH & CH_2\!-\!COOH & CH_2\!-\!COOH \\
\text{(XVIII.)} & \text{(XVII.)} & \text{(XVI.)} \\
\text{Oxalessigsaure.} & &
\end{array}
$$

$$
\begin{array}{ll}
HO\!-\!C\!-\!COOH & \\
CH\!-\!COOH & \longrightarrow \quad 2\ HOOC\!-\!COOH \\
\text{(XIX.)} &
\end{array}
$$

Formelubersicht 1.

(XIX) hinweg bei quantitativem Ablauf 4 Mole Oxalsäure entstehen müßten. Zudem zeigen Modellversuche an Äpfelsäure und Oxalessigsäure (XVIII), daß die Ausbeute je Stufe nur 85—90% d. Th. beträgt, was mit den gefundenen 2,58 Molen Oxalsäure befriedigend übereinstimmt.

Die synthetische Verbindung (VII) wird von Permanganat viel träger angegriffen. Sie verhält sich in dieser Hinsicht ganz wie Zitronensäure (XI), die unter den gleichen Bedingungen nur 0,24 Mole Oxalsäure liefert, während Iso-zitronensäure (XII) 2,32 Mole ergibt (ganz entsprechend der Verlustrechnung). Da die Formeln (VI) und (VII) denen der Iso-zitronensäure(XII) und Zitronensäure (XI) weitgehend analog sind, und die Verbindungen dasselbe unterschiedliche Verhalten gegen Permanganat zeigen wie das Vergleichspaar, kann Formel (VII) ausgeschlossen und der Abbausäure $C_8H_8O_8$ die Formel (VI) eines *n-Butanol(1)-tetracarbonsäure(1,2,3,4)-lactons* zuerteilt werden.

Schließlich wurde die Formel (VI) durch die Synthese wie folgt bewiesen [Schweiger (*103*)]:

Racemat 1.   Racemat 2.

(XX.) 3-Acetoxy-Δ^4-tetrahydrophthalsäure-dimethylester.

Acetoxy-butadien und Fumarsäure-dimethylester.   Racemat 2′. (XXI.) Butanol(1)-tetracarbonsäure.

Durch Diensynthese lassen sich Fumarsäure-dimethylester und Acetoxy-butadien miteinander vereinigen. Die entstehenden beiden Racemate des 3-Acetoxy-Δ^4-tetrahydrophthalsäure-dimethylesters (XX) (Rac. 1 und Rac. 2) lassen sich durch fraktionierte Kristallisation trennen. Rac. 1 und Rac. 2 werden durch Ozonspaltung und Oxydation in zwei racemische Butanol(1)-tetracarbonsäuren überführt. Die Trennung des einen Racemats (Rac. 2′) (XXI) mit Strychnin oder Brucin führte zu einer mit der natürlichen Abbausäure identischen Säure und deren Spiegelbild*.

Somit ist also Formel (VI) als Konstitution der Abbausäure $C_8H_8O_8$ mit Sicherheit bewiesen.

---

* Mit dieser Synthese wäre zugleich die Konfiguration der Abbausäure und damit der Chebulsäure geklärt, wenn es noch gelingt, die absolute Konfiguration zu bestimmen. Jedoch besitzt die Synthese für die Konstitution der Abbausäure volle Beweiskraft.

Wenn nun die Teilformel (V) (S. 74) für die *Chebulsäure* ergänzt werden soll, muß man berücksichtigen, daß die OH-Gruppe am C-Atom 1 der Butanol-tetracarbonsäure mit der Carboxylgruppe des aromatischen Teils zum Lacton geschlossen sein muß, und daß eine Carboxylgruppe der Säure $C_8H_8O_8$ aus dem Benzolring stammt. Auch dann bleiben noch zwei, aber nur zwei Möglichkeiten, nämlich die Formeln (XXII) und (XXIII). [In beiden Formeln sind die C-Atome der Seitenkette so beziffert, wie sie der Butanol-tetracarbonsäure (vgl. VI, S. 75) zugehören.]

(XXII.)

(XXIII.) Chebulsäure.

(XXIV.) Trimethoxy-phthalsäure.

(XXV.) Ellagsäure.

Die Entscheidung zwischen den Formeln (XXII) und (XXIII) lieferte der Abbau des Triamids der Trimethyl-chebulsäure mit Hypochlorit und Alkali. Wie zuerst WEERMANN (*106*) und später HIRST und Mitarbeiter (*42, 4*) gezeigt haben, entsteht bei dieser Reaktion nur aus den Amiden von α-Oxysäuren Natriumcyanat, das mit Semicarbazid als Hydrazodicarbonamid isoliert werden kann, während aus den Amiden von β-Oxysäuren kein Cyanat gebildet wird. Der Abbau des Trimethyl-chebulsäure-triamids ergab 0,83 Mole Kaliumcyanat, isoliert als Hydrazodicarbonamid. Das beweist eindeutig Formel (XXIII), denn nur in ihr liegt — nach Öffnen des Lactonringes — eine α-Oxysäure vor.

Die somit abgeleitete Formel (XXIII) für Chebulsäure fand eine Bestätigung durch R. D. HAWORTH und DE SILVA (*36*), denen der oxydative Abbau der Trimethyl-chebulsäure mit $K_3Fe(CN)_6$ und KOH zu Trimethoxy-phthalsäure (XXIV) gelang, wodurch die Stellung der Carboxylgruppe im aromatischen Kern der Chebulsäure experimentell bewiesen wurde.

Die Aufklärung der Konstitution der Chebulsäure ergab den ersten Hinweis auf einen überraschenden Zusammenhang, der im Kapitel III, S. 89 ausführlich behandelt werden wird. Ellagsäure (XXV), eine lange bekannte Verbindung, die in vielen Gerbextrakten beobachtet wird, besitzt die Bruttoformel $C_{14}H_6O_8$. Chebulsäure die Formel $C_{14}H_{12}O_{11}$. Zwischen den beiden Formeln besteht eine Differenz von 3 H_2O, und die C-Atome beider Verbindungen haben die gleiche Anordnung.

b) 3,6-Digalloyl-glucose.

Die Annahme FREUDENBERGS (23), daß die Digalloyl-glucose, die bei der Hydrolyse der Chebulinsäure mit kochendem Wasser entsteht, 3,6-Digalloyl-glucose sei, hat sich bestätigen lassen. Die Konstitutionsaufklärung ist von O. TH. SCHMIDT, BERG und BAER (76) durchgeführt worden.

Schon aus Versuchen von RICHTER (73), der Chebulinsäure methyliert und bei der Hydrolyse nur Trimethylgallussäure und keine Dimethylgallussäure erhalten hatte, ging hervor, daß die beiden Gallussäuren der Digalloyl-glucose einzeln und nicht als m-Digallussäure gebunden sind. Es ließ sich leicht zeigen, daß die 1-Stellung der Glucose nicht galloyliert ist. Digalloyl-glucose läßt sich ohne Verlust von Gallussäure glykosidieren.

Zunächst wurde das Enneaacetat der Digalloyl-glucose (25) über den Acetobromgerbstoff (Octaacetyl-digalloyl-1-brom-glucose) in Octaacetyl-digalloyl-β-methyl-glucosid überführt. Die Konfiguration am C-Atom 1 und die Ringweite wurde durch alkalische Hydrolyse, die zum β-Methylgluco-pyranosid führte, bestätigt. Durch direkte Glykosidierung der Digalloyl-glucose mit alkoholischer Chlorwasserstoffsäure wurden die nachstehenden Glucoside A, B und C dargestellt. Entsprechend den Erfahrungen der Zuckerchemie wurde diesen Glucosiden die pyranoide Struktur zuerteilt, wenn in der Hitze, und die furanoide, wenn in der Kälte glykosidiert worden war.

A. Digalloyl-β-äthyl-gluco-furanosid, Schmp. 203—204°, $[\alpha] = -46,5°$.
B. Digalloyl-α-(?)äthyl-gluco-furanosid, Schmp. 172°, $[\alpha] = +7°$.
C. Digalloyl-β-(?)äthyl-gluco-pyranosid, Schmp. 183—185°, $[\alpha] = -18,5°$.

Bei Verbindung A wurde die furanoide Struktur bestätigt durch die alkalische Hydrolyse, die zu dem erwarteten β-Äthyl-gluco-furanosid [W. N. HAWORTH und PORTER (37)] führte, das als Tetra-p-azobenzolcarbonsäureester, der sich vorzüglich hierzu eignet, identifiziert wurde.

Insgesamt zeigen die Glykoside *A—C*, daß die Hydroxylgruppen 1, 4 und 5 nicht mit Gallussäure besetzt sein können. Somit blieb noch zu entscheiden zwischen drei möglichen Konstitutionen, nämlich einer 2,3-, einer 2,6- und einer 3,6-Digalloyl-glucose. Versuche zur Osazonbildung verliefen widerspruchsvoll. Mit Phenylhydrazin wurde kein definiertes Reaktionsprodukt erhalten. *p*-Nitrophenylhydrazin lieferte ein rotes, amorphes Präparat, dessen Analysen auf ein *p*-Nitrophenyl-osazon der Digalloyl-glucose stimmen. p-Bromphenylhydrazin indessen führte zu einem kristallisierten Osazon einer Monogalloyl-glucose. Es wäre unrichtig, aus dieser Verbindung zu schließen, daß eine Gallussäure in 2-Stellung gebunden gewesen sein müßte, um von dem Hydrazin verdrängt zu werden. FISCHER und BERGMANN (*18*) erhielten nämlich beim Umsatz von synthetischer 3-Galloyl-glucose mit Phenylhydrazin, unter Abspaltung der Gallussäure aus 3-Stellung, gewöhnliches Glucose-phenyl-osazon. Es hat sich später gezeigt, daß das kristallisierte *p*-Brom-phenyl-osazon ein Abkömmling der 6-Galloyl-glucose ist.

Eine Entscheidung zwischen den drei möglichen Konstitutionen brachte das Studium der Glykolspaltung. Digalloyl-α-äthyl-gluco-furanosid (*B*, S. 80; XXVI) wurde mit Diazomethan zum Hexamethyl-äther methyliert, dann mit $NaJO_4$ behandelt. Da die Verbindung weder Perjodat verbraucht, noch Formaldehyd abspaltet, muß eine Gallussäure in 6-Stellung, die andere in 2- oder 3-Stellung stehen. Schließlich wurde

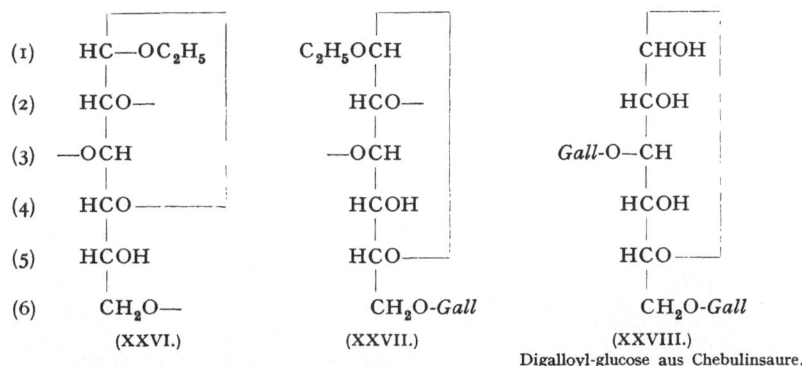

(1)	$HC—OC_2H_5$	C_2H_5OCH	$CHOH$
(2)	$HCO—$	$HCO—$	$HCOH$
(3)	$—OCH$	$—OCH$	*Gall*-$O—CH$
(4)	HCO	$HCOH$	$HCOH$
(5)	$HCOH$	HCO	HCO
(6)	$CH_2O—$	CH_2O-*Gall*	CH_2O-*Gall*
	(XXVI.)	(XXVII.)	(XXVIII.)

Digalloyl-glucose aus Chebulinsaure.

auch Digalloyl-β-äthyl-glucopyranosid (*C*, XXVII) mit Diazomethan methyliert und der Hexamethyläther mit Perjodat geprüft. Da auch diese Verbindung keine Glykolspaltung erleidet, muß die zweite Gallussäure in 3-Stellung gebunden sein. Somit ergibt sich Formel (XXVIII) als *Konstitution der Digalloyl-glucose aus Chebulinsäure.*

Dieses Ergebnis wird aufs beste bestätigt durch die Synthese [O. TH. SCHMIDT und SCHACH (*95*)]. Der Weg der Synthese der 3,6-Digalloyl-glucose hat die Gelegenheit geboten, gleichzeitig die schon von Fischer und Bergmann (*18*) auf andere Weise aufgebaute Mono-galloyl-glucose und die 6-Galloyl-glucose synthetisch zu gewinnen. Deshalb werden im folgenden alle drei Synthesen beschrieben.

Grundsätzlich sind die Wege zu galloylierten Zuckern durch die Gedankengänge der Zuckerchemie vorgeschrieben, und die von Schmidt und Schach durchgeführte Wiederholung der von Fischer und Bergmann schon beschriebenen Darstellung der Monogalloyl-glucose — es ist 3-Galloyl-glucose — gleicht dem Vorbild weitgehend. Indessen verwenden Schmidt und Schach nicht wie die früheren Autoren acylierte Gallussäure-chloride, sondern das von Cavallito und Buck beschriebene Tribenzyl-gallussäurechlorid (*13*).

Dieses Verfahren bringt wichtige Vorteile. Alle Zwischenprodukte der Synthesen sind wohlkristallisierte, gut definierte Verbindungen, was bei der Verwendung acylierter Gallussäuren keineswegs immer der Fall war. Zudem wird die bei den früheren Verfahren immer problematische,

$TbgCl \longrightarrow$

(XXIX.) 1,2,5,6-Diacetonglucose.   (XXX.) 3-Tribenzylgalloyl-diacetonglucose.

(XXXIII.)
1,2-Aceton-3-tribenzylgalloyl-glucose.   (XXXII.)
3-Galloyl-glucose.   (XXXI.)
3-Tribenzylgalloyl-glucose.

partielle, alkalische Abspaltung der phenolisch gebundenen Acyle vermieden. Und schließlich erfolgt die Entfernung der Benzylreste durch katalytische Hydrierung als letzter Schritt der Synthese in neutralem Medium und führt daher immer sofort zu reinen Endprodukten.

Zur Darstellung von *3-Galloyl-glucose* (XXXII) wird die Natriumverbindung von 1,2,5,6-Diacetonglucose (XXIX) mit Tribenzyl-galloyl-chlorid („*TbgCl*") zu 3-Tribenzylgalloyl-diaceton-glucose (XXX) umgesetzt.

Aus Verbindung (XXX) werden in einem Gemisch von 2 *n*-Salzsäure und Aceton in sieben Tagen bei 40° beide Acetongruppen abgespalten. Es entsteht 3-Tribenzylgalloyl-glucose (XXXI), deren hydrierende Entbenzylierung zu 3-Galloyl-glucose (XXXII) führt (Gesamtausbeute, bezogen auf Diacetonglucose = 51%). Obgleich die Zwischenstufen (XXX) und (XXXI) kristallisierte und sicher einheitliche Verbindungen sind, und obgleich die so gewonnene 3-Galloyl-glucose analytisch (auch im Papierchromatogramm) einwandfrei ist, zeigt sie ebensowenig Neigung zur Kristallisation wie das FISCHERsche Präparat, mit dem sie in der Drehung (+47,2°) übereinstimmt. Das Heptaacetat der 3-Galloyl-glucose ist kristallisiert und eignet sich zur Charakterisierung der Verbindung.

Zur Synthese der *3,6-Digalloyl-glucose* (XXXV a) wird Verbindung (XXX) in einem Gemisch von Aceton und 2 *n*-Schwefelsäure 2 Tage bei 20° behandelt, wobei nur 1 Acetongruppe abgespalten und 1,2-Aceton-3-tribenzylgalloyl-glucose (XXXIII) erhalten wird. Zur Einführung der zweiten Tbg-gruppe, wird Verbindung (XXXIII) in chloroformischer Lösung mit knapp einem Mol. Tribenzylgalloyl-chlorid in Gegenwart von Chinolin umgesetzt. Daß die neue Acylgruppe in 6-Stellung eintreten würde, war zu erwarten, nachdem 1,2-Aceton-3,6-dibenzoyl-glucose ebenfalls durch partielle Benzylierung von 1,2-Aceton-3-benzoyl-glucose

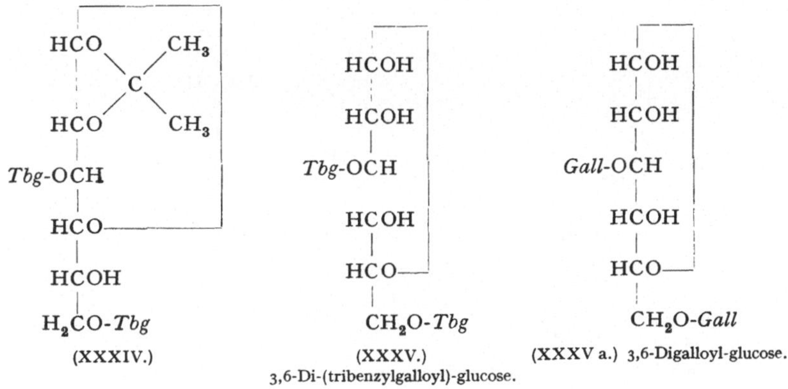

(XXXIV.)   (XXXV.)   (XXXV a.) 3,6-Digalloyl-glucose.

3,6-Di-(tribenzylgalloyl)-glucose.

6*

gewonnen wird [Brigl und Grüner (*11*)]. So entstand aus Verbindung (XXXIII) 1,2-Aceton-3,6-di-tribenzylgalloyl-glucose (XXXIV), deren Hydrolyse (in Tetrahydrofuran + 4 *n*-H_2SO_4, 37°, 33 Tage) zu 3,6-Di-(tribenzylgalloyl)-glucose (XXXV) führte. Die katalytische Hydrierung ergab 3,6-Digalloyl-glucose, die sofort kristallisierte. Die synthetische 3,6-Digalloyl-glucose ist identisch mit der Digalloyl-glucose aus Chebulinsäure. Beide besitzen die gleiche Drehung, den gleichen R_f-Wert im Papierchromatogramm (*89*) und übereinstimmende Löslichkeiten. Auch die Enneaacetate stimmen nach Schmelzp., Misch-schmelzp. und Drehung miteinander überein. Die Gesamtausbeute an Digalloyl-glucose, bezogen auf eingesetzte Diaceton-glucose, betrug 10% d. Th.

Einerseits zur Nachprüfung, ob bei der oben beschriebenen Synthese der 3-Galloyl-glucose vielleicht infolge von Acylwanderung ein Irrtum entstanden sei, anderseits aus Interesse an der Verbindung selbst, wurde auch *6-Galloyl-glucose* (XXXVIII) synthetisiert. Hierzu wurde 1,2-Aceton-3,5-benzal-glucose (XXXVI) (*10, 27*) in chloroformischer Lösung und Tribenzylgalloyl-chlorid und Chinolin zur 1,2-Aceton-3,5-benzal-6-tribenzylgalloyl-glucose (XXXVI a) umgesetzt. Die Abspaltung von Aceton

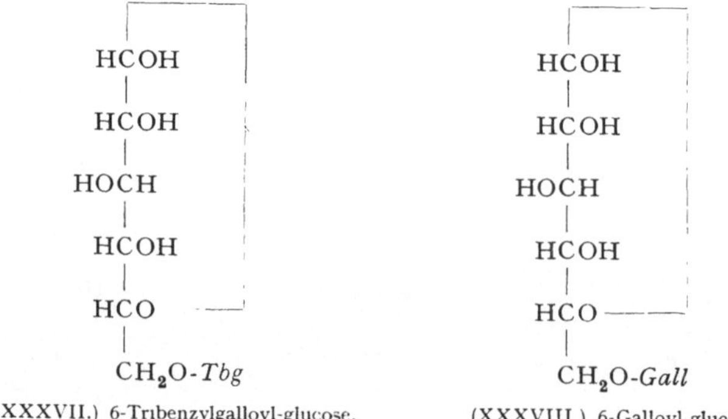

(XXXVI.) 1,2-Aceton-3,5-benzal-glucose.   (XXXVI a.) 1,2-Aceton-3,5-benzal-6-tribenzylgalloyl-glucose

(XXXVII.) 6-Tribenzylgalloyl-glucose.   (XXXVIII.) 6-Galloyl-glucose.

und Benzaldehyd durch 4 *n*-Schwefelsäure in Tetrahydrofuran erforderte bei 40° 9 Tage und führte zur 6-Tribenzylgalloyl-glucose (XXXVII).

Die katalytische Entbenzylierung ergab sofort die schön kristallisierende 6-Galloyl-glucose (XXXVIII). Die Gesamtausbeute, bezogen auf eingesetzte Acetonbenzal-glucose, betrug 54% der theoretisch möglichen. Auch hier wurde das Heptaacetat dargestellt. Bei 6-Galloyl-glucose gelang die Darstellung des *p*-Bromphenylosazons, wenn auch in schlechter Ausbeute. Es zeigte sich, daß dieses Osazon identisch ist mit dem aus der Digalloyl-glucose der Chebulinsäure erhaltenen, oben erwähnten Monogalloyl-glucose-*p*-bromphenylosazon.

Die beiden Monogalloyl-glucosen unterscheiden sich in ihrem Verhalten von der Digalloyl-glucose u. a. dadurch, daß sie in Wasser leichter und in Essigester sehr viel schwerer löslich sind als diese. Die Gerbstoffreaktionen treten bei den einfach galloylierten Verbindungen noch mehr zurück als bei der Digalloylverbindung. So wird z. B. Gelatine-lösung von den Monogalloyl-glucosen (im Gegensatz zur Digalloyl-glucose) nicht gefällt.

c) Neochebulinsäure und 1,3,6-Trigalloyl-glucose.

Die in diesem Abschnitt beschriebenen Ergebnisse sind noch unveröffentlicht, aber größtenteils in Dissertationen und Diplomarbeiten festgelegt.

Der oben beschriebenen Entstehung der 3,6-Digalloyl-glucose bei der Hydrolyse der Chebulinsäure mit kochendem Wasser gehen zwei interessante Reaktionsstufen voraus, die man polarimetrisch und papierchromatographisch beobachten und präparativ erfassen kann (*15, 8*). Wenn man Chebulinsäure in wäßrigem Aceton bei 60° aufbewahrt, so nimmt die optische Drehung allmählich ab und nach 2 Tagen erscheint im Papierchromatogramm ein neuer Fleck mit kleinerem R_f-Wert als Chebulinsäure. Der neue Fleck ergibt mit Anilin-phthalat keine Braunfärbung, zeigt also keine freie reduzierende Gruppe am Zucker (*64, 64a*) an. Während er mit fortschreitender Hydrolysenzeit zunimmt, beobachtet man zunächst noch keines der Spaltstücke, in welche Chebulinsäure zerfallen kann, also keine Chebulsäure, keine Gallussäure, keine Digalloyl- oder Monogalloyl-glucose und keine Glucose selbst, aber natürlich ein Schwächerwerden des Chebulinfleckes. Die neue Verbindung mit dem kleineren R_f-Wert enthält also noch alle Bausteine der Chebulinsäure. Sie wurde „*Neochebulinsäure*" genannt und wird unten näher beschrieben.

Das erste Spaltstück, das dann auftritt, ist Chebulsäure. Gleichzeitig mit ihr entsteht eine neue Verbindung, eine Trigalloyl-glucose, die ebenfalls nicht mit Anilinphthalat reagiert. Als nächstes erst tritt Gallussäure im Chromatogramm auf und gleichzeitig 3,6-Digalloyl-glucose, die durch die Reaktion mit Anilinphthalat zu erkennen ist. Die drei Reaktionsstufen überschneiden einander stärker und die papierchromatographische Verfolgung ist ungünstiger als bei der im Kapitel III (S. ҫ6) beschriebenen, zeitlich früher aufgeklärten Hydrolyse der Chebulagsäure. Aber die Indizien im Verlauf der Hydrolyse genügen, um sowohl Neochebulinsäure wie die Trigalloyl-glucose präparativ darzustellen.

Zur *Darstellung der Neochebulinsäure* bricht man die Hydrolyse nach $4^1/_2$ Tagen ab. Das Aceton wird im Vakuum abgedampft und aus der wäßrigen Lösung kristallisiert die unveränderte (schwerlösliche) Chebulinsäure aus. Das Filtrat wird eingeengt und scheidet beim Stehen im Kühlschrank Neochebulinsäure aus. Durch erneute Hydrolyse der unveränderten Chebulinsäure kommt man zu einer Ausbeute von zirka 58%.

Neochebulinsäure ist in kaltem Wasser schwer, dagegen schon bei 35—40° sehr leicht löslich, so daß sie gut von den begleitenden Stoffen getrennt werden kann. Sie ist schön kristallisiert (fünfeckige Tafeln), hat einen vergleichsweise scharfen Zersetzungspunkt (193—195°) und dreht $+12,7°$ (Äthanol, $c = 2,5$) (Chebulinsäure: $+65°$ in Äthanol $+$ Wasser). Die Kristalle sind wasserhaltig (6 H_2O) und die wasserfreie Substanz ist sehr hygroskopisch.

Die Formel der Neochebulinsäure ist $C_{41}H_{34}O_{28}$ im Gegensatz zu Chebulinsäure $C_{41}H_{32}O_{27}$. Während Chebulinsäure eine einbasische Säure ist, verbraucht Neochebulinsäure zwei Äquivalente Lauge. Mit Diazomethan erhält man aus Chebulinsäure eine amorphe Tridekamethylchebulinsäure [FISCHER und BERGMANN (*18*), vgl. (*85*)], aus Neochebulinsäure dagegen eine amorphe Tetradekamethyl-verbindung, die mit Methyljodid und Silberoxyd zu einer (ebenfalls amorphen) Pentadekamethylneochebulinsäure umgewandelt wird.

Die Umwandlung von Chebulinsäure in Neochebulinsäure vollzieht sich also unter Aufnahme eines Mols Wasser, wobei eine Carboxylgruppe und eine aliphatische Hydroxylgruppe frei werden, d. h. die Umwandlung muß in der Aufspaltung eines Esters oder eines Lactons bestehen. Über Deutungsversuche wird nach Besprechung der Trigalloyl-glucose berichtet.

Zur *Darstellung der Trigalloyl-glucose* hydrolysiert man Chebulinsäure etwas länger (bis zum Auftreten der ersten Spuren Gallussäure im Chromatogramm), oder man setzt Neochebulinsäure ein. Nach Abtrennung von unveränderter Chebulinsäure und Neochebulinsäure durch Kristallisation bringt man die Lösung auf p_H 6,0—6,2 und extrahiert die Trigalloyl-glucose mit Essigester im Vakuum. Die Verbindung kristallisiert nicht, sondern fällt aus wäßriger Lösung als Gallerte. Sie ist in kaltem Wasser recht schwer, in der Wärme auch nur mäßig löslich. Weder der Ennea-methyläther noch das Ennea-acetat sind kristallisiert.

Da die Verbindung mit Anilinphthalat keine Färbung gibt, muß die dritte Gallussäure glykosidisch gebunden sein, d. h. Trigalloyl-glucose aus Chebulinsäure ist *1,3,6-Trigalloyl-glucose*. Die Bestimmung der Konfiguration am C-Atom 1 wurde in der gleichen Weise untersucht, wie es bei Corilagin (S. ҫ4) beschrieben ist, und erwies sich ebenfalls als β-glykosidisch.

Die 1,3,6-Trigalloyl-glucose wurde auf folgendem Wege *synthetisiert* [KLINGER (*45*)]:

2,4-Dibenzyl-lävoglucosan (XXXIX) (*109*) wird mit Tribenzylgalloyl-chlorid in Pyridin in 7 Tagen bei 40° umgesetzt zu 3-Tribenzylgalloyl-2,4-dibenzyl-lävoglucosan (XL). Die Öffnung des 1,6-Ringes zu 3-Tri-benzylgalloyl-2,4-dibenzyl-glucose (XLI) gelingt, wenn Verbindung (XL) in benzolischer Lösung, die wenig einer 10 vol.-proz. Lösung von konz. Schwefelsäure in Dioxan und viel Trifluoressigsäure-anhydrid enthält, 24 Stunden bei 20° aufbewahrt wird, mit etwa 25%iger Ausbeute. Nun wird Verbindung (XLI) mit Tribenzylgalloyl-chlorid und Pyridin umge-setzt, wozu 50 Tage bei 60° erforderlich sind, und (XLII) erhalten. Durch Hydrogenolyse werden alle 11 Benzylgruppen abgespalten. Obgleich die Verbindungen (XXXIX)—(XLII) schon kristallisiert sind, ist das End-produkt, die synthetische 1,3,6-Trigalloyl-glucose (XLIII) wie die natür-liche nur als Gallerte zu erhalten. Analyse, Drehung und R_f-Wert stimmen überein.

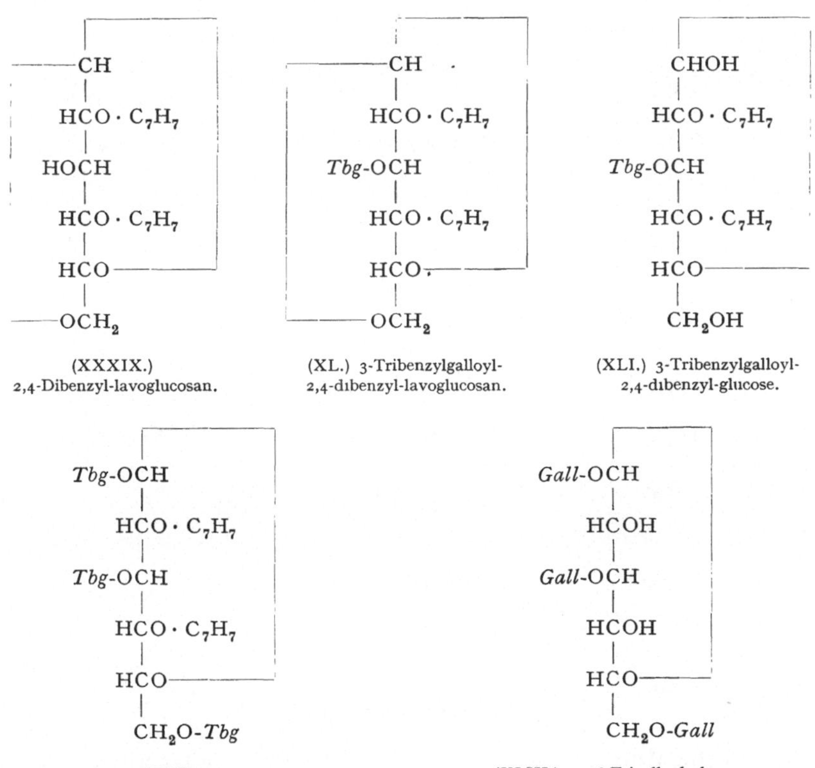

(XXXIX.)
2,4-Dibenzyl-lavoglucosan.

(XL.) 3-Tribenzylgalloyl-
2,4-dibenzyl-lavoglucosan.

(XLI.) 3-Tribenzylgalloyl-
2,4-dibenzyl-glucose.

(XLII.)
1,3,6-Tri-(tribenzylgalloyl)-2,4-dibenzyl-glucose.

(XLIII.) 1,3,6-Trigalloyl-glucose.

Die *Gesamtkonstitution der Chebulinsäure* ist damit noch nicht aufgeklärt. Als Arbeitshypothese hat O. Th. Schmidt (75) die Formel (XLIV) aufgestellt. In ihr ist gesichert die 1,3,6-Trigalloyl-β-D-glucose. Als Haftstelle oder Haftstellen der Chebulsäure an den Zucker kommen nur die Hydroxyle 2 und 4 in Betracht. Da Chebulinsäure eine Monocarbonsäure ist [Freudenberg und Frank (26)], da sie ferner mit Diazomethan 13 und nur 13 Methoxylgruppen aufnimmt, sollte die Chebulsäure mit 2 Carboxylgruppen an 2 Hydroxylgruppen des Zuckers gebunden sein.

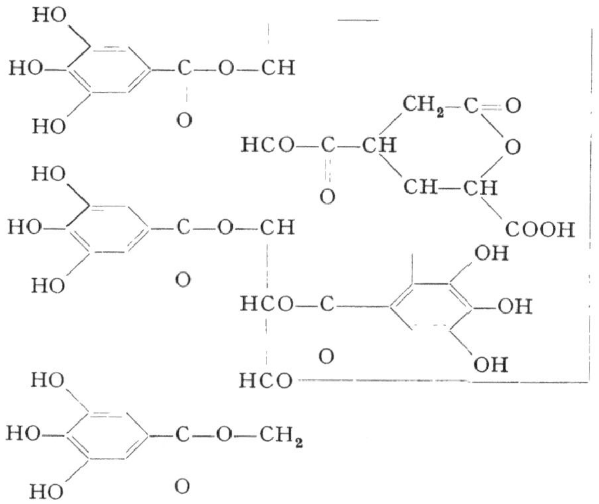

(XLIV.) Chebulinsaure (?).

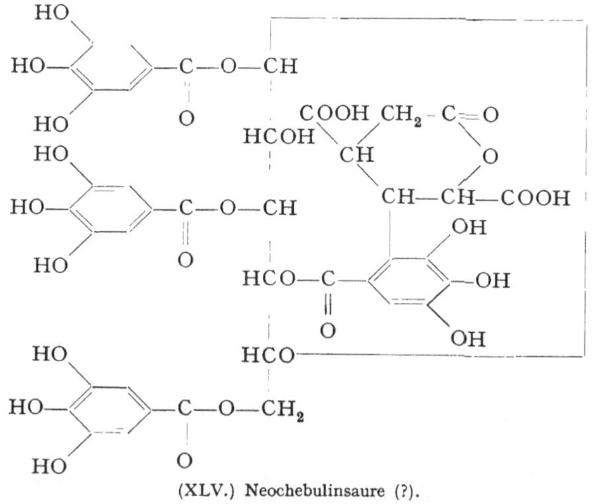

(XLV.) Neochebulinsaure (?).

Die Chebulsäure ist in Formel (XLIV) anders angeordnet als in isoliertem Zustand [Formel (XXIII), S. 79]. Der Grund hierfür sind

Überlegungen über mögliche genetische Beziehungen, die im Kapitel IV behandelt werden (S. 127). Aber auch dann noch stellt die Anordnung und Bindung der Chebulsäure nur eine unter mehreren möglichen Formeln dar.

Da bei der Umwandlung der Chebulinsäure in Neochebulinsäure lediglich die Elemente eines Mols Wasser aufgenommen werden, aber kein Baustein (ganz) abgespalten wird, muß sich diese Umwandlung im Bereich der Chebulsäure abspielen. Am wahrscheinlichsten ist es, daß eine der Esterbindungen zu den Hydroxylgruppen 2 oder 4 der Glucose gelöst wird. Und hier ist es wieder wahrscheinlicher, daß die aliphatische Carboxylgruppe frei wird, denn die aromatische sollte sich grundsätzlich gleich verhalten wie die in 3 und 6 gebundenen Gallussäure-carboxylgruppen. Weniger wahrscheinlich ist es, daß lediglich die Lactongruppe im aliphatischen Teil der Chebulsäure geöffnet wird, denn das Lacton müßte sich teilweise wieder schließen, wobei eine Drehungsänderung zu erwarten wäre, d. h. in diesem Falle sollte die Neochebulinsäure nach der Auflösung ihre Drehung ändern.

Aber alle drei Möglichkeiten stehen in Übereinstimmung mit der Tatsache, daß in Neochebulinsäure eine neue Carboxylgruppe frei ist, die für den Eintritt einer weiteren Methoxylgruppe mit Diazomethan (Tetradekamethyl-neochebulinsäure) zur Verfügung steht, und daß nun auch eine aliphatische Hydroxylgruppe der Methylierung mit Methyljodid und Silberoxyd zugänglich geworden ist. So kann man mit allen angedeuteten Vorbehalten eine wahrscheinliche Formel für Neochebulinsäure (XLV) aufstellen.

III. Ellagen-gerbstoffe.

1. Übersicht.

Als Ellagen-gerbstoffe bezeichnet man solche pflanzlichen Gerbstoffe, die, sei es durch Hydrolyse mit Säuren, sei es auch nur beim Stehen der Extrakte, *Ellagsäure* ausscheiden. Mitunter erfolgt diese Abscheidung auch beim Gerben mit solchen Stoffen auf den Häuten. Es bilden sich Kristallhäufchen, an deren Stelle das rohe Leder dann Flecke aufweist, die der Gerber als „Blume" bezeichnet. Ellagen-gerbstoffe werden von durchaus verschiedenen Pflanzenarten hervorgebracht. Die wichtigsten Gerbmittel dieser Klasse sind:

Valonea (die Fruchtbecher von *Quercus valonea*, einer kleinasiatischen Eiche, oder von *Qu. macrolepis*),

Myrobalanen (die getrockneten, in frischem Zustand pflaumenartigen Früchte von *Terminalia chebula* aus Indien) und

Dividivi und Algarobilla (die getrockneten Fruchtschotten von *Caesalpinia coriaria* und *C. brevifolia*, zweier Akazienarten aus Mittel- und Südamerika).

Aber auch die Gerbstoffe aus unserer europäischen Edelkastanie *(Castanea vesca)* und Eichenrinde spalten bei der Hydrolyse Ellagsäure ab. Ein wenig Ellagsäure wird auch bei der Spaltung des türkischen Gallotannins (aus den Zweig-gallen von *Quercus infectoria*, den sogenannten „Aleppogallen") erhalten. Reicher an Ellagsäure-bildenden Gerbstoffen sind die Schalen der Granatäpfel sowie die Wurzel- und Zweigrinde des Granatbaumes (*Punica granatum*, einer Myrtacea). Nach einer Angabe von Nierenstein (*61*) aus dem Jahre 1919 scheinen auch die Knoppern (gallenähnliche Auswüchse, die sich an jungen Früchten der Stieleichen, *Qu. pedunculata*, durch den Stich der Knoppernwespe bilden) Ellagen-gerbstoffe zu enthalten.

Wie sich im Verlaufe dieses Kapitels ergeben wird, kann man sicher einen Teil der Ellagen-gerbstoffe zur Klasse der Gallotannine rechnen.

Es ist immer wieder versucht worden, solche Verbindungen, die neben anderen Bausteinen (Zucker, Gallussäure) auch Ellagsäure enthalten, aus den Gerbstoffgemischen der Pflanzenextrakte abzutrennen und in reiner Form darzustellen. Dies ist bis vor einigen Jahren in keinem Falle mit Sicherheit gelungen, und alle analytischen Daten über solche Ellagen-gerbstoffe bedürfen der Nachprüfung, sobald die Stoffe einmal in ein-heitlicher Form erhalten werden.

Über die Gerbstoffe der Myrobalanen liegen alte Arbeiten von Löwe (*50*) und Zölffel (*110*) vor, in welchen die Gewinnung amorpher, Ellagsäure enthaltender Produkte beschrieben wird, aus welchen später Nierenstein (*60*) durch mehr-fache Carbäthoxylierung und Verseifung einen kristallisierten Ellagen-gerbstoff der Formel $C_{26}H_{28}O_{19}$ erhalten zu haben angibt, in welchem 1 Mol Ellagsäure mit 2 Molen Glucose verbunden sein sollen. Die Angaben Nierensteins konnten nicht reproduziert werden [O. Th. Schmidt und Nieswandt (*94*)]. Auch die spätere Mitteilung Nierensteins (*62*), er habe dieselbe Verbindung ohne den Umweg über die Carbäthoxylierung aus einer großen Zahl von Pflanzen isoliert, bedarf der Nachprüfung. Reichel, der vorübergehend den Ellagen-gerbstoffen die Konstitution von galloylierten Ellagsäuren zugesprochen hatte (*72*), hat 1941 in einer kurzen Mitteilung (*70*) von der Isolierung einer ganzen Reihe optisch aktiver, offenbar amorpher Ellagen-gerbstoffe berichtet, doch fehlen hierzu experimentelle Angaben.

Den ersten einheitlichen und kristallisierten Ellagen-gerbstoff, der bekannt geworden ist, isolierten 1948 O. Th. Schmidt und Nieswandt (*93, 94*). Er wurde „Chebulagsäure" genannt. Zu seiner Darstellung wurden ursprünglich noch ausschließlich die klassischen Arbeitsmethoden verwandt: Extraktionen bei verschiedenem p_H, fraktionierte Fällungen der Zinkverbindungen, fraktionierte Kristallisation. Zwei Jahre später fanden Schmidt und Lademann (*88*) denselben Gerbstoff auch in Dividivi und entdeckten kurz darauf einen zweiten kristallisierten Ellagen-gerb-stoff in Dividivi, den sie „Corilagin" nannten (*90*). Schließlich konnten O. Th. Schmidt und D. M. Schmidt (*97*) Corilagin auch in Myrobalanen nachweisen. Bei der Isolierung der Gerbstoffe aus Dividivi hat die Methode der Gegenstromverteilung gute Dienste geleistet.

Aus den jungen Trieben der Edelkastanie hat Mayer (*51*) 1952 eine interessante neue Phenolcarbonsäure, die „Dehydro-digallussäure" isoliert und aufgeklärt, die sehr wahrscheinlich ebenfalls an Zucker gebunden

vorkommt, und somit bei den Gallotanninen zu behandeln ist. 1954 haben
O. Th. Schmidt und Bernauer (77) aus *Algarobilla* einen neuen kristallisierten Gerbstoff entdeckt und „Brevilagin" genannt, der gelb gefärbt
und optisch aktiv ist ($[\alpha]_D = + 146{,}6°$, in Wasser), und der bei der Hydrolyse eine bisher unbekannte Phenolcarbonsäure abspaltet, die „Brevifolin-
carbonsäure" genannt wurde. Während die eingehende Untersuchung
des Brevilagins noch aussteht, ist die Konstitution der Brevifolin-carbonsäure weitgehend, und die ihres Decarboxylierungsprodukts „Brevifolin"
völlig geklärt worden.

Aus Valonea schließlich ist es noch nicht gelungen, einen einheitlichen
Gesamtgerbstoff zu isolieren. Indessen haben Schmidt und Komarek (87)
wiederum einen neuen Gerbstoffbaustein, die Valoneasäure aufgefunden
und aufgeklärt.

Zur *Gewinnung* und *Reindarstellung* der neuen Verbindungen wurde allgemein
folgendes Verfahren angewendet: Gereinigte Perkolate oder Extrakte der pflanzlichen Materialien werden, wie das 1936 Phillips (66) in ähnlicher Weise beschrieben
hat, zuerst bei p_H 6—6,2 im Vakuum mit Essigester erschöpfend extrahiert, dann
auf p_H 2—3 gebracht und wiederum erschöpfend extrahiert. Zuerst erhält man
die „phenolsauren" Gerbstoffe, dann die „carboxylsauren". Diese Fraktionen
werden mit Blei oder Zink fraktioniert gefällt und einzelne Fraktionen aus diesen
Fällungen · durch fraktionierte Kristallisation oder durch Gegenstromverteilung
weiter gereinigt. Der Gang der Reinigung kann häufig polarimetrisch und immer
papierchromatographisch überwacht und gesteuert werden.

2. Corilagin.

Die Verbindung wird aus der phenol-sauren Fraktion von Dividivi-
Extrakten über fraktionierte Bleifällung und Gegenstromverteilung mit
Wasser/Essigester gewonnen. Corilagin bildet lange, farblose Prismen,
die sich bei 204—205° zersetzen. Es ist in Wasser ziemlich schwer löslich
und besitzt die auffallend hohe spezifische Drehung von —246° (in
Äthanol), die die Auffindung und Herausarbeitung aus dem Gerbstoff-
gemisch begünstigt hat. Die Bruttoformel ist $C_{27}H_{22}O_{18}$. Corilagin besitzt
keine freie Carboxylgruppe. Mit Diazomethan wird das kristallisierte
Enneamethyl-corilagin erhalten, dessen Weitermethylierung mit Methyljodid und Silberoxyd zum ebenfalls kristallisierten Hendekamethyl-
corilagin führt [O. Th. Schmidt und Lademann (47, 90)]. Auch das
Hendekaacetyl-corilagin ist kristallisiert [O. Th. Schmidt, D. M. Schmidt
und Herok (98)].

Die *Hydrolyse* des Corilagins mit verdünnter Schwefelsäure führt
zu je einem Mol. Glucose, Gallussäure und Ellagsäure (90). Hydrolysiert
man indes Enneamethyl-corilagin mit Alkali [O. Th. Schmidt, Blinn
und Lademann (79)], so erhält man unter Zerstörung des Zuckers zwar
Trimethyl-gallussäure, aber keine methylierte Ellagsäure, sondern
Hexamethoxy-diphensäure.

In den beiden aromatischen Spaltstücken sind alle neun Methoxylgruppen des Enneamethyl-corilagins enthalten. Sie erweisen sich als
Äther-methoxylgruppen, was damit übereinstimmt, daß Corilagin kein

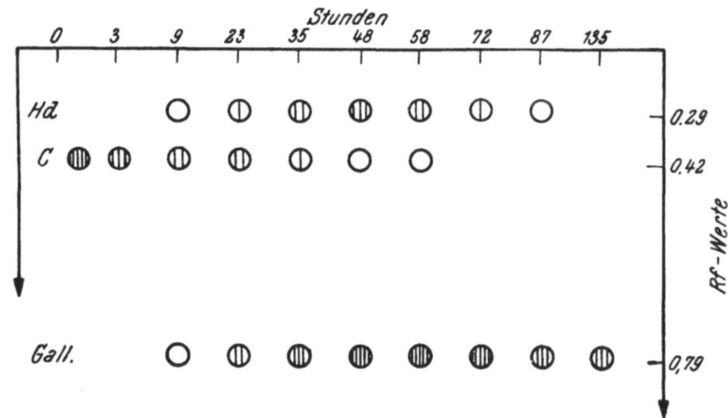

Abb. 2a. Hydrolyse des Corilagins. Indikator $FeCl_3$. Die Intensitat der Flecke ist durch die Schraffur angedeutet. *Hd* = Hexaoxy-diphenoyl-glucose; *C* = Corilagin; *Gall* = Gallussaure.

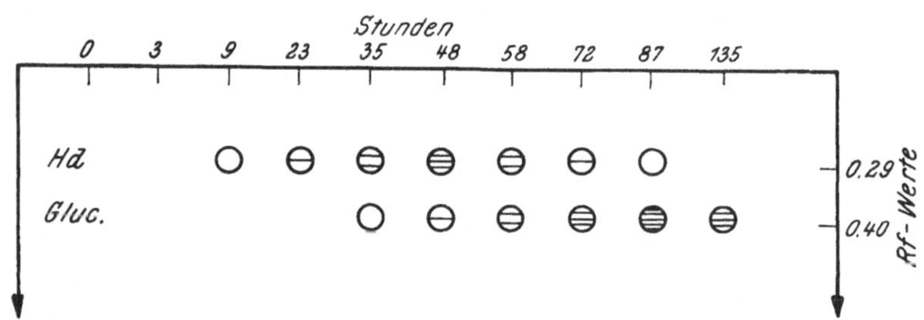

Abb. 2b. Hydrolyse des Corilagins. Indikator Anilinphthalat. *Hd* = Hexaoxy-diphenoyl-glucose; *Gluc* = Glucose.

freies Carboxyl besitzt. Die bei der Hydrolyse des Corilagins auftretende
Ellagsäure ist also nicht als solche, sondern als Hexaoxy-diphensäure,
deren doppeltes Lacton sie ist, an den Zucker gebunden. Da alle neun
aromatischen Hydroxylgruppen im Corilagin frei sind, können Gallussäure und Hexaoxy-diphensäure nicht nach Art der *m*-Digallussäure
miteinander verbunden sein, sondern jede der Säuren ist mit Hydroxylgruppen des Zuckers verestert. Dabei muß notwendigerweise die Hexaoxy-diphensäure mit ihren beiden Carboxylgruppen an zwei Hydroxylgruppen des Zuckers haften.

Die sehr unwahrscheinliche Annahme, daß die Hexaoxy-diphensäure nur mit
einer Carboxylgruppe an den Zucker gebunden sei und mit der anderen ein ge-

mischtes Anhydrid mit der Gallussäure bilde, kann schon auf Grund der Brutto-formel des Corilagins ausgeschlossen werden.

Corilagin wird von Wasser bei 100° allmählich hydrolysiert. Die papierchromatographische Verfolgung dieses Vorganges (*Abb. 2a und 2b*) zeigt, daß nach 9 Stunden zuerst Gallussäure abgespalten wird.

Gleichzeitig tritt, langsamer wandernd als Corilagin, eine neue Ver-bindung auf, die eine freie reduzierende Gruppe (angezeigt mit Anilin-phthalat) (*64, 64a*) besitzt. Es handelt sich hier um eine Hexaoxy-diphenoyl-glucose, deren Menge im weiteren Hydrolysenverlauf nach Maß-gabe abgespaltener Gallussäure zunächst anwächst, um dann wieder ab-zunehmen, da auch Hexaoxy-diphensäure, wenn auch viel langsamer, abgespalten wird und, zuerst nach 35 Stunden, als Ellagsäure ausgeschie-den wird.

(XLVI.) (*R* = H.) Corilagin.
(XLVII.) (*R* = CH₃.)
(LI.) (*R* = CO·CH₃.) Hendekaacetyl-corilagin.

(XLVIII.) (*R* = H.)
(IL.) (*R* = CH₃.)

(La.) 2,4-Dimethyl-α-methylglucosid.

(Lb.) 2,4-Dimethyl-β-methylglucosid.

Der Versuch zeigt, daß die Gallussäure im Corilagin glucosidisch, die Hexaoxy-diphensäure an andere Hydroxylgruppen des Zuckers gebunden ist. Welche Hydroxylgruppen dies sind, zeigt der Abbau des Hendekamethyl-corilagins zu dem zugrunde liegenden methylierten Zucker [Formeln (XLVII) bis (L)] [O. Th. Schmidt, D. M. Schmidt und Herok (98)]. Aus dem völlig methylierten Corilagin (XLVII) wird selbst mit überschüssigem Kaliummethylat in Chloroform nur die Trimethyl-gallussäure in 1-Stellung (als Methylester) abgespalten. Das verbleibende Zweierstück, eine amorphe Octamethyl-(hexaoxy-diphenoyl)-glucose (XLVIII), wurde mit methanolischer Chlorwasserstoffsäure glucosidiert und das erhaltene Octamethyl-(hexaoxy-diphenoyl)-α- und -β-methyl-glucosid-Gemisch (IL) nun mit wäßrigem Alkali hydrolysiert. Nach Abtrennung der Hexamethoxy-diphensäure wurde ein zunächst sirupöses Gemisch von Dimethyl-α- und -β-methylglucosid (L a) und (L b) isoliert, aus dem eine Komponente leicht und rasch auskristallisierte. Sie erwies sich als identisch mit dem 2,4-Dimethyl-β-methylglucosid (L b) von Reeves, Adams und Goebel (69, 69 a).

Somit war bewiesen, daß im Corilagin (XLVI) die Hydroxylgruppen 2 und 4 der Glucose allein frei sind, daß die Hexaoxy-diphensäure in 3- und 6-Stellung gebunden ist und daß ein Pyranose-Ring vorliegt.

Es verblieb noch die Frage nach der Konfiguration am C-Atom 1 der Glucose.

Zu dieser Konfigurationsbestimmung am glykosidischen C-Atom wendeten O. Th. Schmidt, D. M. Schmidt und Herok (98) erstmalig das Verfahren der präparativ schon länger bekannten Umwandlung von β- in α-Glykoside an, nachdem O. Th. Schmidt und Herok (86) es zuvor an Glucogallin erprobt hatten. Das Heptaacetat des Glucogallins (1-Galloyl-β-glucose) wird unter Einwirkung von Bortrifluorid (48) in Chloroform in das Heptaacetat der 1-Galloyl-α-glucose umgewandelt. Auf diesem Wege ist das Heptaacetyl-α-glucogallin selbst aus synthetisch hergestelltem β-Glucogallin leichter zugänglich geworden als durch die direkte Synthese von Fischer und Bergmann (19), und zudem konnte das bisher nur amorph und unrein bekannte α-Glucogallin selbst in reiner kristallisierter Form dargestellt werden. Die Umwandlung des Hepta-acetyl-β-glucogallins ($[\alpha]_D = -24°$, in Wasser) in Heptaacetyl-α-gluco-gallin ($[\alpha]_D = +104°$) ist mit einer kräftigen Rechtsverschiebung der Drehung verbunden. Eine vergleichbare Rechtsverschiebung wurde festgestellt, als das Hendekaacetyl-corilagin (Formel LI, S. 93, $[\alpha]_D = -27,2°$) mit Bortrifluorid in Chloroform in ein neues (amorphes) „Hendekaacetyl-α-corilagin" ($[\alpha]_D = +59,5°$) umgewandelt wurde.

Somit ist gezeigt, daß Corilagin am C-Atom 1 β-Konfiguration besitzt. Diese Methode der Konfigurationsbestimmung dürfte allgemein anwendbar sein. Im vorliegenden Fall beruht der Beweis auf dem Vergleich zweier Substanzen, die in ihrer Konstitution sehr nahe miteinander verwandt sind: Corilagin ist 3,6-Hexaoxy-diphenoyl-glucogallin.

Die aus dem methylierten Corilagin abgespaltene Hexamethoxy-diphensäure unterscheidet sich von der Hexamethoxy-diphensäure, die HERZIG und POLAK (40) durch Methylierung der Tetramethyl-ellagsäure mit Methyljodid und Alkali erhalten haben, zunächst durch den Schmelzpunkt.

Sie schmilzt bei 158—161°, wird bei weiterem Erhitzen zwischen 200—220° wieder fest und schmilzt dann zum zweiten Male wie die HERZIGsche Säure bei 240°. Läßt man nun erkalten und erhitzt erneut, so wird nur der Schmelzp. von 240° erhalten. Dieses Verhalten wies auf Racemisation hin. In der Tat ist die Hexamethoxy-diphensäure aus Corilagin optisch aktiv und dreht +25,8°.

Die Ursache der optischen Aktivität der Hexamethoxy-diphensäure ist die Behinderung der freien Drehbarkeit um die Diphenyl-bindung. Dies ist vermutlich der erste Fall von „Atropisomerie" bei einem Naturstoff.

Es ist auch neuartig, daß eine Dicarbonsäure mit zwei Carboxylgruppen an ein Zuckermolekül gebunden ist*. Die Betrachtung des Modells (Stuart-kalotten) zeigt, daß die Bindung der Hexaoxy-diphensäure in 3,6-Stellung (12-Ring) nicht nur spannungsfrei ist, sondern noch eine gewisse Beweglichkeit zuläßt. Das Modell zeigt aber auch, daß die beiden Benzolringe der gebundenen Diphensäure unmöglich komplanar sein können. Das bedeutet, daß die Hexaoxy-diphensäure in situ optisch aktiv sein muß und diese Eigenschaft nicht erst bei der Methylierung erhält. Bei der Hydrolyse des Corilagins wird die abgespaltene Hexaoxy-diphensäure sehr rasch in Ellagsäure umgewandelt und verliert ihre Asymmetrie. Über die Darstellung und die Eigenschaften der optisch aktiven Hexaoxy-diphensäure und einiger ihrer Abkömmlinge wird in Abschnitt 4, S. 99, berichtet.

Mit der Aufklärung der Konstitution und Konfiguration des Corilagins ist der erste natürliche Gerbstoff der Gallotannin-Klasse vollständig aufgeklärt worden.

3. Chebulagsäure.

a) Beschreibung und Analyse.

Die Chebulagsäure wird aus der carboxylsauren Fraktion der von Chebulinsäure schon weitgehend (durch Kristallisation) befreiten Myrobalanen-perkolate durch fraktionierte Fällung mit Zinkacetat und an-

---

* Bei der Hydrolyse der methylierten Chebulagsäure (vgl. S. 96) ist die gleiche optisch aktive Hexamethoxy-diphensäure von O. TH. SCHMIDT und BLINN (78) schon etwas früher isoliert und ihre „beidarmige" Bindung an die Glucose festgestellt worden. Zur übersichtlicheren Darstellung wurde die Beschreibung dieser Beobachtungen hier vorweggenommen.

schließende fraktionierte Kristallisation [O. Th. Schmidt und Nieswandt (*94*)] oder durch Gegenstromverteilung mit Essigester/Wasser [Blinn (*9*)] gewonnen. Aus der carboxylsauren Fraktion der Dividiviperkolate wird mit Blei gefällt, dann das Material der Gegenstromverteilung unterworfen [O. Th. Schmidt und Lademann (*88*)].

Chebulagsäure bildet meist sehr schön ausgebildete Rhomben, die beim Erhitzen ohne zu schmelzen oberhalb 240° dunkel werden. Sie ist nahezu ebenso schwer löslich wie Chebulinsäure und bei ungünstigem Mischungsverhältnis schwer durch Kristallisation von dieser zu trennen. Die Drehung beträgt —57,2° (in Äthanol), die Bruttoformel $C_{41}H_{30}O_{27}$.

Die Titration zeigt eine freie Carboxylgruppe an (*107*). Bei der durchgreifenden Hydrolyse mit verdünnter Schwefelsäure entstehen je ein Mol. Glucose, Gallussäure, Chebulsäure und Ellagsäure (*94, 88*).

Mit Diazomethan nimmt Chebulagsäure ebenso wie Chebulinsäure 13 Methylgruppen auf. Neben einer kleinen Menge einer kristallisierten Dodekamethyl-chebulagsäure entsteht als Hauptprodukt eine amorphe Tridekamethyl-verbindung. Beide Methylierungsprodukte geben bei der Spaltung mit Kaliummethylat Glucose, Trimethylgallussäure und (+)-Hexamethoxy-diphensäure; die Tridekamethyl-verbindung liefert außerdem Trimethylchebulsäure, die Dodekamethyl-verbindung eine methoxylärmere Chebulsäure (*79*).

Wie bei Corilagin dargelegt, beweist das Auftreten von Hexamethoxy-diphensäure, daß die Ellagsäure, die bei der sauren Hydrolyse der Chebulagsäure erhalten wird, nicht als solche, sondern als Hexaoxy-diphensäure gebunden sein muß und erst nach erfolgter Hydrolyse durch doppelten Lactonschluß entsteht.

Indes ließ sich der Beweis, daß die Hexaoxy-diphensäure „beidarmig" an den Zucker gebunden ist, bei der komplizierteren Chebulagsäure nicht durch einfache Methoxylbilanz führen. Wenn aber Chebulagsäure in geeigneter Weise mit heißem Wasser hydrolysiert wird, so gelingt die Darstellung einer (amorphen) Hexaoxy-diphenoyl-glucose. Diese nimmt mit Diazomethan 6 Methylgruppen auf, wodurch die beidarmige Bindung der Hexaoxy-diphensäure an den Zucker auch für die Chebulagsäure bewiesen ist (*79*). Diese Feststellung ließ es möglich erscheinen, daß — abgesehen von der Chebulsäure — Chebulagsäure und Corilagin in ihrem Bau miteinander übereinstimmen. Das ist in der Tat der Fall.

b) *Hydrolyse mit Wasser; Neochebulagsäure.*

Der schrittweise *Abbau* der Chebulagsäure mit Wasser bei 60° (oder 95°) [O. Th. Schmidt und D. M. Schmidt (*96*)] läßt sich polarimetrisch und papierchromatographisch sehr schön verfolgen und ist ein Musterbeispiel für die Verwendbarkeit der Papierchromatographie zum Studium von Reaktionsabläufen.

Die erste Veränderung der Chebulagsäure mit Wasser bei 60° vollzieht sich, ähnlich wie beim Abbau der Chebulinsäure, ohne Abspaltung eines der vier Bausteine. Schon nach etwa 6 Stunden tritt im Chromatogramm ein neuer Fleck kleineren R_f-Wertes auf, der „Neochebulagsäure" anzeigt. Während die optische Drehung von —57° auf etwa —100° bis —110° ansteigt, nimmt die Umwandlung der Chebulagsäure in Neochebulagsäure zu. Als erstes Spaltstück tritt dann die Chebulsäure in Erscheinung, deren Menge nach 8 Tagen ihren größten Wert erreicht hat. Das gleichzeitig mit Chebulsäure auftretende „Dreierstück" läßt sich nicht nachweisen, weil es in den drei verwendeten Chromatographiergemischen von Chebulsäure überlagert wird. Als nächstes wird nach 8—8$^1/_2$ Tagen die beginnende Abspaltung der Gallussäure sichtbar. Unterbricht man in diesem Zeitpunkt die Hydrolyse, neutralisiert bei $p_H = 6{,}2$ und extrahiert kontinuierlich mit Essigester, so erhält man im Essigester-extrakt den einzigen phenolsauren Anteil im Reaktionsgemisch, das „Dreierstück". Es erweist sich als Corilagin und kann auf diese Weise auch präparativ aus Chebulagsäure mit etwa 52%iger Ausbeute dargestellt werden.

Damit ist gezeigt, daß in der Chebulagsäure die Gallussäure β-glucosidisch in 1-Stellung der Glucose und die Hexaoxy-diphensäure in 3- und 6-Stellung gebunden sind.

Auch die Neochebulagsäure läßt sich leicht darstellen [HENSLER (*39*)], wenn man die Hydrolyse der Chebulagsäure in Wasser bei 55° nach dem ersten Auftreten der Chebulsäure (80—90 Stunden) abbricht, den Hauptteil der unveränderten Chebulagsäure aus der eingeengten Lösung durch Kristallisation abtrennt und den Rest der Gegenstromverteilung (2 Essigester/1 Wasser) unterwirft. Die Neochebulagsäure $C_{41}H_{32}O_{28}$ bildet lange, farblose Prismen vom Schmelzp. 196° (Zers.) und dreht —119° (Äthanol) und —163° (Wasser).

Während Chebulagsäure mit 10 Molen Wasser kristallisiert, enthält Neochebulagsäure lufttrocken 9 Mole Wasser, so daß die Bruttoformeln der beiden Säuren ($C_{41}H_{30}O_{27}$, 10 H_2O und $C_{41}H_{32}O_{28}$, 9 H_2O) miteinander übereinstimmen.

Neochebulagsäure enthält zwei freie Carboxylgruppen.

Mit Diazomethan wird aus Neochebulagsäure die kristallisierte Tetradekamethyl-neochebulagsäure $C_{55}H_{60}O_{28}$, Schmelzp. 246° (Zers.), $[\alpha]_D = —83{,}2°$ (in Aceton) erhalten (*39*).

c) Zur Gesamtkonstitution.

Für die gesamte Konstitution der Chebulagsäure ergibt sich Formel (LII), deren linker Teil, das Corilagin darstellend, gesichert ist, während für die Anordnung und Bindung der Chebulsäure und für die aus For-

mel (LII) abgeleitete Formel (LIII) der Neochebulagsäure die gleichen
Vorbehalte gelten, die bei der Chebulinsäure (S. 88 ff.) besprochen wurden.

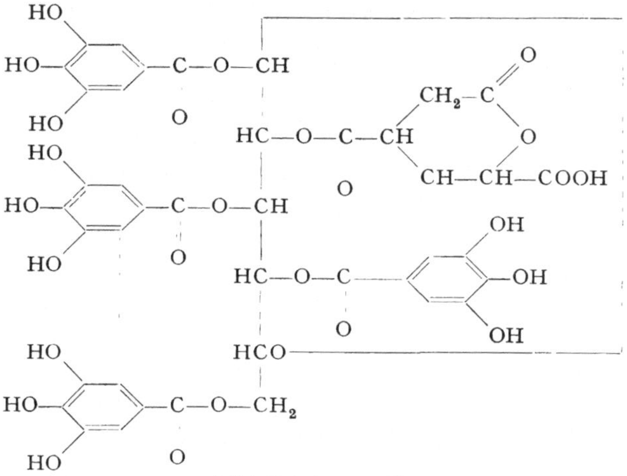

(LII.) Chebulagsaure (?).

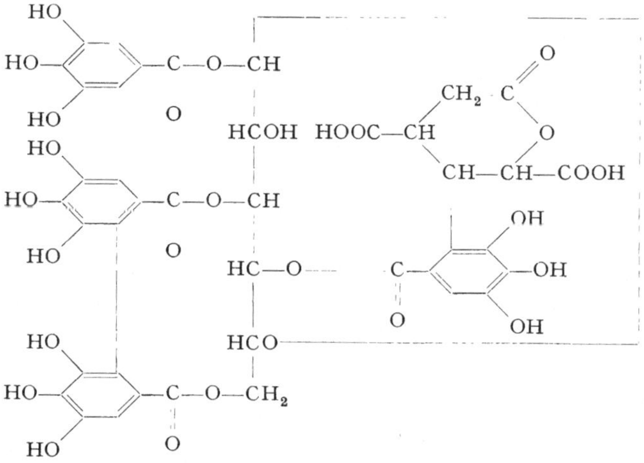

(LIII.) Neochebulagsaure.

Wenn die Annahme zutrifft, daß die Chebulsäure in der Chebulin-
und Chebulagsäure gleich gebunden ist, so würde sich die interessante
Tatsache ergeben, daß die beiden Säuren sich nur dadurch voneinander
unterscheiden, daß zwischen den beiden Gallussäuren der Chebulinsäure
in 3- und 6-Stellung eine Dehydrierung zur Diphensäure stattgefunden
hat. Versuche, Chebulinsäure durch Dehydrierung in Chebulagsäure
überzuführen, sind indes noch nicht gelungen.

Es ist in diesem Zusammenhang erwähnenswert, daß zwar in den Myrobalanen Chebulinsäure und Chebulagsäure nebeneinander vorkommen, daß aber in Dividivi keine Chebulinsäure gefunden worden ist. Es verdient ferner der Erwähnung, daß Corilagin ursprünglich sowohl in Dividivi wie in Myrobalanen neben Chebulagsäure enthalten ist und nicht erst bei der Aufarbeitung aus dieser entsteht [O. TH. SCHMIDT und D. M. SCHMIDT (*97*)]. Da aber beide Früchte vor dem Versand getrocknet werden, ist die Frage nach der eigentlichen Nativität des Corilagins nicht entschieden.

4. Hexaoxy-diphensäure.

a) Darstellung der optisch aktiven Hexamethoxy- und Hexabenzoxy-diphensäuren sowie der racemischen und aktiven Hexaoxy-diphensäuren.

Die Tatsache, daß in den beiden einzigen bisher bekannten Ellagengerbstoffen Corilagin und Chebulagsäure nicht die Ellagsäure selbst, sondern die *Hexaoxy-diphensäure* gebunden enthalten ist, machte es wahrscheinlich, daß dieser Säure für die Chemie dieser Gerbstoffklasse eine allgemeine Bedeutung zukommt. Daher ist die Hexaoxy-diphensäure eingehender untersucht worden. Zunächst stellten O. TH. SCHMIDT und DEMMLER (*80*) die racemische Hexamethoxy-diphensäure (*41*) (LIVa) durch Methylierung von Tetramethyl-ellagsäure mit Natronlauge und Dimethylsulfat dar und trennten das Racemat sowohl über die Chinidin- wie über die Strychninsalze. Die rechtsdrehende Form (Schmelzp. 161°, $[\alpha] = +25,8°$, in Äthanol) erwies sich als identisch mit der aus den methylierten Gerbstoffen erhaltenen (+)-Hexamethoxy-diphensäure. Beide Formen wandeln sich beim Erhitzen in das Racemat um: sie schmelzen bei 161°, erstarren bei 210—220° und schmelzen zum zweiten Mal bei 240°, dem Schmelzp. der rac. Säure. Die Racemisierungs-Geschwindigkeit der aktiven Säure wurde gemessen (*Tabelle 2*) und die Aktivierungswärme der Racemisierung in $n/10$-NaOH zu 25000 cal. bestimmt. Sodann stellten O. TH. SCHMIDT, VOIGT, PUFF und KÖSTER (*99*) durch die Einwirkung von Phenyldiazomethan oder von Benzylchlorid und Kaliumcarbonat die Tetrabenzyl-ellagsäure und daraus durch weitere Benzylierung mit Benzylchlorid und Kaliumhydroxyd die Hexabenzoxy-diphensäure dar. O. TH. SCHMIDT und DEMMLER (*81*) zerlegten das Racemat mit Cinchonin oder Chinidin und erhielten so die beiden aktiven Hexabenzoxy-diphensäuren (Schmelzp. 147—148°, $[\alpha]_D = -63,6°$, bzw. $+66,2°$, in Chloroform). Schließlich spalteten sie sowohl aus der rac. Hexabenzoxy-diphensäure wie aus den aktiven Formen durch katalytische Hydrierung die Benzylgruppen ab und erhielten die rac. Hexaoxy-diphensäure in kristallisiertem Zustand und die aktiven Formen in Lösung. In der nachstehenden *Tabelle 1* sind die zur *D*-Reihe gehörenden Säuren, Dimethylester und Diamide der Hexaoxy-, Hexamethoxy- und Hexabenzoxy-diphensäure zusammengestellt und ihre experimentelle Verknüpfung angegeben.

Tabelle 1. Drehungswerte in der D-Reihe der Hexaoxy-, Hexamethoxyund Hexabenzoxy-diphensäure.

D-Hexabenzoxy-diphensäure-diamid [M] = —1138° (in Äthanol)	←	D-Hexabenzoxy-diphensäure [M] = —381° (in Äthanol) [M] = —423° (in Methanol)	→	D-Hexabenzoxy-diphensäure-dimethylester [M] = —301° (in Aceton)
		↓ H₂		↓ H₂
D-Hexaoxy-diphensäure-diamid [M] = —113° (in Methanol)		D-Hexaoxy-diphensäure [M] = +92,7° (in Methanol)		D-Hexaoxy-diphensäure-dimethylester [M] = +187° (in Methanol)
				↓ CH₂N₂
D-Hexamethoxy-diphensäure-diamid [M] = —432° (in Äthanol)	→	D-Hexamethoxy-diphensäure [M] = +116° (in Methanol) [M] = +109° (in Äthanol)	←	D-Hexamethoxy-diphensäure-dimethylester [M] = +167° (in Äthanol)

Es ist auffallend, daß in der D-Reihe eine Verschiebung der molaren Drehungen von Dimethylester über die Säure zum Diamid von stärker positiven (schwächer negativen) zu schwächer positiven oder negativen (stärker negativen) Werten stattfindet. Dieselbe Erscheinung zeigt auch die 2,2'-Dimethoxy-diphenyl-6,6'-dicarbonsäure von Stanley, McMahon und Adams (104). Die Zuordnung zur D-Reihe ist nach der beobachteten Drehung der (+)-Hexaoxy-diphensäure erfolgt. Die Bestimmung der absoluten Konfiguration war nicht durchführbar. Es hätte sein können, daß am Kalottenmodell der Chebulagsäure oder des Corilagins nur eine Form der *Hexaoxy-diphensäure* in 3-, 6-Stellung „paßt". Diese hätte man abbilden und als D-Form definieren können. Es zeigte sich aber, daß beide Formen gleich gut „passen" [O. Th. Schmidt, D. M. Schmidt und Herok (98)].

Zwei Fragen interessierten besonders beim Studium der *Hexaoxydiphensäure*, nämlich ihre Stabilität in sterischer Hinsicht und vom Standpunkt der Gerbstoff-chemie mehr noch ihre Stabilität im Hinblick auf die Umwandlung in Ellagsäure.

b) Sterische Stabilität der aktiven Formen.

In *Tabelle 2* sind die beobachteten Drehungen und die Halbwertszeiten der Racemisation zusammengestellt.

Tabelle 2. Drehungswerte und Racemisierungsgeschwindigkeit in der Diphensäure-Reihe.

Nr.	Substanz	$[\alpha]_D$	Halbwertszeit der Racemisierung
1	*D*-Hexaoxy-diphensäure .	+29,1° (Methanol)	281′ (in Methanol, 20°)
2	Monoatrium-salz von Nr. 1...............	—56,8° (Methanol)	88,5′ (Methanol, 20°)
3	Dinatrium-salz von Nr. 1	+132,4° (Methanol + Wasser)	342′ (Methanol + Wasser, 20°)
4	*D*-Hexaoxy-diphensäure-dimethylester	+51,1° (Methanol)	215′ (Methanol, 20°)
5	*D*-Hexaoxy-diphensäure-diamid	—33,7° (Methanol)	21 h (Methanol, 20°)
6	*L*-2,2′-Dimethoxy-diphen-säure (*104*).........	—135,5° (Äthanol)	7 h 50′ (siedende Natron-lauge)
7	*D*-Hexamethoxy-diphen-säure (*80*)...........	+25,9° (Äthanol)	14 h 45′ (siedende Natron-lauge)
8	*D*-Hexabenzoxy-diphen-säure	—43,4° (Äthanol)	racemisiert sich nicht

Tabelle 2 zeigt, daß zwischen den Oxy-diphensäuren, den Methoxy-diphensäuren und den Benzoxy-diphensäuren (LIV) bedeutende Unter-schiede in der sterischen Stabilität bestehen. Während Hexabenzoxy-diphensäure weder durch Erhitzen über den Schmelzp. noch in Lösungs-mitteln racemisiert werden kann, vollzieht sich dieser Vorgang bei der *Hexaoxy-diphensäure* schon in der Kälte so leicht, daß während der Hydrierung der Hexabenzoxy-diphensäure zu *Hexaoxy-diphensäure* immer schon ein Teil der gerade gebildeten aktiven Säure racemisiert wird. Deshalb und wegen der manchmal rasch einsetzenden Bildung (und Abscheidung) von Ellagsäure ist die optisch aktive *Hexaoxy-diphensäure* nicht isoliert, sondern nur in Lösung untersucht worden, und die an-gegebenen Drehwerte sind extrapoliert.

Man erkennt, daß die Substituenten am Hydroxyl-Sauerstoff 2 und 2′ (H, CH_3, $CH_2 \cdot C_6H_5$) einen sehr großen Einfluß auf die Behinderung der freien Drehbarkeit um die Diphenyl-achse ausüben. Für die Beurteilung der Stabilität oder Racemisierbarkeit von Diphenyl-verbindungen hat es sich in vielen Fällen bewährt, die Abstände der an den der Diphenyl-bindung

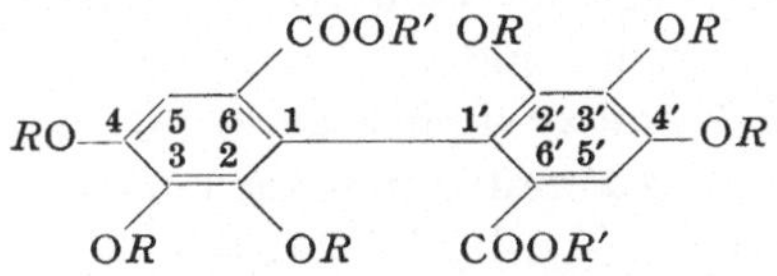

(LIV.) $R = R' = $ H. Oxydiphensaure.
(LIVa.) $R = CH_3$; $R' = $ H. Methoxy-diphensaure.
(LIVb.) $R = CH_2 \cdot C_6H_5$; $R' = $ H. Benzoxy-diphensaure.

benachbarten C-Atome stehenden Atome oder Atomgruppen als Maß zu nehmen (*41*). Obgleich über 2,2'-Dioxy-diphenyl-verbindungen bisher keine Versuche vorlagen, wurden die Abstände C—OH und C—OCH$_3$ in den Tabellen zur Berechnung der Stabilität mit dem gleichen Wert von 1,45 Å eingesetzt. Wollte man annehmen, daß der Substituent R am Hydroxyl-Sauerstoff deshalb keine größere Bedeutung besitzt, weil er ja seinerseits am Sauerstoff gewinkelt und frei drehbar angeordnet ist und nur in besonderen Lagen die freie Drehbarkeit um die Diphenyl-bindung beeinflussen würde, so müßte man erwarten, daß keine oder nur kleinere Unterschiede in der Stabilität der oben beschriebenen Verbindungen bestehen. Dem widerspricht das experimentelle Ergebnis.

Interessant ist der Vergleich der Verbindungen Nr. 4 und 7 der *Tabelle 2*. Bei Hexaoxy-diphensäure-dimethylester (Nr. 4; LIV. $R = $ H; $R' = $ CH$_3$) ist in 2,2'-Stellung OH und in 6,6'-Stellung COOCH$_3$ gebunden. Bei der Hexamethoxy-diphensäure (Nr. 7; LIVa. $R = $ CH$_3$; $R' = $ H) sind an den entscheidenden Stellen mit OCH$_3$ und COOH die gleichen Atome, nur etwas anders gruppiert, angeordnet. Der Unterschied ist groß; die Methoxy-säure ist sehr viel stabiler als der Oxysäure-methylester. Bemerkenswert ist auch die größenordnungsmäßig ähnliche Halbwertszeit von Hexaoxy-diphensäure (Nr. 1; LIV. $R = R' = $ H) und ihrem Dimethylester (Nr. 4; LIV. $R = $ H; $R' = $ CH$_3$). Hier macht also die Substitution der Carboxylgruppe durch Methyl gar nichts aus.

c) Umwandlung in Ellagsäure; „Blume"-Bildung.

Racemische *Hexaoxy-diphensäure*, C$_{14}$H$_{10}$O$_{10}$, unterscheidet sich von Ellagsäure, C$_{14}$H$_6$O$_8$, 2 H$_2$., durch ihre gute Löslichkeit in Methanol, Dioxan, Tetrahydrofuran, Wasser und Aceton, wobei die vollständige Löslichkeit in Aceton das sicherste und schnellste Kriterium für die Freiheit von Ellagsäure ist, da Ellagsäure in diesem Lösungsmittel ganz unlöslich ist. Kristallisierte *Hexaoxy-diphensäure* ist, im Gegensatz zu Ellagsäure, rein weiß und besitzt keinen Schmelzpunkt.

Schon beim Stehen methanolischer Lösungen von Hexaoxy-diphensäure tritt allmähliche Umwandlung in Ellagsäure ein; beim Erwärmen auf 50° ist nach einer Stunde, beim Kochen schon nach 5 Minuten Ellagsäure-Abscheidung zu beobachten. Zusatz von Mineralsäuren beschleunigt die Abscheidung stark. Auch durch Zusatz von Wasser zur methanolischen Lösung wird die Ellagsäure-bildung beschleunigt. Wird eine methanolische Lösung von Hexaoxy-diphensäure mit viel Wasser vermischt, so trübt sie sich nach 1—2 Minuten und nach etwa einer weiteren Minute kristallisiert Ellagsäure aus. In der Lösung läßt sich dann aber noch nach Tagen Hexaoxy-diphensäure papierchromatographisch nachweisen. Andererseits gelingt es jedoch nicht, von Ellagsäure aus, etwa durch tagelanges Schütteln mit Methanol und Wasser, auch nur Spuren von Hexaoxy-diphensäure nachweisbar zu machen.

Wäßrige Lösungen von Hexaoxy-diphensäure enthalten bei 20° nach 4 Wochen neben auskristallisierter Ellagsäure noch unveränderte Säure; auch bei 100° läßt sich nach 7 Stunden noch Hexaoxy-diphensäure nachweisen, aber nicht mehr nach 11 Stunden. Durch Zugabe der gleichen Menge 2 n-Salzsäure zu einer methanolischen Lösung bei 20° ist schon nach 5 Minuten keine Hexaoxy-diphensäure mehr nachzuweisen. Wird trockene Hexaoxy-diphensäure auf 100° erhitzt, so geht sie in Ellagsäure über, doch ist die Umwandlung nach 8 Tagen noch unvollständig.

Aus den beschriebenen Eigenschaften wird verständlich, daß Hexaoxy-diphensäure bei der Hydrolyse von Ellagen-gerbstoffen durch Kochen mit Mineralsäuren nicht gewonnen oder auch nur nachgewiesen werden kann.

Die Kenntnisse der Eigenschaften der racemischen und optisch aktiven Säure machte es möglich, zu prüfen, ob diese wichtige Substanz in Gerbextrakten, die Ellagen-gerbstoffe enthalten, frei vorkommt und welche Rolle sie spielt.

Die rasche Racemisierung macht es unwahrscheinlich, daß in den Extrakten freie, optisch aktive Hexaoxy-diphensäure anzutreffen ist. Aber wie steht es mit der racemischen Säure? Um diese Frage zu klären, untersuchten O. Th. Schmidt und Demmler (*82*) wäßrige, methanolische und acetonische Extrakte von Myrobalanen, Dividivi, Algarobilla, Valonea, Fichtenrinde, Eichenrinde und jungen Kastanientrieben (papierchromatographisch).

In den Fichten-, Eichen- und Kastanien-extrakten war keine Hexaoxy-diphensäure zu sehen. In wäßrigen und manchmal auch in methanolischen Myrobalanen-extrakten wurde sie eindeutig nachgewiesen. Weder Valonea- noch Dividivi- noch Algarobilla-extrakte scheinen die Säure zu enthalten, doch ist in diesen Fällen der Nachweis wegen überdeckender Substanzen gleichen R_f-Wertes unsicher.

Die Hexaoxy-diphensäure ist in den Myrobalanen-extrakten nicht unbegrenzt haltbar. 12 Stunden nach Herstellung eines methanolischen Extrakts ist die Säure manchmal noch nachweisbar, nach 4 Tagen nicht mehr. Setzt man sie einem wäßrigen Myrobalanen-extrakt zu und vermischt die Lösung mit dem gleichen Volumen 2 n-Salzsäure, so ist sie schon nach 5 Minuten nicht mehr nachweisbar.

Zur Prüfung der Frage nach der Herkunft der Hexaoxy-diphensäure in den Extrakten wurden methanolische, wäßrige und acetonische Extrakte von nicht vorbehandeltem Myrobalanen-Mehl mit solchen verglichen, die fermentfrei (sterilisiertes Myrobalanen-Mehl) gewonnen und gehalten waren. Das Ergebnis zeigt *Tabelle 3*.

Tabelle 3. Vergleich von Extrakten von fermentfreiem und unbehandeltem Myrobalanen-Mehl („HODS" = Hexaoxydiphensäure).

Zeit (Stunden)	Extraktionsmittel					
	Methanol		Wasser		Aceton	
	fermentfrei	unbehandelt	fermentfrei	unbehandelt	ferment-frei	unbe-handelt
0	—	HODS (nicht immer)	—	HODS	—	—
24	—	—	—	starke Abscheidung	—	—
48	—	—	Spur Abscheidung	,,	—	—
96	—	Spur Abscheidung	,,	,,	—	—
120	Spur Abscheidung	,,	,,	,,	—	—

Die acetonischen Extrakte enthielten keine Hexaoxy-diphensäure, obgleich diese Säure gerade in Aceton leicht löslich ist. Aus ihnen fiel selbst nach 14 Tagen keine Ellagsäure aus, während sich in den fermentfreien, methanolischen Extrakten nach 5 Tagen ein leichter Niederschlag zeigte. Bemerkenswert ist das Vorhandensein von Hexaoxy-diphensäure in wäßrigen Extrakten unbehandelter Myrobalanen und das frühzeitige Auftreten eines kräftigen Niederschlags im Gegensatz zum fermentfreien wäßrigen Extrakt. Das deutet auf fermentative Spaltung von Ellagen-gerbstoffen as Ursache für das Auftreten der Hexaoxy-diphensäure und anschließende Bildung und Abscheidung der Ellagsäure hin. Das konnte auch dadurch bewiesen werden, daß fermentfreien, wäßrigen Extrakten Tannase-lösungen zugesetzt wurden, worauf sich nach 8 Stunden die Säure nachweisen ließ und nach 24 Stunden eine kristallisierte Abscheidung entstand.

Damit ist gezeigt, daß Fermente die Hauptursache der „Blume"-bildung sind, da sie aus den Ellagen-gerbstoffen Hexaoxy-diphensäure abspalten, die dann in Ellagsäure übergeht. Eine fermentative Spaltung von Gerbstoffen als Ursache der Ellagsäureabscheidung hatte schon PHILLIPS (66) angenommen.

Eine, wenn auch geringe oder langsame hydrolytische Spaltung der Ellagen-gerbstoffe, verursacht durch die eigene Acidität der Gerbextrakte, muß aber außerdem stattfinden, da sich auch im fermentfreien, wäßrigen Extrakt nach 48 Stunden ein Niederschlag abzusetzen beginnt. In methanolischen Extrakten tritt diese Spaltung stark zurück und bleibt in den acetonischen Extrakten praktisch vollständig aus.

Wie besondere Versuche ergeben haben, wird die aus abgespaltener oder zugefügter Hexaoxy-diphensäure in den Gerbextrakten durch Zusatz von Säuren erzeugte Ellagsäure nicht sofort quantitativ abgeschieden. Vielmehr wird ein Teil der Ellagsäure von den Lösungsgenossen in Lösung gehalten („peptisiert") und die Abscheidung erfolgt erst allmählich.

Obgleich bisher nur über Corilagin und Chebulagsäure eingehende Untersuchungen vorliegen, ist es doch wahrscheinlich, daß auch in anderen Ellagen-gerbstoffen nicht die Ellagsäure selbst, sondern Hexaoxy-diphensäure gebunden ist. Bei Valonea jedenfalls konnte dies von O. Th. Schmidt und Grünewald (*83*) bewiesen werden. Wird Valoneaextrakt mit Diazomethan methyliert, dann alkalisch hydrolysiert, so läßt sich Hexamethoxy-diphensäure isolieren. Bemerkenswerterweise handelt es sich hier aber um die *L*-Hexamethoxy-diphensäure.

5. Brevifolin und Brevifolin-carbonsäure.

a) *Beschreibung und Analyse. Konstitution.*

Die wenigen, meist älteren Arbeiten über Algarobilla (getrocknete Fruchtschoten von *Caesalpinia brevifolia*) (*110, 59*), weisen übereinstimmend Gallussäure, Ellagsäure und Glucose (diese nicht sicher) nach.

Ein gewisses Interesse beanspruchen zwei Patentanmeldungen der Fa. Heinemann (*38*) aus dem Jahre 1901, in welchen die Gewinnung von Ellagsäure aus Algarobilla (und Dividivi) durch Kochen der wäßrigen Extrakte beschrieben wird. Auf diese Weise kann nämlich, wie unten ausgeführt wird, ein neuer Gerbstoff-baustein isoliert werden, der wegen seiner Ähnlichkeit mit Ellagsäure von den früheren Bearbeitern übersehen worden war. Die nachstehend beschriebenen Untersuchungen wurden von O. Th. Schmidt und Bernauer (*77*) durchgeführt.

Die Gesamtgerbstoffe von Algarobilla wurden in phenolsaure und carboxylsaure Fraktionen getrennt und beide durch Gegenstromverteilung weiter zerlegt. Der Verlauf der Verteilung wurde papierchromatographisch und polarimetrisch verfolgt.

Trotz der nahen botanischen Verwandtschaft von *Caesalpinia brevifolia* mit *C. coriaria* (Dividivi) ließ sich in Algarobilla weder Corilagin noch Chebulagsäure nachweisen. Es gelang jedoch, aus dem phenolsauren Anteil einen neuen, kristallisierten Gerbstoff zu isolieren, für den der Name „*Brevilagin*" vorgeschlagen wurde. Die Verbindung, die noch der eingehenden Untersuchung bedarf, ist gelb, dreht $+146,6°$ (in Äthanol) und liefert bei der Hydrolyse neben Glucose und Ellagsäure ein neues, kristallisiertes Spaltstück, die „*Brevifolin-carbonsäure*". Diese Säure wurde auch aus dem carboxylsauren Anteil der Algarobilla-extrakte gewonnen.

Brevifolin-carbonsäure ist ebenfalls gelb und gibt bei der Gries-meyer-Reichel-Probe (*71*) genau die gleiche rotviolette Färbung wie Ellagsäure. Die Summenformel ist $C_{13}H_8O_8$.

Mit Diazomethan wird eine kristallisierte Tetramethyl-verbindung erhalten, mit 2,4-Dinitro-phenylhydrazin ein Dinitro-phenylhydrazon. Beim Erhitzen in Wasser, 2 n-H_2SO_4 oder Dimethylanilin gibt Brevifolin-carbonsäure 1 Mol CO_2 ab und geht in eine neue kristallisierte Verbindung $C_{12}H_8O_6$ über, die ,,*Brevifolin*'' genannt wurde. Das Brevifolin läßt sich auch aus acetonischen Algarobilla-extrakten durch längeres Erhitzen mit Wasser gewinnen. Brevifolin ist ebenfalls gelb, gibt die gleiche Farb-reaktion nach Griesmeyer-Reichel wie Ellagsäure und Brevifolin-carbonsäure, ist nahezu ebenso schwer löslich wie Ellagsäure und liefert mit Pyridin eine mikroskopisch ähnliche Additionsverbindung wie Ellagsäure. Die Eisenchlorid-Färbung ist grünstichig blau.

Brevifolin liefert mit Diazomethan einen Trimethyl-äther, mit Acetan-hydrid ein Triacetat. Zusammen mit der Eisenchlorid-reaktion ergibt sich der Schluß, daß drei benachbarte Hydroxylgruppen die einzigen sauren Gruppen des Moleküls sind. Trimethyl-brevifolin, in Wasser unlös-lich, wird von warmer Natronlauge gelöst und beim Ansäuern wieder unverändert ausgeschieden. Mit Dimethylsulfat und Alkali wird aus Trimethyl-brevifolin, $C_{15}H_{14}O_6$, eine farblose Tetramethyl-verbindung, $C_{16}H_{18}O_7$, mit freier Carboxylgruppe erhalten. Diazomethan liefert hier-aus den Methylester $C_{17}H_{20}O_7$. Es ist also eine Lactongruppierung im Brevifolin vorhanden.

Die Tetramethylverbindung wird durch längeres Erwärmen mit ver-dünnter Salzsäure in Trimethyl-brevifolin zurückverwandelt. Diese leichte partielle Entmethylierung deutet auf eine Enoläther-gruppierung in der Tetramethyl-verbindung und eine Enol-lacton-gruppierung im Tri-methyl-brevifolin hin. Trimethyl-brevifolin und die Tetramethyl-ver-bindung geben beide (letztere unter partieller Entmethylierung) in salz-saurem Methanol das gleiche Dinitro-phenylhydrazon. Mit diesem Nach-weis einer Carbonylgruppe sind die Funktionen aller sechs Sauerstoff-atome des Brevifolins festgelegt.

Das Kohlenstoffgerüst des Brevifolins erhellt aus einer Reihe von *Abbauprodukten*. Unterwirft man Trimethyl-brevifolin der Einwirkung von alkalischem Ferricyanid nach R. D. Haworth (*35, 36*), so erhält man wie aus Trimethyl-chebulsäure 3,4,5-Trimethoxy-phthalsäure. Nimmt man die Carboxylgruppe des Lactons in Gallussäurestellung am aroma-tischen Kern an, was — wie weiter unten gezeigt wird — berechtigt ist, so kann die Konstitution des Brevifolins durch die provisorische Formel (LV) ausgedrückt werden, wobei der Rest $C_5H_4O_2$ die Gruppierungen

$$-CO- \quad \text{und} \quad -\overset{\overset{\textstyle O}{|}}{C}=\overset{|}{C}- \quad \text{enthalten muß.}$$

Wollte man diesen Rest als offene Kette formulieren, so müßte außer der Enoldoppelbindung noch eine zweite C—C-Doppelbindung vorhanden sein. Dem widerspricht die Tatsache, daß sich die Tetramethyl-verbindung mit Palladium-Bariumsulfat als Katalysator nicht hydrieren läßt. Völlig unvereinbar mit einer offenen Seitenkette ist aber das Abbauprodukt Bernsteinsäure, das bei der Oxydation des Trimethyl-brevifolins mit Kaliumpermanganat resultiert; und zwar, weil neben den bereits nachgewiesenen Gruppierungen in einer Kette —$C_5H_4O_2$— eine Methylengruppe keinen Platz mehr findet. Daß aber mindestens eine Methylengruppe im Molekül vorhanden ist, überdies α-ständig zum Carbonyl, läßt sich auch durch die Reaktion von Brevifolin und Benzaldehyd in salzsaurem Methanol zeigen, bei welcher Kondensation zu einem tiefgelben Benzalbrevifolin eintritt. Somit kommt man zwangsläufig zu dem Schluß, daß im Brevifolin ein Fünfring mit einem Gallussäurerest verknüpft ist (LVI).

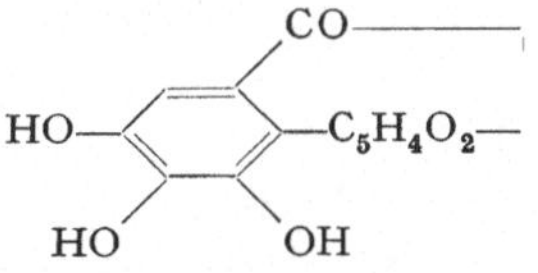

(LV.) 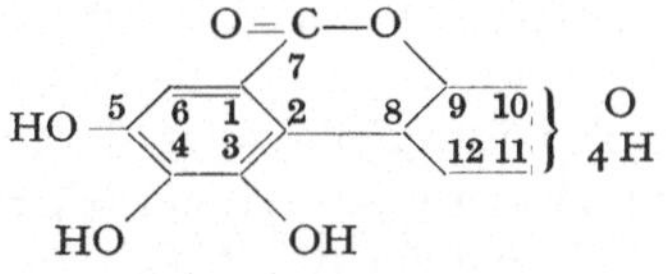   (LVI.)

Auf den Fünfring sind vier Wasserstoff-atome, ein Keto-sauerstoff und eine Enol-doppelbindung zu verteilen. Die Enolisierung muß in der Lage der beiden Sauerstoff-atome zueinander begründet sein. Möglicherweise erfährt sie durch Konjugation mit dem aromatischen System eine zusätzliche Stabilisierung. Der Fünfring hat demgemäß entweder die Anordnung eines enolisierten α-Diketons mit dem Keto-sauerstoff am C-Atom 10 oder eines enolisierten β-Diketons mit dem Keto-sauerstoff am C-Atom 12. In beiden Fällen kann die Enol-doppelbindung nur zwischen den C-Atomen 8 und 9 liegen. Von cyclischen α-Diketonen ist bekannt (*102*), daß sie eine ausgeprägte Enolisierungstendenz haben; das würde der Stabilität des Enol-lactons entsprechen. Deshalb und auf Grund genetischer Überlegungen war von vornherein die α-Anordnung beider Sauerstoff-atome am Fünfring wahrscheinlicher. Um

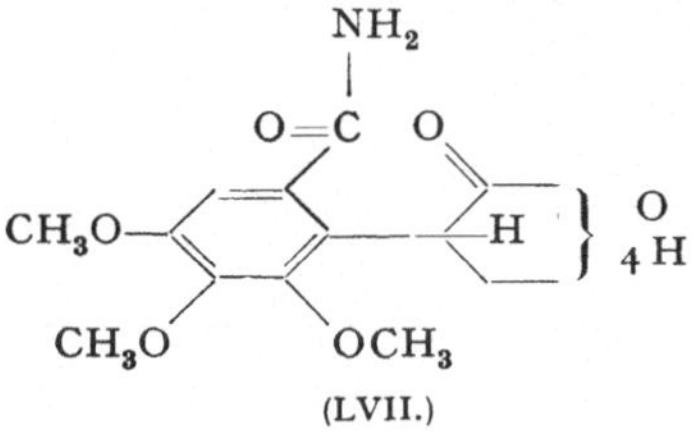

(LVII.) 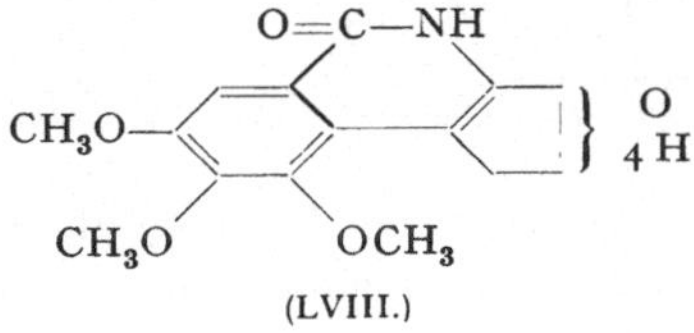   (LVIII.)

sie nachzuweisen, wurde das Amid (LVII) hergestellt, das jedoch mit
o-Phenylen-diamin nicht zu einem Glyoxalin-derivat kondensiert werden
konnte, obgleich es sicher als echtes α-Diketon vorliegt, wie mit Hilfe
des Ultraviolett-Spektrogramms gezeigt werden konnte (vgl. unten).
Mit dem spektroskopischen Befund stimmt überein, daß das Amid kein
Brom verbraucht. Die Reaktion des Amids mit p-Nitrophenylhydrazin
führt nicht zu dem erwarteten Osazon, sondern zum p-Nitrophenyl-
hydrazon des Lactams (LVIII). Offenbar sind die sterischen Verhältnisse
für die bevorzugte intramolekulare Reaktion sehr günstig.

Der Nachweis der α-Anordnung wurde schließlich durch die Oxydation
der Trimethyl-brevifolins mit alkalischem Wasserstoffperoxyd geführt.
α-Diketone reagieren mit diesem Oxydationsmittel nach folgendem
Schema (57):

$$R\text{—}CO\text{—}CO\text{—}R' + H_2O_2 \rightarrow R\text{—}COOH + R'\text{—}COOH.$$

Trimethyl-brevifolin $C_{15}H_{14}O_6$ liefert die erwartete Tricarbonsäure $C_{15}H_{18}O_9$ (1 Mol H_2O wird zur Lactonöffnung aufgenommen), der
die Konstitution (LIX) zukommen muß, mit vorzüglicher Ausbeute.

Für das Brevifolin folgt daraus Formel (LX).

Dem Trimethyl-brevifolin ($C_{15}H_{14}O_6$) kommt die Formel (LXI), dem
Triacetyl-brevifolin ($C_{18}H_{14}O_9$) die Formel (LXII) zu. Das Tetramethyl-
derivat ($C_{16}H_{18}O_7$; Formel LXIII) wurde „Tetramethyl-brevifolsäure"

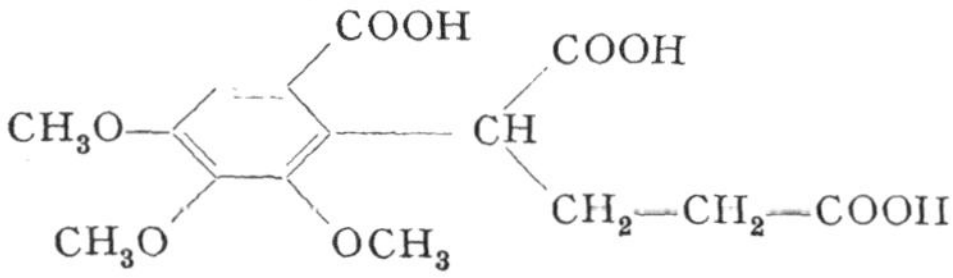

(LIX.)

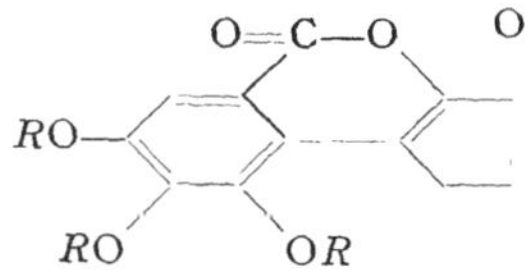

(LX.) $R =$ H. Brevifolin.
(LXI.) $R =$ CH_3. Trimethyl-brevifolin.
(LXII.) $R =$ OC·CH_3. Triacetyl-brevifolin.

genannt, und dessen Ester ($C_{17}H_{20}O_7$; Formel LXIV) „Tetramethyl-
brevifolsäure-methylester". Dem Benzal-brevifolin entspricht die
Formel (LXV).

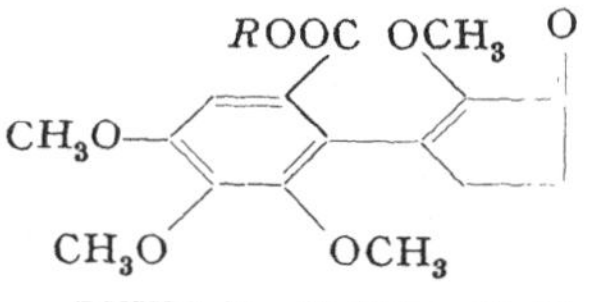

(LXIII.) $R =$ H. Tetramethyl-brevifolsäure.
(LXIV.) $R =$ CH_3. Tetramethyl-brevifolsaure-methylester.

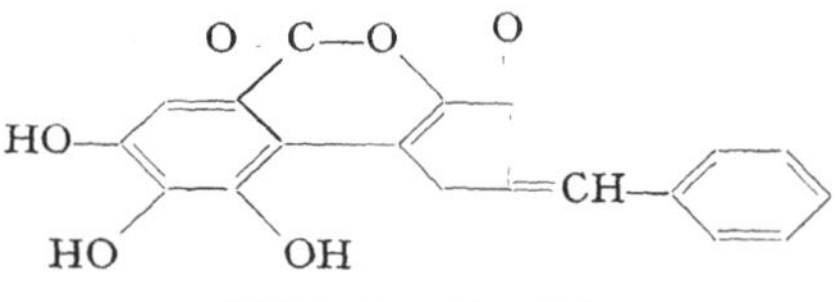

(LXV.) Benzal-brevifolin.

Der noch offene Beweis für die Stellung der Carboxylgruppe am
aromatischen Kern wurde durch die nachstehende Reaktionsfolge erbracht.

Brevifolin (LX) wurde mit Benzylchlorid und wasserfreiem Kalium-carbonat in seinen Tribenzyläther (LXVI) verwandelt. Dieser wurde mit Dimethylsulfat und Lauge unter Lactonöffnung zur Tribenzyl-methyl-brevifolsäure (LXVII) umgesetzt. Dann wurde hydrierend entbenzyliert. Die dabei gebildete Methyl-brevifolsäure (LXVIII) ließ sich durch Erhitzen in Dimethylanilin mit Naturkupfer „C" leicht decarboxylieren. Die decarboxylierte Verbindung (LXIX) („Decarboxy-methyl-brevifol-säure") wurde mit Diazomethan aufmethyliert zur Verbindung (LXX) („Decarboxy-tetramethyl-brevifolsäure") und diese schließlich nach R. D. HAWORTH (*36, 35*) mit alkalischem Kaliumferricyanid oxydiert. Da es nicht von vornherein sicher schien, daß diese Reaktion auch beim Vorliegen einer Methyläthergruppe in dem zu oxydierenden Teil bis zur Carbonsäure führen würde, wurde ihr zuerst Tetramethyl-brevifolsäure (LXIII) unterworfen, und es wurde ebenso wie aus Trimethyl-brevi-folin (LXI) Trimethoxy-phthalsäure erhalten. Auch mit Verbin-dung (LXX) verlief der Abbau glatt und führte zu 2,3,4-Trimethoxy-benzoesäure (LXXI).

Verbindung (LXIX) kristallisiert vorzüglich und verhält sich gegen Ferrichlorid wie Pyrogallol. Die Farbe schlägt sehr rasch von Blaugrün nach Rotbraun um. Die methylierte Verbindung (LXX) konnte nicht zur Kristallisation gebracht werden. Sie ließ sich jedoch in ein 2,4-Dinitro-phenylhydrazon überführen; durch längere Behandlung mit stark salzsaurem 2,4-Dinitrophenyl-hydrazin in Äthanol wurde auch das entsprechende Osazon gewonnen.

(LXVI.) Brevifolin-tribenzylather.

(LXVII.) $R = C_7H_7$. Tribenzyl-methyl-brevifolsaure.
(LXVIII.) $R = H$. Methyl-brevifolsaure.

(LXIX.) $R = H$. Decarboxy-methyl-brevifolsaure.
(LXX.) $R = CH_3$. Decarboxy-tetramethyl-brevifolsaure.

(LXXI.) 2,3,4-Trimethoxy-benzoesäure.

Es bleibt noch die Frage nach der Stellung der freien Carboxylgruppe in der Brevifolin-carbonsäure. Wegen der leichten Decarboxylierbarkeit nehmen O. TH. SCHMIDT und BERNAUER (*77*) an, daß die Brevifolin-carbonsäure eine β-Keto-carbonsäure ist, woraus Konstitution (LXXII) folgt. Dem Trimethyl-brevifolin-carbonsäure-methylester kommt dann Formel (LXXIII) zu.

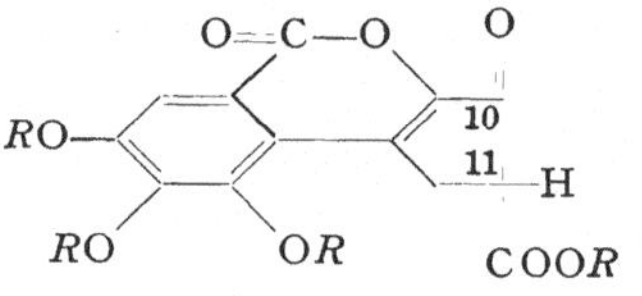

(LXXII.) $R = H$. Brevifolin-carbonsaure.
(LXXIII.) $R = CH_3$. Trimethyl-brevifolin-carbonsäure-methylester.

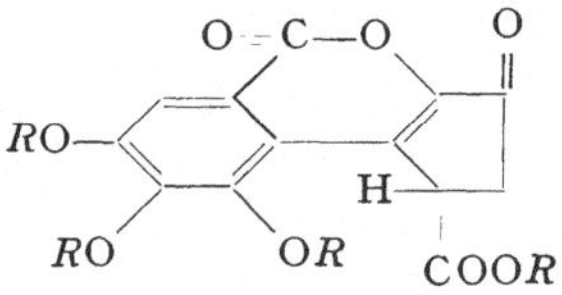

(LXXIV.) $R = H$.
(LXXV.) $R = CH_3$.

Für Formel (LXXII) spricht noch die Tatsache, daß Brevifolin-carbonsäure trotz des asymmetrischen C-Atoms 11 optisch inaktiv ist. Das läßt sich verstehen, da die Nachbarschaft der C=O-Gruppe (C-Atom 10) Enolisierung und Racemisierung ermöglicht, was bei den Formeln (LXXIV) und (LXXV) nicht möglich ist.

Aber auch Verbindung (LXXIV) als vinyloge β-Carboxylsäure würde die Decarboxylierbarkeit verständlich erscheinen lassen und die Wahl zwischen Formel (LXXII) und (LXXIV) bedarf noch der experimentellen Entscheidung, da die genaue Kenntnis der Konstitution der Brevifolin-carbonsäure für genetische Überlegungen von Bedeutung ist.

b) Synthesen des Trimethyl-brevifolins.

Die durch Analyse und Abbaureaktionen abgeleitete Formel (LX, S. 108) des Brevifolins ist durch zwei *Synthesen* des Trimethyl-brevifolins bestätigt worden. Bernauer und O. Th. Schmidt (6) setzten nach einem Verfahren von Meerwein (55) diazotierten Trimethyläther-amino-gallussäure-methylester (LXXVI) in Gegenwart von Cupriacetat mit Cyclopenten-(1)-ol-(2)-on-(3)-methyläther (LXXVII) (5) um. Unter Abspaltung von Chlorwasserstoff, Stickstoff und Methanol (Hydrolyse der Ester- und Enoläther-gruppe) erhielten sie direkt Trimethyl-brevifolin (LXI).

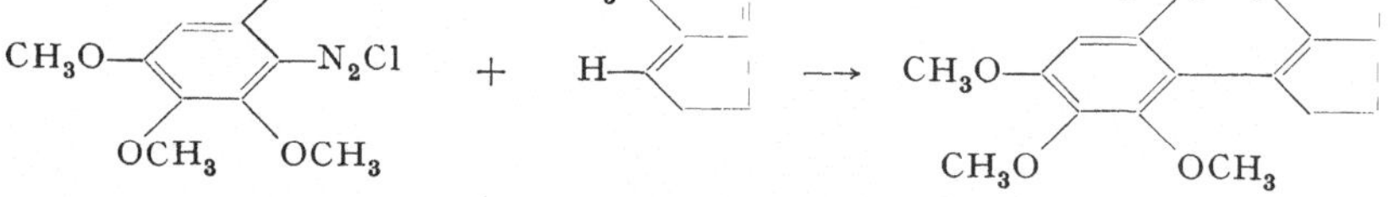

(LXXVI.)
Diazotierter Trimethyläther-aminogallussäure-methylester.

(LXXVII.)
Cyclopenten-(1)-ol-(2)-on-(3)-methyläther.

(LXI.) Trimethyl-brevifolin.

Auf ganz anderem Wege kamen R. D. Haworth und Grimshaw (34) zu Trimethyl-brevifolin. Flavellagsäure (LXXVIII) wurde mit Peroxyd oxydiert zu (LXXIX), das durch Methylierung und Kettenverlängerung in die homologe Säure (LXXX) umgewandelt wurde. Von hier aus führte der Ringschluß durch Kochen mit Diphosphorpentoxyd zu Trimethyl-brevifolin (LXI).

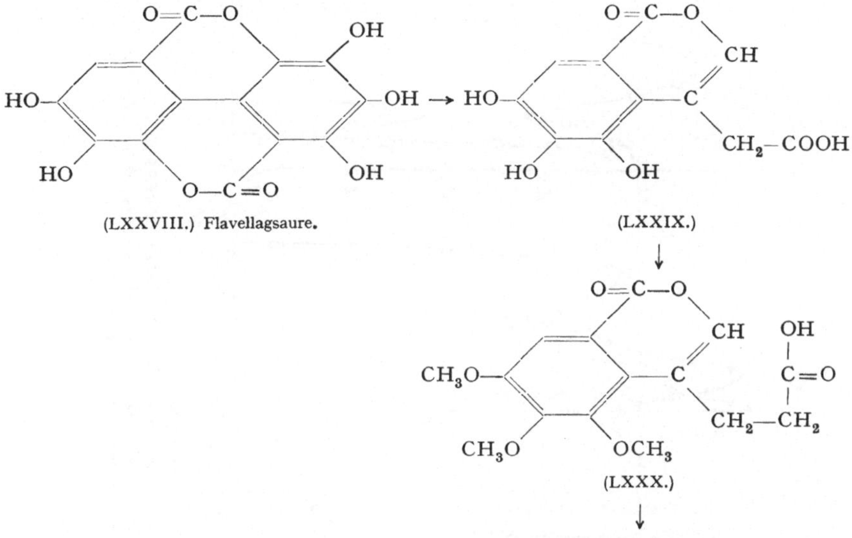

(LXXVIII.) Flavellagsäure.

(LXXIX.)

(LXXX.)

(LXI.) Trimethyl-brevifolin.

c) Zusammenhänge zwischen Konstitution, sterischem Bau und Ultraviolett-Spektrum.

Die oben geschilderte Konstitutionsaufklärung des Brevifolins ist durch U. V.-spektroskopische Untersuchungen unterstützt worden [BERNAUER (5)], die allgemeine Bedeutung haben und deshalb hier besprochen seien.

Alle in *Abb. 3* (S. 112) erwähnten Substanzen, mit Ausnahme von Brevifolin und Hexamethoxy-diphensäure sind in wasserfreiem Dioxan gemessen worden. Hieraus erklärt sich in *Abb. 3 A* die Verschiedenheit zwischen Brevifolin (LX, S. 108) und Brevifolin-trimethyläther (LXI, S. 108), da Brevifolin mit Wasser ein tiefer gelb gefärbtes Hydrat bildet. Aufschlußreich ist in *Abb. 3 A* der Vergleich von Brevifolin-trimethyläther (LXI) mit Tetramethyl-brevifolsäure (LXIII, S. 108), deren Spektren sich grundlegend unterscheiden. Das Hauptmaximum des Brevifolintrimethyläthers bei 264 mμ deutet auf das Vorliegen eines Gallussäure-Gerüsts (Maximum der Trimethyläther-gallussäure bei 266 mμ). Das in den sichtbaren Bereich übergreifende Bandensystem ist durch die Resonanz von aromatischem Kern, Enol-doppelbindung und Keton-doppelbindung verursacht. Die Feinstruktur zusammen mit dem steilen, linearen Abfall der Kurve bei 370—380 mμ weist auf ebenen Bau des gesamten Moleküls (56). Die Tetramethyl-brevifolsäure (LXIII) ist im Gegensatz zu der blaßgelben Trimethyl-verbindung (LXI) völlig farblos. Dementsprechend besitzt sie keine langwellige Absorptionsbande. Das Maxi-

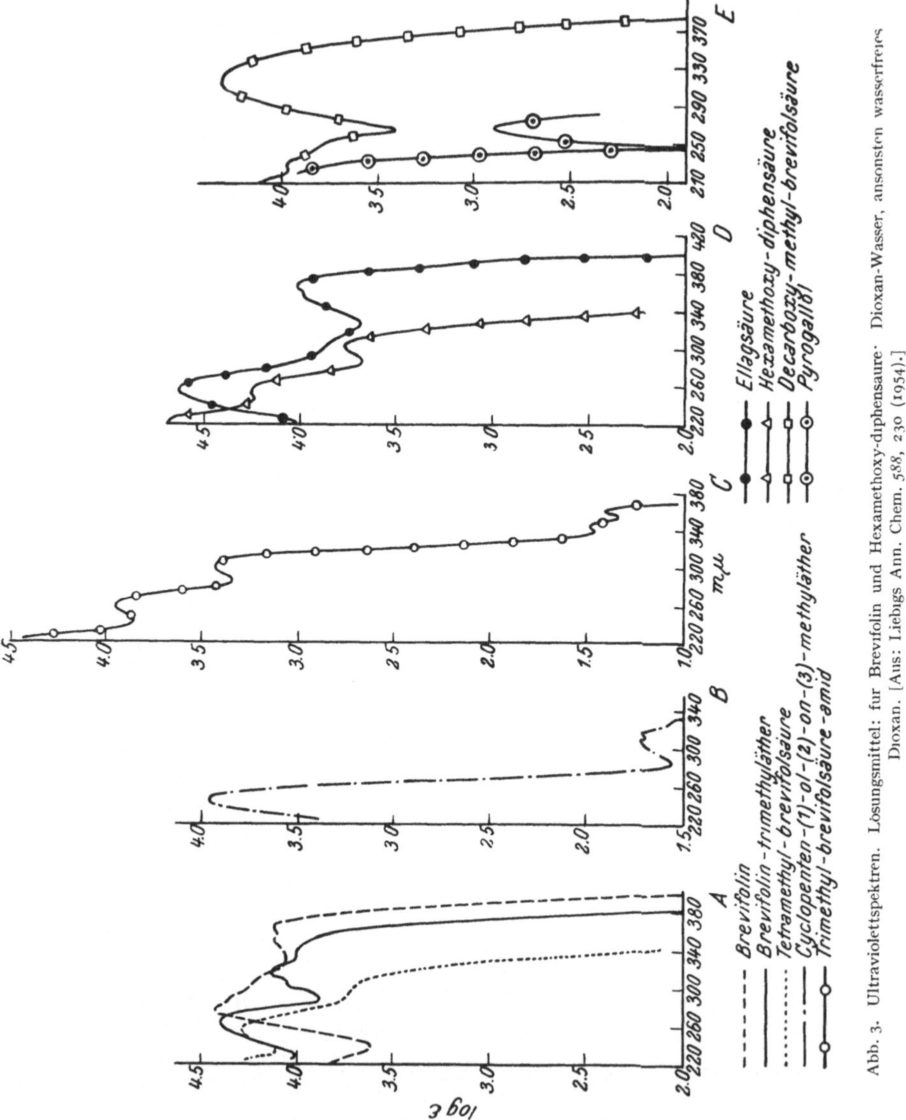

Abb. 3. Ultraviolettspektren. Lösungsmittel: für Brevifolin und Hexamethoxy-diphensäure· Dioxan-Wasser, ansonsten wasserfreies Dioxan. [Aus: Liebigs Ann. Chem. 588, 230 (1954).]

mum bei 255 mμ scheint dem der Gallussäure zu entsprechen, besitzt aber, verglichen mit der Trimethyläther-gallussäure, eine zu hohe Extinktion (log ε = 4,294, statt 3,983). Die „Schulter" bei 300 mμ wird durch die Substitution in 2-Stellung der methylierten Gallussäure verursacht, wofür weiter unten (LXXXIV) noch ein weiteres Beispiel gegeben wird. — Der auffallende Unterschied zwischen den Spektren von Brevifolin-trimethyläther und Tetramethyl-brevifolsäure hat offenbar sterische Ursache. In der Tetramethyl-brevifolsäure (Formel LXIII, S. 108) liegen die beiden Ringe wahrscheinlich in verschiedenen Ebenen, weil sich die Liganden —COOH, phenol-OCH$_3$ und enol-OCH$_3$ gegenseitig behindern. Diese Vorstellung ist von der Atropisomerie bekannt.

CALVIN (*12*) hat angeregt, die Spektren als Kriterium für behinderte freie Drehbarkeit heranzuziehen. Verschiedene Autoren haben an Diphenyl-derivaten den Zusammenhang zwischen räumlicher Lagerung der Ringe und Lichtabsorption nachgewiesen (*67*, *62a*, *7*). Im Falle der Tetramethyl-brevifolsäure (LXIII) hat das System: aromatischer Kern — Enol-doppelbindung — Carbonyl-gruppe mit der Beseitigung der ebenen Anordnung die Möglichkeit, seine π-Elektronen zu einem wirksamen Resonanzsystem zu verschmelzen, und damit auch seinen Charakter als Chromophor verloren*. Die, verglichen mit Trimethyläther-gallussäure, erhöhte Extinktion von Tetramethyl-brevifolsäure erklärt sich durch die Überlagerung des Cyclopentenolon-methyläther-Anteils. Der aus Cyclopentandion mit Diazomethan hergestellte Cyclopenten-(1)-ol-(2)-on-(3)-methyläther (LXXXI) besitzt eine ausgeprägte Bande bei 246 mμ (Abb. 3 *B*).

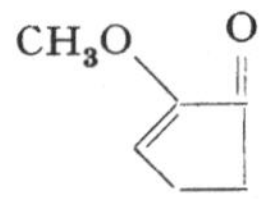

(LXXXI.) Cyclopenten-(1)-ol-(2)-on-(3)-methylather. 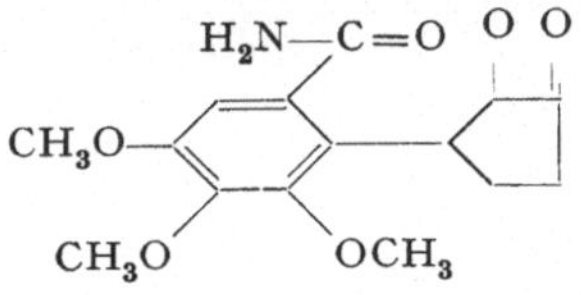   (LXXXII.) Trimethyl-brevifolsaure-amid.

Im Trimethyl-brevifolsäure-amid (LXXXII) liegt ein weiteres lacton-offenes Derivat des Brevifolins vor. Sein Spektrum (*Abb. 3C*) zeigt nicht die erwartete Übereinstimmung mit der Tetramethyl-brevifolsäure. Wie weiter unten aus dem Vergleich mit Hexamethoxy-diphensäure deutlich wird, besitzt das Amid das reine Spektrum einer in 2-Stellung substituierten Trimethyl-äther-gallussäure (abgesehen von zwei schwachen Banden im langwelligen U. V.; s. unten). Der optische Einfluß des Fünfringsystems im Bereich der Gallussäure-Bande muß also beseitigt sein. Dies ist beweisend für die α-Diketon-Anordnung, denn die Extinktion echter α-Diketone ist in dem fraglichen Spektralgebiet um Zehnerpotenzen geringer als die der zugehörigen Enole (*46*). Im Falle

---

* Es ist indessen nicht gelungen, Tetramethyl-brevifolsäure in optische Antipoden zu zerlegen (*49*).

des Amids (LXXXII) fällt sie gegenüber der starken Gallussäure-Bande nicht ins Gewicht. Die Annahme echter α-Diketon-Anordnung für das Amid wird durch dessen Absorption bei langen Wellen bestätigt. Es zeigt bei 344 mμ und 360 mμ zwei Maxima geringer Intensität. Bei echten α-Diketonen der Sterinreihe ist ein Maximum bei 347 mμ gefunden worden (*46*).

Für die optische „Manifestierung" der mit der Lacton-öffnung verbundenen Änderung der sterischen Verhältnisse ist der Vergleich der Ellagsäure und Hexamethoxy-diphensäure ein schönes Beispiel. Die Ellagsäure (LXXXIII) hat zweifellos weitgehend ebenen Bau. Das bedeutet starke Resonanz beider aromatischer Systeme, was im Spektrum (*Abb. 3D*) durch eine intensive Bande bei 386 mμ (mit angedeuteter Feinstruktur) und eine „überhöhte" Gallussäure-Bande zum Ausdruck kommt. Die lacton-offene Hexamethoxy-diphensäure (LXXXIV) besitzt nur die Gallussäure-Bande ($\lambda_{max} = 254$ mμ) und eine weitere Bande

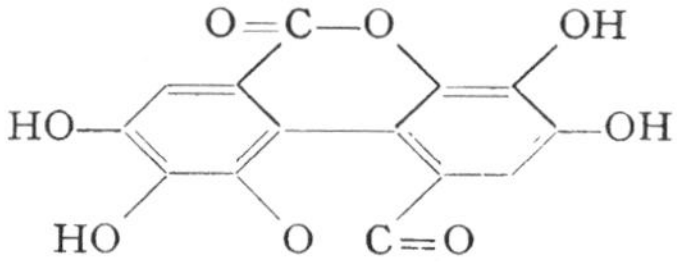

(LXXXIII.) Ellagsaure.

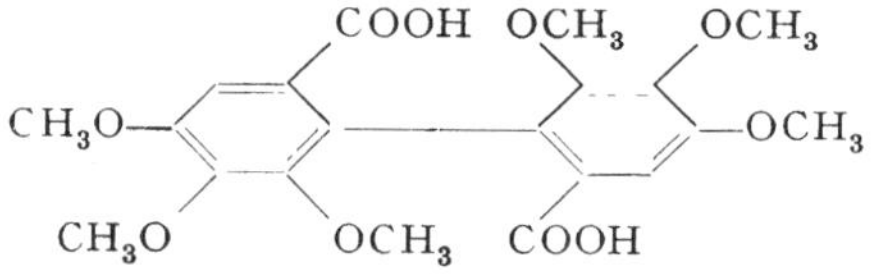

(LXXXIV.) Hexamethoxy-diphensaure.

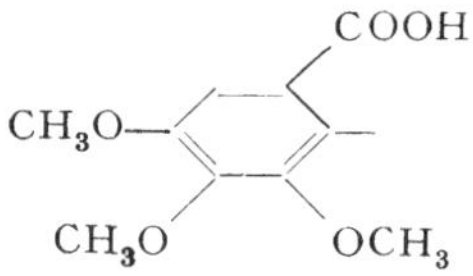

(LXXXV.)

bei etwa 300 mμ, die — wie oben schon angedeutet — immer dann auftritt, wenn eine verätherte Gallussäure in 2-Stellung eine Seitenkette trägt. Die Behinderung der freien Drehbarkeit um die Diphenyl-bindung, welche die Anordnung beider Ringe in einer gemeinsamen Ebene unmöglich macht, ist durch die Isolierung atropisomerer Antipoden der Hexaoxy-, Hexamethoxy- und Hexabenzoxy-diphensäure bewiesen (vgl. S. 95).

Interessant ist der quantitative Vergleich zwischen Trimethyl-brevifolsäure-amid (LXXXII) und Hexamethoxy-diphensäure (LXXXIV). Die Extinktionen einander entsprechender Banden liegen bei dieser doppelt so hoch wie beim Amid. Das System (LXXXV) ist also als Chromophor im Sinne der „Additionsregel" (*46*) auffaßbar.

Eine kräftige Änderung des Spektrums tritt mit der Entfernung der kernständigen Carboxylgruppe ein, wie die Kurve der Decarboxymethyl-brevifolsäure (LXXXVI) erkennen läßt (Abb. 3*E*). Dies entspricht nicht einfach dem Übergang von einer substituierten Gallussäure zu einem entsprechenden Pyrogallol; denn Pyrogallol (*Abb. 3E*) absorbiert, wenn auch viel schwächer, bei etwa der gleichen Wellenlänge wie Gallussäure. Im Spektrum der Verbindung (LXXXVI) erkennt man

(LXXXVI.) Decarboxy-methyl-brevifolsäure.

jedoch eine sehr stark ausgeprägte Bande bei 318 mμ (log $\varepsilon = 4{,}316$). In dieser Verbindung sind offenbar wie beim lacton-geschlossenen Brevifolin-trimethyläther (LXI, S. 108) die Elektronensysteme von Sechs- und Fünfring in Resonanz, was möglich ist, weil die hindernde Carboxylgruppe fehlt. Man kann sogar sagen, daß die Tendenz, ein energiearmes, mesomeres System auszubilden, die komplanare Anordnung beider Ringe verursacht. Für eine beträchtliche Starrheit der Molekel (LXXXVI) spricht auch der hohe Schmelzp. 211—212°. Wahrscheinlich besteht auch ein Zusammenhang zwischen der Leichtigkeit, mit der sich Lactone aus ring-offenen Derivaten des Brevifolins bzw. der Hexaoxy-diphensäure bilden, und dem damit verbundenen Gewinn an Resonanzenergie.

6. Dehydro-digallussäure.

Bei der Untersuchung der Gerbstoffe der Edelkastanie (*Castanea vesca*) entdeckte W. Mayer (*51*) eine neue, schön kristallisierte Phenol-carbonsäure, die den Namen „Dehydro-digallussäure" erhielt.

Junge Blätter und Triebe der Edelkastanie werden extrahiert, der Extrakt mit Blei gefällt, die Fällung zerlegt, und die so gereinigte Gerbstofflösung erschöpfend mit Äther extrahiert. Aus dem Ätherextrakt wird die Dehydro-digallussäure mit 8% Ausbeute, bezogen auf die Gesamtgerbstoffe, gewonnen.

Dehydro-digallussäure ist farblos, gibt eine rasch verblassende tief-blaue Eisenchlorid-Reaktion und verhält sich bei der Cyankali-Probe wie freie Gallussäure. In allen gebräuchlichen Lösungsmitteln ist sie sehr schwer löslich und dürfte früher mit Ellagsäure verwechselt worden sein. Gegen Säuren ist die Verbindung weitgehend stabil, selbst nach zwei-stündigem Kochen mit 48%iger Bromwasserstoffsäure wird sie unver-ändert zurückerhalten. Bei der Einwirkung von heißer verdünnter Natronlauge tritt jedoch Aufspaltung ein und es lassen sich nahezu 50% Gallussäure isolieren.

Die Formel ist $C_{14}H_{10}O_{10}$. Mit Diazomethan wird eine Heptamethyl-verbindung $C_{21}H_{24}O_{10}$ erhalten, die sich zu einer Pentamethyl-verbindung $C_{19}H_{20}O_{10}$ verseifen läßt. Die letztere ist eine Dicarbonsäure, von der ein Di-*p*-bromphenacyl-ester dargestellt wurde. Im Gegensatz zur nicht methylierten Verbindung ist die Pentamethyl-verbindung gegen Alkali völlig beständig.

Von den zehn Sauerstoffatomen der Dehydro-digallussäure liegen fünf als phenolische Hydroxylgruppen vor, und vier gehören Carboxylgruppen an. Für das zehnte Sauerstoffatom bleibt nur die Äther-Funktion. Damit wird der neuen Phenol-carbonsäure ein Diphenyläther-Gerüst zugrunde gelegt. Über die Verteilung der Substituenten liefert einen Hinweis die blaue Eisenchlorid-reaktion und die Cyankali-reaktion. Danach sollte ein Teil der Molekel Gallussäureanordnung haben

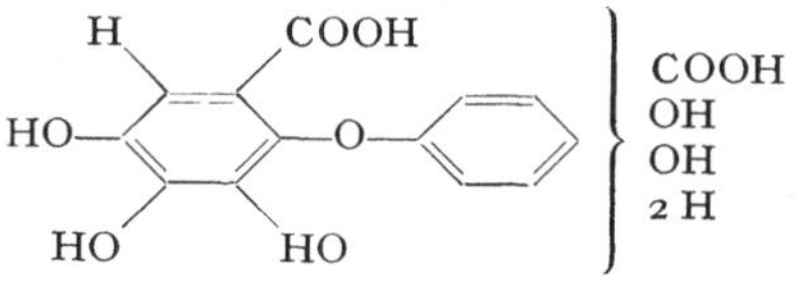

(LXXXVII.)

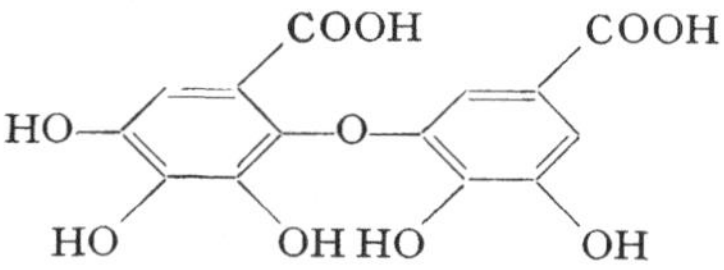

(LXXXVIII.) Dehydro-digallussaure.

(LXXXVII). Aber auch dem anderen Teil könnte Gallussäure zugrunde liegen (Formel LXXXVIII), da die Spaltung mit Alkali zwar weniger als 50%, aber doch nur Gallussäure geliefert hatte. Diese Spaltung ist im übrigen keine Hydrolyse; von dem neben Gallussäure zu erwartenden zweiten Spaltstück, der (damals noch unbekannten) Tetraoxy-benzoesäure oder dem Decarboxylierungsprodukt Tetraoxy-benzol ließ sich keine Spur nachweisen*.

Wenn Dehydro-digallussäure einige Minuten mit konz. Schwefelsäure auf $100°$ erhitzt wird, entsteht ein Xanthon, dem die Richtigkeit der Stellung der Substituenten in (LXXXVIII) vorausgesetzt, die Konstitution (LXXXIX) zukommen muß. Mit Diazomethan und anschließender Verseifung wird der Pentamethyläther des Xanthons (XC) erhalten, der auch aus der Penta- oder der Hepta-methyl-verbindung der Dehydro-

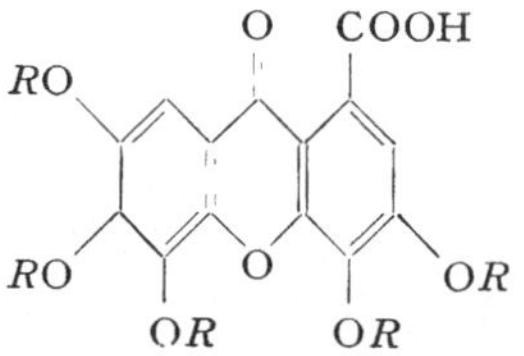

(LXXXIX.) $R = H.$ (XC.) $R = CH_3.$

digallussäure direkt mit konz. Schwefelsäure dargestellt werden kann. Jedenfalls beweist die Xanthon-bildung die Diphenyläther-anordnung und die o-Stellung einer Carboxylgruppe zur Ätherbrücke.

---

* Privatmitteilung von Dr. W. Mayer.

Den endgültigen Beweis der Konstitution entsprechend (LXXXVIII) erbrachte MAYER durch die *Synthese* des Pentamethyl-dehydro-digallus-säure-dimethylesters (XCIII). Hierzu wurde Brom-trimethyläther-gallussäure-methylester (XCI) mit dem Kaliumphenolat des 3,4-Di-methyläther-gallussäure-methylesters (XCII) nach ULLMANN kondensiert. Das synthetische Produkt war mit dem aus Dehydro-digallussäure mit Diazomethan erhaltenen identisch.

(XCI.)
Brom-trimethylather-gallussaure-methylester.

(XCII.)
3,4-Dimethyläther-gallussaure-methylester (K-phenolat).

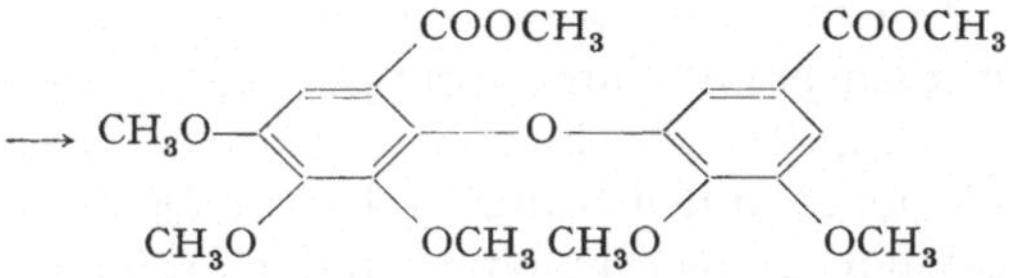

(XCIII.) Pentamethyl-dehydro-digallussaure-dimethylester.

Aber auch die freie Dehydro-digallussäure konnte von FIKENTSCHER (*17*) auf folgendem Weg synthetisiert werden: Brom-trimethyläther-gallussaures Kalium (XCIV) wurde mit dem Kaliumphenolat des 3,4-(Diphenyl-methylen-dioxy)-5-oxy-benzoesäure-methylester (XCV) in

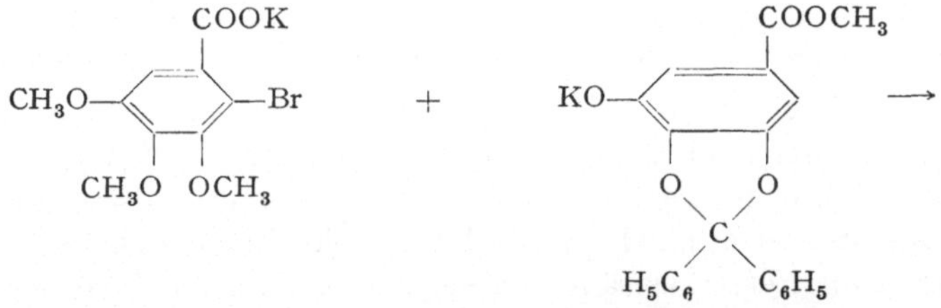

(XCIV.) Brom-trimethylather-gallussaures K.   (XCV.) 3,4-(Diphenyl-methylendioxy)-5-oxy-benzoesaure-methylester (K-phenolat).

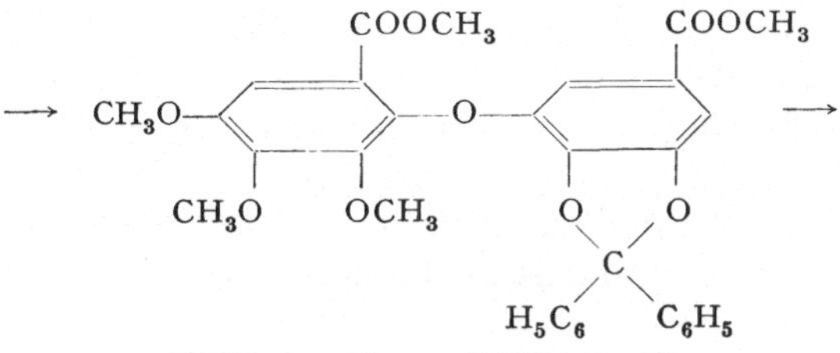

(XCVI.) *R* = CH₃.   (XCVII.) *R* = H.

Gegenwart von Kupferpulver erhitzt. Es entstand Verbindung (XCVI), die mit Alkali zur Dicarbonsäure (XCVII) verseift und dann mit Aluminiumbromid unter Abspaltung der Methylgruppen und der Benzhydryliden-gruppe in Dehydro-digallussäure (LXXXVIII) umgewandelt wurde.

Dehydro-digallussäure fällt Gelatine, hat also Gerbstoff-eigenschaften. Vor kurzem ist sie in der Kastanienrinde an Zucker gebunden aufgefunden worden*.

7. Valoneasäure.

a) Beschreibung und Analyse; Ultraviolett-Spektrum.

Die Fruchtbecher einiger Eichenarten, die in den östlichen Mittelmeerländern wachsen, werden unter der Bezeichnung „Valonea" als eines der besten pflanzlichen Gerbmitteln verwendet. Die Stammpflanzen sind *Quercus valonea* in Kleinasien und *Qu. macrolepis* (großschuppig) in Griechenland. Die Fruchtbecher dieser Eichen sind größer als die der mitteleuropäischen Eichen und ihre Schuppen sind bedeutend stärker entwickelt. Diese Schuppen sind besonders gerbstoffreich und werden als „Trillo" neben Valonea in den Handel gebracht. Es ist bis jetzt noch nicht gelungen, aus Valonea einen einheitlichen Gerbstoff rein darzustellen. Indessen berichten O. Th. Schmidt und Komarek (*87*) über die Isolierung und Aufklärung eines neuartigen Gerbstoff-bausteins aus Valonea.

Die Verbindung wurde entdeckt, als acetonische Extrakte von Valonea nach Reinigung über Bleisalzfällung und fraktionierte Essigester-Extraktion durch Erhitzen mit 5proz Schwefelsäure hydrolysiert wurden. Zur präparativen Darstellung erwies es sich als bedeutend vorteilhafter, käufliche Valonea-Extrakte mit Aceton zu perkolieren und die Perkolate ohne weitere Reinigung zu hydrolysieren. Dabei fällt die neue Verbindung zusammen mit Ellagsäure aus, von der sie dank ihrer etwas größeren Löslichkeit in Aceton und 90proz Dioxan getrennt werden kann.

Die neue Verbindung bildet farblose Kristalle, die keinen Schmelzp. haben, ist optisch inaktiv und besitzt die Formel $C_{21}H_{10}O_{13}$. Sie liefert in Methanol mit (methanolischem) Ferrichlorid eine dunkelgrüne, auf Zusatz von Wasser dunkelblaue Färbung. Die Griesmeyer-Reichelsche Probe (*71*) auf freie Ellagsäure ist negativ, ebenso die Reaktion auf gebundene Ellagsäure (*63, 68*), dagegen ist die Cyankali-Reaktion auf Gallussäure (*108*) positiv. Mit Gelatine und Brucinacetat entstehen Fällungen.

Die Verbindung ist in Wasser unlöslich, löst sich aber leicht in Alkalien, wobei — unter Luftabschluß oder in Wasserstoff-Atmosphäre — in der Kälte keine Veränderungen eintreten. Gegen verdünnte oder halbkonzentrierte Säuren ist die Verbindung auch in der Hitze beständig. Konzentrierte Schwefelsäure oder alkalische Lösungen unter Luftausschluß

---

* Privatmitteilung von Dr. W. Mayer.

führen beim Erhitzen zu Veränderungen, die weiter unten beschrieben werden.

Mit Diazomethan entsteht eine Heptamethyl-verbindung $C_{28}H_{24}O_{13}$, die sich zu einer Hexamethyl-verbindung $C_{27}H_{22}O_{13}$ verseifen läßt. Das Vorliegen einer Carboxylgruppe wird bestätigt durch die Darstellung eines Hexamethyl-amids $C_{27}H_{23}O_{12}N$ aus der Heptamethyl-verbindung. Dank ihrer freien Carboxylgruppe löst sich die Hexamethyl-verbindung in kalter, verdünnter Natronlauge leicht. Erhitzt man die Verbindung mit überschüssiger $n/10$-NaOH, so werden bei der Rücktitration drei Carboxylgruppen angezeigt. Aus der neutralen oder angesäuerten Lösung scheidet sich dann aber die ursprüngliche Hexamethyl-verbindung wieder aus. Da dies Verhalten zwei Lactongruppierungen anzeigt, wurde die neue Verbindung $C_{21}H_{10}O_{13}$ „*Valoneasäure-dilacton*" genannt. Zur weiteren Charakterisierung wurde ein Hexaacetat, $C_{33}H_{22}O_{19}$, dargestellt.

Die positive Cyankali-Reaktion (die sich in allen bisher bekannten Fällen als spezifisch für Gallussäure erwies, solange deren funktionelle Gruppen nicht verändert sind) des Valoneasäure-dilactons legte es nahe, bei dieser Verbindung einen Decarboxylierungsversuch vorzunehmen, in der Erwartung, daß unter ähnlichen Bedingungen wie bei der Gallussäure Kohlendioxyd abgespalten würde. Dies ist in der Tat der Fall. In Chinolin (oder Glycerin) spaltet die Verbindung oberhalb 160—180° in 3 Stunden 1 Mol CO_2 ab. Das „Decarboxyvaloneasäure-dilacton" $C_{20}H_{10}O_{11}$ ist grünstichig hellgelb, hat keinen Schmelzp. und zeigt keine positive Cyankali-Reaktion mehr. Neben diesem Decarboxylierungs-produkt scheint in geringer Menge Ellagsäure zu entstehen.

Ein Versuch, unter gleichen Bedingungen aus dem Hexamethyl-valoneasäure-dilacton Kohlendioxyd abzuspalten, verlief in Überein-stimmung zum Verhalten des Gallussäure-trimethyläthers negativ.

Ein überraschendes Ergebnis lieferte der Versuch, Valoneasäure-dilacton mit Benzylchlorid und Kaliumcarbonat in Acetophenon zu benzylieren (*14*). Es wurde nicht das der Hexamethyl-verbindung ent-sprechende Hexabenzyl-valoneasäure-dilacton erhalten, sondern Tetra-benzyl-ellagsäure und Mono-C-benzyl-tetra-O-benzyl-ellagsäure, also die-selben Reaktionsprodukte wie bei der Benzylierung der Ellagsäure selbst (*99*). Der Versuch zeigt, daß im Valoneasäure-dilacton 1 Mol. Ellagsäure eingebaut sein muß, das unter dem Einfluß von Kaliumcarbonat ab-gespalten wird.

Wird Hexamethyl-valoneasäure-dilacton oder sein Methylester mit Dimethylsulfat und Alkali behandelt, so treten unter gleichzeitiger Öffnung der beiden Lacton-gruppierungen zwei neue Methoxylgruppen auf. Es entsteht die Octamethyl-valoneasäure, $C_{29}H_{30}O_{15}$, deren Triester dank seiner guten Löslichkeit eine Molekulargewichtsbestimmung er-laubte, die die angegebene Bruttoformel bestätigte. Auch ein schön kristallisierter Tri-*p*-brom-phenacylester der Octamethyl-valoneasäure, $C_{53}H_{45}O_{18}$, wurde dargestellt.

Aus den bisherigen analytischen Ergebnissen geht hervor, daß die — in freier Form noch nicht bekannte — Valoneasäure die Formel $C_{21}H_{14}O_{15}$ haben muß, und daß acht Sauerstoff-atome in acht Hydroxylgruppen

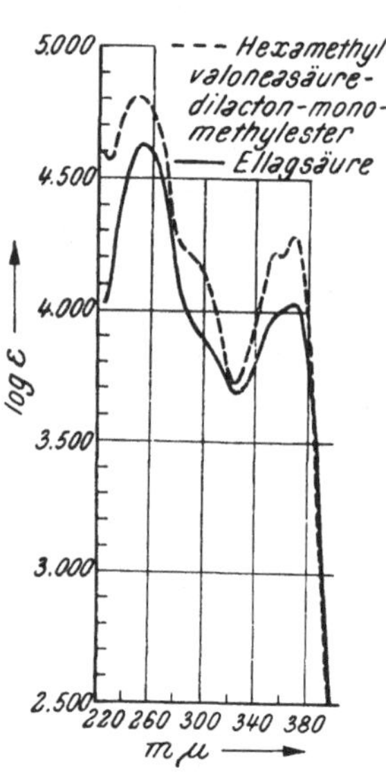

Abb 4. Ultraviolett-Spektren von Ellagsaure und Hexamethylvaloneasaure-dilacton-monomethylester, in Dioxan (Ellagsaure in Dioxan-Wasser). [Aus. Liebigs Ann. Chem. *591*, 153 (1955).]

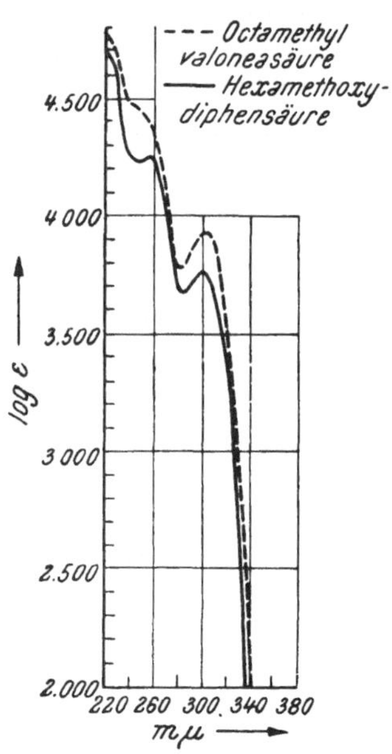

Abb. 5. Ultraviolett-Spektren von Octamethylvaloneasaure und Hexamethoxy-diphensaure in Dioxan. [Aus: Liebigs Ann. Chem. *591*, 153 (1955).]

und sechs Sauerstoff-atome in drei Carboxylgruppen festgelegt sind. Das 15. Sauerstoff-atom muß Ätherfunktion haben, es gelang nämlich nicht, eine Carbonylgruppe nachzuweisen.

Der geringe Wasserstoffgehalt der Valoneasäure oder ihres Dilactons, ebenso die Eisenchlorid-Reaktion weisen auf das Vorliegen aromatischer Ringe hin. Die Zahl der Kohlenstoff-atome (21) legt die Vermutung nahe, daß der Valoneasäure drei Gallussäuren (je C_7) zugrunde liegen, von denen zwei in Form von Ellagsäure bzw. Hexaoxy-diphensäure miteinander verknüpft sind, wofür das Auftreten der Tetrabenzyl-ellagsäure und der Mono-C-benzyl-tetra-O-benzyl-ellagsäure bei der Benzylierungs-reaktion spricht. Die dritte Gallussäure, deren Anwesenheit durch die positive Cyankali-Reaktion belegt ist, müßte dann ätherartig mit der Ellagsäure bzw. Hexaoxy-diphensäure verbunden sein.

Eine gute Stütze dieser zunächst summarischen Ableitung bot die vergleichende U. V.-Spektroskopie. *Abb. 4* zeigt die Absorptionskurve des Hexamethyl-valoneasäure-dilacton-monomethylesters verglichen mit der Absorptionskurve der Ellagsäure (5). Beide Kurven zeigen in ihrem Verlauf gute Übereinstimmung und unterscheiden sich lediglich durch die höheren Extinktionswerte der ersten. Dies kann so gedeutet werden, daß in dem Hexamethyl-valoneasäure-dilacton-monomethylester die Ellagsäure als Chromophor vorliegt, und daß die höheren Extinktionswerte einem zusätzlich vorhandenen Chromophor zuzuschreiben sind, der der Ellagsäure konstitutiv sehr nahe verwandt sein muß, da andernfalls stärkere Veränderungen der Absorptionskurve zu erwarten wären. Dieser zusätzliche Chromophor kann in einem Mol. Gallussäure bestehen, das mit der Ellagsäure verknüpft ist. Die Richtigkeit dieser Deutung wurde durch den Vergleich der Absorptionskurven von Octamethyl-valoneasäure und Hexamethoxy-diphensäure nachgeprüft (*Abb. 5*). In diesen beiden lacton-geöffneten Verbindungen ist das der Ellagsäure und dem Valoneasäure-dilacton eigene Resonanzsystem der koplanaren Benzolringe (5) aufgehoben, womit das langwellige Maximum bei 370 mμ verschwindet. Es war also folgendes zu erwarten:

1. Die Absorptionskurven der Octamethyl-valoneasäure und der Hexamethoxy-diphensäure müssen einen sehr ähnlichen Verlauf haben.

2. Liegen bei der Octamethyl-valoneasäure drei Gallussäure-Chromophore vor, so sollten bei sonst gleichem Kurvenverlauf die Extinktionswerte in den Maxima um etwa 50% höher liegen als bei der Hexamethoxy-diphensäure, die den Gallussäure-Chromophor nur zweimal enthält. Abb. 5 bestätigt diese Erwartung: Die beiden Kurven haben sehr ähnlichen Verlauf, und die Extinktionen bei 254 mμ verhalten sich mit 17900 (Hexamethoxy-diphensäure) und 25600 (Octamethyl-valoneasäure) nahezu wie 2 : 3; ähnlich ist das Verhältnis der Extinktionen mit 5930 : 8530 bei 302 mμ bzw. 304 mμ.

b) Alkali-Spaltung.

Mit der Erkenntnis, daß im Valoneasäure-dilacton 1 Mol. Ellagsäure ätherartig mit 1 Mol. Gallussäure verknüpft ist, verblieben für seine Struktur noch vier Möglichkeiten, die durch die Formeln (XCVIII), (CI), (CII) und (CV) wiedergegeben sind (*Formelübersicht 2*).

Der Benzylierungsversuch hatte den Hinweis ergeben, daß Valoneasäure-dilacton unter dem Einfluß alkalischer Mittel aufgespalten werden kann. Wenn es sich dabei um eine hydrolytische Spaltung der Ätherbindung handelte, so müßte eine ad hoc durchgeführte Alkalispaltung bei Formel (XCVIII) und (CI) zu Oxy-ellagsäure („Flavellagsäure", IC) und Gallussäure (C), bei Formel (CII) und (CV) zu Ellagsäure (CIII) und Tetraoxy-benzoesäure (CIV) führen:

(XCVIII.)

$+ H_2O$

(IC.) Oxy-ellagsäure. $C_{14}H_6O_9$.

$\uparrow + H_2O$

(C.) Gallussäure. $C_7H_6O_5$.

(CI.) $C_{21}H_{10}O_{13}$.

(CII.)

$+ H_2O$

(CIII.) Ellagsäure. $C_{14}H_6O_8$.

$\uparrow + H_2O$

(CIV.) Tetraoxy-benzoesäure. $C_7H_6O_6$.

(CV.) Valoneasäure-dilacton. $C_{21}H_{10}O_{13}$.

Formelübersicht 2.

Die Spaltung des freien Valoneasäure-dilactons verläuft leicht. Sie wurde durch 2stündiges Kochen mit 2 n-Natronlauge durchgeführt, wird aber auch schon durch kochende 1proz. Sodalösung bewirkt. Als einzige definierte Spaltstücke wurden Ellagsäure und Gallussäure isoliert. Für das Auftreten von Oxy-ellagsäure oder Tetraoxy-benzoesäure wurde kein Hinweis gefunden. Die Ausbeute an den isolierten Spaltstücken betrug stets weniger als 50% der eingesetzten Substanz. Es kann sich also bei der Alkalispaltung des Valoneasäure-dilactons nicht um eine einfache Hydrolyse handeln, denn es sind nicht die Bestandteile des Wassers, die an der Spaltstelle eintreten, sondern nur zwei Wasserstoffatome.

Unter den drastischeren Bedingungen der Alkalischmelze sind solche Ätherspaltungen von Asahina (*2, 3*) in der Gruppe der Depsidone beschrieben worden. So lieferte die Physodsäure (CVI) bei dieser Reaktion Orcin (CVII) und 5-n-Amyl-resorcin (CVIII) (*3*), während bei rein hydrolytischer Spaltung n-Amyl-trioxy-benzol (CIX) zu erwarten wäre. Als weiteres Beispiel sei die Alkalischmelze des Hyposalacinols (CX) angeführt, die zu β-Orcin (CXI) und 3,5-Dioxy-p-toluylsäure (CXII) geführt hat.

(CVI.) Physodsaure.

$C_5H_{11}COOH + 2\,CO_2 +$

(CVII.) Orcin.

(CVIII.) n-Amyl-resorcin.

(CIX.) n-Amyl-trioxy-benzol.

(CX.) Hyposalacinol.

(CXI.) β-Orcin.

(XCII.) 3,5-Dioxy-p-toluylsäure.

Ein näher gelegenes Analogie-beispiel bietet die im vorigen Abschnitt (S. 115) behandelte Dehydro-digallussäure, deren Alkalispaltung auch nur Gallussäure in nahezu 50% der theoretischen Ausbeute ergibt.

Über den Mechanismus dieser bemerkenswerten Phenolätherspaltung können vorerst keine Aussagen gemacht werden. Soviel läßt sich jedoch feststellen, daß freie Hydroxylgruppen der Ausgangsverbindungen an der Reaktion beteiligt sein müssen, denn nur das freie Valoneasäure-dilacton läßt sich so spalten; die Hexamethyl-verbindung ist selbst gegen heiße 40proz. Kalilauge resistent. Auch Dehydro-digallussäure verliert durch Verätherung der Hydroxylgruppen ihre Alkali-spaltbarkeit (51).

Die Alkalispaltung des Valoneasäure-dilactons hat die aus den früheren Versuchen wahrscheinlich gemachte Annahme einer Verbindung aus Ellagsäure und Gallussäure bestätigt. Eine Entscheidung zwischen den Formeln (XCVIII), (CI), (CII) und (CV) (S. 122) hat jedoch erst die Xanthonbildung erbracht.

c) Valonea-xanthon.

Erhitzt man Valoneasäure-dilacton, $C_{21}H_{10}O_{13}$, 15 Minuten lang mit konz. Schwefelsäure, so bildet sich unter Verlust eines Mols Wasser in vorzüglicher Ausbeute eine Verbindung $C_{21}H_8O_{12}$, welcher die Formel (CXIII) zukommen muß. Sie enthält noch alle C-Atome des Valoneasäure-dilactons und, wie aus den nachstehend beschriebenen Umsetzungen hervorgeht, noch alle sechs Hydroxylgruppen sowie die beiden Lacton-gruppierung der Ausgangssubstanz, während eine Carboxylfunktion verschwunden ist. Es ergibt sich hieraus, daß die Wasserabspaltung erfolgt sein muß aus der Carboxylgruppe des Gallussäure-anteils des Valoneasäure-dilactons und einem Kernwasserstoff des Ellagsäure-anteils, wobei ein Xanthon-abkömmling entstehen muß.

Die neue Verbindung, das „Valonea-xanthon", ist gelbgrün und gibt eine blaugrüne Eisenchlorid-reaktion. Mit Diazomethan wird das gelbe Hexamethyl-valonea-xanthon $C_{27}H_{20}O_{12}$ (CXIV), mit Essiganhydrid das blaßgelbe Hexaacetyl-valonea-xanthon $C_{33}H_{20}O_{18}$ (CXV) erhalten.

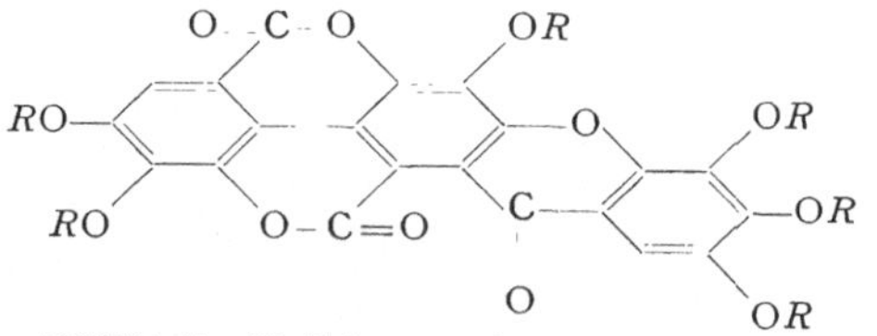

(CXIII.) R = H. Valonea-xanthon.
(CXIV.) R = CH_3. Hexamethyl-valonea-xanthon.
(CXV.) R = CO · CH_3. Hexaacetyl-valonea-xanthon.

Wird das Hexamethyl-valonea-xanthon mit Dimethylsulfat und Alkali methyliert, so entsteht unter Öffnung der beiden Lactonringe eine farblose, in Alkali unlösliche Verbindung $C_{31}H_{32}O_{14}$. Ihr kommt die Formel (CXVI) eines 2,3,4,5,6-Pentamethoxy-8-carboxy-7-(2′,3′,4′-tri-methoxy-6′-carboxy-phenyl)-xanthon-dimethylesters zu, welcher der Einfachheit halber „Octamethyl-valonea-xanthon-dimethylester“ genannt wurde. Von den beiden Estergruppen wird eine normal, die zweite, vermutlich an der Carboxylgruppe an $C_{(8)}$, von Alkali schwer verseift.

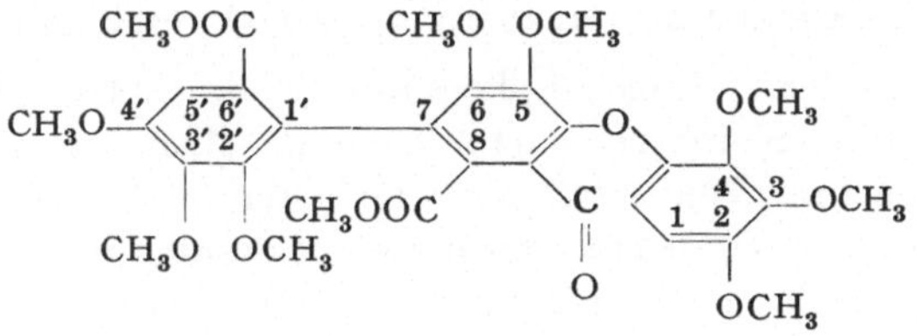

(CXVI.) Octamethyl-valonea-xanthon-dimethylester.

Octamethyl-valonea-xanthon-dimethylester gibt mit konz. Schwefelsäure eine gelbe Lösung. Die gleiche Farberscheinung wird beobachtet, wenn man eine Lösung dieser Verbindung in Eisessig mit etwas konz. Schwefelsäure versetzt. In beiden Fällen verschwindet die Farbe auf Zusatz von Wasser. Ähnliche Gelbverfärbungen unter dem Einfluß von konz. Salz- oder Schwefelsäure zeigen auch unsubstituiertes Xanthon sowie Xanthonderivate, z. B. 3,4-Dimethoxy-xanthon oder 3,4,5,6,7-Pentamethoxy-xanthon-1-carbonsäure-methylester (*51*). Auch in diesen Fällen wird die Farbe durch Zusatz von Wasser wieder aufgehoben.

Die Entstehung des Valonea-xanthons aus Valoneasäure-dilacton beweist, daß diesem die Formel (CV, S. 122) zukommen muß. Für die Valoneasäure selbst, die in freier Form nicht dargestellt wurde, ergibt sich die Formel (CXVII).

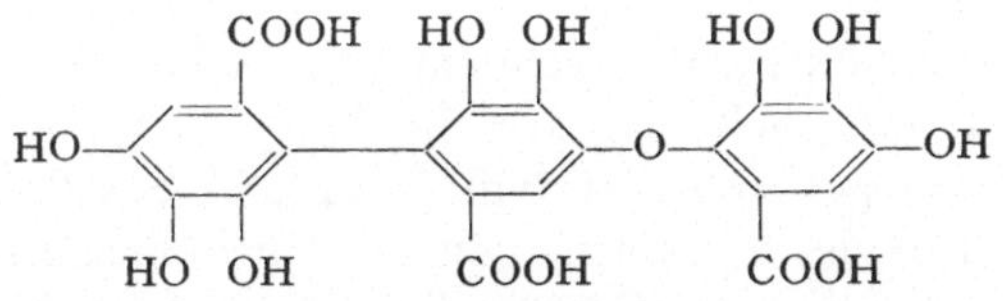

(CXVII.) Valoneasaure.

Zweifellos wird auch Valoneasäure in der Pflanze aus Gallussäure gebildet. Sie vereinigt in sich die beiden Dehydrierungstypen: die „Kerndehydrierung“, die zur Bildung der Ellagsäure oder Hexaoxy-diphensäure führt, und die „Phenoldehydrierung“, durch welche die Dehydro-digallussäure der Edelkastanie entstanden sein muß. Ebenso wie die Hexaoxy-diphensäure mit ihren beiden Carboxylgruppen an Glucose gebunden ist und bei der Abspaltung in Ellagsäure übergeht, scheint auch Valoneasäure in der lacton-offenen Form gebunden zu sein und nach der Abspaltung in Valoneasäure-dilacton überzugehen. O. Th. Schmidt und Grünewald (*83*) methylierten Valonea-extrakt mit Diazomethan, hydrolysierten alkalisch und isolierten aus dem Hydrolysat optisch aktive Octamethyl-valoneasäure.

IV. Mögliche genetische Beziehungen.

1. Zur Bildung der Hexaoxy-diphenoyl-verbindungen.

Das Studium der Hexaoxy-diphensäure (S. 99) hatte unter anderem ergeben, daß diese Säure zwar sehr leicht in Ellagsäure übergeht, daß aber der umgekehrte Weg nicht ohne bedeutende Umwege möglich ist. O. Th. Schmidt und Demmler (*82*) halten es deshalb für ausgeschlossen, daß bei der Entstehung der Ellagen-gerbstoffe Corilagin und Chebulagsäure Glucose und Ellagsäure miteinander reagieren unter Bildung einer Hexaoxy-diphenyl-glucose. Sie halten es aber auch für unwahrscheinlich, daß Hexaoxy-diphensäure, deren Stabilität ja begrenzt ist, vorgebildet ist und in fertigem Zustand mit dem Zucker in („beidarmige") Bindung tritt. (Die geringe sterische Stabilität der optisch aktiven Formen spielt für diese Betrachtung keine Rolle, da racemische Hexaoxy-diphensäure durch die Bindung an die optisch aktive Glucose bevorzugt eine optisch aktive Form annehmen könnte.) Vielmehr glauben sie, daß in Poly- oder Oligo-galloyl-glucosen durch die Wirkung dehydrierender Fermente aus zwei in geeigneter Stellung gebundenen Molekülen Gallussäure die Hexaoxy-diphensäure erst am Zucker gebildet wird.

Eine Stütze dieser Hypothese bildet die Feststellung Erdtmans (*16*), daß in vitro Ellagsäure durch Dehydrierung aus Gallussäure durch Luftsauerstoff in alkalischer Lösung nur dann gebildet wird, wenn man veresterte Gallussäure verwendet. Vor allem erhielt die hier entwickelte Hypothese von der Entstehung der Hexaoxy-diphensäure einen hohen Grad von Wahrscheinlichkeit, nachdem mit Sicherheit feststand, daß in Corilagin und der Chebulagsäure die Hexaoxy-diphensäure in 3,6-Stellung der Glucose gebunden ist (vgl. S. 91 und S. 98). Das sind dieselben Zuckerhydroxyle, an welchen in der Chebulinsäure zwei Gallussäuren gebunden sind (vgl. S. 88). Die Vorstellung ist naheliegend, daß Chebulagsäure, $C_{41}H_{30}O_{27}$, durch einfache (fermentative) Dehydrierung zweier Gallussäuren in situ aus Chebulinsäure, $C_{41}H_{32}O_{27}$, entsteht. Dabei ist freilich zu berücksichtigen, daß es noch nicht völlig sicher ist, daß in den beiden Gerbstoffen die Chebulsäure gleichartig gebunden ist. Jedenfalls vermochten O. Th. Schmidt und Günther (*32*) zwar aus Gallussäure-methylester durch geeignete Dehydrierung Hexaoxy-diphensäure-dimethylester herzustellen, doch ist die Überführung von Chebulinsäure in Chebulagsäure oder von 1,3,6-Trigalloyl-glucose in Corilagin in vitro noch nicht gelungen.

2. Zur Entstehung der Chebulsäure und Brevifolin-carbonsäure.

Schon bevor die Konstitution der Chebulsäure völlig aufgeklärt war, war die Tatsache aufgefallen, daß sich die Bruttoformeln der Chebulsäure, $C_{14}H_{12}O_{11}$, und Ellagsäure, $C_{14}H_6O_8$, gerade um $3\ H_2O$ unterscheiden. Die Vorstellung, daß zwischen den beiden Säuren eine konstitutiver Zusammenhang bestehen könnte, hat zur Aufklärung der Chebulsäure beigetragen, und O. Th. SCHMIDT und MAYER (*92*) sprachen die Vermutung aus, daß ein genetischer Zusammenhang zwischen den beiden Säuren bestehen müsse. Man konnte sich damals, bevor die Hexaoxydiphensäure bekannt war, vorstellen, daß Ellagsäure (CXVIII) in der Pflanze durch oxydierende und reduzierende Fermente so umgewandelt wird, daß unter Aufnahme von 3 Molen Wasser, also in summarischer

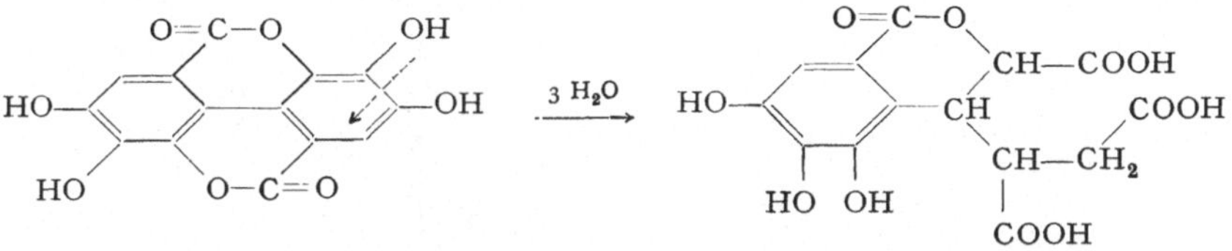

(CXVIII.) Ellagsaure.   (CXIX.) Chebulsäure.

„Hydrolyse", ein aromatischer Kern der Ellagsäure aufgespalten und Chebulsäure (CXIX) gebildet würde. Es zeigte sich später, daß dieser Zusammenhang wahrscheinlich nicht zwischen Ellagsäure, sondern Hexaoxydiphensäure und Chebulsäure besteht. In einem wichtigen Experiment

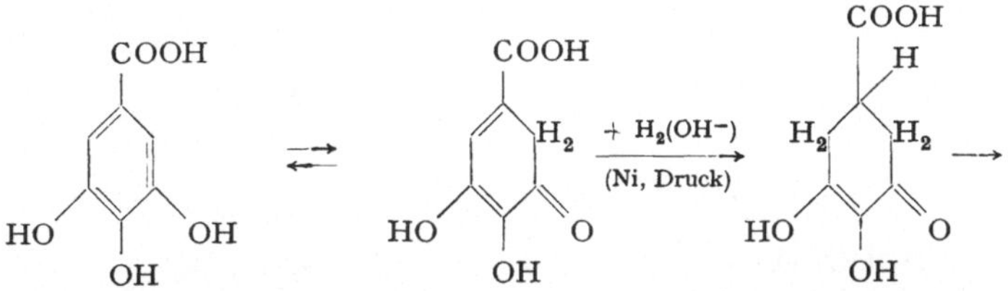

(CXXa.) Gallussäure (CXXb).   (CXXI.) Dihydrogallussaure.

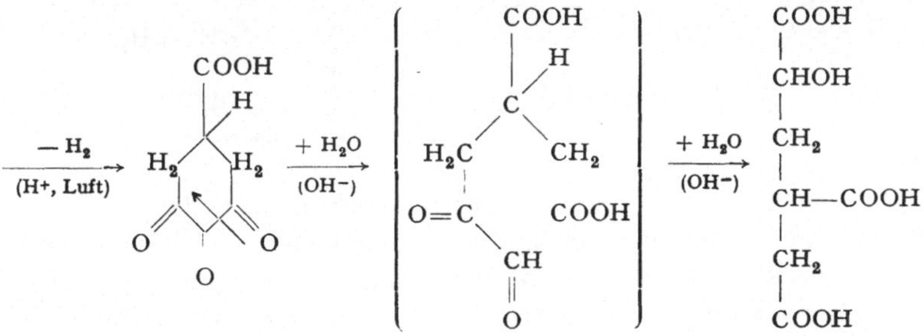

(CXXII.) Cyclohexantrion-carbonsäure.   (CXXIII.)   (CXXIV.) 2-Oxy-4-carboxy-adipinsäure.

haben Mayer, Bachmann und Kraus (53) eine Modellreaktion durchgeführt, welche diese Hypothese sehr stark stützt: Gallussäure (CXXa $\rightleftharpoons$ $\rightleftharpoons$ CXXb) wird in alkalischer Lösung zur Dihydrogallussäure (CXXI) umgewandelt. Diese wird (als cyclisches Redukton) leicht dehydriert und bildet Cyclohexantrion-carbonsäure (CXXII). Mit bemerkenswerter Leichtigkeit wird (CXXII) von verdünnten, wäßrigen Alkalien aufgespalten. Über die Zwischenverbindung (CXXIII) hinweg entsteht 2-Oxy-4-carboxy-adipinsäure (CXXIV). Im Effekt gleicht diese Ringöffnung von Gallussäure, $C_7H_6O_5$, zu 2-Oxy-4-carboxy-adipinsäure, $C_7H_{10}O_7$, durch die Aufnahme von 2 Molen Wasser einer Hydrolyse.

Diese Reaktion läßt sich (auf dem Papier) genau übertragen auf den Übergang von Hexaoxy-diphensäure, $C_{14}H_{10}O_{10}$, zu Chebulsäure, $C_{14}H_{12}O_{11}$, wie in den Formeln (CXXV) bis (CXXIX) zum Ausdruck kommt.

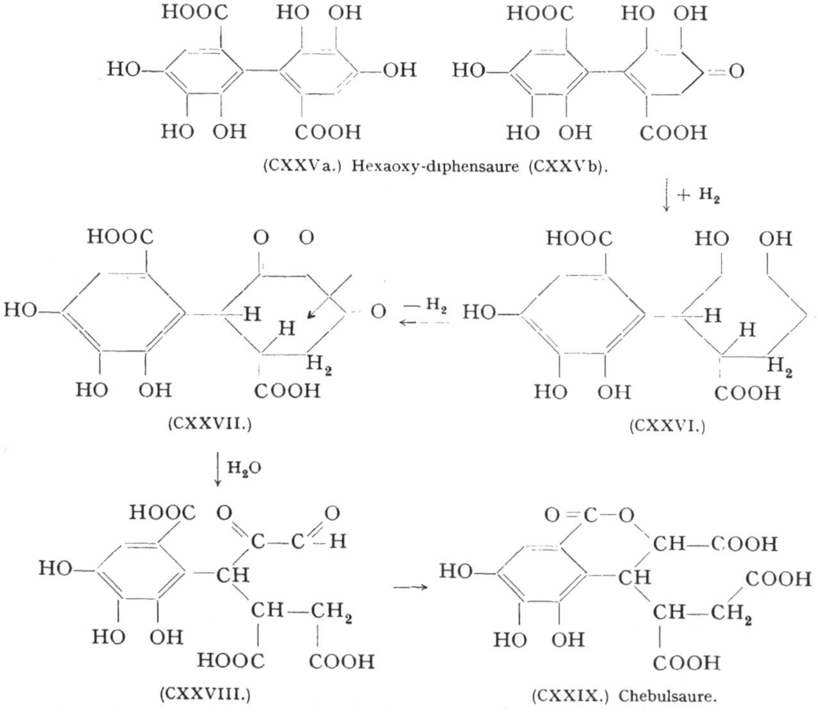

(CXXVa.) Hexaoxy-diphensäure (CXXVb).

(CXXVII.) (CXXVI.)

(CXXVIII.) (CXXIX.) Chebulsäure.

Es läßt sich aber auch eine genetische Beziehung von der Hexaoxy-diphensäure zur Brevifolin-carbonsäure „konstruieren". Dabei muß allerdings berücksichtigt werden, daß bei dem gedachten Übergang neben der Abspaltung von CO_2 eine Dehydrierung stattfinden muß (vgl. Tabelle 4, S. 131). Aber auch dann bleiben noch zwei Reaktionswege zu diskutieren,

je nach der Konstitution, die sich für Brevifolin-carbonsäure als zutreffend
herausstellen wird.

Hat die Säure die Konstitution (CXXXII), so müßte Verbindung
(CXXVII) zunächst decarboxyliert werden (CXXX). Eine solche De-
carboxylierung sollte leicht möglich sein, da die Cyclohexantrion-carbonsäure von MAYER, BACHMANN und KRAUS (53) (CXXII) in der Tat
leicht CO_2 verliert. Verbindung (CXXX) würde dann zu (CXXXI) aufgespalten und schließlich unter Ringschluß und Dehydrierung in (CXXXII)
übergehen.

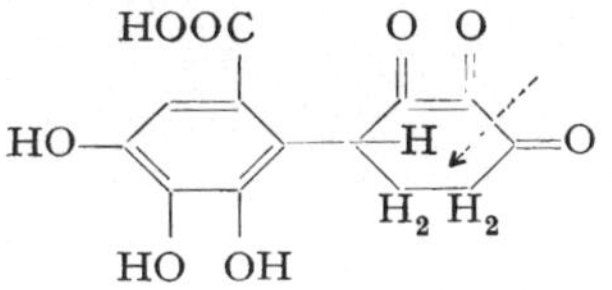

(CXXX.) 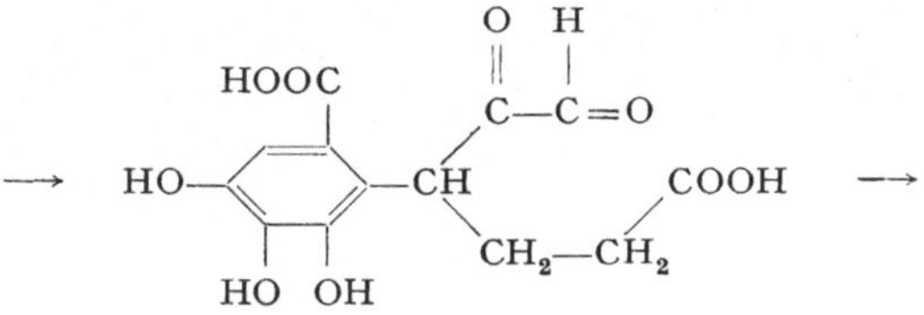   (CXXXI.)

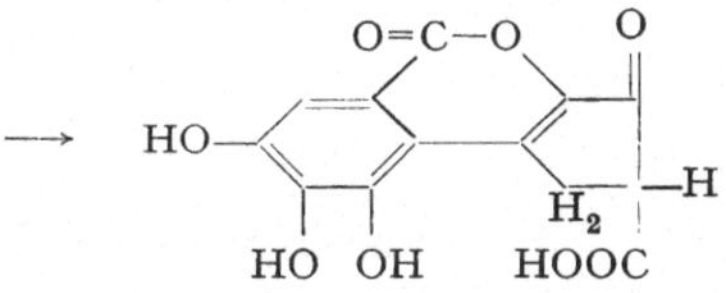

(CXXXII.)

Besitzt Brevifolin-carbonsäure die Konstitution (CXXXIV), so
müßte die obige Zwischenverbindung (CXXVIII) unter Ringschluß und
Dehydrierung zu (CXXXIII) führen, welche die zur Carbonylgruppe
β-ständige Carboxylgruppe verlieren würde.

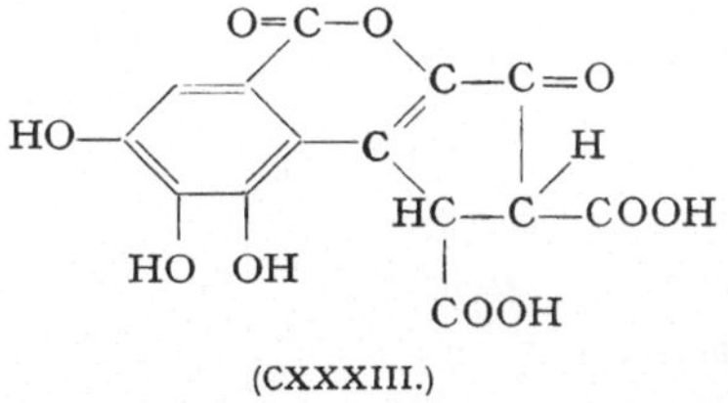

(CXXXIII.) 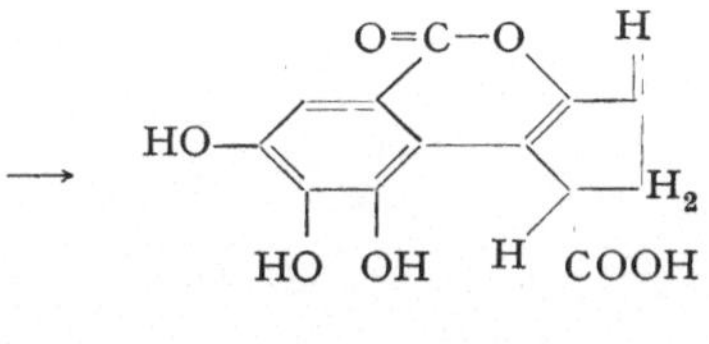   (CXXXIV.)

3. Zur Entstehung der verschiedenen Typen von hydrolysierbaren Gerbstoffen.

In der nachstehenden *Formelübersicht 3* sind alle bisher in Gallotanninen aufgefundenen Phenol-carbonsäuren zusammengestellt. Es
ergibt sich, daß alle auf Gallussäure zurückgehen.

Diese Säuren stehen alle in einer formelmäßig einfachen Beziehung
zur Gallussäure (*Tabelle 4*).

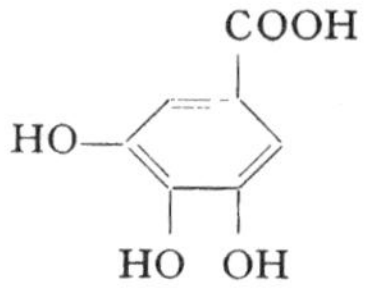

(C.) Gallussaure.

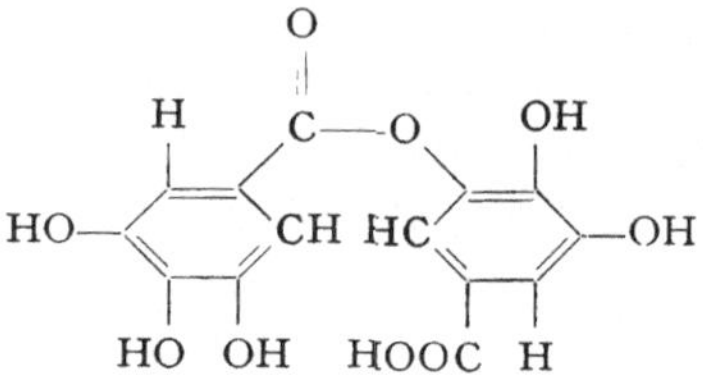

(CXXXV.) *m*-Digallussaure.

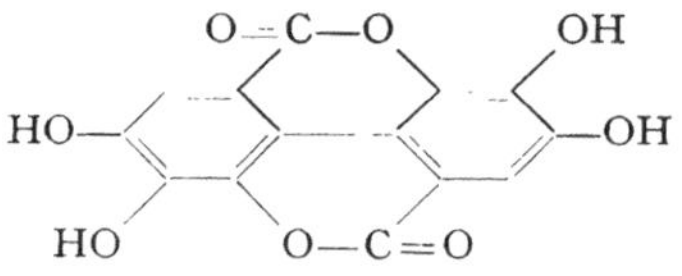

(XXV.) Ellagsaure.

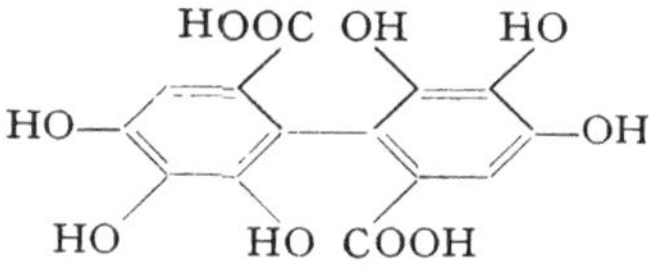

(LIV.) Hexaoxy-diphensäure.

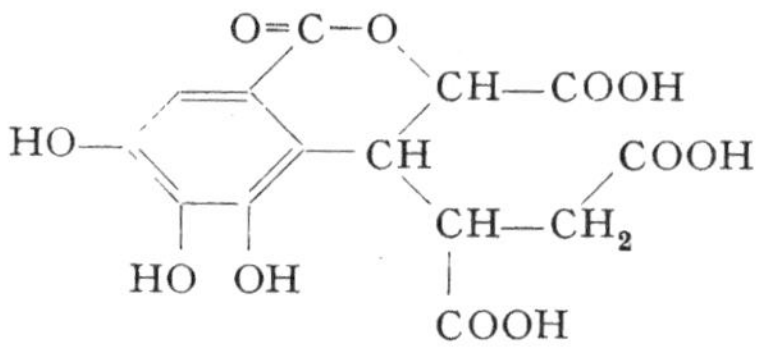

(XXIII.) Chebulsaure.

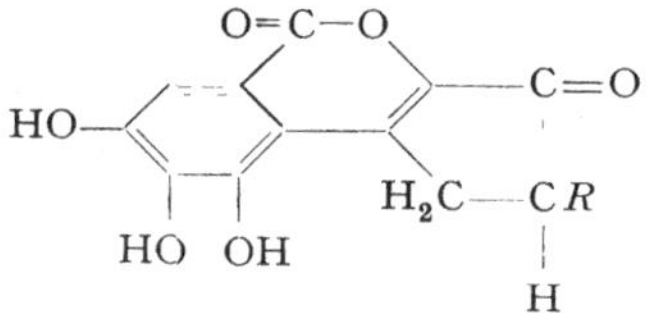

(LXXII.) R = H. Brevifolin.
R = COOH. Brevifolin-carbonsaure.

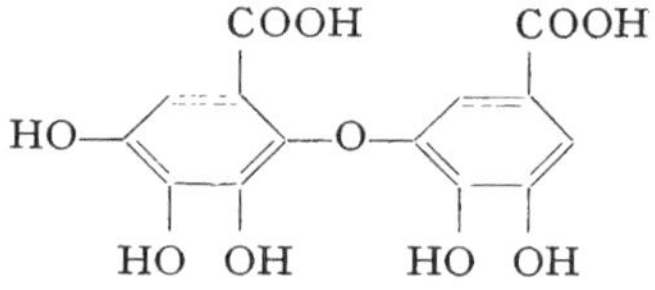

(LXXXVIII.) Dehydro-digallussaure.

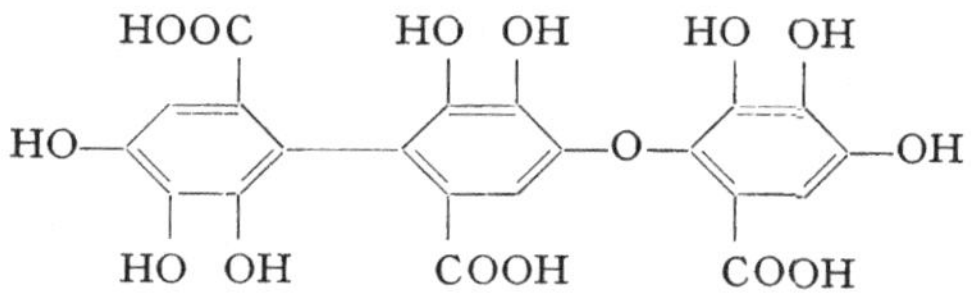

(CXVII.) Valoneasaure.

Formelübersicht 3.

Von O. Th. Schmidt und Mayer (*75, 52*) wurde die Vorstellung entwickelt, daß diese Übergänge sich auch in der Pflanze unter dem Einfluß von Fermenten so einfach abspielen, und daß alle diese Umwandlungen in situ, das heißt an den an Zucker gebundenen Molekülen vor sich gehen könnten, daß also z. B. die Chebulsäure sich aus einer an Zucker gebundenen Hexaoxy-diphensäure bilden könnte. Diese Über-

Tabelle 4. Beziehung der in Gallotanninen vorkommenden Phenol-carbonsäuren zu Gallussäure.

$$2 \text{ Gallussäure} - H_2O \rightarrow \text{m-Digallussäure}$$
$$2 \text{ Gallussäure} - H_2 \rightarrow \text{Hexaoxy-diphensäure}$$
$$\downarrow - 2 H_2O$$
$$2 \text{ Gallussäure} - H_2 \quad - 2 H_2O \rightarrow \text{Ellagsäure}$$
$$2 \text{ Gallussäure} - H_2 \rightarrow \text{Dehydro-digallussäure} \quad + H_2O$$
$$2 \text{ Gallussäure} - H_2 \quad + \quad H_2O \rightarrow \text{Chebulsäure} \leftarrow$$
$$- H_2 - CO_2$$
$$2 \text{ Gallussäure} - 2 H_2 \quad - \quad CO_2 \rightarrow \text{Brevifolin-carbonsäure} \leftarrow$$
$$3 \text{ Gallussäure} - 2 H_2 \rightarrow \text{Valoneasäure}$$

legung hat bei der Aufstellung der vorläufigen Formeln für Chebulin- und Chebulagsäure (SS. 88 und 98) mitgesprochen. Schließlich halten es O. TH. SCHMIDT und MAYER für möglich, daß aus vorgegebenen Oligo- oder Polygalloyl-glucosen gemäß *Formelübersicht 4* die verschiedenen Typen von Gerbstoffen dieser Klasse entstehen könnten.

In einfacher oder doppelter Dehydrierung würde z. B. eine Penta-galloyl-glucose in Ellagen-gerbstoffe übergehen, die durch Umwandlung der (gebundenen) Hexaoxy-diphensäure in Chebulsäure zu Chebulin- und Chebulagsäure führen könnten. Liese Zusammenhänge sind nahe-liegend; der Übergang von Chebulagsäure zum Corilagin ist in vitro

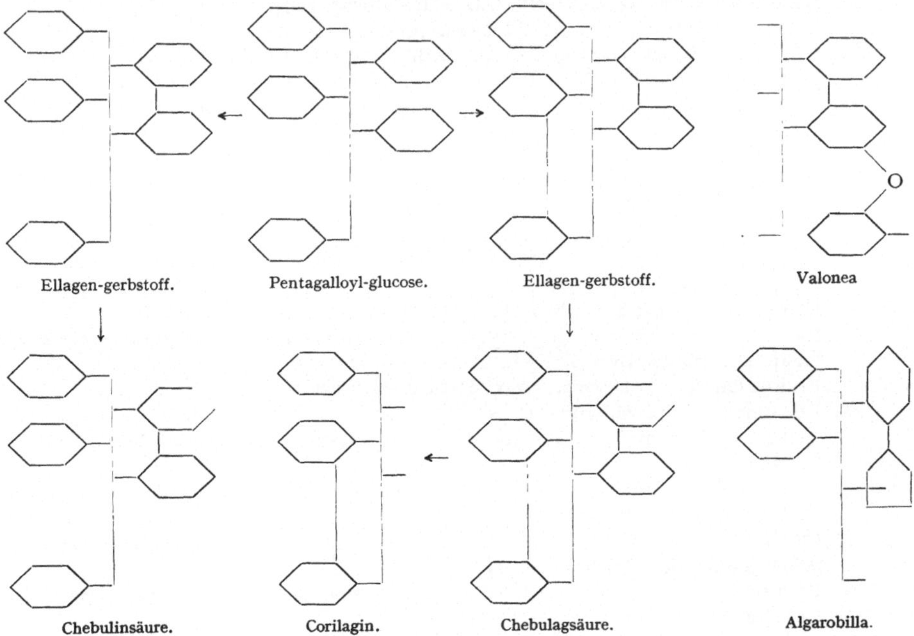

Formelübersicht 4.

9*

realisiert. Ähnliche genetische Beziehungen mögen zwischen galloylierten Zuckern und den (noch weniger bekannten) Gerbstoffen von Valonea und Algarobilla bestehen.

Literaturverzeichnis.

1. Adolphi, W.: Ein Beitrag zur Kenntnis der Chebulinsäure. Arch. Pharmaz. **230**, 684 (1892).

2. Asahina, Y. und J. Asano: Untersuchungen über Flechtenstoffe, XXIII. Mitt.: Über Salazinsäure (II). Ber. dtsch. chem. Ges. **66**, 893 (1933).

3. Asahina, Y. und H. Nogami: Untersuchungen über Flechtenstoffe, XLI. Mitt.: Über die Konstitution der Physodsäure (I). Ber. dtsch. chem. Ges. **67**, 805 (1934).

4. Ault, R. G., W. N. Haworth and E. L. Hirst: The Constitution of Ascorbic Acid. Action of Sodium Hypochlorite on α-Methoxy-acid Amides. J. Chem. Soc. (London) **1934**, 1722.

5. Bernauer, K.: Zusammenhänge zwischen Konstitution, sterischen Verhältnissen und Ultraviolett-Spektren bei Derivaten des Brevifolins. Liebigs Ann. Chem. **588**, 230 (1954).

6. Bernauer, K. und O. Th. Schmidt: Die Synthese des Brevifolin-trimethyläthers. XXII. Mitt. über natürliche Gerbstoffe. Liebigs Ann. Chem. **591**, 153 (1955).

7. Bilbo, A. J. and G. M. Wyman: Steric Hindrance to Coplanarity in *o*-Fluorobenzidines. J. Amer. Chem. Soc. **75**, 5312 (1953).

8. Bittermann, H.: Dipl.-Arbeit, Univ. Heidelberg, 1954.

9. Blinn, F.: Dissert., Univ. Heidelberg, 1951.

10. Brigl, P. und H. Grüner: Kohlenhydrate, XIII. Mitt.: Neue Benzal- und Benzoyl-Derivate der Glucose. Ber. dtsch. chem. Ges. **65**, 1428 (1932).

11. — — Kohlenhydrate, XVIII. Mitt.: Über Benzoate der Glucofuranose. Ber. dtsch. chem. Ges. **66**, 1977 (1933).

12. Calvin, M.: Resonance and Some Physical and Chemical Properties of Biphenyl Types. J. Organ. Chem. (USA) **4**, 256 (1939).

13. Cavallito, C. J. and J. S. Buck: Synthesis of Phenolic Acid Esters. I. Depsides. J. Amer. Chem. Soc. **65**, 2140 (1943).

14. Clinton, R. O. and T. A. Geissman: Gallaldehyde Tribenzyl Ether. J. Amer. Chem. Soc. **65**, 85 (1943).

15. Demmler, K.: Dissert., Univ. Heidelberg, 1952.

16. Erdtman, H.: Phenoldehydrierungen VI. Dehydrierende Kupplung einiger Guajakolderivate. Svensk Kem. Tidskr. **47**, 223 (1935).

17. Fikentscher, R.: Dissert., Univ. Heidelberg, 1956.

18. Fischer, E. und M. Bergmann: Über neue Galloylderivate des Traubenzuckers und ihren Vergleich mit der Chebulinsäure. Ber. dtsch. chem. Ges. **51**, 298 (1918).

19. — — Über das Tannin und die Synthese ähnlicher Stoffe. VI. Ber. dtsch. chem. Ges. **52**, 829 (1919).

20. Fischer, E. und K. Freudenberg: Über das Tannin und die Synthese ähnlicher Stoffe. Ber. dtsch. chem. Ges. **45**, 915 (1912).

21. Freudenberg, K.: Über Gerbstoffe. I.: Hamameli-Tannin. Ber. dtsch. chem. Ges. **52**, 177 (1919).

22. — Über Gerbstoffe. II.: Chebulinsäure. Ber. dtsch. chem. Ges. **52**, 1238 (1919).

23. — Tannin, Cellulose, Lignin, S. 33. Berlin: J. Springer. 1933.

24. FREUDENBERG, K. und F. BLÜMMEL: Hamameli-tannin. III. 17. Mitt. über Gerbstoffe und ähnliche Verbindungen. Liebigs Ann. Chem. **440**, 45 (1924).

25. FREUDENBERG, K. und B. FICK: Über Gerbstoffe. 6.: Chebulinsäure (II). Ber. dtsch. chem. Ges. **53**, 1728 (1920).

26. FREUDENBERG, K. und TH. FRANK: Chebulinsäure. III. 24. Mitt. über Gerbstoffe und ähnliche Verbindungen. Liebigs Ann. Chem. **452**, 303 (1927).

27. FREUDENBERG, K. und G. HÜLL: 2,6-Dimethyl-glucose. Ber. dtsch. chem. Ges. **74**, 237 (1941).

28. FREUDENBERG, K. und D. PETERS: Über Gerbstoffe. 4.: Hamameli-Tannin (II). Ber. dtsch. chem. Ges. **53**, 953 (1920).

29. FRIDOLIN, A.: Dissert., Dorpat, 1884.

30. GILSON, E.: Contribution à l'étude des tannoïdes. Sur les tannoïdes de la rhubarbe de Chine. Bull. Acad. méd. Belgique **16**, 827 (1902).

30a. — Sur deux nouveaux glucotannoïdes. C. R. hebd. Séances Acad. Sci. **136**, 385 (1903).

31. GRÜTTNER, F.: Beiträge zur Chemie der Rinde von *Hamamelis virginica* L. Arch. Pharmaz. **236**, 278 (1898).

32. GÜNTHER, W.: Dipl.-Arbeit, Univ. Heidelberg, 1955.

33. GUYOT, A.: Synthèses au moyen de tétrachlorure de phtalyle fondant à 88⁰ (II). 1. partie: étude de quelques homologues de la diphenylanthrone. Bull. soc. chim. France (3) **17**, 966 (1897).

34. HAWORTH, R. D. and J. GRIMSHAW: Synthesis of Trimethylbrevifolin. Chem. and Ind. **1955**, 199.

35. HAWORTH, R. D., B. P. MOORE and P. L. PAUSON: Purpurogallin. II. Synthesis of Purpurogallone. J. Chem. Soc. (London) **1949**, 3271.

36. HAWORTH, R. D. and L. B. DE SILVA: Chebulinic Acid. J. Chem. Soc. (London) **1951**, 3511.

37. HAWORTH, W. N. and C. R. PORTER: Isolation of Crystalline α- and β-Ethyl-glucofuranosides (γ-Ethylglucosides) and Other Crystalline Derivatives of Glucofuranose. J. Chem. Soc. (London) **1929**, 2796.

38. HEINEMANN, Firma A.: Verfahren zur Gewinnung von Ellagsäure. [Chem. Zbl. **1903**, I, 11].

39. HENSLER, R. H.: Dipl.-Arbeit, Univ. Heidelberg, 1954.

40. HERZIG, J. und J. POLAK: Zur Konstitution der Ellagsäure. III. Mitt. über Laktonfarbstoffe. Monatsh. Chem. **29**, 263 (1908).

41. HILLEMANN, H.: Molekulare Asymmetrie. Angew. Chem. **50**, 435 (1937).

42. HIRST, E. L., E. G. V. PERCIVAL and F. SMITH: Constitution of Ascorbic Acid. Nature (London) **131**, 617 (1933).

43. KARRER, P., H. R. SALOMON und J. PEYER: Das chinesische Tannin. II. Mitt. über Gerbstoffe. Helv. Chim. Acta **6**, 3 (1923).

44. KARRER, P., R. WIDMER und M. STAUB: Über türkisches Tannin. III. Mitt. über Gerbstoffe. Liebigs Ann. Chem. **433**, 288 (1923).

45. KLINGER, G.: Dissert., Univ. Heidelberg, 1955.

46. KORTE, F.: Das U. V.-Spektrum in Abhängigkeit von der Konstitution organischer Verbindungen. Angew. Chem. **63**, 370 (1951).

47. LADEMANN, R.: Dissert , Univ. Heidelberg, 1951.

48. LINDBERG, B.: Action of Strong Acides on Acetylated Glycosides. I. Transformation of some Aliphatic Tetraacetyl-β-glucosides to the α-Form. II. Synthesis of Tetraacetyl-α-tertiary Butylglucoside. Acta Chem. Scand. **2**, 426, 534 (1948).

49. LOHMANN, A.: Dipl. Arbeit, Univ. Heidelberg, 1953.

50. Löwe, J.: Über die Gerbsäure der Dividivi-Schoten und deren Beziehung zur Gallussäure. Z. analyt. Chem. **14**, 35 (1875); Über die Gerbsäure der Myrobalanen und ihre Identität mit der Ellaggengerbsäure. Z. analyt. Chem. **14**, 44 (1875).

51. Mayer, W.: Dehydro-digallussäure. Liebigs Ann. Chem. **578**, 34 (1952).

52. — Die Chemie der Phenolcarbonsäuren der hydrolysierbaren Gerbstoffe. Habilitationsschrift, Univ. Heidelberg, 1952.

53. Mayer., W., R. Bachmann und F. Kraus: Über carbocyclische Reduktone. Dihydropyrogallol und Dihydrogallussäure. Chem. Ber. **88**, 316 (1955).

54. Mayer, W., K. Wick und O. Th. Schmidt: Die Synthese der n-Butanol-(2)-tetracarbonsäure-(1,2,3,4) (3-Oxy-3,4-dicarboxy-adipinsäure). Chem. Ber. **86**, 488 (1953).

55. Meerwein, H., E. Büchner und K. van Emster: Über die Einwirkung aromatischer Diazoverbindungen auf α,β-ungesättigte Carbonylverbindungen. J. prakt. Chem. **152**, 237 (1939).

56. Merkel, E. und Chr. Wiegand: Ultraviolettabsorption und Molekülbau von Diphenyl in den drei Aggregatzuständen und von einigen mit ihm verwandten Verbindungen. Z. Naturforsch. **3B**, 93 (1948).

57. Meyer, H.: Analyse und Konstitutionsermittlung organischer Verbindungen, S. 576. Wien: J. Springer. 1938.

58. Münz, W.: Gerbstoffe der Edelkastanie und des sizilianischen Sumachs. Collegium **714**, 499 (1929).

59. Nierenstein, M.: Zur näheren Kenntnis einiger „Blume" gebender Gerbstoffe. II. Mitt. Collegium **162**, 197 (1905).

60. — Beitrag zur Kenntnis der Gerbstoffe. III. Über Ellagen-gerbsäure. Ber. dtsch. chem. Ges. **43**, 1267 (1910).

61. — The Tannin of the Knopper Gall. J. Chem. Soc. (London) **115**, 1174 (1919).

62. Nierenstein, M. and J. Potter: The Distribution of Myrobalanitannin. Biochemic. J. **39**, 390 (1945).

62a. O'Shaughnessy, M. T. and W. H. Rodebush: Ultraviolet Absorption Spectra of Organic Molecules: The Dependence upon Restricted Rotation and Resonance. J. Amer. Chem. Soc. **62**, 2906 (1940).

63. Paech, K. und M. V. Tracey: Moderne Methoden der Pflanzenanalyse O. Th. Schmidt: Natürliche Gerbstoffe. Bd. III, S. 526. Berlin-Göttingen-Heidelberg: Springer-Verlag. 1955.

64. Partridge, S. M.: Filter-paper Partition Chromatography of Sugars. 1. General Description and Application to the Qualitative Analysis of Sugars in Apple Juice, Egg White and Foetal Blood of Sheep. Biochemic. J. **42**, 238 (1948).

64a. — Aniline Hydrogen Phthalate as a Spraying Reagent for Chromatography of Sugars. Nature (London) **164**, 443 (1949).

65. Perkin, A. G. and Y. Uyeda: Occurrence of a Crystalline Tannin in the Leaves of the *Acer ginnala*. J. Chem. Soc. (London) **121**, 66 (1922).

66. Phillips, H.: Chemistry of Myrobalans. I. Extraction of Myrobalans Nuts and Liquor with Organic Solvents. J. Internat. Soc. Leather Trades Chem. **20**, 230 (1936).

67. Pickett, L. W., G. F. Walter and H. France: The Ultraviolet Absorption Spectra of Substituted Biphenyls. J. Amer. Chem. Soc. **58**, 2296 (1936).

68. Procter, H. R. und J. Paessler: Leitfaden für gerbereichemische Untersuchungen, S. 78. Berlin: J. Springer. 1901.

69. Reeves, R. E., M. H. Adams and W. F. Goebel: The Synthesis of a New Dimethyl-β-methylglucoside. J. Amer. Chem. Soc. **62**, 2881 (1940).

69a. REEVES, R. E. and W. F. GOEBEL: Chemoimmunological Studies on the Soluble Specific Substance of Pneumococcus. J. Biol. Chem. **139**, 511 (1941).

70. REICHEL, L.: Zur Kenntnis der Ellagengerbstoffe. Naturwiss. **29**, 16 (1941).

71. REICHEL, L. und A. SCHWAB: Über Galloyl-ellagsäuren. Chemie und Biochemie der Pflanzenstoffe, VIII. Mitt. Liebigs Ann. Chem. **550**, 152 (1942).

72. REICHEL, L. und E. ULSPERGER: Über Ellagengerbstoffe. Naturwiss. **27**, 628 (1939).

73. RICHTER, W.: Dissert., Univ. Erlangen, 1911. Zur Kenntnis des Eutannins. Arb. Pharm. Inst. Univ. Berlin **9**, 85 (1912).

74. SCHMIDT, O. TH.: Die Konstitution des Zuckers aus Hamameli-tannin. Über Zucker mit verzweigter Kohlenstoffkette. I. Liebigs Ann. Chem. **476**, 250 (1929).

75. — Zur Chemie der natürlichen Gerbstoffe. Z. Pflanzenernährg., Düngung, Bodenkde. **63** (114), 15 (1955).

76. SCHMIDT, O. TH., S. BERG und H. H. BAER: Die Konstitution der Digalloyl-glucose aus Chebulinsäure. VII. Mitt. über natürliche Gerbstoffe. Liebigs Ann. Chem. **571**, 19 (1951).

77. SCHMIDT, O. TH. und K. BERNAUER: Brevifolin und Brevifolin-carbonsäure. XXI. Mitt. über natürliche Gerbstoffe. Liebigs Ann. Chem. **588**, 211 (1954).

78. SCHMIDT, O. TH. und F. BLINN: Über einen Fall von Atropisomerie bei einem Naturstoff. Naturwiss. **38**, 72 (1951).

79. SCHMIDT, O. TH., F. BLINN und R. LADEMANN: Über die Bindung der Ellagsäure in Corilagin und Chebulagsäure. XII. Mitt. über natürliche Gerbstoffe. Liebigs Ann. Chem. **576**, 75 (1952).

80. SCHMIDT, O. TH. und K. DEMMLER: Optisch aktive 2,3,4,2′,3′,4′-Hexamethoxy-diphenyl-dicarbonsäure-6,6′. XII. Mitt. über natürliche Gerbstoffe. Liebigs Ann. Chem. **576**, 85 (1952).

81. — — Racemische und optisch aktive 2,3,4,2′,3′,4′-Hexaoxy-diphenyl-6,6′-dicarbonsäure. XVII. Mitt. über natürliche Gerbstoffe. Liebigs Ann. Chem. **586**, 179 (1954).

82. — — Über das Verhalten und die Rolle der Hexaoxy-diphensäure in Gerbextrakten. XX. Mitt. über natürliche Gerbstoffe. Liebigs Ann. Chem. **587**, 75 (1954).

83. SCHMIDT, O. TH. und H. GRÜNEWALD: unveröffentlicht.

84. SCHMIDT, O. TH. und K. HEINTZ: Über Zucker mit verzweigter Kohlenstoffkette. V. Die Synthese der Hamamelonsäure. Liebigs Ann. Chem. **515**, 77 (1934).

85. SCHMIDT, O. TH., M. HEINTZELER und W. MAYER: Chebulinsäure. I. Mitt. Chem. Ber. **80**, 510 (1947).

86. SCHMIDT, O. TH. und J. HEROK: α-Glucogallin. XVIII. Mitt. über natürliche Gerbstoffe. Liebigs Ann. Chem. **587**, 63 (1954).

87. SCHMIDT, O. TH. und E. KOMAREK: Valoneasäure. XXIII. Mitt. über natürliche Gerbstoffe. Liebigs Ann. Chem. **591**, 156 (1955).

88. SCHMIDT, O. TH. und R. LADEMANN: Chebulagsäure aus Dividivi. IV. Mitt. über natürliche Gerbstoffe. Liebigs Ann. Chem. **569**, 149 (1950).

89. — — Papierchromatographie der Gallotannine. IX. Mitt. über natürliche Gerbstoffe. Liebigs Ann. Chem. **571**, 41 (1951).

90. — — Corilagin, ein weiterer kristallisierter Gerbstoff aus Dividivi. X. Mitt. über natürliche Gerbstoffe. Liebigs Ann. Chem. **571**, 232 (1951).

91. SCHMIDT, O. TH., R. LADEMANN und W. HIMMELE: Notiz über die Spaltsäure $C_{14}H_{12}O_{11}$ aus Chebulin- und Chebulagsäure. XI. Mitt. über natürliche Gerbstoffe. Chem. Ber. **85**, 408 (1952).

92. Schmidt, O. Th. und W. Mayer: Die Konstitution der Spaltsäure $C_{14}H_{12}O_{11}$ aus Chebulin- und Chebulagsäure. V. Mitt. über natürliche Gerbstoffe. Liebigs Ann. Chem. **571**, 1 (1951).

93. Schmidt, O. Th. und W. Nieswandt: Über einen krystallisierten, Ellagsäure enthaltenden Gerbstoff aus Myrobalanen. Naturwiss. **35**, 191 (1948).

94. — — Chebulagsäure, ein krystallisierter Ellagen-Gerbstoff aus Myrobalanen. III. Mitt. über natürliche Gerbstoffe. Liebigs Ann. Chem. **568**, 165 (1950).

95. Schmidt, O. Th. und A. Schach: Synthese der 3-Galloyl-glucose, 6-Galloyl-glucose und 3,6-Digalloyl-glucose. VIII. Mitt. über natürliche Gerbstoffe. Liebigs Ann. Chem. **571**, 29 (1951).

96. Schmidt, O. Th. und D. M. Schmidt: Die Umwandlung von Chebulagsäure in Corilagin. XIV. Mitt. über natürliche Gerbstoffe. Liebigs Ann. Chem. **578**, 25 (1952).

97. — — Über das Vorkommen von Corilagin in Myrobalanen. XV. Mitt. über naturliche Gerbstoffe. Liebigs Ann. Chem. **578**, 31 (1952).

98. Schmidt, O. Th., D. M. Schmidt und J. Herok: Die Konstitution und Konfiguration des Corilagins. XIX. Mitt. über natürliche Gerbstoffe. Liebigs Ann. Chem. **587**, 67 (1954).

99. Schmidt, O. Th., H. Voigt, W. Puff und R. Köster: Benzyläther der Ellagsäure und Hexaoxy-diphensäure. XVI. Mitt. über natürliche Gerbstoffe. Liebigs Ann. Chem. **586**, 165 (1954).

100. Schmidt, O. Th. und C. C. Weber-Molster: Über Zucker mit verzweigter Kohlenstoffkette. III. Die Konfiguration der beiden Fructoheptonsäuren und der Hamamelose. Liebigs Ann. Chem. **515**, 43 (1934).

101. — — Über Zucker mit verzweigter Kohlenstoffkette. IV. Das optische Verhalten der Aldonsäuren in Gegenwart von Natriummolybdat. Liebigs Ann. Chem. **515**, 65 (1934).

102. Schwarzenbach, G. und Ch. Wittwer: Über das Keto-Enol-Gleichgewicht bei cyclischen α-Diketonen. Helv. Chim. Acta **30**, 663 (1947).

103. Schweiger, R.: Dipl.-Arbeit, Univ. Heidelberg, 1954.

104. Stanley, W. M., E. McMahon and R. Adams: Stereochemistry of Diphenyls. XXVII. Comparison of the Racemization of 2,2′-Difluoro-6,6′-dicarboxy-diphenyl and 2,2′-Dimethoxy-6,6′-dicarboxydiphenyl. J. Amer. Chem. Soc. **55**, 706 (1933).

105. Thoms, H.: Zur Gerbstoff-forschung. II. Mitt. über das Eutannin. Apoth.-Ztg. **21**, 354 (1906). Über das Eutannin. Arb. Pharm. Inst. Univ. Berlin **9**, 78 (1912).

106. Weermann, R. A.: L'action de l'hypochlorite de sodium sur les amides d'α-hydroxy-acides et de polyhydroxy-acides, ayant un groupe hydroxyle à la place α. Rec. trav. chim. Pays-Bas **37**, 16 (1918).

107. Will, H.: Dipl.-Arbeit, Univ. Heidelberg, 1951.

108. Young, S.: A Distinctive Test for Gallic Acid. Chem. News **48**, 31 (1883).

109. Zemplén, G., Z. Csürös und S. Angyal: Über benzylierte Derivate des Lävoglucosans und der Glucose. Ber. dtsch. chem. Ges. **70**, 1848 (1937).

110. Zölffel, G.: Über die Gerbstoffe der Algarobilla und der Myrobalanen. Arch. Pharmaz. **229**, 223 (1891).

(Eingelaufen am 22. Dezember 1955.)

Neuere Ergebnisse auf dem Gebiete der glykosidischen Herzgifte: Grundlagen und die Aglykone.

Von **CH. TAMM**, Basel.

Inhaltsübersicht.

I. Einleitung.

Die vergangenen Jahre sind durch eine beachtliche Intensivierung
der Arbeiten auf dem Gebiete der glykosidischen Herzgifte gekennzeichnet.
Eine große Anzahl verschiedenster Pflanzen sind auf den Gehalt an
herzwirksamen Stoffen geprüft worden. Neben bereits bekannten Herzglykosiden konnten zahlreiche neue Wirkstoffe isoliert werden. Bei
vielen von ihnen ist es gelungen, die chemische Konstitution aufzuklären.

In letzter Zeit sind deshalb eine Reihe von zusammenfassenden Publikationen über die Chemie der herzaktiven Glykoside erschienen, (vgl. *52, 92, 218, 269, 287, 293, 164, 356*). Die vorliegende Übersicht will sich nicht nur die Aufzählung und Beschreibung der verschiedenen Eigenschaften der isolierten Herzglykoside zum Ziele setzen, sondern versuchen, vor allem einen Einblick in die Methoden, die zu ihrer Erforschung benützt werden, zu geben. Im Kapitel über die Konstitutionsermittlung liegt das Hauptgewicht auf den Abbaureaktionen und der Struktur der Aglykone. Die Zucker und Glykoside sowie eine Betrachtung über das Auftreten der glykosidischen Herzgifte in der Natur werden Gegenstand eines weiteren Artikels sein.

II. Die Isolierung von herzaktiven Glykosiden.

Die Gewinnung von chemisch reinen Herzglykosiden aus Pflanzen stößt auf mannigfaltige Schwierigkeiten. Die Beschaffung der Pflanzen, welche die gesuchten Stoffe enthalten, ist nicht einfach, da sie oft in den Tropen heimisch, schwer zugänglich und relativ selten sind. Der Gehalt an wirksamer Substanz ist meist gering. In vielen Fällen liegen komplizierte Gemische von verschiedenen Herzglykosiden vor, die ihrerseits mit den chemisch nahe verwandten, aber herzunwirksamen Saponinen vergesellschaftet sind. Die Saponine sind sehr weit verbreitet. Sie verändern die Löslichkeitseigenschaften der Herzgifte, da sie kolloidale Lösungen bilden und die Oberflächenspannung der Pflanzensäfte stark beeinflussen und dadurch die Isolierung erschweren.

Die Herzgifte sind gegen chemische Eingriffe sehr empfindlich. Durch Alkalien werden die Isoverbindungen (14,21- und 16,21-Oxido-cardanolide), die physiologisch unwirksam sind, gebildet und Acetylgruppen verseift. Durch Säuren werden Glykoside, die 2-Desoxyzucker enthalten, leicht hydrolysiert, tertiäre Oxygruppen und Acetoxygruppen in den Geninen abgespalten. Bei starkem Erwärmen tritt Zersetzung der Wirkstoffe ein, und durch den Luftsauerstoff werden autoxydable Aldehydgruppen oxydiert. In gewissen Samen, z. B. von Strophanthus, lagern die sog. allomerisierenden Enzyme (*102*) die Glykoside in die Alloglykoside (17α-Cardenolide) um. Aus diesen Einwirkungen resultieren physiologisch unwirksame Stoffe.

Mit den heute zur Verfügung stehenden Methoden sind die genannten Schwierigkeiten weitgehend gemeistert worden. Sie erlauben die Isolierung von reinen, einheitlichen Glykosiden, ihren Aglykonen und Zuckern.

Die Wahl der Methode richtet sich dabei nach der Art des pflanzlichen Ausgangsmaterials, d. h. ob Samen, Wurzeln, Stengel, Blüten oder Blätter zu untersuchen sind; ferner nach dem Typus des gewünschten Stoffes, d. h. ob die Isolierung der zuckerreichen genuinen Glykoside oder der zuckerärmeren Spaltprodukte oder deren Aglykone angestrebt wird. In allen Fällen werden zunächst Rohextrakte hergestellt, in welchen die gesuchten Stoffe stark angereichert sind. Eine orientierende biologische Prüfung dieser Extrakte auf Herzwirksamkeit und eine orientierende Untersuchung im Papierchromatogramm sind oft von Nutzen. In den wenigsten Fällen gelingt aber die Isolierung von reinen kristallisierten Stoffen ohne die Anwendung weiterer Reinigungs- und Trennungsverfahren.

1. Isolierung von reinen Glykosiden und Aglykonen.

a) Herstellung der Rohextrakte.

Die meisten frischen Pflanzenteile enthalten Enzyme, die in wäßriger Lösung die genuinen zuckerreichen Glykoside zu zuckerärmeren Stoffen abzubauen vermögen.

Wird die Gewinnung der genuinen Glykoside gewünscht, so muß bei der Bereitung der Rohextrakte für die Inaktivierung dieser Enzyme gesorgt werden. STOLL und KREIS (*294, 296*) haben als erste ein derartiges Verfahren ausgearbeitet, das ihnen die Isolierung der Lanatoside A, B und C aus den Blättern von *Digitalis lanata* und der Purpureaglykoside A und B aus den Blättern von *D. purpurea* erlaubte. Sie verhindern die Wirkung der Enzyme durch Zusatz von neutralen Mineralsalzen (z. B. Ammoniumsulfat) zu den frisch gemahlenen Blättern. Die Isolierung der genuinen Glykoside gelingt nicht, wenn die Blätter lange und trocken aufbewahrt worden sind. Die Inaktivierung der Enzyme bzw. Stabilisierung der Droge kann auch durch Behandlung mit Alkoholdämpfen erreicht werden. Haben die Enzyme Gelegenheit, ihre Wirkung zu entfalten, so erhält man statt der Lanatoside die um 1 Mol. Glucose ärmeren Glykoside Acetyldigitoxin-α und -β, Acetylgitoxin-α und -β und Acetyldigoxin-α und -β bzw. Digitoxin, Gitoxin und Digoxin.

Ganz analog kann bei der Gewinnung der genuinen Glykoside aus Zwiebeln [z. B. von Scillaren A aus *Scilla maritima* (*314*)] und aus Samen [z. B. von k-Strophanthosid aus *Strophanthus kombé* (*313*)] verfahren werden.

Isolierung und Trennung der zuckerreichen genuinen Glykoside sind meistens sehr schwierig, da sie häufig schlecht kristallisieren, und die weiter unten besprochenen Trennungsverfahren, die zur Gewinnung von einheitlichen Kristallisaten dienen, wegen ihres stark polaren Charakters oft versagen. Demgegenüber gestaltet sich die Trennung der zuckerärmeren Spaltprodukte, wobei es sich vorwiegend um Monoglykoside, seltener um Aglykone handelt, meist weniger kompliziert. Steht deshalb nicht die Gewinnung der genuinen Glykoside im Vordergrund, so werden mit Vorteil Rohextrakte hergestellt und weiterverarbeitet, wie sie nach

Einwirkung der pflanzeneigenen, zuckerabspaltenden Enzyme erhalten werden. Dieses Verfahren ist besonders von T. REICHSTEIN und seiner Schule entwickelt und bei ihren meisten Isolierungsarbeiten benützt worden.

Das zerkleinerte Pflanzenmaterial (Samen, Blätter, Stengel, Wurzeln, Zwiebeln) wird ohne Zusatz von Mineralsalzen zuerst mit Wasser unter Zusatz von etwas Toluol zur Konservierung mehrere Tage bei 25—37° geweicht und dann mit Alkohol-Wasser-Gemischen von steigendem Alkoholgehalt erschöpfend extrahiert. Samen werden vorher durch Perkolation mit Petroläther entfettet. Die Extrakte werden im Vakuum bei 40° zu einem kleinen Volumen konzentriert, das wäßrige Konzentrat mit der gleichen Menge Alkohol versetzt, mit $Pb(OH)_2$ gereinigt, durch eine Schicht Kieselgur (Hyflo-Super Cel) filtriert, der Alkohol im Vakuum entfernt und die wäßrige Lösung (etwa p_H 6) nacheinander gründlich mit Äther, Chloroform und Chloroform-Alkohol (2 : 1) ausgeschüttelt, diese Auszüge getrocknet und eingedampft. Zur feineren Abstufung kann vor der Extraktion mit Chloroform-Alkohol (2 : 1) noch mit Chloroform-Alkohol (9 : 1)- und (4 : 1)-Gemischen ausgeschüttelt werden.

Für die Prüfung und Isolierung von besonders stark wasserlöslichen Glykosiden wie Ouabain und ähnlichen Stoffen wird die nach dem Ausschütteln verbleibende wäßrige Phase eingedampft, der Rückstand mit Alkohol extrahiert, die Extrakte eingedampft, mit halbgesättigter Na_2SO_4-Lösung versetzt und mit Chloroform-Alkohol (3 : 2) (*303*) ausgeschüttelt (*8, 39*).

Als Beispiele für die Extraktion von Zwiebeln vgl. KATZ (*133*), von Blättern vgl. AEBI und REICHSTEIN (*1*) oder HUNGER (*94*) und von Wurzeln vgl. HUBER et al. (*93*) oder SCHINDLER und REICHSTEIN (*250*). Bei der Extraktion der Samen von Strophanthus muß zusätzlich für die Inaktivierung der unerwünschten allomerisierenden Enzyme gesorgt werden.

Nach dem Verfahren von REICHSTEIN und Mitarb. [detaillierte Vorschrift (*39*)] wird aus den mit Petroläther entfetteten Samen zunächst bei 0° ein Wasserauszug bereitet, der die gewünschten hydrolysierenden Enzyme [Strophanthobiase (*119*)] enthält. Die allomerisierenden Enzyme (*102*) sind unter diesen Bedingungen in Wasser schwer löslich und verbleiben im Samenrückstand. Sie werden durch Zusatz von Alkohol inaktiviert. Der Wasserextrakt mit den hydrolysierenden Enzymen (Strophanthobiase) wird nachträglich in den wäßrigen Glykosidextrakt zur Fermentierung zurückgegeben.

Nach einem Verfahren der Firma C. F. Boehringer & Söhne, G. m. b. H., Mannheim (*19*), lassen sich Monoglykoside aus Strophanthussamen dadurch gewinnen, daß die entfetteten Samen gleichzeitig der Einwirkung von Wasser und halogenierten Kohlenwasserstoffen (z. B. Chloroform) während zirka 14 Tagen bei 25° ausgesetzt werden. Die Chloroformauszüge liefern im Falle der Samen von *Strophanthus kombé* in guter Ausbeute die Monoglykoside (Cymarin, Cymarol usw.) ohne die entsprechenden Allo-Glykoside in nachweisbaren Mengen zu enthalten. Werden die Samen unter den gleichen Bedingungen, aber ohne Anwesenheit von Chloroform aufbewahrt, so ergibt die anschließende Extraktion hauptsächlich die Allo-Glykoside.

DE GRAEVE (*70*) hat bei der Extraktion der Samen von *S. kombé* festgestellt, daß die hydrolysierenden Enzyme (Strophanthobiase) viel rascher wirken als die allomerisierenden Enzyme, wenn die entfetteten Samen mit Wasser bei 37° stehengelassen werden. So lassen sich die Monoglykoside leicht, ohne mit den Allo-Glykosiden verunreinigt zu sein, gewinnen. Bleidiacetat hemmt die allomerisierenden Enzyme spezifisch.

Für die routinemäßige *Analyse* von Strophanthussamen unterwarfen BUSH und TAYLOR (*21*) wie auch HEFTMANN et al. (*78*) die nach dem üblichen Verfahren gewonnenen Rohextrakte direkt einer sauren Hydrolyse und untersuchten die chloroformlöslichen Hydrolysenprodukte im Papierchromatogramm. Die Methode hat den Nachteil, daß sie neben den Aglykonen oft zusätzliche, schwer zu identifizierende Artefakte liefert.

b) Trennung von Substanzgemischen.

Die direkte Kristallisation eines reinen und einheitlichen Glykosids aus den Rohextrakten gelingt selten, da meistens ein kompliziertes Substanzgemisch vorliegt

Für die Auftrennung der Gemische, die sich in den Äther-, Chloroform- und Chloroform-Alkohol (9 : 1)-Extrakten finden, also der weniger polaren Stoffe, bewährt sich gewöhnlich die *Adsorptionschromatographie* an alkalifreiem Al_2O_3 nach der Durchlaufmethode von REICHSTEIN und SHOPPEE (*41, 221*).

Die stärker polaren Glykosidgemische der Chloroform-Alkohol (4 : 1)-, (2 : 1)- und (3 : 2)-Extrakte lassen sich an Al_2O_3 von der gleichen Aktivitätsstufe erst nach Acetylierung oder Benzoylierung chromatographieren (*233*). Milde alkalische Verseifung mit wäßrig-methanolischem $KHCO_3$ bei 20° gibt in den meisten Fällen die freien Glykoside. Unter diesen Bedingungen wird weder der Lactonring geöffnet, noch die Bildung von Isoverbindungen (14,21-Oxido-cardanolide) bewirkt. In den gut wasserlöslichen Lanatosiden ließ sich die Acetylgruppe mit wäßrigem $Ca(OH)_2$ verseifen (*295*). Einige acetylierte Glykoside, die sich von der Digitalose oder Thevetose ableiten [z. B. Monoacetyl-neriifolin (*64*), Digitalinum verum-monoacetat (*228*), Strospesid-monoacetat (*228*), Sarmentosid A-acetat (*216*)], besitzen Acetylgruppen, die mit $KHCO_3$ in wäßrigem Methanol bei 20° nur unvollständig verseift werden. Die Entacetylierung solcher Glykosidacetate gelingt manchmal durch das Enzymgemisch des Hepatopankreassafts der Weinbergschnecke (*64, 228*). FRÈREJAQUE (*58*) hat eine Farbreaktion auf Papier ausgearbeitet, die es erlaubt, Acetylgruppen schon in geringsten Glykosidmengen nachzuweisen. Es ist ferner zu beachten, daß 16-Acetoxygruppen durch den Kontakt mit Al_2O_3 leicht eliminiert werden, z. B. in Oleandrin und seinen Derivaten (*1, 95*).

Diese unerwünschte Nebenreaktion tritt bei der Chromatographie an *Kieselgur-Mg-Silikat-Mischung* (*228*) oder *Silicagel* nicht ein.

Das erstere Adsorbent kann auch für die Chromatographie der stärker polaren Glykoside bzw. Rohextrakte benützt werden. Allerdings ist der Trenneffekt bedeutend schlechter als bei Al_2O_3.

In neuester Zeit ist auch in die Chemie der Herzglykoside die *Verteilungschromatographie* (*169—171*) eingeführt worden. Die Methode eignet sich besonders für die präparative Trennung von Glykosiden und Aglykonen, bei denen die Adsorptionschromatographie an Al_2O_3 und Kieselgur-Mg-Silikat-Mischung versagen oder ungeeignet sind, also auch für stärker polare Glykosidgemische. Sie wurde auf die Herzglykoside zuerst von STOLL et al. (*297*) übertragen.

Sie verwendeten als stationäre Phase Wasser, wobei Baumwoll-Linters (bestehend aus fast reiner Cellulose) oder Diatomitstein als Träger dienten, und als mobile Phase Äthylacetat und Äthylacetat-Butanol-Gemische verwendet wurden. Auf diese Weise gelang es, die amorphe „Mutterlauge" des Scillarens A, das sog. Scillaren B, in acht neue kristallisierte einheitliche Stoffe aufzutrennen. Für eine Variante vgl. (*290*). Eine weitere Ausführungsform von REICHSTEIN und Mitarb. (*80*) benützt ebenfalls Wasser als stationäre Phase, wobei speziell gereinigte Kieselgur als Träger verwendet wird. Als mobile Phase dienen Petroläther-Benzol-Gemische, reines Benzol, Benzol-Chloroform-Gemische, reines Chloroform, sowie Chloroform-*n*-Butanol- oder Benzol-*n*-Butanol-Gemische.

Für die Trennung von Geninen (z. B. Bufadienoliden) haben sich als stationäre Phase Propylenglykol-Wasser (4 : 1) mit Kieselgur als Träger und als mobile Phase Gemische von Benzol und Chloroform bewährt (*358*).

Die Verteilungschromatographie hat den Nachteil, daß sie im Gegensatz zu den Adsorptionschromatographien sehr viel Zeit benötigt; doch sind die Trenneffekte im allgemeinen sehr gut. Sie gewährleistet auch ein konstantes pH von 7, so daß an den herzwirksamen Stoffen bisher nie chemische Veränderungen infolge der Trennungsoperation beobachtet worden sind. Für die Grundlagen der Verteilungschromatographie von Steroiden siehe (*77 b*).

TSCHESCHE und Mitarb. haben die Trennung von Glykosiden durch *Gegenstromverteilung* versucht.

Während die Glykoside der Uzara-Wurzel in einem Apparat mit 18 Stufen nicht erfolgreich war (*341*), ließen sich Extrakte von *Digitalis lanata* durch wiederholte Verteilung über 24 Stufen weitgehend auftrennen (*344*).

Für die Trennung von Gemischen zuckerreicher Glykoside, bei denen die erwähnten Verfahren versagen, besteht die Möglichkeit, diese zunächst mit einer Glucosidase zu behandeln. Das resultierende Gemisch von zuckerärmeren Stoffen läßt sich oft ohne Mühe trennen (*93, 224, 247*).

Glykoside und Aglykone, die eine reaktionsfähige Carbonylfunktion besitzen, wie Aldehyde vom Typus des Strophanthidins, können durch Überführung in wasserlösliche GIRARD-Verbindungen mit Reagens T (*69*) von carbonylfreien Anteilen abgetrennt werden (*141*), wobei auch die Betainhydrazone der Aldehyde unter bestimmten Bedingungen wieder gespalten werden können (*252, 255*).

2. Farbreaktionen.

a) Allgemeine Farbreaktionen.

Für den qualitativen Nachweis kleiner Mengen von Herzglykosiden und Aglykonen sind eine Reihe von Farbreaktionen aufgefunden worden.

Für den quantitativen Nachweis (als Grundlage einer kolorimetrischen Methode) eignen sich nur wenige Reagenzien.

Glykoside werden an ihrer Zuckerkomponente, die durch saure Hydrolyse freigesetzt wird, erkannt. Die Spaltung wird mit Vorteil in einem Gemisch von Salzsäure-Eisessig-Wasser nach KILIANI (*150*) durchgeführt, wenn der Zucker eine gewöhnliche Hexose oder Hexamethylose ist. Glykoside, die einen 2-Desoxyzucker enthalten, lassen sich bereits mit 0,05 n methanolisch-wäßriger H_2SO_4 spalten. Unter den stärkeren Bedingungen von KILIANI werden die freigesetzten 2-Desoxyzucker zerstört. Die freien Zucker werden mit Fehlingscher Lösung nachgewiesen. Zur Ausführung vgl. (*7*).

Glykoside mit 2-Desoxyzucker geben eine positive KELLER-KILIANI-Reaktion, eine für 2-Desoxyzucker charakteristische Farbreaktion (*146, 148*), sofern der 2-Desoxyzucker nicht mit einer Hexose verknüpft ist. Ausführung nach (*43*). Zum Nachweis auf Papier vgl. (*237*). Während z. B. Cymarose und das Monoglykosid Cymarin eine positive Reaktion zeigen, fällt diese bei den entsprechenden Di- und Triglykosiden k-Strophanthin-β und k-Strophanthosid negativ aus. NAGATA, TAMM und REICHSTEIN (*190*) haben deshalb eine empfindliche Reaktion ausgearbeitet, die erlaubt, 2-Desoxyzucker und deren Glykoside auch in den Fällen zu erkennen, wo die KELLER-KILIANI-Reaktion versagt. (Hydrolyse der Glykoside mit Trichloressigsäure. Die freien 2-Desoxyzucker geben mit p-Nitrophenylhydrazin und Alkali eine Blaufärbung.) Alle Glykoside und Aglykone geben mit konz. H_2SO_4 oder 84proz. H_2SO_4 charakteristische Färbungen (*43*), die sich oft als gute Hilfe bei der Identifizierung erweisen. Sowohl die Cardenolide als auch die Bufadienolide geben mit Antimontrichlorid (*163, 378*) und mit Trichloressigsäure (*232, 291*) (ROSENHEIM-Reaktion) Färbungen, die im Ultraviolett-Licht manchmal stark fluoreszieren. Beide Farbreaktionen sind aber unspezifisch für Glykoside und Aglykone.

b) Farbreaktionen für Cardenolide.

Cardenolide geben eine positive LEGAL-Reaktion (Rotfärbung) (*25, 222*). Die Reaktion ist negativ bei den Isoverbindungen (14,21-Oxido-cardanoliden sowie 16,21-Oxido-cardanoliden) und wenn der Lactonring hydriert ist (Cardanolide). Das Reduktionsvermögen des Butenolidrings dient zur Sichtbarmachung der Flecken im Papierchromatogramm mit dem Reagens von TOLLENS (*327—329*). Empfindlicher und spezifischer ist die Reaktion nach RAYMOND (*214, 215, 251*) (*m*-Dinitrobenzol in methanolischem Alkali). Sie wurde auch für die kolorimetrische Bestimmung von Glykosiden herangezogen (*76*). Manchmal günstiger und bequemer ist die Modifikation nach KEDDE (*144, 202*) (3,5-Dinitrobenzoesäure in methanolischem KOH). Glykoside und Aglykone geben nach

BALJET (*5, 6, 12, 13*) mit alkalischer Pikrinsäure, nach WARREN et al. (*362*) mit Naphtochinonsulfonsäure typische Färbungen und nach PETIT et al. (*200, 201*) mit H_3PO_4 im UV.-Licht eine intensive Fluoreszenz. Die erwähnten Farbreaktionen werden von BELLET (*14*) ausführlich diskutiert. TSCHESCHE et al. (*345*) benützten das Xanthydrol-Reagens zur kolorimetrischen Bestimmung der Lanatoside A, B und C und der Purpureaglykoside A und B. Sie haben auch die Empfindlichkeit der anderen Reagenzien bestimmt und miteinander verglichen.

Farbreaktionen auf Papier für 2-Desoxyzucker-glykoside, die eine positive KELLER-KILIANI-Reaktion geben, haben kürzlich FRÈREJACQUE und DURGEAT (*59*) mitgeteilt. Alkoholischer *p*-Dimethyl-aminobenzaldehyd mit H_3PO_4 und alkoholisches Vanillin mit H_3PO_4 geben blaue Färbungen. *p*-Nitranilin mit alkoholischer HCl und Na-perjodat soll nur mit 3-Desoxyzucker-glykosiden eine goldgelbe Färbung liefern.

Nach KRAUS (*155*) lassen sich die normalen Aglykone von den biologisch inaktiven Allo-Aglykonen (17 α-Cardenoliden) unterscheiden, indem die letzteren nach Behandlung mit Alkali eine positive Reaktion mit 2,4-Dinitrophenylhydrazin geben.

c) Farbreaktionen für Bufadienolide.

Für die Glykoside und Aglykone dieser Gruppe ist die LIEBERMANNsche Reaktion (Eisessig-Schwefelsäure-Acetanhydrid) spezifisch (*176, 291*). KATZ verwendet für den Nachweis auf Papier das Gemisch Furfurol-Eisessig-Schwefelsäure (*134, 378*) oder Dimethylaminobenzaldehyd (*137*) und TSCHESCHE konz. H_2SO_4 (*351*). Oft bewähren sich das Antimontrichlorid-Reagens (*358*) und Trichloressigsäure (ROSENHEIM-Reaktion). Die Aglykone zeigen mit Antimontrichlorid im UV.-Licht oft andere Fluoreszenzfarben als im Tageslicht.

Doch sind alle diese Reaktionen für die Charakterisierung des Cumalinrings nicht genügend spezifisch. Spezifisch und sehr empfindlich ist hingegen der Nachweis von Glykosiden und Aglykonen dieses Typs auf Papier durch direkte Photokopie mit gefiltertem UV.-Licht nach BERNASCONI et al. (*15*). Da der Cumalinring durch eine breite Absorptionsbande bei 300 mμ (log $\varepsilon = 3{,}70$—$3{,}72$) charakterisiert ist, verwenden diese Autoren Licht, das hauptsächlich die Wellenlängen um 300 mμ umfaßt.

d) Quantitative Bestimmungsmethoden.

Die quantitative Bestimmung von Herzglykosiden und Aglykonen ist nur an reinen Stoffen zuverlässig. Gemische von reinen Substanzen, wie auch Stoffgemische in galenischen Präparaten und Drogen müssen zuerst getrennt werden. STOLL et al. (*290*) haben bei den Lanatosiden eine Trennung durch Verteilungschromatographie mit einer Genauigkeit von $\pm$ 3% erzielt. Dafür werden 50—100 mg Substanzgemisch benötigt.

Tschesche et al. (*345*) trennten 0,1 mg Lanatosidgemisch durch Papierchromatographie. Die erhaltenen Substanzflecken werden mit einem gemessenen Volumen von Xanthydrol-Reagens aus dem Papier herausgelöst und ihre Menge direkt kolorimetrisch bestimmt. Für weitere Literatur vgl. (*32, 157, 293, 365a*).

3. Papierchromatographie.

Zur Trennung, Identifizierung und Reinheitsprüfung kleiner Mengen von herzwirksamen Glykosiden und Aglykonen ist in den letzten Jahren die Papierchromatographie herangezogen worden (*31*). Die Methode eignet sich auch für einen orientierenden Nachweis dieser Stoffe in Rohextrakten. Bei der orientierenden Untersuchung der Glykoside und Aglykone von Strophanthussamen, von denen nur einzelne Samen zur Untersuchung vorlagen, hat die Methode gute Dienste geleistet (*9, 21, 78, 86, 253*). In der folgenden Übersicht werden die Methoden für schwach polare und stark polare Glykoside und Aglykone separat besprochen.

Abgesehen von einer Ausnahme wurde stets Whatman-Papier No. I verwendet.

a) *Schwach polare Glykoside und Aglykone.*

Die erste Ausführungsform wurde von Schindler und Reichstein (*251*) angegeben [vgl. auch (*123, 124*)].

Sie benützten das von Zaffaroni et al. (*376, 377*) empfohlene Formamid als stationäre Phase. Es kann auch eine Lösung von 25proz. Formamid in Aceton benützt werden, wodurch etwas höhere Laufgeschwindigkeiten und bessere Trennung erzielt werden (*80*). Als mobile Phase dienen Benzol-Chloroform-Gemische und reines Chloroform. Zur Entwicklung der Flecken dient das Raymond-Reagens oder Kedde-Reagens, wobei das letztere im allgemeinen etwas dauerhaftere Färbungen gibt. Die Anwesenheit von Formamid stört die Farbreaktion nicht. Oft geben schon Substanzmengen von 20 γ sichtbare Flecken. Für die Chromatographie von Acetaten benützt man Benzol-Petroläther- und Benzol-Formamid-Gemische.

Bush und Taylor (*21*) empfehlen ein System, das sich bereits bei der Chromatographie von Steroidhormonen (*20*) bewährt hat, nämlich 50proz. wäßriges Methanol als stationäre Phase und Benzol als mobile Phase.

Eine weitere Variante wurde von Heftmann und Levant (*79a*) in der Verwendung von Propylenglykol als stationäre Phase und Toluol als mobile Phase gefunden. Die Laufgeschwindigkeiten der Substanzen sind nicht sehr groß. Zur Erhöhung darf aber kein Chloroform zugesetzt werden, da es mit Propylenglykol mischbar ist. Diese Systeme eignen sich auch für die Trennung der Acetylverbindungen [vgl. auch (*79*)].

Für die Trennung von Aglykonen, insbesondere der Bufadienolide, haben sich als stationäre Phase Propylenglykol-Wasser (4 : 1) und als bewegliche Phase Benzol-Chloroform-Gemische bewährt (*358*). Kaiser chromatographiert mit Xylol-Methyläthylketon (1 : 1) gesättigt mit Formamid (*126*).

b) Stark polare Glykoside und Aglykone.

Diese Methoden wurden besonders für die Trennung und Reinheitsprüfung der stark polaren Glykoside von *Digitalis lanata* und *D. purpurea* entwickelt.

SVENDSEN und JENSEN (*123, 319*) erzielten gute Trennungen bei der Verwendung von Chloroform-Methanol-Wasser (10:2:5), (10:4:5) und (10:8:5), wobei Chloroform die mobile und Wasser die stationäre Phase war. Die gleichen Autoren verwenden auch das Formamid-Chloroform-System (*124*). Es werden aber sehr lange Laufzeiten benötigt. HASSALL und MARTIN (*77*) verwenden als stationäre Phase Wasser oder eine 2,55 *n*-wäßrige Natriumbenzoatlösung, als mobile Phase Chloroform, Äthylacetat, Äthylmethylketon, Äthyloxalat und *n*-Butanol. Äthylacetat-Chloroform-Wasser (10:8:5) und (10:10:15) und Schleicher-Schüll-Papier Nr. 1574 wurden von VASTAGH und TUZSON (*359*) zur Trennung der Digitalis-Glykoside Digitoxin, Gitoxin und der Lanatoside benützt.

Eine Anordnung, die umgekehrt als die übliche ist, wird von TSCHESCHE et al. (*345*) zur Trennung der Glykoside aus *D. purpurea* und *D. lanata* empfohlen. Als stationäre Phase werden mit Wasser gesättigte, organische Lösungsmittel, wie aliphatische Alkohole (C_5—C_8) oder Ester von Malonsäure oder Oxalsäure, und als bewegliche Phase Wasser mit Zusatz von Formamid benützt. Folgende Mischungen erwiesen sich als günstig: *n*-Octanol-Pentanol-Wasser-Formamid (6:2:8:2) und (6:2:1:4) und Pentanol-Wasser (1:1). Filtrierpapier: Schleicher und Schüll Nr. 2043a. Diese Lösungsmittelsysteme wurden bei der erwähnten quantitativen Bestimmungsmethode der Digitalisglykoside verwendet.

SILBERMAN und THORP (*273*) haben die Lanatoside A, B und C im System Äthylacetat-Benzol (86:14), das 1—7,5% Äthanol enthält, getrennt. KAISER verwendet dafür ein Gemisch von Chloroform-Tetrahydrofuran-Formamid (*126*). Zur Trennung der Digitalisglykoside in verschiedenen Lösungsmittelsystemen vgl. auch (*125, 244, 245*).

Eine allgemeine Ausführungsform für Glykoside und Aglykone von sehr unterschiedlicher Polarität haben SCHENKER et al. (*246*) angegeben. Sie verwenden Wasser als stationäre und *n*-Butanol oder *n*-Butanol-Toluol- bzw. *n*-Butanol-Chloroform-Gemische als bewegliche Phase. Im Gegensatz zu dem sonst üblichen Wasser-Butanol-System wird nicht feuchtes *n*-Butanol auf unbehandeltem Papier verwendet, sondern letzteres mit Wasser getränkt. *n*-Butanol- oder *n*- bzw. Iso-amylalkohol-Wasser verwenden HABERMANN et al. (*72*) zur Trennung der Digitalisglykoside.

III. Die Konstitutionsermittlung.

Herzaktive Glykoside sind gemischte Cycloacetale von Zuckern mit einem Alkohol, der als Genin oder Aglykon bezeichnet wird. Sie sind nach folgendem Schema aufgebaut: Aglykon-Z_1-Z_2-Z_3-Z_4.

Ihre Konstitutionsermittlung hat sich sowohl mit der Struktur der Zucker und der Aglykone als auch mit der Verknüpfungsart dieser beiden Bausteine zu befassen. Liegen mehrere Zucker vor, so ist auch ihre Verknüpfung untereinander abzuklären.

A. Abbaureaktionen.

a) *Glykosidspaltung.*

Die Spaltung der Herzglykoside läßt sich entweder auf rein chemischem Wege oder mit Hilfe von Enzymen durchführen.

1. Chemische Methoden.

Die glykosidischen Bindungen lassen sich durch Säuren hydrolytisch spalten. Gegen Basen sind sie beständig. Während die Zucker sich dabei meistens ohne Schwierigkeiten unversehrt fassen lassen, hängt dies bei den Aglykonen von den Bedingungen der Hydrolyse ab. Diese richten sich nach dem Charakter des am Aglykon direkt haftenden Zuckers Z_1. Ist dieser ein *2-Desoxyzucker* (d. h. die Oxygruppe an $C_{(2)}$ eines normalen Zuckers ist durch Wasserstoff ersetzt), so werden seine Glykoside viel rascher hydrolysiert als die der normalen Zucker. Dabei spielt es praktisch keine Rolle, ob das Glykosid des 2-Desoxyzuckers in pyranoider oder furanoider Form vorliegt. Die Geschwindigkeit der Hydrolyse ist bei der furanoiden Form am größten.

Auch bei Glykosiden mit gewöhnlichen Zuckern werden die furanoiden rascher als die pyranoiden Formen gespalten, doch sehr viel langsamer als die pyranoide Form der 2-Desoxyzucker-Glykoside (*218*). OVEREND et al. (*196*) haben diese Geschwindigkeiten bestimmt und miteinander verglichen. Zum Mechanismus der Spaltung von O-Glykosiden vgl. (*55, 225*).

So werden Herzglykoside, in denen Z_1 ein 2-Desoxyzucker ist, bereits mit 0,05-*n* H_2SO_4 bei 90° nach 25 Minuten vollständig gespalten, wobei das Aglykon in quantitativer Ausbeute intakt isoliert werden kann. Zur Ausführung vgl. Spaltung von Odorosid A nach RANGASWAMI und REICHSTEIN (*213*) sowie (*234*). Auf diese Weise gelang es, aus Odorosid B zum ersten Male Uzarigenin intakt zu fassen. Über den Nachweis der 2-Desoxyzucker in Herzglykosiden s. S. 145.

Sind in einem zuckerreichen Glykosid mehrere 2-Desoxyzucker miteinander glykosidisch verknüpft, so tritt unter den genannten Bedingungen zwischen ihnen allen Spaltung ein. Dies ist z. B. bei Digitoxin der Fall [Aglykon = Digitoxigenin, $Z_1=Z_2=Z_3=D$-Digitoxoserest (*294*)]. Viel häufiger ist aber in den genuinen Glykosiden nur der Zuckerrest Z_1 ein 2-Desoxyzucker. Als Z_2 und Z_3 tritt in allen solchen bisher bekannten Fällen der D-Glucoserest auf. Triglykoside, die nach diesem Schema gebaut sind, liefern somit bei dieser milden sauren Hydrolyse das Aglykon und ein Trisaccharid, Diglykoside das Aglykon und ein Disaccharid. Vgl. als Beispiel die Spaltung von k-Strophanthosid und k-Strophanthin-β in Strophanthidin und Strophanthotriose bzw. Strophanthobiose

(*313*), die von Echujin in Digitoxigenin und Strophanthotriose (*88*) oder die von Odorosid K in Uzarigenin und Odorotriose (*231*).

Die Glykosidspaltung ist bedeutend schwieriger, wenn der mit dem Aglykon verknüpfte Zucker normal gebaut ist (Oxygruppe an $C_{(2)}$). Die Hydrolyse muß dann unter so energischen Bedingungen ausgeführt werden, daß zwar der Zucker intakt erhalten wird, das Aglykon aber meist nicht mehr unversehrt gefaßt werden kann (Bildung von Anhydro-Geninen).

Interessiert nur die Isolierung der Zucker, oder liegen keine säureempfindlichen Gruppen vor, so wird für diese Hydrolyse am besten das Gemisch von konz. HCl-Eisessig-Wasser (10 : 3,5 : 5,5) nach KILIANI (*150*) gewählt. Zur Ausführung vgl. (*224*). Diese Reaktion wird, wie früher erwähnt, auch zum Nachweis von Zucker in kleinen Glykosidmengen verwendet (*7*).

Das bekannte Verfahren von MANNICH und SIEWERT (*168*), nach dem solche schwer spaltbaren Glykoside in Acetonlösung mit konz. HCl bei 2.° behandelt werden, gestattet es, ihre Aglykone zum Teil intakt zu gewinnen. Die Ausbeuten sind zwar oft gering und die Tendenz zur Bildung von Dehydratisierungsprodukten groß. Zur Ausführung vgl. (*44, 235*).

Tanghinin und Desacetyl-tanghinin, Glykoside aus den Samen von *Tanghinia venenifera* POIR. lassen sich mit HCl in Aceton nach MANNICH und SIEWERT nicht hydrolysieren. REICHSTEIN und Mitarb. fanden, daß diese Glykoside in Chloroform, das etwa 2proz. HCl-Gas enthält, bei 20° glatt spaltbar sind (*271*). Voraussetzung für das Gelingen der Spaltung ist die Löslichkeit des Glykosids in Chloroform und wahrscheinlich eine erschwerte Abspaltbarkeit der tertiären Oxygruppen im Geninteil, wie dies bei Tanghinigenin tatsächlich der Fall ist (*272*).

Verlaufen alle Spaltversuche negativ, so kann auf das Verfahren von STEINEGGER und KATZ (*286*) zurückgegriffen werden, bei dem direkt das Glykosid einer erschöpfenden CrO_3-Oxydation und Hydrolyse unterworfen wird. Der Zucker wird dabei zerstört. Statt des Aglykons wird das entsprechende 3-Keton in zirka 10% Ausbeute erhalten. Die Methode leistete bei der Konstitutionsermittlung von Desacetyl-neriifolin (I) (*84*), ferner bei Acovenosid A (*322*), Desacetyl-tanghinin (*85,271*) und Gofrusid (*99*) gute Dienste. Die resultierende Ketogruppe an $C_{(3)}$ kann zur gewünschten 3β-Oxygruppe, solange der ungesättigte Lactonring vorhanden ist, nur nach MEERWEIN-PONNDORF oder mit $NaBH_4$ reduziert werden. Erstere Methode liefert ein Gemisch der beiden epimeren Alkohole, wobei derjenige mit axialer Konfiguration vorwiegen dürfte (*10*). Die Reduktion mit $NaBH_4$ gibt fast ausschließlich das äquatoriale Epimere, bei Digitoxigenon (II) also vor allem 3-Epi-digitoxigenin (III) (*270*).

Glykoside mit normalem Zucker sind schon unter sehr milden Bedingungen spaltbar, wenn ihr Genin eine Doppelbindung in 4- oder 5-Stellung besitzt. Dies ist der Fall bei Proscillaridin A (IV) (*311*) und bei Acofriosid L (VII) (*189*). Die nach Abspaltung des Zuckers gebildete Oxygruppe an $C_{(3)}$ in Allylstellung wird leicht abgespalten, und es bildet

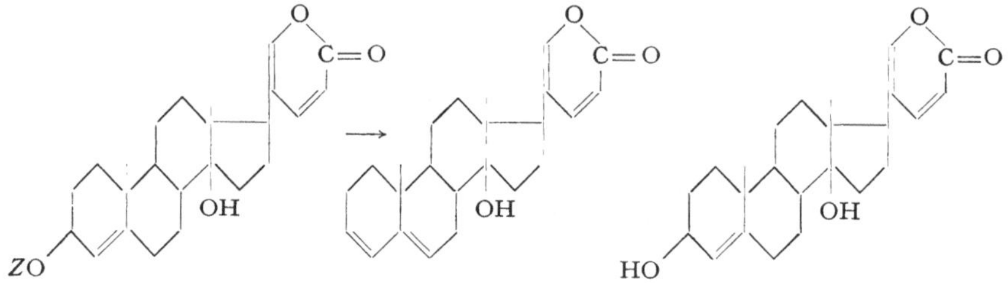

(IV.) Proscillaridin A.
(*Z* = Rhamnoserest.)
Scillaren A (*Z* = Glucosido-rhamnoserest).

(V.) Scillaridin A.

(VI.) Scillarenin.

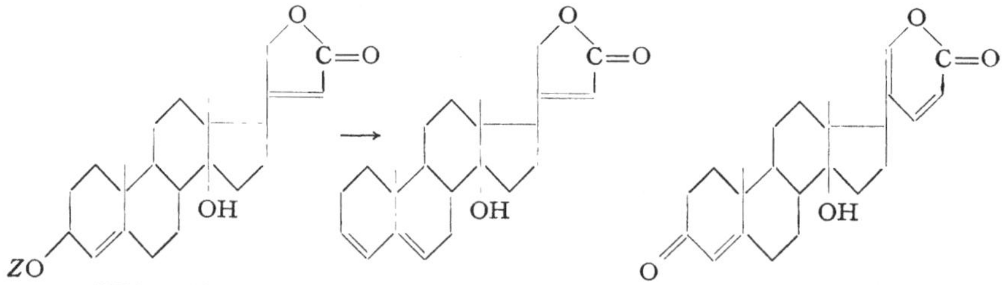

(VII.) Acofriosid L.
(*Z* = Acofrioserest.)

(VIII.) 3,5-Anhydro-periplogenin.

(IX.) Scillarenon.

sich ein 3,5-Dien. Wird die Δ^4-Doppelbindung zuerst hydriert, so sind zur Hydrolyse energischere Bedingungen nötig. Die Verbindung (IV) gibt daher nicht Scillarenin (VI), sondern Scillaridin A (V), und (VII) gibt 3,5-Anhydro-periplogenin (VIII). Im Falle des Scilliglaucosids (XI) konnten STOLL et al. die *D*-Glucose bereits mit 1proz. H_2SO_4 bei 0° abspalten. Es gelang ihnen das intakte Genin Scilliglaucosidin (X) zu fassen (*316*). Daß dabei die Oxygruppe an $C_{(3)}$ nicht eliminiert wird, führen diese Autoren auf die Anwesenheit der Sauerstoffunktion an $C_{(10)}$

(X.) Scilliglaucosidin.   (XI.) Scilliglaucosid.   (XII.)

zurück. Versuche an Δ^4-5-Anhydro-strophanthidin-19-säure als Modell bestätigen ihre Auffassung; denn die 3-ständige Oxygruppe ließ sich mit 1 % H_2SO_4 erst beim Erwärmen entfernen unter Bildung des 3,5-Diens (*318*). Mit methanolischer HCl gibt (XI) das 3,5-Dien (XII).

Es sei noch auf die Stoffe des Milchsafts von *Calotropis procera* hingewiesen, deren Genine statt eines Zuckers Methylreduktinsäure und Oxymethylreduktinsäure tragen. Diese Stoffe lassen sich nicht hydrolysieren. Die Methylreduktinsäure wird in schlechter Ausbeute durch eine Brenzreaktion bei 240° und 0,05 Torr gewonnen. Das Genin wird dabei zerstört (*90*).

2. Enzymatische Methoden.

Für die Strukturaufklärung von Di-, Tri- oder Tetraglykosiden ist es oft erwünscht, die Zucker *stufenweise* abzuspalten. Chemische Spaltverfahren sind nur bedingt selektiv. Hingegen werden durch die Verwendung von *Enzymen* gute Erfolge erzielt. Enzyme, die für die Spaltung von glykosidreichen Bindungen spezifisch sind, finden sich oft zusammen mit den Glykosiden in der Pflanze und sind für die häufig beobachtete Veränderung der pharmakologischen Wirksamkeit von Drogen im Verlaufe der Isolierung verantwortlich.

1926 ist es JACOBS und HOFFMANN (*119*) zum ersten Male gelungen, aus den Samen von *Strophanthus Courmontii* ein Enzympräparat, das die sog. *Strophanthobiase* stark angereichert enthielt, zu gewinnen. Strophanthobiasepräparate wurden dann auch aus den Samen anderer Strophanthusarten gewonnen, wobei sich *S. kombé* und *S. Eminii* als besonders enzymreich erwiesen (*260*). Strophanthobiase spaltet in Glykosiden endständige Glucosereste ab. Die Bindung zwischen Zucker und Aglykon wird nicht gelöst, sofern dieser Zucker nicht *D*-Glucose ist. Strophanthobiase spaltet z. B. das Triglykosid k-Strophanthosid in das Monoglykosid Cymarin und 2 Mol. *D*-Glucose, k-Strophanthin-β in

Cymarin und 1 Mol. *D*-Glucose (*313*), und Strophanthidin-*β*-*D*-glucosid- ⟨1,5⟩ weitgehend in Strophanthidin und *D*-Glucose (*172*). Zur Arbeitsmethodik vgl. (*88, 224*).

Enzyme mit ähnlichen Eigenschaften sind oft auch in glykosidhaltigen Blättern und Zwiebeln enthalten. Die Blätter von *Digitalis lanata* enthalten die sog. Digilanidase (*294*) und die von *D. purpurea* die Digipurpidase (*296*), die Zwiebeln von *Scilla maritima* die Scillarenase (*314*). Glucosidasen aus Hefe (*88, 313*), bitteren Mandeln (*300*), Takadiastase und „Luizym" aus *Aspergillus oryzae* (*64, 88, 310, 351*), ferner Enzympräparate aus dem Hepatopankreassaft der Weinbergschnecke (*Helix pomatia*) (*57, 93*), den Samen von *Adenium multiflorum* (*96*), den Samen von *Coronilla glauca* (*302*) und aus Luzernesamen (*302*) haben sich ebenfalls als außerordentlich nützliche Hilfsmittel für die Glykosidspaltung erwiesen.

Das Handelspräparat Takadiastase ist mit dem Enzympräparat „Luizym" aus *Aspergillus oryzae* nahe verwandt, wird aber vor Erreichen des Mycelstadiums gewonnen, während zur Gewinnung des letzteren die Pilze bis zur Sporenbildung gezüchtet werden.

Durch Züchtung eines speziellen Penicilliumstammes in einer Nährlösung, die als Kohlenstoffquelle nur Rhamnose enthielt, gewannen Stoll et al. ein adaptives Enzympräparat, das in Proscillaridin A (IV) die Bindung zwischen Aglykon und dem Rhanmoserest zu spalten vermochte (*315*). Damit gelang es zum ersten Male, Scillarenin (VI) intakt zu fassen. Durch Modifizierung des enzymatischen Spaltverfahrens kann die Glykosidspaltung durch eine Dehydrierung der Oxygruppe an $C_{(3)}$ des Genins ergänzt werden. Aus Scillaren A (IV, $Z =$ Glucosido-rhamnoserest) und Proscillaridin A (IV) wurde in dieser Weise direkt Scillarenon (IX) erhalten (*311*). Auch Enzympräparate aus tierischen Organen wurden zur Glykosidspaltung herangezogen (*307*).

Schließlich sei erwähnt, daß das Enzympräparat aus dem Hepatopankreassaft der Weinbergschnecke nicht nur glykosidische Bindungen zu lösen vermag, sondern auch eine Esterase enthält, die Acetylgruppen abspaltet. Diese Eigenschaft ist für Glykoside, die schwer verseifbare Acetylgruppen enthalten, wertvoll (*64, 224, 228*).

b) Die Konstitution der Aglykone.

Die zuckerfreien Bausteine der herzwirksamen Glykoside, die Aglykone oder Genine, unterscheiden sich von den anderen bekannten natürlichen Steroiden durch die *cis*-ständige Verknüpfung der Ringe *C* und *D*. Die Ringe *A* und *B* sind meist *cis*-ständig, manchmal aber *trans*-ständig miteinander verbunden. Die Konfiguration an den asymmetrischen C-Atomen Nr. 8, 9, 10 und 13 ist gleich wie bei den Gallensäuren. Alle Aglykone

tragen als Seitenkette an $C_{(17)}$ einen ungesättigten Lactonring in
β-Stellung, ferner an $C_{(3)}$ eine β-ständige sekundäre und an $C_{(14)}$ eine
β-ständige tertiäre Hydroxylgruppe. (Eine α-ständige Oxygruppe an $C_{(3)}$
in einem natürlichen Aglykon ist bisher nur in Urezigenin nachgewiesen
worden.)

Es gibt zwei *Grundtypen von Aglykonen*, die sich im Bau des Lacton-
rings unterscheiden: 1. Der Digitalis-Strophanthus-Typ, der einen
Butenolidring (α,β-ungesättigtes γ-Lacton) trägt. Nach den Vorschlägen
zur Nomenklatur der Steroide* wird dieser Typ mit dem Grundnamen
„Cardenolid" bezeichnet. 2. Der Scilla-Bufo-Typ, der einen Cumalinring
oder α-Pyronring (doppelt ungesättigtes δ-Lacton) aufweist und nach den
Nomenklaturvorschlägen* als *„Bufadienolid"* bezeichnet wird. Die einfachsten Repräsentanten der beiden Grundtypen sind Digitoxigenin
(XIII) und Bufalin (XIV).

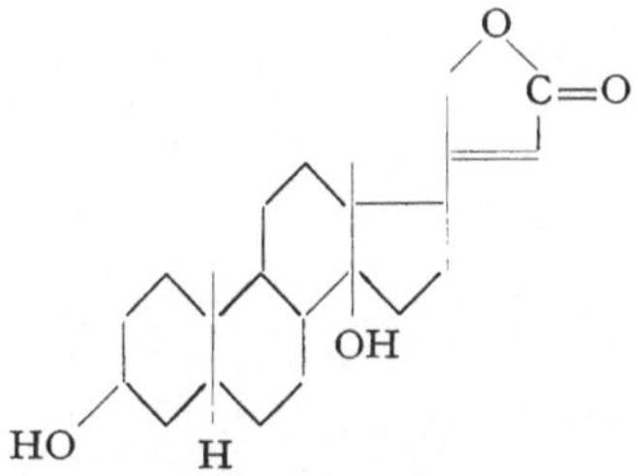

(XIII.) Digitoxigenin.

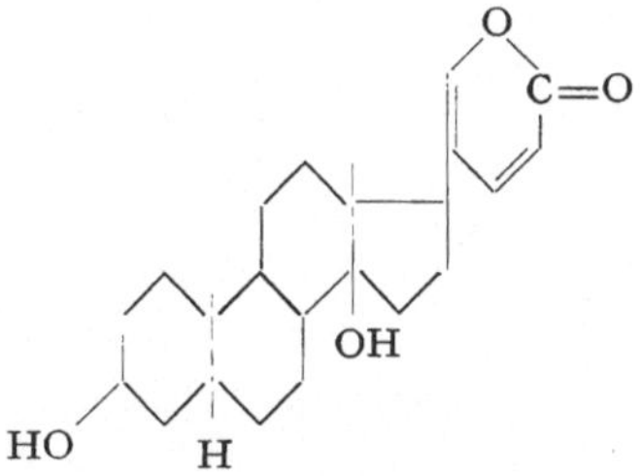

(XIV.) Bufalin.

Im folgenden werden zuerst die Methoden dargelegt, die für den
Konstitutionsbeweis der beiden Grundtypen (XIII) und (XIV) zur Verfügung stehen. Später werden die weiteren Reaktionen besprochen, die
zum Beweis von zusätzlichen Strukturelementen herangezogen werden
müssen.

1. Cardenolide.

Die Cardenolide enthalten 23 C-Atome. Der Butenolidring wird an
den charakteristischen Farbreaktionen (S. 144) und an seinem typischen
UV.-Maximum bei 217 mμ; log ε = 4,21 (in Alkohol) erkannt. Sein
IR.-Spektrum ist durch ν (C=O)-Schwingungen bei 5,59 μ (1790 cm^{-1})
(Vorbande) und 5,71—5,73 μ (1750—1747 cm^{-1}) (Hauptbande) und durch

---

* Vgl. Helv. chim. Acta **34**, 1680 (1951) (Konferenz an der „Ciba Foundation"
in London, 1950). Für die vollständig gesättigten Systeme werden die Grundnamen
„Cardanolid" bzw. „Bufanolid" verwendet. STOLL (*288*) hat vorgeschlagen, den
Grundnamen Cardenolid bzw. Cardanolid und Bufadienolid bzw. Bufanolid aus
historischen Gründen durch Digenolid bzw. Diganolid und durch Scilladienolid bzw.
Scillanolid zu ersetzen. Da über die Nomenklatur noch keine endgültige Entscheidung getroffen worden ist, schließen wir uns den Vorschlägen der Londoner Konferenz an.

eine ν (C=C)-Schwingung bei 6,16—6,18 μ (1624--1618 cm^{-1}) (in Methylenchlorid oder Chloroform) charakterisiert (125a).

Die richtige Formulierung geht auf Elderfield et al. (197) zurück.

α) *Beweis des Kohlenstoffskeletts.* Der direkteste Beweis der Steroidnatur der Aglykone wurde durch ihre Überführung in den Dielsschen Kohlenwasserstoff erbracht, was zuerst bei Anhydro-uzarigenin (347) durch Dehydrierung mit Selen gelang. In gleicher Weise wurde auch Strophanthidin zu demselben Kohlenwasserstoff dehydriert.

Der Abbau der Aglykone zu Ätiansäuren und Alloätiansäuren ist eine weitere Stütze für ihre Steroidnatur und wichtig für die Festlegung der Stereochemie. Die Konstitution eines unbekannten Aglykons läßt sich entweder durch diesen Abbau ermitteln oder durch Verknüpfung mit einem Geninderivat, dessen Struktur durch einen solchen Abbau gesichert worden ist. Eine derartige Reaktionsfolge wurde zum ersten Male

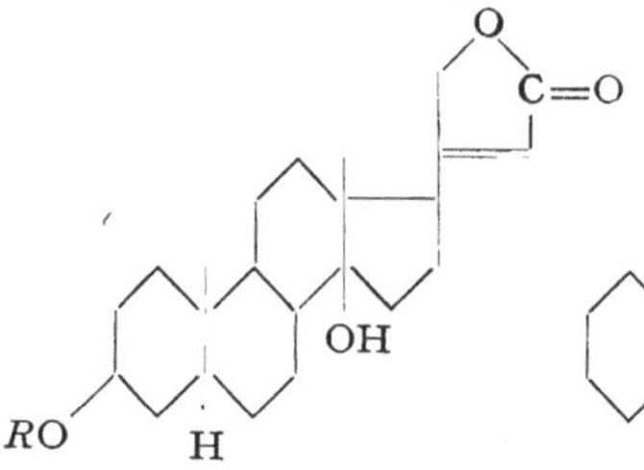

(XV.) Uzarin (R = Glucosido-glucoserest).   (XVII.)   (XVIII.)
(XVI.) Uzarigenin (R = H).

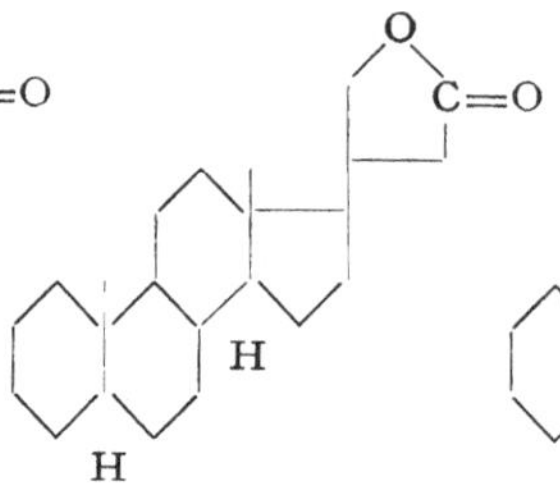

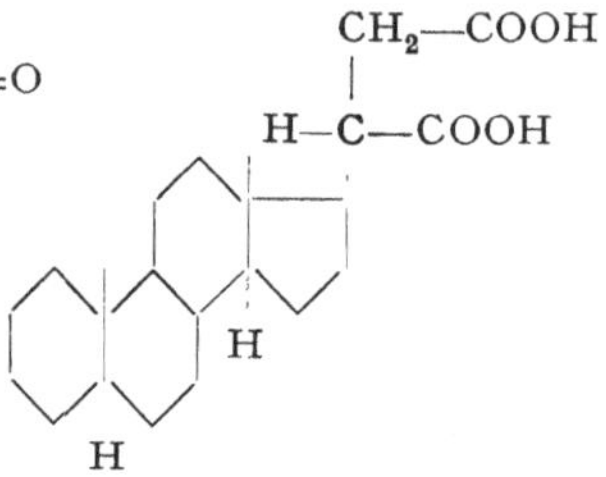

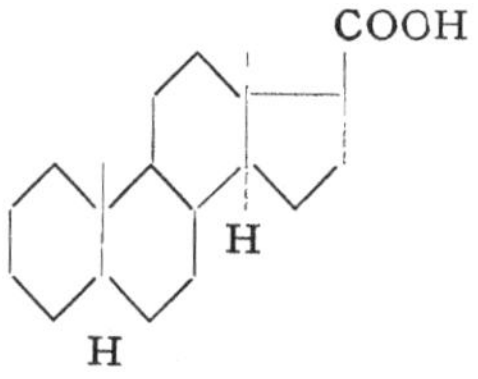

(XIX.) 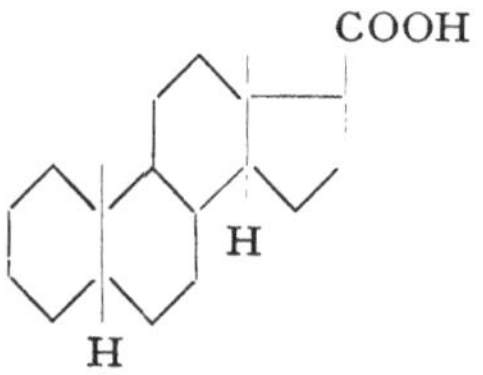   (XX.)

von Tschesche (330—333) an Uzarin (XV) durchgeführt. Behandlung von (XV) mit starken Säuren lieferte ein Gemisch von Anhydrogeninen. Nach Hydrierung der Doppelbindungen, Dehydrierung der Oxygruppe an C$_{(3)}$ und Entfernen der erhaltenen Ketogruppe nach Clemmensen wurde (XVII) gebildet. Die durch Oxydation daraus bereitete zweibasische Säure (XVIII) wurde verestert und mit Phenylmagnesiumbromid in das Bis-diphenylcarbinol übergeführt. Letzteres wurde mit CrO$_3$ zur Alloätiansäure (XIX) abgebaut.

Etwa gleichzeitig gelang es Jacobs und Elderfield (107, 107a) in ähnlicher Weise auch Digitoxigenin (XIII) über Isodigitoxigenin (XXXIX,

S. 158) zur Ätiansäure (XX) abzubauen. Beide Umwandlungen waren ohne Anwendung pyrolytischer Reaktionen durchgeführt worden. Da keine Möglichkeit für eine Inversion an $C_{(5)}$ bestand, ist somit erwiesen, daß Digitoxigenin (XIII) zur Koprostanreihe und Uzarigenin (XVI) zur Cholestanreihe gehören und daß ihre Seitenkette derjenigen der Norcholansäure entspricht.

Die beiden Umwandlungen wurden vor allem aus historischen Gründen erwähnt. Die zum Teil heute üblichen Methoden zur Überführung der Aglykone in Ätiansäuren sind die folgenden.

β) Abbau der Aglykone zu Ätiansäuren.

Der Abbau mit *alkalischem* $KMnO_4$, der an der Strophanthidinsäure zuerst von JACOBS (*101*), dann von ELDERFIELD (*36*) studiert und von FIESER (*52*) richtig formuliert wurde, findet heute kaum mehr Verwendung, da er Inversion an $C_{(17)}$ einschließt und zu den unerwünschten 17α-Ätiansäuren führt.

Durch *Abbau* der acetylierten Genine mit *neutralem* $KMnO_4$ in Aceton bei 20° werden die entsprechenden Ätiansäuren ohne Inversion an $C_{(17)}$ erhalten. Zum Beispiel O-Acetyl-digitoxigenin (XXI) liefert nach Behandlung mit $KMnO_4$ und anschließender Methylierung mit Diazomethan den 3β-Acetoxy-14-oxy-14β-ätiansäure-methylester (XXII) in 42% Ausbeute (*100*). Die neutrale $KMnO_4$-Oxydation ist besonders auch für den Abbau des Cumalinrings der Bufadienolide geeignet.

(XXI.) O-Acetyl-digitoxigenin (Ac = CH_3CO).   (XXII.) 3β-Acetoxy-14-oxy-14β-ätiansäure-methylester.

Der von MEYER und REICHSTEIN (*183*) eingeführte *Ozonabbau* von Acetyl-geninen ist noch schonender als der mit $KMnO_4$ und liefert im

(XXIII.)   (XXIV.)   (XXII.)

allgemeinen etwas bessere Ausbeuten an Ätiansäure. So wird O-Acetyl-digitoxigenin (XXI) bei zirka —40° mit Ozon behandelt und das rohe Ozonid mit Zn in Eisessig reduktiv gespalten. Der entstandene Glyoxylsäureester (XXIII) gibt mit $KHCO_3$ in wäßrigem Methanol das Ketol (XXIV), das nach Oxydation mit HJO_4 oder $NaJO_4$ (*274*, *331*) und Methylierung mit Diazomethan den Methylester (XXII) liefert. Dieser Ester ist von RUZICKA und Mitarbeiter teilsynthetisch bereitet worden, so daß seine Konstitution gesichert ist (*240*). (Für diesen Abbau dürfen die Acetyl-genine keine gegen HJO_4 bzw. $NaJO_4$ empfindliche Gruppierung aufweisen.) Unter den gleichen Voraussetzungen lassen sich auch Acetyl-glykoside abbauen. Die Bildung der Ketole vom Typus (XXIV) gibt die Möglichkeit, Cardenolide in die natürlichen *Hormone der Nebennierenrinde* überzuführen. Acetyliertes (XXIV) gab nach Abspaltung der Oxygruppe an $C_{(14)}$, Hydrierung und anschließender

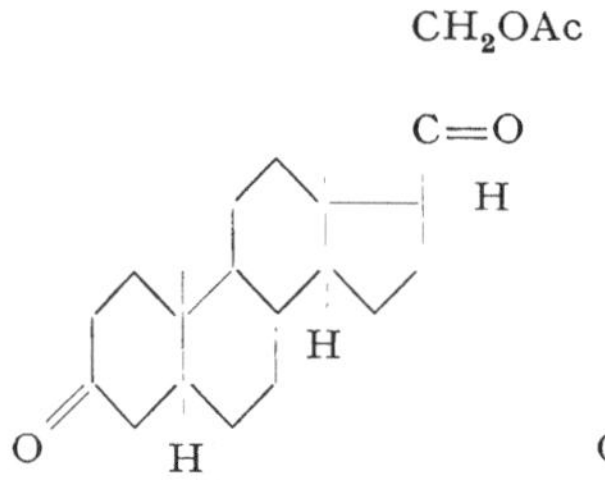

(XXV.)

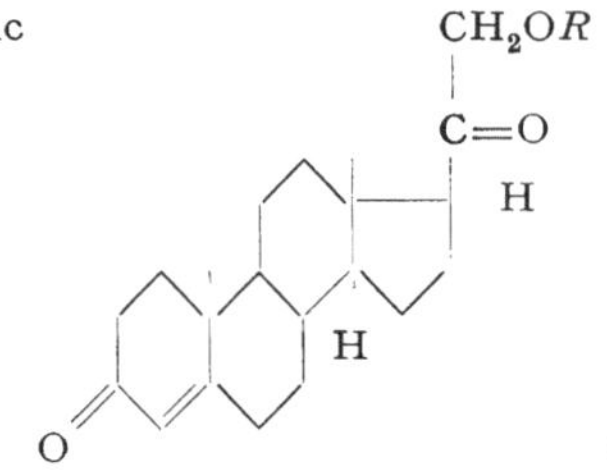

(XXVI.) (*R* = H) Desoxy-corticosteron.
(XXVII.) (*R* = Ac.)

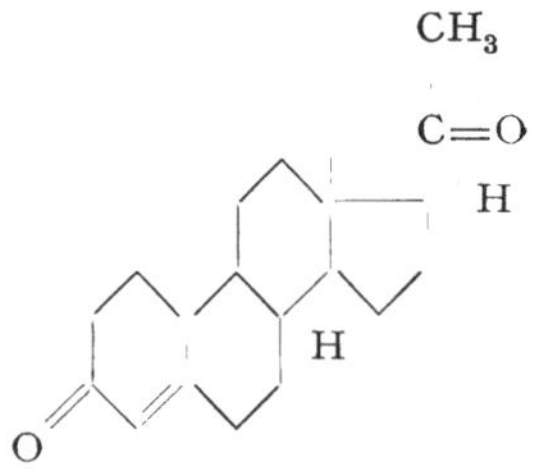

(XXVIII.) Progesteron.

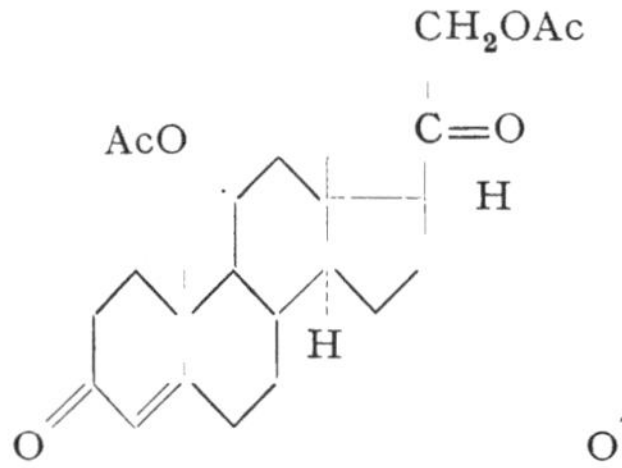

(XXIX.) Di-O-acetyl-11-epi-corticosteron.

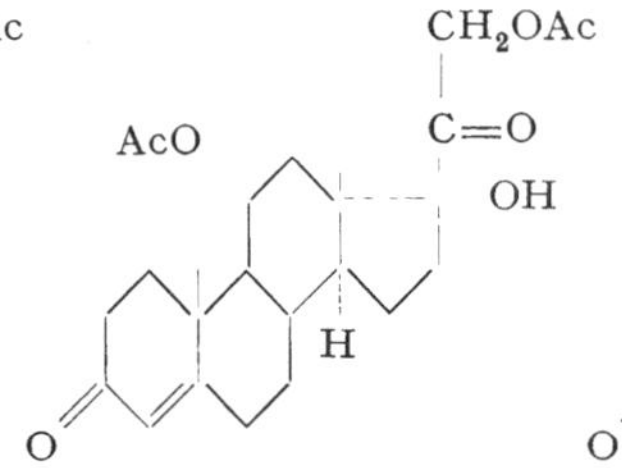

(XXX.) Di-O-acetyl-11-epi-17-oxy-corticosteron.

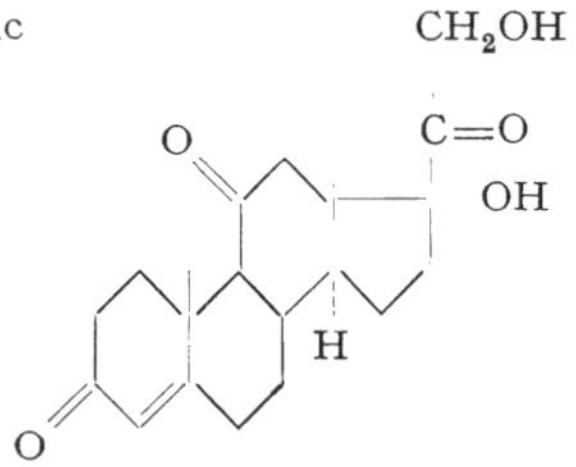

(XXXI.) Cortison.

CrO_3-Oxydation (XXV) (*183*), das in bekannter Weise (*219*) in Acetyl-desoxy-corticosteron (XXVII) umgewandelt wurde. Auf ähnliche Art wurden aus Periplogenin (LIX, S. 163) Desoxy-corticosteron (XXVI) und dessen Acetat (XXVII) (*159*) und Progesteron (XXVIII) (*159*), und aus Sarmentogenin Di-O-acetyl-11-epi-corticosteron (XXIX), Di-O-acetyl-11-epi-17-oxy-corticosteron (XXX) und Cortison (XXXI) bereitet (*162*).

γ) Abspaltung der Hydroxylgruppe an $C_{(14)}$. Die HO-Gruppe an $C_{(14)}$ ist unter normalen Bedingungen nicht acetylierbar und nicht dehydrierbar, woraus sich ihre tertiäre Natur ergibt. Der Beweis der 14β-Stellung stützt sich vor allem auf die Bildung der Isoverbindungen (14,21-Oxido-cardanolide) (siehe unten). Die Abspaltung der HO-Gruppe ermöglicht zunächst die Bildung von zwei isomeren Anhydroprodukten. Weitere Isomere können durch Wanderung der Doppelbindung entstehen.

O-Acetyl-digitoxigenin (XXI) gibt mit verdünnter Mineralsäure ein Gemisch von „α"-Anhydro-(XXXII) und „β"-Anhydro-O-acetyl-digitoxigenin (XXXIII).

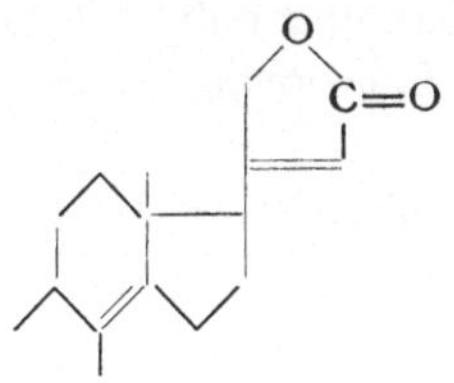

(XXXII.) „α"-Anhydro-O-acetyl-digitoxigenin-$\Delta^{8,14}$. 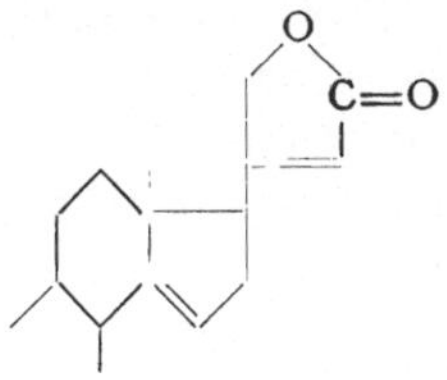   (XXXIII.) „β"-Anhydro-O-acetyl-digitoxigenin-Δ^{14}.

Stärkere Mineralsäure (z. B. konz. HCl) gibt vorwiegend die „α"-Form, die offenbar das stabilere Isomere ist; denn (XXXIII) läßt sich unter diesen Bedingungen zu (XXXII) isomerisieren (*26, 108*), während POCl$_3$ in Pyridin und SOCl$_2$ in Pyridin (*259*) vorwiegend das erwünschte „β"-Derivat liefern (*100*). Beide Anhydroprodukte bilden keine Isoverbindungen mehr, was ein weiterer Beweis für die 14β-Stellung dieser ursprünglichen Oxygruppe ist. Das „β"-Isomere ist stärker linksdrehend als „α" und läßt sich mit Pt in Eisessig glatt hydrieren, während „α" nicht hydrierbar ist. Zum molekularen Drehungsbeitrag der Doppelbindung in 8,14-Stellung vgl. auch (*272*). Das UV.-Spektrum zeigt im Bereich von 206—220 mμ das typische Maximum der tetrasubstituierten Doppelbindung (*17*). Die Isomeren lassen sich auch im IR.-Spektrum voneinander unterscheiden (*16, 26*).

Analoge Verhältnisse werden beim diesbezüglich näher untersuchten Digoxigenin (*26, 206, 277, 337*) und bei den 14-Oxyätiansäuren angetroffen.

Befindet sich ein weiterer Substituent in der Nähe der Oxygruppe an C$_{(14)}$, so läßt sie sich mit POCl$_3$ und Pyridin schwer oder manchmal gar nicht abspalten (*272*).

δ) *Stereochemie der Substituenten des Digitoxigenins (Grundtyp XIII) und seiner Isomeren* (vgl. Tabelle 1, S. 196). Der erwähnte, stereochemisch eindeutig verlaufende Abbau von *O-Acetyl-digitoxigenin* (XXI) zum 3-β Acetoxy-ätiansäure-methylester (XXII) beweist die *cis*-ständige Verknüpfung der Ringe *A* und *B*, die 3β-Stellung der sekundären Hydroxylgruppe und die 17β-Stellung des Butenolidringes in Digitoxigenin (XIII, S. 153). Trotz der 3β-Stellung der Oxygruppe geben die Aglykone im

Gegensatz zu anderen Steroiden im allgemeinen keine schwerlöslichen Digitonide (*74*); vgl. Digitoxigenin (*100*), Sarmentogenin (*132*), Digoxigenin und Strophanthidin (*336*). Die Konfiguration der tertiären Oxygruppe an $C_{(14)}$ wurde zunächst von den Isogeninen (14,21-Oxidocardanoliden) abgeleitet, die sich aus den Geninen unter dem Einfluß von starkem Alkali in irreversibler Reaktion bilden. Die *Isogenine* geben keine positive Legal-Probe und weisen bei 217 mμ keine selektive Absorption auf. Ihre HO-Gruppe an $C_{(14)}$ läßt sich nicht mehr anhydrisieren, während die HO-Gruppe an $C_{(3)}$ acetylierbar und dehydrierbar geblieben ist. Die Oxygruppe an $C_{(14)}$ ist mit dem Lactonring unter Ringbildung in Reaktion getreten.

Die Reaktion dürfte nach folgendem Mechanismus verlaufen [vgl. auch (*36, 110, 111, 197*)]:

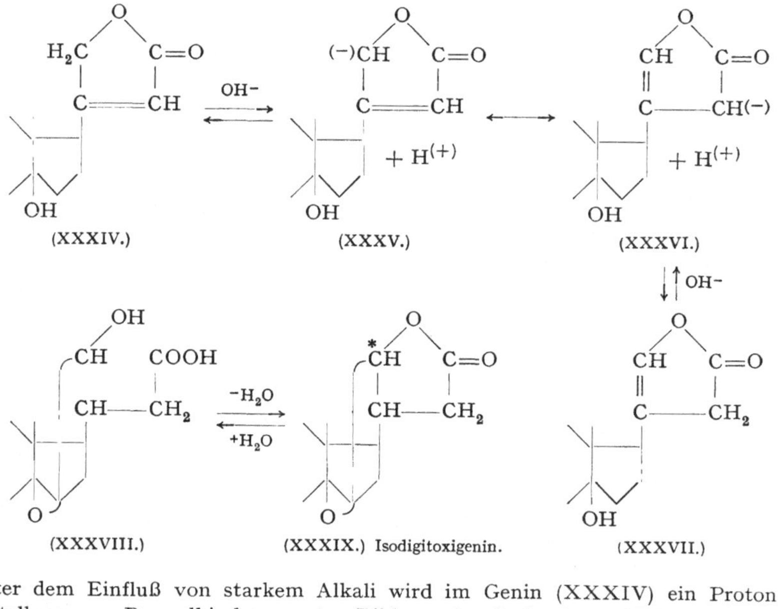

(XXXIV.)   (XXXV.)   (XXXVI.)

(XXXVIII.)   (XXXIX.) Isodigitoxigenin.   (XXXVII.)

Unter dem Einfluß von starkem Alkali wird im Genin (XXXIV) ein Proton in α-Stellung zur Doppelbindung unter Bildung des Carbanions (XXXV) losgelöst. Eine weitere Grenzform von (XXXV) ist (XXXVI), die im Gleichgewicht mit (XXXVII) steht. Durch Alkali katalysierte Gleichgewichte zwischen α,β- und β,γ-ungesättigten Carbonylverbindungen sind bekannt und untersucht worden (*3*). Im nächsten Schritt addiert sich die HO-Gruppe von $C_{(14)}$ an die Doppelbindung des Lactonringes, wobei ausgehend von (XXXVII) die Bildung der 14,21-Oxido-Verbindung (XXXIX) begünstigt ist, da ein Sechsring entsteht. Andernfalls entsteht ein gespannteres Fünfringsystem. Die Reaktion kann mit der Anlagerung von Alkoholen an Dihydropyran oder Dihydrofuran verglichen werden. (XXXIX) steht mit der cyclischen Halbacetalform (XXXVIII) im Gleichgewicht.

Die Isogenine bilden Methylester und Semicarbazone (*109, 110*). An $C_{(21)}$ entsteht ein neues Asymmetriezentrum. Die Isoverbindungen

Glykosidische Herzgifte. **159**

beweisen, daß die Substituenten an $C_{(14)}$ und $C_{(17)}$ sich auf der gleichen Seite des Ringsystems befinden.

Die Möglichkeit der Umkehrung der beiden Asymmetriezentren während der Isomerisierung ist aber a priori nicht auszuschließen. Der früher erwähnte Abbau von Digitoxigenin (XIII, S. 153) über Isodigitoxigenin (XXXIX) zur Ätiansäure (XX, S. 154) zeigte, daß dies nicht der Fall ist. Ein weiterer Beweis ergab sich an dem aus Digitoxigenin (XIII) bereiteten Ketol (XXIV, S. 155), das bei der Oxydation mit CrO_3 nicht in die Ätiansäure, sondern in das Ketolacton (XL) übergeht. Derartige Ketolactone zeigen im UV.-Spektrum ein typisches Maximum bei 353—355 mμ (log $\varepsilon = 1{,}62$) und im IR.-Spektrum eine charakteristische Bande bei 5,71 μ (1750 cm^{-1}); die Oxygruppe ist nicht mehr sichtbar (272). Ihre Bildung ist nur möglich, wenn sich die beiden beteiligten Substituenten auf der gleichen Seite des Ringsystems befinden. Erst nach Behandlung von (XL) mit H_2O_2 und Methylierung der entstandenen Ätiansäure erhält man den Ester (XXII, S. **155**).

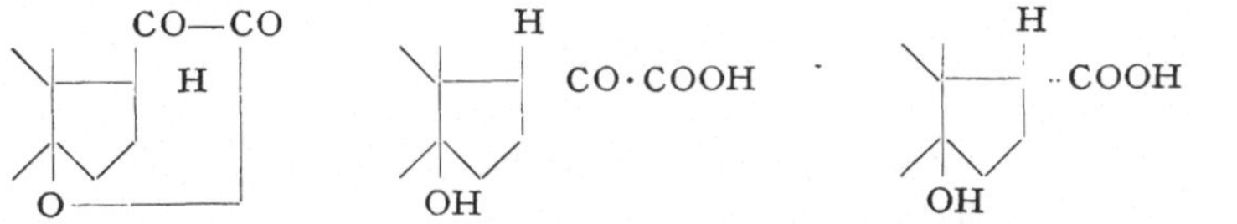

(XL.)   (XLI.)   (XLII.) 17α-Ätiansäure.

Mit warmem Alkali entsteht aus dem Ketolacton (XL) unter gleichzeitiger Epimerisierung an $C_{(17)}$ die Ketosäure (XLI), die kein Lacton mehr bildet und mit H_2O_2 in die 17α-Ätiansäure (XLII) übergeht. Die 14β-Stellung der tertiären HO-Gruppe wurde schließlich noch, wie bereits erwähnt, durch die Synthese des Methylesters (XXII) endgültig bewiesen (240). SPEISER und REICHSTEIN (282, 283) bauten in ähnlicher Weise Periplogenin (LIX) zur entsprechenden Ätiansäure ab.

Durch Dehydrierung der 3-Oxygruppe mit CrO_3 geht Digitoxigenin (XIII) in Digitoxigenon (II) über. Die Ketogruppe von (II) läßt sich glatt mit $NaBH_4$ wieder reduzieren, ohne daß der Lactonring angegriffen wird (vgl. S. 150). Es bildet sich dabei praktisch kein Digi-

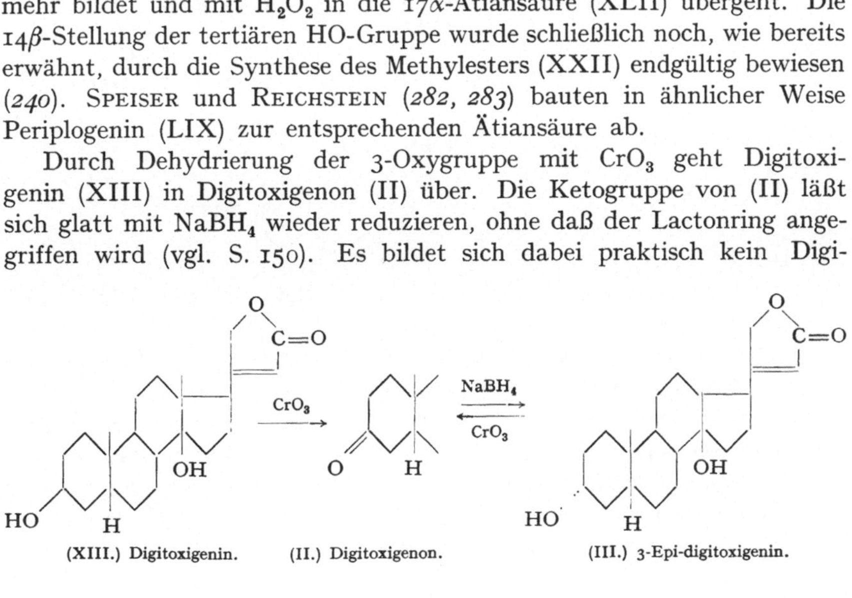

(XIII.) Digitoxigenin.   (II.) Digitoxigenon.   (III.) 3-Epi-digitoxigenin.

toxigenin (XIII), sondern, wie zu erwarten, fast nur das Epimere mit äqua-
torialem OH, nämlich *3-Epi-digitoxigenin* (III) (*270*). Da bei der Rückoxydation von (III) wieder Digitoxigenon (II) erhalten wird, ist bewiesen, daß nur Epimerisierung an $C_{(3)}$ eingetreten ist. 3-Epi-digitoxi-
genin ist physiologisch unwirksam (*270*). In der Natur ist dieses Aglykon
bisher nicht gefunden worden.

Uzarigenin (XVI), ein natürliches Aglykon, ist ebenfalls mit Digitoxigenin (XIII) isomer. Es besitzt die Struktur eines $3\beta,14$-Oxy-5α-
cardenolids, was durch den früher erwähnten Abbau von Uzarin (XV,
S. 154) zur Alloätiansäure (XIX) [Tschesche (*330—333*)] und denjeni-
gen von reinem, aus Odorosid B gewonnenen Uzarigenin in den 3β-Acet-
oxy-14-oxy-14β-allo-ätiansäure-methylester (XLIII) nach Rangaswami
und Reichstein (*213*) bewiesen worden ist. Die beiden an $C_{(20)}$ isomeren
Dihydro-uzarigenine sind von Plattner und Ruzicka (*207*) teilsynthetisch
bereitet worden, was ein weiterer Beweis für die Konstitution des
Uzarigenins ist.

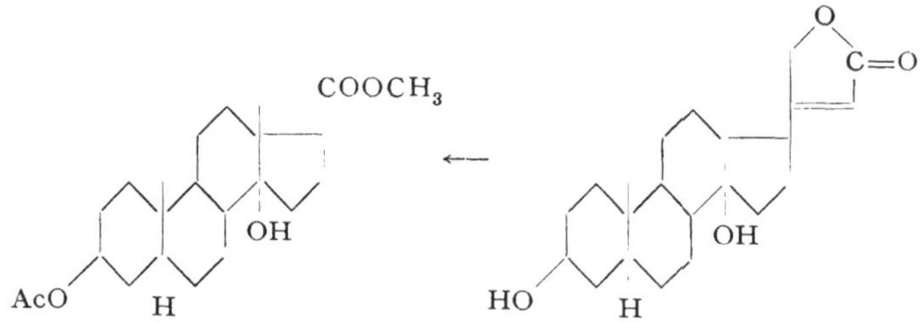

(XLIII.) 3β-Acetoxy-14-oxy-14β-allo-atiansaure-methylester.   (XVI.) Uzarigenin.

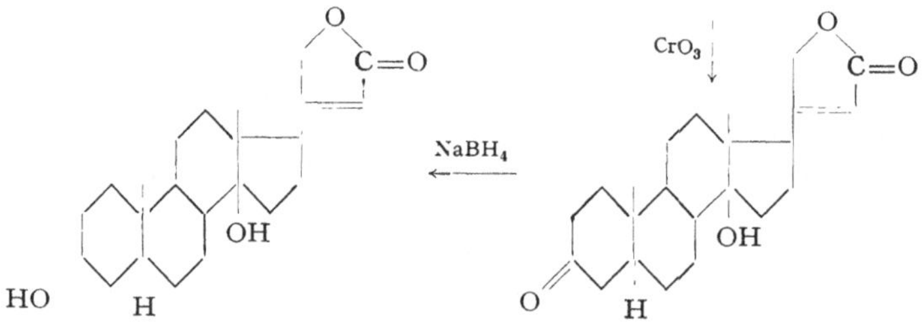

(XLV.) Urezigenin.   (XLIV.) Uzarigenon.

Die Dehydrierung von Uzarigenin (XVI) mit CrO_3 gibt Uzarigenon
(XLIV). Die Reduktion des Ketons (XLIV) mit $NaBH_4$ liefert ein
Gemisch von isomeren Alkoholen, aus welchem Tschesche trotz des
weitaus überwiegenden Anteiles an Uzarigenin (äquatoriale 3β-Oxygruppe)
auch das Epimere (XLV) isolieren konnte (*349*). Dieses war mit natür-
lichem *Urezigenin* identisch. Urezigenin ist ein natürliches Aglykon,
das bisher einzige mit einer α-ständigen 3-Oxygruppe.

ε) Stellung und Konfiguration von zusätzlichen funktionellen Gruppen. Weitere Aglykone bekannter Konstitution (vgl. Tabelle 1, S. 196). Eine zusätzliche *Hydroxylgruppe an* $C_{(1)}$ besitzt *Acovenosigenin A*. SCHLEGEL, TAMM und REICHSTEIN (*259*) bewiesen seine Konstitution als 1β,3β,14-Trioxy-cardenolid (XLIX). Schon in früheren Versuchen hatten TAMM und REICHSTEIN (*322*) Hinweise für eine Sauerstoffunktion an $C_{(1)}$ gewonnen. Der direkte Abbau von Acovenosid A (XLVII), ein Monoglykosid des Acovenosigenins A (XLIX) mit CrO_3 nach STEINEGGER und KATZ (*286*), gab ein Butenolid, das eine Δ^2-ungesättigte 1-Ketongruppierung aufwies (XLVI) [UV.-Spektrum in Alkohol: Maxima bei 217 mμ (log ε = 4,40) und 330 mμ (log ε = 1,93)]. Acovenosid A (XLVII) gibt mit Acetanhydrid und Pyridin bei 20° ein schwer trennbares Stoffgemisch, das u. a. Tri-O-acetyl-acovenosid A und Di-O-acetyl-acovenosid A (XLVII, Z =

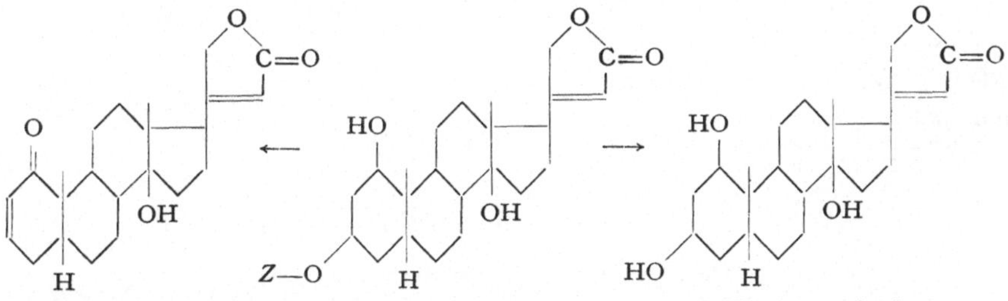

(XLVI.) (XLVII.) Acovenosid A. (Z = Acovenoserest.) (XLIX.) Acovenosigenin A.

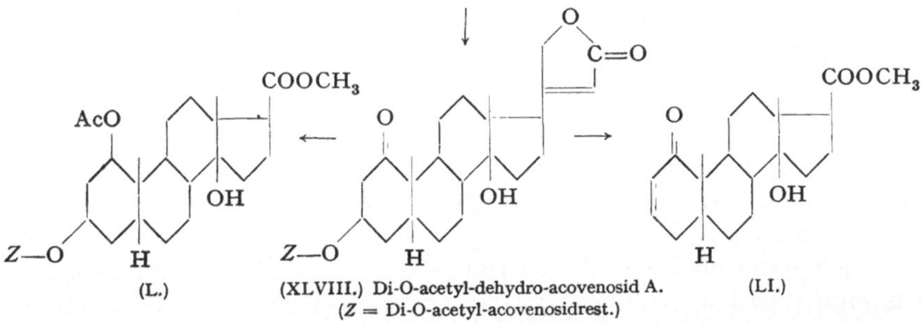

(L.) (XLVIII.) Di-O-acetyl-dehydro-acovenosid A. (LI.)
(Z = Di-O-acetyl-acovenosidrest.)

= Di-O-acetyl-acovenosidrest) enthält. (XLVII) besitzt im Geninteil noch eine freie, offensichtlich schwer acetylierbare sekundäre Oxygruppe. Sie läßt sich mit CrO_3 glatt dehydrieren. Das resultierende Di-O-acetyl-dehydro-acovenosid A (XLVIII) liefert nach Abbau mit O_3 usw. und Methylierung neben dem normalen zuckerhaltigen Ätiansäuremethyl-ester (L, Z = Di-O-acetyl-acovenoserest) (aus dem ursprünglich beigemeng-ten Tri-O-acetyl-acovenosid A stammend) einen zuckerfreien, sauerstoff-ärmeren Ätiansäure-methylester, der nach dem UV.-Spektrum [Maxima bei 225 mμ (log ε = 3,91) und 333 mμ (log ε = 1,91) in Alkohol] durch eine

Δ^2-ungesättigte Ketongruppierung entsprechend (LI) gekennzeichnet ist. Acovenosigenin A (XLIX) selbst läßt sich vollständig acetylieren und durch Abbau mit Ozon, Methylierung und nach Entfernen der tert. Oxygruppe an $C_{(14)}$, in den $1\beta,3\beta$-Diacetoxy-ätiansäure-methylester (LII) und schließlich in den Dioxyester (LIII) überführen. Die Hydroxylgruppe an $C_{(3)}$ kann mit N-Bromacetamid selektiv dehydriert werden. Acetylierung und Chromatographie an Al_2O_3 des erhaltenen Ketoalkohols (LIV) führt zum α,β-ungesättigten Ketoester (LVI). Nach Absättigung der Doppelbindung wird der bekannte 3-Keto-ätiansäure-methylester (LVIII) erhalten. Damit ist das Steringerüst, die Haftstelle

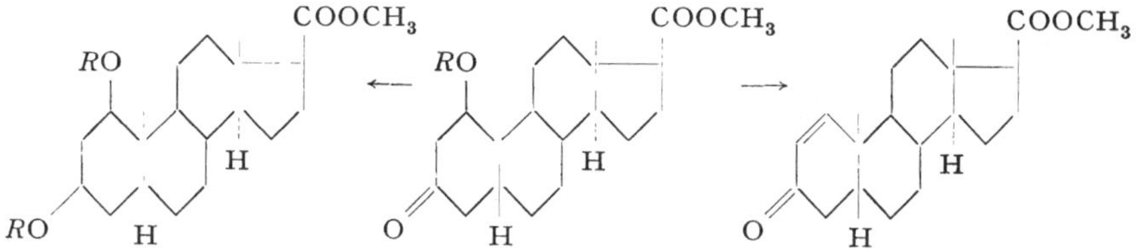

COOCH₃   COOCH₃   COOCH₃

(LII.) $(R = Ac.)$ $1\beta,3\beta$-Diacetoxy-atiansaure-methylester.
(LIII.) $(R = H.)$

(LIV.) $(R = H.)$
(LV.) $(R = Ac.)$

(LVI.)

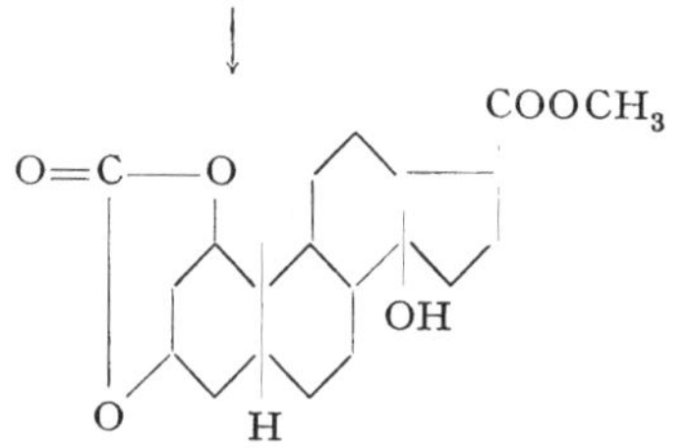

COOCH₃   COOCH₃

(LVII.)

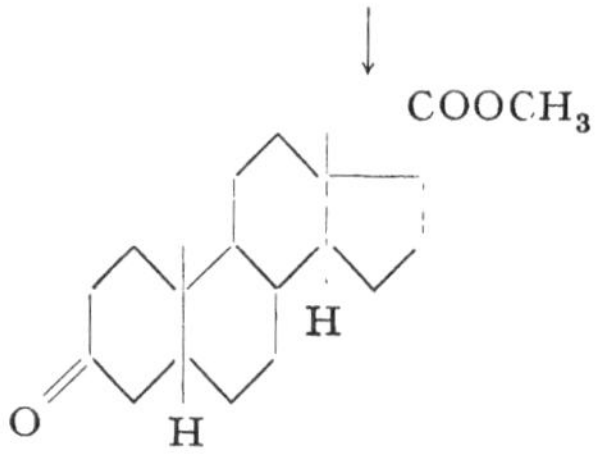

(LVIII.) 3-Keto-atiansäure-methylester.

des Butenolidringes in 17β-Stellung und einer HO-Gruppe in 3-Stellung bewiesen. Der ungesättigte Ketoester (LVIII) muß aus einem β-Oxy-Keton bzw. 1,3-Diol (entsprechend LIII) entstanden sein, da letzteres, wie auch Acovenosigenin A (XLIX) selbst, gegen $NaJO_4$ beständig ist. Die räumliche Lage der beiden Hydroxylgruppen im Ester (LIII) ergibt sich aus der glatten Bildung des cyclischen Kohlensäureesters (LVII), was nur bei bis-axialer Anordnung $(1\beta,3\beta)$ dieser beiden Oxygruppen möglich ist. Acovenosigenin A ist das erste natürliche Steroid, in dem eine Oxygruppe in 1-Stellung eindeutig bewiesen ist.

Eine tertiäre, nicht acetylierbare und nicht dehydrierbare Hydroxylgruppe an $C_{(5)}$ tritt in *Periplogenin* (LIX) auf. Oxydation von (LIX) mit CrO_3 in Eisessig gibt Anhydro-periplogenon (LX), das eine am UV.-Spektrum leicht erkennbare zusätzliche α,β-ungesättigte Ketongruppierung besitzt, was die Stellung der ursprünglichen HO-Gruppe an $C_{(5)}$ beweist (*85*).

(LIX.) Periplogenin.   (LX.) Anhydro-periplogenon.   (LXI.) 3β-Acetoxy-allo-ätiansäure-methylester.

Der Abbau von (LIX) zum 3β-Acetoxy-allo-ätiansäure-methylester (LXI) beweist die β-Konfiguration für die HO-Gruppe an $C_{(3)}$ und den Butenolidring. Da in Derivaten von (LIX) 3- und 5-ständige HO-Gruppen zur Bildung eines cyclischen Carbonats (*282*) mit Phosgen und eines cyclischen Sulfits (*208*) befähigt sind, ergibt sich auch für die HO-Gruppe an $C_{(5)}$ die β-Konfiguration. Die Verschiedenheit von Tetrahydro-14-anhydroperiplogenin mit dem teilsynthetisch bereiteten Isomeren mit α-ständiger HO-Gruppe an $C_{(5)}$ ist ein weiterer Konstitutionsbeweis (*208*).

Eine *Sauerstoffunktion an* $C_{(19)}$ weisen *Corotoxigenin* (LXII) und *Coroglaucigenin* (LXIII), ferner *Strophanthidin* (LXIV) und *Strophanthidol* (LXV) auf.

(LXII.) (*R* = —CHO.) Corotoxigenin.   (LXIV.) (*R* = —CHO.) Strophanthidin.
(LXIII.) (*R* = —CH₂OH.) Coroglaucigenin.   (LXV.) (*R* = —CH₂OH.) Strophanthidol.

Die anguläre Aldehydgruppe wird im Spektrum in Alkohol durch ein weiteres Maximum bei 310 mμ; log ε = 1,49 für (LXII) (*255*), und bei 303 mμ; log ε = 1,45 (*99*) für (LXIV) erkannt [Cymarin: 308 mμ; log ε = zirka 1,45 (*68, 197*)]. Die Aldehydgruppe in (LXII) ist im allgemeinen viel reaktionsträger als in (LXIV), doch geben beide Typen Oxime und Semicarbazone. Ihre Verknüpfung an $C_{(19)}$ und Beziehung zur HO-Gruppe an $C_{(3)}$ ergab sich bei Strophanthidin und bei Corotoxigenin vor allem durch die Bildung der cyclischen Lactole (LXVI) (*106*) bzw. (LXVIII) (*302*) mit alkoholischem HCl, die weder Aldehydreaktionen, noch die Reaktionen eines sekundären Alkohols gaben.

Die β-Konfiguration der Oxygruppe an $C_{(5)}$ und ihre *cis*-Stellung zur Aldehydgruppe in (LXV) ergab sich aus der in einer Cyanhydrinsynthese,

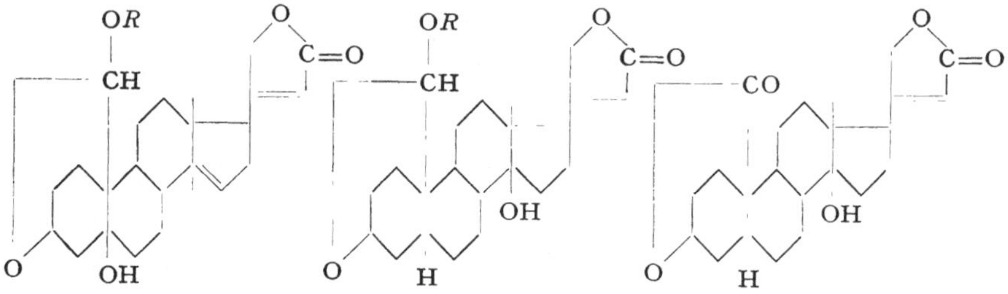

$$OR \quad CH \quad C=O \quad O \quad OH \qquad OR \quad CH \quad C=O \quad O \quad OH \quad H \qquad CO \quad C=O \quad O \quad OH \quad H$$

(LXVI.) (R = Alkyl.)   **(LXVII.)** (R = H.)   **(LXIX.)**

((LXVIII.) (R = CH_3 oder C_2H_5.)

ausgehend von Dihydro-strophanthidin (LXXI), gebildeten Oxysäure. Sie bildete mit der HO-Gruppe an $C_{(5)}$ sehr leicht ein Lacton, während die HO-Gruppe an $C_{(3)}$ unverändert blieb. Anguläre Aldehyd- und Oxymethylgruppen lassen sich mit CrO_3 leicht zu den entsprechenden Carbonsäuren oxydieren, wenn die 3β-ständige HO-Gruppe durch Acetylierung geschützt wird [Beispiele: Carbonsäure aus O-Acetyl-strophanthidin (22); Carbonsäure aus O-Acetyl-corotoxigenin (302)]. Mit $KMnO_4$ läßt sich auch freies Strophanthidin zur Carbonsäure oxydieren; doch ist die Ausbeute merklich schlechter (22). Die Aldehyde mit freier 3β-Oxygruppe bilden bei der CrO_3-Oxydation leicht neutrale Dilactone vom Typus (LXIX) (99). Möglicherweise liegt in Lösung ein Teil von Aldehyden vom Typus (LXII) bereits in der Lactolform (LXVII) vor, was die leichte Bildung von (LXIX) begünstigt. Deshalb sind Aldehyde vom Typus (LXII) ohne 5-ständige HO-Gruppe autoxydierbar (99, 136).

Durch Reduktion der Aldehydgruppe zur Oxymethyl- oder Methylgruppe werden diese Aglykone mit bekannten Stoffen verknüpft. Dafür ist die katalytische Hydrierung unbrauchbar, da der Butenolidring rascher reduziert wird. Selektive Reduktion der Aldehydgruppe zur Oxymethylgruppe gelingt hingegen mit Al-Isopropylat nach MEERWEIN-PONNDORF oder mit Al-Amalgam nach RABALD und KRAUS (210) [Beispiele: (LXV) aus (LXIV); Antiarol aus Antiarin (34)]. Der Cumalinring wird von Al-Amalgam angegriffen; hingegen bleibt er bei der Reduktion nach MEERWEIN-PONNDORF unverändert [vgl. z. B. Scilliglaucosidin (316)]. Die Reduktion von (LXII) gelingt zwar mit Al-Amalgam nicht, da dessen Aldehydgruppe zu träge ist (99), aber mit $NaBH_4$ nach HUNGER und REICHSTEIN (98). Im Gegensatz zu $LiAlH_4$ greift $NaBH_4$ den Butenolidring und den Cumalinring nicht an, da dieses Reagens Carbonylgruppen reduziert, ohne Estergruppen zu spalten (27). Durch Reduktion mit $NaBH_4$ gelang es z. B., Corotoxigenin (LXII) in Coroglaucigenin (LXIII) überzuführen. (LXIII) ließ sich durch Umwandlung in den 3β-19-Diacetoxy-allo-ätiansäure-methylester (LXX) (151) mit Strophanthidol (LXV) verknüpfen, was die Konfiguration der Substituenten an $C_{(3)}$, $C_{(17)}$ und $C_{(19)}$ bewies (99).

Wegen der leichten Autoxydierbarkeit gewisser angulärer Aldehydgruppen werden diese mit Vorteil zuerst zur Oxymethylgruppe reduziert, da bei den weiteren Abbaureaktionen weniger Komplikationen eintreten. Die vollständige Reduktion der angulären Aldehydgruppe zur Methylgruppe gelingt nach WOLFF-KISHNER, wobei von den Dihydrogeninen ausgegangen wird. Der Umsatz mit den Aglykonen direkt würde Iso-genine liefern. Die Hydrierung der Doppelbindung des Lactonringes

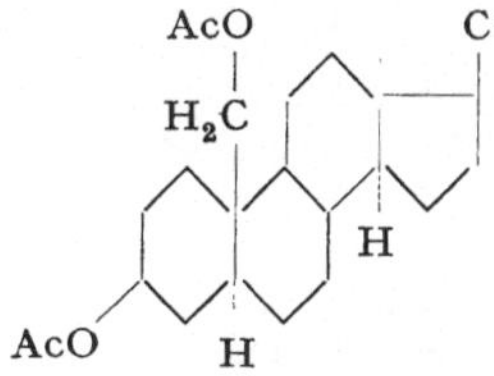

(LXX.) 3β-19-Diacetoxy-allo-atiansäure-methylester.

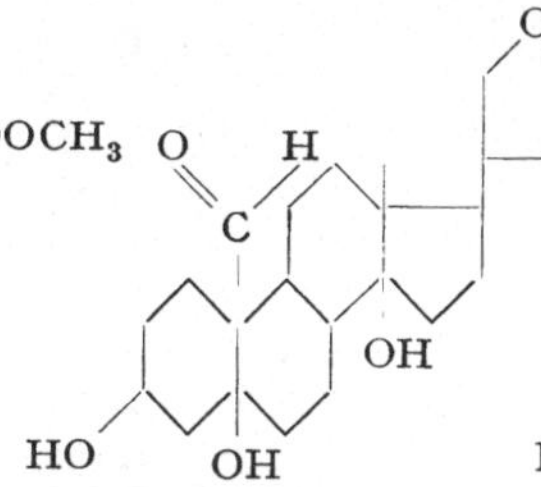

(LXXI.) Dihydro-strophanthidin.

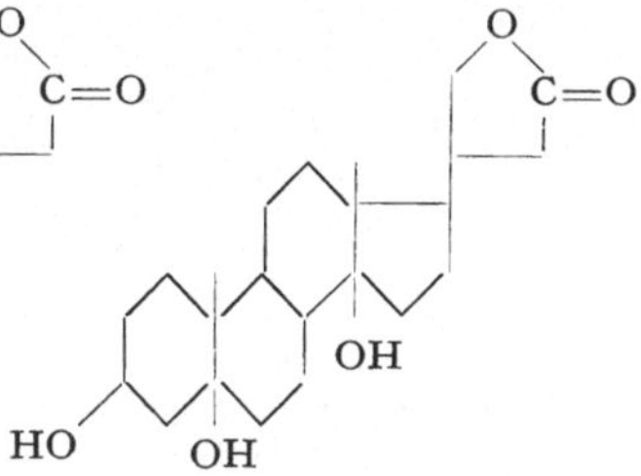

(LXXII.) Dihydro-periplogenin.

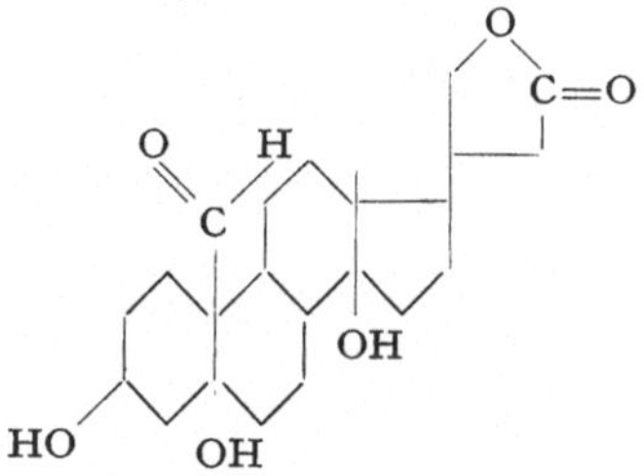

(LXXIII.) Dihydro-corotoxigenin.

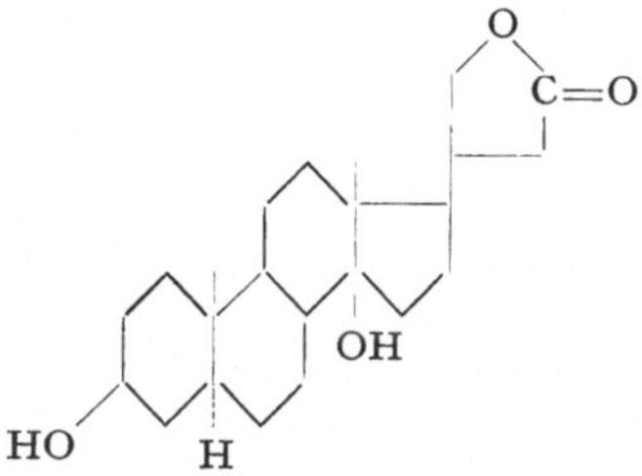

(LXXIV.) Dihydro-uzarigenin.

kann zu zwei an $C_{(20)}$ raumisomeren Dihydrogeninen führen. Es bildet sich vorwiegend das eine der Isomeren (*208, 255*). So wurden Dihydro-strophanthidin (LXXI) in Dihydro-periplogenin (LXXII) (*208*), Dihydro-corotoxigenin (LXXIII) in Dihydro-uzarigenin (LXXIV) (*255*) übergeführt. Der Übergang von (LXXI) in (LXXII) beweist die räumliche Stellung der Aldehydgruppe an $C_{(10)}$ und der HO-Gruppe an $C_{(5)}$. Für (LXXIII) wurde so die Konfiguration an $C_{(5)}$ und der tertiären HO-Gruppe an $C_{(14)}$ bewiesen. Daß unter den Bedingungen der WOLFF-KISHNER-Reaktion keine Isomerisierung eingetreten war, wurde dadurch bewiesen, daß Strophanthidin (LXIV, S. 163) mit Äthandithiol in das cyclische Mercaptal übergeführt wurde, das nach vorsichtiger reduktiver Entschwefelung mit RANEY-Nickel Periplogenin (LIX) lieferte (*281*). In ähnlicher Weise wurde Adonitoxigenin (XCIV, S. 169) mit Gitoxigenin (LXXXV, S. 168) verknüpft (*348*).

Der Versuch, die Oxymethylgruppe von Strophanthidol (LXV) über das Mesylat und Jodid in die Methylgruppe überzuführen, schlug fehl, da das Neopentyl-jodid Anlaß zu Umlagerungen gab (*151*).

Eine *Sauerstoffunktion an* $C_{(11)}$ wird in *Sarmentogenin* (LXXVI), *11-Epi-sarmentogenin* (LXXIX) und *Desarogenin* (= 11-Dehydro-sarmentogenin) (LXXVIII) angetroffen. Von diesen drei Geninen ist 11-Epi-sarmentogenin (LXXIX) bisher nicht in der Natur gefunden worden. Die drei Genine (LXXVI), (LXXIX) und (LXXVIII) liefern bei der

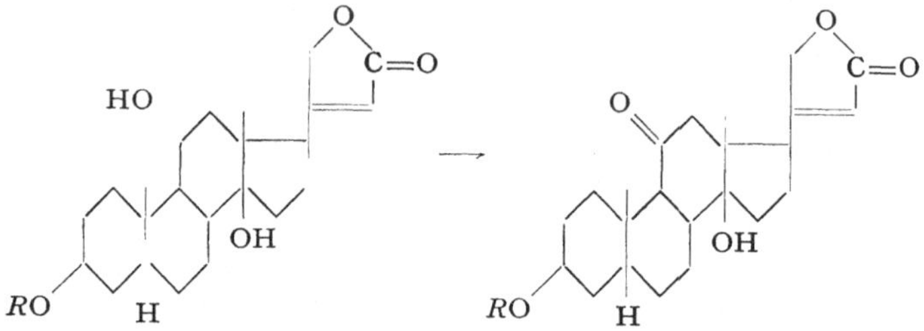

(LXXV.) (*R* = Sarmentoserest.) Sarmentocymarin.
(LXXVI.) (*R* = H.) Sarmentogenin.

(LXXVII.) (*R* = Sarmentoserest.)
11-Dehydro-sarmentocymarin.
(LXXVIII.) (*R* = H.) Desarogenin.

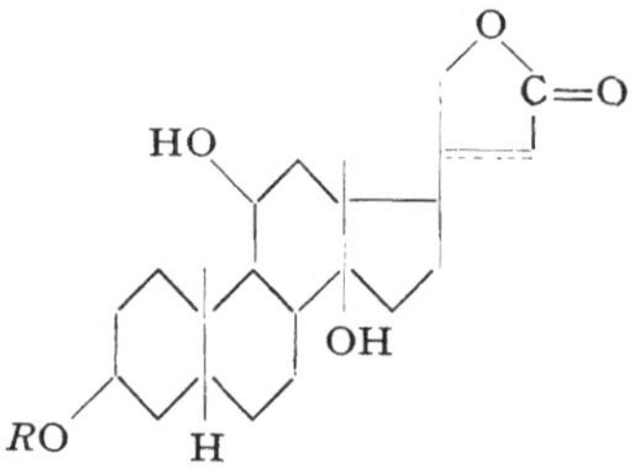

(LXXIX.) (*R* = H). 11-Epi-sarmentogenin.

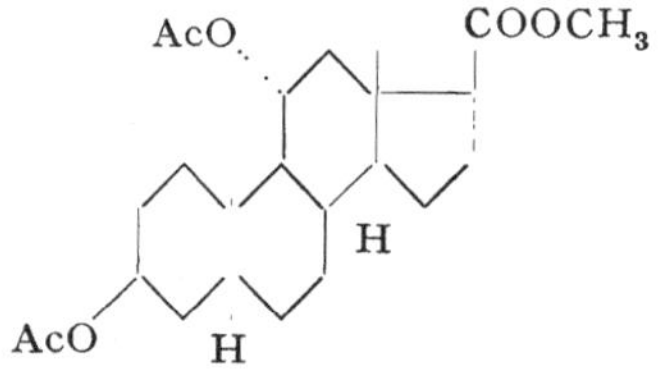

(LXXX.) 3β,11α-Diacetoxy-atiansaure-methylester.

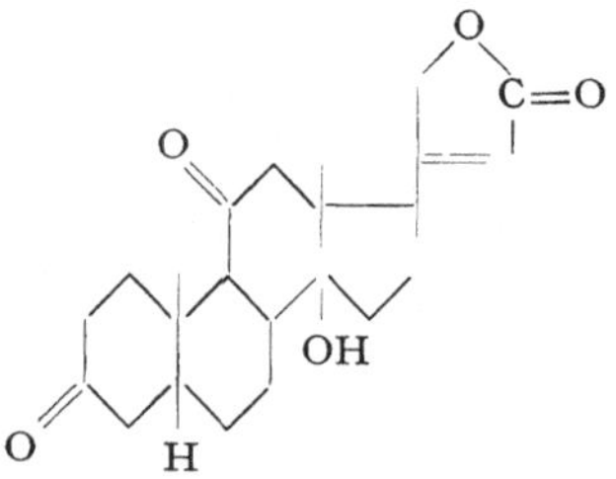

(LXXXI.) 3,11-Dehydro-sarmentogenin.

Dehydrierung mit CrO_3 dasselbe Diketon, 3,11-Dehydro-sarmentogenin (LXXXI). Dieses gibt aber nur ein Monosemicarbazon und ein Monoxim (*117, 141, 276*). Die Ketogruppe ist wegen sterischer Hinderung durch die anguläre Methylgruppe an $C_{(10)}$ reaktionsträge. Bei katalytischer Hydrierung der 11-Ketogruppe entsteht vorwiegend die axiale 11β-Oxygruppe, die im Gegensatz zur äquatorialen 11α-Oxygruppe nicht mehr acetylierbar ist (*161, 217, 284, 338*). Da (LXXVI) ein Diacetylderivat gibt, kommt der Oxygruppe an $C_{(11)}$ die α-Konfiguration zu. Der schlüssige Beweis wurde von K**ATZ** (*132*) durch den Abbau von (LXXVI) in den $3\beta,11\alpha$-Diacetoxy-ätiansäure-methylester (LXXX) (*131*) erbracht. Durch vorsichtige Dehydrierung von Sarmentocymarin (LXXV) mit CrO_3 in Eisessig kann man 11-Dehydro-sarmentocymarin (LXXVII) gewinnen (*49*). Die Zuckerkomponente wird unter diesen Bedingungen nicht angegriffen. Milde saure Hydrolyse von (LXXVII) liefert 11-Dehydro-sarmentogenin (LXXVIII) (*49*). Durch Reduktion von (LXXVIII) mit $NaBH_4$ gelangte S**CHINDLER** (*249*) zum 11-Epi-sarmentogenin (LXXIX). In analoger Weise wurden auch Desarosid (11-Dehydro-sarnovid) und 11-Episarnovid bereitet [(LXXVII) bzw. (LXXIX), mit $R =$ Digitaloserest].

Eine *Hydroxylgruppe an* $C_{(12)}$ tritt in *Digoxigenin* (LXXXII) auf. Auch dieses Genin läßt sich mit Chromtrioxyd zu einem Diketon dehydrieren, das nur ein Monosemicarbazon bzw. Monoxim liefert (*117, 141*). Die Ketogruppe in 12-Stellung ist reaktionsträge, was wahrscheinlich durch die Anwesenheit der 14β-ständigen Hydroxylgruppe hervorgerufen wird, da sowohl Anhydrodigitoxigenon (*276*) und 12-Keto-cholansäurederivate mit Ketonreagenzien reagieren.

(LXXXII.) Digoxigenin.   (LXXXIII.) 3,12-Diketo-atiansaure-methylester.   (LXXXIV.) $3\beta,12\beta$-Dioxy-atiansaure-methylester.

Der Abbau von Digoxigenin (LXXXII) zur 3,12-Diketo-ätiansäure (vgl. LXXXIII) (*285*) sicherte die Haftstellen der beiden ursprünglichen sekundären Hydroxylgruppen, und die Überführung in den $3\beta,12\beta$-Dioxy-ätiansäure-methylester (LXXXIV) ihre Konfiguration (*199, 326, 364*, vgl. auch *26*). Digoxigenin (LXXXII) ist somit ein $3\beta,12\beta,14$-Trioxycardenolid.

Substituenten an $C_{(16)}$ besitzen *Gitoxigenin* (LXXXV), *Oleandrigenin* (16-O-Acetyl-gitoxigenin) (LXXXVII), *Gitaloxigenin* (16-O-Formyl-gitoxigenin) (LXXXVIIa) und *Adonitoxigenin* (XCIV). Energische Hydrolyse von (LXXXV) mit Mineralsäure (konz. HCl bei 0°) gibt Dianhydrogitoxigenin (LXXXVIII), dessen UV.-Spektrum Maxima bei 222,5 mμ; log $\varepsilon = 4{,}05$ und 337,5 mμ; log $\varepsilon = 4{,}30$ in Alkohol aufweist [vgl. auch (*335, 344, 348, 374*)].

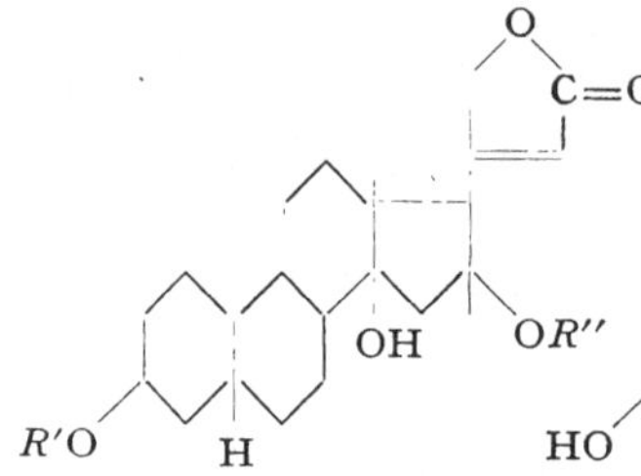

(LXXXV.)
(R′ = R″ = H.) Gitoxigenin.
(LXXXVI.)
(R′ = R″ = Ac.) Di-O-Acetyl-gitoxigenin.
(LXXXVII.)
(R′ = H; R″ = Ac.) Oleandrigenin.
(LXXXVIIa.)
(R′ = H; R″ = —CHO.) Gitaloxigenin.

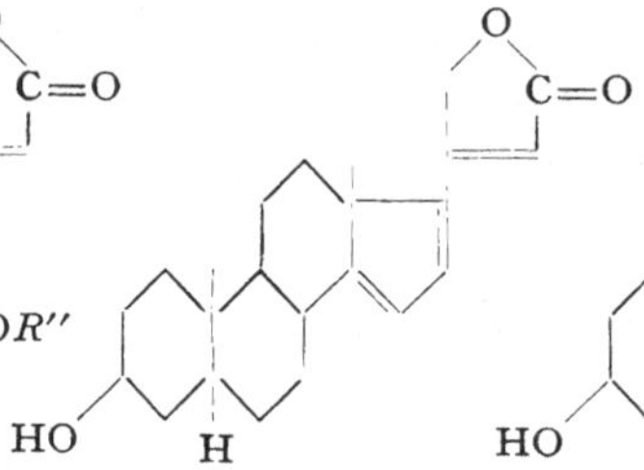

(LXXXVIII.)
Dianhydro-gitoxigenin.

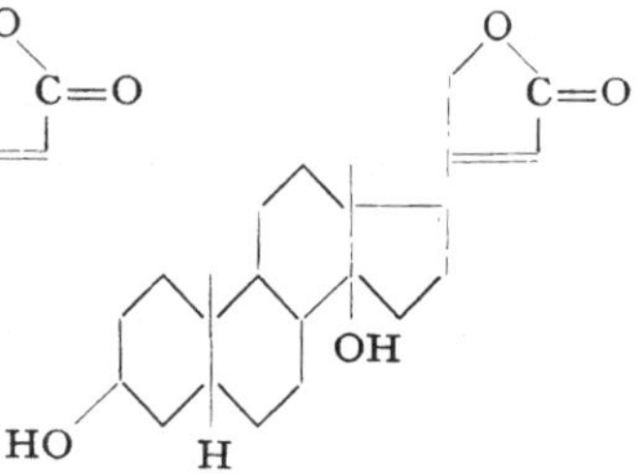

(LXXXIX.)
16-Monoanhydro-gitoxigenin.

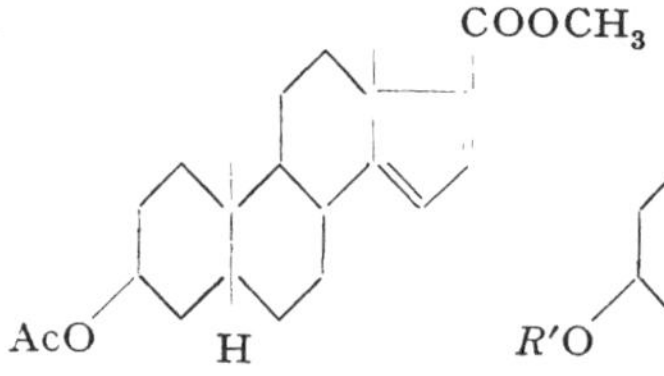

(XC.)
3β-Acetoxy-atia-dien-
(14,16)-säure-methylester.

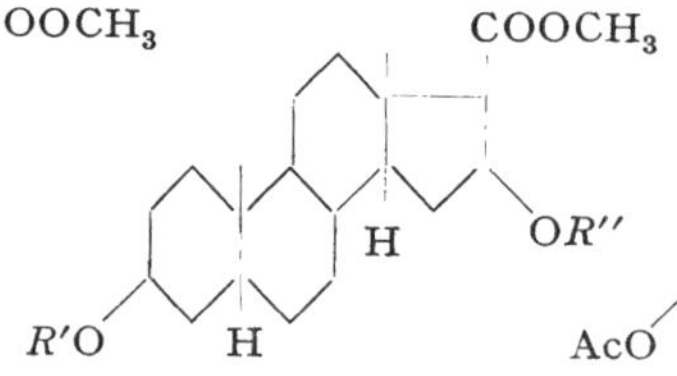

(XCI.)
(R′ = R″ = Ac.) $3\beta,16\beta$-Diacetoxy-
atiansaure.
(XCII.) (R′ = Ac; R″ = H.)

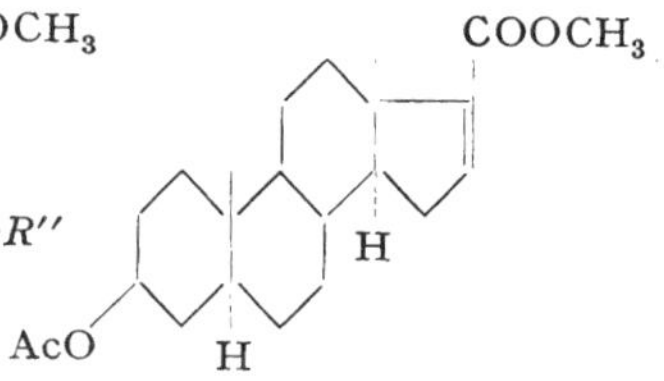

(XCIII.)
3β-Acetoxy-ätien-(16)-säure-
methylester.

Butenolide mit einer 16-Acetoxy-gruppe oder 16-Formyloxy-gruppe gehen durch Pyrolyse (*89*) oder beim Kontakt mit Al_2O_3 (*173*) in 16-Mono-anhydro-gitoxigenin-derivate über. Zum Beispiel entsteht aus Oleandrigenin (LXXXVII) *16-Monoanhydro-gitoxigenin* (LXXXIX). Gitaloxigenin ist das 16-Formyl-gitoxigenin (*71*). (LXXXIX) wurde aus O-Acetyl-digitoxigenin durch Behandeln mit N-Bromsuccinimid und anschließendem Kochen mit Pyridin gewonnen (*239, 241*). Das UV.-Spektrum von (LXXXIX) in Alkohol zeigt ein Maximum bei 273 mμ; log ε = 4,4 (*173*). An diesen Reaktionen und dem UV.-Sepktrum der Reaktionsprodukte können 16-ständige Oxy- bzw. Acetoxygruppen rasch identifiziert werden (*1, 95*). Die 16-Stellung dieser Substituenten war zuerst von JACOBS (*114*) erkannt worden. Der Abbau von Di-O-acetyl-gitoxigenin (LXXXVI) zur $3\beta,16\beta$-Diacetoxy-ätiansäure (XCI), die partiell zu (XCII) verseift wurde, lieferte dafür eine weitere Stütze (*174*). (XCII) gab bei der Behandlung mit $POCl_3$ in Pyridin ein Gemisch von 3β-Acetoxy-ätien-(16)-säure-methylester (XCIII) und 3β-Acetoxy-ätia-dien-(14,16)-säure-methylester (XC) (*174, 175*). (XCIII) war identisch mit einem teilsynthetisch bereiteten Präparat (*174*) und zeigte im UV.-Spektrum ein Maximum bei 225 mμ, log ε = 4,1 (in Alkohol). Das UV.-Spektrum von (XC) zeigt ein Maximum bei zirka 294 mμ, log ε = zirka 4,17.

Für die β-Konfiguration der 16-Oxygruppe spricht, daß sie beim Übergang von Gitoxigenin (LXXXV) in Isogitoxigenin und in β-Dihydrogitoxigeninsäure mit $C_{(21)}$ der Seitenkette unter Ringbildung reagiert (*112 bis 115, 108*). Ferner bereitete MOORE (*187*), ausgehend von 3β-Acetoxypregnen-(16)-on-(20), den 3β,16α-Diacetoxy-ätiansäure-methylester. Dieser war verschieden vom entsprechenden 3β,16-Diacetoxy-ester, der aus Di-O-acetyl-gitoxigenin (LXXXVI) erhalten worden war.

Ein weiterer Vertreter mit einer Hydroxylgruppe an $C_{(16)}$ ist *Adonitoxigenin* (XCIV) (*143*). Das Aglykon besitzt noch eine zusätzliche anguläre Aldehydgruppe, vermutlich an $C_{(10)}$. Der Beweis dafür wurde von TSCHESCHE und PETERSEN (*348*) durch die Überführung von (XCIV) in Gitoxigenin (LXXXV) erbracht. (Reduktive Entschwefelung des Äthylenmercaptals mit RANEY-Nickel.)

(XCIV.) Adonitoxigenin.

Die aus Strophanthussamen gewonnenen physiologisch inaktiven *Allo-Glykoside* (17α-Cardenolide) liefern bei der sauren Hydrolyse die gleichen Zucker wie die aktiven Formen, aber ein isomeres Aglykon. Das UV.-Spektrum und die funktionellen Gruppen der Allo-Aglykone sind unverändert. Sie bilden aber keine Isoverbindungen (14,21-Oxidocardanolide) mehr. Durch Abbau von Allo-periplogenin zu 3β-Acetoxy-17α-allo-ätiansäure-methylester wurde bewiesen, daß in den Allo-Geninen ein 17α-ständiger Butenolidring vorliegt (*282, 283*). Bekannt sind *Allo-strophanthidin* und *Allo-periplogenin*.

ζ) Aglykone mit teilweise bekannter Konstitution. Im folgenden werden nur Aglykone besprochen, deren Konstitution schon weitgehend aufgeklärt worden ist (vgl. Tabelle 1, S. 196).

Adynerigenin, $C_{23}H_{32}O_4$, hat in der letzten Zeit die Aufmerksamkeit auf sich gezogen. Es stehen die Formeln (XCV) und (XCIX), die beide von TSCHESCHE (*339, 340, 352*) vorgeschlagen worden sind, und die Formel (XCVIII), die von CARDWELL (*26*) stammt, zur Diskussion. Auf Grund der biologischen Unwirksamkeit von Adynerin und Δ^7-Adynerin hatte TSCHESCHE zunächst in (XCVI) und (XCIX) die 3-Oxygruppe α-ständig formuliert (*342*). Die β-Stellung dieser Oxygruppen an $C_{(3)}$ ist aber bewiesen (siehe unten). CARDWELL formuliert in (XCVII) und

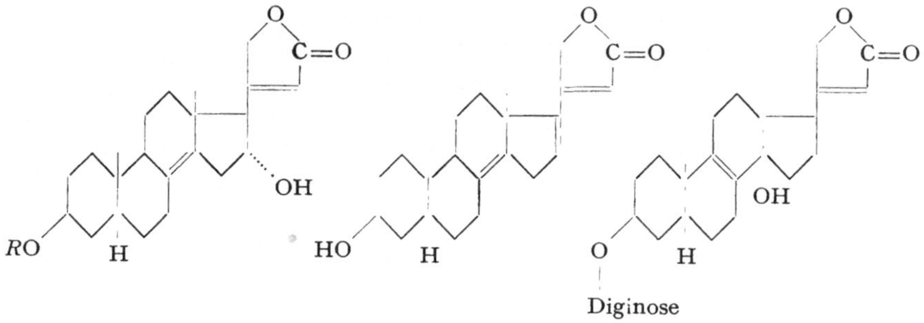

(XCVII.) (R = Oleandrose-rest.) (CVII.) (XCVI.) Adynerin [TSCHESCHE (*352*)].
(XCVIII.) (R = H) Adynerigenin [CARDWELL (*26*)].

Diginose

verd. HCl   konz. HCl

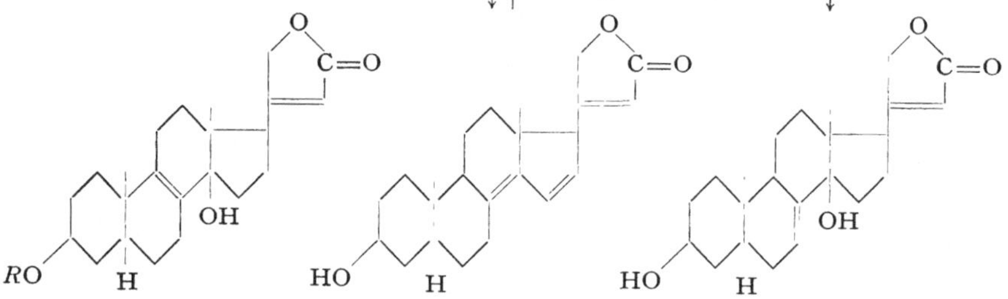

(XCIV.) (R = Oleandrose-rest?) (CII.) (XCIX.) Δ^7-Adynerigenin [TSCHESCHE (*352*)].
(XCV.) (R = H) Adynerigenin
  [TSCHESCHE (*339, 340*)].

verd. HCl H_2, Pt   verd. HCl

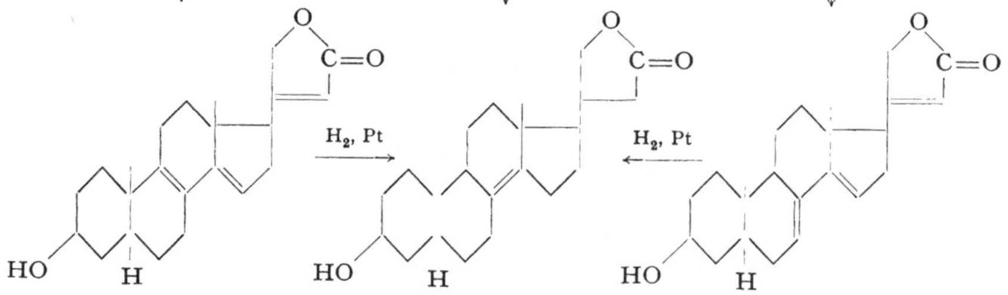

H_2, Pt H_2, Pt

(C.)   (CIII.) Dihydro-x-anhydro-digitoxigenin.   (CI.)

konz. HCl   konz. HCl

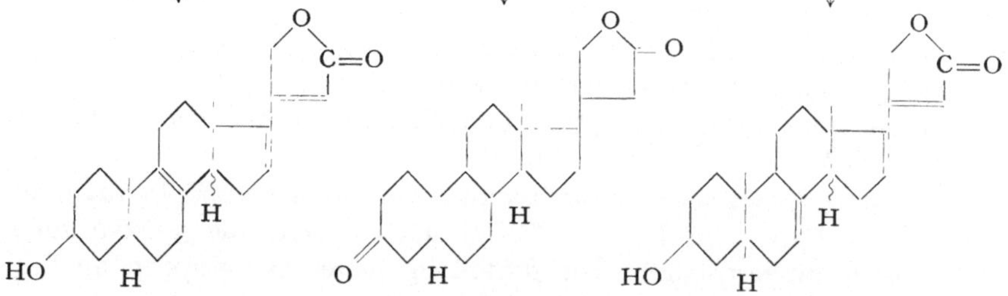

(CV.)   (CIV.)   (CVI.)

(XCVIII) die 16-Oxygruppe in Analogie zu Gitoxigenin als α-ständig. Es ist aber anzunehmen, daß sie β-ständig ist, da für Gitoxigenin die α-Stellung mit ziemlicher Sicherheit ausgeschlossen werden konnte (*187*). In allen drei Formeln, (XCV), (XCIX) und (XCVIII), wird zum Ausdruck gebracht, daß neben den zwei Sauerstoffatomen des Butenolidringes die beiden anderen O-Atome als HO-Gruppen vorliegen, von denen sich die eine bei der Bildung einer Isoverbindung mit dem Butenolidring beteiligen kann. Ferner liegt eine schwer hydrierbare, isolierte Doppelbindung vor, die nach den IR.-Spektren von Adynerin und Iso-adynerin ditertiär sein dürfte.

Nach Tschesche (*342, 352*) nimmt diese Doppelbindung in (XCVI) die 8,9-Stellung ein. Wird Adynerin mit Säure zum Genin hydrolysiert, was schon unter äußerst milden Bedingungen gelingt, so wandert sie nach Tschesche (*352*) in die 7-Stellung. Demnach ist das eigentliche Adynerigenin nicht bekannt, da bei der Hydrolyse nur das sog. Δ^7-Adynerigenin (XCIX) erhalten wird. Diese Annahme stützt sich auf die Farbreaktion nach Tortelli-Jaffé, die bei (XCVI) positiv, bei (XCIX) aber negativ ausfällt, ferner auf die IR.-Spektren von (XCIX) und (CI), die auf das Vorliegen eines trisubstituierten Äthylens deuten.

Adynerigenin geht nach Behandeln mit verdünnter HCl in Anhydro-adynerigenin über. Nach dem Spektrum (neues Maximum bei 247 mμ, log $\varepsilon = 4{,}1$) ist die neue Doppelbindung in Konjugation mit der ursprünglichen getreten. Alle drei Formulierungen für Anhydro-adynerigenin, nämlich (C), (CI) und (CII), wie sie sich aus den ursprünglichen Formelvorschlägen ergeben, sind damit vereinbar. Wird Anhydro-adynerigenin anschließend in der Kälte mit konz. HCl behandelt, so wandert die eine der Doppelbindungen in Konjugation zum Butenolidring, was im UV. das dafür typische Maximum bei 280 mμ (log $\varepsilon = 4{,}4$) hervorruft. Es entstehen die Isomeren (CV), (CVI) bzw. (CVII). Dieses Resultat scheint am ehesten für die Richtigkeit der Formel (CVII) zu sprechen, da sich die Doppelbindung in diesem Falle nur um 1 C-Atom verschieben muß.

Die katalytische Hydrierung von Anhydro-adynerigenin (C), (CI) oder (CII) mit Pt gibt Tetrahydro-anhydro-adynerigenin. Dieses ist sowohl nach Cardwell (*26*) wie auch nach Tschesche (*342*) mit Dihydro-α-anhydro-digitoxigenin (CIII) identisch. Damit ist das Steringerüst und die 3β-Stellung der einen Hydroxylgruppe für Adynerigenin bzw. Adynerin endgültig bewiesen, nachdem schon früher (*340*) durch die Überführung von Tetrahydro-anhydro-adynerigenin in Tetrahydro-anhydro-digitoxigenon (CIV) ein Hinweis für das Vorliegen des normalen Steringerüsts erhalten worden war. [Die Absättigung der Doppelbindung in 8 : 14-Stellung gelang durch katalytische Hydrierung mit Pt und HCl nach Windaus (*373*)]. Die Identität von Tetrahydro-anhydro-adynerigenin mit (CIII) scheint ebenfalls für die Richtigkeit der Formel (CII) für Anhydro-adynerigenin und somit auch für die Cardwellsche Adynerigeninformel (XCVIII) zu sprechen. Doch dürfen die Strukturen (C) und (CI) für Anhydro-adynerigenin auf Grund des Hydrierungsresultats nicht a priori ausgeschlossen werden. Es ist nämlich bekannt, daß bei solchen Hydrierungen die Doppelbindungen sich verschieben können *(272)*. Gegen eine Oxygruppe in 16-Stellung in Adynerigenin bzw. Adynerin (XCVIII) bzw. (XCVII) spricht der Befund von TSCHESCHE *(352)*, daß in Adynerin und Adynerigenin nur die eine der beiden HO-Gruppen acetylierbar ist. Die andere scheint eine tertiäre (eventuell eine schwer acetylierbare sekundäre) Hydroxylgruppe zu sein.

Adynerin wurde, wohl wegen des gemeinsamen Auftretens mit Oleandrin, früher willkürlich als Oleandrosid formuliert *(26, 339, 340)*. Doch TSCHESCHE *(342)* zeigte, daß die Zuckerkomponente dieses Glykosids die *D*-Diginose ist.

Aus den bisherigen Untersuchungen über Adynerin und Adynerigenin geht hervor, daß die Strukturvorschläge (XCVII) und (XCVIII) *(26)* kaum zutreffend sein dürften, daß aber auch die jüngsten Vorschläge *(352)* einer weiteren experimentellen Stütze bedürfen, insbesondere was die Lage der isolierten Doppelbindung betrifft.

Neriantogenin. Isomer zu Adynerigenin ist Neriantogenin $C_{23}H_{32}O_4$. Dieses Aglykon besitzt zwei acetylierbare Oxygruppen und eine zusätzliche isolierte Doppelbindung im Kern. Durch Wasserabspaltung mit Säure geht Neriantogenin in Dianhydro-gitoxigenin (LXXXVIII) über *(339, 340, 352)*. Damit sind das Steringerüst und eine 3β-ständige Hydroxylgruppe gesichert. Auf Grund dieser Tatsache schlug TSCHESCHE *(339, 340)* zuerst Formel (CVIII) vor. Die Vermutung von CARDWELL *(26)*,

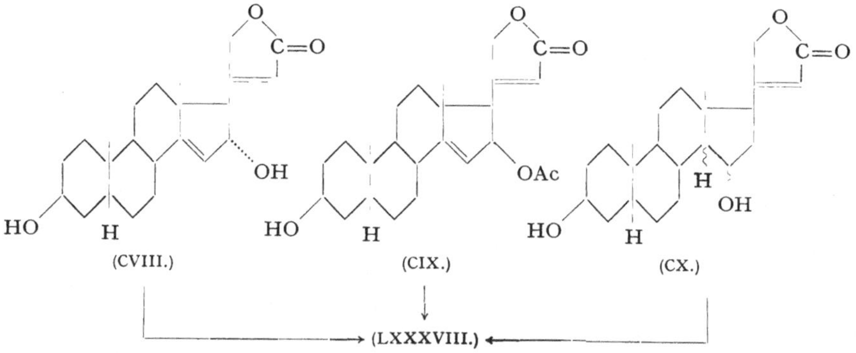

daß Neriantogenin mit Δ^{14}-Anhydro-oleandrigenin (CIX) identisch sei, erwies sich als unrichtig, da nach dem UV.-Spektrum [Maxima bei 221 mμ (log ε = 3,95) und 270 mμ (log ε = 4,31)] die Doppelbindung die 16-Stellung einnimmt (wie in Δ^{16}-Anhydro-gitoxigenin). (Interessanterweise verschwindet die kurzwellige Bande, wenn Neriantogenin acetyliert wird.) TSCHESCHE vermutet deshalb, daß die zweite Oxygruppe die 15-Stellung einnimmt, und schlägt für Neriantogenin die vorläufige Formel (CX) vor.

Aus dem acetylierten Gemisch der Aglykone von „Uzaron" erhielt
TSCHESCHE (*349*) u. a. ein neues Geninacetat $C_{25}H_{34}O_5$, das er *O-Acetyl-smalogenin* nannte.

Das freie Genin, dem die Formel $C_{23}H_{32}O_5$ zukommen müßte, ist noch
nicht bekannt. Auf Grund von orientierenden Versuchen schlägt
TSCHESCHE für O-Acetyl-smalogenin die Formel (CXI) vor. Die Anwesenheit eines Butenolid-rings stützt sich auf das UV.-Spektrum. Es

(CXI.)   (CXII.)

läßt keine weitere Carbonylfunktionen erkennen. Mit KOH liefert (CXI)
kein gewöhnliches 14,21-Cardanolid, sondern ein Lacton, dessen Spektrum
wahrscheinlich ein Maximum bei 267 mμ aufweist, und das mit Tri-
phenyl-tetrazoliumchlorid ein tiefrotes Triphenylformazan liefert (reaktive
Aldehydgruppe). Dem Lacton wird deshalb die provisorische Strukturformel (CXII) zuerteilt. Die Formeln (CXI) und (CXII) bedürfen aber
noch weiterer Beweise.

Xysmalogenin $C_{23}H_{34}O_4$, läßt sich ebenfalls aus der Uzarawurzel
isolieren (*341*). Es besitzt zwei sekundäre Hydroxylgruppen, die beide
mit CrO_3 dehydrierbar sind. Offenbar fehlt die übliche tertiäre Oxygruppe
in 14-Stellung. Da nur die eine der beiden HO-Gruppen acetylierbar
ist, nimmt TSCHESCHE $C_{(11)}$ als Haftstelle für die andere, schwer
acetylierbare an und schlägt für Xysmalogenin die Strukturformel (CXIII)
vor. Die Dehydrierung von (CXIII) würde demnach zum 3,11-Dehydro-
genin (CXV) und die des Monoacetats (CXIV) zu (CXVI) führen. Diese
Formeln sind aber keineswegs als gesichert zu betrachten. O-Monoacetyl-

(CXIII.) (*R* = H.)   (CXV.)   (CXVI.)
(CXIV.) (*R* = Ac.)

xysmalogenin (CXIV) spaltet mit H_2SO_4 oder $POCl_3$ in Pyridin Wasser ab. Es entsteht ein Gemisch von Monoanhydro-xysmalogenin-acetaten, in denen die Lage der neuen Doppelbindung nicht festgestellt ist.

(CXVII.) ($R = $ H.) Tanghinigenin.
(CXVIII.) ($R = $ Ac.)

(CXXII.)

(CXXIII.)

Desacetyl-tanghinin

(CXIX.) Tanghinigenon.

(CXX.) ($R = $ H.) 3-Epi-tanghinigenin.
(CXXI.) ($R = $ Ac.)

(CXXXII.) 3α-Acetoxy-ätien-(7)-säure-methylester.

(CXXVI.)

(CXXIV.)

(CXXXIII.)

(CXXV.)

(CXXVII.)

(CXXX.)

$$\text{(CXXVIII.)} \qquad \text{(CXXIX.)} \qquad \text{(CXXXI.)}$$

Tanghinigenin und *3-Epi-tanghinigenin* $C_{23}H_{32}O_5$ besitzen nach
SIGG, TAMM und REICHSTEIN (*272*) die Formeln (CXVII) bzw. (CXX). In
diesen Formeln ist das Vorliegen des Steringerüsts, die Lage der beiden
Hydroxylgruppen sowie der Seitenkette sicher bewiesen. Für die Lage
des Oxydrings fehlt noch der endgültige Beweis. Die Gewinnung des
intakten Tanghinigenins war anfänglich durch die ungewöhnlich schwere
Spaltbarkeit des Tanghinins bzw. Desacetyl-tanghinins verunmöglicht
(*63, 85*). Durch oxydativen Abbau des Zuckerrests (*286*) in Desacetyl-
tanghinin gelang es, Tanghinigenon (CXIX) zu gewinnen (*85*). Später
wurde das intakte Genin (CXVII) durch die energische Spaltung des
Glykosids mit HCl-Gas in Chloroform gut zugänglich (*271*). Die Dehydrierung des Genins (CXVII) mit CrO_3 gibt Tanghinigenon (CXIX).
Durch Reduktion von (CXIX) mit $NaBH_4$ entsteht fast ausschließlich
3-Epi-tanghinigenin (CXX), das sich mit CrO_3 wieder zu Tanghinigenon
(CXIX) zurückoxydieren läßt (*271*). Desacetyl-tanghinin liefert mit
KOH keine Isoverbindung (*61*), sondern eine Säure (*271*), da die Oxygruppe an $C_{(14)}$ durch den vermutlichen Oxydring gehindert ist. Der Abbau von O-Acetyl-tanghinigenin (CXVIII) (*272*) mit Ozon usw. führt zum
Ketol (CXXII), das mit CrO_3 das Ketolacton (CXXIII) liefert. Die
Richtigkeit der Struktur von (CXXIII) wird durch die UV.- und IR.-Spek-
tren erhärtet. Die Bildung des Ketolactons beweist, daß der Lactonring
und die tertiäre Oxygruppe sich auf derselben Seite des Ringsystems
(in β-Stellung) befinden. Der weitere Abbau des Ketols (CXXII) führt
zum Ätiansäure-ester (CXXV). Der analoge Abbau des Epimeren (CXXI)
gibt den Ätiansäure-ester (CXXIV) (*272*). Die beiden Ester (CXXIV)
und (CXXV) liefern nach Verseifung, Remethylierung und Dehydrierung
mit CrO_3 denselben Ketoester (CXXVI), wodurch bewiesen ist, daß sie
sich nur durch Epimerie an $C_{(3)}$ unterscheiden. $LiAlH_4$ greift den Oxydring nicht an.

Die Reaktion von $SOCl_2$ in Pyridin auf den Ester (CXXIV) hat eine
Wasserabspaltung und Öffnung des Oxydrings zur Folge. Es entsteht
das konjugierte heteroannulare Dienol (CXXVII) ($\lambda_{max} = 250\,m\mu$,
$\log \varepsilon = 4{,}4$). Durch Dehydrierung mit CrO_3 in Pyridin geht (CXXVII)
in das Dienon (CXXX) über ($\lambda_{max} = 300\,m\mu$, $\log \varepsilon = 4{,}36$). Wird nach
der $SOCl_2$-Behandlung des Esters (CXXXIV) das resultierende Roh-

produkt direkt katalytisch mit Pt hydriert, so entsteht ein 3α-Acetoxy-ätiensäure-methylester, dessen Doppelbindung auf Grund des molekularen Drehungsbeitrags höchstwahrscheinlich die $8:9$-Stellung einnimmt (CXXXIII). Der gleiche Ätiensäure-ester (CXXXIII) entsteht bei der katalytischen Hydrierung des 3α-Acetoxy-ätien-(7)-säure-methylesters (CXXXII) mit Pt in Eisessig. Durch diese Identität sind, unabhängig von der wirklichen Lage der Doppelbindung, das Steringerüst, die 3β-Oxy-5β-Konfiguration und die 17β-Stellung des Butenolidrings für Tanghinigenin (CXVII) eindeutig bewiesen.

Der epimere 3β-Acetoxy-ätiansäure-methylester (CXXV) gibt nach Behandlung mit $SOCl_2$ und Hydrierung den der Formel (CXXXII) entsprechenden Ätiensäure-ester. Daneben entsteht ein Ketosäure-ester, dessen Ketongruppe nach den UV.- und IR.-Spektren die 7-Stellung einnimmt. Der Ester ist aber mit dem bekannten 3β-Acetoxy-7-keto-ätiansäure-methylester nicht identisch. Die Konfiguration an $C_{(8)}$, $C_{(9)}$ oder $C_{(14)}$ dürfte somit „anormal" sein, wie es durch die Formel (CXXVIII) zum Ausdruck kommt.

Wird der 3β-Oxy-ätiansäure-methylester (CXXV, aber mit freier HO-Gruppe) mesyliert, so entsteht bei der Umsetzung des Mesylats mit NaJ und Reduktion des Jodids mit Zn in Äthanol-Eisessig in einer sehr merkwürdigen, noch nicht abgeklärten Reaktion der um zwei O-Atome ärmere Ester (CXXIX). Die Lage der α,β-ungesättigten Ketongruppierung ergibt sich aus dem UV.-Spektrum (Maxima bei 260 mμ, $\log \varepsilon = 4{,}00$ und 318 mμ, $\log \varepsilon = 1{,}99$). Die Abspaltung der tertiären HO-Gruppe und die Aufsprengung des Oxydrings müssen mit der Mesylierung der 3-Oxygruppe gekoppelt sein, denn die entsprechenden 3-Acetoxy-verbindungen verändern sich bei diesen Reaktionen nicht. Verseifung von (CXXIX), Reduktion der entstandenen Säure mit Li in NH_3 und n-Propanol, Remethylierung und Rückoxydation mit CrO_3 ergeben den 7-Ketosäureester (CXXXI), der anormale Konfiguration an $C_{(8)}$ und $C_{(9)}$ besitzen muß, da er vom 7-Keto-ätiansäure-methylester verschieden ist.

Das fünfte Sauerstoffatom des Tanghinigenins (CXVII) muß ätherartig gebunden sein, da das Ketolacton (CXXIII) weder eine Hydroxylgruppe noch eine Ketogruppe enthält. Die $7\alpha,15\alpha$-Stellung scheint am wahrscheinlichsten, da sie die schwere Abspaltbarkeit der tertiären Oxygruppe an $C_{(14)}$ erklärt. Sie ist aber noch nicht bewiesen.

Abogenin (CXXXV) wird durch milde saure Hydrolyse des Abomonosids (CXXXIV) erhalten (*88*). Es besitzt höchstwahrscheinlich die Bruttoformel $C_{25}H_{36}O_6$. Bei der Acetylierung entsteht das Acetylderivat (CXXXVI). Neben Abogenin (CXXXV) entsteht bei der Spaltung des Glykosids (CXXXIV) Anhydro-abogenin als Nebenprodukt, dem mit Vorbehalt die Formel $C_{25}H_{32}O_5$ zuerteilt wird. Da Anhydro-abogenin nicht acetylierbar ist und nach dem UV.-Spektrum und den molekularen

(CXXXIV.) (R = Cymaroserest.) Abomonosid.
(CXXXV.) (R = H.) Abogenin.
(CXXXVI.) (R = Ac.)

(CXXXVII.) Anhydro-abogenin (?).

Drehungswerten eine 3,5-Diengruppierung aufweist, wird ihm die vorläufige Strukturformel (CXXXVII) zuerteilt. Dies läßt vermuten, daß Abogenin eine isolierte Doppelbindung in 4- oder 5-Stellung besessen haben muß.

Allo-glaucotoxigenin (CXXXVIII) besitzt nach den Untersuchungen von Stoll (*302*) die Bruttoformel $C_{23}H_{32}O_6$ und die vorläufige Strukturformel (CXXXVIII). Es bildet ein Monoxim und das Diacetylderivat (CXXXIX), das bei der Oxydation mit CrO_3 eine Monocarbonsäure liefert. Bei der Einwirkung von methanolischer HCl wird das Cyclo-halbacetal (CXL) erhalten, was in Analogie zu Strophanthidin (LXIV, S. 163) als

(CXXXVIII.) (R = H.) Allo-glaucotoxigenin.
(CXXXIX.) (R = Ac.)

(CXL.)

Beweis für die 3β-ständige Hydroxylgruppe und die Aldehydgruppe an $C_{(10)}$ dient. Allo-glaucotoxigenin (CXXXVIII) liefert mit Alkali eine Isoverbindung, in der nur noch eine der Hydroxylgruppen acetylierbar ist. Daraus wird geschlossen, daß (CXXXVIII) nicht an $C_{(14)}$, sondern an $C_{(16)}$ eine Hydroxylgruppe besitzt, die reagiert hat.

Da das Genin biologisch unwirksam ist, wird der Butenolidring 17α-ständig und somit auch die Oxygruppe an $C_{(16)}$ α-ständig formuliert. Ein Abbau, um diese Annahmen zu beweisen, ist bisher nicht durchgeführt worden.

Von dem mit (CXXXVIII) isomeren *Glaucorigenin* (*302*) ist lediglich bekannt, daß es ein Monoxim und ein Diacetylderivat bildet.

Sarmutogenin $C_{23}H_{32}O_6$ und *Caudogenin* $C_{23}H_{32}O_6$. Schindler und Reichstein konnten für Sarmutogenin die Strukturformel (CXLVI) (*227*, *258*) und für Caudogenin (CXLII) (*258*) wahrscheinlich machen. Caudogenin und Sarmutogenin zeigen in ihrem UV.-Spektrum neben der intensiven Bande des Butenolidrings bei 217 mμ eine weitere bei 295 mμ (log ε = 1,76) bzw. 287 mμ (log ε = 1,89), die von einer isolierten Keto-gruppe herrührt. Diese Gruppe liegt in beiden Geninen als Ketol-gruppierung in der 11,12-Stellung vor, denn sowohl (CXLII) wie auch (CXLVI) geben bei der Dehydrierung mit CrO_3 Sarmutogenon (CXLVII), einen hellgelben Neutralstoff. Der Beweis für das Vorliegen einer 11,12-Diketogruppierung in (CXLVII) stützt sich zunächst auf das Spektrum (Maxima der schwachen Ketobanden bei 285 mμ, log ε = 2,35 und 370 mμ, log ε = 1,32), das gut mit demjenigen des 11,12-Diketo-cholan-säure-methylesters übereinstimmt.

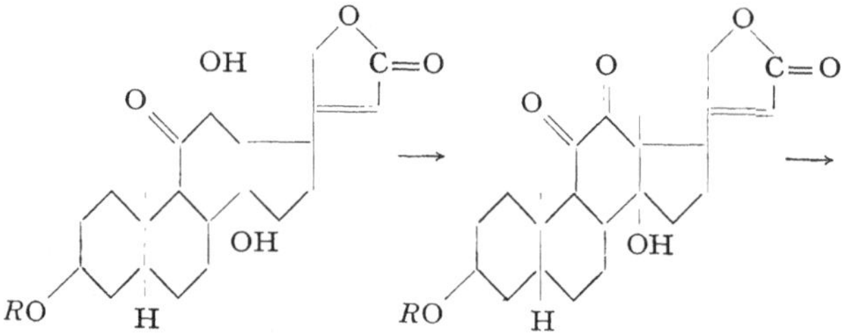

(CXLV.)
Acetyl-caulutogenin

(CXLI.) (R = Oleandroserest.) Caudosid.
(CXLII.) (R = H) Caudogenin.

(CXLIII.) (R = Oleandroserest.) Decosid.
(CXLIV.) (R = H.) Decogenin.

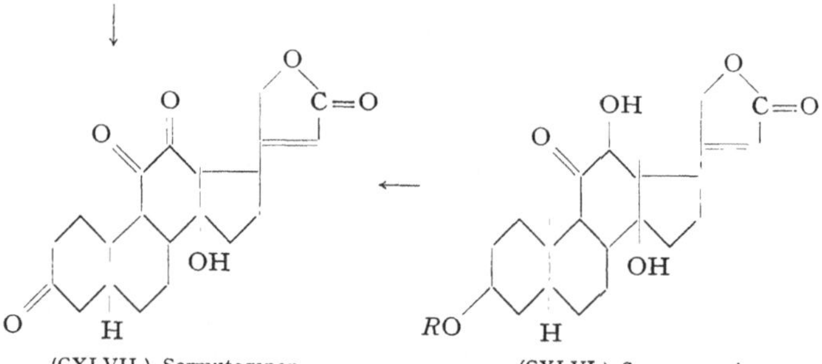

(CXLVII.) Sarmutogenon.

(CXLVI.) Sarmutogenin.

Durch vorsichtige Dehydrierung von Caudosid (CXLI) mit CrO_3 oder Cupriacetat werden Desocid (CXLIII) (Zuckerrest intakt) und durch milde saure Hydrolyse von (CXLIII) *Decogenin* (CXLIV) erhalten (*248*). (CXLIII) und (CXLIV) zeigen im UV. eine sehr ähnliche Absorption wie Sarmutogenon (CXLVII). Zu erwähnen ist, daß Decogenin (CXLIV) ein Monoacetylderivat liefert, dessen UV.-Spektrum stark von demjenigen von (CXLIV) abweicht und das deshalb als *Acetyl-caulutogenin* (CXLV) be-zeichnet wird. (CXLVII) lagert sich erst in Gegenwart von Basen in die

Enolform um (Nachweis mit $FeCl_3$), was für 11,12-Diketo-steroide typisch ist, während sich hexacyclische o-Diketone sonst spontan, ohne Anwesenheit von Basen umlagern (Diosphenole). Damit ist gezeigt, daß Caudogenin (CXLII) und Sarmutogenin (CXLVI) sich nur durch Isomerie der Substituenten in der 11,12-Stellung unterscheiden, wofür vier Möglichkeiten bestehen.

Die Ketogruppe beider Genine ist reaktionsträge, wie dies für 11-Keto- und 12-Keto-14β-oxy-steroide zu erwarten ist. Die beiden sekundären Hydroxylgruppen lassen sich in beiden Geninen leicht acetylieren, womit die 11β-Oxy-12-keto-Gruppierung ausscheidet. Aus dem Vergleich der schwachen Ketobande der Genine und ihrer Acetylderivate mit den UV.-Spektren der vier isomeren 3α-Oxy-cholansäuren mit 11,12-Ketol-Gruppierung resp. 11,12-Ketolacetat-Gruppierung (*11*) ergibt sich, daß mit großer Wahrscheinlichkeit Caudogenin (CXLII) die 11-Keto-12α-oxy-Gruppierung und Sarmutogenin (CXLVI) die 11-Keto-12β-oxy-Gruppierung besitzen. Doch bedürfen diese Zuordnungen wie auch das Steringerüst noch eines sicheren Beweises*.

Sarverogenin, *Inertogenin* und *Leptogenin* besitzen die Bruttoformel $C_{23}H_{30}O_7$ und unterscheiden sich wie Caudogenin (CXLII) und Sarmutogenin (CXLVI) durch Isomerie in der 11,12-Ketol-Gruppierung. Das Vorliegen einer isolierten Ketogruppe ergibt sich aus den UV.-Spektren (*82*), die neben der üblichen Bande des Butenolidrings bei 217 mμ noch solche im Bereich von 279—295 mμ aufweisen. Die von TAYLOR (*324, 325*) für Sarverogenin vorgeschlagene Formel (CXLIX) wird durch HEGEDÜS, TAMM und REICHSTEIN (*82*) gesichert, was die 11,12-Ketolgruppierung betrifft. Für Inertogenin und Leptogenin schlagen die gleichen Autoren die Teilformel (CLI) und (CLIII) vor.

Diese Vorschläge stützen sich auf die folgenden Befunde.

Von den sieben O-Atomen sind zwei im Butenolidring festgelegt; ein weiteres liegt in Form einer reaktionsträgen Ketogruppe vor. Alle drei Genine liefern Diacetylderivate, die gegen CrO_3 beständig sind. Sarverogenin gibt auch ein kristallisiertes Dibenzoat (*23*). Dies zeigt gleichzeitig, daß zwei weitere acetylierbare sekundäre Hydroxylgruppen anwesend sind. Für die verbleibenden beiden Hydroxylgruppen kommt zunächst aus Analogiegründen eine tertiäre Oxygruppe an $C_{(14)}$ in Frage. Eine zweite tertiäre Oxygruppe an $C_{(5)}$ konnte ausgeschlossen werden, da das aus Di-O-acetyl-intermediosid durch saure Hydrolyse gewonnene 11-Monoacetyl-sarverogenin mit freier Oxygruppe an $C_{(3)}$ bei der Dehydrierung mit CrO_3 und Behandlung mit Eisessig keine Δ^4-ungesättigte

---

* *Anmerkung bei der Korrektur:* KÜNDIG-HEGEDÜS und SCHINDLER (*156 a*) bewiesen die Konstitution des Sarmutogenins (CXLVI) und des Caudogenins (CXLII) durch den Abbau von (CXLVI) zum $3\beta,12\beta$-Diacetoxy-11-keto-ätiansäuremethylester.

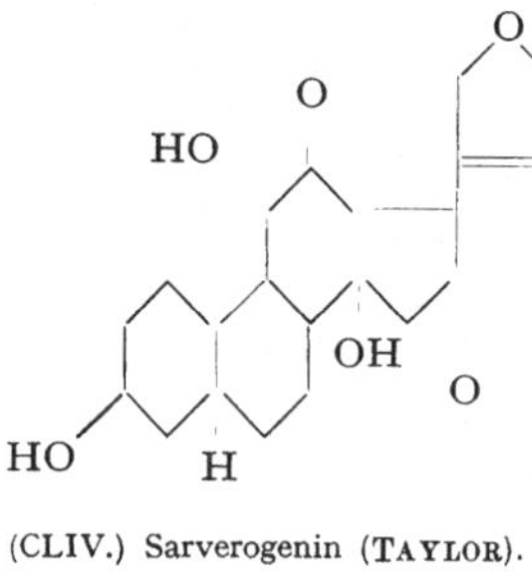

(CLIV.) Sarverogenin (TAYLOR).

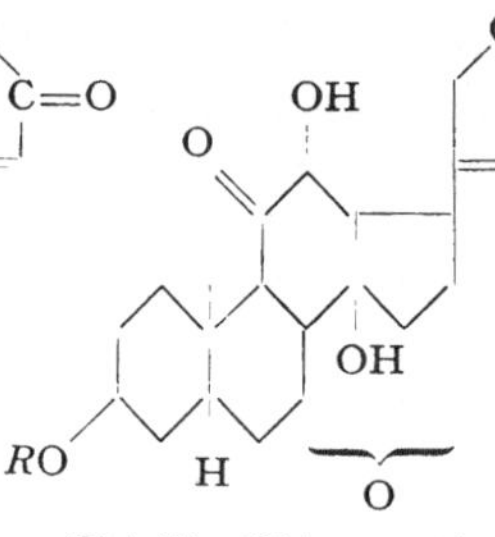

(CL.) (R = Diginose-rest.)
(CLI.) (R = H.) Inertogenin.

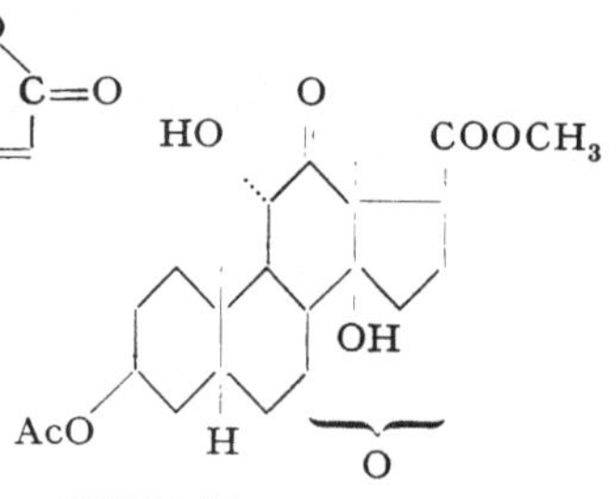

(CLVII.) Ätiansäure-methylester.

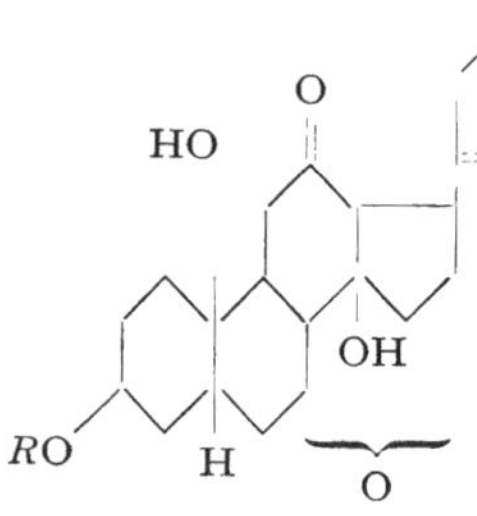

(CXLVIII.) (R = Diginose-rest.) Intermediosid.
(CXLIX.) (R = H.) Sarverogenin.

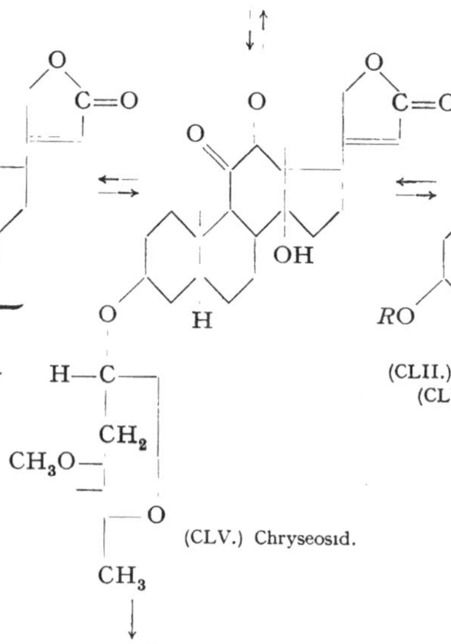

(CLV.) Chryseosid.

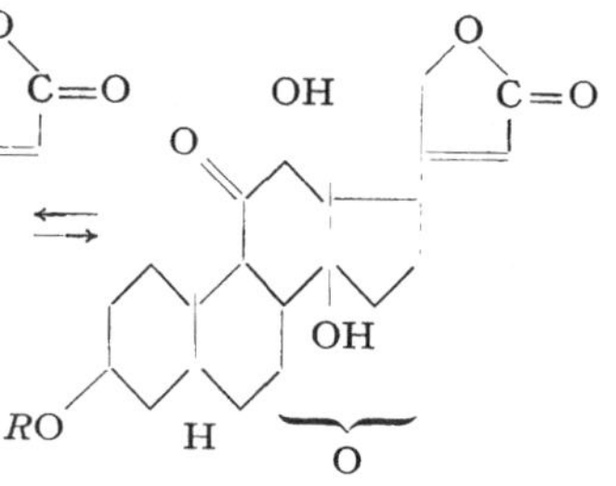

(CLII.) (R = Diginose-rest.) Leptosid.
(CLIII.) (R = H.) Leptogenin.

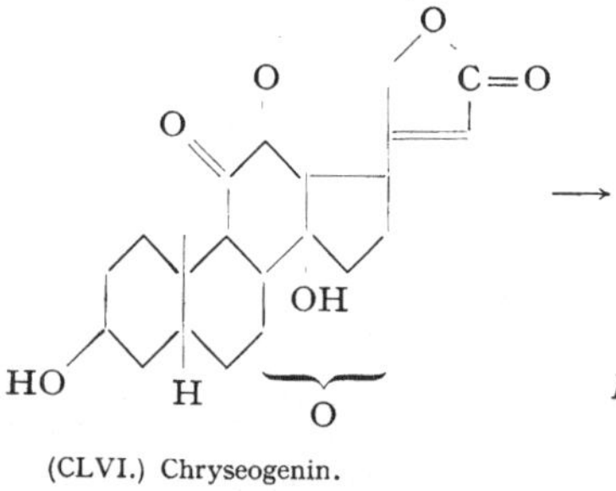

(CLVI.) Chryseogenin.

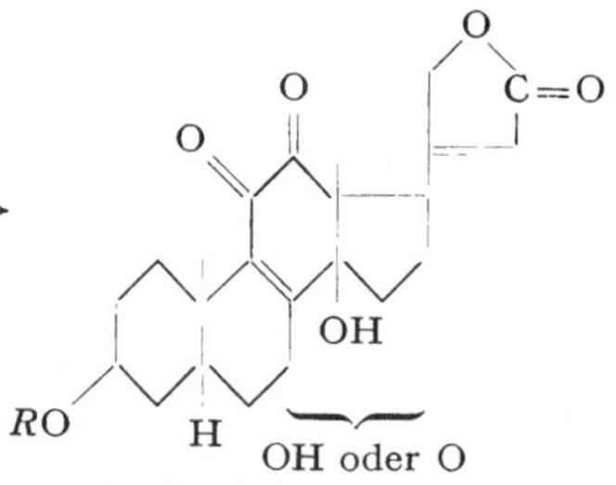

(CLVIII.) (R = H.) Flavogenin.
(CLVIX.) (R = Ac.) O-Acetyl-flavogenin.

3-Ketongruppierung ausbildete (*82*). Das siebente Sauerstoffatom scheint am ehesten ätherartig gebunden als inerter Oxydring vorzuliegen, wie es z. B. in Formel (CLIV) angedeutet ist. Dies wäre eine Erklärung für die große Stabilität von Sarverogenin (CXLIX) gegenüber Mineralsäuren (*324*).

Für den Nachweis der Ketolgruppierung in den Geninen (CXLIX), (CLI) und (CLIII) eignet sich die direkte Dehydrierung mit CrO_3 nicht, die bei Caudogenin (CXLII) und Sarmutogenin (CXLVI) zu Sarmutogenon (CXLVII) führt, da dabei nur in Spuren ein neutrales, kristallisiertes Oxydationsprodukt (Sarverogenon) entsteht (*23*). Hingegen gelingt es, die 11,12-Ketolgruppierung in den freien Glykosiden partiell mit CrO_3 oder Cupriacetat zu dehydrieren (*82*). Intermediosid (CXLVIII), Inertosid (CL) und Leptosid (CLII) geben alle das gleiche gelbe Dehydrierungsprodukt *Chryseosid* (CLV). Die Diketogruppierung enolisiert erst in Gegenwart von Alkali. (CLV) besitzt somit eine 11,12-Diketogruppierung. Damit steht auch das UV.-Spektrum im Einklang. Intermediosid (CXLVIII), Inertosid (CL) und Leptosid (CLII) und somit auch die Genine dieser Glykoside unterscheiden sich nur durch Isomerie in der 11,12-Ketolgruppierung. Diese Schlußfolgerung wird weiter durch die Tatsache gestützt, daß (CLV) bei der Reduktion mit Al-Amalgam ein Gemisch der drei ursprünglichen Glykoside (CXLVIII), (CL) und (CLII) entsteht, das auch präparativ durch Verteilungschromatographie getrennt worden ist.

Aus dem Ergebnis der Acetylierung, dem Vergleich der molekularen Drehungen und der UV.-Spektren der Genine und ihrer Acetylderivate (Maxima bei 215 mμ, log ε = 4,08; Schulter bei 265—275 mμ, log ε = = 2,04—2,0) mit den Werten ähnlich gebauter Cholansäurederivate (*11*), folgt mit großer Wahrscheinlichkeit, daß Sarverogenin (CXLIX) ein 11α-Oxy-12-keto-cardenolid, Inertogenin (CLI) ein 11-Keto-12β-oxy-cardenolid und Leptogenin (CLIII) ein 11-Keto-12α-oxy-cardenolid ist (*82*). Der Vergleich des Verhaltens dieser Genine gegenüber $NaJO_4$ mit den entsprechenden 11,12-Ketolen der Cholansäurereihe unterstützt diese Zuordnung (*82*).

Die Abspaltung des Zuckerrests in Chryseosid (CLV) liefert *Chryseogenin* (CLVI) (*82*), das auch durch Dehydrierung von Sarverogenin (CXLIX) mit Cupriacetat gewonnen werden kann. (CLVI) scheint nicht sehr stabil zu sein, da es schon bei der Acetylierung in Pyridin-Eisessig bei 20° ein Mono-acetylderivat liefert, dessen UV.-Spektrum sich stark geändert hat und dessen Entstehung am besten durch Überlagerung einer —CO—CO—C=C—Gruppierung mit dem Butenolidring erklärt werden kann, wie sie in den Formeln (CLVIII) und (CLIX) zum Ausdruck gebracht ist. (CLVIII) wird als *Flavogenin* bezeichnet. (CLIX) ist gegenüber Eisessig beständig und bildet in Gegenwart von Basen kein Enol mehr.

Zu erwähnen ist, daß O-Diacetyl-sarverogenin mit Ozon usw. zum Ätiansäure-methylester (CLVII) abgebaut worden ist (*321*). Bisher ist es aber noch nicht gelungen, das Steringerüst, eine 14β-Oxygruppe und den Oxydring in den drei Geninen sicher zu beweisen*.

Antiarigenin und al-Dihydro-antiarigenin. Die Hauptglykoside des Milchsafts von *Antiaris toxicaria* LESCH. sind α-Antiarin (CLX) und β-Antiarin (CLXI). Die beiden Glykoside unterscheiden sich nur im Zuckeranteil voneinander (*34, 35, 147, 149*); sie besitzen somit das gleiche Genin (CLXIV). Antiarigenin (CLXIV) ist aber nicht mit Sicherheit bekannt (*35*). Ihm kommt nach den vorliegenden Untersuchungen die Bruttoformel $C_{23}H_{32}O_7$ zu.

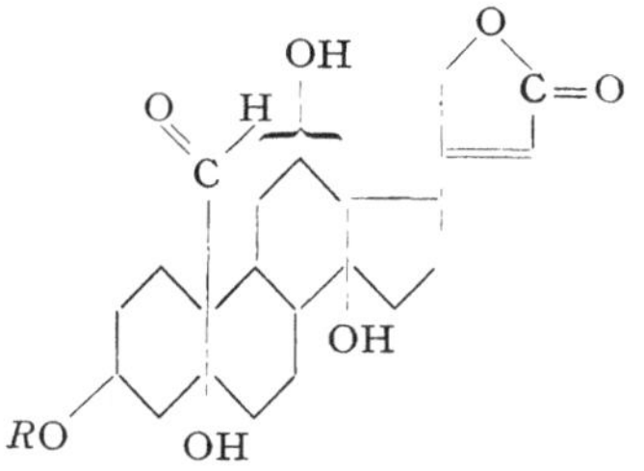

(CLX.) (R = Rhamnoserest.) α-Antiarin.
(CLXI.) (R = Gulomethyloserest.) β-Antiarin.

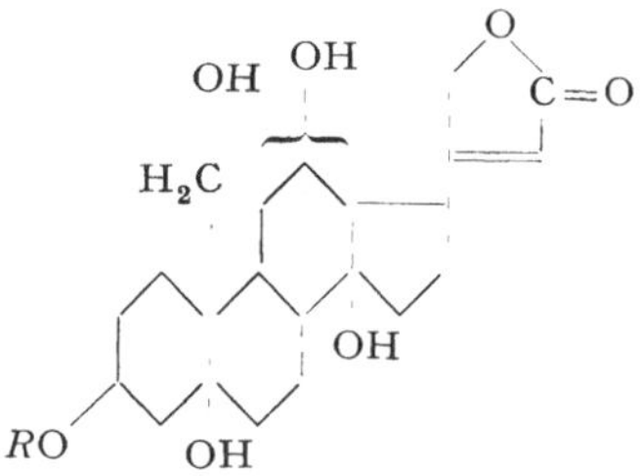

(CLXII.) (R = Rhamnoserest.) al-Dihydro-α-antiarin.
(CLXIII.) (R = Gulomethyloserest.) al-Dihydro-β-antiarin.

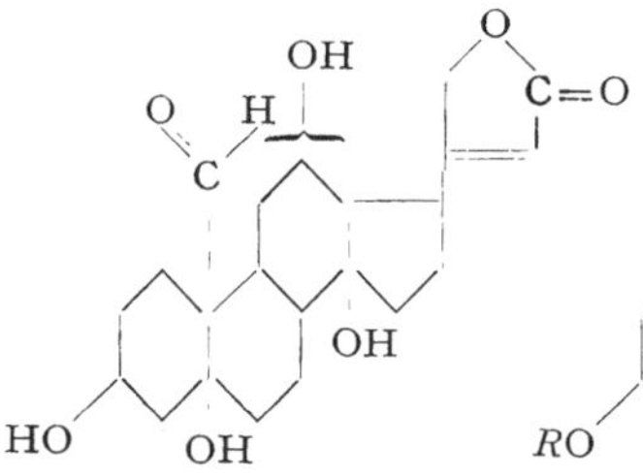

(CLXIV.) Antiarigenin.

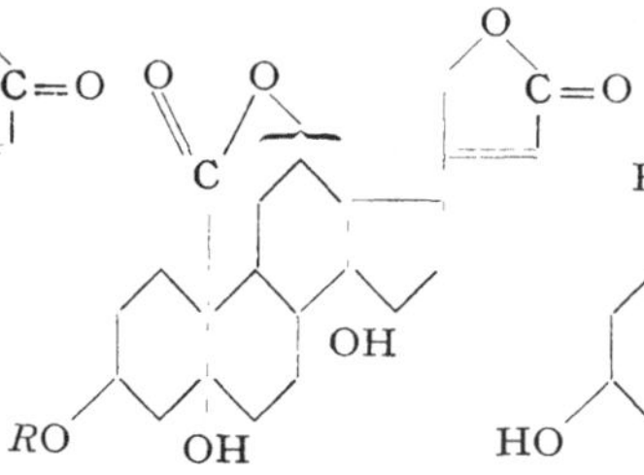

(CLXV.)

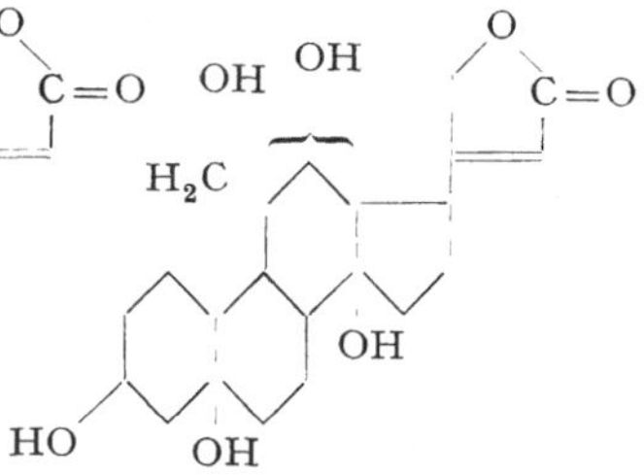

(CLXVI.) al-Dihydro-antiarigenin.

Wegen der schweren Spaltbarkeit der Glykoside sind die meisten Untersuchungen nicht am Genin selbst durchgeführt worden. Antiarigenin besitzt eine Oxogruppe, denn α-Antiarin liefert ein Oxim (*147*). Beide Antiarine weisen im Spektrum neben dem Maximum des Butenolidrings ein weiteres bei 301 mμ (log ε = 1,46) auf (*34, 35*). Am ehesten dürfte eine Aldehydgruppe an $C_{(10)}$ vorliegen. Durch Benzoylierung von (CLX) und (CLXI) werden in beiden Fällen Tri-O-benzoyl-derivate erhalten, die noch eine freie sekundäre Hydroxylgruppe im Geninteil besitzen. Mit CrO_3 in Eisessig liefern diese beiden Derivate keine Säure, sondern einen

---

* *Anmerkung bei der Korrektur:* SCHINDLER (*249 a*) bewies das Vorliegen der tertiären HO-Gruppe an $C_{(14)}$. Sie ist *cis*-ständig zum Butenolidring angeordnet. Ferner wird das Vorliegen eines Tetrahydrofuranringes, der an $C_{(15)}$ haftet, als sehr wahrscheinlich angenommen.

Neutralstoff. Offenbar liegt die Aldehydgruppe nicht frei vor, sondern bildet mit der nicht benzoylierten Oxygruppe ein cyclisches Lactol, das bei der Dehydrierung in ein Lacton übergeht, entsprechend (CLXV). Die beteiligte Oxygruppe muß sich deshalb in der Nähe der ursprünglichen Aldehydgruppe befinden. Die 16-Stellung konnte mit Sicherheit ausgeschlossen werden (35). Die Existenz der Aldehydgruppe findet eine weitere Stütze in der glatten Reduktion von α-Antiarin (CLX) zu al-Dihydro-α-antiarin (CLXII) mit Al-Amalgam (34) und in der analogen, noch glatter verlaufenden Reduktion von β-Antiarin (CLXI) zu al-Dihydro-β-antiarin (CLXIII) mit $NaBH_4$ (35). (CLXIII) gibt erwartungsgemäß ein Pentacetyl- und ein Pentabenzoylderivat, die beide gegen CrO_3 beständig sind. Aus al-Dihydro-β-antiarin (CLXIII) läßt sich im Gegensatz zu (CLXI) durch Spaltung mit HCl in Aceton relativ glatt al-Dihydro-antiarigenin (CLXVI) gewinnen. (CLXVI) liefert ein Triacetyl- und Tribenzoylderivat, die beide gegen CrO_3 beständig sind.

· Das Vorliegen einer 14β-Oxygruppe ergibt sich aus der Bildung eines Ketolactons beim Abbau von Penta-O-acetyl-al-dihydro-β-antiarin mit $KMnO_4$ (35). Die siebente Sauerstoffunktion dürfte am ehesten eine tertiäre Oxygruppe an $C_{(5)}$ sein. Doch sind alle Strukturformeln provisorisch, da die Verknüpfung der Antiarine mit einem bekannten Steroid noch nicht gelungen ist.

Nigrescigenin $C_{23}H_{32}O_7$ (247). Nach dem UV.-Spektrum besitzt dieses Genin eine isolierte Carbonylgruppe. Es liefert ein Diacetyl- und ein Dibenzoylderivat. Das Diacetylderivat gibt mit CrO_3 in Eisessig vorwiegend saure Anteile. Die neutralen Anteile sind vom Ausgangsmaterial verschieden. Weitere Untersuchungen sind bisher nicht durchgeführt worden.

Ouabagenin $C_{23}H_{34}O_8$ (103, 105). Es ist das sauerstoffreichste der bisher bekannten Cardenolide. Nach einem Vorschlag von MANNICH und SIEWERT (168) besitzt es die Strukturformel (CLXXI), wobei die im Formelbilde nicht festgelegte Oxygruppe die 11-Stellung einnimmt. Von den acht Sauerstoffatomen sind zwei im Butenolidring festgelegt, was aus der Bildung von Iso-ouabain und Dihydro-ouabain, aus Ouabain, dem Rhamnosid des Ouabagenins, hervorgeht (105). Ouabagenin und seine Derivate zeigen das für Cardenolide typische UV.-Spektrum (185, 211). Aus der Bildung der Isoverbindung wird auf das Vorliegen einer tertiären Oxygruppe an $C_{(14)}$ geschlossen (103, 105). Die weiteren fünf O-Atome liegen ebenfalls als alkoholische Hydroxylgruppen vor. Vier von ihnen sind acetylierbar; denn Ouabagenin gibt ein Tetracetyl-derivat (CLXXIII), das gegen CrO_3 in Eisessig beständig ist (185). Daneben entsteht das Triacetylderivat (CLXXII) (Diskussion der Struktur siehe unten). Die fünfte und letzte Oxygruppe ist tertiärer Natur. Ihre Haftstelle dürfte in Analogie zu Strophanthidin $C_{(5)}$ sein. Da Ouabagenin weder von

Pb-tetracetat (*168*) noch von Perjodsäure (*278*) angegriffen wird, ist eine 1,2-Glykolgruppierung unwahrscheinlich [vgl. auch (*33, 320*)]. Die Resultate neuerer Untersuchungen von verschiedenen Arbeitskreisen sprechen stark dafür, daß von den sechs Hydroxylgruppen fünf an den C-Atomen 1, 3, 5, 14 und 19 und die sechste höchstwahrscheinlich am $C_{(11)}$ haften, wie es zum Teil FIESER (*53*) und besonders MANNICH (*168*), ohne es weiter zu begründen, vermutet hatten.

Ouabagenin (CLXXI) wird aus Ouabain (CLXVII) durch Abspaltung des Rhamnoserestes mit HCl in Aceton gewonnen (*168*). Es entsteht dabei zunächst Ouabagenin-monoacetonid. SNEEDEN und TURNER (*278*) zeigten, daß das als Nebenprodukt entstehende sog. Anhydro-ouabagenin mit Ouabagenin-acetonid identisch ist; ebenso ist das Acetylierungsprodukt „Anhydro-ouabagenin-diacetat" mit Di-O-acetyl-ouabagenin-monoacetonid identisch. Die Acetongruppe ist mit verdünnter Essigsäure abspaltbar, so daß Ouabagenin sehr leicht aus dem Monoacetonid erhältlich ist. Aus Di-O-acetyl-ouabagenin-monoacetonid (CLXIX) entsteht „Ouabagenin-β-diacetat" (*211*). Nach Einwirkung von Mineralsäure erhält man aber wegen Acylwanderung vorwiegend das isomere „Ouabagenin-α-diacetat" (*211, 278*). In Ouabagenin-monoacetonid nimmt die Acetongruppe nach TAMM (*320*) die 1,19-Stellung ein entsprechend (CLXVIII),

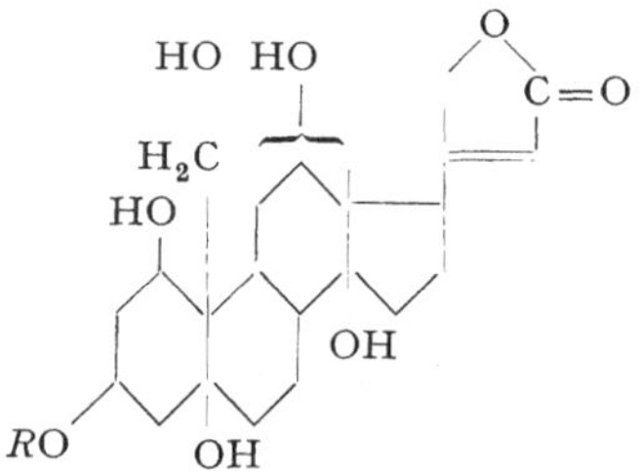

(CLXVII.) (R = Rhamnoserest.) Ouabain.

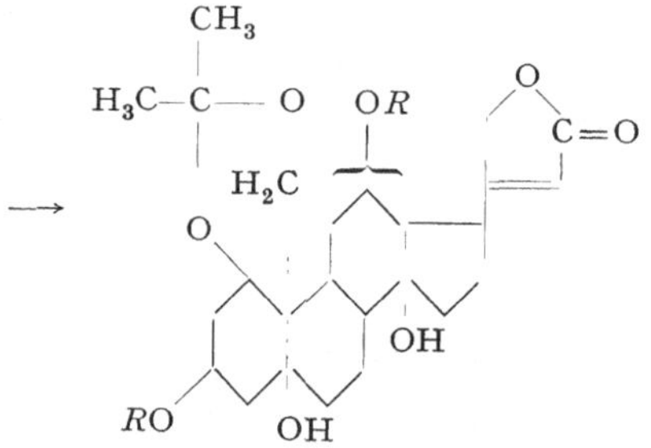

(CLXVIII.) (R = H.)
(CLXIX.) (R = Ac.) Di-O-acetyl-ouabagenin-monoacetonid.

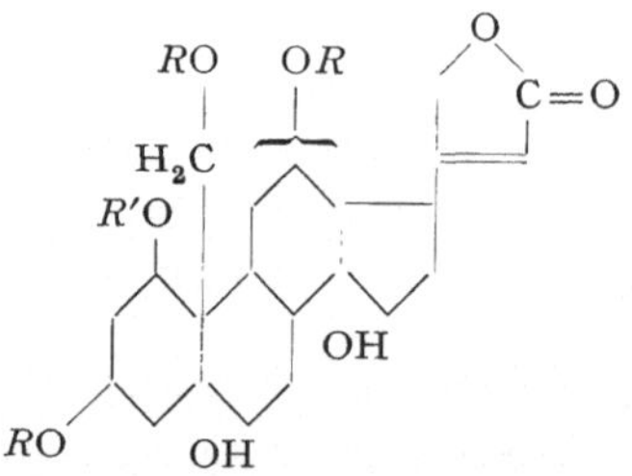

(CLXXI.) (R = R′ = H.) Ouabagenin.
(CLXXII.) (R = Ac; R′ = H.) Tri-O-acetyl-ouabagenin.
(CLXXIII.) (R = R′ = Ac.) Tetra-O-acetyl-ouabagenin.

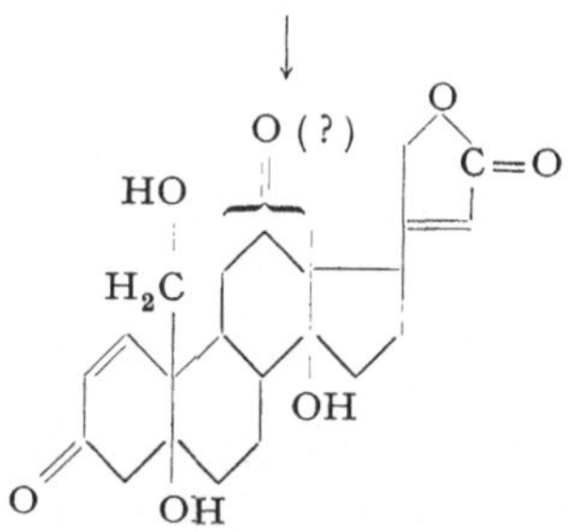

(CLXX.)

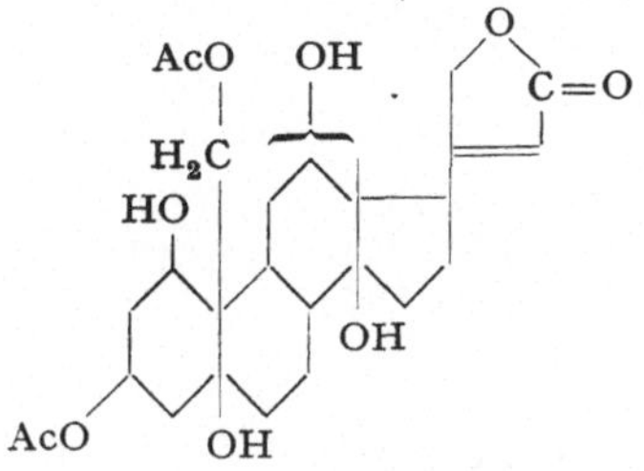

„α-Ouabagenin-diacetat", „β-Ouaba-genin-diacetat".

(CLXXIV.)

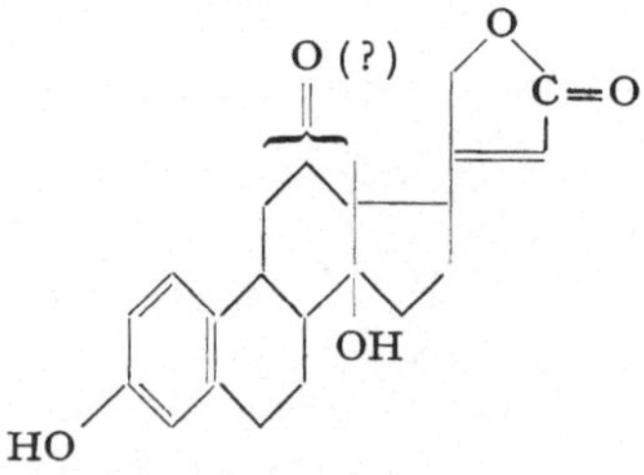

(CLXXV.)

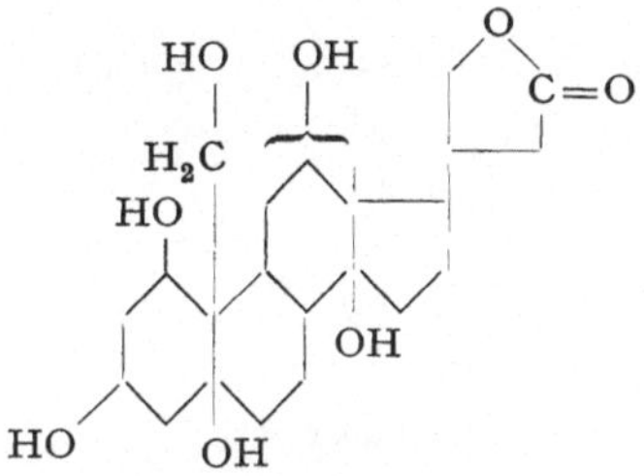

(CLXXVI.) Dihydro-ouabagenin.

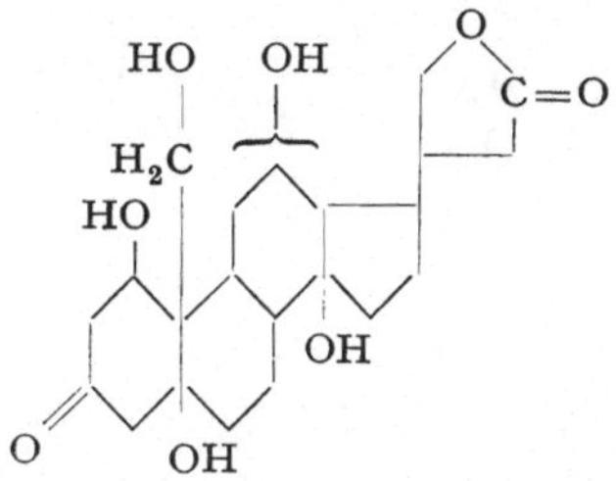

(CLXXVII.)

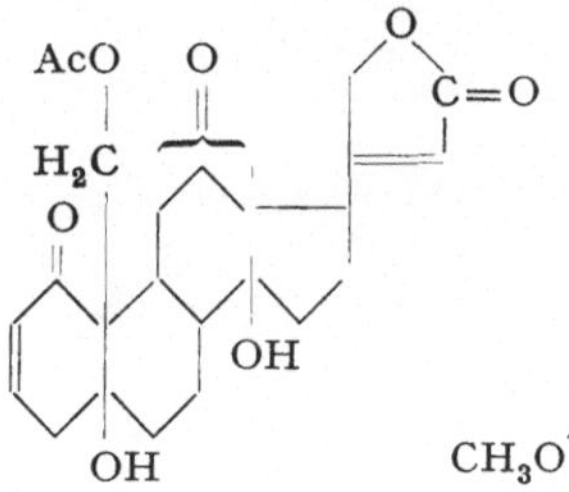

(CLXXVIII.)

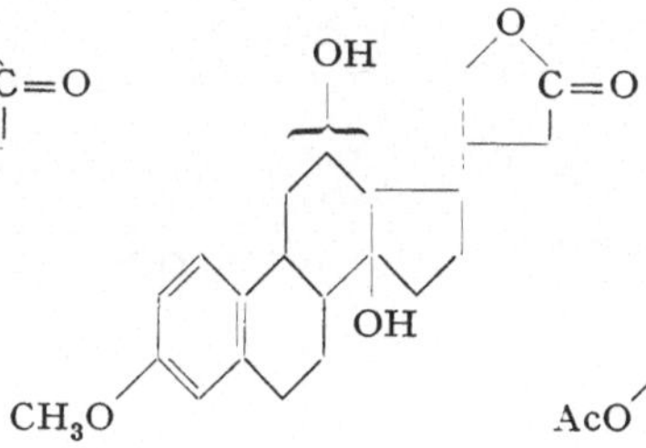

(CLXXIX.)

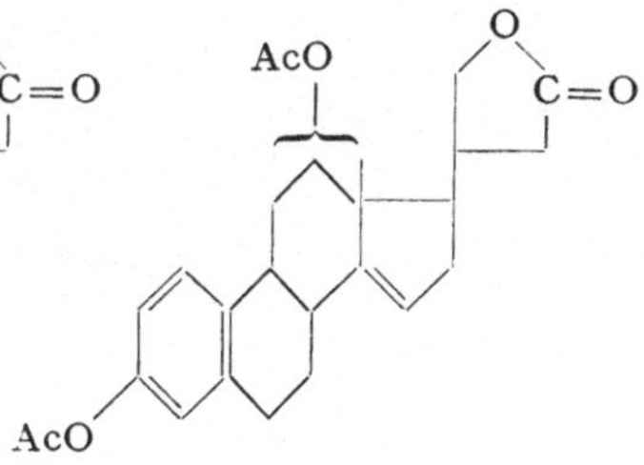

(CLXXX.)

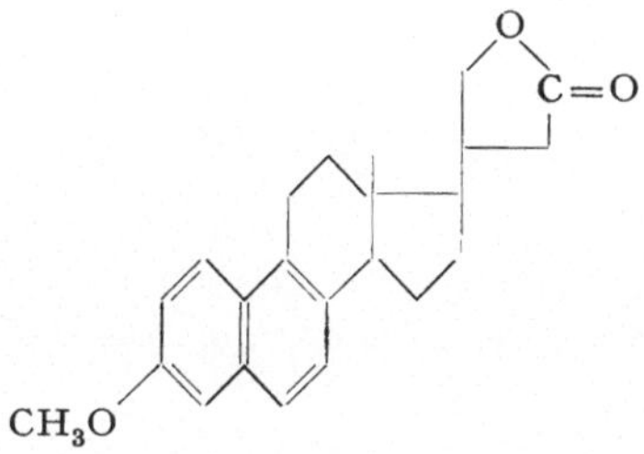

(CLXXXI.)

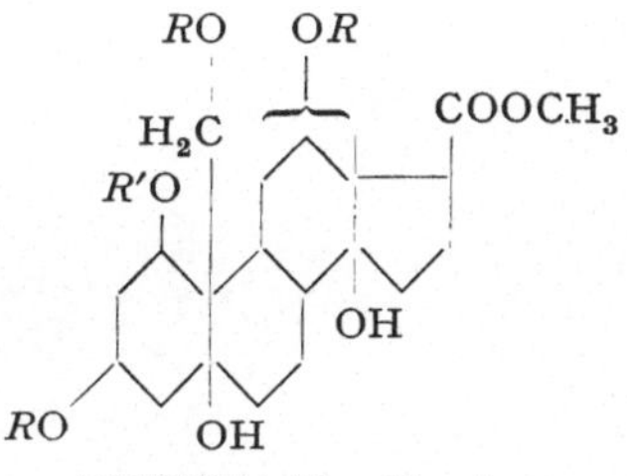

(CLXXXII.) $(R = R' = Ac.)$
(CLXXXIII.) $(R = Ac;\ R' = H.)$
(CLXXXIV.) $(R = R' = H.)$

während früher die $1\beta,3\beta$-Stellung willkürlich angenommen wurde [vgl. z. B. (*278*)]. (CLXVIII) geht nämlich nach Behandlung mit CrO_3 in Pyridin und anschließender Chromatographie an Al_2O_3 in den Neutral-

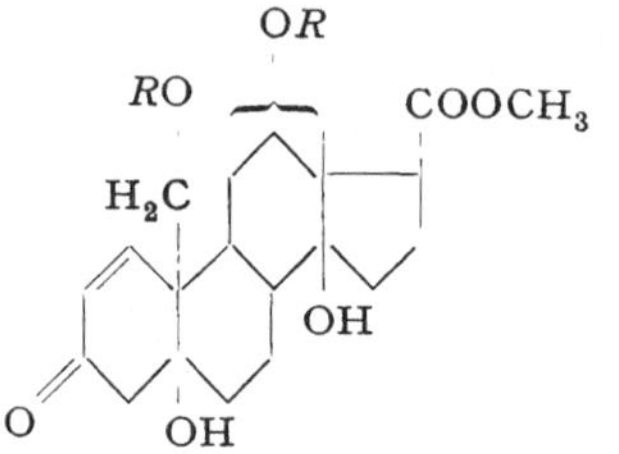

(CLXXXV.) (R = H.) (CLXXXVI.) (R = Ac.)

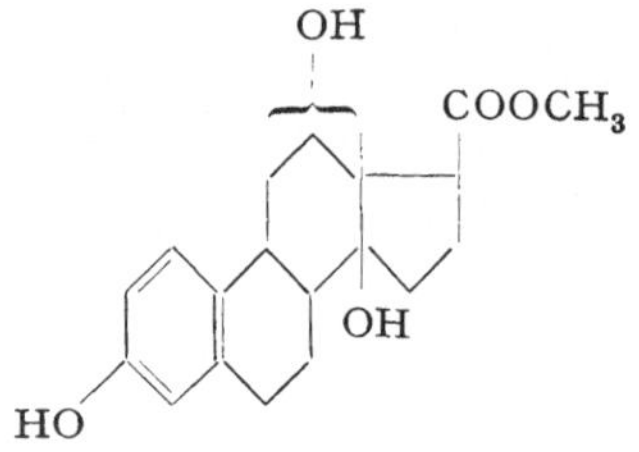

(CLXXXVII.)

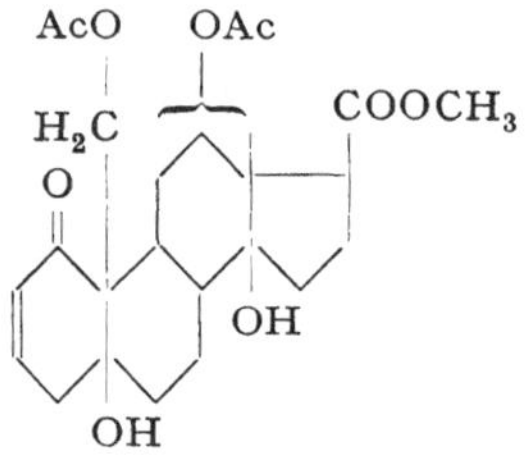

(CLXXXVIII.)

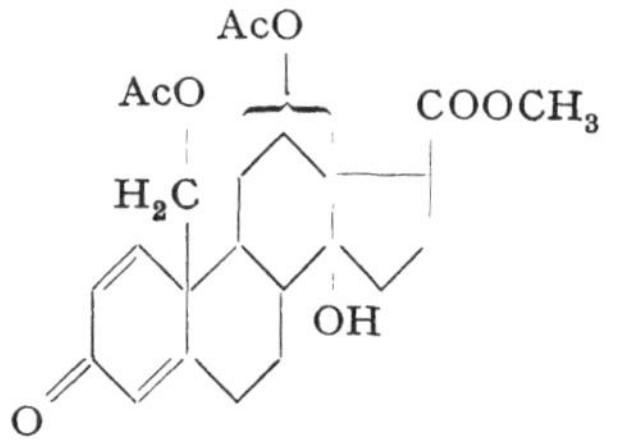

(CLXXXIX.)

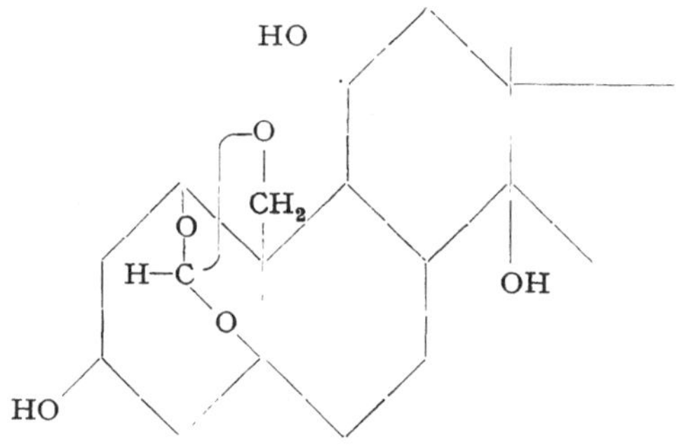

(CXC.)

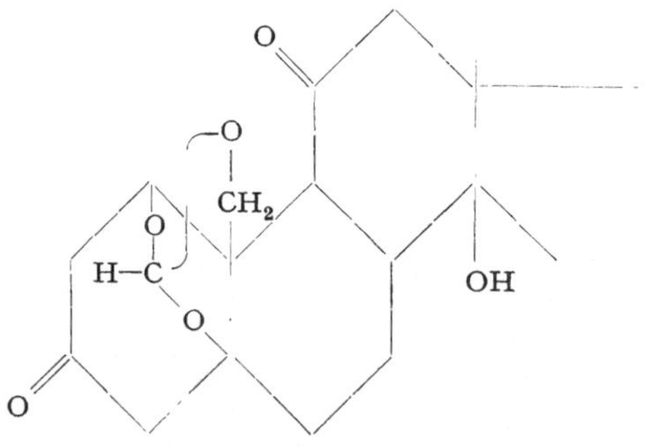

(CXCI.)

stoff (CLXX) über, der nach dem UV.-Spektrum eine Δ^1-3-Ketongruppie-
rung besitzt und nach Behandlung mit Eisessig bei 125° das kristallisierte

$$(CXCII.)$$

Phenol (CLXXV) liefert. [Durch Abbau des Butenolidringes wurde in analoger Reaktionsfolge der (CLXX) entsprechende Ätiansäure-methyl-ester erhalten (*321*).] Es hat Eliminierung der angulären Oxymethyl-gruppe und Aromatisierung des Ringes A stattgefunden. Ausgehend von Dihydro-ouabagenin (CLXXVI) gelang es SNEEDEN und TURNER (*279, 280*), mit Pt und Sauerstoff die 3-Oxygruppe partiell zu dehydrieren. Das entstandene Monoketon (CLXXVII) lieferte nach Behandlung mit Base und nach Methylierung 1 Mol. Formaldehyd und den Phenoläther (CLXXIX). Nach Acetylierung von (CLXXIX) erhielten diese Autoren durch Abspaltung der 14-Oxygruppe mit $POCl_3$ in Pyridin das einfach ungesättigte Dihydrogenin (CLXXX). In Gegenwart von Säure tritt Wanderung der Δ^{14}-Doppelbindung und Eliminierung der intermediär entstandenen allylischen Acetoxygruppe ein, und es bildet sich das β-Methoxy-naphtalin-derivat (CLXXXI).

REICHSTEIN und Mitarb. (*185, 211*) hatten Tetra-O-acetyl-ouaba-genin (CLXXIII) und Tri-O-acetyl-ouabagenin (CLXXII), dessen richtige Struktur erst von FLOREY und EHRENSTEIN (*54*) erkannt worden ist, zu den entsprechenden Ätiansäure-methylestern (CLXXII) und (CLXXIII) abgebaut. Die Verseifung der Acetoxylgruppen der beiden Ester führte zum selben (CLXXXIV). Es gelang EHRENSTEIN (*54*), die 3-Oxygruppe mit N-Bromacetamid partiell zu dehydrieren und nach Chromatographie an Al_2O_3 zum Ester (CLXXXV) zu gelangen, der nach dem UV.-Spektrum eine Δ^1-ungesättigte 3-Ketongruppierung enthält. Der Ester ist im Ring A ähnlich wie (CLXX) gebaut. Mit Eisessig trat auch bei diesem Stoff leicht Aromatisierung ein unter Bildung des Phenols (CLXXXVII), das allerdings amorph blieb. Die Acetylierung von (CLXXXV) gab das Diacetat (CLXXXVI), das nach analoger Behandlung mit Eisessig das amorphe $\Delta^{1,4}$-Dien-on-(3) (CLXXXIX) lieferte. Die Struktur von (CLXXXVII) und (CLXXXIX) stützt sich auf die UV.-Spektren.

Aus dem durch Abbau von Tri-O-acetyl-ouabagenin (CLXXII) ge-wonnenen Ätiansäureester (CLXXXIII) bereiteten FLOREY und EHREN-

STEIN (*54*) durch Dehydrierung mit CrO_3 in Eisessig und Chromatographie an Al_2O_3 den Ester (CLXXXVIII), der nach dem UV.-Spektrum eine Δ^2-ungesättigte 1-Ketongruppierung aufweist. Einen im Ring A gleich gebauten Stoff (CLXXVIII) erhielt TAMM (*320*) bei der Dehydrierung von ,,Ouabagenin-β-diacetat'' mit CrO_3 in Pyridin. Demnach muß ,,Ouabagenin-β-diacetat'' entweder a priori die Struktur (CLXXIV) besitzen, was aber im Gegensatz zu den erwähnten Befunden von SNEEDEN und TURNER (*278*) steht, oder im Verlaufe der Reaktion eine solche intermediär bilden.

Diese Ergebnisse sind, wie gesagt, eine starke Stütze für die Hydroxyle an $C_{(1)}$, $C_{(3)}$, $C_{(5)}$ und $C_{(19)}$ und lassen nur die Haftstelle der letzten leicht acetylierbaren, sekundären Hydroxylgruppe offen. TSCHESCHE und SNATZKE (*353*) bereiteten aus Ouabagenin (CLXXI) ein Orthoformiat, dem sie die Struktur (CXC) zuerteilen. Die drei Oxygruppen in 1-, 5- und 19-Stellung sind dadurch geschützt, während diejenigen in 3- und der vermutlichen 11-Stellung noch frei sind. Sie sind mit CrO_3 in Pyridin dehydrierbar; es entsteht das Diketon (CXCI). Nach der Reduktion der beiden Ketogruppen mit $NaBH_4$ und Acetylierung des erhaltenen Triolgemisches läßt sich als einziges Produkt ein Monoacetylderivat fassen. Die eine der beiden ursprünglichen Oxygruppen ist in dieser Reaktionsfolge offenbar epimerisiert worden, da sie nicht mehr acetylierbar ist. Dieses Resultat spricht für die 11-Stellung dieser Sauerstoffunktion, wobei sie in Ouabagenin (CLXXI) α-ständig und im obigen Monoacetylderivat β-ständig, entsprechend (CXCII), angeordnet ist.

Alle diese Versuche setzen voraus, daß Ouabagenin ein normales Steringerüst besitzt. Der Beweis dafür ist aber noch nicht geleistet worden.

2. Bufadienolide.

Die Bufadienolide enthalten 24 C-Atome. Der Cumalinring wird an Farbreaktionen (S. 145) und an seinem typischen Maximum bei 300 mμ, $\log \varepsilon = 3{,}74$ (in Alkohol) erkannt. Sein IR.-Spektrum ist durch eine ν (C=O)-Schwingung bei 5,81—5,82 μ (1720—1718 cm^{-1}) und durch ν (C=C)-Schwingungen bei 6,10—6,12 μ (1639—1634 cm^{-1}) und 6,24 μ (1603 cm^{-1}) (in Methylenchlorid oder Chloroform) gekennzeichnet (*125 a*). Die erste richtige Formulierung stammt von STOLL (*289, 292*).

α) *Beweis des Kohlenstoffskeletts.* Das Vorliegen des Steringerüstes wurde zuerst an Scillaren A (CXCIII) bewiesen (*292*). Durch Behandeln mit methanolischer HCl ging (CXCIII) in Anhydroscillaridin A (CXCIV) über, das nach katalytischer Hydrierung unter Hydrogenolyse Allocholansäure (CXCV) lieferte. Wird Scillaren A zuerst katalytisch hydriert, dann erst mit methanolischer HCl behandelt und nachhydriert, so ent-

Glucose-Glucose

(CXCIII.) Scillaren A.   (CXCIV.) Anhydro-scillaridin A.   (CXCV.) (R = H.) Allocholansaure.
(CXCVI.) (R = OH.) 3β-Oxy-allocholansaure.

steht die 3β-Oxy-allocholansäure (CXCVI). Damit sind Stellung und Konfiguration der sekundären HO-Gruppe als 3β-ständig bewiesen. Mit dieser HO-Gruppe ist in (CXCIII) der Zucker verknüpft (*304*). Die von WIELAND et al. (*369*) aus Bufotalin erhaltene Isobufocholansäure ist wahrscheinlich die 14β,17α-Cholansäure. Die Wasserabspaltung bei Bufotalin hatte zum $\Delta^{3,14,16}$-ungesättigten Bufotalinderivat geführt, das bei der Hydrierung vorwiegend die Produkte mit 14β,17α-Konfiguration lieferte [vgl. auch PLATTNER et al. (*207*)]. Die Isolierung von 3β-Acetoxy-isobufocholansäure und Isobufocholansäure wurde von K. MEYER wiederholt (*177, 180*).

β) Abbau zu Ätiansäuren und einige Besonderheiten: Aglykone mit bekannter Konstitution (vgl. Tabelle 2, S. 200). Der Abbau der Bufa-dienolide zu Ätiansäuren gelingt mit neutralem $KMnO_4$ in Aceton. Als neutrales Nebenprodukt entsteht ein 14β-Oxy-20-keto-pregnan-21-säurelacton-(21 → 14) (CXCIX), was ein weiterer Beweis ist, daß der Cumalinring und die HO-Gruppe an $C_{(14)}$ sich auf der gleichen Seite des Ringsystems befinden [vgl. die Bildung analoger Ketolactone beim Abbau von Digitoxigenin, Periplogenin und Gitoxigenin, sowie (*198*)]. Der Ozonabbau führt nicht zu den Ätiansäuren, sondern zu Gemischen, deren Zusammensetzung noch unbekannt ist (*177*). Durch den oxydativen Abbau zu den Ätiansäuren gelang es, folgende Bufadienolide mit Cardenoliden von gesicherter Konstitution zu verknüpfen: *Bufalin* (XIV, S. 153) mit Digitoxigenin (XIII, S. 153) (*177*), *Hellebrigenin* (= *Bufotalidin*) (CXCVII) mit Strophanthidin (LXIV, S. 163) (*261*), *Telocinobufagin* (CXCVIII) mit Periplogenin (LIX, S. 163) (*178*), *Gama-bufotalin* (CC) mit Sarmentogenin (LXXVI, S. 166) (*179*), *Bufotalin* (CCI) mit Oleandrigenin (LXXXVII, S. 168) bzw. Gitoxigenin (LXXXV, S. 168) (*180*), *Bovogenin A* (CCIII, S. 190) mit Corotoxigenin (LXII, S. 163) und *Bovogenol A* (CCIIIa) mit Coroglaucigenin (LXIII, S. 163) (*139*). URSCHELER, TAMM und REICHSTEIN (*358*) zeigten, daß Hellebrigenin mit Bufotalidin identisch ist. Es ist das erste Genin, das sowohl

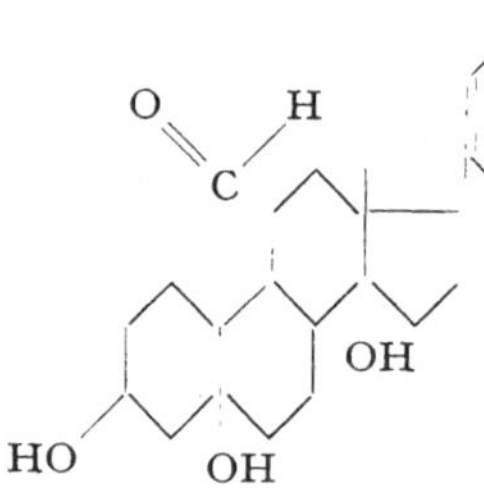

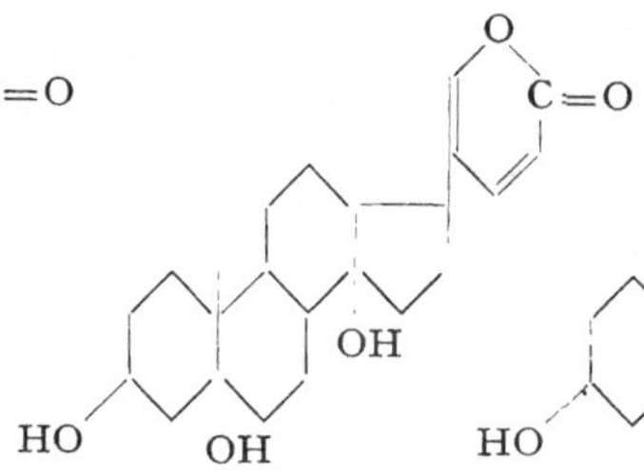

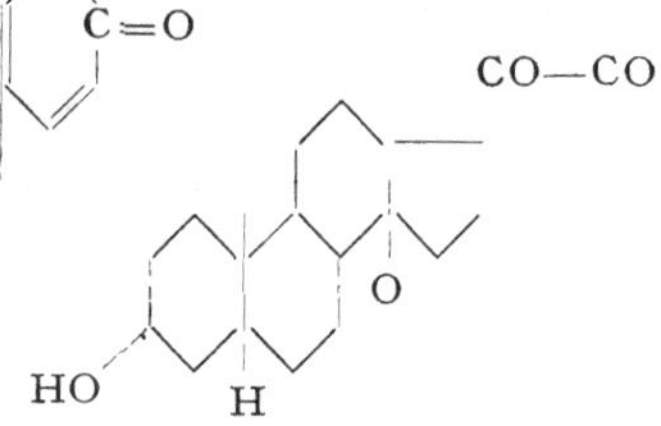

(CXCVII.) Hellebrigenin (Bufotalidin). (CXCVIII.) Telocinobufagin. (CXCIX.)

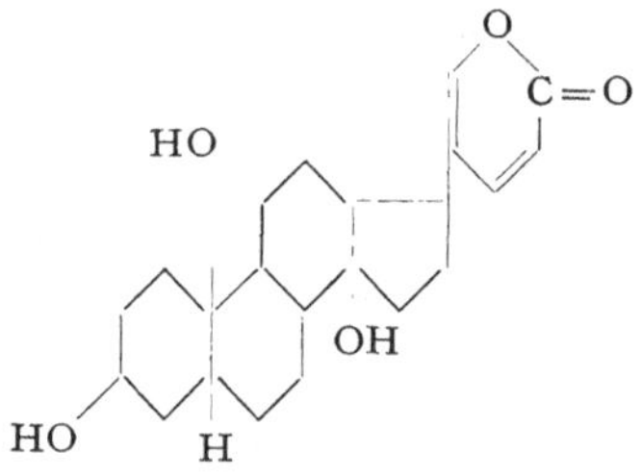

(CC.) Gamabufotalin. 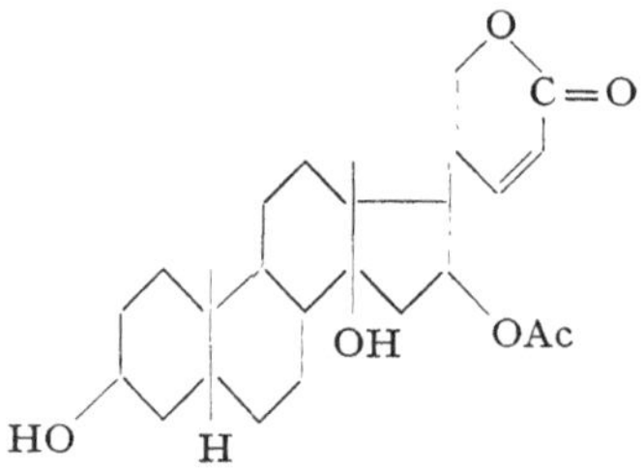 (CCI.) Bufotalin.

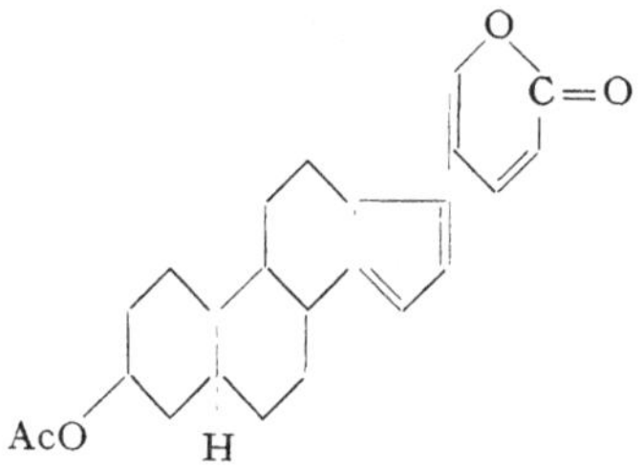

(CCII.)

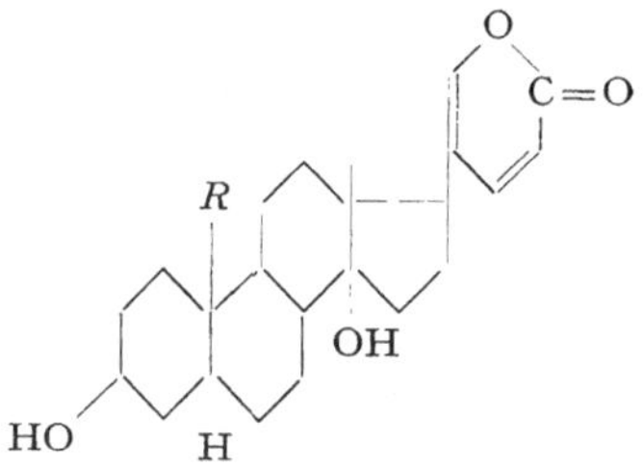

(CCIII.) $(R = CHO.)$ Bovogenin A.
(CCIII a.) $(R = CH_2OH.)$ Bovogenol A.

aus pflanzlichem wie aus tierischem Material isoliert werden konnte. O-Acetyl-bufotalin geht mit 3proz. äthanolischer HCl in das $\Delta^{14,16}$-Di-anhydro-derivat (CCII) über, dessen Spektrum ein intensives Maximum bei 300 mμ, log $\varepsilon = 4{,}22$ und eine Schulter bei 340—380 mμ, log $\varepsilon = 3{,}2$ bis 2,9 (in Alkohol) aufweist. Mit $POCl_3$ in Pyridin wird nur das einfach ungesättigte 14-Monoanhydro-Derivat erhalten.

Scillarenin, $C_{24}H_{32}O_4$ (CCIV) und *Scilliglaucosidin* (CCVI) besitzen eine zusätzliche Doppelbindung im Ring *A*. (CCIV) gibt bei der Dehydrierung Scillarenon (CCV), das eine Δ^4-ungesättigte 3-Ketongruppierung besitzt. In Analogie zu den Sterinen legten STOLL und Mitarbeiter diese Doppelbindung zunächst in die 5-Stellung, bis sie später die 4-Stellung beweisen konnten (*311*). Scillaren A (CXCIII) lieferte nämlich bei der Spaltung mit einem adaptiven Enzym aus einem Penicillium-Stamm neben wenig Scillarenin hauptsächlich Scillarenon (CCV), das sich auf Grund seines

(CCIV.) Scillarenin.   (CCV.) Scillarenon.   (CCVI.) Scilliglaucosidin.

UV.-Spektrums und des Vergleiches molekularer Drehungen als Δ^4-3-Keton erwies und bei der Reduktion nach MEERWEIN-PONNDORF Scillarenin (CCIV) lieferte. Eine Wanderung der Doppelbindung dürfte bei dieser Reaktion kaum eingetreten sein. Eine weitere Stütze für die Konstitution des Scillarenins ergibt sich aus dem Vergleich von molekularen Drehungswerten.

Scilliglaucosidin, $C_{24}H_{30}O_5$ (CCVI), das Genin des Scilliglaucosids, besitzt neben einer isolierten 4-Doppelbindung eine anguläre Aldehydgruppe an $C_{(10)}$ (*316*). Diese Sauerstoffunktion ist dafür verantwortlich, daß die Abspaltung der HO-Gruppe an $C_{(3)}$ und die Bildung des 3,5-Diens im Vergleich zu Scillarenin erheblich erschwert sind, wie durch den Vergleich mit Δ^4-5-Anhydro-strophanthidin-19-säure-methylester gezeigt wurde (*318*). Die Farbintensität bei der ROSENHEIM-Reaktion wird dadurch stark vermindert (*317*).

Es ist auffallend, daß in (CCVI) die Aldehydgruppe an $C_{(10)}$ kein cyclisches Halbacetal mit der 3β-Oxygruppe bildet. Sie läßt sich nach MEERWEIN-PONNDORF zur Oxymethylgruppe reduzieren und mit CrO_3 zur Carboxylgruppe oxydieren (*316*).

γ) Aglykone mit teilweise bekannter Konstitution (vgl. Tabelle 2, S. 200). Für *Artebufogenin* wurde von K. MEYER (*182*), der dieses Genin aus der chinesischen Droge Ch'an Su isolierte, die Bruttoformel $C_{24}H_{32}O_4$ aufgestellt und die provisorische Strukturformel (CCVII) abgeleitet. (CCVII) läßt sich mit CrO_3 in Eisessig zu einem Monoketon dehydrieren.

(CCVII.) Artebufogenin.   (CCVIII.)   (CCIX.)

Bei der Acetylierung liefert (CCVII) die zwei isomeren Acetylderivate A
und B. Acetat A (CCVIII) ist ein umgelagertes Produkt. Dessen Abbau
mit $KMnO_4$ und Methylierung führten zum Ätiansäureester (CCIX), der
nach dem UV.-Spektrum eine isolierte Ketogruppe besitzt, die sich bei
der Hydrierung mit Pt in Eisessig nicht verändert. Es ist zu vermuten,
daß sie bereits im Acetat A enthalten ist. Bei der Verseifung des Esters
(CCIX) mit Alkali tritt keine Umlagerung ein, da er nach der Remethylierung und Reacetylierung unverändert zurückgewonnen werden kann.
Artebufogenin ist physiologisch unwirksam. Möglicherweise ist es ein
Umlagerungsprodukt von Resibufogenin.

Resibufogenin (*182*) stammt ebenfalls aus dem chinesischen Ch'an
Su und ist bisher nur in amorpher Form erhalten worden. Hingegen
ist Mono-O-acetyl-resibufogenin kristallisiert. Meyer (*182*) leitete die
Formel $C_{26}H_{34}O_5$ ab und erteilte ihm die vorläufige Struktur (CCXI).
Diese Verbindung ist gegen CrO_3 in Eisessig beständig. Das freie Resibufogenin (CCX) bildet ein kristallisiertes Hydrochlorid (CCXII), das
acetylierbar und mit CrO_3 in Eisessig dehydrierbar ist, ohne HCl zu verlieren, so daß man annehmen kann, daß die vermutlich an $C_{(3)}$ sitzende

(CCX.) (R = H.) Resibufogenin.
(CCXI.) (R = Ac.)

(CCXII.)

(CCXIII.)

Oxygruppe reagiert hat. Der Abbau von (CCXI) mit $KMnO_4$ und die Methylierung der entstandenen Ätiansäure geben den Ätiansäure-ester (CCXIII),
der sich mit Alkali in noch nicht abgeklärter Weise zum Teil isomerisiert.
Eine Oxygruppe läßt sich weder in (CCXI) noch in (CCXIII) nachweisen,
so daß die vierte Sauerstoffunktion vermutlich als Oxydring vorliegt.

Marinobufagin, $C_{24}H_{32}O_5$, besitzt nach K. MEYER (*181, 198*) die Strukturformel (CCXIV). Von den fünf Sauerstoffatomen sind zwei im Cumalinring festgelegt. Eine weitere liegt als acetylierbare sekundäre

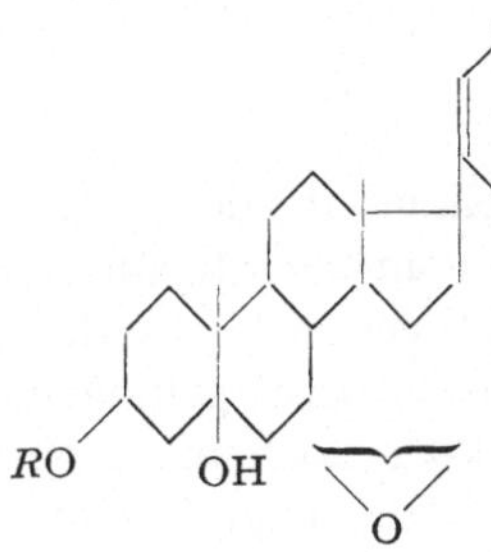

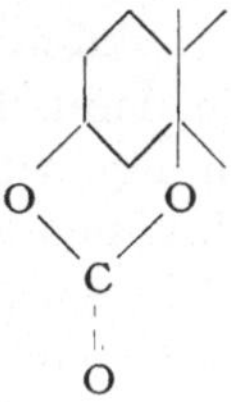

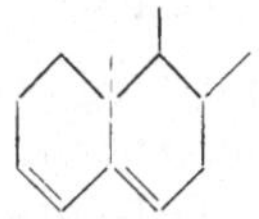

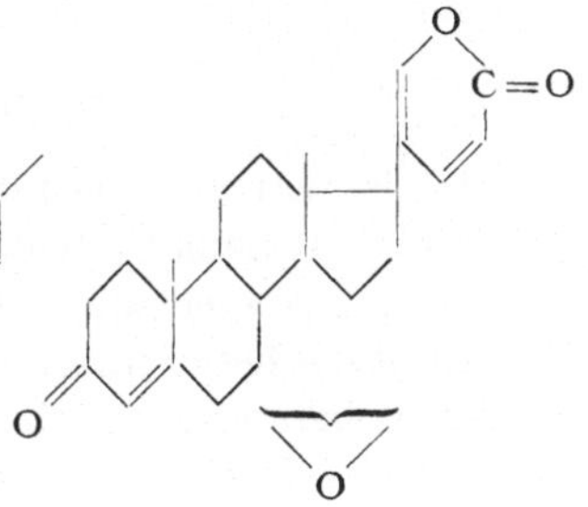

(CCXIV.) (*R* = H.) Marinobufagin. (CCXVI.) (CCXVII.) (CCXVIII.)
(CCXV.) (*R* = Ac.)

HO-Gruppe an $C_{(3)}$ vor, die vierte als 5β-ständige tertiäre Oxygruppe, denn Marinobufagin gibt mit CrO_3 in Eisessig das Δ^4-ungesättigte 3-Keton (CCXVIII). Durch Pyrolyse von (CCXV) wird Dianhydro-marino-bufagin (CCXVII) erhalten, das eine 3,5-Dien-gruppierung enthält. Die Konstitution von (CCXVII) und (CCXVIII) stützt sich auf die UV.- und IR.-Spektren und den Vergleich molekularer Drehungswerte. Die IR.-Spektren von (CCXVII) und (CCXVIII) lassen keine Hydroxyl-banden erkennen, so daß das fünfte Sauerstoffatom in diesen genannten Stoffen wie auch in Marinobufagin (CCXIV) als Oxydring vorliegen dürfte. (CCXIV) bildet den cyclischen Kohlensäureester (CCXVI), was nur bei *cis*-ständiger Anordnung der beteiligten Hydroxyle möglich ist. Sie sind beide β-ständig angeordnet. Der Abbau von (CCXV) mit $KMnO_4$ wie auch die katalytische Hydrierung von (CCXIV) verlaufen unübersichtlich.

Scillirosidin, $C_{26}H_{34}O_7$, besitzt nach STOLL (*305, 306, 312*) die Struktur (CCXIX). Auffallend sind die Acetoxygruppe im Cumalinring, die isolierte, schwer hydrierbare Doppelbindung in 8,9-Stellung und die 12-Oxygruppe. Die umfangreichen Untersuchungen der Jahre 1942/43,

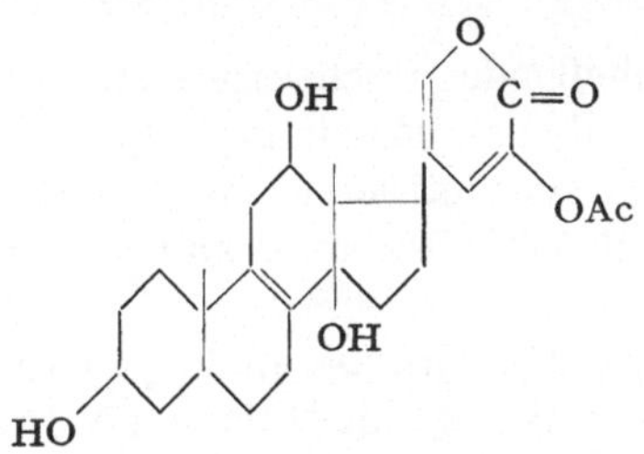

(CCXIX.) Scillirosidin (STOLL). 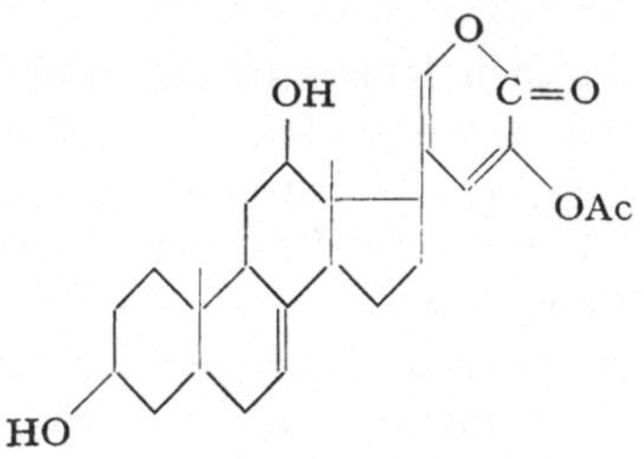 (CCXX.) Scillirosidin (TSCHESCHE).

die zu dieser Formel geführt haben [vgl. Tschesche (*164*)], sind einzig durch Tschesche (*342*) ergänzt worden, der die Doppelbindung auf Grund von IR.-Spektren in die 7-Stellung verlegte, entsprechend der Formel (CCXX). Doch sind beide Formeln noch nicht endgültig gesichert.

B. Teilsynthese der Aglykone.

Die Konstitutionsermittlung der Glykoside gelingt in der Regel durch den systematischen Abbau. Die dadurch gewonnenen Resultate können durch geeignete Teilsynthesen ergänzt und gesichert werden. Die Totalsynthese eines natürlichen Glykosids ist bisher nicht gelungen, da der vollständige Aufbau eines natürlichen Aglykons noch nicht geglückt ist. Nur die Einführung von einigen, für die Genine charakteristischen funktionellen Gruppen in das Steringerüst ist erfolgreich durchgeführt worden. Hingegen liegen Synthesen für die Zucker vor. Ferner sind Methoden zur Verknüpfung der beiden Komponenten zu Glykosiden ausgearbeitet worden.

Für die Synthese des Butenolidringes der Cardenolide stehen mehrere Wege offen. Die richtige Formulierung der Cardenolide als β-substituierte α,β-ungesättigte γ-Lactone an Stelle der ursprünglichen Annahme eines β,γ-ungesättigten Lactonringes ist erst durch diese Synthesen und den Vergleich mit den Angelica-Lactonen als Modellen ermöglicht worden (*197*). Als Ausgangsmaterial dienen entweder die gut zugänglichen $C_{(21)}$-Methylketone vom Typus (CCXXI), $C_{(21)}$-Diazoketone vom Typus (CCXXII), oder Ketolacetate vom Typus (CCXXIII), die sich mit Bromessigester

(CCXXI.)   (CCXXII.)   (CCXXIII.)   (CCXXIV.) 17α-Uzarigenin.

und Zn nach Reformatzky und anschließender Cyclisierung zu Stoffen mit dem gewünschten Butenolidring aufbauen lassen [vgl. (*356*)]. So gelang es, 17α-Uzarigenin (CCXXIV) zu synthetisieren (*206*), wobei bei der Reformatzky-Reaktion die empfindliche HO-Gruppe an $C_{(14)}$ bereits eingeführt war.

Für die Einführung einer β-ständigen HO-Gruppe an $C_{(14)}$ wird ein Ätiansäure-methylester oder ein $C_{(21)}$-Methylketon mit Hilfe von N-Bromsuccinimid und Pyridin in den doppelt ungesättigten Ester (CCXXV) bzw. in das Methylketon (CCXXVI) übergeführt. (CCXXV) und (CCXXVI) lassen sich mit einer Persäure leicht zu einfach ungesättigten 14,15β-Oxidoverbindungen (CCXXVII) oxydieren. Bei der anschließenden

$$\text{COOCH}_3 \qquad \overset{\overset{\displaystyle CH_3}{|}}{C}=O \qquad R \qquad R \qquad R$$

$$\text{(CCXXV.)} \qquad \text{(CCXXVI.)} \qquad \underset{O}{\text{(CCXXVII.)}} \qquad \underset{OH}{\text{(CCXXVIII.)}} \qquad \underset{OH}{\text{(CCXXIX.)}}$$

katalytischen Hydrierung mit Pt entsteht aber nur zum geringsten Teile das gewünschte Reduktionsprodukt (CCXXVIII) mit 17β-ständigem Substituenten R, sondern zur Hauptsache das isomere (CCXXIX) mit 17α-ständigem R (*207, 205*). Deshalb gelang es bisher nur, Allo-uzarigenin (CCXXIV) mit 17α-ständigem Butenolidring zu bereiten. Die Einführung einer β-ständigen HO-Gruppe an $C_{(5)}$ erfolgt am einfachsten durch reduktive Spaltung eines 4,5β-Oxyds mit $LiAlH_4$ nach PLATTNER et al. (*203, 204*).

Der Aufbau des Cumalinringes zur Bereitung von Bufadienoliden ist bisher an Steroidderivaten nicht geglückt. Es sind nur einfache Modelle synthetisiert worden (*66, 67*).

IV. Tabellen.

Vorbemerkung zu den Tabellen 1—4.

Nur einheitliche und analysierte Stoffe sind berücksichtigt. Bei den $[\alpha]_D$-Werten bedeuten: Al = Alkohol; An = Aceton; Chf = Chloroform; D = Dioxan; Me = Methanol; Py = Pyridin und W = Wasser.

Die Werte der „mittleren letalen Dosis" wurden an der Katze durch den HATCHER-Test ermittelt und sind als mg Substanz pro kg Körpergewicht angegeben.

Bei der Glucose handelt es sich immer um die *D*-Glucose.

Bei der Angabe der funktionellen Gruppen bedeuten: Ac = $CH_3 \cdot CO—$, Oxo eine Carbonylfunktion und $\varDelta$ eine Doppelbindung. Die Verknüpfung der Ringe A und B ist, wo nichts anderes erwähnt, *cis*-ständig. Alle Cardenolide zeigen im UV.-Spektrum eine Hauptbande bei 217 mμ (log ε = 4,22). Nur zusätzlich auftretende Maxima sind vermerkt, wobei die in Klammern zugefügten Zahlen die log ε-Werte sind. Das bei 300 mμ (log ε = 3,7) auftretende Maximum der Bufadienolide ist nicht angegeben.

Tabelle 1.

Aglykon	Formel	Schmp.	$[\alpha]_D$
1. Aglykone mit bekannter Konstitution.			
16-Mono-anhydro-gitoxigenin	$C_{23}H_{32}O_4$	227—235°/247—248°	+92,7° (Me)
Digitoxigenin	$C_{23}H_{34}O_4$	248—250°	+19,1° (Me)
3-Epi-digitoxigenin	$C_{23}H_{34}O_4$	274—282°	+26,8° (Me)
Uzarigenin	$C_{23}H_{34}O_4$	230—246°/240—256°	+14,0° (Al)
Urezigenin	$C_{23}H_{34}O_4$	270—275°	+4,1° (Chf)
Desarogenin(= 11-Dehydro-sarmentogenin)	$C_{23}H_{32}O_5$	252—253°/285—295°	+10,6° (Me)
Acovenosigenin A	$C_{23}H_{34}O_5$	272—274°	+2,0° (Me)
Periplogenin	$C_{23}H_{34}O_5$	135—140°/168—169°/ 185—190°/200—205° 235—240°	{ +29,8° (Me) } { +29.1° (Chf) }
Allo-periplogenin	$C_{23}H_{34}O_5$	220—250°	+40,6° (Me)
Sarmentogenin	$C_{23}H_{34}O_5$	272—274°	{ +21,1° (Me) } { +18,9° (An) }
11-Epi-sarmentogenin	$C_{23}H_{34}O_5$	252—258°	+29,2° (Me)
Digoxigenin	$C_{23}H_{34}O_5$	206—209°/220—222°	+61,3° (Me)
Gitoxigenin	$C_{23}H_{34}O_5$	224—225°	+32,6° (Me)
Gitaloxigenin	$C_{24}H_{34}O_6$	215—217°	—
Oleandrigenin	$C_{25}H_{36}O_6$	110—115°/225—228°	—9,8° (Me)
Corotoxigenin	$C_{23}H_{32}O_5$	221°/216—225°	+43° (Me)
Coroglaucigenin	$C_{23}H_{34}O_5$	249—250°/250—255°	+23° (Me)
Adonitoxigenin	$C_{23}H_{32}O_6$	172—178°	—
Strophanthidin	$C_{23}H_{32}O_6$	136—138°/177—178° 220—230°	{ +39,5° (Chf) } { +40,8° (Al) } +44,3° (Me)
Allo-strophanthidin	$C_{23}H_{32}O_6$	248—250°	+37° (Al)
Strophanthidol	$C_{23}H_{34}O_6$	138—142°/150°→222°	+37° (Me)
2. Aglykone mit teilweise bekannter und unbekannter Konstitution.			
Adynerigenin	$C_{23}H_{32}O_4$	238—242°	+18° (Py) (Me)
Neriantogenin	$C_{23}H_{32}O_4$	255—259°	—
Subst. B_1 (= Subst. B_2) aus *Xysmalobium undulatum*	$C_{23}H_{32}O_4$	254—258°	+11,6° (Me)

Cardenolide.

Funktionelle Gruppen	UV.-Spektrum λ max zusätzlich zu 217 mμ (4,22) in mμ	Mittlere letale Dosis mg/kg (Katze)	Literatur
3 β-OH, 14 β-OH, Δ^{16}	273 (4,4)	—	(89, 95)
3 β-OH, 14 β-OH	—	0,424—0,459	(100, 375)
3 α-OH, 14 β-OH	—	unwirksam	(270)
3 β-OH, 14 β-OH, A/B trans	—	—	(213, 341)
3 α-OH, 14 β-OH, A/B trans	—	—	(341, 349)
3 β-OH, 11-Oxo, 14 β-OH	297 (1,74)	—	(49)
1 β-OH, 3 β-OH, 14 β-OH	—	—	(44, 259, 322)
3 β-OH, 5 β-OH, 14 β-OH	—	0,719	(120, 142, 160, 282, 283, 303)
3 β-OH, 5 β-OH, 14 β-OH, 17 α-Butenolidring	—	—	(142, 282, 283)
3 β-OH, 11 α-OH, 14 β-OH	—	—	(117, 132)
3 β-OH, 11 β-OH, 14 β-OH	—	—	(249)
3 β-OH, 12 β-OH, 14 β-OH	—	0,253	(199, 275, 325)
3 β-OH, 14 β-OH, 16 β-OH	—	1,85	(187, 374)
3 β-OH, 14 β-OH, 16 β-O · CHO	—	—	(71)
3 β-OH, 14 β-OH, 16 β-OAc	—	—	(187, 193)
3 β-OH, 14 β-OH, 19-Oxo, A/B trans	310 (1,51)	0,699	(254, 302)
3 β-OH, 14 β-OH, 19-OH, A/B trans	—	0,479	(99, 302)
3 β-OH, 14 β-OH, 16 β-OH, 19-Oxo	zirka 305 (1,5)	—	(143, 348)
3 β-OH, 5 β-OH, 14 β-OH, 19-Oxo	303 (1,45)	0,274	(50, 116, 220)
3 β-OH, 5 β-OH, 14 β-OH, 19-Oxo, 17 α-Butenolidring	303 (1,45)	—	(102, 141)
3 β-OH, 5 β-OH, 14 β-OH, 19-OH	—	—	(18, 210)
3 β-OH, 14 β-OH, $\Delta^{7\,(?)}$	—	unwirksam	(193, 339, 340, 342, 352)
—	—	unwirksam	(340, 352)
—	295 (2,66)	—	(93)

Fortsetzung der Tabelle 1

Aglykon	Formel	Schmp.	$[\alpha]_D$
Subst. D aus *Xysmalobium undulatum*	$C_{23}H_{32}O_4$	227—248°	+16,6° (Me)
Tanghiferigenin	$C_{23}H_{32}O_4$	278—280°	+61,9° (Chf)
Xysmalogenin	$C_{23}H_{34}O_4$	230—248°	+19° (Al)
Aglykon a aus *Xysmalobium undulatum*	$C_{23}H_{34}O_4$	225—240°	+3,8° (Me)
Subst. D aus *Acokanthera longiflora*	$C_{23}H_{30}O_5$	226—227°	+86,6° (Me)
Subst. F aus *Acokanthera longiflora*	$C_{23}H_{30}O_5$	305—308°	+12,6° (Py-Me 1 : 1)
Decogenin	$C_{23}H_{30}O_5$	230—235°	+66,1° (Me)
Tanghinigenin	$C_{23}H_{32}O_5$	187—188°/195—196°	+14,1° (Chf)
3-Epi-tanghinigenin	$C_{23}H_{32}O_5$	230—235°	+28,4° (Chf)
Abogenin	$C_{23}H_{34}O_{5-6}$	210—217°	+11,0° (Me)
Flavogenin (= Subst. Nr. 782)	$C_{23}H_{28}O_6$	240—243°	+34,2° (Me)
Caudogenin	$C_{23}H_{32}O_6$	213—220°	—82,2° (Me)
Sarmutogenin	$C_{23}H_{32}O_6$	258—262°	+48,9° (Me)
Erysimidin	$C_{23}H_{32}O_6$	161—164°/227—229°	—
Corchortoxin	$C_{23}H_{32}O_6$	247°	+67,9° (Al)
Calotropagenin	$C_{23}H_{32}O_6$	240°	+42°
Allo-glaucotoxigenin	$C_{23}H_{32}O_6$	235—236°	+27° (Me)
Corchorgenin	$C_{23}H_{32}O_6$	226—227°	+90° (Al)
Genin H. M. 15 aus *Acokanthera friesiorum*	$C_{23}H_{32}O_6$	190—200°	+83,6° (Me)
Glaucorigenin	$C_{23}H_{32}O_6$	228°	—86,3° (Me)
Quilengenin	$C_{23}H_{32-34}O_6$	238—242°/264—267°	+12,5° (Me)
Genin H. 15 aus *Strophanthus amboënsis*	$C_{23}H_{32-34}O_6$	238—242°/250—254°	+13,6° (Me)
Sarverogenin	$C_{23}H_{32}O_7$	126—130°/140—150°/ 162—170°/182—188°/ 227—230°	+44,7° (Me)
Inertogenin	$C_{23}H_{32}O_7$	227—232°	—56,4° (Me)
Leptogenin	$C_{23}H_{32}O_7$	244—246°	+78,4° (Me)
Nigrescigenin	$C_{23}H_{32}O_7$	238—241°/162—168° 243—246°	+24,8° (Me)
Antiarigenin	$C_{23}H_{32}O_7$	242—248°	+43,2° (Me)
al-Dihydro-antiarigenin	$C_{23}H_{34}O_7$	161—162°	+37,4° (Me)
E. Sche. 12 aus *Periploca nigrescens*	$C_{25}H_{34}O_8$	190°/230—234°	+29,8° (Chf)

Funktionelle Gruppen	UV.-Spektrum λ max zusätzlich zu 217 mμ (4,22) in mμ	Mittlere letale Dosis mg/kg (Katze)	Literatur
—	—	—	*(93)*
—	295 (1,50)	—	*(62, 65a)*
—	—	—	*(341, 349)*
—	—	—	*(93)*
—	—	—	*(7)*
—	—	—	*(7)*
3 β-OH, 11,12-Dioxo, 14 β-OH	370 (1,50)	—	*(248)*
3 β-OH, 14 β-OH, Äther-O	—	1,016	*(271, 272)*
3 α-OH, 14 β-OH, Äther-O	—	unwirksam	*(271, 272)*
—	—	—	*(88)*
3 β-OH, 11,12-Dioxo, $\Delta^{8,9}$, 14 β-OH	440 (1,43)	—	*(23, 82)*
3 β-OH, 11-Oxo, 12 α-OH, 14 β-OH	295 (1,76)	—	*(258)*
3 β-OH, 11-Oxo, 12 β-OH, 14 β-OH	287 (1,89)	—	*(227, 258)*
—	300	—	*(51)*
—	—	0,39	*(127)*
—	—	—	*(91)*
—	—	> 5,0	*(302)*
—	—	0,266	*(28)*
—	—	—	*(189)*
—	—	0,607	*(302)*
—	—	—	*(37)*
—	zirka 280 (1,71)	—	*(37)*
3 β-OH, 11 α-OH, 12-Oxo, 14 β-OH, Äther-O	279 (1,85)	—	*(23, 82, 325)*
3 β-OH, 11-Oxo, 12 α-OH, 14 β-OH, Äther-O	292 (1,70)	—	*(82)*
3 β-OH, 11-Oxo, 12 β-OH, 14 β-OH, Äther-O	290 (1,90)	—	*(82)*
—	295 (1,95)	0,229	*(247)*
—	zirka 305 (1,8) für α-Antiarin	—	*(34, 35)*
—	—	0,595	*(35)*
—	275 (2,08) Inflexion; 325 (1,90)	—	*(247)*

Fortsetzung der Tabelle 1

Aglykon	Formel	Schmp.	$[\alpha]_D$
Ouabagenin	$C_{23}H_{34}O_8$	235—238°/255—256° 262—268°	+11,3° (W)
O-Acetyl-smalogenin	$C_{25}H_{34}O_5$	265—267°	—
Nebenprodukt A aus *Strophanthus Eminii*	$C_{25}H_{36}O_6$	257—261°	+53,1° (Chf)
E. Sche. 16 aus *Periploca nigrescens*	$C_{25-27}H_{34-36}O_{7-8}$	236—239°	+54,8° (Chf)
Diginigenin............	$C_{21}H_{28}O_4$	115°	—226° (An)

Tabelle 2.

Aglykon	Formel	Schmp.	$[\alpha]_D$
1. Aglykone mit bekannter Konstitution.			
Scillarenin	$C_{24}H_{32}O_4$	225—238°	{ —16,8° (Me) / +17,9° (Chf) }
Bufalin	$C_{24}H_{34}O_4$	244—248°	—8,7° (Chf)
Scilliglaucosidin.........	$C_{24}H_{30}O_5$	245—248°	{ +49,5° (Me) / +78° (Chf) }
Scilliglaucosidin-19-ol	$C_{24}H_{32}O_5$	233—234°	{ —13° (Me) / +0,8° (Chf) }
Bovogenin A (= Bowieasubst. G)	$C_{24}H_{32}O_5$	247—262°	0° (Me)
Bovogenol A (= al-Di-hydro-Bowieasubst. A)..	$C_{24}H_{34}O_5$	260—270°	—15 4° (Me)
Telocinobufogenin (= Telocinobufagin)	$C_{24}H_{34}O_5$	160—175°/210—211°	+4,4° (Chf)
Hellebrigenin (= Bufo-talidin)	$C_{24}H_{32}O_6$	150—153°/237—240° 250—253°	+17,8° (An)
Hellebrigenol..........	$C_{24}H_{34}O_6$	146—153°	+4,1 (Chf-Me 1 · 1)
Bufotalin	$C_{26}H_{36}O_6$	148—156°/222—232°	+5,4° (Chf)
Gamabufotalin (= Gama-bufagin oder Gamabufo-genin)	$C_{24}H_{36}O_6$	253—255°/261—263°	—
2. Aglykone mit teilweise bekannter oder unbekannter Konstitution.			
Arenobufagin..........	$C_{24}H_{32}O_4$	220°/233°	—
Artebufogenin	$C_{24}H_{32}O_4$	275—283°	+29,1° (Chf)
Nabogenin	$C_{24}H_{30-32}O_5$	250—266°	+34,1° (Chf-Me 1 : 1)

Funktionelle Gruppen	UV.-Spektrum λ max zusätzlich zu 217 mμ (4,22) in mμ	Mittlere letale Dosis mg/kg (Katze)	Literatur
1 β-OH, 3 β-OH, 5 β-OH, 14 β-OH, 19-OH, 11 α(?)-OH	—	—	(*168, 211*)
—	—	—	(*349*)
—	—	—	(*160*)
—	zirka 300 (1,48)	—	(*247*)
—	—	unwirksam	(*265—268*)

Bufadienolide.

Funktionelle Gruppen	UV.-Spektrum λ max zusätzlich zu 300 mμ (3,7) in mμ	Mittlere letale Dosis mg/kg (Katze)	Literatur
3 β-OH, 14 β-OH, Δ^4	—	0,125	(*298, 311*)
3 β-OH, 14 β-OH	—	0,137	(*154, 176, 177*)
3 β-OH, 14 β-OH, 19-Oxo, Δ^4	—	0,064	(*300, 316*)
3 β-OH, 14 β-OH, 19-OH, Δ^4	—	—	(*316*)
3 β-OH, 14 β-OH, 19-Oxo, *A/B trans*	—	—	(*138*)
3 β-OH, 14 β-OH, 19-OH, *A/B trans*	—	—	(*138, 139*)
3 β-OH, 5 β-OH, 14 β-OH	—	0,101	(*176, 178*)
3 β-OH, 5 β-OH, 14 β-OH, 19-Oxo	—	0,077	(*260, 261, 358, 368*)
3 β-OH, 5 β-OH, 14 β-OH, 19 β-OH	—	—	(*140 a*)
3 β-OH, 14 β-OH, 16 β-OAc	—	0,137	(*176, 180, 370, 365*)
3 β-OH, 11 α-OH, 14 β-OH	—	0,100	(*152, 179*)
—	—	0,093	(*30*)
—	—	—	(*182*)
—	—	—	(*140*)

Fortsetzung der Tabelle 2

Aglykon	Formel	Schmp.	$[\alpha]_D$
Marinobufogenin (= Marinobufagin)	$C_{24}H_{32}O_5$	224—225°	+10,0° (Chf)
Kilimandscharogenin A ...	$C_{24}H_{34}O_5$	277—280°	+28,0° (Me)
Bufotalinin	$C_{24}H_{30-32}O_6$	195—201°	+19,2° (Chf)
Arenobufogenin	$C_{24}H_{32}O_6$	252°	—
Regularobufogenin	$C_{24}H_{34}O_6$	235—236°	—
Scillicoelosidin	$C_{24}H_{30-32}O_6$	233—234°	+84,9° (Me)
O-Acetyl-resinobufogenin .	$C_{26}H_{34}O_5$	218—230°	—1,1° (Chf)
Cinobufagin	$C_{26}H_{34}O_6$	216—217°	—3,6° (Chf)
Cinobufotalin	$C_{26}H_{34}O_7$	259—262°	+10,7° (Chf)
Scillirosidin	$C_{26}H_{34}O_7$	173—175°	—23° (Me)

Tabelle 3.

Aglykon	Glykosid	Formel	Schmp.
1. Aglykone mit bekannter Konstitution.			
Digitoxigenin $C_{23}H_{34}O_4$	Evatromonosid	$C_{29}H_{44}O_7$	181—186°
	Evomonosid	$C_{29}H_{44}O_8$	238—240°
	Somalin (= Honghelo-sid G)	$C_{30}H_{46}O_7$	133—136°/197—198°
	Odorosid A	$C_{30}H_{46}O_7$	183—184°/198—206°
	Digistrosid	$C_{30}H_{46}O_7$	173—175°/211—212°
	Neriifolin	$C_{30}H_{46}O_8$	218—225°
	Honghelin.............	$C_{30}H_{46}O_8$	133—136°
	Odorosid H............	$C_{30}H_{46}O_8$	235—238°
	Monoacetyl-neriifolin (= Cerberin = Veneni-ferin)	$C_{32}H_{48}O_9$	212—215°
	Odorosid H-monoacetat..	$C_{32}H_{48}O_9$	227—228°
	Evobiosid	$C_{35}H_{44}O_{13}$	195—198°
	Echubiosid	$C_{36}H_{56}O_{12}$	231—234°
	Odorosid D	$C_{36}H_{56}O_{12}$	219°/254—262°
	Gracilosid = Odorosid F .	$C_{36}H_{56}O_{13}$	298—302°
	Odorobiosid G	$C_{36}H_{56}O_{13}$	243—245°
	Thevebiosid	$C_{36}H_{56}O_{13}$	208—210°

Funktionelle Gruppen	UV.-Spektrum λ max zusätzlich zu 300 mμ (3,7) in mμ	Mittlere letale Dosis mg/kg (Katze)	Literatur
3 β-OH, 5 β-OH, Äther-O	—	1,489—1,552	(*122, 181, 198, 358*)
—	—	—	(*38*)
—	—	0,619	(*358, 368*)
—	—	—	(*366, 367*)
—	—	—	(*121*)
—	—	—	(*300*)
—	—	unwirksam	(*182*)
—	—	0,2085	(*153, 176*)
—	—	0,199	(*153, 176*)
—	—	0,057	(*308*)

Cardenolid - Glykoside.

$[\alpha]_D$	Mittlere letale Dosis mg/kg (Katze)	Zucker	Literatur
—14,6° (Me)	—	*D*-Digitoxose	(*354*)
—30,6° (Me)	0,278	*L*-Rhamnose	(*77a, 184, 242, 323*)
{ +9,5° (Al) +13,4° (Chf) }	0,372	*D*-Cymarose	(*73, 87, 88, 95*)
+7,0° (Chf)	0,186	*D*-Diginose	(*212, 213*)
—	—	Sarmentose	(*166*)
{ —50,2° (Me) —60° (Chf) —77° (Py) }	0,196	*L*-Thevetose	(*56, 83, 84*)
—11,2° (Me)	—	*D*-Thevetose	(*65, 223*)
+6,2° (Me), +8,7° (Chf)	0,200	*D*-Digitalose	(*212, 224, 229*)
—86° (Me)	0,370	*L*-Thevetose	(*83—85*)
+11,1° (Chf)	—	*D*-Digitalose + CH_3COOH	(*224*)
—28,2° (Me)	—	*L*-Rhamnose + Glucose	(*184, 242*)
+4,2° (Me)	0,290	*D*-Cymarose + Glucose	(*88*)
—19,2° (D)	0,594	*D*-Diginose + Glucose	(*212, 230*)
—15,3° (Me)	—	*D*-Digitalose + Glucose	(*2, 230, 236*)
—7,4° (Me)	0,273	*D*-Digitalose + Glucose + + CH_3COOH	(*224, 230*)
—62,9° (Me)	1,004	*L*-Thevetose + Glucose	(*83, 84*)

Fortsetzung der Tabelle 3

Aglykon	Glykosid	Formel	Schmp.
	Digitoxin	$C_{41}H_{64}O_{13}$	263—275°
	Evonosid	$C_{41}H_{64}O_{18}$	202—208°
	Echujin	$C_{42}H_{66}O_{17}$	165—172°
	Thevetin	$C_{42}H_{66}O_{18}$	189—192°
	Odorotriosid G-mono-acetat	$C_{42}H_{66}O_{18}$	265—266°/282—284°
	Acetyl-digitoxin-α	$C_{43}H_{66}O_{14}$	170—190°/217—221°
	Acetyl-digitoxin-β	$C_{43}H_{66}O_{14}$	221—224°
	Purpureaglykosid A (= Des-acetyl-digilanid A)	$C_{47}H_{74}O_{18}$	275—280°
	Lanatosid A (= Digi-lanid A)	$C_{49}H_{76}O_{19}$	245—248° (Zers.)
Uzarigenin $C_{23}H_{34}O_4$	Desgluco-cheirosid A	$C_{29}H_{44}O_8$	239—244°
	Odorosid B	$C_{30}H_{46}O_7$	150°/200—201°
	Cheirosid A (= Cheirosid H)	$C_{35}H_{54}O_{13}$	293—295°
	Uzarin	$C_{35}H_{54}O_{14}$	266—270°
	Odorobiosid K	$C_{36}H_{56}O_{12}$	173—176°/222—269°
	Odorosid K	$C_{42}H_{66}O_{17}$	195°/243—267°
Urezigenin $C_{23}H_{34}O_4$	Urezin	$C_{35}H_{54}O_{14}$	165°/185—192°
Desarogenin (= 11-Dehydro-sar-mentogenin) $C_{23}H_{32}O_5$	11-Dehydro-sarmento-cymarin	$C_{30}H_{44}O_8$	130—137°
	11-Dehydro-divaricosid	$C_{30}H_{44}O_8$	217—219°
	Desarosid	$C_{30}H_{44}O_9$	250—252°/265—268°
Acovenosigenin A $C_{23}H_{34}O_5$	Acovenosid A	$C_{30}H_{46}O_9$	223—226°
	Acovenosid C	$C_{42}H_{66}O_{19}$	188—190°
Periplogenin $C_{23}H_{34}O_5$	Periplocymarin	$C_{30}H_{46}O_8$	136—140°/210—213°
	Emicymarin	$C_{30}H_{46}O_9$	159—164°/200—205°
	Periplocin	$C_{36}H_{56}O_{13}$	207—208°
	Ledienosid	$C_{29}H_{44}O_9$	169—175°
	Vanderosid	$C_{30}H_{46}O_8$	171—175°
	Emicin	$C_{36}H_{56}O_{14}$	253—255°
Allo-periplogenin $C_{23}H_{34}O_5$	Allo-periplocymarin	$C_{30}H_{46}O_8$	128—131°
	Allo-emicymarin	$C_{30}H_{46}O_9$	162°/260—264°

$[\alpha]_D$	Mittlere letale Dosis mg/kg (Katze)	Zucker	Literatur
$+4,8°$ (D) $+16,7°$ (Chf)	0,325—0,420	3 *D*-Digitoxose	(*294, 301*)
$-35,1°$ (Me)	0,839	*L*-Rhamnose $+$ 2 Glucose	(*184, 242*)
$-9,8°$ (W) $-6,5°$ (Me)	0,304	*D*-Cymarose $+$ 2 Glucose	(*88*)
$-66,9°$ (Me)	0,889	*L*-Thevetose $+$ 2 Glucose	(*29, 56, 83, 84, 334*)
$-20,1°$ (Me)	0,622	*D*-Digitalose $+$ 2 Glucose $+$ $+$ CH_3COOH	(*212, 224*)
$+4,8°$ (Py), $+24°$ (Me)	0,447	3 *D*-Digitoxose $+$ CH_3COOH	(*299*)
$+16,2°$ (Py), $+26,3°$ (Me)	—	3 *D*-Digitoxose $+$ CH_3COOH	(*295, 299*)
$+10,8°$ (75% Al) $-5,4°$ (Py)	0,337	3 *D*-Digitoxose $+$ Glucose	(*294, 296, 301*)
$+31°$ (75% Al) $+2°$ (Py)	0,380	3 Digitoxose $+$ Glucose $+$ $+$ CH_3COOH	(*294, 295, 309, 372*)
$-12,9°$ (Me)	1,332	*D*-Fucose	(*188, 264*)
$-19,5°$ (Chf)	2,102	*D*-Diginose	(*212, 213, 231*)
$-23,8°$ (Py)	0,683	*D*-Fucose $+$ Glucose	(*188, 263, 264*)
$-27°$ (Py)	4,58	2 Glucose	(*93, 336, 341*)
$-36,0°$ (D) $-31,6°$ (Chf-Me 9:1)	2,290	*D*-Digitalose $+$ Glucose	(*231*)
$-38,7°$ (Me)	4,735	*D*-Digitalose $+$ 2 Glucose	(*229, 231*)
$-4,8°$ (Al)	3,611	2 Glucose	(*341*)
$-7,2°$ (An)	—	*D*-Sarmentose	(*49*)
$-42,6°$ (Me)	—	*L*-Oleandrose	(*248*)
$+6,7°$ (Me) $-10,7°$ (Chf)	0,596	*D*-Digitalose	(*249*)
$-65,0°$ (An)	0,240	*L*-Acovenose	(*44, 195, 259*)
$-63,0°$ (Me)	0,247	*L*-Acovenose $+$ 2 Glucose	(*186, 259*)
$+27,6°$ (Me) $+32,6°$ (Chf)	0,154	*D*-Cymarose	(*43, 46, 48, 120, 160*)
$+11,7°$ (Me) $+10,3°$ (Chf)	0,152	*D*-Digitalose	(*43, 46, 48, 104, 141, 160*)
$+22,9°$ (Me)	0,121	*D*-Cymarose $+$ Glucose	(*303*)
$+5,2°$ (Me)	0,141	Fucose	(*165*)
$+7,8°$ (Chf)	0,229	*D*-Diginose	(*166*)
$+14,7°$ (Chf)	—	*D*-Digitalose $+$ Glucose	(*238*)
$+48,3°$ (Me)	unwirksam	*D*-Cymarose	(*141, 142*)
$+28,7°$ (Me)	unwirksam	*D*-Digitalose	(*104, 141, 142, 158*)

Fortsetzung der Tabelle 3

Aglykon	Glykosid	Formel	Schmp.
Sarmentogenin $C_{23}H_{34}O_5$	Rhodexin A	$C_{29}H_{44}O_9$	250°
	Divaricosid	$C_{30}H_{46}O_8$	220—223°
	Sarmentocymarin	$C_{30}H_{46}O_8$	129—132°/149—155°/ 209—211°
	Kwangosid	$C_{30}H_{46}O_8$	212—217°
	Sarnovid	$C_{30}H_{46}O_9$	149—154°/223—225°
	Sargenosid-diacetat (= ,,Sarmentosid B'') . .	$C_{40}H_{60}O_{16}$	210—215°/260—265°
11-Epi-sarmento-genin $C_{23}H_{34}O_5$	11-Epi-sarnovid	$C_{30}H_{46}O_9$	156—159°
Digoxigenin $C_{23}H_{34}O_5$	Digoxin	$C_{41}H_{64}O_{14}$	265°
	Acetyl-digoxin-α (= Digorid B)	$C_{43}H_{66}O_{15}$	230° (Zers.)
	Acetyl-digoxin-β (= Digorid A)	$C_{43}H_{66}O_{15}$	170°/258° (Zers.)
	Desacetyl-digilanid C	$C_{47}H_{74}O_{19}$	265—268°
	Lanatosid C (= Digi-lanid C)	$C_{49}H_{76}O_{20}$	245—248° (Zers.)
Gitoxigenin $C_{23}H_{34}O_5$	Rhodexin B	$C_{29}H_{44}O_9$	262°
	Gitorin	$C_{29}H_{44}O_{10}$	205—212°
	16-Desacetyl-hon-ghelosid A	$C_{30}H_{46}O_8$	208—210°
	16-Desacetyl-oleandrin . . .	$C_{30}H_{46}O_8$	238—240°
	16-Desacetyl-cryptograndosid A	$C_{30}H_{46}O_8$	198—199°
	Strospesid (= Desgluco-digitalinum verum)	$C_{30}H_{46}O_9$	251—253°
	Strospesid-monoacetat . . .	$C_{32}H_{48}O_{10}$	232—234°
	Oridigin	$C_{35}H_{54}O_{13}$	255° (Zers.)
	Rhodexin C	$C_{35}H_{54}O_{14}$	275°
	Digitalinum verum	$C_{36}H_{56}O_{14}$	241—244°
	Digitalinum verum-monoacetat	$C_{38}H_{58}O_{15}$	257—262°
	Gitoxin	$C_{41}H_{64}O_{14}$	285°/300°
	Acetyl-gitoxin-α	$C_{43}H_{66}O_{15}$	203—204°
	Acetyl-gitoxin-β	$C_{43}H_{66}O_{15}$	275—276°
	Hexa-O-acetyl-gitofucosid	$C_{47}H_{66}O_{11}$	153—154°
	Purpureaglykosid B (= Desacetyl-digi-lanid B)	$C_{47}H_{74}O_{19}$	240—242°

$[\alpha]_D$	Mittlere letale Dosis mg/kg (Katze)	Zucker	Literatur
—20° (Al)	0,096	*L*-Rhamnose	(*191*)
—46,0° (Me)	0,165	*L*-Oleandrose	(*257, 258*)
—13,2° (Me)	0,202	*D*-Sarmentose	(*23, 42, 117, 166*)
—9,4° (Chf)	—	*D*-Diginose	(*166*)
+6,9° (Me), +5,1° (Chf)	0,149	*D*-Digitalose	(*42, 166*)
+1,5° (An)	unwirksam	*D*-Digitalose + Glucose	(*24, 49, 216*)
+18,8° (Me)	0,398	*D*-Digitalose	(*249*)
+10,4° (Py)	0,280	3 *D*-Digitoxose	(*294, 295*)
+18,0° (Py)	0,36	3 *D*-Digitoxose + CH₃COOH	(*167, 295*)
{+29,2° (Py)} {+30,4° (Al)}	0,375	3 *D*-Digitoxose + CH₃COOH	(*167, 295*)
+12,2° (75% Al)	0,228	3 *D*-Digitoxose + Glucose	(*294*)
{+33,5° (Al)} {+22,6° (D)}	0,280	3 *D*-Digitoxose + Glucose + + CH₃COOH	(*294, 295*)
—39,5° (Al)	0,34	*L*-Rhamnose	(*191*)
+7° (Me)	0,437	*D*-Glucose	(*194*)
+13,6° (Me)	0,669	*D*-Cymarose	(*95*)
—24,9° (Me)	wie Oleandrin	*L*-Oleandrose	(*193*)
—3,4° (Me)	—	*D*-Sarmentose	(*2*)
+15,5° (Me)	0,586; 0,407	*D*-Digitalose	(*228*)
+18,8° (Me)	0,691	*D*-Digitalose + CH₃COOH	(*228*)
+12,5° (Me)	wie Digitoxin	2 Desoxymethylpentose + + Glucose	(*167*)
—17,7° (70% Al)	2,3	Rhamnose + Glucose	(*192*)
+1,6° (Me)	1,332	*D*-Digitalose + Glucose	(*228, 243*)
—2,9° (Me)	3,331; 3,843	*D*-Digitalose + Glucose + + CH₃COOH	(*228*)
{ +3,5° (Py) } {+22° (Chf-Me 1 : 1)}	0,727	3 *D*-Digitoxose	(*294, 299, 301*)
{+16,0° (Py)} {+28,7° (Me)}	0,525	3 *D* Digitoxose + CH₃COOH	(*315*)
+26,9° (Py)	—	3 *D*-Digitoxose + CH₃COOH	(*315*)
—	—	Fucose + Glucose + + 6 CH₃COOH	(*164*)
{ +15,5° (75% Al) } { +2,5° (Py) }	0,369	3 *D*-Digitoxose	(*294, 296, 301, 309*)

Fortsetzung der Tabelle 3

Aglykon	Glykosid	Formel	Schmp.
	Lanatosid B (= Digilanid B).............	$C_{49}H_{76}O_{20}$	$232°, 245—248°$ (Zers.)
	Hepta-O-acetyl-gluco-gitofucosid	$C_{49}H_{68}O_{21}$	$153—154°$
	Octa-O-acetyl-glucogitorin	$C_{51}H_{70}O_{22}$	$142—146°$
Gitaloxigenin (= 16-Formyl-gitoxigenin) $C_{24}H_{34}O_6$	Gitaloxin	$C_{42}H_{64}O_{15}$	$250—253°$
Oleandrigenin (= 16-Acetylgitoxi-genin) $C_{25}H_{36}O_6$	Honghelosid A..........	$C_{32}H_{48}O_9$	$208—211°$
	Oleandrin	$C_{32}H_{48}O_9$	$250°$
	Cryptograndosid A	$C_{32}H_{48}O_9$	$122—124°$
	Honghelosid C	$C_{38}H_{58}O_{14}$	$155—158°$
	Urechitoxin	$C_{38}H_{58}O_{14}$	$157—159°$
	Tetra-O-acetyl-crypto-grandosid B	$C_{46}H_{66}O_{18}$	$123—125°$
16-Anhydro-gitoxigenin $C_{23}H_{32}O_4$	16-Desacetyl-16-anhydro-honghelosid A	$C_{30}H_{44}O_7$	$208—211°/218°$
	16-Desacetyl-16-anhydro-oleandrin	$C_{30}H_{44}O_7$	$230—234°$
	16-Desacetyl-16-anhydro-cryptograndosid A	$C_{30}H_{44}O_7$	$230—232°$
	16-Anhydro-strospesid ...	$C_{30}H_{44}O_8$	$242—246°$
	16-Anhydro-strospesid-monoacetat	$C_{32}H_{46}O_9$	$157—160°/232—243°$
	16-Desacetyl-16-anhydro-cryptograndosid B	$C_{36}H_{54}O_{12}$	$198—202°$
	16-Anhydro-digitalinum verum-monoacetat	$C_{38}H_{56}O_{14}$	$284—289°$
Corotoxigenin $C_{23}H_{32}O_5$	Strobosid	$C_{29}H_{42}O_8$	$204—206°$
	Boistrosid.............	$C_{29}H_{42}O_8$	$213—219°$
	Gofrusid	$C_{29}H_{42}O_9$	$250—257°$
	Millosid..............	$C_{30}H_{44}O_8$	$142—146°$
	Pauliosid.............	$C_{30}H_{44}O_8$	$203—205°$
	Christyosid	$C_{30}H_{44}O_9$	$210—213°$
Coroglaucigenin $C_{23}H_{34}O_5$	Frugosid	$C_{29}H_{44}O_9$	$160—170°/240—242°$
Adonitoxigenin $C_{23}H_{32}O_6$	Adonitoxin.............	$C_{29}H_{42}O_{10}$	$240—255°/262—265°$
Strophanthidin $C_{23}H_{32}O_6$	Corchorosid A	$C_{29}H_{42}O_9$	$167—169°/188—190°$

$[\alpha]_D$	Mittlere letale Dosis mg/kg (Katze)	Zucker	Literatur
$\left\{ \begin{array}{l} +36,7° \text{ (Al)} \\ +6° \quad \text{(Py)} \end{array} \right\}$	0,403	3 D-Digitoxose + Glucose + + CH$_3$COOH	(294, 295, 309)
—2° (Me)	—	Fucose + Glucose + + 7 CH$_3$COOH	(343)
—	—	2 Glucose + 8 CH$_3$COOH	(164)
—7° (Py)	—	3 D-Digitoxose + H · COOH	(71)
—14,0° (Me)	0,387	D-Cymarose	(95)
—52,1° (Me)	0,197	L-Oleandrose	(193)
—32,9° (Me)	0,220	D-Sarmentose	(2)
—9,6° (Me)	0,364	D-Cymarose + Glucose	(95)
—58,4° (Me)	—	L-Oleandrose + Glucose	(75, 372)
—22,7° (Chf)	—	D-Sarmentose + Glucose	(2)
+75,8° (Me)	unwirksam	D-Cymarose	(95)
+18,3° (Me)	unwirksam	L-Oleandrose	(2)
+53,2° (Me)	unwirksam	D-Sarmentose	(2)
+69,4° (Me)	—	D·Digitalose	(96, 229)
+66° (Me)	—	D-Digitalose + CH$_3$COOH	(229)
+28,3° (Me)	—	D-Sarmentose + Glucose	(2)
—11,2° (Py)	—	D-Digitalose + Glucose + + CH$_3$COOH	(229)
—13,5° (Me)	0,255	D-Boivinose	(254)
+5,1° (Me)	—	D-Digitoxose	(254)
—5,1° (Me)	0,191	D-Allomethylose	(99, 145)
—1,4° (Me)	1,330	D-Cymarose	(254)
—10,1° (Me)	0,714	D-Sarmentose	(254)
+10,5° (Me)	1,456	D-Digitalose	(256)
—17,4° (80% Me)	0,161	D-Allomethylose	(97, 99)
—25,7° (50% Al)	0,191	L-Rhamnose	(143, 233, 348)
$\left\{ \begin{array}{l} +11° \quad \text{(Me)} \\ +19,2° \text{ (Me)} \\ +14,5° \text{ (Chf)} \end{array} \right\}$	0,077	D-Boivinose	(60, 156)

Fortsetzung der Tabelle 3

Aglykon	Glykosid	Formel	Schmp.
	Helveticosid	$C_{29}H_{42}O_9$	$151—155°$
	Convallatoxin	$C_{29}H_{42}O_{10}$	$211—212°/243—247°$
	Desgluco-cheirotoxin	$C_{29}H_{42}O_{10}$	$188—189°$
	Strophanthidin-β-D-glucosid —$\langle 1,5 \rangle$	$C_{29}H_{42}O_{11}$	$234—236°$
	Cymarin	$C_{30}H_{44}O_9$	$132—134°/148°$ $185—187°$
	Convallosid.	$C_{35}H_{52}O_{15}$	$201—204°$
	Cheirotoxin	$C_{35}H_{52}O_{15}$	$210—211°$
	k-Strophanthin-β	$C_{36}H_{54}O_{14}$	$195°$
	k-Strophanthosid	$C_{42}H_{64}O_{19}$	$199—200°$
Allostrophanthidin $C_{23}H_{32}O_6$	Allo-cymarin	$C_{30}H_{44}O_9$	$150°$
Strophanthidol $C_{23}H_{34}O_6$	Convallatoxol.	$C_{29}H_{44}O_{10}$	$171—173°$
	Strophanthidol-β-D-glucosid —$\langle 1,5 \rangle$	$C_{29}H_{44}O_{11}$	$187—190°$
	Cymarol	$C_{30}H_{46}O_9$	$222—240°$
	k-Strophanthol-γ.	$C_{42}H_{66}O_{19}$	$195—200°$
2. Aglykone mit teilweise bekannter Konstitution.			
Adynerigenin $C_{23}H_{32}O_4$	Adynerin	$C_{30}H_{44}O_7$	$228—234°$
Neriantogenin $C_{23}H_{32}O_4$	Neriantin	$C_{29}H_{42}O_9$	$206—208°$
Tanghiferigenin $C_{23}H_{32}O_4$	Neotanghiferin.	$C_{32}H_{46}O_9$	$250—252°$
Xysmalogenin $C_{23}H_{34}O_4$	Xysmalorin	$C_{35}H_{54}O_{14}$	$220—224°$
Tanghinigenin $C_{23}H_{32}O_5$	Desacetyl-tanghinin (= Pseudotanghinin) . .	$C_{30}H_{44}O_9$	$217°/239—245°$
	Tanghinin	$C_{32}H_{46}O_{10}$	$127—129°$
	Protanghinin	$C_{38}H_{56}O_{15}$	$162°$
Abogenin $C_{23}H_{34}O_{5-6}$	Abomonosid.	$C_{32}H_{48}O_9$ (ev. $C_{30}H_{46}O_{8-9}$)	$128—131°$
	Abobiosid	$C_{38}H_{58}O_{14}$ (?)	$244—249°$
Decogenin $C_{23}H_{30}O_6$	Decosid	$C_{30}H_{42}O_9$	$195—198°$

$[\alpha]_D$	Mittlere letale Dosis mg/kg (Katze)	Zucker	Literatur
$\begin{cases}+30,6° \text{ (Me)} \\ +26,0° \text{ (Chf)}\end{cases}$	0,080	D-Digitoxose	(190)
$\begin{cases}0° \text{ (Me)} \\ -15° \text{ (Py)}\end{cases}$	0,079	L-Rhamnose	(128, 130, 346)
$+1,3°$ (Me)	0,096	D-Gulomethylose	(188, 264)
$\begin{cases}+21,0° \text{ (W)} \\ +17° \text{ (Me)}\end{cases}$	0,091	Glucose	(172, 357)
$+35°$ (Chf)	0,111	D-Cymarose	(43, 46, 48, 118, 313)
$-10,4°$ (80% Al)	0,215	L-Rhamnose + Glucose	(262)
$-17,2°$ (Me)	0,119	D-Gulomethylose + Glucose	(188, 263, 264)
$+31,8°$ (Me)	0,120	D-Cymarose + Glucose	(118, 119, 313)
$+13,8°$ (Me)	0,126	D-Cymarose + 2 Glucose	(313)
$+43°$ (Me)	unwirksam	D-Cymarose	(102, 141)
$-10,0°$ (Me)	0,087	L-Rhamnose	(98, 350)
$\begin{cases}+5,8° \text{ (Me)} \\ +4,6° \text{ (W)}\end{cases}$	—	Glucose	(158)
$+22,1°$ (80% Me)	0,994	D-Cymarose	(18, 43, 46, 48, 160)
$+8,6°$ (Me)	—	D-Cymarose + 2 Glucose	(210)
$+7,5°$ (Py)	unwirksam	Diginose	(193, 339, 340, 342, 352)
—	unwirksam	—	(339, 340, 352)
$\begin{cases}+49,6° \text{ (Al)} \\ -48,6° \text{ (Chf)}\end{cases}$	0,944	L-Thevetose	(62, 65a, 85)
$-17,8°$ (Py)	unwirksam	2 Glucose	(341)
$-57,9°$ (Al)	0,231	L-Thevetose	(61, 63, 64, 85)
$-83,6°$ (Me)	0,352	L-Thevetose + CH_3COOH	(85)
$-86,0°$ (Me)	—	L-Thevetose + Glucose + CH_3COOH	(64)
$-18,0°$ (Chf)	0,679	D-Cymarose	(88)
$-26,2°$ (Me)	0,699	D-Cymarose + Glucose	(88)
$+3,0°$ (Me)	—	L-Oleandrose	(248)

Fortsetzung der Tabelle 3

Aglykon	Glykosid	Formel	Schmp.
Sarmutogenin $C_{23}H_{32}O_6$	Sarmutosid............	$C_{30}H_{44}O_9$	132—135°/233—244°/ 250—252°
	Ambosid..............	$C_{30}H_{44}O_9$	195—197°/200—201°
	Musarosid.............	$C_{30}H_{44}O_{10}$	229—232°
Caudogenin $C_{23}H_{32}O_6$	Caudosid	$C_{30}H_{44}O_9$	249—252°
Erysimidin $C_{23}H_{32}O_6$	Erysimin..............	$C_{29}H_{42}O_9$	168—172°
Calotropagenin $C_{23}H_{32}O_6$	Calotropin	$C_{29}H_{40}O_9$	221° (Zers.)
Quilengenin $C_{23}H_{32-34}O_6$	Quilengosid (= Subst. Nr. 795) ...	$C_{30}H_{44-46}O_9$	151—152°/151—170°
Chryseogenin $C_{23}H_{32}O_7$	Chryseosid	$C_{30}H_{40}O_{10}$	145—147°/156—158°
Sarverogenin $C_{23}H_{32}O_7$	Sarverosid	$C_{30}H_{44}O_{10}$	125—130°/145°
	Intermediosid	$C_{30}H_{44}O_{10}$	135—140°/187—180°/ 198—200°
	Panstrosid	$C_{30}H_{44}O_{11}$	185°/212—218°/ 230—233°
	Interosid	$C_{36}H_{54}O_{15}$	236—238°
	i-Strophanthosid	$C_{42}H_{64}O_{20}$	195—200°
Inertogenin $C_{23}H_{32}O_7$	Inertosid	$C_{30}H_{44}O_{10}$	155—168°/227—244°
Leptogenin $C_{23}H_{32}O_7$	Leptosid	$C_{30}H_{44}O_{10}$	199—202°
Antiarigenin $C_{23}H_{32}O_7$	α-Antiarin	$C_{29}H_{42}O_{11}$	242—243°
	β-Antiarin.............	$C_{29}H_{42}O_{11}$	206°/233—240°/ 240—245°
al-Dihydro-antiari- genin $C_{23}H_{34}O_7$	al-Dihydro-α-antiarin ...	$C_{29}H_{44}O_{11}$	211—213°
	al-Dihydro-β-antiarin	$C_{29}H_{44}O_{11}$	248—257°
Ouabagenin $C_{23}H_{34}O_8$	Ouabain (= g-Strophanthin) ...	$C_{29}H_{44}O_{12}$	183—187°/241°
3. *Verwandte Derivate.*			
Diginigenin $C_{21}H_{28}O_4$	Diginin	$C_{28}H_{40}O_7$	155—183°/193—196°
Convallamaretin $C_{26}H_{40}O_5$	Convallamarin	$C_{44}H_{70}O_{19}$	amorph

$[\alpha]_D$	Mittlere letale Dosis mg/kg (Katze)	Zucker	Literatur
$+10,4°$ (Me)	0,478	D-Sarmentose	(226, 258)
$+25,0°$ (Me)	0,827	D-Diginose	(81)
$+29,0°$ (Me)	0,481	D-Digitalose	(226)
$-99,8°$ (Me)	0,959	L-Oleandrose	(257, 258)
$+43,5°$ (Al)	—	2-Desoxy-hexamethylose	(51)
$+55,7°$ (Me)	0,12	—	(90)
$-39,1°$ (Me)	2,874 und 7,127	D-Diginose	(37, 39)
$+55,0°$ (Chf)	unwirksam	D-Diginose	(82)
$+12,1°$ (An)	0,403	D-Sarmentose	(23, 40, 45, 47)
$+18,9°$ (An)	1,839	D-Diginose	(39, 45, 47, 82)
$+31,7°$ (Me)	0,997	D-Digitalose	(39, 45, 47)
$+12,7°$ (W)	—	D-Diginose + Glucose	(363)
$\begin{cases}-5,8° \text{ (Me)}\\-9,3° \text{ (W)}\end{cases}$	—	D-Diginose + 2 Glucose	(355)
$-45,8°$ (Me)	unwirksam	D-Diginose	(80, 82)
$+42,9°$ (Me)	1,888	D-Diginose	(80, 82)
$-3,9°$ (Me)	0,116	D-Gulomethylose = Anti-arose	(34, 35, 146)
$\begin{cases}\quad 0° \quad \text{ (Me)}\\+2,2° \text{ (Me-W 1:1)}\end{cases}$	0,100	L-Rhamnose	(35, 147)
$-9,4°$ (Me)	0,120	D-Gulomethylose	(34)
$-6,7°$ (Me)	0,107	L-Rhamnose	(35)
$-44,3°$ (Me)	0,116	L-Rhamnose	(4, 103, 105, 168)
$\begin{cases}-223° \text{ (Chf)}\\-172° \text{ (Me)}\\-176° \text{ (Al)}\end{cases}$	unwirksam	D-Diginose	(129, 265—268, 343)
$-66°$ (Me)	unwirksam	L-Rhamnose + 2 Glucose	(360, 361)

Tabelle 4.

Aglykon	Glykosid	Formel	Schmp.
Scillarenin $C_{24}H_{32}O_4$	Proscillaridin A	$C_{30}H_{42}O_8$	$215—218°$
	Scillaren A	$C_{36}H_{52}O_{13}$	$230—240°/270—273°$
	Glucoscillaren A	$C_{42}H_{62}O_{18}$	$228—232°$
	Transvaalin $=$ Scillaren A		
Scilliglaucosidin $C_{24}H_{30}O_5$	Scilliglaucosid	$C_{30}H_{40}O_{10}$	$164—166°$
Bovogenin A $C_{24}H_{32}O_5$	Bovosid A...............	$C_{31}H_{44}O_9$	$205—235°/240—245°$
Bovogenol A $C_{24}H_{34}O_5$	Bovosidol A	$C_{31}H_{46}O_9$	$220—238°$
Hellebrigenin $C_{24}H_{32}O_6$	Desgluco-hellebrin.......	$C_{30}H_{42}O_{10}$	$237—242°/269—273°$
	Hellebrin	$C_{36}H_{52}O_{15}$	$283—284°$
Hellebrigenol $C_{24}H_{34}O_6$	Desgluco-hellebrol	$C_{30}H_{44}O_{10}$	$170—172°$
Scillirosidin $C_{26}H_{34}O_7$	Scillirosid	$C_{32}H_{46}O_{12}$	$170°$

Bufadienolid-Glykoside.

$[\alpha]_D$	Mittlere letale Dosis mg/kg (Katze)	Zucker	Literatur
—82,6° (Me)	0,157	L-Rhamnose	(*297*)
—73,8° (75% Al)	0,145	L-Rhamnose + Glucose	(*297*)
—66° (Me)	0,172	L-Rhamnose + 2 Glucose	(*297*) (*378*)
+106° (Me)	0,069	Glucose	(*297*)
+73,3° (Me)	0,20	L-Thevetose	(*133—135, 139*)
—87,0° (Me)	—	L-Thevetose	(*135, 139*)
—24,9° (Me)	0,086	L-Rhamnose	(*260*)
—23,4° (50% Me)	0,104	L-Rhamnose + Glucose	(*129, 261*)
—32,7° (Me)	0,092	L-Rhamnose	(*98*)
—59° (Me)	0,120	Glucose	(*305, 312*)

Literaturverzeichnis.

1. AEBI, A. und T. REICHSTEIN: Über die Glykoside der Blätter von *Cryptostegia grandiflora* (ROXB.) R. BR. (Asclepiadaceae). Helv. Chim. Acta **33**, 1013 (1950).

2. — — Die Glykoside von *Strophanthus gracilis* K. SCH. et PAX. Helv. Chim. Acta **34**, 1277 (1951).

3. ALEXANDER, E. R.: Principles of Ionic Organic Reactions, p. 281. New York: J. Wiley and London: Chapman a. Hall. 1950.

4. ARNAUD: Sur la matière cristallisée active des flèches empoisonnées des Çomalis, extraite du bois d'Ouabaïo. C. R. hebd. Séances Acad. Sci. **106**, 1011 (1888).

5. BALJET, H.: Les glucosides digitaliques. Une nouvelle réaction d'identité. Schweiz. Apoth.-Ztg. **56**, 71 (1918).

6. — Les glucosides digitaliques. Une nouvelle réaction d'identité. Schweiz. Apoth.-Ztg. **56**, 84 (1918).

7. BALLY, P. R. O., K. MOHR und T. REICHSTEIN: Die Glykoside von *Acokanthera longiflora* STAPF. Helv. Chim. Acta **34**, 1740 (1951).

8. — — — Die Glykoside der Wurzelrinde von *Acokanthera friesiorum* MARKGR. Helv. Chim. Acta **35**, 45 (1952).

9. BALLY, P. R. O., O. SCHINDLER und T. REICHSTEIN: Die Glykoside von *Strophanthus mirabilis* GILG. Helv. Chim. Acta **35**, 138 (1952).

10. BARTON, D. H. R.: The Stereochemistry of *cyclo*Hexane Derivatives. J. Chem. Soc. (London) **1953**, 1027.

11. BAUMGARTNER, G. und CH. TAMM: $3\alpha,12\alpha$-Dioxy-11-keto-cholansäure. Helv. Chim. Acta **38**, 441 (1955).

12. BELL, F. K. and J. C. KRANTZ, Jr.: Digitalis. VIII. The Baljet Reaction, Digitoxin and Digitoxigenin. J. Amer. Pharmaceut. Assoc., Sci. Ed. **38**, 107 (1949).

13. — — Digitalis. VII. Effect of Various Alkalis on the Sensitivity of the Baljet Reaction for Digitoxin. J. Amer. Pharmaceut. Assoc., Sci. Ed. **37**, 297 (1948).

14. BELLET, P.: Recherches sur les réactions colorées des hétérosides cardiotoniques et, principalement, du digitoxoside et du gitoxoside. Ann. pharm. franç. **8**, 471 (1950).

15. BERNASCONI, R., H. P. SIGG und T. REICHSTEIN: Nachweis von Glykosiden und Aglykonen vom Scilla-Bufo-Typ auf Papierchromatogrammen durch direkte Photokopie. Helv. Chim. Acta **38**, 1767 (1955).

16. BLADON, P., J. M. FABIAN, H. B. HENBEST, H. P. KOCH and G. W. WOOD: Studies in the Sterol Group. LII. Infra-red Absorption of Nuclear Tri- and Tetra-substituted Ethylenic Centres. J. Chem. Soc. (London) **1951**, 2402.

17. BLADON, P., H. B. HENBEST and G. W. WOOD: Studies in the Sterol Group. LV. Ultra-violet Absorption Spectra of Ethylenic Centres. J. Chem. Soc. (London) **1952**, 2737.

18. BLOME, W., A. KATZ und T. REICHSTEIN: Cymarol, ein neues herzaktives Glykosid aus *Strophanthus kombé*. Pharmac. Acta Helv. **21**, 325 (1946).

19. BOEHRINGER, C. F. und Söhne, G. m. b. H.: D. B. P. 18012 (1951).

20. BUSH, I. E.: Methods of Paper Chromatography of Steroids Applicable to the Study of Steroids in Mammalian Blood and Tissues. Biochemic. J. **50**, 370 (1952).

21. BUSH, I. E. and D. A. H. TAYLOR: The Paper-Chromatographic Examination of the Cardiac Aglycones of *Strophanthus* Seeds. Biochemic. J. **52**, 643 (1952).

22. BUZAS, A. und T. REICHSTEIN: Die zwei durch Abbau von Strophanthidin erhältlichen, an C-17 raumisomeren $3\beta,5,14$-Trioxy-14-iso-21-norpregnan-19,20-disäuren. Helv. Chim. Acta **31**, 84 (1948).

23. BUZAS, A., J. v. EUW und T. REICHSTEIN: Die Glykoside der Samen von *Strophanthus sarmentosus* P. DC. 2. Mitt. Helv. Chim. Acta 33, 465 (1950).

24. CALLOW, R. K. and D. A. H. TAYLOR: The Cardio-active Glycosides of *Strophanthus sarmentosus* P. DC. "Sarmentoside B" and its Relation to an Original Sarmentobioside. J. Chem. Soc. (London) 1952, 2299.

25. CANBÄCK, T.: Chemical Assay of Digitalis and *Strophanthus* Glycosides. J. Pharm. Pharmacol. 1, 201 (1949).

26. CARDWELL, H. M. E. and S. SMITH: Digitalis Glucosides. VII. The Structure of the Digitalis Anhydrogenins and the Orientation of the Hydroxy-groups in Digoxigenin and Gitoxigenin. J. Chem. Soc. (London) 1954, 2012.

27. CHAIKIN, S. W. and W. G. BROWN: Reduction of Aldehydes, Ketones and Acid Chlorides by Sodium Borohydride. J. Amer. Chem. Soc. 71, 122 (1949).

28. CHAKRABARTI, J. K. and N. K. SEN: Bitter Constituents of the Seeds of *Corchorus olitorius* L., "Corchorgenin"—A New Cardiac-active Aglycone. J. Amer. Chem. Soc. 76, 2390 (1954).

29. CHEN, K. K. and A. L. CHEN: The Occurrence of Thevetin and Kokilphin in the Nuts of *Thevetia ycottli*. Amer. J. Physiol. 123, 36 (1938).

30. CHEN, K. K., H. JENSEN and A. L. CHEN: The Physiological Action of the Principles Isolated from the Secretion of *Bufo arenarum*. J. Pharmacol. 49, 1 (1933).

31. CONDSEN, R., A. H. GORDON and A. J. P. MARTIN: Qualitative Analysis of Proteins: a Partition Chromatographic Method using Paper. Biochemic. J. 38, 224 (1944).

32. DEQUEKER, R. and M. LOOBUYCK: The Determination of Digitalis Glycosides by Means of Keller-Kiliani and Pesez-Dequeker Reagents. J. Pharm. Pharmacol. 7, 522 (1955).

33. DJERASSI, C. and R. EHRLICH: Lead Tetraacetate Oxidation of Steroidal Glycols. Some Observations on Ouabagenin. J. Organ. Chem. (USA) 19, 1351 (1954).

34. DOEBEL, K., E. SCHLITTLER und T. REICHSTEIN: Beitrag zur Kenntnis des α-Antiarins. Helv. Chim. Acta 31, 688 (1948).

35. DOLDER, F., CH. TAMM und T. REICHSTEIN: Die Glykoside von *Antiaris toxicaria* LESCH. Helv. Chim. Acta 38, 1364 (1955).

36. ELDERFIELD, R. C.: Strophanthin. XXXV. The Nature of "the Acid $C_{23}H_{30}O_8$" from Strophanthidin. J. Biol. Chem. 113, 631 (1936).

37. EUW, J. v., H. HEGEDÜS, CH. TAMM und T. REICHSTEIN: Die Glykoside der Samen von *Strophanthus amboënsis* (SCHINZ) ENGL. et PAX. sowie einer verwandten Form. Helv. Chim. Acta 37, 1493 (1954).

38. EUW, J. v., G. A. O. HEITZ, H. HESS, P. SPEISER und T. REICHSTEIN: Die Glykoside von *Strophanthus Welwitschii* (BAILL.) K. SCHUM. Helv. Chim. Acta 35, 152 (1952).

39. EUW, J. v., H. HESS, P. SPEISER und T. REICHSTEIN: Die Glykoside von *Strophanthus intermedius* PAX. Helv. Chim. Acta 34, 1821 (1951).

40. EUW, J. v., A. KATZ, J. SCHMUTZ und T. REICHSTEIN: Die Glykoside der Samen von *Strophanthus sarmentosus* P. DC. I. Festschrift Prof. P. CASPARIS, S. 178. Zürich: Schweiz. Apothekerverein. 1949.

41. EUW, J. v., A. LARDON und T. REICHSTEIN: Teilsynthese des Corticosterons. Helv. Chim. Acta 27, 1287 (1944) [s. insbes. Fußnote 2, S. 1292].

42. EUW, J. v., F. REBER et T. REICHSTEIN: Sur la provenance de la sarmentocymarine: *Strophanthus* spec. var. *sarmentogenifera* No. MPD 50 (Communication préliminaire). Helv. Chim. Acta 34, 413 (1951).

43. EUW, J. v. und T. REICHSTEIN: Die Glykoside der Samen von *Strophanthus Nicholsonii* HOLM. Helv. Chim. Acta **31**, 883 (1948).

44. — — Acovenosid A und Acovenosid B, zwei Glykoside aus den Samen von *Acokanthera venenata* G. DON. 1. Mitt. Helv. Chim. Acta **33**, 485 (1950).

45. — — Die Glykoside der Samen von *Strophanthus Gerrardi* STAPF. Helv. Chim. Acta **33**, 522 (1950).

46. — — Die Glykoside der Samen von *Strophanthus hypoleucus* STAPF. Helv. Chim. Acta **33**, 544 (1950).

47. — — Die Glykoside der Samen von *Strophanthus Courmontii* SACL. Helv. Chim. Acta **33**, 1006 (1950).

48. — — Die Glykoside der Samen von *Strophanthus hispidus* P. DC. 1. Mitt. Helv. Chim. Acta **33**, 1546 (1950).

49. — — Sargenosid („Sarmentosid B") und einige Derivate des Sarmentogenins. Helv. Chim. Acta **35**, 1560 (1952).

50. FEIST, F.: Strophanthin und Strophanthidin. Ber. dtsch. chem. Ges. **31**, 534 (1898).

51. FEOFILAKTOV, V. V. and P. M. LOSHKAREV: Erysimine-glycoside with Cardiac Action from *Erysimum canescens* ROTH. C. R. (Doklady) Acad. Sci. (USSR) **94**, 709 (1954) [Chem. Abstr. **49**, 6287 (1955)].

52. FIESER, L. F. and M. FIESER: Natural Products Related to Phenanthrene. 3rd Ed., p. 507. New York: Reinhold Publ. Corp. 1949.

53. FIESER, L. F. and M. S. NEWMAN: Ouabain. J. Biol. Chem. **114**, 705 (1936).

54. FLOREY, K. and M. EHRENSTEIN: Investigations on Steroids. XXII. Studies on Ouabagenin. I. J. Organ. Chem. (USA) **19**, 1174 (1954).

55. FOSTER, A. B. and W. G. OVEREND: The Acidic Hydrolysis of *O*-Glycosides. Chem. and Ind. **1955**, 566.

56. FRÈREJACQUE, M.: La nériifoline, nouvel hétéroside digitalique de *Thevetia neriifolia*. C. R. hebd. Séances Acad. Sci. **221**, 645 (1945).

57. — Thévétine, nériifoline et monoacétylnériifoline. C. R. hebd. Séances Acad. Sci. **225**, 695 (1947).

58. — Caractérisation du groupe acétyle dans les hétérosides digitaliques. C. R. hebd. Séances Acad. Sci. **240**, 1804 (1955).

59. FRÈREJACQUE, M. et M. DURGEAT: Détection des glucosides digitaliques. C. R. hebd. Séances Acad. Sci. **236**, 410 (1953).

60. — — Poisons digitaliques des graines de jute. C. R. hebd. Séances Acad. Sci. **238**, 507 (1954).

61. FRÈREJACQUE, M. et V. HASENFRATZ: Sur les principes immédiats des amandes de Tanghin. C. R. hebd. Séances Acad. Sci. **222**, 149 (1946).

62. — — Sur la tanghiférine, nouvel hétéroside des amandes de *Tanghinia venenifera*. Identité de la pseudotanghinine et de la désacétyl-tanghinine. C. R. hebd. Séances Acad. Sci. **223**, 642 (1946).

63. — — Sur les hétérosides digitaliques de *Tanghinia venenifera*. C. R. hebd. Séances Acad. Sci. **222**, 815 (1946).

64. — — Sur le tanghinoside, nouvel hétéroside des amandes fraîches de *Tanghinia venenifera*. C. R. hebd. Séances Acad. Sci. **226**, 268 (1948).

65. — — La hongkeline, nouvel hétéroside digitalique cristallisé de *Adenium Hongkel*. C. R. hebd. Séances Acad. Sci. **229**, 848 (1949).

65a. FRÈREJACQUE, M., H. P. SIGG und T. REICHSTEIN: Neotanghiferin. Helv. Chim. Acta **39** (1956) (im Druck).

66. FRIED, J. and R. C. ELDERFIELD: Studies on Lactones Related to the Cardiac Aglycones. V. Synthesis of 5-Alkyl-α-pyrones. J. Organ. Chem. (USA) **6**, 566 (1941).

67. FRIED, J. and R. C. ELDERFIELD: Studies on Lactones Related to the Cardiac Aglycones. VI. The Action of Diazomethane on Certain Derivates of α-Pyrone. J. Organ. Chem. (USA) **6**, 577 (1941).

68 FRIED, J., R. G. LINVILLE and R. C. ELDERFIELD: Studies on Lactones Related to the Cardiac Aglycones. VII. Synthesis of 3,14-Bisdesoxy-thevetigenin and of 14-Desoxy-thevetigenin. J. Organ. Chem. (USA) **7**, 362 (1942) [cf. especially p. 365].

69. GIRARD, A. et G. SANDULESCO: Sur une nouvelle série de réactifs du groupe carbonyle, leur utilisation à l'extraction des substances cétoniques et à la caractérisation microchimique des aldéhydes et cétones. Helv. Chim. Acta **19**, 1095 (1936).

70. GRAEVE, P. DE: Préparation de la cymarine à partir du *Strophanthus kombé*. Bull. soc. chim. France [5] **21**, 449 (1954).

71. HAACK, E., F. KAISER und H. SPINGLER: Gitaloxin, ein neues Hauptglykosid der *Digitalis purpurea*. Naturwiss. **42**, 441 (1955).

72. HABERMANN, E., W. MÜLLER und A. SCHREGLMANN: Papierchromatographische Charakterisierung von Stoffen und Stoffgemischen der Digitalis-Gruppe. Arzneimittelforsch. **3**, 30 (1953).

73. HARTMANN, M. und E. SCHLITTLER: Über afrikanische Pfeilgiftpflanzen. I. Mitt. *Adenium somalense* BALF. fil. Helv. Chim. Acta **23**, 548 (1940).

74. HASLAM, R. M. and W. KLYNE: The Precipitation of 3β-Hydroxysteroids by Digitonin. Biochemic. J. **55**, 340 (1953).

75. HASSALL, C. H.: The Cardiac Glycosides of *Urechites suberecta*. J. Chem. Soc. (London) **1951**, 3193.

76. HASSALL, C. H. and A. E. LIPPMAN: The Colorimetric Determination of Cardiac Glycosides. Analyst **78**, 126 (1953).

77. HASSALL, C. H. and S. L. MARTIN: Paper Chromatography of the Cardiac Glycosides. J. Chem. Soc. (London) **1951**, 2766.

77a. HAUENSTEIN, H., A. HUNGER und T. REICHSTEIN: Evomonosid aus den Samen von *Evonymus europaea* L. Helv. Chim. Acta **36**, 87 (1953).

77b. HEFTMANN, E.: Partition Chromatography of Steroids. Chem. Rev. **55**, 679 (1955).

78. HEFTMANN, E., P. BERNER, A. L. HAYDEN, H. K. MILLER and E. MOSETTIG: Identification of Cardiac Glycosides and Aglycones in *Strophanthus* Seeds by Paper Chromatography. Arch. Biochem. Biophys. **51**, 329 (1954).

79. HEFTMANN, E. and A. L. HAYDEN: Paper Chromatography of Steroid Sapogenins and their Acetates. J. Biol. Chem. **197**, 47 (1952).

79a. HEFTMANN, E. and A. J. LEVANT: Paper Chromatography of Cardiac Glycosides, Aglycones, and Acetates. J. Biol. Chem. **194**, 703 (1952).

80. HEGEDÜS, H., CH. TAMM und T. REICHSTEIN: Die Glykoside von *Strophanthus intermedius* PAX. 2. Mitt. Trennung des Kristallisats Nr. 790. Helv. Chim. Acta **36**, 357 (1953).

81. — — — Ambosid. Helv. Chim. Acta **37**, 2204 (1954).

82. — — — Inertosid, Leptosid, Chryseosid und Flavogenin. Helv. Chim. Acta **38**, 98 (1955).

83. HELFENBERGER, H. und T. REICHSTEIN: Thevetin. I. Helv. Chim. Acta **31**, 1470 (1948).

84. — — Thevetin. II. Helv. Chim. Acta **31**, 2097 (1948).

85. — — Die Glykoside von *Tanghinia venenifera* POIR. Helv. Chim. Acta **35**, 1503 (1952).

86. HESS, H., P. SPEISER, O. SCHINDLER und T. REICHSTEIN: Die Glykoside von *Strophanthus Ledienii* STEIN. Helv. Chim. Acta **34**, 1854 (1951).

87. HESS, J. C. und A. HUNGER: Identifizierung von Honghelosid G mit Somalin. Helv. Chim. Acta **36**, 85 (1953).

88. HESS, J. C., A. HUNGER und T. REICHSTEIN: Die Glykoside von *Adenium Boehmianum* SCHINZ. Helv. Chim. Acta **35**, 2202 (1952).

89. HESSE, G.: Über das Oleandrin. Ber. dtsch. chem. Ges. **70**, 2264 (1937).

90. HESSE, G., L. J. HEUSER, E. HÜTZ und F. REICHENEDER: Zusammenhänge zwischen den wichtigsten Giftstoffen der *Calotropis procera*. V. Mitt. über afrikanische Pfeilgifte. Liebigs Ann. Chem. **566**, 130 (1950).

91. HESSE, G. und F. REICHENEDER: Über das afrikanische Pfeilgift Calotropin. I. Liebigs Ann. Chem. **526**, 252 (1936).

92. HEUSSER, H.: Konstitution, Konfiguration und Synthese digitaloider Aglykone und Glykoside. Fortschr. Chem. organ. Naturstoffe **7**, 87 (1950).

93. HUBER, H., F. BLINDENBACHER, K. MOHR, P. SPEISER und T. REICHSTEIN: Die Glykoside der Wurzeln von *Xysmalobium undulatum* R. BR. 1. Mitt. Helv. Chim. Acta **34**, 46 (1951).

94. HUNGER, A.: Glykoside aus den Blättern von *Urechites lutea* (L.) BRITTON. Helv. Chim. Acta **34**, 898 (1951).

95. HUNGER, A. und T. REICHSTEIN: Glykoside aus *Adenium Honghel* A. DC. Helv. Chim. Acta **33**, 76 (1950).

96. — — Glykoside aus den Samen von *Adenium multiflorum* KL. Helv. Chim. Acta **33**, 1993 (1950).

97. — — Frugosid, ein zweites kristallisiertes Glykosid aus den Samen von *Gomphocarpus fruticosus* (L.) R. BR. Helv. Chim. Acta **35**, 429 (1952).

98. — — Reduktion von Aldehydgruppen in herzaktiven Glykosiden und Aglykonen mit Natriumborhydrid. Ber. dtsch. chem. Ges. **85**, 635 (1952).

99. — — Die Konstitution von Gofrusid und Frugosid. Helv. Chim. Acta **35**, 1073 (1952).

100. HUNZIKER, F. und T. REICHSTEIN: Abbau von Digitoxigenin zu 3β-Oxy-ätiocholansäure. Helv. Chim. Acta **28**, 1472 (1945).

101. JACOBS, W. A.: The Oxidation of Strophanthidin. J. Biol. Chem. **57**, 553 (1923).

102. — Allocymarin and Allostrophanthidin. An Enzymatic Isomerization of Cymarin and Strophanthidin. J. Biol. Chem. **88**, 519 (1930).

103. JACOBS, W. A. and N. M. BIGELOW: The Degradation of Isoouabain. J. Biol. Chem. **101**, 15 (1933).

104. — — The Strophanthins of *Strophanthus Eminii*. J. Biol. Chem. **99**, 521 (1933).

105. — — Ouabain or g-Strophanthin. J. Biol. Chem. **96**, 647 (1932).

106. JACOBS, W. A. and A. M. COLLINS: Anhydrostrophanthidin and Dianhydrostrophanthidin. J. Biol. Chem. **59**, 713 (1924).

107. JACOBS, W. A. and R. C. ELDERFIELD: Degradation of the Lactone Side Chain of Digitoxigenin. Science (Washington) **80**, 434 (1934).

107a. — — The Structure of the Cardiac Aglucones. J. Biol. Chem. **108**, 497 (1935).

108. — — The Oxidation of Anhydro-aglucon Derivatives. J. Biol. Chem. **113**, 611 (1936).

109. JACOBS, W. A. and E. L. GUSTUS: The Oxidation of Trianhydrostrophanthidin. J. Biol. Chem. **74**, 805 (1927).

110. — — Isostrophanthidin and its Derivatives. J. Biol. Chem. **74**, 811 (1927).

111. — — Digitoxigenin and Isodigitoxigenin. J. Biol. Chem. **78**, 573 (1928).

112. — — Gitoxigenin and Isogitoxigenin. J. Biol. Chem. **79**, 553 (1928).

113. — — Gitoxigenin and Isogitoxigenin. J. Biol. Chem. **82**, 403 (1929).

114. JACOBS, W. A. and E. L. GUSTUS: The Correlation of Gitoxigenin with Digitoxigenin. J. Biol. Chem. **86**, 199 (1930).

115. — — The Oxidation and Isomerization of Gitoxigenin. J. Biol. Chem. **88**, 531 (1930).

116. JACOBS, W. A. and M. HEIDELBERGER: Strophanthidin. J. Biol. Chem. **54**, 253 (1922).

117. — — Sarmentocymarin and Sarmentogenin. J. Biol. Chem. **81**, 765 (1929).

118. JACOBS, W. A. and A. HOFFMANN: On Crystalline Kombé Strophanthin. J. Biol. Chem. **67**, 609 (1926).

119. — — On K-Strophanthin-β and other Kombé Strophanthins. J. Biol. Chem. **69**, 153 (1926).

120. — — Periplocymarin and Periplogenin. J. Biol. Chem. **79**, 519 (1928).

121. JENSEN, H.: Chemical Studies on Toad Poisons. VII. *Bufo arenarum, Bufo regularis* and *Xenopus laevis*. J. Amer. Chem. Soc. **57**, 1765 (1935).

122. JENSEN, H. and K. K. CHEN: Chemical Studies on Toad Poisons. III. The Secretion of the Tropical Toad, *Bufo marinus*. J. Biol. Chem. **87**, 755 (1930).

123. JENSEN, K. B.: Paper Chromatography of Cardiac Glycosides and Aglycones from *Digitalis purpurea*. Acta Pharmacol. Toxicol. **9**, 99 (1953).

124. — Paper-chromatographic Separation and Fluorometric Determination of Cardiac Glycosides and Aglycones from *Digitalis purpurea*. Acta Pharmacol. Toxicol. **10**, 69 (1954).

125. JENSEN, K. B. and K. TENNÖE: Paper Chromatography of Cardiac Glycosides and Aglycones from *Digitalis lanata*. J. Pharm. Pharmacol. **7**, 334 (1955).

125a. JONES, R. N. and F. HERLING: Characteristic Group Frequencies in the Infrared Spectra of Steroids. J. Organ. Chem. (USA) **19**, 1252 (1954).

126. KAISER, F.: Die papierchromatographische Trennung von Herzgiftglykosiden. Chem. Ber. **88**, 556 (1955).

127. KARRER, P. und P. BANERJEA: Corchortoxin, ein herzwirksamer Stoff aus Jute-Samen. Helv. Chim. Acta **32**, 2385 (1949).

128. KARRER, W.: Untersuchungen über herzwirksame Glucoside. Festschrift E. C. BARELL, S. 242. Basel: Selbstverlag. 1936.

129. — Über Hellebrin, ein kristallisiertes Glykosid aus *Radix Hellebori nigri*. Helv. Chim. Acta **26**, 1353 (1943).

130. KATZ, A.: Kristallisiertes Convallatoxin-azetat und Convallatoxin-azetat-säure. Pharm. Acta Helv. **22**, 244 (1947).

131. — 3β,11α-Diacetoxy-ätiocholansäure-methylester. Helv. Chim. Acta **30**, 883 (1947).

132. — Konstitution des Sarmentogenins. Helv. Chim. Acta **31**, 993 (1948).

133. — Über die Glykoside von *Bowiea volubilis* HARVEY. 1. Mitt. Helv. Chim. Acta **33**, 1420 (1950).

134. — Über die Glykoside von *Bowiea volubilis* HARVEY. 2. Mitt. Helv. Chim. Acta **36**, 1344 (1953).

135. — Über die Glykoside von *Bowiea volubilis* HARVEY. 3. Mitt. Helv. Chim. Acta **36**, 1417 (1953).

136. — Über die Glykoside von *Bowiea volubilis* HARVEY. 4. Mitt. Pharm. Acta Helv. **29**, 77 (1954).

137. — Über die Glykoside von *Bowiea kilimandscharica* MILDBR. Pharm. Acta Helv. **29**, 369 (1954).

138. — Über die Glykoside von *Bowiea volubilis* HARVEY. 5. Mitt. Helv. Chim. Acta **37**, 451 (1954).

139. — Über die Glykoside von *Bowiea volubilis* HARVEY. 6. Mitt. Helv. Chim. Acta **37**, 833 (1954).

140. KATZ, A.: Über die Glykoside von *Bowiea volubilis* HARVEY. 7. Mitt. Helv Chim. Acta **38**, 1565 (1955).

140 a. — Personliche Mitteilung.

141. KATZ, A. und T. REICHSTEIN: Untersuchung der Samen von *Strophanthus kombé* und einer als „Semen Strophanthi hispidi“ bezeichneten Handelsdroge sowie Bemerkungen zur Konstitution des Sarmentogenins. Pharm. Acta Helv. **19**, 231 (1944).

142. — — *allo*-Periplocymarin und *allo*-Periplogenin sowie Beitrag zur Konstitution von Emicymarin und *allo*-Emicymarin. Helv. Chim. Acta **28**, 476 (1945).

143. — — Adonitoxin, das zweite stark herzwirksame Glykosid aus *Adonis vernalis*. Pharm. Acta Helv. **22**, 437 (1947).

144. KEDDE, D. L.: The Chemical Investigation of Digitalis Preparations. Pharm. Tijdskr. **1947**, 169.

145. KELLER, M. und T. REICHSTEIN: Gofrusid, ein kristallisiertes Glykosid aus den Samen von *Gomphocarpus fruticosus* (L.) R. BR. Helv. Chim. Acta **32**, 1607 (1949).

146. KILIANI, H.: Über den Milchsaft von *Antiaris toxicaria*. Arch. Pharmaz. **234**, 438 (1896).

147. — Über den Milchsaft von *Antiaris toxicaria*. Ber. dtsch. chem. Ges. **43**, 3574 (1910).

148. — Über Digitoxin und Gitalin. Arch. Pharmaz. **251**, 567 (1913).

149. — Über α- und β-Antiarin und über kristallisiertes Eiweiß aus Antiaris-Saft. Ber. dtsch. chem. Ges. **46**, 667 (1913).

150. — Über Digitalinum verum. Ber. dtsch. chem. Ges. **63**, 2866 (1936).

151. KOECHLIN, H. und T. REICHSTEIN: Abbauversuche an Strophanthidin. Helv Chim. Acta **30**, 167 (1947).

152. KOTAKE, M.: Über das Krötengift. II. Mitt. Die giftigen Bestandteile des Sekretes der japanischen Kröte *(Bufo bufo japonicus)*. Liebigs Ann. Chem. **465**, 11 (1928).

153. KOTAKE, M. und K. KUWADA: Chemische Versuche über Krötengift. VI. Zur Kenntnis der Bestandteile des Ch'an Su (Senso). Scient. Pap. Inst. Physic. Chem. Res. (Japan) **32**, 1 (1937).

154. — — Chemische Versuche uber Krötengift. 10. Mitt. Chemische Konstitution des Bufalins. Scient. Pap. Inst. Physic. Chem. Res. (Japan) **36**, 106 (1939) [Chem. Zbl. **1939** II, 1681].

155. KRAUS, J.: Über die Einwirkung von Alkali auf einige steroide Herzgifte und deren Allo-Isomere. Z. physiol. Chem. (Hoppe-Seyler) **292**, 21 (1953).

156. KREIS, W., CH. TAMM und T. REICHSTEIN: Die Glykoside der Samen von *Corchorus capsularis* L. Helv. Chim. Acta **39** (1956) (im Druck).

156a. KÜNDIG-HEGEDÜS, H. und O. SCHINDLER: Die Konstitution von Sarmutogenin. Helv. Chim. Acta **39**, 904 (1956).

157. KÜSSNER, W., F. REIFF und H. W. VOIGTLÄNDER: Über die Prüfung des Digitoxins. Arch. Pharmaz. **288**, 284 (1955).

158. LAMB, I. D. and S. SMITH: The Glucosides of *Strophanthus Eminii*. J. Chem. Soc. (London) **1936**, 442.

159. LARDON, A.: Progesteron- und Desoxycorticosteron-Derivate aus Periplogenin. Helv. Chim. Acta **32**, 1517 (1949).

160. — Die Glykoside der Samen von *Strophanthus Eminii* ASCH. et PAX. Helv Chim. Acta **33**, 639 (1950).

161. LARDON, A. und T. REICHSTEIN: Ester der 3,11-Diketo-cholansäure, 3α-Oxy-11-keto-cholansäure und 3α,11α-Dioxy-cholansaure. Helv. Chim. Acta **26**, 586 (1943).

162. LARDON, A. und T. REICHSTEIN: Teilsynthese von Cortison und verwandten Stoffen aus Sarmentogenin. Pharm. Acta Helv. **27**, 282 (1952).

163. LAWDAY, D.: Detection of Digitalis Glycosides on Paper Chromatograms. Nature (London) **170**, 415 (1952).

164. LETTRÉ, H., H. H. INHOFFEN und R. TSCHESCHE: Über Sterine, Gallensäuren und verwandte Naturstoffe, 2. Aufl., S. 287. Stuttgart: F. Enke. 1954.

165. LICHTI, H., CH. TAMM und T. REICHSTEIN: Die Glykoside der Samen von *Strophanthus Ledienii* STEIN, 2. Mitt. Helv. Chim. Acta **39** (1956) (im Druck).

166. — — — Die Glykoside der Samen von *Strophanthus Vanderijstii* STANER. Helv. Chim. Acta **39** (1956) (im Druck).

167. MANNICH, C. und W. SCHNEIDER: Über die Glykoside von *Digitalis orientalis* L. Arch. Pharmaz. **279**, 223 (1941).

168. MANNICH, C. und G. SIEWERT: Über g-Strophanthin (Ouabain) und g-Strophanthidin. Ber. dtsch. chem. Ges. **75**, 737 (1942).

169. MARTIN, A. J. P.: Partition Chromatography. Annu. Rep. Chem. Soc. (London) **45**, 276 (1948).

170. MARTIN, A. J. P. and R. L. M. SYNGE: Separation of the Higher Monoamino-acids by Counter-current Liquid-Liquid Extraction: The Amino-acid Composition of Wool. Biochemic. J. **35**, 91 (1941).

171. — — A New Form of Chromatogram Employing Two Liquid Phases. 1. A Theory of Chromatography. 2. Application to the Micro-Determination of the Higher Monoamino-acids in Proteins. Biochemic. J. **35**, 1358 (1941).

172. MAULI, R., CH. TAMM und T. REICHSTEIN: Die Glykoside von *Glossostelma spathulatum* (K. SCHUM) BULLOCK. Helv. Chim. Acta **39** (1956) (im Druck).

173. MEYER, K.: Ein neuer Abbau des Gitoxigenins. Helv. Chim. Acta **29**, 718 (1946).

174. — 3β-Acetoxy-ätio-cholen-(16)-säure-methylester aus Gitoxigenin. Helv. Chim. Acta **29**, 1580 (1946).

175. — Nachtrag zum Abbau des Gitoxigenins mit Kaliumpermanganat. Helv. Chim. Acta **29**, 1908 (1946).

176. — Über herzaktive Krötengifte (Bufogenine). Pharm. Acta Helv. **24**, 222 (1949).

177. — Konstitution des Bufalins. Helv. Chim. Acta **32**, 1238 (1949).

178. — Konstitution des Telocinobufagins. Helv. Chim. Acta **32**, 1593 (1949).

179. — Konstitution des Gamabufotalins. Helv. Chim. Acta **32**, 1599 (1949).

180. — Konstitution des Bufotalins. Helv. Chim. Acta **32**, 1993 (1949).

181. — Telocinobufagin aus *Bufo marinus* (L.) SCHNEID. (= *Bufo agua* SEBA) und einige Bemerkungen zur Konstitution des Marinobufagins. Helv. Chim. Acta **34**, 2147 (1951).

182. — Resibufogenin und Artebufogenin aus Ch'an Su. Helv. Chim. Acta **35**, 2444 (1952).

183. MEYER, K. und T. REICHSTEIN: Überführung digitaloider Lactone in Ketole vom Typus der Corticosteroide. Helv. Chim. Acta **30**, 1508 (1947).

184. MEYRAT, A. und T. REICHSTEIN: Evonosid, ein herzwirksames Glykosid aus den Samen des Pfaffenhütchens, *Evonymus europaea* (L.). Pharm. Acta Helv. **23**, 135 (1948).

185. — — Abbau des Ouabagenins. I. Helv. Chim. Acta **31**, 2104 (1948).

186. MOHR, K. und T. REICHSTEIN: Acovenosid C. Die Glykoside der Samen von *Acokanthera venenata* G. DON. 2. Mitt. Helv. Chim. Acta **34**, 1239 (1951).

187. MOORE, J. A.: Synthese der $3\beta,16\alpha$-Dioxy-ätiansäure. Beitrag zur Kenntnis der Konfiguration der Hydroxylgruppe an C-16 in Gitoxigenin. Helv. Chim. Acta **37**, 659 (1954).

188. Moore, J. A., Ch. Tamm und T. Reichstein: Die Glykoside der Goldlack-samen, *Cheiranthus Cheiri* L. 3. Mitt. Helv. Chim. Acta 37, 755 (1954).

189. Muhr, H., A. Hunger und T. Reichstein: Die Glykoside der Samen von *Acokanthera friesiorum* Markgr. Helv. Chim. Acta 37, 403 (1954).

190. Nagata, W., Ch. Tamm und T. Reichstein: Die Glykoside von *Erysimum crepidifolium* H. G. L. Reichenbach und *E. helveticum* (Jacquin) A. P. DC. Helv. Chim. Acta 39 (1956) (im Druck).

191. Nawa, H.: Rhodexin A and B, Cardiac Glycosides of *Rhodea japonica*. Proc. Japan Acad. 27, 436 (1951).

192. — Components of *Rhodea japonica*. I. Separation of New Cardioactive Gluco-sides, Rhodexin A and B. J. pharmac. Soc. Japan 72, 404 (1952).

193. Neumann, W.: Über Glykoside des Oleanders. Ber. dtsch. chem. Ges. 70, 1547 (1937).

194. Okada, M.: Components of *Nerium odorum* Leaves. III. J. pharmac. Soc. Japan 73, 86 (1953).

195. Okada, M. and A. Yamada: A Component of *Digitalis purpurea* L. Seeds. J. pharmac. Soc. Japan 73, 525 (1953).

196. Overend, W. G., M. Stacey and J. Staněk: Deoxy-sugars. VII. A Study of the Reactions of Some Derivatives of 2-Deoxy-D-Glucose. J. Chem. Soc. (London) 1949, 2841.

197. Paist, W. D., E. R. Blout, F. C. Uhle and R. C. Elderfield: Studies on Lactones Related to the Cardiac Aglycones. III. The Properties of β-Substituted $\Delta^{\alpha,\beta}$-Butenolides and a Suggested Revision of the Structure of the Side Chain of the Digitalis-Strophanthus Aglycones. J. Organ. Chem. (USA) 6, 273 (1941).

198. Pataki, S. und K. Meyer: Zur Konstitution des Marinobufagins, II. Mitt. Helv. Chim. Acta 38, 1631 (1955).

199. Pataki, S., K. Meyer und T. Reichstein: Die Konfiguration des Dioxigenins. (Teilsynthese des $3\beta,12\alpha$- und des $3\beta,12\beta$-Dioxy-ätiansäure-methylesters.) Helv. Chim. Acta 36, 1295 (1953).

200. Pesez, M.: Le xanthydrol, réactif des désoses. Ann. pharm. franç. 10 104 (1952).

201. Petit, A., M. Pesez, P. Bellet et G. Amiard: Sur le digitoxoside pur. Bull. soc. chim. France [5] 17, 288 (1950).

202. Pinxteren, J. A. C. van and A. L. O. M. Smithuis: The Colour Reaction between 3,5-Dinitrobenzoic Acid on Digitalis and *Strophanthus kombé* Gly-cosides. Pharmac. Weekbl. 89 741 (1954).

203. Plattner, Pl. A., H. Heusser und A. B. Kulkarni: Über die reduktive Auf-spaltung von Steroidepoxyden mit Lithiumaluminiumhydrid. I. Synthese von $3\beta,5$-Dioxy-koprostan. Helv. Chim. Acta. 31, 1885 (1948).

204. — — — Über die reduktive Aufspaltung von Steroidepoxyden mit Lithium-aluminiumhydrid. III. Vereinfachte Synthesen von Derivaten des 5-Oxy-koprostans. Helv. Chim. Acta 32, 265 (1949).

205. Plattner, Pl. A., L. Ruzicka, H. Heusser und E. Angliker: Synthese von 14-Oxy-Steroiden. II. Verbindungen der Allo-pregnanolon-Reihe. Helv. Chim. Acta 30, 385 (1947).

206. — — — — Synthese von *allo*-Uzarigenin. Beitrag zur Konstitutionsaufklärung der *allo*-Aglykone. Helv. Chim. Acta 30, 1073 (1947).

207. Plattner, Pl. A., L. Ruzicka, H. Heusser, J. Pataki und Kd. Meier: Über die Synthese von 14-Oxy-Steroiden. Helv. Chim. Acta 29, 942 (1946).

208. Plattner, Pl. A., A. Segre und O. Ernst: Über die Stereochemie des Stro-phanthidins und des Periplogenins. Helv. Chim. Acta 30, 1432 (1947).

209. PRINS, D. A.: Synthese von *d*-Cymarose. Helv. Chim. Acta **29**, 378 (1946).

210. RABALD, E. und J. KRAUS: Zur Kenntnis des Strophanthins. Z. physiol. Chem. (Hoppe-Seyler) **265**, 39 (1940).

211. RAFFAUF, R. F. und T. REICHSTEIN: Weitere Abbauversuche mit Ouabagenin (II). Helv. Chim. Acta **31**, 2111 (1948).

212. RANGASWAMI, S. und T. REICHSTEIN: Die Glykoside von *Nerium odorum* SOL. I. Pharm. Acta Helv. **24**, 159 (1949).

213. — — Konstitution von Odorosid A und Odorosid B. Die Glykoside von *Nerium odorum* SOL., 2. Mitt. Helv. Chim. Acta **32**, 930 (1949).

214. RAYMOND, W. D.: The Detection and Estimation of Ouabain and Strophanthin. Analyst **63**, 478 (1938).

215. — The *m*-Dinitrobenzene Reaction of Ouabain and its Application to the Examination of East African Arrow Poison. Analyst **64**, 113 (1939).

216. REBER, F. und T. REICHSTEIN: Trennung der Sarmentoside A, C, D und E. Pharm. Acta Helv. **28**, 1 (1953).

217. REICH, H. und T. REICHSTEIN: 11 α-Keto- und 11 α-Oxy-cholansäure. Helv. Chim. Acta **26**, 562 (1943).

218. REICHSTEIN, T.: Chemie der herzaktiven Glykoside. Angew. Chem. **63**, 412 (1951).

219. REICHSTEIN, T. und H. G. FUCHS: Desoxy-corticosteron und andere Pregnanderivate aus Ätio-lithocholsäure und 3β-Oxy-ätiocholansäure. Helv. Chim. Acta **23**, 658 (1940).

220. REICHSTEIN, T. und H. ROSENMUND: Isolierung eines kristallisierten, stark herzwirksamen Glykosides aus *Adonis vernalis* und seine Identifizierung als Cymarin. Pharm. Acta Helv. **15**, 150 (1940).

221. REICHSTEIN, T. and C. W. SHOPPEE: Chromatography of Steroids and Other Colourless Substances by the Method of Fractional Elution. Discuss. Faraday Soc. **7**, 305 (1949).

222. REYLE, K. und T. REICHSTEIN: Partialsynthese des Strophanthidin-α-*D*-lyxosids. Helv. Chim. Acta **35**, 98 (1952).

223. — — Teilsynthese des Digitoxigenin-β-*D*-thevetosids, seine Identifizierung mit Honghelin und Versuche zur Teilsynthese eines Digitoxigenin-*L*-thevetosids. Helv. Chim. Acta **35**, 195 (1952).

224. RHEINER, A., A. HUNGER und T. REICHSTEIN: Die Glykoside von *Nerium odorum* SOL., 3. Mitt. Die Konstitution von Odorosid G. Helv. Chim. Acta **35**, 687 (1952).

225. RICHARDS, G. N.: Hydrolysis of Glycosides and Cyclic Acetals. Chem. and Ind. **1955**, 228.

226. RICHTER, R., K. MOHR und T. REICHSTEIN: Sarmutosid und Musarosid. Glykoside der Samen von *Strophanthus sarmentosus* A. P. DC. 4. Mitt. Helv. Chim. Acta **36**, 1073 (1953).

227. RICHTER, R., O. SCHINDLER und T. REICHSTEIN: Zur Konstitution von Sarmutosid und Musarosid. Glykoside von *Strophanthus sarmentosus* A. P. DC. 6. Mitt. Helv. Chim. Acta **37**, 76 (1954).

228. RITTEL, W., A. HUNGER und T. REICHSTEIN: Digitalinum verum und Strospesid (= Desgluco-digitalinum verum). Berichtigung früherer Angaben. Helv. Chim. Acta **35**, 434 (1952).

229. — — — „Odorosid E", Odorosid H, Odorosid-K-acetat und „Kristallisat J". Die Glykoside von *Nerium odorum* SOL. 4. Mitt. Helv. Chim. Acta **36**, 434 (1953).

230. RITTEL, W. und T. REICHSTEIN: Odorosid D und Odorosid F. Die Glykoside von *Nerium odorum* SOL. 5. Mitt. Helv. Chim. Acta **36**, 554 (1953).

231. — — Odorosid K und Odorobiosid K. Die Glykoside von *Nerium odorum* SOL. 6. Mitt. Helv. Chim. Acta **37**, 1361 (1954).

232. Rosenheim, O.: A Specific Colour Reaction for Ergosterol. Biochemic. J. **23**, 47 (1929).

233. Rosenmund, H. und T. Reichstein: Isolierung eines weiteren krystallisierten stark herzwirksamen Glykosids aus *Adonis vernalis*. Pharm. Acta Helv. **17**, 176 (1942).

234. Rosselet, J. P. und A. Hunger: Intermediosid (Substanz Nr. 761). Helv. Chim. Acta **34**, 1036 (1951).

235. Rosselet, J. P., A. Hunger und T. Reichstein: Panstrosid (Substanz Nr. 762). Helv. Chim. Acta **34**, 2143 (1951).

236. Rosselet, J. P. und T. Reichstein: Die Glykoside von *Strophanthus gracilis* K. Schum. et Pax. 2. Mitt. Odorosid H und Gracilosid. Helv. Chim. Acta **36**, 787 (1953).

237. Ruppol, E. et I. Turkovic: Contribution à l'étude des Strophanthus. I. Note sur les glucosides du *Strophanthus arnoldianus*. II. Tableau récapitulatif des hétérosides de divers strophanthus. J. pharmac. Belgique **36**, 95 (1954).

238. — — Les hétérosides du *Strophanthus Preussii* (Engl. et Pax) type G. 4666. J. pharmac. Belgique **37**, 221 (1955).

239. Ruzicka, L., Pl. A. Plattner und H. Heusser: Die Einwirkung von N-Brom-succinimid auf digitaloide Aglykone. Helv. Chim. Acta **29**, 473 (1946).

240. Ruzicka, L., Pl. A. Plattner, H. Heusser und Kd. Meier: Synthese des 3β-Acetoxy-14-oxy-14-allo-ätio-cholansäure-methylesters, eines Abbauproduktes aus Digitoxigenin, und zweier Umwandlungsprodukte des Gitoxigenins. 14-Oxy-Steroide. IV. Helv. Chim. Acta **30**, 1342 (1947).

241. Ruzicka, L., Pl. A. Plattner und J. Pataki: Über die Einwirkung von N-Bromsuccinimid auf $\Delta^{20,22}$-3β-Acetoxy-nor-allo-cholensäure-methylester. Helv. Chim. Acta **28**, 1360 (1945).

242. Šantavý, F. und T. Reichstein: Evonosid, ein kristallisiertes, herzwirksames Glykosid aus *Evonymus europaea* L. II. Mitt. Helv. Chim. Acta **31**, 1655 (1948).

243. Sasakawa, Y.: Isolation of Digitalinum verum from *Digitalis purpurea* Leaves. J. pharmac. Soc. Japan **74**, 474 (1954).

244. — Paper Partition Chromatography of *Digitalis purpurea* Glycosides. J. pharmac. Soc. Japan **74**, 721 (1954).

245. — Paper Partition Chromatography of *Digitalis purpurea* Glycosides (2). J. pharmac. Soc. Japan **75**, 946 (1955).

246. Schenker, E., A. Hunger und T. Reichstein: Zur Papierchromatographie von stark polaren herzwirksamen Glykosiden und Aglykonen. Helv. Chim. Acta **37**, 680 (1954).

247. — — — Die Glykoside von *Periploca nigrescens* Afzel. Helv. Chim. Acta **37**, 1004 (1954).

248. Schindler, O.: 11-Dehydro-divaricosid und Decosid (C-Dehydro-caudosid). Helv. Chim. Acta **38**, 140 (1955).

249. — 11-Epi-sarmentogenin, Desarosid (11-Dehydro-sarnovid) und 11-Epi-sarnovid. Helv. Chim. Acta **38**, 538 (1955).

249 a. Schindler, O.: Zur Konstitution von Sarverogenin. Helv. Chim. Acta **39**, 375 (1956).

250. Schindler, O. und T. Reichstein: Die Glykoside der Wurzeln von *Adenium Honghel* A. DC. Helv. Chim. Acta **34**, 18 (1951).

251. — — Nachweis und Trennung digitaloider Glykoside und Aglykone durch Chromatographie auf imprägniertem Filterpapier. Helv. Chim. Acta **34**, 108 (1951).

252. — — Isolierung von Strophanthidin und Strophanthidin-glykosiden mit Girards Reagens T. Helv. Chim. Acta **34**, 521 (1951).

253. SCHINDLER, O. und T. REICHSTEIN: Vergleich dreier Samenproben von Sarmentogenin liefernden Strophanthus-Varianten mit Hilfe von Papier-chromatographie. Helv. Chim. Acta 34, 608 (1951).

254. — — Die Glykoside der Samen von *Strophanthus Boivinii* BAILL. Helv. Chim. Acta 35, 673 (1952).

255. — — Millosid, Pauliosid, Strobosid und Boistrosid. Die Glykoside von *Strophanthus Boivinii* BAILL. 2. Mitt. Helv. Chim. Acta 35, 730 (1952).

256. — — Christyosid (Substanz Nr. 764). Helv. Chim. Acta 36, 370 (1953).

257. — — Die Glykoside der Samen von *Strophanthus divaricatus* (LOUR.) HOOK. et ARN. Helv. Chim. Acta 36, 1007 (1953).

258. — — Divaricosid und Caudosid. Helv. Chim. Acta 37, 667 (1954).

259. SCHLEGEL, W., CH. TAMM und T. REICHSTEIN: Die Konstitution von Aco-venosid A. 2. Mitt. Helv. Chim. Acta 38, 1013 (1955).

260. SCHMUTZ, J.: Hellebrin. Pharm. Acta Helv. 22, 373 (1947).

261. — Die Konstitution des Hellebrigenins. Helv. Chim. Acta 32, 1442 (1949).

262. SCHMUTZ, J. und T. REICHSTEIN: Convallosid, ein stark herzwirksames Gly-kosid aus Samen *Convallariae majalis* L. Pharm. Acta Helv. 22, 359 (1947) [s. insbes. S. 370, Herstellung der Strophanthobiase].

263. SCHWARZ, H., A. KATZ und T. REICHSTEIN: „Cheirotoxin", ein herzwirksames Glykosid, sowie andere Inhaltsstoffe von Goldlacksamen (*Cheiranthus Cheiri* L.). Pharm. Acta Helv. 21, 250 (1946).

264. SHAH, N., K. MEYER und T. REICHSTEIN: Glykoside aus Goldlacksamen, *Cheiranthus Cheiri* L. II. Pharm. Acta Helv. 24, 113 (1949).

265. SHOPPEE, C. W.: Diginin. 3. Mitt. Abbau des Diginigenins zu einem Kohlen-wasserstoff Diginan $C_{21}H_{36}$. Helv. Chim. Acta 27, 246 (1944).

266. — Diginin und Diginigenin, 4. Mitt. Helv. Chim. Acta 27, 426 (1944).

267. SHOPPEE, C. W. und T. REICHSTEIN: Diginin. 1. Mitt. Helv. Chim. Acta 23, 975 (1940).

268. — — Diginin. 2. Mitt. Zur Konstitution der Diginose. Helv. Chim. Acta 25, 1611 (1942).

269. SHOPPEE, C. W. und E. SHOPPEE: Steroids: Cardiotonic Glycosides and Agly-cones, Toad Poisons: Steroid Saponins and Sapogenins. In: E. H. RODD, Chemistry of Carbon Compounds, Vol. II B, p. 983. Amsterdam-Houston-London-New York: Elsevier Publ. Co. 1953.

270. SIGG, H. P., CH. TAMM und T. REICHSTEIN: 3-Epi-digitoxigenin. Helv. Chim. Acta 36, 985 (1953).

271. — — — Tanghinigenin und 3-Epitanghinigenin. Helv. Chim. Acta 38, 166 (1955).

272. — — — Abbau des Tanghinigenins. Helv. Chim. Acta 38, 1721 (1955).

273. SILBERMAN, H. und R. H. THORP: The Estimation of the Component Cardiac Glycosides in Digitalis Plant Samples Using Paper Chromatography and Fluorescence Photography. J. Pharm. Pharmacol. 5, 438 (1953).

274. SIMPSON, S. A., J. F. TAIT, A. WETTSTEIN, R. NEHER, J. v. EUW, O. SCHINDLER und T. REICHSTEIN: Die Konstitution des Aldosterons. Helv. Chim. Acta 37, 1200 (1954).

275. SMITH, S.: Digoxin, a New Digitalis Glucoside. J. Chem. Soc. (London) 1930, 508.

276. — Digitalis Glucosides. V. On the Constitution of Digoxigenin. J. Chem. Soc. (London) 1935, 1305.

277. — Digitalis Glucosides. VI. The Existence of Two Anhydrodigoxigenins. J. Chem. Soc. (London) 1936, 354.

278. Sneeden, R. P. A. and R. B. Turner: Ouabagenin. I. The Relationship between Ouabagenin Monoacetonide and "Anhydroöuabagenin". J. Amer. Chem. Soc. **75**, 3510 (1953).

279. — — Observations on the Structure of Ouabagenin. Chem. and Ind. **1954**, 1235.

280. — — Ouabagenin. II. The Hydroxyl Groups of the A/B Ring System. J. Amer. Chem. Soc. **77**, 130 (1955).

281. Speiser, P.: Periplogenin aus Strophanthidin. Helv. Chim. Acta **32**, 1368 (1949).

282. Speiser, P. und T. Reichstein: Konfiguration des Periplogenins und des *allo*-Periplogenins. Helv. Chim. Acta **30**, 2143 (1947).

283. — — Konfiguration des Periplogenins und *allo*-Periplogenins. II. Helv. Chim. Acta **31**, 622 (1948).

284. Steiger, M. und T. Reichstein: Die Funktion des letzten Sauerstoffatoms. Helv. Chim. Acta **20**, 817 (1937).

285. — — Ein neuer Abbau des Digoxigenins. Helv. Chim. Acta **21**, 828 (1938).

286. Steinegger, E. und A. Katz: Abbau von Glucosiden mit Chromsäure. Pharm. Acta Helv. **22**, 1 (1947).

287. Stoll, A.: Über die herzwirksamen Glykoside der Digitalisgruppe. Die Pharmazie **5**, 328 (1950).

288. — Bemerkungen zu den „Vorschlägen zur Nomenklatur der Steroide". Helv. Chim. Acta **34**, 2141 (1951).

289. — Sur les substances cardiotoniques de la scille maritime (*Scilla maritima* L.). Experientia **10**, 282 (1954).

290. Stoll, A., E. Angliker, F. Barfuss, W. Kussmaul und J. Renz: Trennung und Bestimmung herzwirksamer Glykoside und deren Aglykone durch Chromatographie an Silicagel. Helv. Chim. Acta **34**, 1460 (1951).

291. Stoll, A. und A. Hofmann: Die Hydrierung des Scillarens A und die physiologische Prüfung einiger Scillarenderivate. Helv. Chim. Acta **18**, 401 (1935).

292. Stoll, A., A. Hofmann und A. Helfenstein: Die Identität der α-Scillansäure mit Allocholansäure. Helv. Chim. Acta **18**, 644 (1935).

293. Stoll, A. und E. Jucker: Herzglykoside. In: K. Paech und M. V. Tracey, Moderne Methoden der Pflanzenanalyse, Bd. III, S. 205. Berlin-Gottingen-Heidelberg: Springer-Verlag. 1955.

294. Stoll, A. und W. Kreis: Die genuinen Glykoside der *Digitalis lanata*, die Digilanide A, B und C. Helv. Chim. Acta **16**, 1049 (1933).

295. — — Acetyldigitoxin, Acetylgitoxin und Acetyldigoxin. Helv. Chim. Acta **17**, 592 (1934).

296. — — Genuine Glucoside der *Digitalis purpurea*, die Purpureaglucoside A und B. Helv. Chim. Acta **18**, 120 (1935).

297. — — Neue herzwirksame Glykoside aus der weißen Meerzwiebel. Helv. Chim. Acta **34**, 1431 (1951).

298. — — Darstellung von Scillarenin, dem primären Aglykon von Scillaren A, mit Hilfe eines adaptiven Enzyms. Helv. Chim. Acta **34**, 2301 (1951).

299. — — Acetyl-digitoxin-α und Acetyl-digitoxin-β. Helv. Chim. Acta **35**, 1318 (1952).

300. Stoll, A., W. Kreis und A. v. Wartburg: Die hydrolytische Spaltung von herzwirksamen Glykosiden der weißen Meerzwiebel. Helv. Chim. Acta **35**, 2495 (1952).

301. — — — Die Kristallisation der Desacetyl-lanatoside A und B (Purpurea Glykoside A und B). Helv. Chim. Acta **37**, 1134 (1954).

302. Stoll, A., A. Pereira und J. Renz: Über herzwirksame Glykoside und Aglykone der Samen von *Coronilla glauca*. Helv. Chim. Acta **32**, 293 (1949).

303. STOLL, A. und J. RENZ: Über Periplocin, das genuine herzwirksame Glykosid der *Periploca graeca*. Helv. Chim. Acta **22**, 1193 (1939).

304. — — Die Überführung von Scillaren A in Epi-allo-lithocholsäure (3β-Oxy-allo-cholansäure). Helv. Chim. Acta **24**, 1380 (1941).

305. — — Über Scillirosid, ein gegen Nager spezifisch wirksames Gift der roten Meerzwiebel. Helv. Chim. Acta **25**, 43 (1942).

306. — — Der Lactonring des Scillirosids. Helv. Chim. Acta **25**, 377 (1942).

307. — — Spaltung von Herzglykosiden mit Enzympräparaten aus tierischen Organen. Helv. Chim. Acta **34**, 782 (1951).

308. — — Der enzymatische Abbau des Scillirosids zum Scillirosidin. Helv. Chim. Acta **33**, 286 (1950).

309. — — Herzglykoside der *Digitalis ferruginea* L. Helv. Chim. Acta **35**, 1310 (1952).

310. STOLL, A., J. RENZ und A. BRACK: Spaltung von Herzglykosiden durch Pilzenzyme. Helv. Chim. Acta **34**, 397 (1951).

311. — — — Die Lage der Kerndoppelbindung im Scillarenin. Helv. Chim. Acta **35**, 1934 (1952).

312. STOLL, A., J. RENZ und A. HELFENSTEIN: Über die Struktur des Scillirosids. Helv. Chim. Acta **26**, 648 (1943).

313. STOLL, A., J. RENZ und W. KREIS: k-Strophanthosid, das Hauptglucosid der Samen von *Strophanthus kombé*. Helv. Chim. Acta **20**, 1484 (1937).

314. STOLL, A., E. SUTER, W. KREIS, B. B. BUSSEMAKER und A. HOFMANN: Die herzaktiven Substanzen der Meerzwiebel. Scillaren A. Helv. Chim. Acta **16**, 703 (1933).

315. STOLL, A., A. v. WARTBURG und W. KREIS: Acetyl-gitoxin-α und Acetyl-gitoxin-β. Helv. Chim. Acta **35**, 1324 (1952).

316. STOLL, A., A. v. WARTBURG und J. RENZ: Die Konstitution des Scilliglaucosidins. Helv. Chim. Acta **36**, 1531 (1953).

317. — — — Das Verhalten von Scilliglaucosidin und anderen ungesättigten Steroiden bei der Rosenheim-Reaktion. Helv. Chim. Acta **36**, 1565 (1953).

318. — — — Partialsynthese und Eigenschaften von Δ^4-5-Anhydrostrophanthidin-19-säure-methylester, einer Modellsubstanz für Scilliglaucosidin-19-säure-methylester. Helv. Chim. Acta **36**, 1557 (1953).

319. SVENDSEN, A. B. und K. B. JENSEN: Papierchromatographische Untersuchung von Digitalisglykosiden. Pharm. Acta Helv. **25**, 241 (1950).

320. TAMM, CH.: Die Konstitution des Ouabagenin-monoacetonids. Helv. Chim. Acta **38**, 147 (1955).

321. TAMM, CH. und G. BAUMGARTNER: Abbauversuche an Ouabagenin (unveröffentlicht).

322. TAMM, CH. und T. REICHSTEIN: Abbau von Acovenosid A. I. Mitt. Helv. Chim. Acta **34**, 1224 (1951).

323. TAMM, CH. und J. P. ROSSELET: Die Konstitution von Evomonosid. Helv. Chim. Acta **36**, 1309 (1953).

324. TAYLOR, D. A. H.: Some Reactions of Sarverogenin. J. Chem. Soc. (London) **1952**, 4832.

325. — The Structure of Sarverogenin. Chem. and Ind. **1953**, 62.

326. — The Degradation of Digoxigenin. J. Chem. Soc. (London) **1953**, 3325.

327. TOLLENS, B.: Über ammon-alkalische Silberlösung als Reagens auf Aldehyd. Ber. dtsch. chem. Ges. **15**, 1635 (1882).

328. — Über ammon-alkalische Silberlösung als Reagens auf Formaldehyd. Ber. dtsch. chem. Ges. **15**, 1828 (1882).

329. — Über die Circular-Polarisation des Traubenzuckers (Dextrose). III. Ber. dtsch. chem. Ges. **17**, 2234 (1884).

330. Tschesche, R.: Zur Konstitution des Uzarins. Z. physiol. Chem. (Hoppe-Seyler) **222**, 50 (1933).

331. — Die Konstitution der pflanzlichen Herzgifte. Angew. Chem. **47**, 729 (1934).

332. — Abbau eines Genins der pflanzlichen Herzgifte zu einem Gallensäure-derivat. Z. physiol. Chem. (Hoppe-Seyler) **229**, 219 (1934).

333. — Ätio-allocholansäure und ihre Identifizierung mit der Abbausäure aus Uzarigenin. Ber. dtsch. chem. Ges. **68**, 7 (1935).

334. — Die Konstitution des Thevetins. Ber. dtsch. chem. Ges. **69**, 2368 (1936).

335. — Zur Kenntnis des Oleandrins. Ber. dtsch. chem. Ges. **70**, 1554 (1937).

336. Tschesche, R. und K. Bohle: Die Konstitution des Uzarigenins. Ber. dtsch. chem. Ges. **68**, 2252 (1935).

337. — — Die Konstitution des Digoxigenins. Ber. dtsch. chem. Ges. **69**, 793 (1936).

338. — — Die Konstitution des Sarmentogenins. Ber. dtsch. chem. Ges. **69**, 2497 (1936).

339. — — Zur Konstitution des Adynerins. Ber. dtsch. chem. Ges. **71**, 654 (1938).

340. Tschesche, R., K. Bohle und W. Neumann: Die Nebenglykoside des Oleanders. Ber. dtsch. chem. Ges. **71**, 1927 (1938).

341. Tschesche, R. und K.-H. Brathge: Die Glykoside der Uzara-Wurzel. Chem. Ber. **85**, 1042 (1952).

342. Tschesche, R. und G. Grimmer: Zur Konstitution des Adynerins und ein Beitrag zur Stellung der Doppelbindung im Scillirosid. Chem. Ber. **87**, 418 (1954).

343. — — Neue Glykoside aus den Blättern von *Digitalis purpurea* und *Digitalis lanata.* Chem. Ber. **68**, 1569 (1955).

344. Tschesche, R., G. Grimmer und F. Neuwald: Gitorin, ein neues Glykosid aus *Digitalis lanata* und über ein quantitatives Mikrobestimmungsverfahren für Aglykone und Zucker in Glykosiden vom Typ der Fünfringlactone. Chem. Ber. **85**, 1103 (1952).

345. Tschesche, R., G. Grimmer und F. Seehofer: Die quantitative Trennung und Identifizierung von Herzglykosiden aus *Digitalis purpurea* und *lanata* durch „echte Verteilungschromatographie an Papier". Chem. Ber. **86**, 1235 (1953).

346. Tschesche, R. und W. Haupt: Über das Convallatoxin. Ber. dtsch. chem. Ges. **69**, 459 (1936).

347. Tschesche, R. und H. Knick: Die Dehydrierung des Uzarigenins mit Selen (vorl. Mitt.). Z. physiol. Chem. (Hoppe-Seyler) **222**, 58 (1933).

348. Tschesche, R. und R. Petersen: Die Konstitution des Adonitoxigenins. Chem. Ber. **86**, 574 (1953).

349. Tschesche, R., M. E. Rühsen und G. Snatzke: Über die Aglykone der Nebenglykoside der Uzara-Wurzel. Chem. Ber. **88**, 686 (1955).

350. Tschesche, R. und F. Seehofer: Die Cardenolid-Glykoside der Blätter von *Convallaria majalis.* Chem. Ber. **87**, 1108 (1954).

351. Tschesche, R., K. Sellhorn und K.-H. Brathge: Zur Spaltung von Herzglykosiden mit Luizym. Chem. Ber. **84**, 576 (1951).

352. Tschesche, R. und G. Snatzke: Über Adynerin und Neriantin. Chem. Ber. **88**, 511 (1955).

353. — — Zur Konstitution des Ouabagenins. Chem. Ber. **88**, 1558 (1955).

354. Tschesche, R., S. Wirtz und G. Snatzke: Über die herzwirksamen Glykoside der Wurzeln von *Evonymus atropurpurea* (Jacq). Chem. Ber. **88**, 1619 (1955).

355. Turkovic, I.: Extraction et identification d'un trioside des graines de *Strophanthus Intermedius* Pax: l'„i" Strophanthoside. Bull. acad. royale méd. Belgique [6] **19**, 55 (1954).

356. TURNER, R. B.: Structure and Synthesis of Cardiac Genins. Chem. Rev. 43, 1 (1948).

356a. TUZSON, J. und G. VASTAGH: Über die quantitative papierchromatographische Trennung von Digitalis-Glykosiden. Pharm. Acta Helv. 30, 444 (1955).

357. UHLE, F. C. and R. C. ELDERFIELD: Synthetic Glycosides of Strophanthidin. J. Organ. Chem. (USA) 8, 162 (1943).

358. URSCHELER, H. R., CH. TAMM und T. REICHSTEIN: Die Giftstoffe der europäischen Erdkröte Bufo bufo bufo L. Helv. Chim. Acta 38, 883 (1955).

359. VASTAGH, G. und J. TUZSON: Papierchromatographische Trennung der Digitalis-Glucoside. Pharm. Zentralhalle 92, 88 (1953).

360. VOSS, W. und G. VOGT: Über Convallamarin. Ber. dtsch. chem. Ges. 69, 2333 (1936).

361. VOSS, W. und W. WACHS: Über die Alkoholyse von Phenolglucosiden. Liebigs Ann. Chem. 522, 240 (1936).

362. WARREN, A. T., F. O. HOWLAND and L. W. GREEN: Colorimetric Assay of Digitoxin. J. Amer. Pharmaceut. Assoc., Sci. Ed. 37, 186 (1948).

363. WATTIEZ, M. N.: Contribution à l'étude des Strophanthus Africains. (Strophanthus intermedius, PAX) 2e comm. Bull. acad. royale méd. Belgique [6] 18, 147 (1953).

364. WENNER, V. und T. REICHSTEIN: Vereinfachte Methode zur Bereitung von 3α-,12α-Dioxy-ätiocholansäure. Helv. Chim. Acta 27, 965 (1944).

365. WIELAND, H.: Über den Giftstoff der Kröte. Sitzungsber. Bayr. Akad. Wiss. (math.-physik. Kl.) 1920, 329.

366. WIELAND, H. und H. BEHRINGER: Zur Konstitution des Bufotalins. Liebigs Ann. Chem. 549, 209 (1941).

367. — — Zur Konstitution des Bufotalins. Liebigs Ann. Chem. 549, 209 (1941) [s. insbes. S. 234, Anhang].

368. WIELAND, H., G. HESSE und R. HÜTTEL: Weiteres zur Konstitutionsfrage. Liebigs Ann. Chem. 524, 203 (1936).

369. WIELAND, H., G. HESSE und H. MEYER: Der Abbau des Bufotalins zu einer Cholansäure. Liebigs Ann. Chem. 493, 272 (1932).

370. WIELAND, H. und F. J. WEIL: Über das Krötengift. Ber. dtsch. chem. Ges. 46, 3315 (1913).

371. WINDAUS, A. und E. HAACK: Über die Formel des Digitalinum verum. Ber. dtsch. chem. Ges. 62, 475 (1929).

372. WINDAUS, A. und L. HERMANNS: Über Cymarin, den wirksamen Bestandteil aus Apocynum cannabinum. Ber. dtsch. chem. Ges. 48, 979 (1915).

373. WINDAUS, A., O. LINSERT und H. J. ECKHARDT: Über Iso-dehydrocholesterin (Δ^6,Δ^8-Cholestadien-3-ol). Liebigs Ann. Chem. 534, 22 (1938).

374. WINDAUS, A. und G. SCHWARTE: Über ein in Chloroform unlösliches Glykosid aus Digitalisblättern, das Gitoxin. Ber. dtsch. chem. Ges. 58, 1515 (1925).

375. WINDAUS, A. und G. STEIN: Über die Formel des Digitoxins. Ber. dtsch. chem. Ges. 61, 2436 (1928).

376. ZAFFARONI, A., R. B. BURTON and E. H. KEUTMANN: The Application of Paper Partition Chromatography to Steroid Analysis. J. Biol. Chem. 177, 109 (1949).

377. — — — Adrenal Cortical Hormones: Analysis by Paper Partition Chromatography and Occurrence in the Urine of Normal Persons. Science (Washington) 111, 6 (1950).

378. ZOLLER, P. und CH. TAMM: Die Konstitution des Transvaalins. Helv. Chim. Acta 36, 1744 (1953).

(Eingelaufen am 27. Dezember 1955.)

Natural Tropolones
and Some Related Troponoids.

By TETSUO NOZOE, Sendai, Japan.

Contents.

I. Introduction.

Cycloheptatriene, a seven-membered ring polyene, was obtained by LADENBURG (*118*) in 1881, and by MERLING (*122*) in 1891, by the systematic degradation of an alkaloid, tropine. Its structure was established by WILLSTÄTTER (*241*) by synthesis. This substance, called *tropilidene*, was a very unstable yellow oil and did not show any aromatic character, in contrast to the corresponding six-membered "polyene", *viz.* benzene. At the same time MERLING had obtained a yellow, crystalline product by heating tropilidene with bromine but was unable to establish its structure.

In 1945, DEWAR (*62–64*) proposed a seven-membered *enolone* structure–a novel aromatic system–for the mold product, stipitatic acid, and the alkaloid, colchicine, and he gave the name of *tropolone* to the parental structure of such a system. This was the incentive that led to the discovery of some twenty natural tropolones during the next few years.

Independently of these studies, *hinokitiol*, a terpenoid compound, was found to contain such an unsaturated, seven-membered ring, to yield various substitution products, and to possess some very peculiar characteristics quite different from all existing benzenoid compounds [NOZOE (*134, 136*)].

Since 1949 various related compounds, the *"troponoids"* containing the skeleton of tropone (II), tropolone (III), and troponeimine (IV), have been synthesized. There is a fair amount of common characteristics among these troponoid compounds but some marked differences were also found to be caused by the presence or absence of a side-chain or other substituent, and, especially, of a fused ring, or by the position of such groups. Tropolones undergo facile electrophilic substitution, as in the case of hinokitiol; and troponoid compounds in general submit more easily to nucleophilic substitution as well as to rearrangement to benzenoid compounds.

NOZOE and his collaborators (*7, 8*) have also prepared various new derivatives of azulene (V), as well as some hitherto unknown heterocyclic azulenoid compounds (VI: *X*, *Y*, *Z* = hetero atom), starting from a troponoid. Thus, the number of new types of troponoid, azulenoid, and related substances has continued to increase.

(I.) Tropylium cation.   (II.) Tropone.   (III.) Tropolone.   (IV.) Troponeimine.

(V.) Azulene.   (VI.) Heterocyclic azulenoid.   (VII.) Benzo[b]tropazine.

Recently, DOERING and KNOX (*73*) have reexamined MERLING's yellow crystalline product mentioned and found that this substance was the bromide of the cycloheptatrienium or tropylium cation (I), which constitutes the fundamental skeleton of such unsaturated seven-membered aromatic systems. Its detailed examination by physical means has secured the basis of the non-benzenoid, seven-membered aromatic system (*83*). The present writer (*7, 8*) proposed the generic term, "*tropoid* compounds" for all those substances that possess the fundamental structure (I), including troponoids, azulenoids, and other polycyclic compounds, such as (VII); and the term, "*tropoid system*" is used for this seven-membered aromatic system.

When compared to such an extensive and still growing class of compounds the number of the naturally occurring tropolones found so far is still small. However, these compounds are distributed widely in terpenoids, fungal metabolites, alkaloids, glucosides, and pigments. To prove the presence of the tropolone ring is comparatively simple nowadays but great difficulties were encountered earlier in the elucidation of the structure of natural tropolones and in most instances the final conclusions were reached after many detours. There is as yet no certain routine method available for the clarification of the structure of unknown compounds containing a complicated tropoid system. For this reason and since our extensive knowledge in this field has developed from the study of natural tropolones, the progress leading to the structural determination of the natural tropolones will be described in the present survey. Furthermore, some general methods of the synthesis of various types of troponoids will be outlined and the properties of the compounds described. It is hoped that this will facilitate further research.

The terms tropone (cycloheptatrien-1-one) and tropolone (2-hydroxy-cycloheptatrien-1-one) have now been generally accepted. The positions of substituents are designated by numbers as shown in the symbols (VIII) to (X) (9).

    (VIII.)   (IX.)   (X.)   (XI.)

In the case of tropolones [such as (IX) and (X): $R' = H$] substituted in positions other than 5, only a single isomer is usually found in free form because of the mobility of the proton between two oxygen atoms at $C_{(1)}$ and $C_{(2)}$ (8, 9); hence, such compounds are not numbered as in (X) but as in (IX). In ethers and esters, however, a pair of isomers corresponding to two forms is obtained. Since the electrophilic substitution of tropolones was assumed to be very simular to that of phenols, Nozoe (134, 136) and Dewar (65), in the early period of tropolone chemistry, made use of the prefixes *ortho*, *meta* and *para* for substituents of the tropolone ring, while most workers in England and in the United States (2) used α, β, and γ for such designations (XI). However, since such prefixes have caused some misconceptions, a numbering system is used at the present time.

Several review articles on troponoid compounds have been published (1–10), and in order to save space, the present survey will frequently refer to these reviews, instead of mentioning each original paper.

II. Naturally Occurring Tropolones.

1. Terpenoid Tropolones.

a) Occurrence.

It is well known that woods originating from Cupressaceae are generally resistant to decay (78) and, as shown in *Table 1*, the heartwood of these trees or the essential oil thereof has yielded several antibiotic substances such as α-, β- (hinokitiol), and γ-thujaplicins, and nootkatin. From the structural point of view and considering their occurrence, these compounds can be regarded as tropolones of terpenoid origin. Not only the tropolone type but also the content in these trees is dependent on the geographical location, even when the same species is considered. Japanese Hinoki (*Chamaecyparis obtusa* Sieb. et Zucc.) is devoid of tropolones (133, 134) in contrast to specimens originating from Formosa.

Table 1. List of Plants Containing Terpenoid Tropolones.

Botanical name	Common name	Geographical location	Type of tropolone**	References
Chamaecyparis taiwanensis MASAMUNE et SUZUKI	Taiwan-hinoki*	Formosa (Taiwan)	α, β	(*133, 134*)
C. formosensis MATSUM.	Benihi*	Formosa	(trace)	(*133*)
C. nootkatensis SPACH.	Alaskan Yellow cedar	U. S. A.	*Nk*	(*37*)
Thuja plicata D. DON	Western Red cedar	U. S. A.	α, β, γ	(*14, 15, 80, 246*)
Thuja plicata D. DON		Sweden	α, γ	(*80, 92*)
T. occidentalis L.	Northern White cedar	Sweden	α, γ	(*93*)
T. occidentalis L.	Nioi-hiba*	Japan	α, β	(*126*)
T. Standishii CARR.	Nezuko*	Japan	β	(*127*)
Thujopsis dolabrata ZIEB. et ZUCC.	Hiba*	Japan	α, β	(*196*)
Libocedrus decurrens TORREY	Incense cedar	U. S. A.	β, γ	(*245, 247*)
Cupressus macrocarpa HARTW.	Monterey cypress	New Zealand	β, *Nk*	(*57, 246*)
Juniperus chinensis L.	Byakushin*	Japan	β, *Nk*	(*128*)

* Japanese common name. ** α, β, γ: Thujaplicins; *Nk*: Nootkatin.

Quite recently, ZAVARIN and ANDERSON (*246, 247*) have developed a chromatographic procedure for the separation of Cupressaceae heartwood extracts and the identification of the tropolonic constituents.

b) Hinokitiol or β-Thujaplicin, $C_{10}H_{12}O_2$ (m. p. 51–52°), and Hinokitin, $C_{30}H_{33}O_6Fe$ (m. p. 251–252°).

Studies on Hinokitiol (134). The essential oil of Taiwan Hinoki, different from that of Japanese Hinoki, is dark red and its acid component has been known to give a deep red coloration with ferric chloride. In 1926, HIRAO found that by treating this acid component with ferric chloride a red pigment, hinokitin, m. p. 251°, was obtained. He assumed that this substance, also found in the Hinoki oil, was a natural product and the source of the red tint of Taiwan Hinoki heartwood.

In 1936 NOZOE (*133*) found unexpectedly that hinokitin contained iron and that its formula agreed with $C_{30}H_{33}O_6Fe$. The acid substance, $C_{10}H_{12}O_2$, obtained by heating hinokitin with strong alkali, was named hinokitiol. Hinokitiol showed an acidity intermediate between that of acetic acid and phenol and it formed stable complex salts not only with iron but also with almost any metal ion. NOZOE assumed that hinokitiol was an enolized form of a 1,2- or 1,3-diketone containing two double bonds, however, its structure could not be clarified at that time.

Later, Nozoe and Katsura (*143*) obtained hinokitiol in pure, crystalline form and carried out extensive structural studies. They established the following facts. Catalytic hydrogenation of hinokitiol results in absorption of 4 to 4.5 moles of hydrogen to yield chiefly a saturated diol, $C_{10}H_{20}O_2$, besides a monohydric alcohol, $C_{10}H_{20}O$ and a ketone, $C_{10}H_{18}O$. In general hinokitiol does not react with ketonic reagents but forms a neutral monoacetate so that one of the two oxygen atoms must be in an enolic hydroxyl while the other forms an inactive carbonyl group. The number of double bonds in hinokitiol was assumed to be three but the exact number remained undetermined because of various difficulties.

Since hinokitiol affords acetone (besides carbon dioxide and oxalic acid) on oxidation it must possess an isopropyl group. The oxidation of the diol or of the dialdehyde, obtained by treatment of the diol with lead tetraacetate, affords a dicarboxylic acid, $C_{10}H_{18}O_4$. Since the distillation of this acid with acetic anhydride yields a ketone, $C_9H_{16}O$, hinokitiol certainly represents an α-enolone of a seven-membered ring. Various properties of hinokitiol have shown that it is not a benzenoid compound. Because the oxidation of hinokitiol with alkaline hydrogen peroxide afforded β-isopropyl-levulinic acid, besides an unsaturated dibasic acid, it was interpreted erroneously to be the isopropyl homolog of cycloheptadien-2-ol-1-one corresponding to $C_{10}H_{14}O_2$. Evidently, too much emphasis had been laid on the presence of the saturated keto acid, $C_8H_{14}O_3$ (*134*).

Subsequent studies by Nozoe, Sebe, and others (cf. *134*) during 1940–1947 revealed that hinokitiol possessed the following characteristic properties: 1. Hinokitiol, $C_{10}H_{12}O_2$, is extremely stable towards both fused alkali and concentrated or fuming sulfuric acid. 2. It withstands the Bouveault-Blanc reduction but its methyl ether is reduced to the saturated diol, $C_{10}H_{20}O_2$. 3. In spite of being an α-enolone, hinokitiol shows a fair degree of acidity and like β-enolones forms inner complex salts. 4. It yields various substitution products such as chloro, bromo, iodo, nitro and azo compounds, but does not submit to direct sulfonation or the Friedel-Crafts reaction. 5. Hinokitiol shows a certain amount of basicity; it dissolves in conc. mineral acids but is precipitated on dilution with water. It is not nitrated in the presence of conc. sulfuric acid. 6. Dinitro-hinokitiol (XII: $X' = H$, $X'' = X''' = NO_2$) is easily converted to benzoic acid derivatives or their esters when warmed with aqueous acetic acid or alcohol (*134*).

In short, hinokitiol was found to possess marked aromatic character in spite of being an unsaturated, seven-membered ring compound.

On the basis of the experimental results just mentioned the structure (XIV) was assigned to hinokitiol but the exact position of the isopropyl side-chain could not be established at that time. Nevertheless, considering the similarity of its electrophilic substitution and that of

(XII.)   (XIII.)

phenols, the positions substituted on the seven-membered ring were assumed to be at 3, 5, and 7 [cf. (XII)]; and, based on the number of mono-, di-, and especially tri-halogen derivatives actually obtained, it seemed certain that the side-chain of hinokitiol occupied the 4-position (*134*).

5,7-Dinitro-hinokitiol affords condensation products with ammonia, amines, and some ketonic reagents and a highly colored such product (XIII) with *o*-phenylenediamine (*134*). The orange colored azo compounds of hinokitiol are easily converted into reddish-purple pigments (hinopurpurin), when heated with acetic acid or ethanol, especially in the presence of a trace of hydrochloric acid (see p. *271*).

Since 1948, the structures of various substitution and condensation products of hinokitiol were determined by NOZOE and others (*6, 8, 135, 136*).

Structure of β-Thujaplicin. In 1933, ANDERSON and SHERRARD (*15*) discovered a toxic "phenolic" substance, $C_{10}H_{12}O_2$, m. p. 82°, in western red cedar, and later an isomer (m. p. 52–52.5°) that crystallized from the mother liquor on standing for a long period of time (*14*).

In 1948, ERDTMAN and GRIPENBERG (*80*) obtained from the heartwood of the same tree cultivated in Sweden, besides the substance of m. p. 82°, a third isomer of m. p. 34°, and termed these antibiotic substances α-, β-, and γ-thujaplicin in the increasing order of their melting points. They all give the same green ferric chloride reaction and form green copper complex salts soluble in chloroform. Since they all show identical ultraviolet spectra, they were assumed to possess the same conjugated system. The workers mentioned carried out the same degradation reactions with the three thujaplicins and clarified their respective structures. This was the first published report on the structural determination of natural tropolones.

The oxidation of the stereoisomeric mixture of the octahydro compound (XV), obtained by the catalytic reduction of β-thujaplicin, afforded

β-isopropyl-pimelic acid (XVI) and the pyrolysis of its barium salt gave 3-isopropyl-cyclohexanone (XVII).

(XIV.)   (XV.)   (XVI.)   (XVII.)

Hinokitiol (β-thujaplicin).   Octahydro-β-thujaplicin. β-Isopropyl-pimelic acid.   3-Isopropyl-cyclohexanone.

Considering these conversions, the structure of 4-isopropyl-tropolone (XIV) was assigned to β-thujaplicin (*14*). Later, hinokitiol was found to be identical with β-thujaplicin (*134*). NOZOE as well as COOK and their collaborators, synthesized independently the three isomeric thujaplicins (p. 257).

c) γ-Thujaplicin, $C_{10}H_{12}O_2$ (m. p. 82°).

ERDTMAN and GRIPENBERG (*79*) obtained γ-isopropyl-pimelic acid and 4-isopropyl-cyclohexanone from octahydro-γ-thujaplicin, as in the case of the β-isomer mentioned, and assigned the structure (XVIII) to γ-thujaplicin.

(XVIII.) γ-Thujaplicin.   (XIX.) α-Thujaplicin.

d) α-Thujaplicin, $C_{10}H_{12}O_2$ (m. p. 26° and 34°).

GRIPENBERG (*92*) proved that α-thujaplicin is 3-isopropyl-tropolone (XIX) in the same manner as in the case of β- and γ-thujaplicin. It is interesting to note that α-thujaplicin crystallizes in two forms melting at 26° and 34°, but it has not been determined whether this is due to dimorphism or to the formation of tautomers (*136, 160*). It should also be noted that α-thujaplicin is different in some respects from the β- and γ-compounds. Thus, the boiling point of the α-isomer is considerably lower than those of the β- or γ-isomers, and its copper complex is less soluble and higher melting than those of the two other isomers. Although electrophilic substitution (*160, 244*) of α-thujaplicin is similar to that of the β- and γ-compounds, it is not sulfonated when heated with sulfamic acid at 150–160°. These differences are probably caused by the steric effect of the isopropyl group adjacent to the hydroxyl in (XIX) [NOZOE et al. (*136, 160*)].

e) Nootkatin, $C_{15}H_{20}O_2$ (m. p. 95°).

ERDTMAN and his co-workers (*37, 81*) assumed that nootkatin (XX) had a tropolone structure because of its typical coloration with ferric chloride and the formation of a green copper complex; and the same assumption

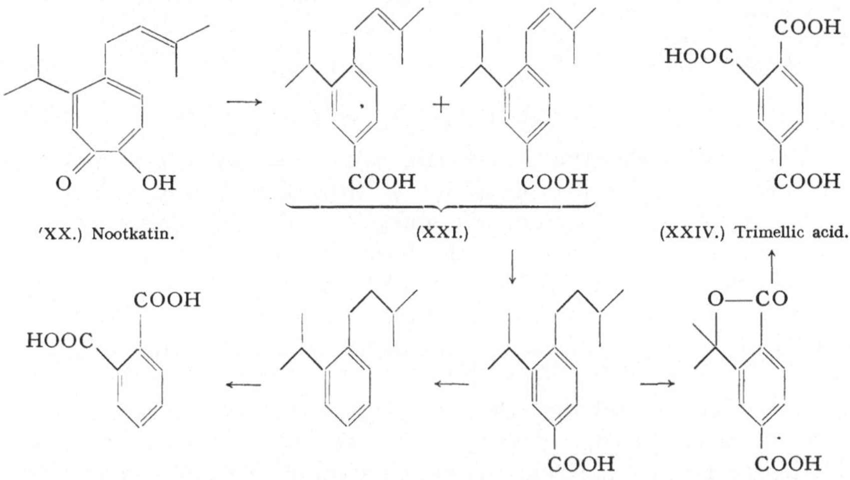

'XX.) Nootkatin. (XXI.) (XXIV.) Trimellic acid.

(XXVI.) (XXV.) (XXII.) (XXIII.)

Chart 1. Structural Determination of Nootkatin.

was made by AULIN-ERDTMAN (*18, 21*) on the basis of the spectrum. The presence of a 3-methylbut-2-enyl side-chain in nootkatin was postulated from the formation of acetone and isobutyric acid by the oxidation of an α-glycol, $C_{15}H_{22}O_4$, which is derived from nootkatin (*81*). CAMPBELL and ROBERTSON (*36*) concluded from X-ray data that this side-chain must be located at the 5-position in β-thujaplicin. Recently, DUFF and ERDTMAN (*76*) treated nootkatin methyl ether with sodium methoxide and submitted the acid mixture (XXI) thereby obtained to reduction. The dihydro acid (XXII) formed was oxidized with dilute nitric acid to give a lactone. The latter yielded trimellic acid (XXIV) by further oxidation (*Chart* 1) and the structure of the lactone was established as (XXIII) by synthesis. From these results and from the fact that the hydrocarbon (XXV), obtained by the decarboxylation of (XXII), yields on oxidation phthalic acid (XXVI), ERDTMAN and his co-workers established the formula (XX) for nootkatin.

2. Hydroxytropolone-carboxylic Acids.

a) Occurrence as Mold Metabolites.

As shown in *Table 2*, three types of carboxylic acids belonging to the tropolone series have been found as metabolites of molds (next page).

Table 2. Occurrence of Tropolones as Mold Metabolites.

Mold Species	Metabolite tropolone	References
Penicillium puberulum BAINIER	Puberulic acid and Puberulonic acid	(*30*)
P. aurantio-virens BIOURGE		(*22, 30, 54*)
P. Johannioli ZALESKI		(*29*)
P. cyclopium viridicatum		(*29*)
P. stipitatum THOM.	Stipitatic acid	(*29*)

b) *Stipitatic Acid,* $C_8H_6O_5$ *(m. p. 302–304°).*

According to RAISTRICK et al. (*29, 30*), the stipitatic acid molecule contains three active hydrogens but is titrated as a dibasic acid. It forms a deep yellow disodium salt and gives a deep red coloration with ferric chloride. Methylation with diazomethane yields two neutral, isomeric methyl ethers (XXVII and XXVIII: $R' = R'' = R''' = CH_3$), a weakly acidic dimethyl compound (XXVII or XXVIII: $R' = R''' = = CH_3$; $R'' = H$) with methanolic hydrogen chloride, and a dibasic monomethyl derivative (XXVII: $R' = R''' = H$; $R'' = CH_3$) with alkaline dimethyl sulfate. Decarboxylation of stipitatic acid affords a monobasic acid, $C_7H_6O_3$, that gives the same red coloration with ferric chloride as the original acid. These observations suggest the presence of two enolic or phenolic hydroxyls and probably of one carboxylic group. Although stipitatic acid does not react with ketonic reagents, its reduction product forms a 2,4-dinitrophenylhydrazone so that the fifth oxygen atom seems to be in a masked carbonyl. It seemed surprising at the time that stipitatic acid when fused with alkali gave 5-hydroxy-isophthalic acid (XXIX) in good yields. Stipitatic acid is stable against concentrated mineral acids, and does not give an addition product but a mono-substitution product when treated with bromine in 80% acetic acid.

Although the composition of stipitatic acid is rather simple, it shows many characteristics that differentiated it from all known compounds, and hence, a satisfactory structure could not be secured for some time.

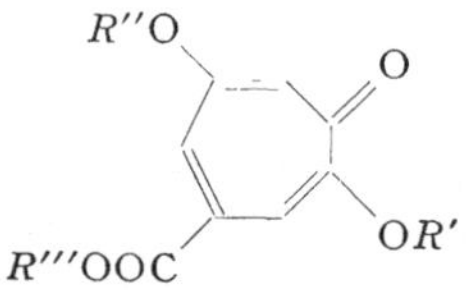

(XXVII.)

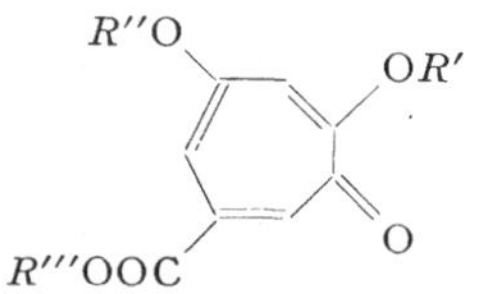

(XXVIII.)

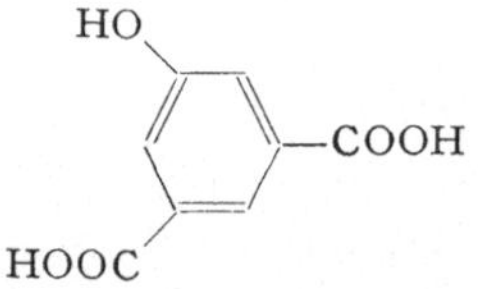

(XXIX.) 5-Hydroxy-isophthalic acid.

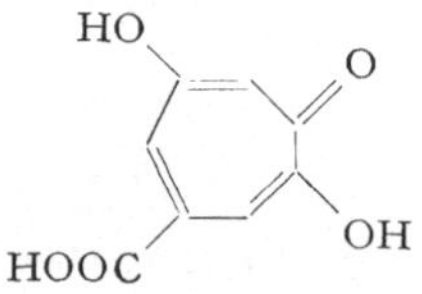

(XXX.) Stipitatic acid.

As a result of theoretical evaluation of all experimental data, DEWAR (*62*) proposed a tropolone structure for the first time (1945) and assigned the formula (XXX) to stipitatic acid. This suggestion became the starting point of the extensive development of troponoid chemistry.

TODD et al. (*56*) oxidized stipitatic acid (XXX) with cold alkaline hydrogen peroxide and obtained aconitic acid (XXXIII) and malonic acid. Later JOHNSON et al. (*26*) proved the correctness of structure (XXX) by total synthesis of stipitatic acid (p. 257).

c) Puberulic Acid, $C_8H_6O_6$ (m. p. 318–320°).

In 1932, BIRKINSHAW and RAISTRICK (*30*) found in the culture medium of *Penicillium puberulum* an acid, $C_8H_6O_6$, which colored deep reddish brown with ferric chloride; it was termed puberulic acid. They isolated this acid from other fractions of the extract in the form of its sparingly soluble, deep-red nickel salt and liberated the acid component with mineral acid. However, it was found that this preparation was a mixture of puberulic acid and a yellow acid, termed puberulonic acid. Puberulic acid could be isolated and purified *via* its diacetate, since the yellow acid does not form a diacetate. Puberulic acid is titrated as a dibasic acid and gives a dimethyl ether with alkaline dimethyl sulfate. A treatment with diazomethane affords a neutral tetramethyl ether, easily soluble in cold water.

The foregoing observations seemed to indicate that the puberulic acid molecule contained two phenolic hydroxyls and probably two carboxyl groups; however, none of the known or newly synthesized dihydroxybenzene-dicarboxylic acids of the same composition showed the properties of puberulic acid.

BARGER and DORRER (*22*) found that puberulic acid easily underwent decarboxylation when heated to about its melting point and they obtained a substance, $C_7H_6O_4$, that gave the same coloration with ferric chloride as the original acid. RAISTRICK et al. (*30*) assumed that puberulic acid was somehow related to stipitatic acid, and DEWAR (*64*) postulated that it represented a monohydroxy-stipitatic acid but he failed to elucidate its structure.

Based on the observation that the oxidation of puberulic acid with alkaline hydrogen peroxide (as in the case of stipitatic acid) gave aconitic acid (XXXIII) in good yield, and on some experimental results obtained by other workers, TODD et al. (*55*) assigned the structure (XXXI), or its tautomer, to puberulic acid. However, in contrast to stipitatic acid and γ-thujaplicin, puberulic acid is stable to alkali fusion up to 320°.

JOHNSON et al. (*104*) were able to synthesize puberulic acid by hydrolyzing bromostipitatic acid.

The formation of aconitic from puberulic acid may be understood by assuming that the latter reacts in its tautomeric form (XXXII).

(XXXI.) Puberulic acid.   (XXXII.)   (XXXIII.) Aconitic acid.

d) Puberulonic Acid, $C_9H_4O_7$ (m. p. 298°; decomp.).

The yellow puberulonic acid, found as mentioned in some culture media together with puberulic acid, does not submit to ordinary acetylation or decarboxylation, and does not form a methyl ester (*30*). BARGER and DORRER (*22*) found that, when this acid is titrated with alkali, the yellow color changes to deep pink at the neutralization point, and then the color fades out completely in the alkaline pH range. Although this colorless solution turns yellow when acidified, further addition of alkali fails to recover the pink color. From this "abnormal" behavior the authors concluded that a lactone or a pseudo-acid group is present in puberulonic acid.

On the basis of the analytical data of some puberulonic acid derivatives and especially of the quantitative conversion of puberulonic acid into puberulic acid, $C_8H_6O_6$ and CO_2, when heated in dilute sulfuric acid or even in water, TODD et al. (*54*) corrected the formula (*30*), $C_8H_4O_6$ to $C_9H_4O_7$. Considering the experimental results mentioned and the results of the potentiometric titration, the structure (XXXIV) or that of the isomeric γ-lactone was proposed for puberulonic acid (*55*).

On the other hand, AULIN-ERDTMAN (*19, 20*) determined the ultraviolet spectrum of puberulonic acid in various solvents and at various pH and, assuming that it had a true tropolone structure and not the triketo-lactonic acid structure (XXXIV) mentioned, suggested the acidanhydride structure (XXXV). The latter formulation is in agreement with the infrared spectral data reported by AULIN-ERDTMAN and THEORELL (*21*) as well as by TODD and his co-workers (*106*).

(XXXIV.)   (XXXV.) Puberulonic acid.

3. Purpurogallin.

a) Possible Occurrence in Nature.

NIERENSTEIN (*131*) isolated a red pigment, "dryophantin", $C_{23}H_{28}O_{15}$, from the "red pea gall" produced by *Dryophanta divisa* ADL. on the leaves of *Quercus pedunculata* EHRL., and found that its hydrolysis yielded two moles of glucose and one mole of purpurogallin. Later, NIERENSTEIN and SWANTON (*132*) reported that they had obtained several pigments which were also believed to be purpurogallin glucosides occurring in various galls, and produced either by insects or fungi on the leaves, buds or shoots of *Quercus*, *Genista*, *Tilia*, and *Potentilla* species. Recently, some workers have denied the presence of purpurogallin derivatives as natural products (cf. *9*). However, purpurogallin that has been known since old times, is produced easily from pyrogallol by the action of various oxidation agents. Since its tropolonoid structure was established fairy early, some significant features of the study of this substance will be described below.

b) The Structure of Purpurogallin, $C_{11}H_8O_5$ (m. p. 276°, decomp.).

In 1869, GIRARD termed a red pigment purpurogallin, prepared from pyrogallol by oxidation with silver nitrate and potassium permanganate. Subsequently, PERKIN et al. found that four out of the five oxygen atoms of purpurogallin are in phenolic (enolic) hydroxyl groups that form di-, tri-, and tetramethyl ethers (*2, 3, 5*). Purpurogallin is substituted by bromine and can be converted in good yields into the isomeric carboxylic acid "purpurogallone", when heated with concentrated alkali. PERKIN (*198*) assigned the formula (XXXVI) to purpurogallone, and a later synthesis carried out by HAWORTH et al. (*99*) proved this assumption to be correct. NIERENSTEIN et al. (*61*) favored the structure (XXXVII) for purpurogallin, while WILLSTÄTTER et al. (*242*) proposed (XXXVIII), although supporting data were not available.

(XXXVI.) Purpurogallone.   (XXXVII.)   (XXXVIII.)

In 1948, however, BARLTROP et al. (*23*) obtained 3,4,5-trimethoxyphthalic acid (XLI) by the permanganate oxidation of the tetramethyl ether (XL) and proposed a new formula (XXXIX) for purpurogallin. Independently and simultaneously, HAWORTH et al. (*98*) carried out the catalytic reduction of the purpurogallin tetramethyl ether (XL) using Adams' catalyst. The tetrahydro derivative (XLII) thus prepared was then oxidized with alkaline hydrogen peroxide and the product identified with the synthetic dicarboxylic acid (XLIII). These authors have thus established the presence of the seven-membered ring in (XLII) and proposed the corresponding 3,4-benzotropolone structure (XXXIX) for purpurogallin. Moreover, these experiments determined the relative positions of the three hydroxyl groups in the benzotropolone ring.

Finally, the structure of purpurogallin was unequivocally established by total synthesis [Haworth et al. (*38, 39*)] and by X-ray diffraction studies [Dunitz (*77*)].

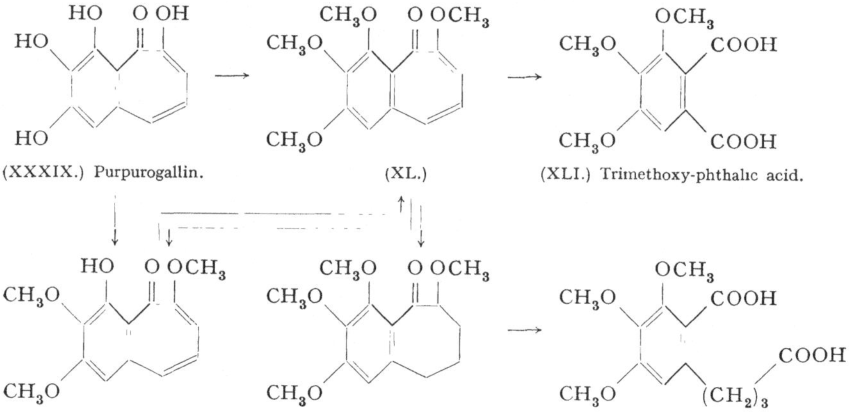

(XXXIX.) Purpurogallin.   (XL.)   (XLI.) Trimethoxy-phthalic acid.

(XLIV.)   (XLII.)   (XLIII.)

A treatment of the tetramethyl ether (XL) with dilute acid results in the hydrolysis of one methoxyl located in the benzene ring to the tri-methyl ether (XLIV) and the same trimethyl ether is also formed directly when purpurogallin is methylated. It is interesting to note that Haworth et al. prepared the dicarboxylic acid (XLV) by autoxidation of purpurogallin in alkaline medium or by hydrogen peroxide oxidation. Furthermore, they obtained a compound assumed to be 4-methyl-tropolone (XLVI) by thermal decarboxylation of the acid (XLV) (*96, 98*). The structure of this substance was later confirmed by synthesis [Nozoe et al. (*165*)].

(XXXIX.) Purpurogallin. ⟶   (XLV.)   (XLVI.) 4-Methyltropolone.

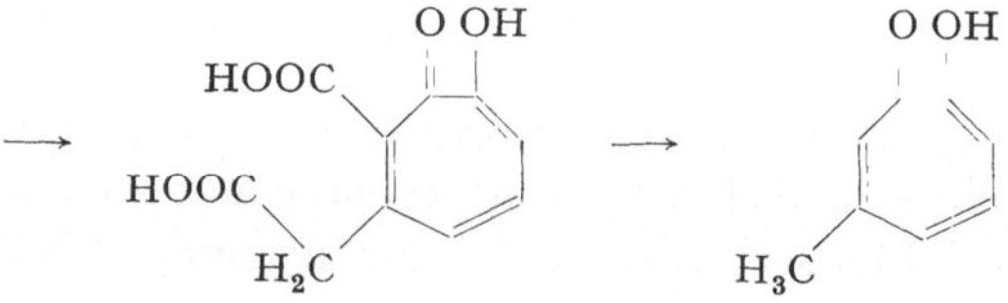

4. Alkaloidal Tropolones.

a) Occurrence.

Colchicine and related compounds are found in the corms, seeds, flowers, and pericarps of some plants belonging to the Liliaceae such as *Colchicum autumnale* L., *C. speciosum* STEO., *Gloriosa superba* L., and other species of *Colchicum, Gloriosa, Androcymbium, Bulbocodium*, and *Merendera* (*9, 48*). In most instances the separation and purification of the alkaloids are conveniently effected by chromatography (*17, 211, 212, 214a, 215*); and for the isolation and detection of alkaloids in a small

Table 3. Colchicum Alkaloids (*214, 214a*).

Name of compound	Formula	M. p. (°)	$[\alpha]_D$ in chloroform	Number of groups OCH$_3$	CH$_3$CO
Colchicine	$C_{22}H_{25}O_6N$	154–156	—119°	4	1
Colchiceine	$C_{21}H_{23}O_6N$	175–177	—256°	3	1
Substance B.............	$C_{21}H_{23}O_6N$	256–257	—172°	4	1**
C.............	$C_{21}H_{23}O_6N$	130–180	—135°	3	1
E$_1$	$C_{21}H_{25}O_5N$				
F.............	$C_{21}H_{25}O_5N$	184–186	—127°	4	—
S.............	$C_{22}H_{25}O_6N$	136–138	—117°	3	—
U.............	$C_{19}H_{21}O_5N$		?	3	—
R............	?	196–198	?	?	?
H-3 (?)	?	183–185	?	?	?
Ta...........	?	133–135	—211°	3	—
To...........	?	236–238	—65°***	4	1
Glucoside (M)............	$C_{27}H_{33}O_{11}N$	310–314		3	1
Substance I*	$C_{22}H_{25}O_6N$	184–186	+307°	4	1
D*............	$C_{21}H_{23}O_5N$	234–236	+294°	3	1
J*............	$C_{22}H_{25}O_6N$	274–278	—445°	4	1
P*............	?	228–229	—226°	4	1

* No tropolone ring. ** Formyl group. *** In pyridine.

sample of plant material, paper chromatography can be applied (*121, 199*). Polarographic methods are sometimes used for the detection and determination of Colchicum alkaloids (*209, 215*). As shown in *Table 3*, more than ten minor alkaloids, besides colchicine, have been found in detailed studies by ŠANTAVÝ et al. (*214, 214a*). Most of these alkaloids are known to contain the tropolone ring, and their hydrolysates, obtained with hydrochloric acid, turn green with ferric chloride; however, the alkaloids marked with* in *Table 3* do not give such coloration and are supposed to represent rearrangement products which do not contain the tropolone ring.

b) Colchicine, $C_{22}H_{25}O_6N$ (m. p. 155–157°), and Colchiceine, $C_{21}H_{23}O_6N$ (m. p. 172°).

Meadow saffron or autumn crocus (*Colchicum autumnale* L.) has been used since old times as a remedy for gout. In 1833, Geiger and Hesse discovered a new alkaloid in the seeds and corms of this plant and named it colchicine.

Colchicine, which has the most complicated structure among the known natural tropolones, was the subject of extensive studies because of its peculiar chemical properties as well as its highly interesting biological effects. Although numerous pertinent papers have been published, in the present article only the evidence necessary for the establishment of the colchicine structure will be given, especially concerning the tropolonic nature of ring *C*. Further details will be found in the reviews written by Cook and Loudon (*48*), and by Lettré (*119*).

The Windaus Formula. Zeisel (cf. *248*) found that one of the four methoxyls in the colchicine molecule, $C_{22}H_{25}O_6N$, is very reactive and one mole of methanol is liberated easily when the alkaloid is heated in water containing a small amount of hydrochloric acid. Thus colchiceine, $C_{21}H_{23}O_6N$, is formed. When hot strong acid is applied, one mole of acetic acid is set free and the so-called "trimethylcolchicinic acid", $C_{19}H_{21}O_5N$, is obtained. Its acetylation yields colchiceine, whose methylation reverts to the original colchicine (*5*).

In contrast to colchicine, colchiceine and trimethylcolchicinic acid (desacetyl-colchiceine) have acid character; they give dark green coloration with ferric chloride and form green copper complexes.

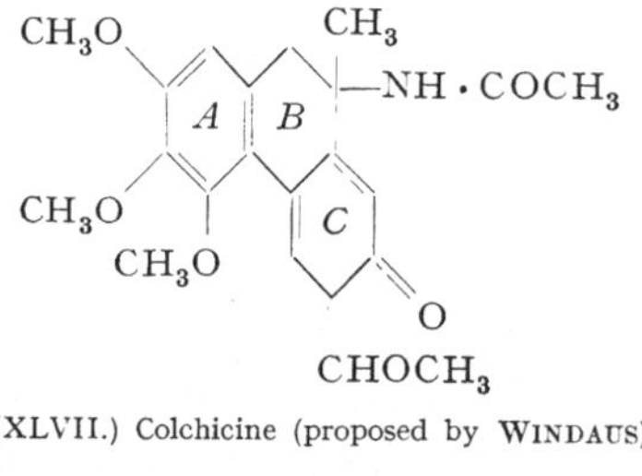

(XLVII.) Colchicine (proposed by Windaus). 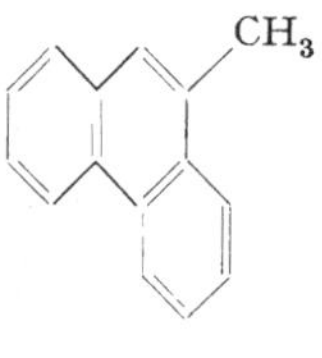   (LI.) 9-Methylphenanthrene.

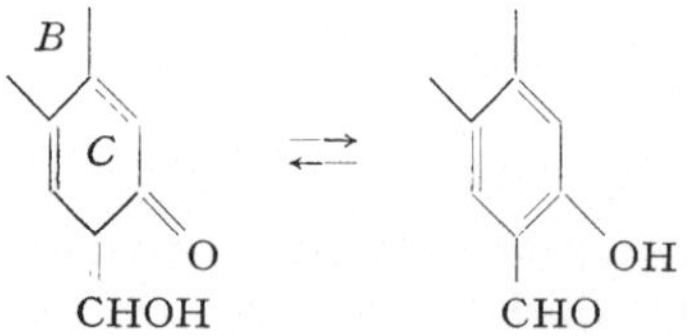

(XLVIIIa.) Colchiceine.   (XLVIIIb.) 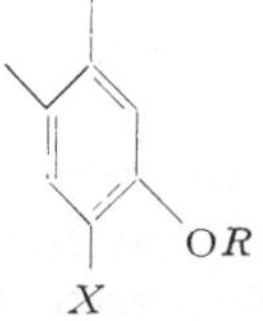   (XLIX.) 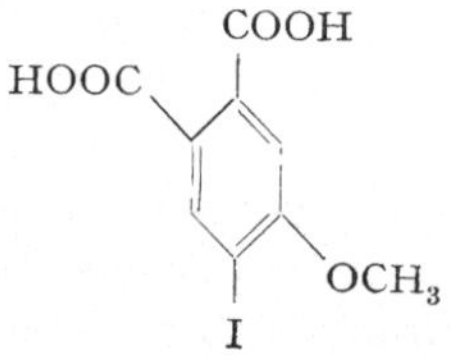   (L.) 4-Iodo-5-methoxy-phthalic acid.

During the period 1910–1923 extensive studies of colchicine were carried out by Windaus (*243*) who proposed the tricyclic structure (XLVII). The main arguments in favor of this structure follow.

Drastic oxidation of colchicine with potassium permanganate affords 3,4,5-trimethoxy-phthalic acid (XLI, p. 246) which secures the structure of ring *A*. Benzoylation of trimethylcolchicinic acid gives a dibenzoate whose mild hydrolysis yields a mono-N-benzoate. The bis-benzenesulfonate of this acid appears in two isomeric forms, whose cautious hydrolysis affords the same mono-N-benzenesulfonate. These N-acyl derivatives show dark green coloration with ferric chloride, in contrast to the O,N-diacyl derivatives. It was assumed on the basis of these observations that colchicine is not a methyl ester but a methyl ether (XLVII) of a hydroxymethylene ketone [colchiceine (XLVIII a)], and that the two isomeric O-acyl derivatives represent stereoisomers. However, colchiceine does not react with ketonic reagents, thus differing from salicylaldehyde (cf. XLVIII b).

A treatment of colchicine with iodine and alkali results in facile elimination of one carbon atom and the formation of N-acetyl-iodo-colchinol (XLIX: $X = I$, $R = H$), $C_{20}H_{22}O_5NI$, which is no longer an enolone but a phenol, and is converted by hot permanganate oxidation of its methyl ether (XLIX: $X = I$, $R = CH_3$) into an iodomethoxy-phthalic acid (L) (*243*). The latter is 4-iodo-5-methoxy-phthalic acid as found by synthesis [Grewe (*90*)].

Windaus (*243*) obtained deaminocolchinol methyl ether both by Hofmann degradation of colchinol methyl ether and by dehydration of the carbinol derived from the former. Demethylation of the deamino compound by boiling with hydriodic acid followed by zinc dust distillation gave 9-methylphenanthrene (LI). Consequently, Windaus favored the formula (LII: $R = NHCOCH_3$) for N-acetylcolchinol methyl ether, and (XLVII) for colchicine itself.

(LII.)

(LIII.) Deamino-colchinol methyl ether (Cook).

Further Experimental Evidence. In 1934, Dustin found that colchicine has an inhibiting effect on the mitosis of animal cells at an early stage; and in 1937, Blakeslee reported that colchicine effected doubling of the chromosomes in the cell, thereby giving a polyploid. Moreover, its use in tumor therapy seemed possible (cf. *119*). Because of these interesting and important biological aspects the study of colchicine was taken up again and its various reactions were reinvestigated. In 1940, Cook et al. (*42*) pointed out the incorrectness of ring *B* as it appears in the Windaus formula (XLVII).

Judging from the stability of the free amine and carbinol derived from the N-acetylcolchinol methyl ether mentioned above, these compounds could not possibly represent dihydrophenanthrene derivatives as suggested by Windaus (*243*). Hence, Cook et al. (*5, 27, 34*) compared deaminocolchinol methyl ether with various synthetic compounds with the result that it should be represented by formula (LIII). Thus, it came to light that under the drastic conditions applied by Windaus, during the demethylation of the deamino compound a rearrangement of ring *B* and the formation of 9-methylphenanthrene took place. From these results Cook (cf. *65*) assumed that ring *B* also might be seven-membered in colchicine itself.

Dewar's Colchicine Formula. At about the same time, Dewar (*63, 64*) came to the conclusion that colchicine must contain a tropolone ring, considering the observations mentioned, especially the failure of colchicine and colchiceine to react with ketonic reagents and the similarity of their behavior to that of stipitatic acid (*62*), rather than salicylaldehyde. Furthermore, almost identical ultraviolet spectra of colchicine and colchiceine were reported by Bursian (*35*). Dewar proposed the structure (LIV) for colchicine ($R = CH_3$) and for colchiceine ($R = H$);

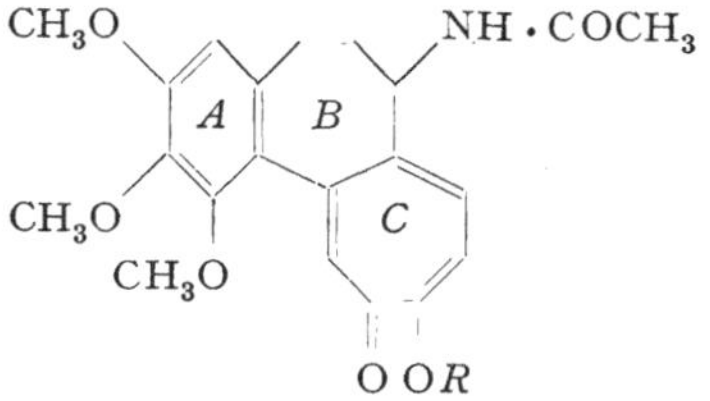

(LIV.) Colchicine ($R = CH_3$).
Colchiceine ($R = H$) (Dewar).

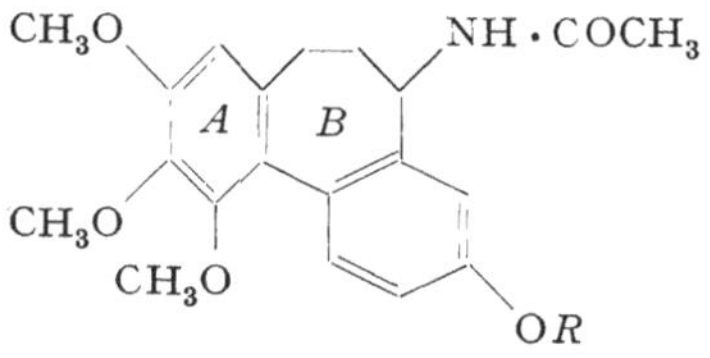

(LV.) Colchinol methyl ether ($R = CH_3$).
N-Acetyl-colchinol ($R = H$).

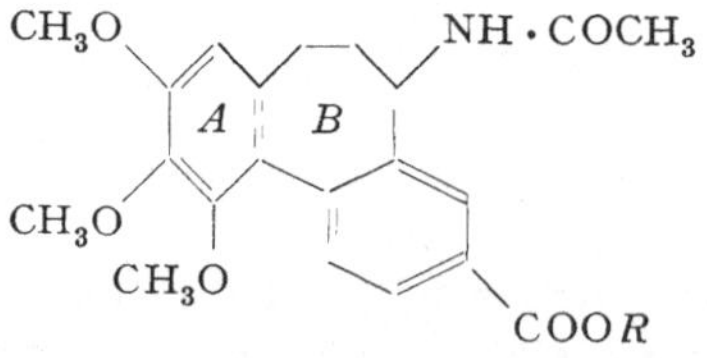

(LVI.)

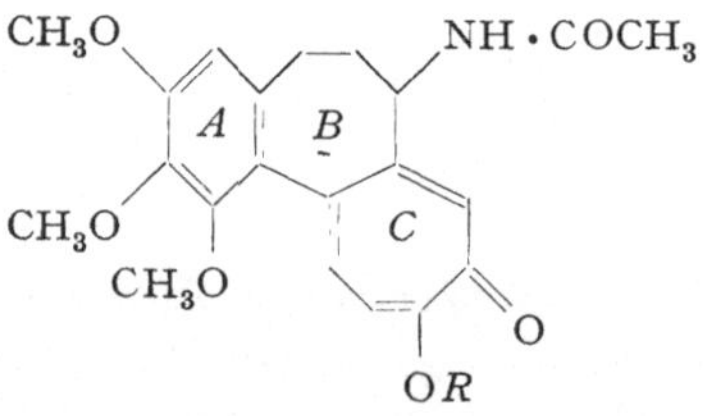

(LVII.) Colchicine ($R = CH_3$).
Colchiceine ($R = H$) (Čech, Šantavý et al.).

he also stated that the carbonyl and methoxyl groups may have to be interchanged or ring C rotated about its junction with ring B.

Although DEWARS formulation explained all chemical properties of colchicine, the experimental proof concerning rings B and C was given by later investigators.

Structure of Ring B. Besides COOK, TARBELL et al. (*232*) as well as HORNING, ULLYOT et al. (*102*) contributed to the clarification of the seven-membered structure of ring B. This structure was unequivocally established when COOK et al. (*47*) and RAPOPORT et al. (*204*) succeeded in the total synthesis of colchinol methyl ether (LV; $R = CH_3$).

The Tropolonic Nature of Ring C. TARBELL et al. (*16, 108*) carried out a detailed examination of the hydrogenation of colchicine and colchiceine, reported earlier by BURSIAN (*35*), and they obtained a number of reduction products. One of these, hexahydro-colchiceine (LVIII), was oxidized with periodic acid to the monoaldehyde (LX) and the course of this conversion was formulated as follows.

(LVIII.) Hexahydro-colchiceine.　　　　(LIX.)　　　　(LX.)

The similarity between colchicine and γ-thujaplicin was also revealed in polarographic (*31, 210*) and infrared spectroscopic studies (*220*). ŠANTAVÝ (*208*) and FERNHOLZ (*84*) obtained a carboxylic acid ester (colchic acid or allo-colchiceine) when colchicine was treated with alkoxide. FERNHOLZ (*84*) assigned the structure (LVI: $R = CH_3$) to his "allo-colchicine" since he could prepare it from N-acetylcolchinol (LV: $R = H$), that was also obtained by treating colchiceine with alkaline hydrogen peroxide [ŠANTAVÝ et al. (*40*)].

The ring C of colchicine was found to undergo facile rearrangement to benzenoid structure. Considering the structures of the rearrangement products (LV) and (LVI), it was thought that a formula such as (LVII: $R = CH_3$) or its isomer would be preferable to (LIV) proposed by DEWAR (cf. *5, 40, 47, 84, 215*). PEPINSKY et al. (*109*) have reported that the tricyclic tropolonoid structure proposed by DEWAR could be confirmed by X-ray analysis and that the fine structure of ring C seemed more likely to be represented by the modified structure (LVII: $R = CH_3$) favored by ČECH and ŠANTAVÝ (*40*).

Diazomethane converts colchiceine into colchicine and an isomeric isocolchicine (*230*).

Since the respective biological activities of these isomers were found to differ considerably, various derivatives of colchicine and isocolchicine were synthesized and their spectra and optical rotation were studied (*82, 101, 200*).

After numerous tropolones had been synthesized, colchicine and colchiceine were compared with these compounds; and, although the chemical characteristics of colchiceine were substantially different from those of 4,5-benzotropolone (*231, 232*), colchiceine and colchicine showed the following six features, very similar to those of tropolone and its alkyl or aryl derivatives and methyl ethers: 1. Colchicine is more easily soluble in water than colchiceine (*63*). 2. Colchiceine but not colchicine forms metal complexes. 3. The methoxyl group in ring C is easily substituted by amino (*87, 94, 237*), hydrazino or mercapto groups (*139, 239, 240*). 4. Colchicine easily undergoes condensation with guanidine, thiourea or cyanoacetamide to form heterocyclic azulenoids (*8, 46a, 88, 139*). 5. Colchicine is rearranged readily by alkoxides to a benzenoid carboxylic acid. 6. The action of excess bromine on colchiceine results, *via* the perbromide, in ring contraction (*120*).

Detailed Examination of Ring C. As shown above, the various reactions of colchicine are similar to those of the tropolones, with the exception of the relative ease or difficulty of such conversions. These results leave no doubt about the tropolonoid nature of colchicine. Since there still remained some unclarified structural details, and final proof of the relative positions of carbonyl and methoxyl groups were lacking, more direct degradative and synthetic arguments for the correct structure of ring C were desirable.

In order to establish the size of ring C and the position of the carbonyl group, Rapoport et al. (*202, 203*) have carried out a reliable, stepwise reduction of colchicine through the dimethylamino compound (LXII) and they obtained a tetrahydro-demethoxy derivative (LXIII) and a compound (LXIV) having no oxygen in ring C.

Very recently, Muller and Velluz (*125*) have connected colchicine and isocolchicine, respectively, with thiocolchicine and isothiocolchicine (LXV and its isomer). When comparing the ultraviolet spectra of the unsaturated cyano compounds (LXVIII) and (LXIX) obtained from (LXV) and its isomer *via* (LXVI), (LXVII), or their respective isomers, they came to the conclusion that colchicine should be represented by

(LXX) and isocolchicine by (LVII; $R = CH_3$); thus, they wished to inter-change the two structures proposed earlier.

On the other hand, CORRODI and HARDEGGER (58) have recently carried out drastic ozonolysis of colchicine and, by the reoxidation of the ozonolytic product with performic acid, obtained besides oxalic acid, crystalline N-acetyl-L-glutamic acid (LXXI). This cleavage has established beyond doubt that the sole asymmetric center (cf. LXX) in colchicine and allied alkaloids possesses the configuration of D-glyceraldehyde.

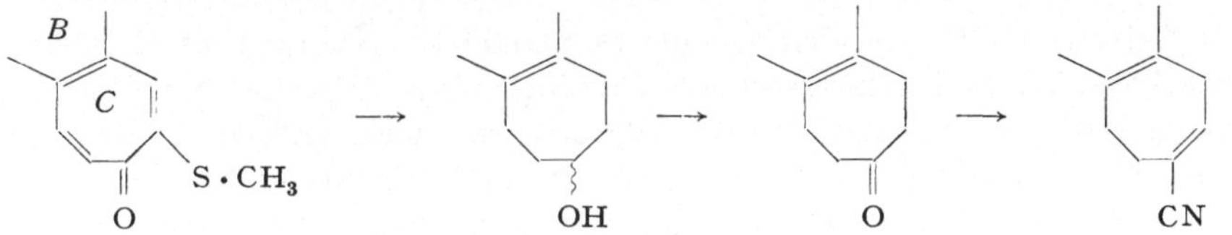

(LXV.)   (LXVI.)   (LXVII.)   (LXVIII.)

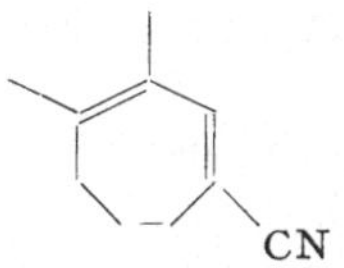

(LXIX.) 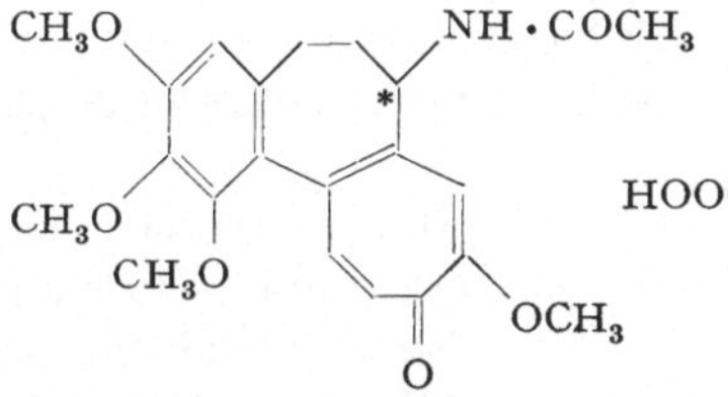   (LXX.) 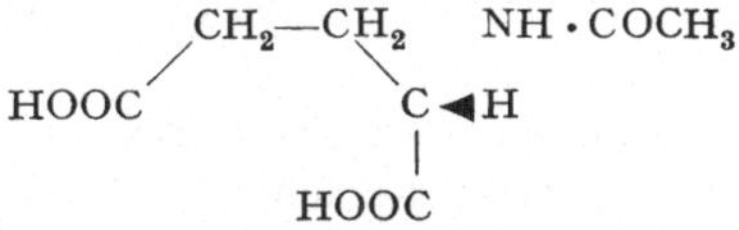   (LXXI.) N-Acetyl-L-glutamic acid.

In summary it may be stated that formulas (LXXI a) and (LXXI b) must be assigned to the compounds of the colchicine and isocolchicine series, but a final choice is not possible at the present time (86 a).

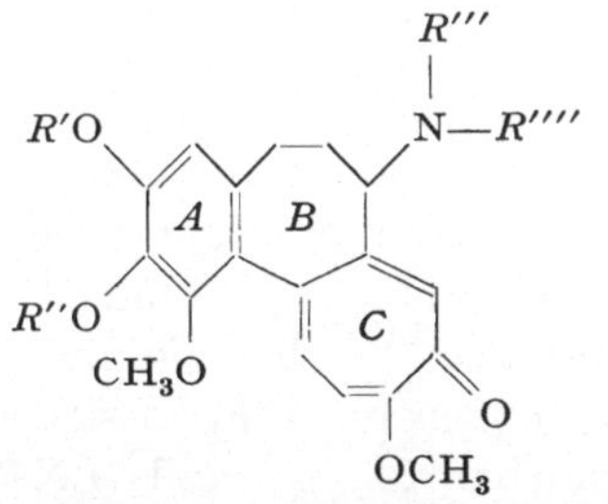

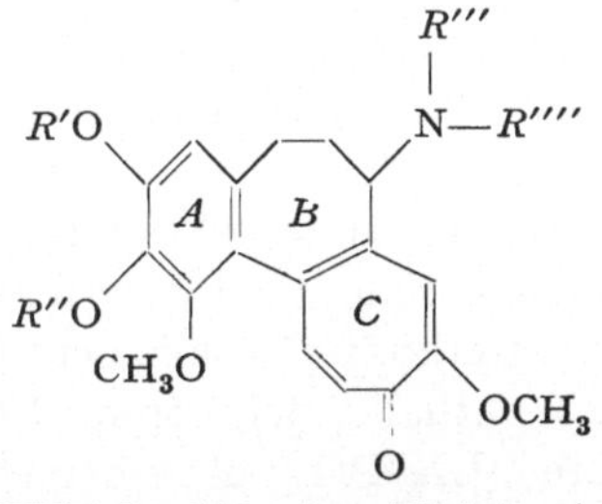

(LXXI a) Colchicine (or isocolchicine) series.   or   (LXXI b.) Isocolchine (or colchicine) series.

The structures of some related compounds of this series follow:

	R'	R''	R'''	R''''
(LXXII.) Desacetyl-colchicine	CH_3	CH_3	H	H
(LXXIII.) Dimethyl-desacetyl-colchicine	CH_3	CH_3	CH_3	CH_3
(LXXIV.) Formyl-desacetyl-colchicine	CH_3	CH_3	HCO	H
(LXXV.) Demecolcine	CH_3	CH_3	CH_3	H
(LXXVI.) 2-Demethyl-colchicine	H	CH_3	CH_3CO	H
(LXXVII.) 3-Demethyl-colchicine	CH_3	H	CH_3CO	H
(LXXVIII.) Colchicoside	$C_6H_{11}O_5$	CH_3	CH_3	H

c) N-Formyl-desacetyl-colchicine, $C_{21}H_{23}O_6N$ (m. p. 257°).

The formylation of desacetyl-colchiceine followed by methylation with diazomethane and chromatographic purification of the product resulted in the isolation of a compound identical with substance "B" (Table 3, p. 247). Considering this result, Šantavý and Reichstein (*215*) assigned the structure (LXXIV) to substance "B".

d) Demecolcine or Colchamine, $C_{21}H_{25}O_5N$ (m. p. 186°).

Substance "F", or demecolcine as named by Šantavý et al. (*217*), is a basic alkaloid with a secondary amino group. It yields methylamine on heating with concentrated hydrochloric acid and is in several respects similar to colchicine (*212, 217, 236*). Ueno (*236*) as well as Šantavý, Reichstein et al. (*238*) have independently established the structure (LXXV) of this alkaloid, based on the observation that di-methyl-desacetyl-colchicine (LXXIII) can be obtained either by mono-methylation of substance "F" or by dimethylation of desacetyl-colchicine (LXXII).

Furthermore, Šantavý et al. (*238*) synthesized demecolcine itself from desacetyl-colchiceine by a reliable method, *via* its anil.

Kiselev and Menshikov (*111*) also assigned the same structural formula (LXXV) to colchamine, an alkaloid isolated from *Colchicum speciosum*. It is believed to be identical with demecolcine (LXXV) (*217, 238*).

e) 2-Demethyl-colchicine and 3-Demethyl-colchicine, $C_{21}H_{23}O_6N$.

The substances "C" and "E₁" [Šantavý (*211, 215*)] have so similar properties that their separation is very difficult. Both are phenolic compounds and are converted into colchicine when methylated with diazomethane. Šantavý et al. (*216*) carried out analogous ethylation experiments, followed by permanganate oxidation, and obtained 3,4-di-methoxy-5-ethoxy-phthalic acid from substance "C", but 3,5-dimethoxy-4-ethoxy-phthalic acid from "E₁". Consequently, they assigned the structure (LXXVI) of 2-demethyl-colchicine to "C" and (LXXVII), i. e. 3-demethyl-colchicine to "E₁" (cf. p. 247).

f) Colchicoside, $C_{27}H_{33}O_{11}N$ (m. p. 216–218°).

Bellet et al. (*28*) isolated colchicoside (LXXVIII, p. 253) from meadow saffron. This compound was found to contain three methoxyl groups and one acetyl group, and its hydrolysis with hydrochloric acid gave glucose and an aglucone $C_{20}H_{21}O_6N$ (m. p. 275–280°). These workers were able to prepare colchicoside from this aglucone which was assumed to be 2-demethyl-colchicine (cf. *82*).

g) Substances "I" and "J" (Lumicolchicines), $C_{22}H_{25}O_6N$, and Substance "D",
$$C_{21}H_{23}O_5N.$$

Šantavý (*213*) found in meadow saffron several minor alkaloids which do not give the ferric chloride reaction and yield an oxime. Of these alkaloids, substances "I" and "J" are identical with lumicolchicines "I" and "II", which are irradiation products of colchicine. Substance "D" is identical with a similar product of substance "E_1".

Grewe et al. (*91*) studied the irradiation products of dilute aqueous colchicine solutions, with exclusion of air, and obtained α-, β-, and γ-lumicolchicines. Of these, β and γ were found to be identical, respectively, with lumicolchicines "II" and "I", and they gave different tetrahydro compounds by catalytic hydrogenation. The ultraviolet spectrum of α-lumicolchicine was found to be different from those of the β- and γ-isomers; furthermore, it could not be hydrogenated in the presence of Adams' catalyst.

Recently, Forbes (*86*) observed that, when heated with HCl, the dihydro compound (LXXX)—obtained by reducing β-lumicolchicine with sodium borohydride—suffered immediate hydrolysis to (LXXXI) which showed the characteristics of an acyloin. The compound (LXXXI) was found on perbenzoic acid titration to contain only one aliphatic double bond, similar to the foregoing tetrahydro compound (*91*); hence its molecules must include a fourth ring. Degradation of the α-glycol (LXXXII), obtained by sodium borohydride reduction of the acyloin just mentioned, yielded with sodium metaperiodate the dialdehyde (LXXXIII). Since the latter did not form a monoaldehyde by internal condensation, thus differing from hexahydro-colchiceine (p. 251), Forbes assumed that a bridging of a tropolone ring (ring C) had occurred during photoisomerization with formation of (LXXIX).

Since the oxidation of β-lumicolchicine or of colchicine affords trimethoxy-phthalic anhydride (p. 246), evidently, there has been no change with reference to ring A; however, it has not been possible to decide whether or not ring B and its asymmetric center had been changed.

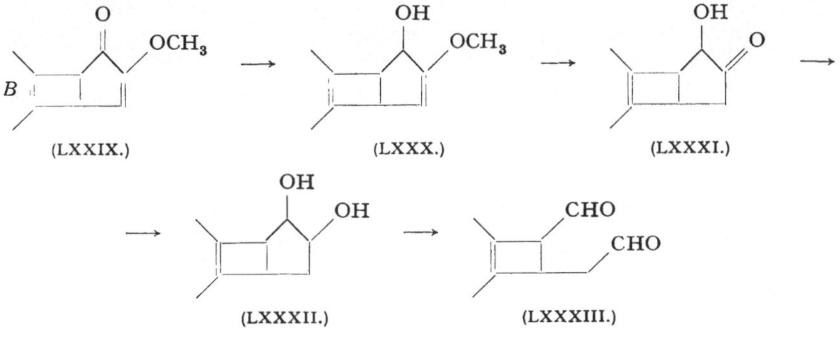

(LXXIX.) → (LXXX.) → (LXXXI.) → (LXXXII.) → (LXXXIII.)

Forbes (*86*) also observed that exactly the same conversion, i. e. (LXXIX) → (LXXXIII), had occurred in the case of γ-lumicolchicine. Considering also the quasi-identity of the ultraviolet and infrared curves of these two isomers in the higher frequency regions, Forbes concluded that they represent either geometric isomers or a pair of enantiomorphs. Assuming that the size of ring *B* has remained the same as in colchicine, then the structure of substance "I" or "J" will be (LXXXIV: $R = CH_3$), and that of substance "D", (LXXXIV; $R = H$).

It is interesting to note that such photoisomerization products do occur in nature.

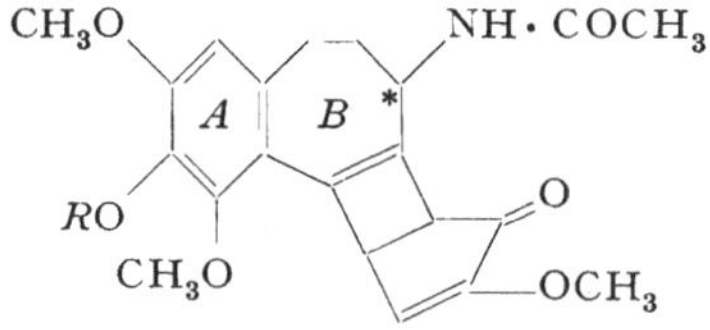

(LXXXIV.)

III. The Synthesis of Troponoids.

1. Tropolones and Tropones.

Tropolone. In 1950, the synthesis of tropolone was reported simultaneously and independently from four laboratories. Doering and Knox (*69, 70*) succeeded in synthesizing tropolone by the permanganate oxidation of cycloheptatriene (tropilidene) (LXXXV) which can be prepared by photochemical ring-widening of benzene in the presence of diazomethane. Later, these authors made use of this method in the synthesis of tropolone derivatives containing the isopropyl, phenyl or cyclohexyl group in the 4 and 5 positions (*72*).

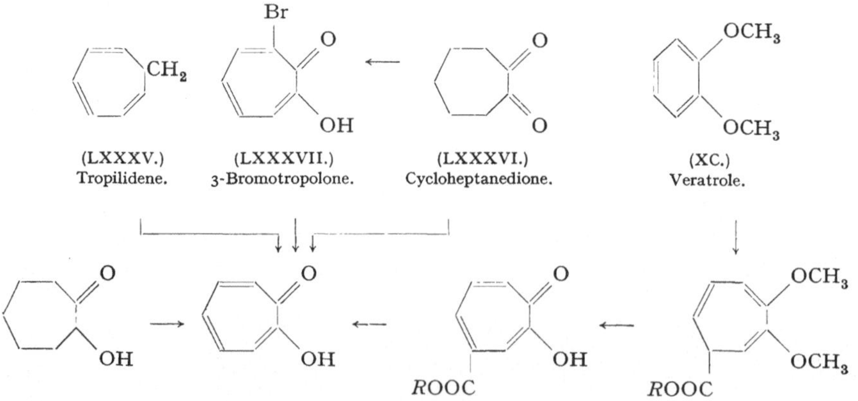

(LXXXV.) Tropilidene.   (LXXXVII.) 3-Bromotropolone.   (LXXXVI.) Cycloheptanedione.   (XC.) Veratrole.

LXXXVIII.)   (III.) Tropolone.   (LXXXIX.) 4-Carboxethyl-tropolone.   (XCI.)

Cook et al. (*44, 45*) and independently, Nozoe et al. (*188*), synthesized 3-bromotropolone (LXXXVII), together with some other bromotropolones, by treating cycloheptanedione (LXXXVI) with bromine in acetic acid. Tropolone itself can be obtained by the dehydrogenation of (LXXXVI) with N-bromosuccinimide (*188*).

This method of dehydrogenating with bromine has relatively wide application and has been used in the synthesis of α-, β-, and γ-thujaplicins (*51, 157, 186, 187*) as well as other alkyltropolones (*136, 165*). Dehydrogenation of the acyloin (LXXXVIII) with bromine also gives tropolone (*114*). Tropolones can be obtained almost quantitatively from bromotropolones by catalytic hydrogenolysis, in the presence of a palladium-carbon catalyst and sodium acetate (*45*).

On the other hand, Haworth et al. (*95, 96*) oxidized 4-methyltropolone methyl ether to the 4-carboxylic acid (LXXXIX; $R = $ H), *via* the 4-aldehyde, and obtained tropolone by decarboxylating the acid.

Tropolone-carboxylic Acid. Johnson et al. (*25*) synthesized 4-ethoxycarbonyl-tropolone (LXXXIX; $R = C_2H_5$) by bromination-dehydrobromination of the cycloheptatriene-carboxylic acid derivative (XCI), obtained by reacting diazoacetic ester with veratrole (XC). This method was useful in the synthesis of stipitatic acid and other tropolone-carboxylic acids (*26, 105*).

Tropone. The synthesis of tropone (II, cf. below) was carried out in 1951 by Doering et al. (*67*) by ring-enlargement of anisole (XCII) with diazomethane to the compound (XCIII) which was then dehydrogenated with bromine. Dauben et al. (*60*) synthesized tropone by treating cycloheptenone with bromine to form 2,4,7-tribromotropone (XCIV), followed by selective reduction by means of a poisoned palladium catalyst. Nozoe et al. (*148*) obtained 2,4,7-tribromotropone by direct bromination of suberone, $(CH_2)_6CO$ (cycloheptanone).

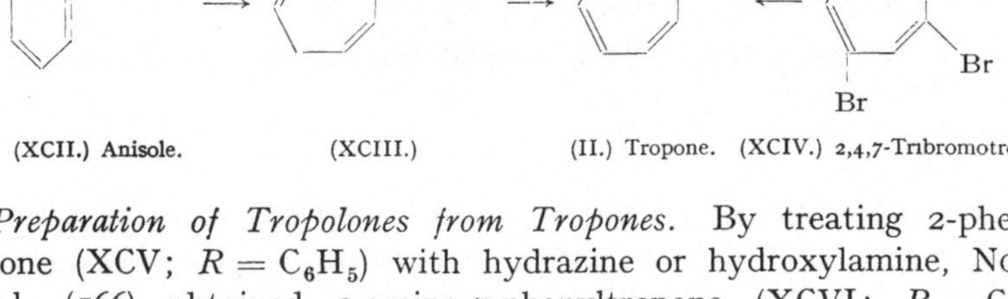

(XCII.) **Anisole.**   (XCIII.)   (II.) Tropone.   (XCIV.) 2,4,7-Tribromotropone.

Preparation of Tropolones from Tropones. By treating 2-phenyltropone (XCV; $R = C_6H_5$) with hydrazine or hydroxylamine, Nozoe et al. (*166*) obtained 2-amino-7-phenyltropone (XCVI; $R = C_6H_5$, $X = NH_2$) which was converted into 3-phenyltropolone (XCVII;

$R = C_6H_5$) by heating with alkali. Numerous derivatives of tropolone, such as 3-methyl-, 3-p-methoxyphenyl-, 3-benzyl-, 3-α-naphthyl-, and 3-β-naphthyltropolones, were synthesized by the same method (*8, 9, 184, 227*). Bromotropones (XCVI; $X = Br$) obtained by bromination of tropones can also be converted into tropolones by hydrolysis with mineral acid (*74, 149*).

(XCV.) 2-Phenyltropone. ($R = C_6H_5$.) (XCVI.) (XCVII.) 3-Phenyltropolone. ($R = C_6H_5$.)

2. Benzotropolones.

3,4-Benzotropolone. Cook and others (*46, 53*) synthesized 3,4-benzo-tropolone (XCIX) from the dione (XCVIII) by dehydrogenation with either palladium-carbon or bromine. N-bromosuccinimide can also be used in this process (*123*). Following the same method, Haworth et al. (*38, 39*) synthesized purpurogallin and its analogs.

(XCVIII.) (XCIX.) 3,4-Benzotropolone. (C.) Dibenzotropolone.

(CI.) Phthalaldehyde. (CII.) 4,5-Benzotropolone. ($R = H$.)

4,5-Benzotropolone. Tarbell et al. (*231, 233*) as well as Fernholz et al. (*85*) synthesized 4,5-benzotropolone (CII; $R = H$) and its homologs by condensing phthalaldehyde (CI) with methoxyacetone or methoxy-methyl ethyl ketone. The formation of the tropolone ring in such single-step condensation is a rare phenomenon.

3,4,5,6-Dibenzotropolone. Sakan and Nakazaki (*207*) prepared di-benzotropolone (C) by the oxidation of dibenzosuberone with selenium dioxide. However, this compound no longer showed tropolonoid charac-teristics.

3. Colchicine Analogs.

a) *Approach to the Synthesis.*

Because of the chemical and biological interest in colchicine several chemists made efforts to synthesize colchicine and its analogs, on the basis of purely chemical and chemotherapeutical considerations. The majority of these workers intended to start the synthesis from compounds containing the rings A and B, and to close the tropolonoid ring C in the last step. So far as is known to the present writer, none of these investigators has been able to reach this final step. On the other hand, some papers describe syntheses of colchicine analogs starting from a tropolone (corresponding to ring C), attaching a benzenoid ring (A) to it, and attempting to close ring B in a final operation.

In the present Section, only the second approach will be discussed. Some syntheses of phenyl-, p-methoxyphenyl-, and benzyl-tropolones have already been presented. Recently, 5-aryltropolones such as (CIII) were synthesized in the writer's laboratory, by decomposing the diazonium salt prepared from 5-aminotropolone in benzene, anisole, or trimethoxybenzene solution (*8*).

b) *Styryl-tropolones.*

Condensation of the dicarboxylic acid (XLV, p. 246), originating from purpurogallin, with benzaldehyde (or derivatives) under the conditions of the PERKIN or KNOEVENAGEL synthesis affords styryl-tropolone-

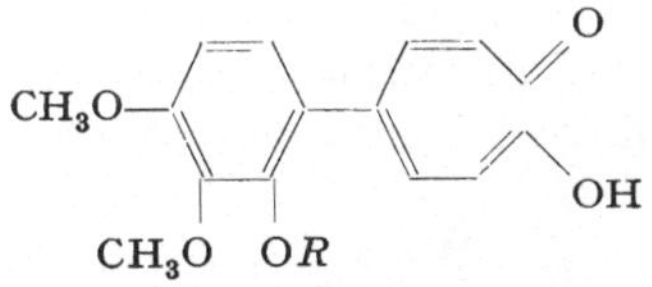

(CIII.) 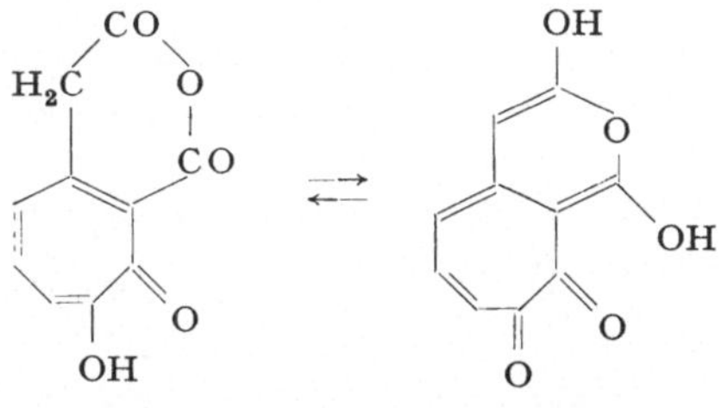　　　(CIVa.)　　　(CIVb.)

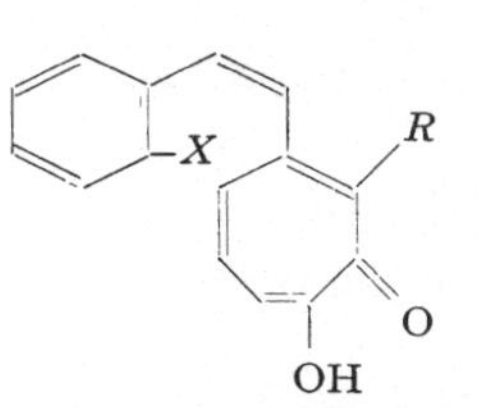

(CVa; X = H, R = COOH.)
(CVb; X = H, R = H.)
(CVc; X = NH₂, R = H.)

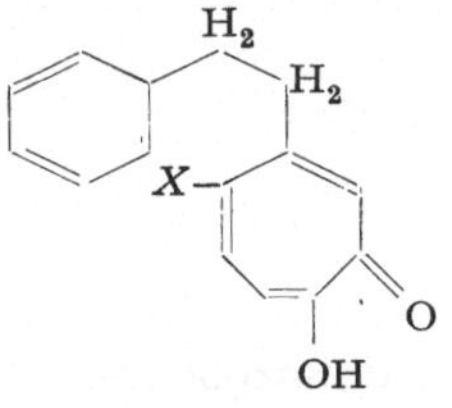

(CVIa; X = H.)
(CVIb; X = azo.)
(CVIc; X = NH₂.)

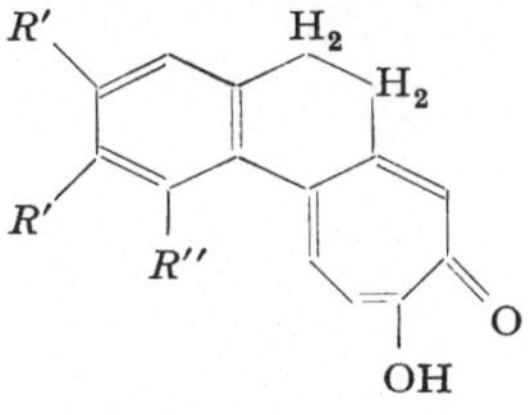

(CVIIa; R' = R'' = H.)
(CVIIb; R' = R'' = OCH₃.)
(CVIIc; R' = OCH₃, R'' = H.)

carboxylic acids of the type (CVa) that easily undergo decarboxylation to styryl-tropolones (CVb) [Haworth et al. (59); Nozoe et al. (155, 156); Tarbell et al. (234)]. In this instance, the dicarboxylic acid is assumed to take part in the reaction in the form of its red-colored anhydride (CIV), the structure of which is thought to be more like (CIVb) than (CIVa) (156).

c) Phenylethyl-tropolones and their Ring Closure.

An attempt was made to cyclize the amino derivative (CVc) of the styryl-tropolone mentioned, under the conditions of the Pschorr reaction but this objective could not be attained (155, 234). This failure was probably due to the extreme lability of the cis-form of styryl-tropolone. Thus only the trans-form was originally present, which made a ring closure virtually impossible (156, 234). Thereupon, Nozoe et al. (8) attempted this ring closure first by preparing a dihydro derivative (CVIa) by the catalytic reduction of styryl-tropolone (CVb) in alkaline medium, converting it to the amino compound (CVIc) via the azo compound (CVIb), and submitting it to the Pschorr reaction. The anticipated ring closure did take place and the tricyclic troponoid (CVIIa) appeared.

Attempt was also made to prepare by this method the trimethoxy compound, desacetylamino-norcolchiceine (CVIIb), but the composition of the reaction product indicated that one of the three methoxyls had been hydrolyzed and the dimethoxy compound (CVIIc) formed (8).

d) Phenylpropyl-tropolone and Derivatives.

Interaction of phenylacetaldehyde with the dicarboxylic acid (XLV) mentioned afforded the compound (CVIII) with three carbon atoms located between benzene and tropolone rings (155). This compound was converted into phenylpropyl-tropolone (CIX) and its various derivatives in order to attempt ring closure (197) as indicated by the dotted line.

(CVIII)

(CIX.) Phenylpropyl-tropolone.

4. Halotropones.

The reaction of thionyl chloride with tropolone and 3-bromotropolone yields the 2-chloro- and 2,7-dichlorotropones (11, 71). Nozoe, Seto et al.

(*194, 223*) found that the decomposition of the hydrogen halide salt of 2-hydrazinotropone, obtained from tropolone methyl ether, affords in the presence of copper salt 2-halotropone in a good yield. This method is very convenient as it leads to various halotropones with well-established structures [SETO (*223–226*)]. For example, 2-bromo-7-chlorotropone (CXI) and 3-bromo-2-chlorotropone (CXIII) are thus obtained from the two isomeric 3-bromotropolone methyl ethers (CX) and (CXII).

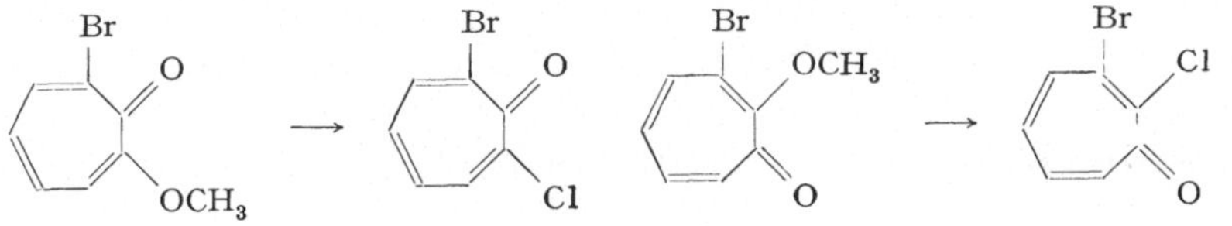

(CX.) (CXI.) (CXII.) (CXIII.)
2-Bromo-7-chlorotropone. 3-Bromo-2-chlorotropone.

Decomposition of 2-hydrazinotropone in the presence of copper sulfate also affords tropone (*171, 223*). The action of bromine on suberone is very complicated and the product includes, depending on the conditions, a tribromo derivative (*75*), tetra-, penta-, and hexabromo-suberones, and 2,4,7-tribromotropone (XCIV, p. 257) as well as 4-bromotropone and 2,5-dibromotropone [NOZOE et al. (*8, 164*)].

These halotropones are cyclic vinylogs of acyl halides and are important in that they may be utilized in syntheses of troponoid and azulenoid compounds.

5. 3- and 4-Hydroxytropones.

In connection with the difference in the chemical characteristics of tropone and tropolone, the synthesis of the tropolone isomers, 3- and 4-hydroxytropone, seemed to be of considerable interest. JOHNSON et al. (*105, 107*) succeeded in preparing 3-hydroxytropone (see below) from a compound obtained by ring enlargement of veratrole or resorcinol dimethyl ether with diazoacetic ester. 4-Hydroxytropone was synthesized by the hydrolysis of 4-bromotropone [NOZOE et al. (*164*)] and by the diazoacetic ester method, starting from hydroquinone ether [JOHNSON et al. (*41*)].

6. Heterocyclic Troponoids.

Several troponoids containing fused heterocyclic rings of various types have been synthesized. Most of them have been obtained by condensation of monocyclic troponoids, and some of them probably exist in the form of tautomeric aza-azulenes (p. 286). We will give the formulas only of the following representatives of this series with the pertinent literature references: Derivatives of furanotropolone (CXIV) and its S-analog (*100a, 235*); anhydride of 3,3'-bitropolonyl and its S-

and N-analogs (CXV; $X = O$, S or NH) (*13, 197*); derivatives of pyrrolo-tropone (cf. CXVI) (*150*); derivatives of oxazolotropones (CXVII and CXVIII) (*107, 197, 129*); imidazolotropone (CXIX) (*197*); and derivatives of pyridotropolone (CXX) (*12, 49, 229*).

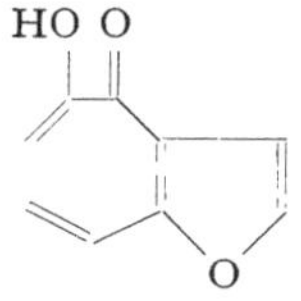

(CXIV.) Furanotropolone. 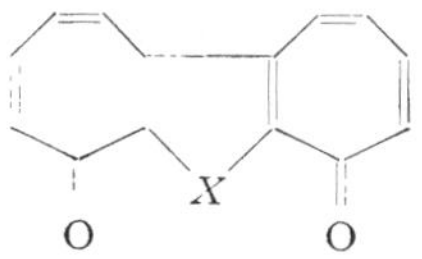　　(CXV.) 3,3′-Bitropolonyl derivatives. 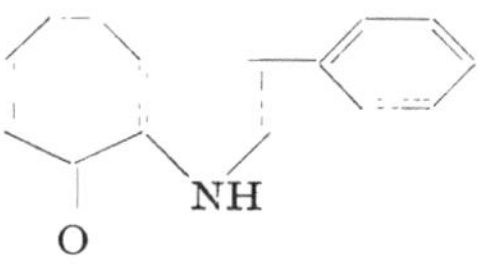　　(CXVI.) Phenylpyrrolotropone.

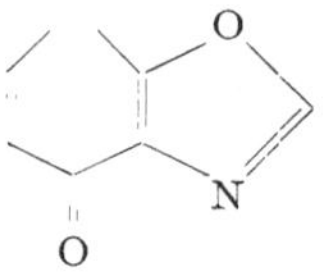

(CXVII.) 　Oxazolotropones. 　(CXVIII.)　　(CXIX.) Imidazolotropone. 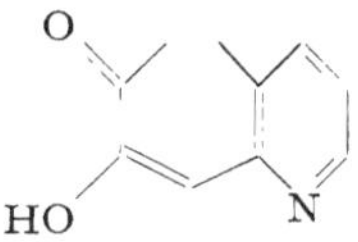 (CXX.) Pyridotropolone.

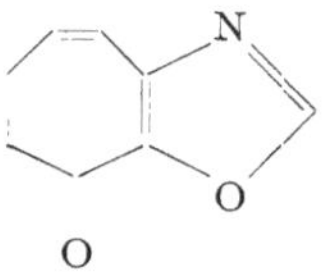

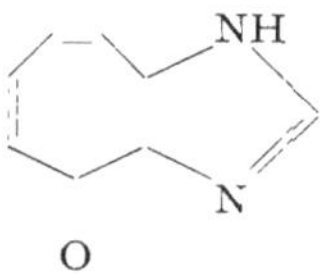

IV. Physical Properties and Fine Structure.

1. General Considerations.

As mentioned, physicochemical studies of tropolone and related compounds have been progressing rapidly, since DEWAR had proposed a then new tropolone ring system (p. 234) in connection with the structures of two natural products, stipitatic acid and colchicine. By means of physical methods, such as spectra, X-ray and electron diffraction, the fine structure of the tropolone ring has been gradually clarified. Recent investigations of the infrared and Raman spectra of the tropylium cation (I) (*83*) have established beyond doubt that an aromatic, planar, seven-membered ring system is the basic structure of all tropoid compounds.

Physical studies in this direction are extending over a wide range, also including dipole moments, magnetic susceptibilities, heats of formation, polarography, solubility and acidity tests, molecular orbital treatment, etc. In the present Section only brief mention will be made of some of these methods that have specially contributed to the clarification of the troponoid ring. The reader is referred to several review articles now available (*2, 8, 9, 115, 116*).

2. Acidity and Complex Formation.

As will be mentioned later, troponoids in general show basic properties due to their characteristic structure (CXXI or CXXIII); furthermore, they form stable salts with mineral acids and yield molecular compounds such as picrates. In contrast, the molecules of tropolones and isomeric hydroxytropones possess enolic hydroxyl groups and hence show some acidity, the strength of which lies between those of phenol and acetic acid. The actual acidity of each individual compound varies considerably depending on the presence or absence of substituents and the character and position of the latter. Tropolones generally form yellow to orange-colored, sparingly water-soluble sodium salts; they also form crystalline salts with ethylenediamine. Halo- and nitro-tropolones give salt-like complexes with p-toluidine and other arylamines, and such complexes in many instances undergo dehydrative condensation to 2-arylamino-tropones (9, *134*). These crystalline salts and complexes are made use of in the identification of tropones and tropolones. However, tropolones containing halogens and/or nitro groups scarcely form stable mineral acid salts or picrates.

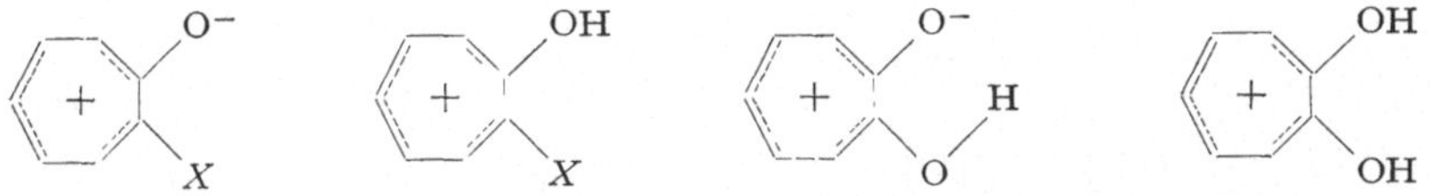

(CXXI.)   (CXXII.)   (CXXIII.)   (CXXIV.)

The pKa values of 3- and 4-hydroxytropones are, 5.4 and 5.1, respectively, showing stronger acidity than that of tropolone (pKa 6.9).

Tropolones very easily form chelated complexes with various metals; and such complexes crystallize readily from organic solvents and show sharp melting points. Especially the copper and iron complexes are used in the isolation and purification of tropolones. As mentioned above, the deep red or green coloration with ferric chloride facilitates their detection.

3. Ultraviolet Spectra.

The spectra of troponoids are similar, to some extent, to those of salicylaldehyde and other benzenoid compounds; however, troponoids and benzenoids can be distinguished when the intensities of the bands are compared (*89*). This procedure has often been used in the study of naturally occurring tropolones.

In general, the ultraviolet spectral curve of a troponoid can be divided into two sections: 1. Very intense extinction in the 200–300 mμ region with ε values ranging from 10^4 – 10^5 ("region A"), and 2. somewhat weaker intensities at 300–400 mμ ($\varepsilon = 10^3$–10^4) ("region B").

The band in region A is common to tropones and tropolones. It is hardly affected by the presence of alkyl groups but shifts slightly to longer wave lengths when halogen, hydroxyl or ester groups are introduced (*Table 4*, next page).

Table 4. Ultraviolet Spectra of Troponoids.

Compound	λ max mμ (log ε)			Reference
	Region A	Region B	*	
Tropone	225 (4.33), 228 (4.34), 231.5 (4.34), 239 (4.10)	312.5 (3.92)	e	(60) (67)
	225 (4.34)	297 (3.74), 310 (3.67)	a	
2-Bromotropone	246 (4.18)	315 (3.79)	b	(223)
2-Phenyltropone	227 (4.40), 270 (3.85)	323 (3.96)	c	(166)
2-Methoxytropone	235 (4.46)	314 (3.87), 319 (3.88), 350 (3.81)	d	(45)
Tropolone	222 (4.37), 232 (4.36), 238 (4.37)	322 (3.84), 340 (3.64), 356 (3.73), 374 (3.74)	b	(45)
	228 (4.36), 237 (4.36)	320 (3.83), 351 (3.76)	e	
3-Hydroxytropone	247 (4.51), 255 (4.41)	298 (3.65), 309 (3.52)	d	(105)
4-Hydroxytropone	228 (4.27)	337 (4.11)	c	(164)
3-Hydroxytropolone	246 (4.60)	328 (3.88), 365 (3.83), 376 (3.92)	c	(105)
2-Aminotropone	268 (3.97)	340 (3.94), 375 (3.84), 385 (3.76), 395 (3.89)	b	(194)
2-Mercaptotropone	237 (4.03), 268 (4.18)	420 (4.08)	b	(178)
2-Methylthiotropone	229 (4.02), 247 (4.06), 283 (3.83)	328 (3.88), 344 (3.93), 376 (3.96), 391 (3.81)	b	(178)
α-Thujaplicin	266 (4.43)	320 (3.81), 357 (3.77)	d	(79)
β-Thujaplicin	236 (4.37)	322 (3.72), 353 (3.66)	a	(72)
γ-Thujaplicin	225 (4.48)	323 (4.04), 358 (3.83), 375 (3.72)	a	(72)
Colchicine	244 (4.00)**	351 (3.79)**	c	cf. (35)
Colchiceine	243 (4.51)**	350 (4.26)**	c	cf. (35)
Stipitatic acid	262 (4.49)	332 (3.71), 360 (3.61)	e	(56)
Puberulic acid	270 (4.55)	350 (3.86)	e	(54)

* Solvents: *a*, isooctane; *b*, cyclohexane; *c*, methanol; *d*, ethanol; and *e*, water.
** The values were determined in the writer's laboratory.

The band in region B is strongly influenced by the structure of the individual troponoid, especially by the presence or absence of such hydroxyl, amino, or mercapto groups which are able to form hydrogen bonds with the carbonyl at $C_{(1)}$. This band can belong to either of the following two types.

The first one is the *tropone type* band, a characteristic of tropone, alkyl- or aryltropones, halotropones, and some other compounds from which such groups are absent that might form intramolecular hydrogen bonds. These compounds show only a single absorption band in the region, 300–340 mμ. Tropolones also show this type of band when converted into their methyl ethers or acetates.

The other spectral type is the *tropolone type*, showing two bands of about equal intensity at 300–350 mμ and 350–400 mμ. These bands may differ somewhat depending on the character and position of substituents; they are altered by conjugation or by change of solvent. The longest wave length region is related to the ionization potential of the substituent. The valley between the two maxima in region B may be distinct or broad and unsharp. However, the curves can be differentiated easily from those of other aromatic substances and are very useful in the detection of troponoids.

The tropolone anions show two very sharp and intense absorption maxima in a still longer wave length region. Especially the tropolone and 3-hydroxytropolone curves have marked fine structure that tends to disappear when a hydroxyl-containing solvent is used. The ultraviolet spectrum of 2-aminotropone is as a whole similar to that of tropolone. The main absorption shifts to considerably longer wave lengths in the case of 2-mercaptotropone.

4. Infrared Spectra.

Several studies have been made on the infrared spectra of troponoid compounds, mostly referring to carbonyl and hydroxyl bands.

Frequency Region of 4000 to 2000 cm.$^{-1}$ The C—H stretching vibration of the tropone ring is similar to that of benzenoid compounds and appears as a band of weak intensity in the region, 3060–3010 cm.$^{-1}$ A detailed study of this band is not yet available. The O—H stretching vibration of dissolved or gaseous tropolone appears at 3140 cm.$^{-1}$ and in the solid state at 3210 cm.$^{-1}$ The majority of tropolones show strong absorption in the region, 3250–3180 cm.$^{-1}$ in the solid state and this band broadens and shifts to slightly lower frequencies in case of a solution. In the spectra of some tropolone derivatives, such as 3,7-dibromo- and 3,5,7-tribromo-tropolones, this band is very weak when solids are investigated (*111*).

Carbonyl Band. The C=O stretching vibration of tropone appears at 1645 cm.$^{-1}$ in carbon tetrachloride and with high intensity at 1651 cm.$^{-1}$

in the gaseous state. In the case of tropolone the carbonyl band is located at 1613, 1620, and 1628 cm.$^{-1}$ (solid, solution, gas, respectively), indicating a displacement of 20–25 cm.$^{-1}$ toward the lower frequency region as compared with tropone. The location of the carbonyl band of tropone in a frequency region lower than that of ordinary $\alpha\beta : \alpha'\beta'$-unsaturated ketones, ($p$-benzoquinone, 1664 cm.$^{-1}$) is probably due to high polarity of tropone and to the ring strain. The carbonyl band of 3-hydroxytropone appears at 1647 cm.$^{-1}$, approximately coinciding with the corresponding tropone band; however, the tropolone band is found at much lower frequencies. This observation as well as the fact that the OH band of tropolone is displaced from the normal hydroxyl frequency (3600 cm.$^{-1}$) to a lower frequency region, indicate the presence of intramolecular hydrogen bonding.

Other Regions. In the region of 1600 to 1200 cm.$^{-1}$, tropone exhibits a band at 1580 cm.$^{-1}$ in liquid state, and at 1613 cm.$^{-1}$ in the gaseous state; and tropolone shows strong absorption at 1548 cm.$^{-1}$ (solid) and at 1573 cm.$^{-1}$ (gas). These frequencies correspond to the C=C stretching vibration, and are analogous to the band in the azulene spectrum at 1570 cm.$^{-1}$ The same band appears in the region, 1600–1550 cm.$^{-1}$ in tropone and at 1570–1540 cm.$^{-1}$ in tropolone curves. Some common absorption bands are found at 1100–700 cm.$^{-1}$ in the spectra of various derivatives, and some of these have been assigned to the out-of-plane vibration of C—H, analogous to those in benzene derivatives. The bands located in this region may be used for the determination of substituted positions in troponoids. A band at 3425 cm.$^{-1}$ has been reported for tropone (*67*). However, it could not be found when tropone gas was studied and was probably caused by moisture in the highly hygroscopic samples (*110*).

5. X-Ray and Electron Diffraction.

Independently of organic-chemical work, X-ray crystallographic studies of tropolone were carried out using the copper chelate complex (*205*). The carbocyclic ring of tropolone was identified as a planar, almost regular heptagon. The carbon-carbon distance of about 1.4 Å offers evidence for the aromatic nature of tropolones as established by organic-chemical methods. Similar results were obtained later with tropolone hydrochloride and sodium tropolonate (*218*). X-Ray measurements were also extended to colchicine, nootkatin, purpurogallin, and two isomeric methyl ethers of 3-bromotropolone.

An important suggestion concerning the true nature of troponoids, especially the problem of tautomerism in tropolone, was offered by the observation that, whereas the lengths of the two carbon-oxygen bonds in tropolone hydrochloride are identical, the analogous value in the tropolone copper complex is different. The electron diffraction studies of tropolone and tribromotropolone have confirmed the model based on X-ray diffraction data.

6. Dipole Moments.

Dipole moment values for tropone, tropolone, and numerous derivatives are now available. The dipole moments of tropone and tropolone are, respectively, 4.17 D and 3.53 D (*115, 116*), and that of 4-hydroxytropone is 5.9 D (*8*). These values indicate that a substantial contribution is made by the ionic structure (CXXI, p. 263) in which the oxygen atom of the carbonyl group carries a negative charge. KUBO and KURITA (*115, 116*) found that the direction of the dipole moment in tropolone (III, p. 235) is in agreement with the line drawn from the center of the ring through the center of two oxygen atoms, with the carbon ring as the positive end. This has made it possible (as in the benzene series) to determine the position of substituents in numerous derivatives of tropolone and hinokitiol by measuring the dipole moments.

7. Polarography.

In general, the relationship between the shape of the reduction wave and p_H in troponoid compounds is to a certain extent similar to that in benzaldehyde. At p_H 4, tropolone and its alkyl derivatives show one wave of single-electron reduction. With the increase of p_H the polarogram becomes more complicated and there appear two waves of two-electron reduction. When p_H is further increased to 8, one wave of two-electron reduction appears (*230a*). Finally when the p_H value is still more increased, this results in the decrease of wave height and the reduction wave does not appear in 0.01 N potassium hydroxide. The half-wave potential is – 1.1 V. at p_H 4 and – 1.46 V. at p_H 8. It is more difficult to reduce dissociated tropolone molecules than non-dissociated ones; and it is more difficult to reduce tropolones than their methyl ethers. It must be stressed that in the polarography of tropolones the use of boric acid as buffer will not show on the acid side the reduction wave of tropolone itself but that of the tropolone-boric acid complex which is more easily reducible than tropolone.

As described in an earlier Section (p. 251), the presence of a tropolone ring in colchicine had been postulated in view of the similarity of the γ-thujaplicin and colchiceine polarograms. Polarographic experiments have also been conducted with various tropones, tropolones, and their derivatives [ŠANTAVÝ et al. (*24, 103*)].

V. Chemical Properties.

1. General Properties of Troponoid Rings.

a) Ketonic Properties.

Tropone itself reacts with ketonic reagents to form a semicarbazone and a phenylhydrazone but 2-aryltropones yield only 2,4-dinitrophenyl-hydrazones. However, these tropones, when treated with hydrazine, form 2-aminotropone in good yield but not hydrazones. With hydroxylamine, they give 2-aminotropone as well as the oxime [Nozoe et al. (*166, 171, 174, 184*)]. This facile amination of the ring is one of the characteristics of the tropone. In contrast, tropolone (*45*) and 3- and 4-hydroxytropone (*107, 164*) as well as 2-aminotropone (*194*) are indifferent to ketonic reagents. However, 3,4-benzotropolone methyl ether (*46, 85*) and 4,5-benzotropolone and its methyl ether (*231*) give 2,4-dinitrophenylhydra-zones. Furthermore, dibenzotropolone mentioned earlier (*207*) forms a dioxime and quinoxaline derivatives. This is very likely caused by the decrease of resonance of the tropolone ring when fused with a benzene ring, whereby ketonic properties are strengthened and the tropolonic character is weakened.

Nozoe et al. (*8, 134, 177*) found that the 5-nitroso, 5-nitro and 5-arylazo derivatives of tropolone undergo condensation in their tauto-meric tropoquinonoid form (CXXV) with *o*-phenylenediamine to give quinoxalotropone derivatives (CXXVI) or their benzo[*b*]tropazine tautomers (VII). Condensation with ethylenediamine affords pyrazino-tropone derivatives (CXXVII).

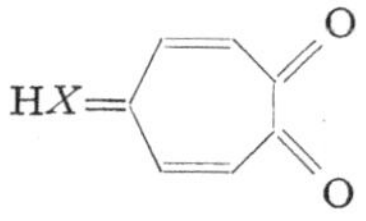

(CXXV.) 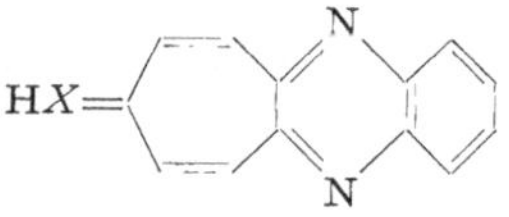   (CXXVI.) 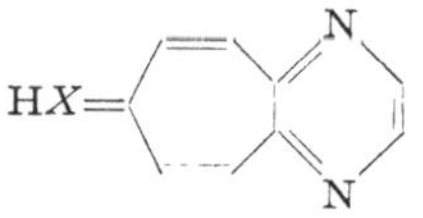   (CXXVII.)

b) Hydroxylic Function and Methyl Ethers.

The enolic hydroxyl in the tropolone ring is the source of its acidity, hence the acidity disappears when a tropolone is esterified or etherified. The methyl ether can be obtained by treating tropolone with diazo-methane or its sodium or silver salt with methyl iodide (*70, 134, 183*).

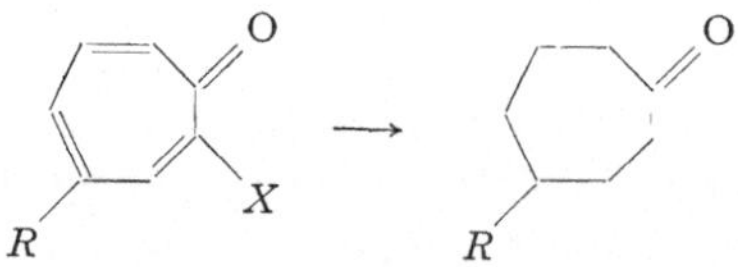

(CXXVIII.)   (CXXIX.) 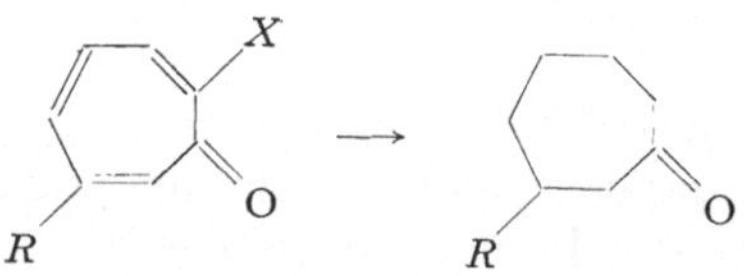   CXXX.)   (CXXXI.)

Ethers are also obtained by treatment with methanol and hydrochloric acid or in some instances with alkaline dimethyl sulfate (*46, 70, 231*). The ethers generally give crystalline hydrochlorides or picrates. The fact that the methyl ether is more soluble in water than the free tropolone is another characteristic of tropolonoids (*30, 63, 183*).

Non-symmetrically substituted tropolones, such as 3-bromo- (*159, 162*) and 4-methyltropolone (*96*), give two isomeric methyl ethers [cf. (CX) and (CXII) p. 261].

HAWORTH et al. (*12, 13*) established the structure of the two methyl ethers, (CXXVIII) and (CXXX) ($R = CH_3$, $X = OCH_3$), of 4-methyltropolone by converting them, *via* hydrazino compounds, to the tropone derivatives (CXXVIII) and (CXXX) ($R = CH_3$, $X = H$) according to STEVENS and McFADYEN or by preparing the corresponding dimethylamino compounds (CXXVIII) and (CXXX) ($R = CH_3$, $X = N(CH_3)_2$), followed by reduction to the structurally clarified 4- and 3-methylsuberones (CXXIX) and (CXXXI) ($R = CH_3$). SETO (*224, 225*) established the structure of the two isomeric methyl ethers of α- and β-thujaplicin by converting them into halotropones, through the hydrazino compound, and then to the 2-, 3-, and 4-isopropylsuberones, the latter having well-known structures.

It should be noted that 3,4- and 4,5-benzotropolones form only one kind of methyl ether because of the resonance inhibition of the tropolone ring and by the effect of the fused ring; moreover, such methyl ethers are not as reactive as those of the corresponding monocyclic tropolones (*46, 231*). In spite of the tricyclic structure, the methoxyl group in the colchicine and isocolchicine molecules is very reactive and undergoes substitution with amino and methylthio groups (p. 252), because ring B fused with the tropolone ring is a non-planar, alicyclic system.

c) Stability and Double Bond Character.

Both tropone and tropolone are resistant to hot concentrated acids, since they are stabilized in form of their conjugated acid cations (CXXII or CXXIV, p. 263). Displaying their basic properties, tropones and tropolones easily form picrates, styphnates, and urea complexes. Tropolone and 4-hydroxytropone are very resistant to alkali and stabilize as anions; however, tropone is highly labile in the presence of alkali or a primary amine (*174*). 2-Phenyltropone (*166*) and 3-hydroxytropone (*105*) are somewhat more stable in alkaline media than tropone itself but tend to resinify when warmed with alkali.

Catalytic reduction of tropolone and its homologs in the presence of platinum oxide results, as in the case of natural tropolones, in the formation of monoalcohol, ketol, and ketones, besides a saturated diol. However, the tropolone ring is not reduced catalytically in the presence of palladium so that the free tropolone can be obtained in good yields from halotropolones, and the corresponding aminotropolones from nitroso-, nitro-, or azotropolones. Nevertheless, the degree of unsaturation of polyhalogen derivatives of 3- and 4-hydroxytropones is higher than that of halotropolones. The dehalogenation by selective hydrogenation of such derivatives is difficult, the compounds being mostly reduced to the saturated ketol (*145, 228*).

As mentioned earlier, tropone and tropolone form stable salts with mineral acids but do not add hydrogen chloride to double bonds. On the other hand, the interaction of bromine and tropone in acetic acid gives first a red complex, which then is converted into a colorless tetra-bromo addition compound. On heating, two moles of hydrogen bromide are liberated and 2,7-dibromotropone is formed (*171, 174*). The same statement is valid for 2-phenyltropone and halotropones which first form adducts and then substitution products (*124*).

Tropones, tropolones, and thujaplicins undergo the Diels-Alder reaction to give maleic anhydride adducts (*174, 180, 221, 222*). 4-Hydroxytropone, when treated with diazomethane, not only forms the methyl ether but also tends to undergo addition at the double bond to yield a pyrazolo-tropone derivative (*197*).

d) Oxidative Degradation of the Tropolone Ring.

As mentioned earlier, tropolone can also be obtained by the alkaline permanganate oxidation of cycloheptatriene; however, the yields are low because the compound has a tendency to decompose when treated with ordinary oxidizing agents.

Oxidation with alkaline hydrogen peroxide gives *cis,cis*-muconic acid from tropolone (*45*) and *o*-carboxycinnamic acid from 3,4- or 4,5-benzo-tropolone (*53,85*). It is interesting to note that, as mentioned before, colchicine is converted by alkaline hydrogen peroxide into N-acetyl-colchinol (p. 250). While such treatment was effective in the structural clarification of stipitatic and puberulic acids, it caused complications, at least for a time, when the hinokitiol structure was being investigated (cf. p. 238). Later studies of Nozoe et al. (*176*) revealed, however, that the unsaturated dibasic acid, obtained by hydrogen peroxide oxidation of hinokitiol, is identical with α-isopropyl-*cis,cis*-muconic acid (CXXXIII), while the liquid acid represents its monolactone (CXXXIV). It was also found that persulfate oxidation of tropolone or hinokitiol results in hydroxylation of the 3- and 5-positions (*185*), while hydrogen peroxide oxidation of tropolone itself gives first 3-hydroxytropolone which is then rapidly cleaved to *cis,cis*-muconic acid (*142*). Consequently, Nozoe and others (*8*) postulated that hydroxyl compounds such as (CXXXII) and

(CXXXV) are formed as intermediates during the oxidation of hino-kitiol with hydrogen peroxide.

The formation of β-isopropyl-levulinic acid (CXXXVIII) may be explained in the following manner. Oxidation of hinokitiol gives the dihydroxy intermediate (CXXXV) which suffers oxidative degradation (in its keto form CXXXVI) to (CXXXVIII), *via* the β-ketoacid (CXXXVII) (see *Chart 2*). It should be noted that the isomer of (CXXXIII), β-iso-propyl-muconic acid, and the isomer of (CXXXVIII), α-isopropyl-levulinic acid, have not been obtained so far.

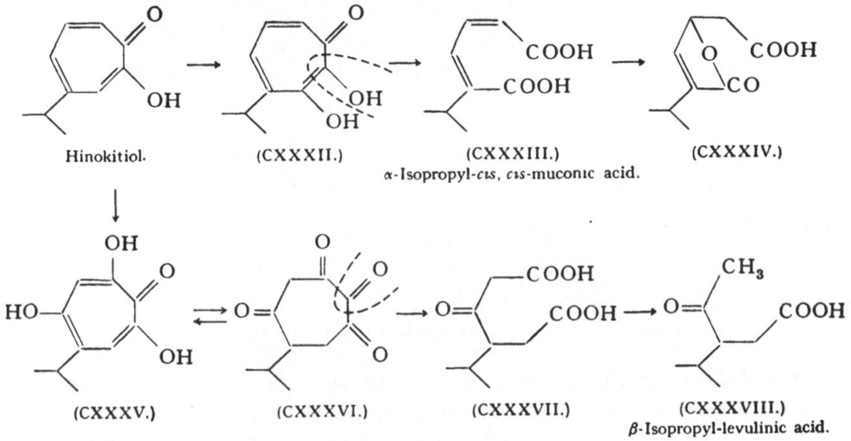

Hinokitiol.   (CXXXII.)   (CXXXIII.)   (CXXXIV.)
α-Isopropyl-*cis*, *cis*-muconic acid.

(CXXXV.)   (CXXXVI.)   (CXXXVII.)   (CXXXVIII.)
β-Isopropyl-levulinic acid.

Chart 2. Oxidation of Hinokitiol with Hydrogen Peroxide.

In the case of α-thujaplicin, only the compound (CXXXIII) is formed under the influence of hydrogen peroxide, while γ-thujaplicin yields (CXXXVIII) and β-isopropyl-muconic acid (probably in its *trans* form) (*142*). Such observations would indicate that this kind of oxidation is a very delicate process.

It was easy to predict that polyhydroxy-tropolones will convert to their alicyclic keto forms as do polyhydroxy-benzenes. From stipitatic and puberulic acids, aconitic and malonic acids are supposed to be formed in a similar manner (p. 243).

The azo compounds (CXXXIX) of hinokitiol undergo easy rearrange-ment to hinopurpurins (p. 239). Recent investigations have shown that hinopurpurin possesses a seven-membered quinonoid, i. e. tropoquinonoid structure; and, as a result of detailed studies concerning the hydrogen peroxide oxidation products of hinopurpurin [NOZOE et al. (*138*, *140*)], this pigment was proved to have structure (CXL) (*Chart 3*, next page).

This sequence of reactions has revealed that in (CXXXIX) a dehydro-genative ring closure had taken place between the tertiary carbon of the

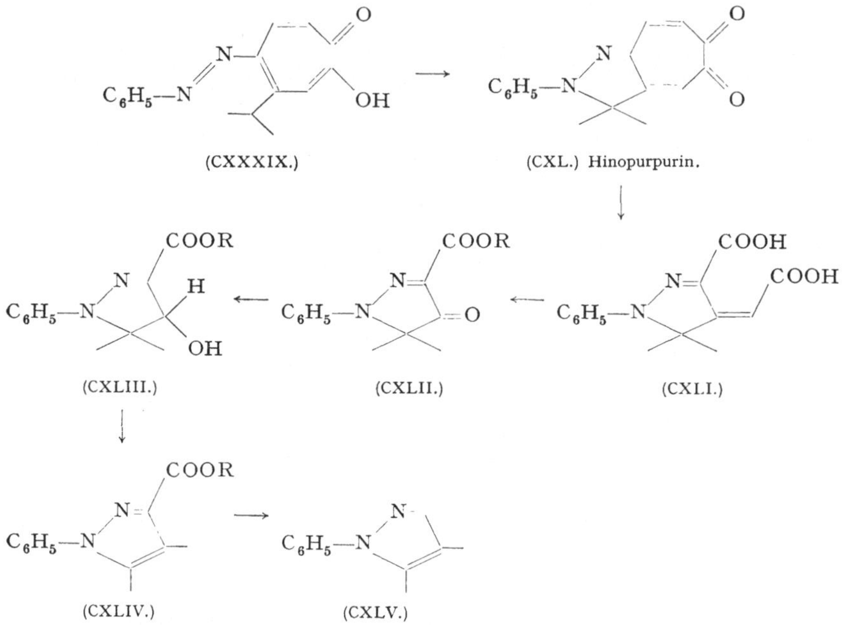

(CXXXIX.)   (CXL.) Hinopurpurin.

(CXLIII.)   (CXLII.)   (CXLI.)

(CXLIV.)   (CXLV.)

Chart 3. Oxidative Degradation of Hinopurpurin.

isopropyl group and a nitrogen atom of the azo group. Furthermore, it was observed that heating of azohinokitiol and benzoquinone in methanol caused almost quantitative formation of hinopurpurin.

e) Reduction of Tropolones.

Tropolone is not reduced in the presence of palladium but is catalytically hydrogenated in the presence of ADAMS' catalyst to give a mixture of cycloheptanediol, cycloheptanone, and cycloheptanol.

Reduction of purpurogallin in the presence of platinum had usually given a mixture of products [HAWORTH et al. (98)]. Recently, however, WALKER (240a) carried out the reduction with 10% palladium-charcoal catalyst under pressure (40 lb.) and obtained in good yield tetrahydropurpurogallin of an acyloin structure. COOK et al. (52) isolated a substance assumed to be cycloheptenedione on the lithium aluminum hydride reduction of tropolone; however, the product was benzaldehyde in the case of tropolone methyl ether. HEILBRONNER and ESCHENMOSER (219) obtained a colorless carbinol base (CXLVI) when reducing purpurogallin tetramethyl ether (XL) with an excess of the hydride and also observed that (CXLVI) was converted into a red benzotropylium cation (CXLVII) when dissolved in dilute acids. These workers also prepared (CXLVIII) from the compounds (CXLVI) and (CXLVII) by oxidation with the pyridine-chromium trioxide complex. The 4,5-benzotropolone derivative (CL) was also obtained by reducing (CXLVII) with zinc dust and sulfuric acid to a ketone (CXLIX) and oxidizing the latter with selenium dioxide.

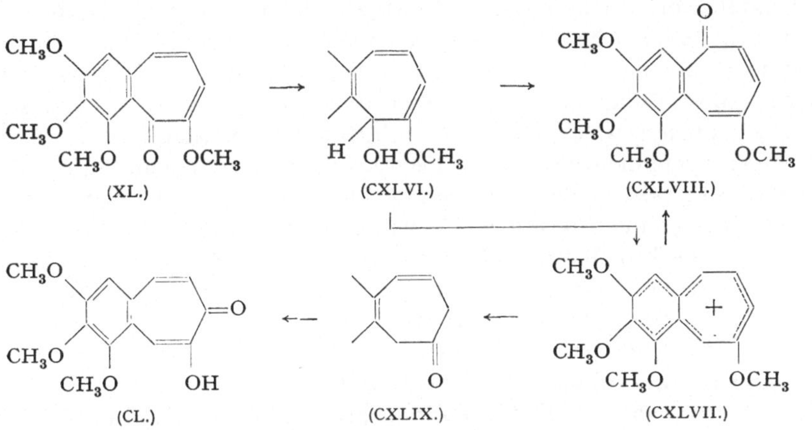

(XL.)   (CXLVI.)   (CXLVIII.)

(CL.)   (CXLIX.)   (CXLVII.)

2. Cationoid and Free Radical Reactions.

a) General Considerations.

After tropolone and its homologs had been synthesized, they (like hinokitiol) were found to undergo easily electrophilic substitution. The positions easily substituted were as expected 3, 5, and 7 in the tropolone ring (9, 135, 136). Predictions made on the basis of molecular orbital calculations (32, 65, 117) were approximately in accordance with the observations as far as tropolone itself was concerned. However, such reactions of various tropolones are of a complicated nature and the result is strongly dependent on structural differences, on the type of cationoid reagent used and even on slight differences in the experimental conditions (8, 136). The mechanisms of electrophilic substitution of tropolones seem to be similar to those of benzenoid compounds, especially phenols. Electrophilic substitutions that take place easily in tropolones are, azo coupling, nitrosation, sulfonation (by sulfamic acid), hydroxymethylation, and halogenation. FRIEDEL-CRAFTS type alkylations and acylations, aldehyde and ketone syntheses according to GATTERMANN, and HOESCH, and FRIES rearrangements, which all are important electrophilic substitutions and typical for benzenoid compounds, could not be realized in the tropolone class (7, 52). Neither are tropolones sulfonated by means of concentrated or fuming sulfuric acids and even nitration, a usually facile reaction, is inhibited in the presence of concentrated sulfuric acid and acetic anhydride (134, 136). This is caused by the amphoteric character of tropolones that assume a conjugate acid (protonated cation) form (CXXIV, p. 263) in strong acid media or they form chelated complexes with the metal salts in the Friedel-Crafts catalyst, thus resisting cationoid substitution (*136*). However, Nozoe, Seto et al. (*181, 182*) succeeded in the sulfonation of tropolones by heating with sulfamic acid at 150°.

On the basis of his theoretical calculations Dewar (*65*) had predicted that in tropolone both electrophilic and radical substitutions will take place almost exclusively at the 5-position. However, the actual substitution process was found to be influenced by various factors (*6, 7*). Of the electrophilic substitutions mentioned, azo coupling and nitrosation are liable to take place exclusively at the 5-position but halogens tend to enter the 3- and 7-positions. As will be shown later, the halogenation, especially the bromination of tropolones yields as a first step a highly colored complex, whose conversion finally affords the substituted compound proper. Thus, the position attacked may be different from that substituted directly, in a simple reaction.

Nitration and sulfonation processes represent an intermediate reaction type. In tropolone itself, the position attacked is mostly 5 and only if this position is already occupied or in case of a steric interference by a group present in the 4-position, will the substitution take place easily at the 3-(or 7-)position. It is understandable that the coplanar, seven-membered troponoid ring is more sensitive to steric hindrance than is the planar, six-membered benzene ring.

b) Location of Substituents.

There are various methods available for the location of substituents in tropolones. Of these, organic chemical procedures include the rearrangement of tropolone derivatives to benzenoid compounds and the conversion of tropolone derivatives to structurally known troponoids by changing the substituents present or by further substitution. Physicochemical methods, such as the measurement of dipole moments, X-ray diffraction, and infrared spectra, are also useful in determining the position of substituents.

For this purpose the use of nitrotropolones is especially convenient since they give numerous isomers, readily undergo aromatization, and are easily converted to, and correlated with, amino, halogen, and some other derivatives. Consequently, we will now briefly discuss such procedures, choosing the nitro compounds of hinokitiol as an example (*134–136*) (*Chart 4*).

Hinokitiol forms two dinitro compounds. The main product (CLIV) of the nitration rearranges very easily to give 4,6-dinitro-*m*-cumic acid (CLVII), hence compound (CLIV) is 5,7-dinitrohinokitiol (*161*). Mononitration of hinokitiol gives five mononitro compounds whose reduction affords only three different aminohinokitiols, thus indicating that four

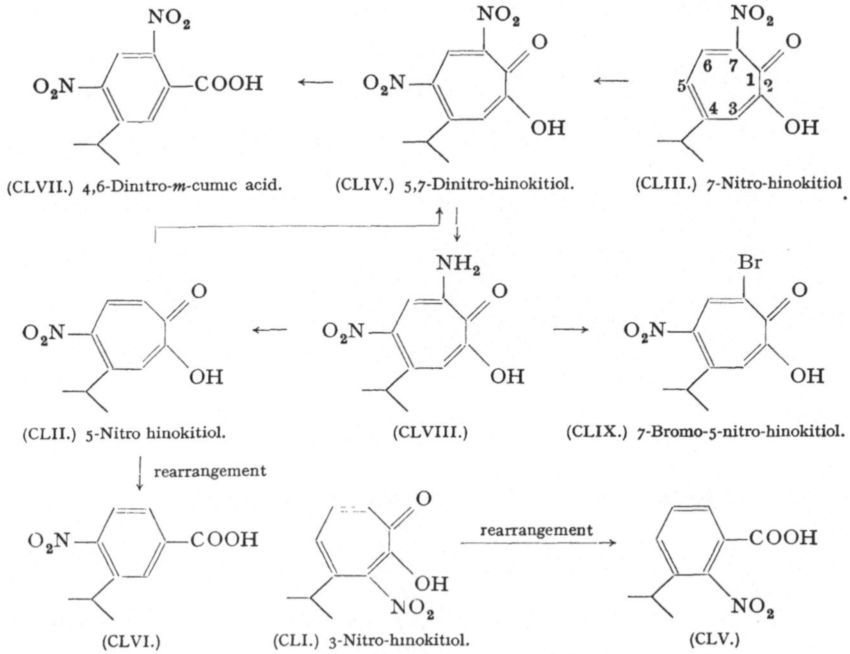

Chart 4. Nitro Derivatives of Hinokitiol.

of the mononitro derivatives must be interpreted as two pairs of isomers containing the nitro group in the same position (*158*). Two of the mononitro-hinokitiols, viz. (CLI) and (CLII) were identified as the 3- and 5-nitro compounds considering the structure of their respective rearrangement products, (CLV) and (CLVI). The remaining mononitro compound (CLIII) must be 7-nitro-hinokitiol since its further nitration gave 5,7-dinitro-hinokitiol (CLIV) which was also obtained by nitrating compound (CLII) (*152, 158*). The compound (CLVIII), obtained by the partial reduction of the dinitro derivative (CLIV) yielded, on deamination 5-nitro-hinokitiol (CLII), and 7-bromo-5-nitrohinokitiol (CLIX) in the SANDMEYER reaction (*137*).

On the other hand, the azohinokitiol could be identified as the 5-azo compound (CLXI) (*Chart 5*), since the amino compound (CLX), obtained by its reduction was found to be identical with the reduction product of the structurally clarified 5-nitro-hinokitiol (CLII) (*152*).

Bromination of hinokitiol affords, besides a tribromo compound, two dibromo and three monobromo derivatives (*172*). The structures of the three monobromo-hinokitiols, viz. (CLXII), (CLXIII), and (CLXIV), were secured by converting the mononitro-hinokitiols (CLII), (CLIII), and (CLI), into the corresponding amino derivatives and then, by the

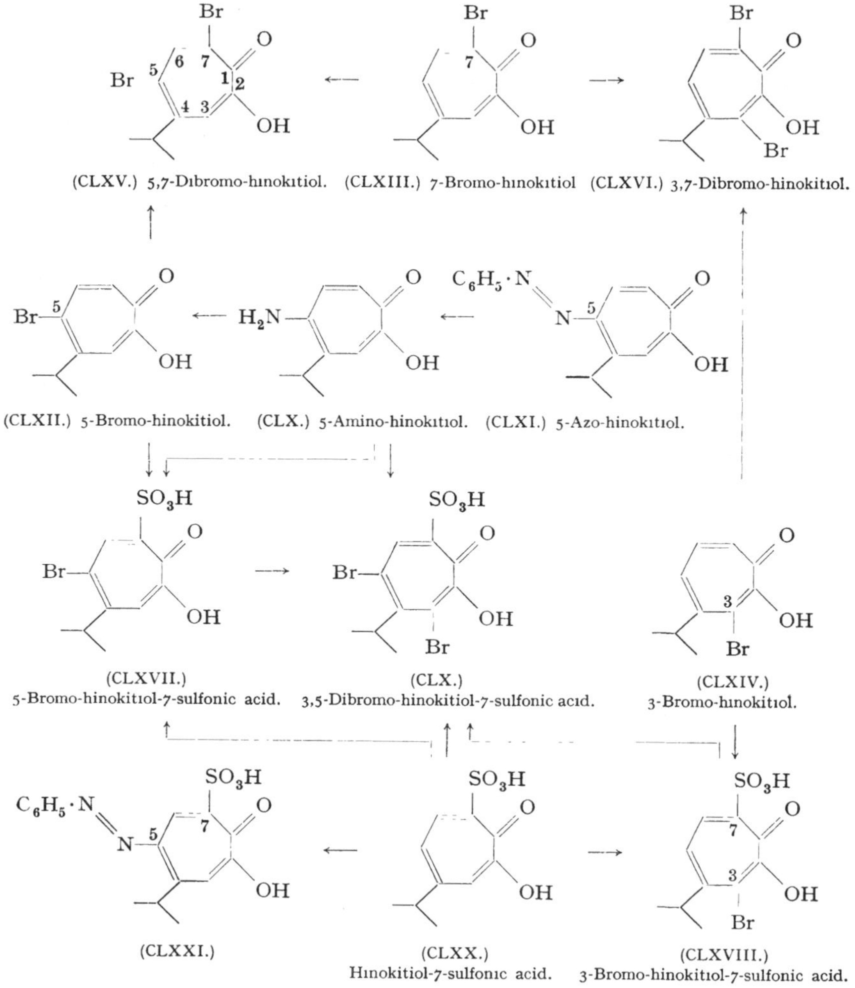

(CLXV.) 5,7-Dibromo-hinokitiol.   (CLXIII.) 7-Bromo-hinokitiol   (CLXVI.) 3,7-Dibromo-hinokitiol.

(CLXII.) 5-Bromo-hinokitiol.   (CLX.) 5-Amino-hinokitiol.   (CLXI.) 5-Azo-hinokitiol.

(CLXVII.)
5-Bromo-hinokitiol-7-sulfonic acid.   (CLX.) 3,5-Dibromo-hinokitiol-7-sulfonic acid.   (CLXIV.) 3-Bromo-hinokitiol.

(CLXXI.)   (CLXX.) Hinokitiol-7-sulfonic acid.   (CLXVIII.) 3-Bromo-hinokitiol-7-sulfonic acid.

Chart 5. Substitution Products of Hinokitiol.

SANDMEYER reaction, into the respective bromo compounds (*152, 154*). The structure of the dibromo-hinokitiol (CLXVI) was determined by its formation by further bromination of both monobromo derivatives, (CLXIII) and (CLXIV). Similarly, the other dibromo-hinokitiol (CLXV) was identified with the compound obtained from the two monobromo derivatives (CLXII) and (CLXIII) (*172*).

Hinokitiol-sulfonic acid (CLXX) is to be formulated as the 7-derivative, since it coupled to give (CLXXI); furthermore, compound (CLXVII) obtained by brominating (CLXX) was identical with the sulfonation product of 5-bromo-hinokitiol (CLXII). Likewise, sulfonation of 3-bromo-hinokitiol (CLXIV) afforded the 7-sulfonic acid (CLXVIII) (*181*) (*Chart 5*).

Subsequently, substituents in tropolone and 4-methyltropolone have been located by means of similar methods (*9, 162, 165, 179*).

c) Steric Effect in Substitution Processes.

As a result of the establishment of the structure of numerous substitution products of tropolones (see p. 276) there seemed no definite evidence that substituents had entered into any position other than 3, 5, and 7 in the tropolone ring*. However, it became increasingly clear that such electrophilic substitution processes are markedly influenced by the steric effect of the neighboring groups in the tropolone series (*8, 136*).

Coupling invariably takes place in the 5-position of the tropolone ring and the reaction does not occur at all if this position is occupied. This conversion may therefore indicate vacancy in the 5-position (*165, 169, 181*). The azo coupling of γ-thujaplicin (*79*) and 5-chloro-hinokitiol (*169*) does afford a reddish brown, amorphous product but the latter is not a true azo compound (*8*). Although the presence of a 4-isopropyl group does not interfere with the coupling of hinokitiol at the 5-position, the reaction product has a tendency to rearrange to hinopurpurin (*138*, p. 271). 4-*tert*-Butyltropolone seems to be completely resistant to azo coupling (*136*).

Nitrosation also affects the 5-position in most instances (*195*) but this process is inhibited by the steric hindrance of the 4-alkyl group and thus the yield of 5-nitroso compound obtained from 4-methyltropolone is markedly diminished (*97, 165*). In the case of hinokitiol (4-isopropyl-tropolone), nitrosation results mainly in the formation of 4-isopropyl-salicylic acid and very little 3-nitro-hinokitiol (*189*).

Nitration of hinokitiol affords mainly 7-nitro-hinokitiol and less of the 3- and 5-derivatives. It is interesting to note that the sterically most hindered 3-position is more easily attacked than the 5-position, apart from the sterically least hindered 7-position (*135, 151, 158*). Nozoe and others (*152*) have obtained from hinokitiol five different mononitro compounds whose reduction gave only three amino derivatives, viz. the 3-, 5-, and 7-amino-hinokitiols. Evidently, when a nitro group enters the 5- or 3-position, the steric interference of the 4-isopropyl group causes the formation of isomers (*135, 158*).

---

* Nozoe et al. (*170*) obtained an azo dye in crystalline form and in good yields by coupling 3,5,7-trihydroxymethyl-tropolone. In this instance, it seems that the azo coupling had taken place at the 4-position but this point needs further investigation.

The nitration of tropolone itself is simpler than that of hinokitiol; mainly the 5-nitro compound is formed (*50*, *147*). The behavior of 4-methyl-tropolone is intermediate between that of tropolone and hinokitiol and the nitration leads to three mononitro compounds (*97*, *165*). The nitration of α- and γ-thujaplicins has also been carried out (*8*, *244*).

Sulfonation of tropolone itself affords mainly the 5-derivative. γ-Thujaplicin and 5-bromotropolone, in which the 5-position is already occupied, as well as compounds containing a 4-isopropyl group yield 7-sulfonic acids (*162*, *165*, *181*, *182*, *187*). It should be stressed that, in spite of its free 5- and 7-positions, α-thujaplicin is not sulfonated (*160*), although in other electrophilic substitutions it behaves in the same manner as do the β- and γ-isomers. This shows anew that a slight structural modification in tropolones may lead to markedly altered results in such reactions.

Hydroxymethylation of tropolone itself affords, depending on the conditions, mono-, di-, or trihydroxymethyl derivatives. The positions thus substituted are 3, 5, and 7, as in many other instances. Hydroxymethylation of 4-methyltropolone or hinokitiol gives mainly the 7-substituted derivative (*170*), indicating the steric effect of alkyl groups. The structural determination of these products is carried out by converting them into methyltropolones.

d) Halogenation of Tropolones and 2-Aminotropones.

As has been mentioned above, halogenation of tropolones in general and especially the bromination or iodination attacks the 3- and 7-positions, rather than the 5-position.

The interaction of bromine and tropolone in solvents such as chloroform and carbon tetrachloride produces first a reddish-orange, crystalline complex which in contact with water or an alcohol decomposes immediately to give a mixture of true substitution products (*43*, *192*). It is also known that similar bromine complexes are formed by the methyl ethers of tropolones and by bromotropolones (*192*) as well as methylthiotropones (*178*). That tropolone itself is more easily attacked in the 3- and 7-positions by bromination is probably due to the formation of an intermediate complex which is converted into the true substitution compound (*7*, *9*).

With reference to electrophilic substitutions of 2-aminotropone, it is noteworthy that the azo coupling attacks the 5-position while the bromination takes place (in contrast to that of tropolone) at 7, 5, and 3, in decreasing order of easiness (cf. CLXXII). The structure of such substituted 2-aminotropone can be determined by converting it, by alkali hydrolysis, into the corresponding tropolone derivative (*194*).

e) Benzotropolones.

Like other monocyclic tropolones, 3,4- and 4,5-benzotropolones easily undergo electrophilic substitution. In the case of 3,4-benzotropolone (XCIX), the halogen tends to enter the 7-position. Coupling and sulfonation are also easy but the preferred positions cannot yet be designated (*46, 146*). With 4,5-benzotropolone (CII; $R = H$, p. 258), both nitration and halogenation, as well as coupling have been carried out and mainly 3-derivatives have been obtained. Bromination gave a 3,7-dibromo compound (*231*).

f) 3- and 4-Hydroxytropones.

Like tropolone itself, hydroxytropones also undergo electrophilic substitution easily, especially coupling, halogenation, and nitration. In the case of 3-hydroxytropone (CLXXIII), both halogenation and azo coupling take place exclusively at the 2-position (*107*), while 4-hydroxytropone (CLXXIV) couples at 5. Bromination and chlorination in the cold give 5,7-disubstituted derivatives and bromination in the heat yields the 2,5,7-tribromo compound (*164, 197*).

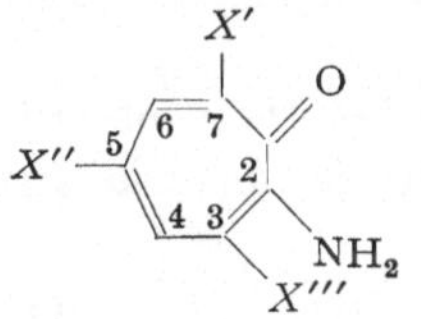

(CLXXII.) 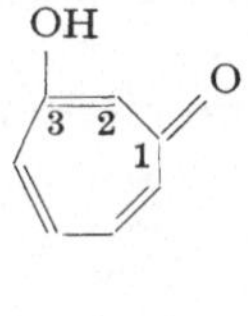   (CLXXIII.) 3-Hydroxy-tropone. 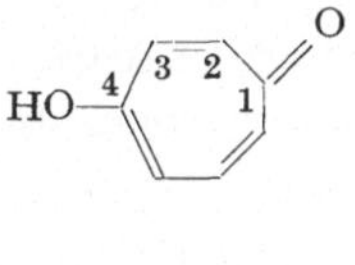 (CLXXIV.) 4-Hydroxy-tropone.

g) Free Radical Reactions.

Examples of free radical reactions in the tropolonoid series are rare. The radical formed by the pyrolysis of the diazonium salt, obtained from 5-aminotropolone, reacts with benzene or its methoxyl derivatives to form 5-aryltropolones (p. 259). This radical also reacts with tropolone to yield 3,5'-bitropolonyl (CLXXV). Heating of 3-iodotropolone in the presence of copper dust gave 3,3'-bitropolonyl and its anhydride (CLXXVI) (*8*).

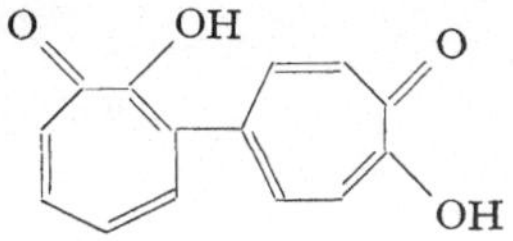

(CLXXV.) 3,5'-Bitropolonyl. 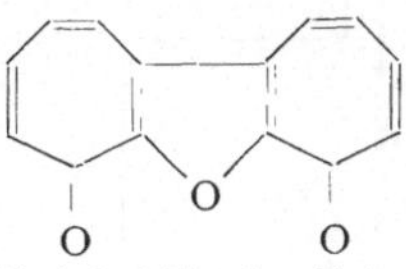   (CLXXVI.) Anhydride of 3,3'-bitropolonyl.

3. Anionoid Substitution and Rearrangements.

a) General Considerations.

As mentioned above, in troponoid compounds the contribution of structure (CXXI) (p. 263), in which each of the carbon atoms in the seven-membered ring is positively charged, is substantial. Hence such molecules are liable to be attacked by anionoid reagents. This is especially marked in troponoids containing such substituents (halogen, methoxyl) that can be easily eliminated in the form of an anion; then either a differently substituted troponoid is formed or rearrangement to benzenoid structures takes place. Evidently, anionoid reactions of troponoid compounds are more complicated than their cationoid reactions. The results will be strongly influenced by the structure of the troponoid compound, the type of anionoid reagent, and the experimental conditions.

In contrast to the behavior of the methyl ethers of tropolones and colchicine, in benzotropolone methyl ethers the substitution of methoxyl with a base encounters great difficulty, due to resonance inhibition (9).

The reagents which are liable to cause nucleophilic substitution in troponoids are, alkalis, alkoxides, sulfides, mercaptides, ammonia, amines, and cyanides, as well as phenyllithium and GRIGNARD reagents. The formation of 2-aminotropones from tropones on treatment with hydroxylamine or hydrazine is a typical example of nucleophilic substitution of a hydrogen atom for which the characteristic tropone structure is responsible (166).

The use of rearrangement processes of tropolones, their methyl ethers, and halotropones by alkali, alkoxide, or ammonia for the structural

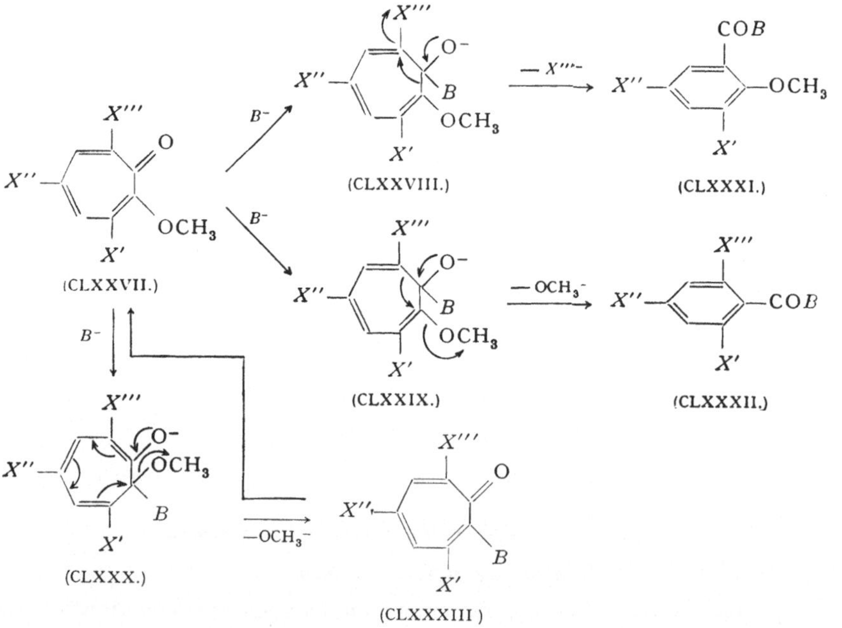

Chart 6. Anionoid Substitution and Rearrangement of Tropolones.

clarification of troponoids has been mentioned above. We may add that such reaction products may represent a complicated mixture. These rearrangement products are not only benzoic acid derivatives but also include derivatives of salicylic acid and of salicylaldehyde. On interaction with GRIGNARD reagent, benzophenone and triphenylcarbinol derivatives are sometimes formed (*100, 167*).

It may be stated, without going into details of the reaction mechanism (*9, 112*), that an attack on carbon atom 1 would afford a rearrangement product, while that on the carbon atom 2 would yield a seven-membered ring derivative. For example (*Chart 6*), methyl ethers of substituted tropolones (CLXXVII) form, when attacked by a base B^-, the rearrangement products (CLXXXI) and (CLXXXII) *via* the respective intermediates (CLXXVIII) and (CLXXIX); furthermore, the displacement product (CLXXXIII) of the substituent group (methoxyl) will be formed, *via* the intermediate (CLXXX) (*149, 159*).

It has been found that in such reactions those substitution and rearrangement products are obtained whose formation requires the assumption that carbon atoms other than $C_{(1)}$ and $C_{(2)}$ (especially those not carrying halogen or methoxyl) were attacked. Recently, DOERING and others (*66*) examined the troponoid-benzenoid rearrangement using a labeled troponoid compound, with ^{14}C in the position 1. They proved that the rearrangement had taken place by the attack on $C_{(1)}$ and proposed a mechanism with an intermediate containing a cyclopropane ring.

More recently, NOZOE and others succeeded in synthesizing various azulene derivatives and heterocyclic azulenoid compounds by the interaction of cyanoacetic, malonic and acetoacetic esters, thiourea, guanidine, or cyanoacetamide and various tropolone methyl ethers and halotropones, in the presence of sodium alkoxide. Thus, numerous tropoid compounds can be prepared starting from relatively simple troponoids.

In some instances the new substituent enters a position different from that of the liberated group, e. g., $C_{(7)}$ instead of $C_{(2)}$.

On the basis of the reaction products, anionoid substitutions of troponoids can be classified into the three following types: 1. Rearrangement reaction induced by an attack on carbonyl-carbon by a base; 2. simple displacement of substituents with a base; and 3. rearrangement and displacement (''abnormal'' displacement) by attack of an unsubstituted carbon atom by a base. Several anionoid reagents will now be discussed.

b) Alkali and Alkoxides.

Tropolones that are not substituted with electronegative groups do not undergo rearrangement unless heated with strong alkali at a fairly high temperature (*70*), since they form stable anions in alkali. When, however, nitro groups are attached to the ring or when the tropolone ring is fused with a benzene ring, then the compounds display a tendency to undergo rearrangement. As mentioned earlier, such rearrangements can be used for structural determinations.

In order to compare the behavior of tropolone methyl ether with that of colchicine, DOERING and KNOX (70) refluxed this ether in methanol in the presence of sodium methoxide and found that it was converted into methyl benzoate, although the reaction rates were low. This method has become important in the structural clarification of such tropolone derivatives that resist rearrangement in aqueous alkali.

Halotropones rearrange easily in cold alkali to form benzoic acid derivatives but in some instances tropolone and 3-hydroxytropone derivatives appear, depending on the structure of the halotropones treated. For example, the two isomeric methyl ethers obtained from 3-bromotropolone behave in entirely different manner towards alkali: one of them (CX) reverts back to the original tropolone compound but the other (CXII) rearranges quantitatively to o-bromobenzoic acid (*159*). When the two isomeric chlorobromotropones, prepared from these two methyl ethers, are treated with ethanolic alkali, the 2-chloro-7-bromo compound (CXI) forms o-chlorobenzoic acid and a minute

(CLXXXIV.) 3-Chloro-salicylaldehyde.

(CLXXXV.) 2-Chloro-3-hydroxytropone $(R = H)$.

amount of 3-bromotropolone as well as 3-chloro-salicylaldehyde (CLXXXIV), while the other isomer, *viz.* the 2-chloro-3-bromo compound (CXIII) yields o-bromobenzoic acid and 2-chloro-3-hydroxytropone and its ethyl ether (CLXXXV; $R = C_2H_5$) (*228*).

In contrast to their methyl ethers, free halotropolones are alkali stable at ordinary temperature but they do react with alkali under drastic conditions. For example, alkali fusion of 3-bromotropolone results in formation of 4-hydroxytropolone (which is formed by ''abnormal'' displacement) and of the 3-hydroxy derivative (*112, 144*).

c) Ammonia and Amines.

Prior to its structural clarification, colchicine was known to give ''colchiceinamide'' when reacted with ammonia (cf. *48*). Likewise, tropolone methyl ethers yield very easily 2-amino- or 2-hydrazinotropone when treated with ammonia, alkyl- or arylamines, or hydrazine (cf. *9*).

In contrast, 2-halotropone derivatives tend to give both amino-substituted and rearranged products. This reaction is a complicated one and the ratio of substituted product to rearrangement product is influenced by various factors including the position of the substituent and the amine used.

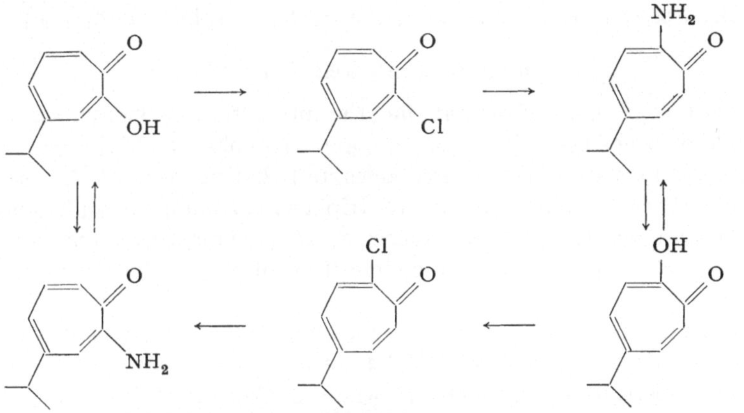

(CLXXXVI.)

Chart 7. Interconversion of 4- and 5-Isopropyltropolones.

By means of the "abnormal" reaction represented in *Chart 7*, NOZOE and his co-workers were able to establish the correlation between β- and γ-thujaplicins by mutual interconversion (*193*).

Halotropolones are ammonia-stable but, when heated with the potassium salt of an arylsulfonamide, they are easily substituted and yield various aminotropolones (*112, 144*).

d) Sulfides, Mercaptides, and Cyanides.

Halotropones, halotropolones and their methyl ethers undergo nucleophilic substitution relatively easily on treatment with alkali sulfides, alkyl or aryl mercaptides or cyanides to yield mercapto or cyano derivatives (cf. *9*). Mercapto-tropone itself is stabilized by the formation of an intramolecular hydrogen bond, but mercapto-tropolone and those derivatives in which the substituent is in a position that cannot chelate with the carbonyl group, undergo facile dehydrogenation to the disulfide (*178*). Cyano-tropolones can be hydrolyzed easily to carboxy-tropolones (*8, 9, 112, 144*).

Alkali cyanides in alcohol solution show marked preference for entering an unsubstituted position in the ring rather than for displacing halogen; e.g., 3-bromotropolone yields 4-cyanotropolone (*112, 144*).

e) Anionic Substitution in Strong Acids.

DOERING et al. (*68, 71*) found that halotropones and halotropolones have the tendency to exchange their halogen atoms in concentrated hydrochloric or hydrobromic acid. On heating 2,4,7-tribromotropone with concentrated sulfuric acid, 2,5,7-tribromo-4-hydroxytropone is formed (*145*); this can be interpreted as a nucleophilic substitution of the 4-hydrogen atom by means of the oxidative effect of the acid.

f) Grignard Reagents and Phenyllithium.

NOZOE et al. (*166*) observed that the interaction of phenylmagnesium bromide or other GRIGNARD reagents and a tropolone methyl ether results in the easy formation of 2-aryltropones, but that in case of halotropolone methyl ethers a rearrangement to triphenylcarbinol, benzophenone, or related compounds also takes place (*8, 167*). This whole reaction is a fairly complicated one and the composition of the product is influenced by various factors.

According to HAWORTH et al. (*13, 100*), the effect of phenyllithium on 4-methyltropolone (CLXXXVII *a*) results in the sole formation of a normal substitution product *b*), whereas the interaction of phenyl-magnesium bromide and either of the two isomeric methyl ethers of *a*), *viz.* *c*) or *d*), results in abnormal substitution and the formation of *e*) and *f*); at the same time rearrangement takes place and triphenyl-carbinol derivatives appear (*Chart 8*).

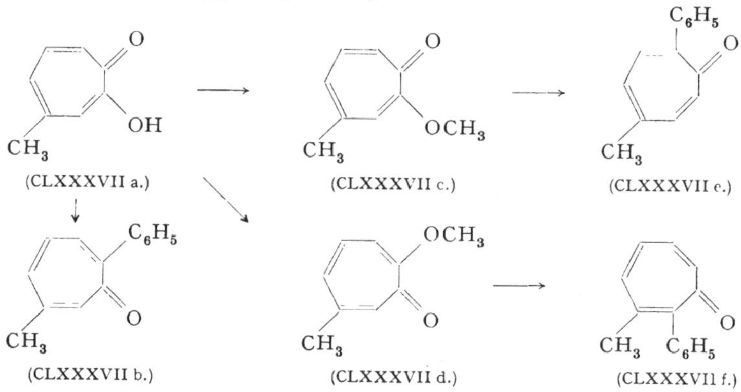

(CLXXXVII a.) → (CLXXXVII c.) → (CLXXXVII e.)
(CLXXXVII b.) (CLXXXVII d.) → (CLXXXVII f.)

Chart 8. Interaction of 4-Methyltropolone and Some Organometallic Compounds.

g) Rearrangements with Alkali Hypohalites or by Perhalogenation.

Alkali hypoiodite converts colchiceine into iodocolchinol; and heating tropolone with excess alkali hypoiodite or hypobromite yields triiodo- or tribromophenol [DOERING et al. (*70*)]. The interaction of iodine and tropolone in the presence of one mole of sodium carbonate results in the formation of 2,4-diiodobenzoic acid, besides iodotropolone (*113*).

FERNHOLZ (*85*) found that, when an excess of bromine reacts with bromotropolone in acetic acid, first the perbromide (CLXXXVIII) is formed which, on treatment with alkali, gives tribromophenol (CXC) *via* the intermediate (CLXXXIX). The so-called "tribromo-colchiceinic acid" prepared by WINDAUS (*243*) by reacting colchiceine with excess bromine in acetic acid was formulated by LETTRÉ (*120*) as (CXCI), corresponding to (CLXXXIX). NOZOE and others (*134*) also found that chlorohinokitiols on further chlorination suffered similar rearrangement.

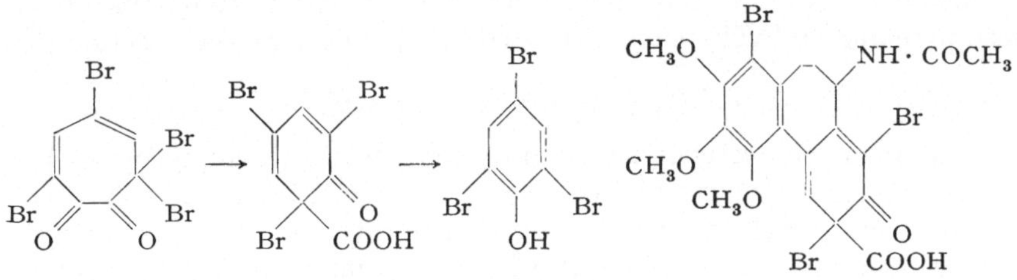

(CLXXXVIII.) (CLXXXIX.) (CXC.) (CXCI.)

h) Some Other Rearrangement Reactions.

Troponoid compounds are rearranged also by some other reagents. For example, the two isomeric amino compounds (CXCII) and (CXCIV), prepared from hinokitiol (152), when diazotized in sulfuric acid and then heated undergo rearrangement (while liberating nitrogen) to the corresponding isomeric isopropylsalicylic acids (CXCIII) and (CXCV) (*153*). Similar phenomena have been observed with the corresponding amino derivatives of tropolone (*50*), 4-methyltropolone (*97*) or colchiceine (*130*), but not with the amino derivative of 4,5-benzotropolone (*130*). The rearrangement of hinokitiol to the salicyclic acid derivative (CXCIII) during attempted nitrosation has been mentioned (*195*).

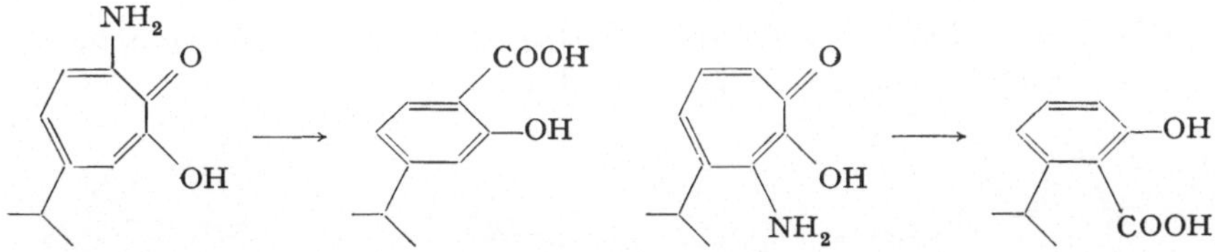

(CXCII.) (CXCIII.) (CXCIV.) (CXCV.)

Excess thionyl chloride cleaves tropolone to give *o*-chlorobenzaldehyde (*11*) while the interaction of lithium aluminum hydride and tropolone results in the formation of benzaldehyde (*52*).

As we have shown above, troponoids are relatively easily rearranged by a number of reagents whereby the composition of the product is influenced by various factors.

4. Formation of Azulenoid Compounds.

Although, as mentioned, the troponoid molecule possesses a seven-membered ring containing three carbon-carbon double bonds, such compounds are fairly stable and derivatives with functional groups in adjacent ring positions can be obtained easily. Consequently, such derivatives can be further transformed to azulenoid compounds by fusion with a second, fivemembered ring. In this manner, numerous azulenoid compounds have been prepared and will be characterized briefly in the present Section (8).

a) 2-Oxo-1,2-dihydro-1-oxa-azulene.

Sodium malonic ester or acetoacetic ester reacts with tropolone methyl ethers or 2-halotropones and gives, respectively, (CXCVIa) and (CXCVIb) in good yields. Sulfuric acid converts them into the parent compound, 2-oxo-1,2-dihydro-1-oxaazulene (CXCVIc). A mild alkali treatment of (CXCVIa) and (CXCVIb) leads, respectively, to troponyl-acetic acid (CXCVIIa) and troponyl-acetone (CXCVIIb) (227).

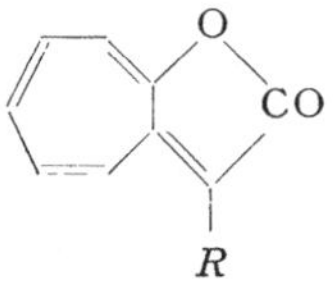

(CXCVIa; $R = COOC_2H_5$.)
(CXCVIb; $R = COCH_3$.)
(CXCVIc; $R = H$.)

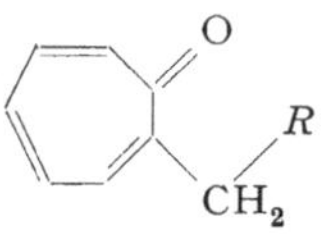

(CXCVIIa; $R = COOH$.)
(CXCVIIb; $R = COCH_3$.)

b) 2-Oxo-1,2-dihydro-1-aza-azulene and 1-Aza-azulene.

Application of malonic ester to 2-aminotropones or of cyanoacetamide to the methyl ethers of tropolones, both in the presence of sodium alkoxide, respectively affords (CXCVIIIa) and (CXCVIIIb). Hydrolysis of (CXCVIIIa) followed by decarboxylation gives the parent compound, 2-oxo-1,2-dihydro-1-aza-azulene (CXCVIIIc).

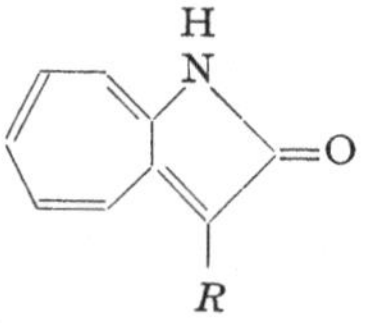

(CXCVIIIa; $R = COOC_2H_5$.)
(CXCVIIIb; $R = CN$.)
(CXCVIIIc; $R = H$.)

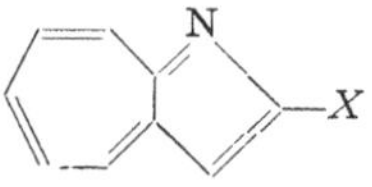

(CXCIXa; $X = Cl$.)
(CXCIXb; $X = H$.)

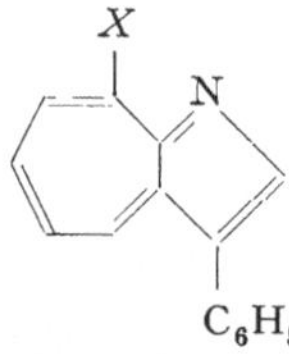

(CCa; $X = Cl$.)
(CCb; $X = H$.)

Treatment of (CXCVIIIc) with phosphoryl chloride yields 2-chloro-1-aza-azulene (CXCIXa) from which the parent compound, 1-aza-azulene (CXCIXb) is obtained through its hydrazino derivative (190, 191).

By the application of the FISCHER indole synthesis to the condensate of phenylacetaldehyde and 2-hydrazinotropone, a pyrrolotropone derivative (CXVI, p. 262) is obtained, whose treatment as in the foregoing case affords 3-phenyl-1-aza-azulene (CCb) through its chloro derivative (CCa) (*150*).

c) 2-Oxo-1,2-dihydro-1-thia-3-aza-azulene.

Thiourea and 2-halotropone give compound (CCIa) that can be hydrolyzed to 2-oxo-1,2-dihydro-1-thia-3-aza-azulene (CCIb) (*197*).

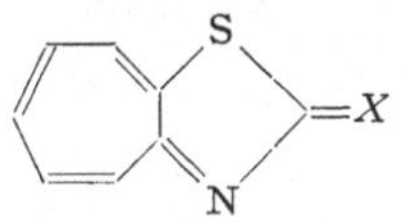

(CCIa; X = NH.)
(CCIb; X = O.)

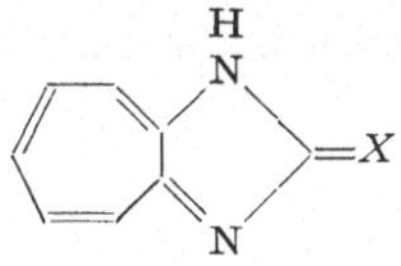

(CCIIa; X = NH.)
(CCIIb; X = O.)

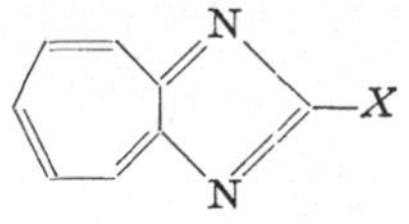

(CCIIIa; X = SH.)
(CCIIIb; X = H.)

d) 2-Oxo-1,2-dihydro-1,3-diaza-azulene and 1,3-Diaza-azulene.

A similar interaction of tropolone methyl ether and thiourea or guanidine in the presence of sodium alkoxide affords, respectively, (CCIIIa) and (CCIIa). Hydrolysis of (CCIIa) with acid or alkali, as well as a treatment of (CCIIIa) with mercuric oxide results in the formation of (CCIIb) (*173*). Compounds corresponding to (CCIIa) and (CCIIIa) can also be obtained from colchicine (*139*). When (CCIIIa) is treated with dilute nitric acid the parent compound, 1,3-diaza-azulene (CCIIIb) is formed (*168*).

e) 4,5-Imidazolo-tropone and 4,5-Triazolo-tropone.

5-Nitroso-tropolone when reacted with hydroxylamine gives tropoquinone trioxime, whose catalytic reduction leads to 2,5-diaminotroponeimine (CCIV). The latter compound is converted into (CCVa) by reaction with formic acid, and into (CCVIa) with nitrous acid. By hydrolyzing (CCVa) and (CCVIa) the respective parent compounds are obtained, *viz.* 4,5-imidazolo-tropone (CCVb) and 4,5-triazolo-tropone (CCVIb) (*141*, *175*).

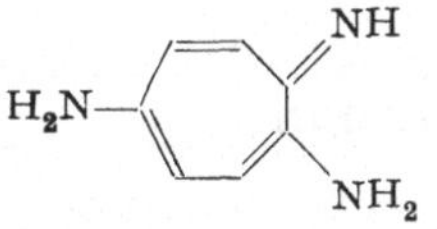

(CCIV.) 2,5-Diamino-troponeimine.

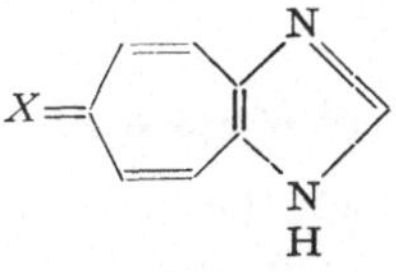

(CCVa; X = NH.)
(CCVb; X = O.)

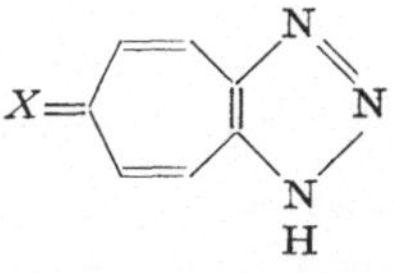

(CCVIa.) (X = NH.)
(CCVIb.) (X = O.) 4,5-Triazolo-tropone.

In these compounds, the tautomeric 6-amino- or 6-hydroxyaza-azulene type may also be considered but the the compounds belong to the troponoid type as shown in the formulas.

f) Azulene.

The action of more than two moles of cyanoacetic ester on 2-halo-tropones or tropolone methyl ethers, in the presence of sodium alkoxide, affords in good yields the homocyclic azulene derivatives (CCVII a), (CCVIII a), and (CCIX a). The use of cyanoacetamide instead of cyanoacetate gives (CCIX b) as well as the afore-mentioned 1-aza-azulene derivative (CXCVIII b); and the use of malononitrile gives (CCVIII b).

(CCVII a; $X = NH_2$.) (CCVIII a; $X = OH$.)   (CCIX a; $X' = COOC_2H_5$, $X'' = NH_2$, $X''' = CN$.)
(CCVII b; $X = H$.)   (CCVIII b; $X = NH_2$.) (CCIX b; $X' = X''' = CONH_2$, $X'' = NH_2$.)
(CCVII c; $X = Cl$.)   (CCIX c; $X' = X''' = H$, $X'' = NH_2$.)
                                                 (CCIX d; $X' = X''' = H$, $X'' = Cl$.)

The compound (CCVII a) yields (CCVII b) by deamination and (CCVII c) by the SANDMEYER reaction. Hydrolysis of (CCVII a, b, and c), followed by decarboxylation gives, respectively, 2-aminoazulene (CCIX c), azulene (V, p. 235), and 2-chloroazulene (CCIX d). The chloro compounds (CCIX d) and (CCVII c) can be further converted into numerous new azulene derivatives by nucleophilic displacement.

Azulenoids can also be prepared by starting from alkyl- or halogen-substituted troponoids, but whereas "normal" condensation products are formed when tropolone methyl ethers are used, "abnormal" reactions are likely to occur when halotropones are reacted. Such reactions are so strongly affected by the steric effect of the side-chain in troponoids that one should be very cautious when determining the structure of the azulenoids thus formed (8).

Electrophilic substitution of azulene itself cannot introduce the substituent in the 2-position; hence, the method mentioned for the formation of 2-substituted derivatives is of great importance (163).

VI. Biogenetical Problems and Conclusion.

As was pointed out at the outset, the number of the known naturally occurring tropolones is still relatively low but they are widespread in nature. In the early stages of the development of tropolone chemistry, it was even thought that tropolone might play an important role in the biogenesis of natural products, considering the following processes: The easy formation of purpurogallin from pyrogallol; the biosynthesis of hydroxytropolone-carboxylic acids from glucose in some molds; and the extremely facile aromatization of certain tropolone derivatives.

Thus, RAISTRICK (*201*) thought that tropolones could constitute a bridge between the aromatic and aliphatic metabolites of molds; and DEWAR (*65*) assumed that 5-aminotropolone might be an *in vivo* precursor of *p*-aminobenzoic acid and that the biosynthesis of other benzene derivatives may pass through tropolone intermediates. ROBINSON (*206*) suggested that the biosynthesis of tropolone could well take place by ring enlargement of polyhydroxybenzenes effected by formaldehyde or its biological equivalent. ROBINSON also presumed that tropolone might be a possible intermediate in the biosyntheses of strychnine, emetine, yohimbine, and other alkaloids, and he formulated a hypothesis explaining the biosynthesis of colchicine.

A recent hypothesis was suggested by BRUCE (*33*), that the oxidative polymerization of 5,6-dihydroxyindole might pass through a troponoid stage such as a pyrrolo-cycloheptindole whose further oxidative polymerization would give rise to melanine.

All these hypotheses are very interesting and could not well be excluded; however, no experimental evidence is available at the present time that would either prove or disprove these theories.

ERDTMAN (*78*) has discussed the biogenetic relationship between thujaplicins and terpenes. Although thujaplicins do not possess the usual isoprenoid structure, they may be considered as of terpenoid origin. For example, tropolones are often found together with terpenoid phenols, such as carvacrol, thymoquinone, and thymohydroquinone, and with terpenoid carboxylic acids such as rhodinic acid and α- and β-chamic acids (*37, 134, 196, 245*). The discovery of nootkatin, that might be termed a sesquiterpenoid tropolone, represents a further argument in favor of the terpenic origin of tropolones, or at least, of the existence of a common precursor. It is hoped that experimental data will be forthcoming soon in support of these hypotheses.

It should be also stressed that much is still to be done in the field of tropoid chemistry. As has been demonstrated above, troponoid compounds submit to both electrophilic and nucleophilic substitution, often accompanied by delicate aromatization and other rearrangement processes. Furthermore, it has become possible to synthesize homocyclic or heterocyclic azulenoids starting from troponoids. Thus, type and number of tropoid compounds are being increased steadily.

An elaborate tropoid system has now been firmly established, including numerous compounds whose characteristics extend from alicyclic to aromatic features and whose structures are either homocyclic or heterocyclic.

Of course, there still exist numerous · unexplained phenomena, unclarified reaction products, and unexplored reaction mechanisms. Several fundamental problems will have to be solved before the road will be clear for the application of tropoids in the fields of industry and pharmaceutics, especially chemotherapeutics. For that reason, it is desirable to find more advantageous methods for the preparation of tropones and tropolones to be used as starting materials in syntheses. Without doubt, new types of troponoids will be detected in natural sources in times to come.

References.

Within the limitations of the present review, it was impossible to present an exhaustive bibliography, especially with reference to older papers, furthermore to contributions in the fields of synthesis and physicochemistry. Hence, it is believed that the following General References (*1*)–(*10*) will be helpful.

General References.

1. CHOPIN, J.: Les Tropolones. Bull. soc. chim. France **1951**, D 57.
2. COOK, J. W. and J. D. LOUDON: The Tropolones. Quart. Rev. Chem. Soc. (London) **5**, 99 (1951).
3. HUBER, G.: Das Tropolon und seine Derivate. Angew. Chem. **63**, 501 (1951).
4. JOHNSON, A. W.: Aromaticity in Seven-Membered Ring Systems. J. Chem. Soc. (London) **1954**, 1331.
5. LOUDON, J. D.: Colchicine and Related Compounds. Annu. Rep. Chem. Soc. (London) **45**, 187 (1948).
6. NOZOE, T.: Chemistry of Tropolone and Its Allied Compounds. Chem. and Chem. Ind. (Japan) **4**, 348 (1951).
7. — Recent Advance in Tropolone Chemistry. Chem. and Chem. Ind. (Japan) **7**, 378, 413 (1954).
8. — General Outline of Tropoid Chemistry (From Troponoids to Azulenoids). Science Repts. Tôhoku Univ. [1] **40**, (1956) (in press).
9. PAUSON, P. L.: Tropones and Tropolones. Chem. Rev. **55**, 9 (1955).
10. ŠANTAVÝ, F.: Tropilidene, Tropone and Tropolone. Chem. Listy **47**, 1534 (1953).

Special References.

11. ABADIR, B. J., J. W. COOK, J. D. LOUDON and D. K. V. STEEL: Tropolones. Part V. Halogeno*cyclo*heptatrienones. J. Chem. Soc. (London) **1952**, 2350.
12. AKROYD, P., R. D. HAWORTH and J. D. HOBSON: Purpurogallin. Part IX. The Structure of the Isomeric O-Methyl Ethers of β-Methyltropolone. J. Chem. Soc. (London) **1951**, 3427.
13. AKROYD, P., R. D. HAWORTH and P. R. JEFFERIES: Purpurogallin. Part XI. Some Further Rearrangement Reactions of the Tropolones. J. Chem. Soc. (London) **1954**, 286.
14. ANDERSON, A. B. and J. GRIPENBERG: Antibiotic Substances from the Heart Wood of *Thuja plicata* D. DON. IV. The Constitution of β-Thujaplicin. Acta Chem. Scand. **2**, 644 (1948).

15. ANDERSON, A. B. and E. C. SHERRARD: Dehydroperillic Acid, an Acid from Western Red Cedar (*Thuja plicata* DON). J. Amer. Chem. Soc. 55, 3813 (1933).

16. ARNSTEIN, H. R. V., D. S. TARBELL, G. P. SCOTT and H. T. HUANG: Studies in the Structure of Colchicine. The Structure of Ring C. J. Amer. Chem. Soc. 71, 2448 (1949).

17. ASHLEY, J. N. and J. O. HARRIES: Purification of Colchicine by Chromatography. J. Chem. Soc. (London) 1944, 677.

18. AULIN-ERDTMAN, G.: Studies in the Tropolone Series. I. Thujaplicins and Nootkatin. Acta Chem. Scand. 4, 1031 (1950).

19. — Studies in the Tropolone Series. II. Puberulonic Acid. Acta Chem. Scand. 4, 1325 (1950).

20. — Studies in the Tropolone Series. IV. Stipitatic, Puberulic and Puberulonic Acids. Acta Chem. Scand. 5, 301 (1951).

21. AULIN-ERDTMAN, G. and H. THEORELL: Studies in the Tropolone Series. III. Infra-red Spectra. Acta Chem. Scand. 4, 1490 (1950).

22. BARGER, G. and O. DORRER: Chemical Properties of Puberulic Acid, $C_8H_6O_6$, and a Yellow Acid, $C_8H_4O_6$. Biochemic. J. 28, 11 (1934).

23. BARLTROP, J. A. and J. S. NICHOLSON: The Oxidation Products of Phenols. I. The Structure of Purpurogallin. J. Chem. Soc. (London) 1948, 116.

24. BARTEK, J., T. MUKAI, T. NOZOE and F. ŠANTAVÝ: Polarography of Tropone and Some of its Derivatives. Collect. Czech. Chem. Communs. 19, 885 (1954).

25. BARTELS-KEITH, J. R. and A. W. JOHNSON: Synthesis of Tropolone Derivatives. Chem. and Ind. 1950, 677.

26. BARTELS-KEITH, J. R., A. W. JOHNSON and W. I. TAYLOR: Synthesis of Stipitatic Acid. Chem. and Ind. 1951, 337.

27. BARTON, N., J. W. COOK and J. D. LOUDON: Colchicine and Related Compounds. Part V. The Structure of Windaus's Deaminocolchinol Methyl Ether. J. Chem. Soc. (London) 1945, 176.

28. BELLET, P., G. AMIARD, M. PESEZ and A. PETIT: Colchicoside. II. Partial Synthesis and Constitution. Ann. pharm. franç. 10, 241 (1952) [Chem. Abstr. 47, 3323 (1953)].

29. BIRKINSHAW, J. H., A. R. CHAMBERS and H. RAISTRICK: Studies in the Biochemistry of Micro-organisms. 70. Stipitatic Acid, $C_8H_6O_5$, a Metabolic Product of *Penicillium stipitatum* THOM. Biochemic. J. 36, 242 (1942).

30. BIRKINSHAW, J. H. and H. RAISTRICK: Studies in the Biochemistry of Micro-organisms. XXIII. Puberulic Acid $C_8H_6O_6$ and an Acid $C_8H_4O_6$, New Products of the Metabolism of Glucose by *Penicillium puberulum* BAINIER and *Penicillium aurantio-virens* BIOURGE. With an Appendix on Certain Dihydroxybenzenedicarboxylic Acids. Biochemic. J. 26, 441 (1932).

31. BRDIČKA, R.: The Polarographic Analogy of γ-Thujaplicin with Colchiceine. Ark. Kemi, Mineral. Geol. 26 B, No. 19 (1948).

32. BROWN, R. D.: *cyclo*Heptatrienone, *cyclo*Pentadienone, and γ-Pyrone. J. Chem. Soc. (London) 1951, 2670.

33. BRUCE, J. M.: The Formation of Melanin from 5 : 6-Dihydroxyindole. Chem. and Ind. 1954, 310.

34. BUCHANAN, G. L., J. W. COOK and J. D. LOUDON: Colchicine and Related Compounds. Part IV. Synthesis of 2 : 3 : 4 : 5-, 2 : 3 : 4 : 6-, and 2 : 3 : 4 : 7-Tetramethoxy-9-methylphenanthrenes. J. Chem. Soc. (London) 1944, 325.

35. BURSIAN, K.: Über das Colchicin. Ber. dtsch. chem. Ges. 71, 245 (1938).

36. CAMPBELL, R. B. and J. M. ROBERTSON: The Structure of Nootkatin. An X-Ray Determination. Chem. and Ind. 1952, 1266.

37. Carlsson, B., H. Erdtman, A. Frank and W. E. Harvey: The Chemistry of the Natural Order *Cuppressales*. VIII. Heartwood Constituents of *Chamae-cyparis nootkatensis*. Carvacrol, Nootkatin, and Chamic Acid. Acta Chem. Scand. 6, 690 (1952).

38. Caunt, D., W. D. Crow and R. D. Haworth: Purpurogallin. Part V. Some Improvements in Synthetical Methods. J. Chem. Soc. (London) 1951, 1313.

39. Caunt, D., W. D. Crow, R. D. Haworth and C. A. Vodoz: Purpurogallin. Part III. Synthesis of Purpurogallin and some Analogues. J. Chem. Soc. (London) 1950, 1631.

40. Čech, J. and F. Šantavý: The Effect of Hydrogen Peroxide in Alkaline Medium on Colchiceine. Collect. Czech. Chem. Communs. 14, 532 (1949).

41. Coffey, R. S., R. B. Johns and A. W. Johnson: Synthesis of 4-Hydroxy-tropone. Chem. and Ind. 1955, 658.

42. Cohen, A., J. W. Cook and E. M. F. Roe: Colchicine and Related Compounds. Part I. Some Observations on the Structure of Colchicine. J. Chem. Soc. (London) 1940, 194.

43. Cook, J. W., A. R. M. Gibb and R. A. Raphael: Tropolones. Part III. Halogenated Derivatives of Tropolone. J. Chem. Soc. (London) 1951, 2244.

44. Cook, J. W., A. R. M. Gibb, R. A. Raphael and A. R. Somerville: The Synthesis of Tropolone. Chem. and Ind. 1950, 427.

45. — — — — Tropolones. Part I. The Preparation and General Characteristics of Tropolone. J. Chem. Soc. (London) 1951, 503.

46. — — — — Tropolones. Part IV. The Preparation of α,β-Benzotropolone. J. Chem. Soc. (London) 1952, 603.

46a. Cook, J. W., W. Graham and (in part) A. Cohen, R. W. Lapsley and C. A. Lawrence. Colchicine and Related Compounds. Part III. J. Chem. Soc. (London) 1944, 322.

47. Cook, J. W., J. Jack, J. D. Loudon, G. L. Buchanan and J. MacMillan: Colchicine and Related Compounds. Part XI. Synthesis of N-Acetylcolchinol Methyl Ether. J. Chem. Soc. (London) 1951, 1397.

48. Cook, J. W. and J. D. Loudon: Colchicine. In: R. H. F. Manske and H. L. Holmes: The Alkaloids. Vol. II, p. 261. New York: Academic Press. 1952.

49. Cook, J. W., J. D. Loudon and D. K. V. Steel: Pyridotropolone. Chem. and Ind. 1952, 562.

50. — — — Tropolones. Part VII. Orientation in Tropolone. J. Chem. Soc. (London) 1954, 530.

51. Cook, J. W., R. A. Raphael and A. I. Scott: Tropolones. Part II. The Synthesis of α-, β-, and γ-Thujaplicins. J. Chem. Soc. (London) 1951, 695.

52. — — — Tropolones. Part VI. Further Reactions of Tropolone. J. Chem. Soc. (London) 1952, 4416.

53. Cook, J. W. and A. R. Somerville: Synthesis of 3 : 4-Benztropolone. Nature (London) 163, 410 (1949).

54. Corbett, R. E., C. H. Hassall, A. W. Johnson and A. R. Todd: Puberulic and Puberulonic Acids. Part I. The Molecular Formula of Puberulonic Acid and Consideration of Possible Benzenoid Structures for the Acids. J. Chem. Soc. (London) 1950, 1.

55. Corbett, R. E., A. W. Johnson and A. R. Todd: Puberulic and Puberulonic Acids. Part II. Structure. J. Chem. Soc. (London) 1950, 6.

56. — — — The Structure of Stipitatic Acid. J. Chem. Soc. (London) 1950, 147.

57. Corbett, R. E. and D. E. Wright: Heartwood Extractives. Part I. The **Extractives** of *Libocedrus Bidwillii* and *Cupressus Macrocarpa*. Chem. and Ind. 1953, 1258.

58. CORRODI, H. und E. HARDEGGER: Die Konfiguration des Colchicins und verwandter Verbindungen. Helv. Chim. Acta **38**, 2030 (1955).

59. CROW, W. D., R. D. HAWORTH and P. R. JEFFERIES: Purpurogallin. Part X. Further Studies on the Oxidation Products of Purpurogallin and Purpurogallincarboxylic Acid. J. Chem. Soc. (London) **1952**, 3705.

60. DAUBEN, H. J., Jr. and H. J. RINGOLD: Synthesis of Tropone. J. Amer. Chem. Soc. **73**, 876 (1951).

61. DEAN, H. F. und M. NIERENSTEIN: Über Purpurogallin. II. Ber. dtsch. chem. Ges. **46**, 3868 (1913).

62. DEWAR, M. J. S.: Structure of Stipitatic Acid. Nature (London) **155**, 50 (1945).

63. — Structure of Colchicine. Nature (London) **155**, 141 (1945).

64. — Structure of Colchicine. Nature (London) **155**, 479 (1945).

65. — Tropolone. Nature (London) **166**, 790 (1950).

66. DOERING, W. v. E. and D. B. DENNEY: The Structural Fate of the Carbonyl Carbon Atom in the Tropolone-Benzoic Acid Rearrangement. J. Amer. Chem. Soc. **77**, 4619 (1955).

67. DOERING, W. v. E. and F. L. DETERT: Cycloheptatrienylium Oxide. J. Amer. Chem. Soc. **73**, 876 (1951).

68. DOERING, W. v. E. and C. F. HISKEY: Reactions of Halotropones and Related Compounds. J. Amer. Chem. Soc. **74**, 5688 (1952).

69. DOERING, W. v. E. and L. H. KNOX: Synthesis of Tropolone. J. Amer. Chem. Soc. **72**, 2305 (1950).

70. — — Tropolone. J. Amer. Chem. Soc. **73**, 828 (1951).

71. — — Rearrangement of Halotropones. Chloride Exchange in Tribromotropolone. J. Amer. Chem. Soc. **74**, 5683 (1952).

72. — — Synthesis of Substituted Tropolones. J. Amer. Chem. Soc. **75**, 297 (1953).

73. — — The Cycloheptatrienylium (Tropylium) Ion. J. Amer. Chem. Soc. **76**, 3203 (1954).

74. DOERING, W. v. E. and J. R. MAYER: The Synthesis of α-Phenyltropolone from Tropolone. J. Amer. Chem. Soc. **75**, 2387 (1953).

75. DOERING, W. v. E. and A. A. SAYIGH: On the Reported Rearrangement of 2,4,7-Tribromotropone to 3,5-Dibromobenzamide. J. Amer. Chem. Soc. **76**, 39 (1954).

76. DUFF, S. R. and H. ERDTMAN: The Chemistry of the Natural Order Cupressales. X. Nootkatin. Chem. and Ind. **1954**, 432.

77. DUNITZ, J. D.: Purpurogallin: by X-Ray Fourier Synthesis. Nature (London) **169**, 1087 (1952).

78. ERDTMAN, H.: Chemistry of some Heartwood Constituents of Conifers and their Physiological and Taxonomic Significance. Progress Organ. Chem. **1**, 22 (1952).

79. ERDTMAN, H. and J. GRIPENBERG: Antibiotic Substances from the Heart Wood of *Thuja plicata* D. DON. II. The Constitution of γ-Thujaplicin. Acta Chem. Scand. **2**, 625 (1948).

80. — — Antibiotic Substances from the Heart Wood of *Thuja plicata* D. DON. Nature (London) **161**, 719 (1948).

81. ERDTMAN, H. and W. E. HARVEY: The Chemistry of the Natural Order Cupressales. IX. Nootkatin. Chem. and Ind. **1952**, 1267.

82. FABIAN, J., V. DELAROFF, P. POIRIER et M. LEGRAND: Étude spectrale ultraviolette et infrarouge des colchicines et apparentés. Bull. soc. chim. France **1955**, 1455.

83. FATELEY, W. G. and E. R. LIPPINCOTT: The Vibrational Spectrum and Structure of the Tropylium Ion. J. Amer. Chem. Soc. **77**, 249 (1955).

84. Fernholz, H.: Über die Umlagerung des Colchicins mit Natriumalkoholat und die Struktur des Ringes C. Liebigs Ann. Chem. **568**, 63 (1950).

85. Fernholz, H., E. Hartwig und J. Ch. Salfeld: Einige Untersuchungen an Tropolonen und Vergleiche mit dem Colchicin. Liebigs Ann. Chem. **576**, 131 (1952).

86. Forbes, E. J.: Colchicine and Related Compounds. Part XIV. Structure of β- and γ-Lumicolchicine. J. Chem. Soc. (London) **1955**, 3864.

86 a. — The Structure of Colchicine and Isocolchicine. Chem. and Ind. **1956**, 192.

87. Fourneau, J.-P.: Dérivés de la colchicine. I. Colchaminones et désacétyl-colchaminones. Bull. soc. chim. France **1955**, 1569.

88. Fourneau, J.-P. et I. Grundland: Dérivés de la colchicine. II. Amino-2 colchimidazoles (imino-2 colchimidazolines). Bull. soc. chim. France **1955**, 1571.

89. Gillam, A. E. and E. S. Stern: An Introduction to Electronic Absorption Spectroscopy in Organic Chemistry. London: Arnold Publ. Ltd. 1954.

90. Grewe, R.: Über die Jod-methoxy-phthalsäure aus Colchicin. Ber. dtsch. chem. Ges. **71**, 907 (1938).

91. Grewe, R. und W. Wulf: Die Umwandlung des Colchicins durch Sonnenlicht. Ber. dtsch. chem. Ges. **84**, 621 (1951).

92. Gripenberg, J.: Antibiotic Substances from the Heart Wood of *Thuja plicata* D. Don. III. The Constitution of α-Thujaplicin. Acta Chem. Scand. **2**, 639 (1948).

93. — The Constituents of the Wood of *Thuja occidentalis* L. Acta Chem. Scand. **3**, 782 (1949).

94. Hartwell, J. L., M. V. Nadkarni and J. Leiter: N-Substituted Colchicein-amides. J. Amer. Chem. Soc. **74**, 3180 (1952).

95. Haworth, R. D. and J. D. Hobson: Synthesis of Tropolone. Chem. and Ind. **1950**, 441.

96. — — Purpurogallin. Part IV. Some Properties of Tropolones. J. Chem. Soc. (London) **1951**, 561.

97. Haworth, R. D. and P. R. Jefferies: Purpurogallin. Part VIII. Nitration and Nitrosation of β-Methyltropolone. J. Chem. Soc. (London) **1951**, 2067.

98. Haworth, R. D., B. P. Moore and P. L. Pauson: Purpurogallin. Part I. J. Chem. Soc. (London) **1948**, 1045.

99. — — — Purpurogallin. Part II. Synthesis of Purpurogallone. J. Chem. Soc. (London) **1949**, 3271.

100. Haworth, R. D. and P. B. Tinker: Purpurogallin. Part XII. The Action of Organometallic Compounds on the Tropolones. J. Chem. Soc. (London) **1955**, 911.

100 a. Heyer, W. und W. Treibs: Über polycyclische Tropolone. II. Das 3,4-Furo-tropolon und 5-Oxy-3,4-thiopheno-tropolon. Liebigs Ann. Chem. **595**, 203 (1955).

101. Horowitz, R. M. and G. E. Ullyot: Colchicine. Some Reactions of Ring C. J. Amer. Chem. Soc. **74**, 587 (1952).

102. Horowitz, R., G. E. Ullyot, E. C. Horning, M. G. Horning, J. Koo, M. S. Fish, J. A. Parker and G. N. Walker: Colchicine. Nature of the B-Ring. J. Amer. Chem. Soc. **72**, 4330 (1950).

103. Jambor, B., J. Bartek und F. Šantavý: Polarographie des β-Oxytropons. Collect. Czech. Chem. Communs. **20**, 1244 (1955).

104. Johns, R. B., A. W. Johnson and J. Murray: Synthetic Experiments in the *cyclo*Heptatrienone Series. Part III. Syntheses of Puberulic Acid and *iso*-Stipitatic Acid. J. Chem. Soc. (London) **1954**, 198.

105. Johns, R. B., A. W. Johnson and M. Tišler: Synthetic Experiments in the *cyclo*Heptatrienone Series. Part IV. 3-Hydroxytropone. J. Chem. Soc. (London) **1954**, 4605.

106. JOHNSON, A. W., N. SHEPPARD and A. R. TODD: Puberulic and Puberulonic Acids. Part III. The Structure of Puberulonic Acid. J. Chem. Soc. (London) **1951**, 1139.

107. JOHNSON, A. W. and M. TIŠLER: Synthetic Experiments in the *cyclo*Heptatrienone Series. Part VI. Further Reactions of 3-Hydroxytropone. J. Chem. Soc. (London) **1955**, 1841.

108. KEMP, A. D. and D. S. TARBELL: Studies on the Structure of Colchicine. Reduction Products from Ring C. J. Amer. Chem. Soc. **72**, 243 (1950).

109. KING, M. V., J. L. DEVRIES and R. PEPINSKY: An X-Ray Diffraction Determination of the Chemical Structure of Colchicine. Acta Crystallogr. **5**, 437 (1952).

110. KINUMAKI, S., K. AIDA and Y. IKEGAMI: The Infrared Spectra of Tropoid Compounds. Part I. Tropone and Tropolone. Science Repts. Research Insts. Tôhoku Univ. **A 8**, (1956) (in press).

111. KISELEV, V. V. and G. P. MENSHIKOV: Chemical Properties of Colchamine. C. R. (Doklady) Acad. Sci. (USSR) **88**, 825 (1953) [Chem. Abstr. **48**, 3952 (1954)].

112. KITAHARA, Y.: Anionoid Reactions of Bromotropolones. Science Repts. Tôhoku Univ. [1] **39** (1955) (in press).

113. KITAHARA, Y. and T. ARAI: Studies on Tropolone Derivatives. IX. Iodination of Tropolone. Proc. Japan Acad. **27**, 423 (1951).

114. KNIGHT, J. D. and D. J. CRAM: Mold Metabolites. VI. The Synthesis of Tropolone. J. Amer. Chem. Soc. **73**, 4136 (1951).

115. KUBO, M., Y. KURITA and M. KIMURA: The Molecular Structure of Tropolone. In: Y. ASAMI and K. HIGASHI, Molecular Structure and Related Problems. (Monograph Ser. Res. Inst. Appl. Electr., No. 4.) Sapporo Japan. 1954.

116. KURITA, Y.: Dipole Moments of Troponoids. Science Repts. Tôhoku Univ. [1] **38**, 85 (1954).

117. KURITA, Y. and M. KUBO: Molecular Orbital Treatment of Tropolone. Bull. Chem. Soc. Japan **24**, 13 (1951).

118. LADENBURG, A.: Die Constitution des Atropins. Liebigs Ann. Chem. **217**, 74 (1883).

119. LETTRÉ, H.: Zur Konstitution des Colchicins. Angew. Chem. **59**, 218 (1947).

120. LETTRÉ, H., H. FERNHOLZ und E. HARTWIG: Zur Kenntnis der Tribromocolchiceinsäure. Liebigs Ann. Chem. **576**, 147 (1952).

121. MAČÁK, V., I. BARTOSOVA et F. ŠANTAVÝ: Chromatographie sur papier de quelques alcaloïdes des colchicées. Substances tirées du colchique et leurs dérivés. Comm. XLI. Ann. pharm. franç. **12**, 555 (1954).

122. MERLING, G.: Über Tropin. Ber. dtsch. chem. Ges. **24**, 3108 (1891).

123. MOROE, T., T. MATSUSHIMA and O. MIYAMOTO: Synthesis of 3,4-Benzotropolone. J. Pharmac. Soc. Japan **72**, 1238 (1952).

124. MUKAI, T.: On the Bromination of 2-Phenyltropone. Science Repts. Tôhoku Univ. [1] **38**, 280 (1954).

125. MULLER, G. et L. VELLUZ: Sur l'emplacement du carbonyle troponique dans la colchicine. Bull. soc. chim. France **1955**, 1452.

126. NAKATSUKA, T.: The Occurrence of α- and β-Thujaplicin in the Essential Oil from the Wood of *Thuja occidentalis* L. J. Japan Forest. Soc. **34**, 248 (1952).

127. — The Occurrence of Hinokitiol in the Wood of *Thuja standishii* CARR. J. Japan Forest. Soc. **35**, 291 (1953).

128. NAKATSUKA, T. and Y. HIROSE: Occurrence of Nootkatin and Hinokitiol in the Wood of *Juniperus chinensis* L. J. Japan Forest. Soc. **37**, 196 (1955).

129. Nicholls, G. A. and D. S. Tarbell: β,γ-Benzotropolone and Related Compounds. J. Amer. Chem. Soc. **74**, 4935 (1952).

130. — — Colchicine and Related Compounds. J. Amer. Chem. Soc. **75**, 1104 (1953).

131. Nierenstein, M.: The Colouring Matter of the Red Pea Gall. J. Chem. Soc. (London) **115**, 1328 (1919).

132. Nierenstein, M. and A. Swanton: The Colouring Matters of Galls. Biochemic. J. **38**, 373 (1944).

133. Nozoe, T.: Über die Farbstoffe im Holzteile des „Hinoki"-Baumes. I. Hinokitin und Hinokitiol. Bull. Chem. Soc. Japan **11**, 295 (1936).

134. — Studies on Hinokitiol. Part III. General Survey. Science of Drugs (Japan) **3**, 174 (1949); Science Repts. Tôhoku Univ. [1] **34**, 199 (1950).

135. — General Considerations on the Substitution Reactions of Hinokitiol. Proc. Japan Acad. **26**, No. 9, 30 (1950).

136. — Substitution Products of Tropolone and Allied Compounds. Nature (London) **167**, 1055 (1951).

137. Nozoe, T., H. Akino and K. Sato: On o'-Amino-p-Nitrohinokitiol. Proc. Japan Acad. **27**, 565 (1951).

138. Nozoe, T., S. Ebine, S. Ito and I. Takasu: On Hinopurpurins, the Rearrangement Products of Azohinokitiols. I. Proc. Japan Acad. **27**, 197 (1951).

139. Nozoe, T., T. Ikemi and S. Ito: Some Derivatives from Colchicine. Proc. Japan Acad. **30**, 609 (1954).

140. Nozoe, T., T. Ikemi and T. Ozeki: Structure of Hinopurpurin. Proc. Japan Acad. **31**, 445 (1955).

141. Nozoe, T., S. Ito and K. Matsui: 1,2,3-Triazazulene Derivatives. Proc. Japan Acad. **30**, 313 (1954).

142. Nozoe, T., S. Ito, K. Matsui and T. Ozeki: Action of Hydrogen Peroxide on Troponoids. II. Tropolone, β-Methyltropolone, and α- and γ-Thujaplicins. Proc. Japan Acad. **30**, 604 (1954).

143. Nozoe, T. and S. Katsura: On the Structure of Hinokitiol. J. Pharmac. Soc. Japan **64**, 181 (1944).

144. Nozoe, T. and Y. Kitahara: Anionoid Substitution of 3-Bromotropolone. Proc. Japan Acad. **30**, 204 (1954).

145. Nozoe, T., Y. Kitahara and H. Abe: Hydroxytropones. I. 2,4,7-Tribromo-5-hydroxytropone. Proc. Japan Acad. **29**, 347 (1953).

146. Nozoe, T., Y. Kitahara and T. Ando: On the Derivatives of 3 : 4-Benzotropolone. Proc. Japan Acad. **27**, 107 (1951).

147. Nozoe, T., Y. Kitahara, T. Ando and E. Kunioka: Studies on Tropolone Derivatives. VIII. Nitro Derivatives of Tropolone. II. Proc. Japan Acad. **27**, 231 (1951).

148. Nozoe, T., Y. Kitahara, T. Ando and S. Masamune: Tropone and Its Derivatives. I. Syntheses of Bromo Derivatives of Tropone and Tropolone by the Direct Bromination of Cycloheptanone. Proc. Japan Acad. **27**, 415 (1951).

149. Nozoe, T., Y. Kitahara, T. Ando, S. Masamune and H. Abe: Nucleophilic Substitution of Troponoid Compounds. I. 2,4,7-Tribromotropone. Science Repts. Tôhoku Univ. [1] **36**, 166 (1952).

150. Nozoe, T., Y. Kitahara and T. Arai: Synthesis of 3-Phenyl-1-azazulene. Proc. Japan Acad. **30**, 478 (1954).

151. Nozoe, T., Y. Kitahara and K. Doi: Determination of Substituent Positions in Hinokitiol Derivatives. I. Proc. Japan Acad. **26**, No. 10, 25 (1950).

152. — — — Determination of Substituent Positions in Hinokitiol Derivatives. II. o- and o'-Monoaminohinokitiol. Proc. Japan Acad. **27**, 156 (1951).

153. NOZOE, T., Y. KITAHARA and K. DOI: On a New Type of Aromatization by the Diazotization of o-Aminotropolone Derivatives. J. Amer. Chem. Soc. **73**, 1895 (1951).

154. — — — Diazo Reaction of o-Aminohinokitiols. Proc. Japan Acad. **27**, 282 (1951).

155. NOZOE, T., Y. KITAHARA, K. DOI and S. MASAMUNE: Syntheses of m-Styryl-tropolones and 1-(m-Tropolonyl)-3-phenylpropenes. Proc. Japan Acad. **28**, 291 (1952).

156. NOZOE, T., Y. KITAHARA, K. DOI, S. MASAMUNE, M. ENDO, M. ISHII and S. JE-GYUN: Styryltropolone and Its Derivatives. Science Repts. Tôhoku Univ. [1] **38**, 257 (1954).

157. NOZOE, T., Y. KITAHARA and S. ITO: On the Synthesis of α-Thujaplicin (o-Iso-propyltropolone). Proc. Japan Acad. **26**, No. 7, 47 (1950).

158. NOZOE, T., Y. KITAHARA, E. KUNIOKA and K. DOI: On Mononitrohinokitiols. Proc. Japan Acad. **26**, No. 9, 38 (1950).

159. NOZOE, T., Y. KITAHARA and S. MASAMUNE: Rearrangement Reactions of Bromotropolones with Basic Reagents. Proc. Japan Acad. **27**, 649 (1951).

160. NOZOE, T., Y. KITAHARA, K. YAMANE and T. IKEMI: On Derivatives of α-Thujaplicin. I. Proc. Japan Acad. **27**, 193 (1951).

161. NOZOE T., Y. KITAHARA, K. YAMANE and K. YAMAKI: On Dinitroisohino-kitiol. Proc. Japan Acad. **26**, No. 8, 14 (1950).

162. NOZOE, T., Y. KITAHARA, K. YAMANE and A. YOSHIKOSHI: Studies of Tropolone Derivatives. I. Substitution Products of α-Monobromotropolone. Proc. Japan Acad. **27**, 18 (1951).

163. NOZOE, T., S. MATSUMURA, Y. MURASE and S. SETO: Synthesis of 2-Amino-azulene Derivatives from 2-Halogenotropones. Chem. and Ind. **1955**, 1257.

164. NOZOE, T., T. MUKAI, Y. IKEGAMI and T. TODA: Synthesis and Properties of 4-Hydroxytropone. Chem. and Ind. **1955**, 66.

165. NOZOE, T., T. MUKAI, M. KUNORI, T. MUROI and K. MATSUI: o-, m- and p-Methyltropolones and their Derivatives. Science Repts. Tôhoku Univ. [1] **35**, 242 (1951).

166. NOZOE, T., T. MUKAI, J. MINEGISHI and T. FUJISAWA: On 2-Phenyltropone and 3-Phenyltropolone. Science Repts. Tôhoku Univ. [1] **37**, 388 (1953).

167. NOZOE, T., T. MUKAI and I. MURATA: Rearrangement Reactions of Tropolone Methyl Ethers by Grignard Reagents. Proc. Japan Acad. **28**, 142 (1952).

168. — — — 1,3-Diazazulene. J. Amer. Chem. Soc. **76**, 3352 (1954).

169. NOZOE, T., T. MUKAI and K. TAKASE: On Azo-coupling of Halogenohino-kitiols. Proc. Japan Acad. **27**, 236 (1951).

170. — — — Hydroxymethyl Compounds of Tropolone, 4-Methyltropolone and 4-Isopropyltropolone, and their Reduction Products. Science Repts. Tôhoku Univ. [1] **36**, 40 (1952).

171. — — — On Tropone. Science Repts. Tôhoku Univ. [1] **40**, (1955) (in press).

172. NOZOE, T., T. MUKAI, K. TAKASE and A. MATSUKUMA: Chlorobromo and Polybromo Substitution Products of Hinokitiol. Proc. Japan Acad. **27**, 152 (1951).

173. NOZOE, T., T. MUKAI, K. TAKASE, I. MURATA and K. MATSUMOTO: Reaction of Guanidine on Tropolone Methyl Ether. Proc. Japan Acad. **29**, 452 (1953).

174. NOZOE, T., T. MUKAI, K. TAKASE and T. NAGASE: Tropone and Its Derivatives. VII. On Tropone (Cycloheptatrienone). Proc. Japan Acad. **28**, 477 (1952).

175. NOZOE, T., M. SATO, S. ITO, K. MATSUI and T. MATSUDA: 6-Amino-1,3-di-azazulene. Proc. Japan Acad. **29**, 565 (1953).

176. Nozoe, T., M. Sato, S. Ito, K. Matsui and T. Ozeki: Action of Hydrogen Peroxide on Tropolones. I. Hinokitiol. Science Repts. Tôhoku Univ. [1] **39,** 190 (1955).

177. Nozoe, T., M. Sato and T. Matsuda: Action of Various Amines and Ketonic Agents on 5-Nitrosotropolone. Science Repts. Tôhoku Univ. [1] **37,** 407 (1953).

178. Nozoe, T., M. Sato and K. Matsui: Nucleophilic Substitution of Troponoid Compounds. II. Mercaptotropones and Mercaptotropolones. Science Repts. Tôhoku Univ. [1] **37,** 211 (1953).

179. Nozoe, T., S. Seto, S. Ebine and S. Ito: On p-Aminotropolone. J. Amer. Chem. Soc. **73,** 1895 (1951).

180. Nozoe, T., S. Seto and T. Ikemi: Diels-Alder Reaction of Tropolones with Maleic Anhydride. Proc. Japan Acad. **27,** 655 (1951).

181. Nozoe, T., S. Seto, T. Ikemi and T. Arai: On Sulfonic Acid Derivatives of Hinokitiol. Proc. Japan Acad. **26,** No. 9, 50 (1950).

182. — — — — Studies on Tropolone Derivatives. II. Sulfonic Acid Derivatives. Proc. Japan Acad. **27,** 24 (1951).

183. — — — — Studies on Tropolone Derivatives. III. On Methyl Ethers of Tropolone and Its Halogen Derivatives. Proc. Japan Acad. **27,** 102 (1951).

184. Nozoe, T., S. Seto, T. Ikemi, T. Sato and K. Watanabe: Synthesis of 3-(p-Methoxyphenyl)-tropolone. Science Repts. Tôhoku Univ. [1] **38,** 130 (1954).

185. Nozoe, T., S. Seto, S. Ito, M. Sato and T. Katono: On Hydroxyl Derivatives of Tropolone, α-Thujaplicin and Hinokitiol, and the Attempted Preparation of the so-called ,,Tropoquinones". Science Repts. Tôhoku Univ. [1] **37,** 191 (1953).

186. Nozoe, T., S. Seto, K. Kikuchi, T. Mukai, S. Matsumoto and M. Murase: On the Synthesis of Hinokitiol (m-Isopropyltropolone). Proc. Japan Acad. **26,** No. 7, 43 (1950).

187. Nozoe, T., S. Seto, K. Kikuchi and H. Takeda: On the Synthesis of γ-Thujaplicin (p-Isopropyltropolone). Proc. Japan Acad. **27,** 146 (1951).

188. Nozoe, T., S. Seto, Y. Kitahara, M. Kunori and Y. Nakayama: On the Synthesis of Tropolone (Cycloheptatrienolone). Proc. Japan Acad. **26,** No. 7, 38 (1950).

189. Nozoe, T., S. Seto, M. Kunori and T. Sato: On the Aromatization Occurring during the Nitrosation of Hinokitiol. Proc. Japan Acad. **28,** 89 (1952).

190. Nozoe, T., S. Seto, S. Matsumura and T. Terasawa: The Synthesis of 1-Aza-azulan-2-one and Its Electrophilic Substitution. Chem. and Ind. **1954,** 1356.

191. — — — — The Synthesis of 1-Aza-azulene and Its Derivatives. Chem. and Ind. **1954,** 1357.

192. Nozoe, T., S. Seto, T. Mukai, K. Yamane and A. Matsukuma: Studies on Tropolone Derivatives. VII. Halogen Derivatives of Tropolone. (II.) Proc. Japan Acad. **27,** 224 (1951).

193. Nozoe, T., S. Seto and T. Sato: Reactions of Isomeric Isopropyl-2-halotropones with Alcoholic Ammonia. Proc. Japan Acad. **30,** 473 (1954).

194. Nozoe, T., S. Seto, H. Takeda, S. Morosawa and K. Matsumoto: On 2-Aminotropones. Science Repts. Tôhoku Univ. [1] **36,** 126 (1952).

195. Nozoe, T., S. Seto, H. Takeda and T. Sato: On p-Nitrosotropolone. Science Repts. Tôhoku Univ. [1] **35,** 274 (1951).

196. Nozoe, T., A. Yasue and K. Yamane: On the Acidic Constituents of the Essential Oil of *Thujopsis dolabrata.* Occurrence of α-Thujaplicin. Proc. Japan Acad. **27,** 15 (1951).

197. Nozoe, T. and co-workers: unpublished data.

198. Perkin, A. G.: Purpurogallin. Part II. J. Chem. Soc. (London) **101**, 803 (1912).

199. Potěšilová, H., I. Bartosova et F. Šantavý: Identification d'alcaloïdes dans quelques colchicées très rares. Substances tirées du colchique et leurs dérivés. Comm. XLII. Ann. pharm. franç. **12**, 616 (1954).

200. Raffauf, R. F., E. E. Bumbier and G. E. Ullyot: The Specific Rotation of Isocolchicine. J. Amer. Chem. Soc. **76** 1707 (1954).

201. Raistrick, H.: A Region of Biosynthesis. Proc. Roy. Soc. (London) **A 199**, 141 (1949).

202. Rapoport, H., J. E. Campion and J. E. Gordon: An Alternative Degradation of Colchicine to Octahydrodemethoxydesoxydesacetamidocolchicine and Hexahydrodemethoxydesacetamidocolchicine. J. Amer. Chem. Soc. **77**, 2389 (1955).

203. Rapoport, H., A. R. Williams, J. E. Campion and E. P. Donald: The Degradation of Colchicine to Octahydrodemethoxydesoxydesacetamidocolchicine. J. Amer. Chem. Soc. **76**, 3693 (1954).

204. Rapoport, H., A. R. Williams and M. E. Cisney: The Synthesis of *dl*-Colchinol Methyl Ether. J. Amer. Chem. Soc. **73**, 1414 (1951).

205. Robertson, J. M.: The Crystal Structure of Cupric Tropolone and the Dimensions of the Tropolone Ring. J. Chem. Soc. (London) **1951**, 1222.

206. Robinson, R.: Biogenesis of Tropolones (Cycloheptatrienolones). Proc. Roy. Soc. (London) **B 138**, 15 (1951).

207. Sakan, T. and M. Nakazaki: The Synthesis of Dibenzotropolone. J. Inst. Polytechnic Osaka City Univ. **1**, 23 (1950).

208. Šantavý, F.: Préparation de l'acide colchicique à partir de la colchicine. Helv. Chim. Acta **31**, 821 (1948).

209. — Polarographic Determination of Colchicine. Pharm. Acta Helv. **23**, 380 (1948).

210. — Polarography and Spectrography of Colchicine, Colchiceine and Similar Substances. Collect. Czech. Chem. Communs. **14**, 145 (1949).

211. — Isolation of New Substances from the Flowers and Pericarps of Meadow Saffron (*Colchicum autumnale* L.). Colchicum Compounds and their Derivatives, 13th Comm. Collect. Czech. Chem. Communs. **15**, 552 (1950).

212. — Isolierung neuer Stoffe aus den Knollen der Herbstzeitlose, *Colchicum autumnale* L. Substanzen der Herbstzeitlose und ihre Derivate, 14. Mitt. Pharm. Acta Helv. **25**, 248 (1950).

213. — Substanzen der Herbstzeitlose und ihre Derivate. XXII. Photochemische Produkte des Colchicins und einige seiner Derivate. Collect. Czech. Chem. Communs. **16**, 665 (1951).

214. Šantavý, F., Z. Hoščálková, R. Podivínský and H. Potěšilová: Substances of *Colchicum autumnale* and their Derivatives. XXXVIII. Isolation of Further Compounds from the Corms of *Colchicum autumnale*. Chem. Listy **48**, 886 (1954).

214a. Šantavý, F. and V. Mačák: Substances of *Colchicum autumnale* and their Derivatives. XXXIV. Isolation of Further Substances from the Flowers of *Colchicum autumnale*. Chem. Listy **47**, 1214 (1953).

215. Šantavý, F. und T. Reichstein: Isolierung neuer Stoffe aus den Samen der Herbstzeitlose *Colchicum autumnale* L. Substanzen der Herbstzeitlose und ihre Derivate, 12. Mitt. Helv. Chim. Acta **33**, 1606 (1950).

216 Šantavý, F., M. Talaš und O. Tělupilová: Substanzen der Herbstzeitlose und ihre Derivate. XXVIIIb. Beitrag zur Konstitution der Substanzen C und E_1. Collect. Czech. Chem. Communs **18**, 710 (1953).

217. Šantavý, F., R. Winkler und T. Reichstein: Zur Konstitution von Deme-
     colcin (Substanz F) aus *Colchicum autumnale* L. Substanzen der Herbstzeitlose
     und ihre Derivate, 35. Mitt. Helv. Chim. Acta 36, 1319 (1953).
218. Sasada, Y., K. Osaki and I. Nitta: The Crystal Structure of Tropolone
     Hydrochloride. Acta Crystallogr. 7, 113 (1954).
219. Schaeppi, W. H., R. W. Schmid, E. Heilbronner und A. Eschenmoser:
     Untersuchungen in der Benztropylium-Reihe. I. Das 2,3,4,4-Tetramethoxy-
     benztropylium-Kation. Helv. Chim. Acta 38, 1874 (1955).
220. Scott, G. P. and D. S. Tarbell: Studies in the Structure of Colchicine. An
     Infrared Study of Colchicine Derivatives and Related Compounds. J. Amer.
     Chem. Soc. 72, 240 (1950).
221. Sebe, E. and Y. Itsuno: Diels-Alder Reaction of Troponoids. II. Reaction
     of Tropolone and Tropolone Methyl Ether. Proc. Japan Acad. 29, 107 (1953).
222. Sebe, E. and C. Osako: Diels-Alder Reaction of Troponoids. I. Reaction of
     Hinokitiol (*m*-Isopropyltropolone). Proc. Japan Acad. 28, 282 (1952).
223. Seto, S.: Improvement of the Synthetic Method for 2-Halotropones and their
     Reactions. I. Improvement of the Synthetic Method of 2-Halotropones and
     the Synthesis of 2,3-Dihalotropones. Science Repts. Tôhoku Univ. [1] 37,
     275 (1953).
224 — Improvement of the Synthetic Method for 2-Halotropones and their
     Reactions. II. Application of the Synthetic Reactions of 2-Halotropones (1).
     Structural Determination of the Hydrazino Compounds derived from Hino-
     kitiol; Comparison with the Stevens-McFadyen-Method. Science Repts.
     Tôhoku Univ. [1] 37, 286 (1953).
225. — Improvement of the Synthetic Method for 2-Halotropones and their
     Reactions. III. Application of the Synthetic Reaction of 2-Halotropones (2).
     Determination of the Structure of α-Thujaplicin Methyl Ethers. Science
     Repts. Tôhoku Univ. [1] 37, 292 (1953).
226. — Improvement of the Synthetic Method for 2-Halotropones and their
     Reactions. IV. Application of the Synthetic Method of 2-Halotropones (3).
     Derivation of γ-Thujaplicin from Hinokitiol. Science Repts. Tôhoku Univ.
     [1] 37, 297 (1953).
227. — Improvement of the Synthetic Method for 2-Halotropones and their Reac-
     tions. V. On the Structure of the Condensates of 2-Chlorotropone with Ethyl
     Acetoacetate or Ethyl Malonate. Science Repts. Tôhoku Univ. [1] 37, 367 (1953).
228. — Improvement of the Synthetic Method for 2-Halotropones and their
     Reactions. VI. Alkaline Reaction of 2,7- and 2,3-Dihalotropones. Science
     Repts. Tôhoku Univ. [1] 37, 377 (1953).
229. Slack, R. and C. F. Attridge: Pyridotropolones. Chem. and Ind. 1952, 471.
230. Sorkin, M.: *iso*-Colchicin. Helv. Chim. Acta 29, 246 (1946).
230a. Tanaka, N., S. Sato and T. Nozoe: Polarography of Troponoids. I. (un-
     published).
231. Tarbell, D. S. and J. C. Bill: The Properties of 4,5-Benztropolone and
     Related Compounds, as Compared to those of Colchicine and other Tropolones.
     J. Amer. Chem. Soc. 74, 1234 (1952).
232. Tarbell, D. S., H. R. Frank and P. E. Fanta: Studies on the Structure of
     Colchicine. J. Amer. Chem. Soc. 68, 502 (1946).
233. Tarbell, D. S., G. P. Scott and A. D. Kemp: 4,5-Benztropolone and Related
     Compounds. J. Amer. Chem. Soc. 72, 379 (1950).
234. Tarbell, D. S., R. F. Smith and V. Boekelheide: Synthetic Studies on the
     Colchicine Problem. The Preparation and Properties of Some Styryltropolones.
     J. Amer. Chem. Soc. 76, 2470 (1954).

235. TREIBS, W. und W. HEYER: Über ein Oxyfuro-tropolon. Chem. Ber. **87**, 1197 (1954).

236. UENO, Y.: On the Alkaloids of a Meadow Saffron. Proc. Japan Acad. **29**, 266 (1953).

237. UFFER, A.: Über Colchiceinamide. Helv. Chim. Acta **35**, 2135 (1952).

238. UFFER, A., O. SCHINDLER, F. ŠANTAVÝ und T. REICHSTEIN: Teilsynthese des Demecolcins und einiger anderer Colchicinderivate. Substanzen der Herbstzeitlose und ihre Derivate, 36. Mitt. Helv. Chim. Acta **37**, 18 (1954).

239. VELLUZ, L. et G. MULLER: La thiocolchicine. Bull. soc. chim. France **1954**, 755.

240. — — La thiocolchicine. IV. Étude de deux isomères. Bull. soc. chim. France **1955**, 198.

240a. WALKER, G. N.: Hydrogenation of Purpurogallin and Its Derivatives. J. Amer. Chem. Soc. **77**, 6699 (1955).

241. WILLSTÄTTER, R.: Synthese des Tropidins (vorl. Mitt.). Ber. dtsch. chem. Ges. **34**, 129 (1901).

242. WILLSTÄTTER, R. und H. HEISS: Über die Konstitution des Purpurogallins. Liebigs Ann. Chem. **433**, 17 (1923).

243. WINDAUS, A.: Untersuchungen über die Konstitution des Colchicins. Liebigs Ann. Chem. **439**, 59 (1924).

244. YAMANE, K. and S. MOROSAWA: On Derivatives of α-Thujaplicin. II. On Nitroderivatives of α-Thujaplicin (o-Isopropyltropolone). Bull. Chem. Soc. Japan **27**, 18 (1954).

245. ZAVARIN, E. and A. B. ANDERSON: Extractive Components from Incense-Cedar Heartwood (*Libocedrus decurrens* TORREY). I. Occurrence of Carvacrol, Hydrothymoquinone, and Thymoquinone. J. Organ. Chem. (USA) **20**, 82 (1955).

246. — — Paper Chromatography of the Tropolones of Cupressaceae. J. Organ. Chem. (USA) **21**, 332 (1956).

247. — — Extrahierbare Bestandteile des Kernholzes der Kalifornischen Fluß-zeder (Incense-Cedar, *Libocedrus decurrens* TORREY). IV. Vorkommen und Chromatographie von Thujaplicinen. Chem. Ber. **89**, 545 (1956).

248. ZEISEL, S. und K. R. STOCKERT: Über einige bromhaltige Abkömmlinge des Colchicins. Monatsh. Chem. **34**, 1339 (1913).

Addendum.

249. KLOSTER-JENSEN, E., N. TARKÖY, A. ESCHENMOSER und E. HEILBRONNER: Untersuchungen in der Benztropylium-Reihe. III. 2,7-Polymethylen-4,5-benz-tropone. Helv. Chim. Acta **39**, 786 (1956).

250. SCHMID, R. W., E. KLOSTER-JENSEN, E. KOVÁTS und E. HEILBRONNER: Untersuchungen in der Benztropylium-Reihe. IV. Die Bildungsenthalpien des 2,7-Dimethyl-4,5-benztropons, des 2,7-Dodecamethylen-4,5-benztropons und des 2,7-Pentamethylen-4,5-benztropons. Helv. Chim. Acta **39**, 806 (1956).

(Received, February 23, 1956.)

Alkaloids Related to Anthranilic Acid.

By J. R. PRICE, Melbourne, Australia.

With 2 Figures.

Contents.

I. Introduction.

Attention has frequently been drawn to the existence of recurring patterns in the molecular architecture of natural products and these patterns have been made the basis for hypothetical schemes of biogénesis of the molecules in which they occur. Some of these schemes are now being put to the test by means of mutant studies and isotopic tracer techniques and, although the results will doubtless reveal differences in detail between the hypothetical and the experimentally determined mechanisms of synthesis, there is no doubt in the minds of most organic chemists that similar structural patterns will be proved to reflect similar metabolic pathways. In the following pages the chemistry of a number of alkaloids having a certain structural feature in common will be discussed. This structural feature, representing sometimes only a small portion of the molecule, has the skeletal pattern (I), and for convenience will be referred to as the anthranilic acid unit.

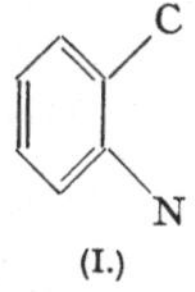

(I.)

Some support for the belief that this anthranilic acid pattern is of biogenetic significance is found in the distribution of these alkaloids, a great many of them occurring in the same plant family, the Rutaceae. It has been shown, however, that anthranilic acid is intimately bound up with tryptophan metabolism so that the substances dealt with in this review are to be regarded — biogenetically — as related to the much larger and more widespread group of indole alkaloids. Nevertheless, they are not indoles* but quinoline and quinazoline derivatives and therefore warrant separate treatment. The subject of *Cinchona* alkaloids, which constitute a biogenetic link between the quinoline and indole groups, is too extensive for inclusion in the present discussion.

II. Anthranilic Acid Derivatives.

In *Neurospora* and other microorganisms it has been established that anthranilic acid is a precursor for the synthesis of tryptophan, with indole as an intermediate. Should a similar mechanism operate in the higher plants it might account for the infrequency with which anthranilic acid, as such, has been found. It is accumulated, however, by a certain maize mutant and has been isolated from the tassels by TEAS and ANDERSON (*109*). Methyl anthranilate — an important component of a number of essential oils used in perfumery — on the other hand, is known to occur in several plant families, notably the Rutaceae *(Citrus, Eriostemon)* but also the Amaryllidaceae (jonquil, tuberose), Anonaceae (ylang ylang oil), Cruciferae *(Cheiranthus)*, Oleaceae (jasmine), Rubiaceae *(Gardenia)*, and possibly Zingiberaceae. It is accompanied in certain of these oils by indole. Methyl N-methylanthranilate is found in some *Citrus* oils and in oil of rue (*53, 113*). Anthranilic acid also occurs in ester combination in *Aconitum* and *Delphinium* alkaloids, the amino group being acylated by aliphatic acids. Controlled hydrolysis of methyl-lycaconitine, for example, gives 1-methyl-succinylanthranilic acid (*51*).

Degradation, as well as synthesis, of tryptophan in lower organisms has been shown to involve anthranilic acid and a similar pattern has been demonstrated in animals (*38*). A number of the intermediates in the degradation, such as kynurenine, come within the scope of this discussion but the only one of particular interest in alkaloid chemistry is 3-hydroxyanthranilic acid. This is formed from hydroxy-kynurenine and evidently undergoes oxidative ring cleavage between carbon atoms 3 and 4 to an aldehydic intermediate, recyclization of which leads to nicotinic acid (*116*).

(II.) Damascenine.

Damascenine. 3-Hydroxyanthranilic acid occurs in *Nigella damascena* in the form of its trimethyl derivative, methyl 3-methoxy-2-methyl-anthranilate, known as damascenine (II).

The structure of damascenine was established by EWINS (*47*) by synthesis from 2-nitro-3-methoxybenzoic acid. LEETE, MARION and SPENSER (*72*) fed tryptophan-3-C^{14} to mature *Nigella damascena* plants but no radioactivity was detected in the damascenine isolated from them.

---

* In some instances such as evodiamine, the anthranilic acid moiety is combined with a separate indole nucleus.

The significance of this result is not clear; it raises doubt as to whether tryptophan metabolism is the same in all higher plants as in the animals and microorganisms which have been studied, and emphasizes the need for more detailed investigation. The excretion of 3-hydroxy-anthranilic acid in small amounts in human urine has recently been reported (*78*).

III. Simple Quinoline Derivatives.

Echinopsine. 1-Methyl-4-quinolone (III) has been isolated by GRESHOFF (*52*) from the seeds of *Echinops ritro* and a number of other *Echinops* species and given the name echinopsine. The structure was established by SPÄTH and KOLBE (*100*). The substance is a weak base, which is converted to 1-methyl-1,2,3,4-tetrahydro-quinoline on reduction with sodium and alcohol or to 4-chloroquinoline by heating with phosphorus pentachloride and phosphorus oxychloride at 150°. SPÄTH and KOLBE synthesized the alkaloid by heating 4-hydroxy-quinoline with methyl iodide. PRICE (*90*) obtained 1-methyl-4-quinolone by decarboxylating 1-methyl-4-quinolone-3-carboxylic acid resulting from oxidation of a number of acridone alkaloids. 1-Methyl-4-quinolone is nitrated in the 3-position by boiling 68% nitric acid; prolonged treatment gives 3,6-dinitro-1-methyl-4-quinolone which, like 1,3,5-trinitrobenzene, is soluble in alkali and is oxidized in alkaline solution to 4-hydroxy-3,6-dinitro-1-methyl-2-quinolone. Further oxidation of this carbostyril derivative yields 5-nitro-N-methyl-anthranilic acid.

(III.)
Echinopsine.

BEATTIE (*18*) examined normal and fasciated plants of *Syndesmon thalictroides* growing within a short distance of each other and found that, while the former were devoid of alkaloids, the latter contained a mixture of 3-methylquinoline-4-carboxylic acid and the methyl and ethyl esters of *iso*carbostyril-3-carboxylic acid. No subsequent examination of this plant has been recorded, though reinvestigation of such an interesting co-occurrence of quinoline and *iso*quinoline derivatives is certainly desirable.

Flindersine, first isolated from the wood of *Flindersia australis* (Rutaceae) by MATTHES and SCHREIBER (*76*) and assigned by them the molecular formula $C_{23}H_{26}O_7N_2$, was later shown by BROWN, HOBBS, HUGHES and RITCHIE (*27*) to be correctly represented as $C_{14}H_{13}O_2N$ and to possess structure (IV). Permanganate oxidation of flindersine gives flindersinic acid $C_{14}H_{13}O_6N$ (V) which is hydrolysed by dilute acid to 4-hydroxy-2-quinolone, α-hydroxy*iso*butyric acid and carbon dioxide. Phosphorus oxychloride converts flindersine to chlorodesoxy-flindersine which can be oxidized to chlorodesoxy-flindersinic acid, $C_{14}H_{12}O_5NCl$ (VI) and this yields the same products as flindersinic acid on acid hydrolysis. Hydrogenation of chlorodesoxy-flindersinic acid removes the chlorine atom to give an acid $C_{14}H_{13}O_5N$ (VII) which is pyrolysed smoothly to 4-hydroxy-quinoline-3-carboxylic acid, thus establishing the angular position for the dimethylpyran ring.

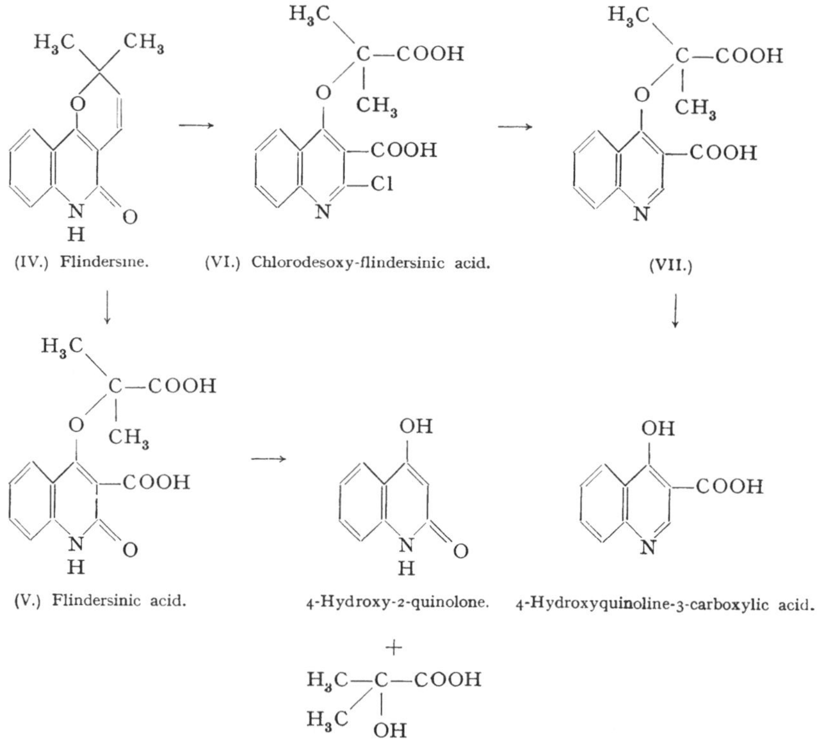

(IV.) Flindersine. (VI.) Chlorodesoxy-flindersinic acid. (VII.)

(V.) Flindersinic acid. 4-Hydroxy-2-quinolone. 4-Hydroxyquinoline-3-carboxylic acid.

This conclusion was confirmed in several ways but only after a number of stumbling blocks had been overcome. For example, crystallization of flindersinic acid from ethanol gives rise to the ethyl ester of 4-hydroxy-2-quinolone-3-carboxylic acid, while the dialdehyde obtained by periodate oxidation of the *cis*-glycol formed by treating flindersine with osmium tetroxide gives, on acid hydrolysis, 4-hydroxy-2-quinolone-3-aldehyde and acetoin, the latter arising by rearrangement of the initially formed α-hydroxy*iso*butyraldehyde.

Flindersine has been synthesized by BROWN, HUGHES and RITCHIE (*28*) as follows. Esterification of 4-hydroxy-2-quinolone with β,β-dimethylacryloyl chloride gives the ester (VIII) which is isomerized by aluminium chloride to the dihydropyrone (IX). Catalytic reduction of (IX) gives the pyranol (X) and this on dehydration is converted to flindersine.

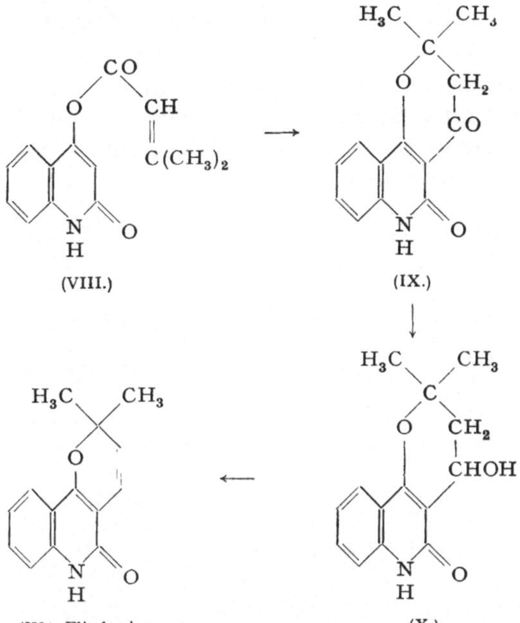

(VIII.)   (IX.)

(IV.) Flindersine.   (X.)

Alkaloids of Angostura. The most extensively investigated group of simple quinoline alkaloids are the Angostura bases, from *Galipea* or *Cusparia* species (Rutaceae), the structures of which are due essentially to SPÄTH and his collaborators (*98, 99, 105—107*). The major alkaloids, *cusparine* and *galipine*, resemble one another closely. Both, like 2- or 4-methoxyquinoline, are isomerized on heating with methyl iodide, losing one O-methyl group and gaining an N-methyl group. Both, in different ways, have been degraded to quinoline and both, by alkali fusion to protocatechuic acid. Oxidation of galipine sulphate by permanganate gives veratric acid and an acid $C_{11}H_9O_3N$, containing one methoxyl group, which has been shown to be 4-methoxyquinoline-2-carboxylic acid. Structure (XI) has been established for galipine by condensing veratric aldehyde with 4-methoxy-2-methylquinoline in the presence of zinc chloride to give the veratrylidene derivative (XII) which undergoes catalytic hydrogenation to (XI), identical with the naturally occurring alkaloid.

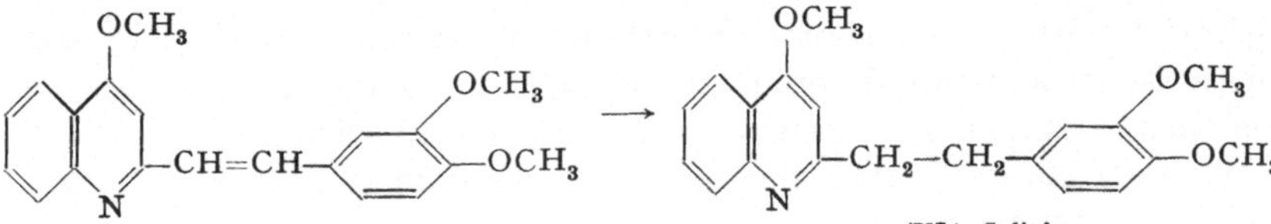

(XII.)   (XI.) Galipine.

A similar synthesis, using piperonal in place of veratric aldehyde, gives cusparine. The phenolic alkaloid, *galipoline* (XIV), which yields galipine on methylation has been synthesized by condensing 4-chloro-2-methyl-quinoline with veratric aldehyde. Replacement of the chlorine atom by the benzyloxy group gives (XIII) which on hydrogenation and acid hydrolysis is converted to galipoline.

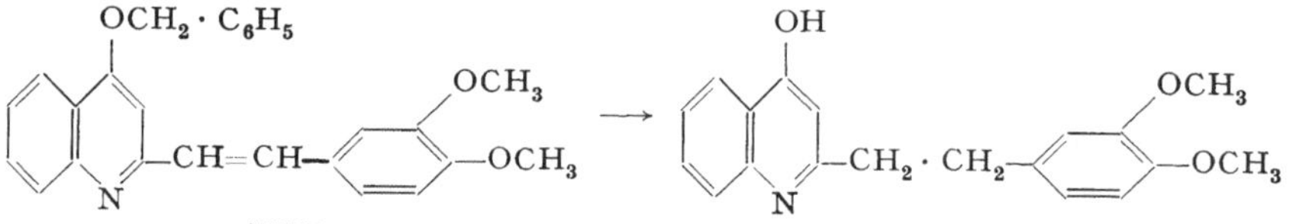

(XIII.)

(XIV.) Galipoline.

A fourth alkaloid, *cuspareine*, has been shown by SCHLÄGER and LEEB (*93*) to belong to the same group. Mild oxidation with permanga-

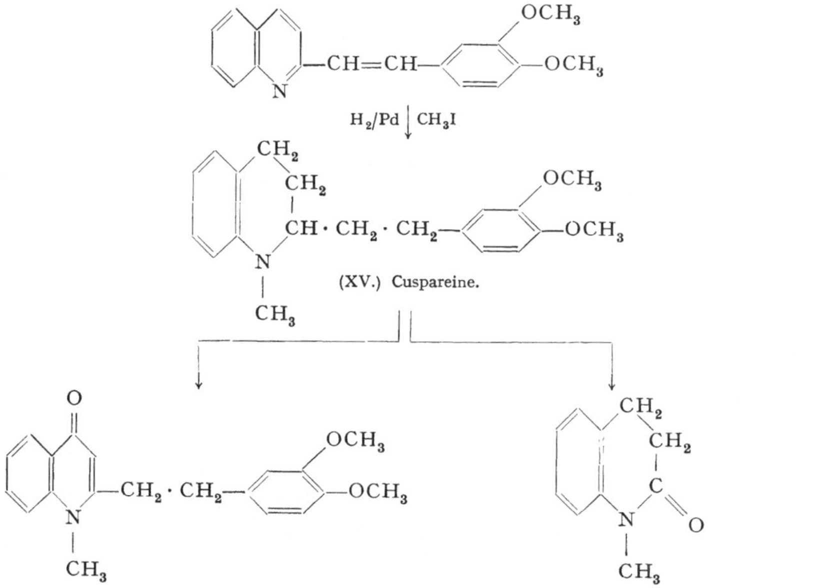

(XV.) Cuspareine.

(XVI.) *iso*Galipine.

(XVII.) 1-Methyl-1,2,3,4-tetrahydroquinol-2-one.

nate in acetone converts cuspareine (XV) to *iso*galipine (XVI), obtained by heating galipine with methyl iodide. More vigorous oxidation of cuspareine with permanganate in hot acetone gives veratric acid and 1-methyl-1,2,3,4-tetrahydroquinol-2-one (XVII). Structure (XV) deduced from these reactions has been confirmed by synthesizing *DL*-cuspareine by condensation of veratric aldehyde and quinaldine followed by hydrogenation and N-methylation.

The synthesis of *nor*cuspareine has also been effected from either *o*-nitrobenzalacetone or veratralacetone by condensing with veratric aldehyde or *o*-nitrobenzaldehyde respectively, followed by hydrogenation of the resulting *o*-nitrobenzalveratral-acetone.

Considerable interest attaches to the *minor alkaloids* isolated from Angostura bark by SPÄTH and his collaborators. These comprise quinoline, 2-methylquinoline, 1-methyl-2-quinolone, 2-*n*-amylquinoline and 4-methoxy-2-*n*-amylquinoline. The two last-named compounds have been synthesized by conventional methods.

The structural relations evident between these alkaloids and the four substituted phenylethylquinoline derivatives led SPÄTH and PIKL (*106, 107*) to suggest that Angostura alkaloids are produced in the plant from a common precursor, probably an anthranilic acid derivative. SCHÖPF and LEHMANN (*94*) have extended this idea, suggesting *o*-aminobenzaldehyde, because of its greater reactivity, as a more probable precursor than anthranilic acid.

In support they have examined the condensation of *o*-aminobenzaldehyde with methyl ketones under so-called "physiological conditions", that is, in dilute aqueous solution at room temperature and at p_H values as near neutrality as possible. It is found that condensation takes place only at high p_H values and that with methyl *n*-amyl ketone the "wrong" methylene group reacts, resulting in the formation of 2-methyl-3-*n*-butylquinoline. But with β-keto acids reaction in the desired sense proceeds smoothly under nearly neutral conditions. Thus, a 75% yield of 2-*n*-amylquinoline results from the reaction of $M/80$ caproylacetic acid at p_H 8 in ten days. SCHÖPF and THIERFELDER (*97*) later suggested that the 4-methoxyquinoline alkaloids may result from condensation of *o*-aminobenzoylacetic acid with an appropriate aldehyde, followed by dehydrogenation and methylation. For example:

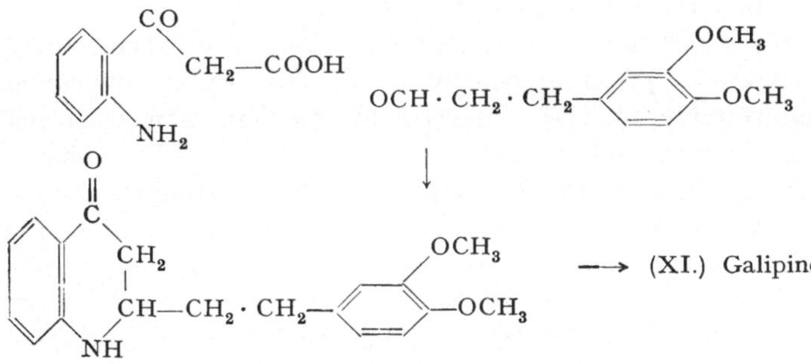

CO—CH$_2$—COOH ... NH$_2$ + OCH·CH$_2$·CH$_2$— (OCH$_3$, OCH$_3$) $\downarrow$ C(=O)—CH$_2$—CH—CH$_2$·CH$_2$— (OCH$_3$, OCH$_3$), NH $\longrightarrow$ (XI.) Galipine (p. 307).

ROBINSON (*92*), suggests the oxidation product (XVIII) — nearer still to the tryptophan precursor — as a more probable intermediate.

Alkaloids of *Lunasia amara*. The alkaloids of *Lunasia amara* (Rutaceae) from the Philippines and Indonesia have been studied by several groups of workers (*57*), and a number of them described but the structures not determined. Recently, JOHNSTONE, PRICE and TODD (*64*) working with a northern Australian form of *L. amara* have found as the principal basic constituent a previously unreported alkaloid $C_{17}H_{15}O_2N$ and shown it to be *1-methyl-2-phenyl-7-methoxy-4-quinolone* (XIX). Nitric acid

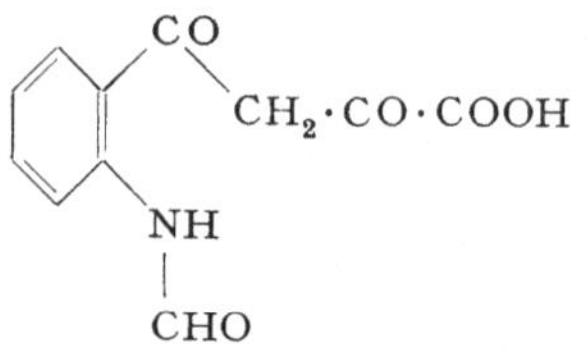

(XVIII.)
(XIX.) 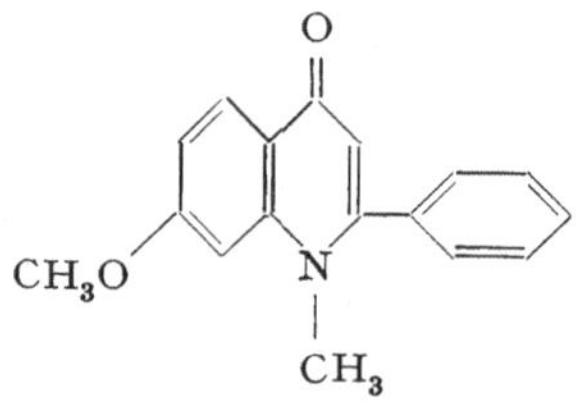
(XX.) 3-Hydroxy-2,4,6-trinitrophenyl-methyl-nitramine. 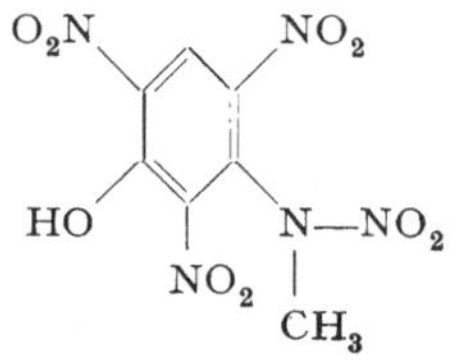

oxidation gives benzoic acid and 3-hydroxy-2,4,6-trinitrophenyl-methyl-nitramine (XX) together with the mono- and dinitro derivatives of the alkaloid. The nitro-derivatives undergo hydrolytic ring cleavage to what are evidently ω-nitroacetophenones and benzoic acid. The hydrochloride of the alkaloid is not demethylated by heating to 250°, consequently the methoxyl group cannot be located *peri* to a quinolone carbonyl group. These data, though not conclusive, suggest that the alkaloid has structure (XIX) and this has been established by synthesis.

7-Methoxy-4-hydroxy-2-phenylquinoline had been synthesized by ELDERFIELD et al. (*45*) from benzoyl-*m*-anisidide (XXI) by conversion to the chloro-imide (XXII), followed by condensation with malonic ester, saponification and decarboxylation. The possibility that cyclization might give rise to the alternative 5-methoxy compound had been eliminated by conversion of the 4-hydroxyquinoline to the 4-chloro compound followed by catalytic dehalogenation to 2-phenyl-7-methoxy-quinoline identical with that prepared as described by BORSCHE and WAGNER-ROEMMICH (*20*) by the Doebner synthesis from *m*-anisidine. Methylation of 7-methoxy-4-hydroxy-2-phenylquinoline with dimethyl sulphate and caustic potash (*64*) gives (XIX) identical with the naturally occurring alkaloid.

(XXI.) Benzoyl-*m*-anisidide.

(XXII.)

(XIX.)

Quinoline Derivatives from Microorganisms. Although not derived from higher plants certain simple quinoline derivatives produced by microorganisms warrant mention here. BRACKEN, POCKER and RAISTRICK (*21*) have isolated *cyclopenin*, $C_{17}H_{14}O_3N_2$, from culture filtrates of *Penicillium cyclopium* and shown it to be hydrolysed by dilute mineral acids to *viridicatin*, $C_{15}H_{11}O_2N$, which had previously been isolated by CUNNINGHAM and FREEMAN (*37*) from the dried mycelium of a strain of *P. viridicatum.* Viridicatin gives on methylation an OO- or ON-dimethyl derivative and on acetylation a monoacetyl derivative. Oxidation of viridicatin in ethanolic potash by oxygen gas yields *o*-aminobenzophenone and oxalic acid, or by permanganate in boiling acetone solution gives oxalyl-*o*-aminobenzophenone (XXIII). 3-Hydroxy-4-phenyl-2-quinolone (XXIV), identical with viridicatin, has been synthesized from *o*-aminobenzophenone by heating with chloroacetic anhydride and cyclizing the resulting chloroacetyl dervative (XXV) by aqueous ethanolic alkali.

(XXV.)

(XXIV.) Viridicatin.

(XXIII.)

Cyclopenin, which is optically active, gives on acid hydrolysis one molecule each of viridicatin, methylamine and carbon dioxide. Structures (XXVI) and (XXVII) have been proposed by BRACKEN, POCKER and RAISTRICK (*21*) but EDWARD (*44*) suggests that (XXVIII), in which complete aromatization is blocked, is also worthy of consideration.

(XXVI.)   (XXVII.)   (XXVIII.)

From the culture filtrates of another microorganism, *Pseudomonas aeruginosa*, DOISY and his collaborators (*56, 114, 115*) have isolated five antibiotic substances and shown three of them to be 2-*n*-heptyl-, 2-*n*-nonyl- and 2-(Δ^1-nonenyl)-4-hydroxyquinolines. A crystalline product with the capacity to antagonize the antibacterial action of dihydrostreptomycin was later isolated from culture filtrates of *Ps. pyocyanea* by LIGHTBOWN (*73, 74*). This antagonist has been shown by CORNFORTH and JAMES (*35*) to be a mixture containing as major components 2-*n*-heptyl- and 2-*n*-nonyl-4-hydroxyquinoline-N-oxides, together with some 2-*n*-undecyl-4-hydroxyquinoline-N-oxide, the nonyl compound (XXIX) having the highest activity. These unusual N-oxides have been synthesized by stannous chloride reduction of the α-(*o*-nitrobenzoyl) derivatives of the appropriate β-keto acids. CORNFORTH and JAMES (*35*) suggest the condensation of anthranilic acid with a β-keto acid as a plausible mode of biosynthesis and claim that, since available β-keto acids should in general be even-numbered, such an origin would explain why the side-chains have odd numbers of carbon atoms.

(XXIX.) 2-*n*-Nonyl-4-hydroxyquinoline-N-oxide.

IV. Acridine Alkaloids.

Eleven acridine alkaloids have so far been described, all derivatives of 5-acridone and all from members of the family Rutaceae. The majority are N-methyl-5-acridones and give 1-methyl-4-quinolone-3-carboxylic acid (XXX) on oxidation (*89*). The simplest, *2,4-dimethoxy-10-methyl-5-acridone* (XXXI) isolated by LAMBERTON and PRICE (*71*) from the leaves of *Acronychia baueri*, has been synthesized by DRUMMOND and LAHEY (*42*) by condensing phloramine dimethyl ether with *o*-chlorobenzoic acid, cyclizing the resulting diphenylamine carboxylic acid (XXXII) by means of phosphorus oxychloride, and methylating by heating the acridone potassium salt with dimethyl sulphate.

(XXXII.)

(XXXI.)
2,4-Dimethoxy-10-methyl-5-acridone.

(XXX.) 1-Methyl-4-quinolone-3-carboxylic acid.

Melicopicine from *Melicope fareana*, shown by CROW and PRICE (*36*) to be 1,2,3,4-tetramethoxy-10-methylacridone (XXXIII, p. 314), has likewise been synthesized by HUGHES, NEILL and RITCHIE (*60*) by way of the diphenylamine carboxylic acid, but in this instance cyclization with phosphorus oxychloride gives 1,2,3,4-tetramethoxy-5-chloroacridine. Sodium methoxide in dry methanol converts the chloro compound to 1,2,3,4,5-pentamethoxy-acridine which is rearranged to melicopicine by heating with methyl iodide. The four homonuclear trimethoxy-10-methylacridones have been synthesized by the same route and one of these, *2,3,4-trimethoxy-10-methylacridone*, was subsequently isolated from the leaves of *Evodia alata* by GELL, HUGHES and RITCHIE (*48*). HUGHES and RITCHIE (*62*) have also synthesized this alkaloid by oxidizing, with alkaline ferricyanide, the methosulphate of 2,3,4-trimethoxy-acridine prepared by condensing *o*-aminobenzaldehyde with 1,2,3,5-tetrahydroxy-benzene.

Melicopicine is oxidized by nitric acid to two quinones, (XXXIV) and (XXXV), each of which can be further demethylated by mild alkaline treatment to the same hydroxymethoxyquinone (XXXVI) and this, by hydrobromic acid to the dihydroxyquinone (XXXVII) (cf. next page).

The electron-attracting character of the acridone oxygen atom facilitates nucleophilic attack at carbon atoms 2 and 4 and this is manifest

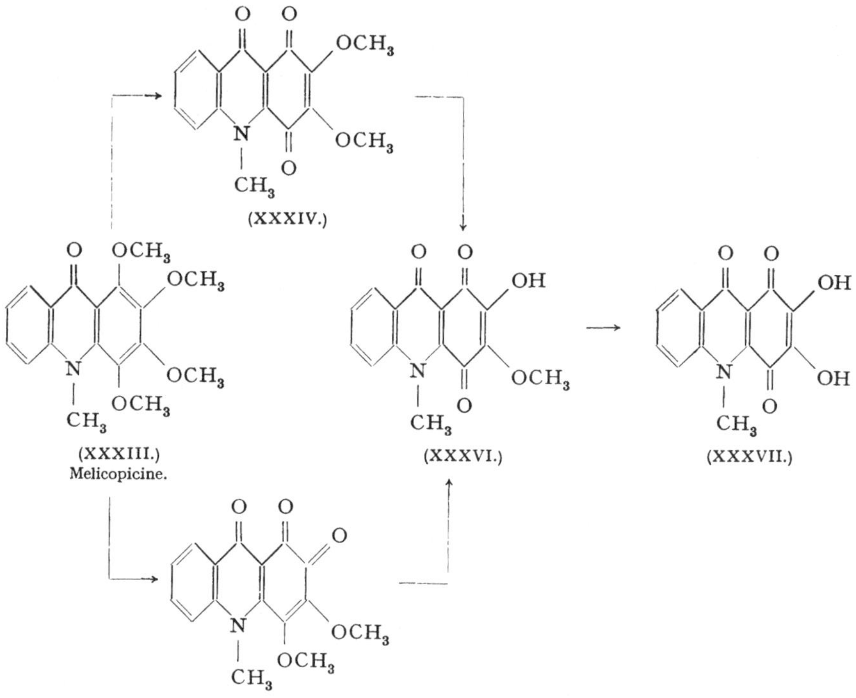

(XXXIV.)

(XXXIII.)
Melicopicine.

(XXXVI.)

(XXXVII.)

(XXXV.)

in the reaction with alcoholic alkali of three of the alkaloids which contain methylenedioxy groups. These alkaloids — *evoxanthine* (2,3-methylenedioxy-4-methoxy-) (59), *melicopine* (1,2-methylenedioxy-3,4-dimethoxy-) (XLI) (36) and *melicopidine* (2,3-methylenedioxy-1,4-dimethoxy-10-methylacridone) (XXXIX) (36) — undergo ether interchange reactions involving fission of the methylenedioxy ring. The products, alkoxyphenols in which the nature of the alkoxy group depends on the alcohol used in the reaction, can be oxidized with nitric acid to quinones. Thus, nitric acid converts the phenol (XXXVIII), obtained from melicopidine (XXXIX) and methanolic potash, to (XXXV), and the corresponding phenol (XL) from melicopine (XLI) and methanolic potash to (XXXIV). The structures

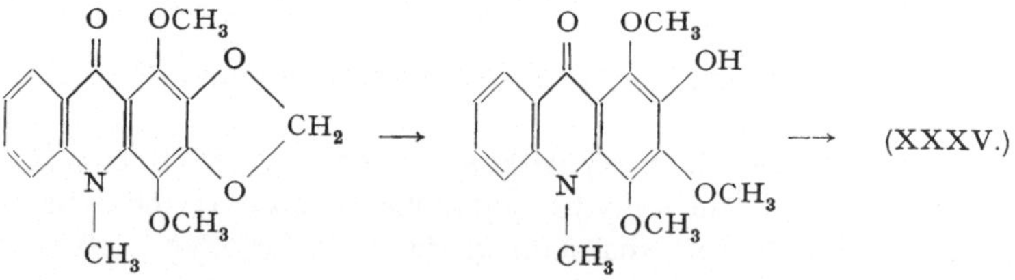

(XXXIX.) Melicopidine.

(XXXVIII.)

(XXXV.)

of melicopine, melicopidine and evoxanthine are established by these reactions and certain modifications of them. Methylation of the dimethoxyphenol resulting from the action of methanolic potash on evoxanthine gave 2,3,4-trimethoxy-10-methylacridone identical with a synthetic specimen.

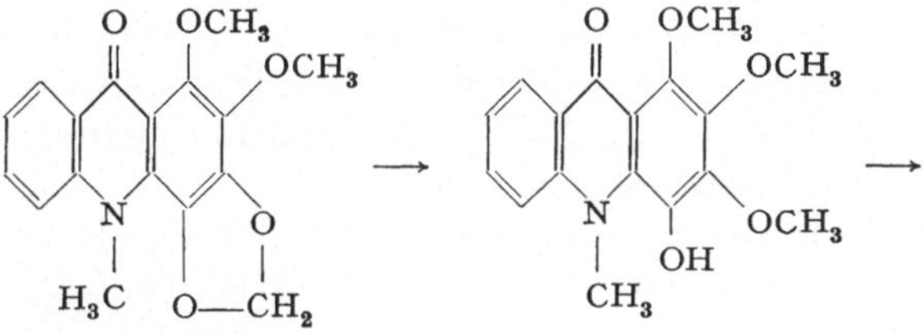

$\longrightarrow$ (XXXIV.)

(XLI.) Melicopine.   (XL.)

Evoxanthidine and *xanthevodine* lack the 10-methyl group and on oxidation give 4-hydroxyquinoline-3-carboxylic acid. Their structures follow from the methylation of evoxanthidine to evoxanthine and of xanthevodine to melicopidine.

Acronycine, $C_{20}H_{19}O_3N$ (*23, 42*), is oxidized by permanganate in acetone to a dibasic acid, $C_{20}H_{19}O_7N$, which readily loses one mole of carbon dioxide. Pyrolysis of the resulting monobasic acid (acronycinic acid) gives α-hydroxy*iso*butyric acid and 2,4-dihydroxy-10-methylacridone, a methoxyl group undergoing demethylation during the reaction. Acronycine, therefore, contains a dimethylpyran ring, a conclusion supported by other reactions. Hydrogenation gives dihydroacronycine; ozonolysis

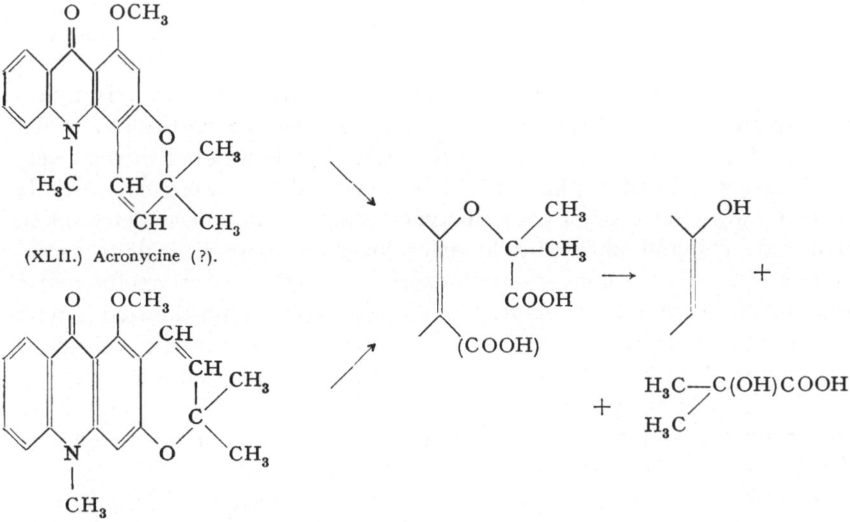

(XLII.) Acronycine (?).

(XLIII.) Acronycine (?).

a phenolic aldehyde, 1- or 3-formyl-2-hydroxy-4-methoxy-10-methyl-acridone; bromination a monobromo-acronycine which is ozonized not to an aldehyde but to an acid showing that the bromine atom is located (as would be expected) on the α-carbon atom. Nitration gives first a mono- then a tri-nitroacronycine which is slowly oxidized by boiling 68% nitric acid to 6-nitro-1-methyl-4-quinolone-3-carboxylic acid. Dihydroacronycine, on the other hand, is rapidly oxidized by nitric acid, under the same conditions as melicopicine etc., to (XXX). A distinction between structures (XLII) and (XLIII) has not yet been made.

Demethylation of the 4-methoxyl group observed during the pyrolysis of acronycinic acid is a characteristic of the acridone alkaloids. In all of them the 4-position is substituted by a methoxyl group, and in all of them this methoxyl group is readily demethylated giving what have been referred to as the "*nor* alkaloids".

The acridones are pale yellow to yellow in colour and are very weak bases. Their mineral acid salts (which are hydrolysed by water) undergo demethylation on heating either at the melting point or in alcoholic solution to give the more highly-coloured 4-hydroxy compounds. These are even weaker bases than the alkaloids and are cryptophenolic in character due to hydrogen bonding between the 4 and 5 oxygen atoms. HUGHES, MATHESON, NORMAN and RITCHIE (*58*), who examined the demethylation of a number of methoxy-acridones and 10-methylacridones, attribute the ease of demethylation to the increased stability of the hydrogen-bonded product.

The two remaining acridones from natural sources are, *4-hydroxy-2,3-dimethoxy-10-methylacridone* and 4-hydroxy-2,3-dimethoxy-acridone *(xanthoxoline)* obtained from the leaves of *Evodia xanthoxyloides* (*29, 61*). Though probably occurring as such in the plant, they could be artifacts caused by the acid treatment involved in the isolation process. 4-Hydroxy-2,3-dimethoxy-10-methylacridone also occurs, together with its methylation product, in the leaves of *Evodia alata*, but is isolated in relatively much larger quantities than would be expected if it were not initially present (*48*). Since 2,3,4-trimethoxyacridone has not been detected in plant extracts, and since 4-methoxyacridones are more difficult to demethylate than 4-methoxy-10-methylacridones (*58*), xanthoxoline also probably does occur in the plant. This alkaloid has been synthesized (*29*) by heating at its melting point the hydrochloride of 2,3,4-trimethoxyacridone obtained by acid hydrolysis of 2,3,4-trimethoxy-5-chloroacridine.

The ultraviolet spectra of a number of the acridone alkaloids have been reported and discussed in detail by BROWN and LAHEY (*24*) [see also LAMBERTON and PRICE (*71*)].

HUGHES and RITCHIE (*62*) have pointed out that in all the known acridine alkaloids, one benzene nucleus is unsubstituted while the second contains alkoxyl substituents in the 2- and 4-positions. These authors investigated the condensation of *o*-aminobenzaldehyde with phenols under "physiological conditions" and found that only phloroglucinol and 1,2,3,5-tetrahydroxybenzene reacted. Pyrogallol, as well as number of mono- and dihydric phenols, were tried without success. The reaction between phloroglucinol and *o*-aminobenzaldehyde, first investigated by ELIASBERG and FRIEDLÄNDER (*46*), proceeds smoothly at room temperatu.e in dilute aqueous solution to give 2,4-dihydroxyacridine (XLIV). Dependence of the yield on the p_H is shown by the following figures:

$$p_H \quad 4 \quad 5 \quad 6 \quad 7 \quad 8 \quad 9 \quad 10 \quad 11 \quad 12 \quad 13$$
$$\text{Yield \%} \quad 0 \quad 0 \quad 5 \quad 28 \quad 90 \quad 87 \quad 87 \quad 83 \quad 81 \quad 43$$

(XLIV.) 2,4-Dihydroxyacridine.

Methylation of 2,4-dihydroxyacridine with diazomethane to the corresponding dimethoxy compound proceeds almost quantitatively. Conversion to methosulphate followed by oxidation with hot alkaline ferricyanide gives 2,4-dimethoxy-10-methylacridone, the yield again being almost quantitative. Nuclear oxidation of 2,4-dihydroxyacridine has not been effected, but 1,2,3,5-tetrahydroxybenzene condenses with *o*-aminobenzaldehyde in hot alkaline solution and the only substance that has been isolated after methylation of the crude reaction product is 1,2,3-trimethoxyacridine, in 40% yield. Oxidation of the methosulphate gives 1,2,3-trimethoxy-10-methylacridone. HUGHES and RITCHIE (*62*) suggest that the biosynthesis of the acridone alkaloids involves the condensation of *o*-aminobenzaldehyde (or its equivalent) with phloroglucinol (or its equivalent) to form 2,4-dihydroxyacridine, and that this is followed by oxidation of the acridine to acridone, nuclear oxidation in the case of tri- or tetra-hydroxy derivatives and N and/or O alkylation. ROBINSON (*92*), however, again points out the possibilities of the tryptophan oxidation product (XVIII, p. 310) or, alternatively, the condensation of anthranilic acid with triacetic acid.

V. Furoquinoline Alkaloids.

Simple Furoquinolines.

The furoquinoline alkaloids, like the acridones, are found only in the Rutaceae, but they are widespread within the family and many occurrences have been reported.

Dictamnine. The simplest member of the group, dictamnine (XLV) was first isolated from *Dictamnus albus* by THOMS (*112*) and later from *Skimmia repens* (*14*) and *Evodia littoralis* (*33*). ASAHINA, OHTA and INUBUSE (*14*) have shown that dictamnine contains one methoxyl group and, like 2- and 4-methoxy quinolines and certain of the Angostura bases, is isomerized by heating with methyl iodide to *iso*dictamnine (XLVI) containing one methylimino but no methoxyl group. The methoxyl group in the alkaloid can be demethylated giving *nor*dictamnine (XLVII), by either acid or alkaline hydrolysis. Methylation, with dimethyl sulphate, converts *nor*dictamnine to *iso*dictamnine; with diazomethane a mixture of dictamnine and *iso*dictamnine is produced (*81*). Dictamnine is oxidized by permanganate in acetone to a mixture of an aldehyde, dictamnal (XLVIII), and an acid, dictamnic acid. Dictamnic acid undergoes demethylation and decarboxylation with boiling hydrochloric acid to 4-hydroxy-2-quinolone. It should, therefore, be either 4-hydroxy-2-methoxy-quinoline-3-carboxylic acid (XLIX) or 4-methoxy-2-quinolone-3-carboxylic acid (L).

ASAHINA, OHTA and INUBUSE synthesized an acid, m. p. 225°, which they believed to be (XLIX), by reduction of dimethyl *o*-nitrobenzoyl malonate with zinc and hydrogen chloride in ethanol, followed by hydrolysis with methanolic potash. They found it to differ from dictamnic acid. They concluded that dictamnic acid has structure (L), and as the second oxygen atom is unreactive and evidently present in an ether linkage, assigned the linear tricyclic structure (XLV) to dictamnine.

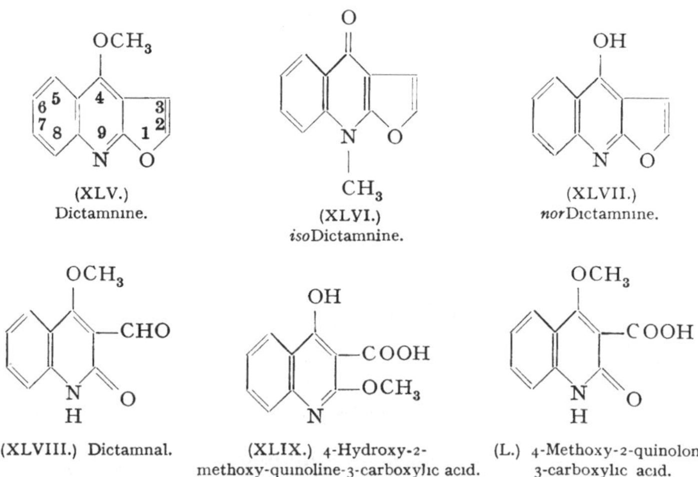

(XLV.) Dictamnine.

(XLVI.) *iso*Dictamnine.

(XLVII.) *nor*Dictamnine.

(XLVIII.) Dictamnal.

(XLIX.) 4-Hydroxy-2-methoxy-quinoline-3-carboxylic acid.

(L.) 4-Methoxy-2-quinolone-3-carboxylic acid.

However, BROWN, HOBBS, HUGHES and RITCHIE (*27*) in connection with the degradation of flindersine (p. 305) have repeated this synthesis using methanol in place of ethanol as solvent. Hydrolysis of the crude product with methanolic potash gave a substance, $C_{11}H_9O_4N$, m. p. 225°,

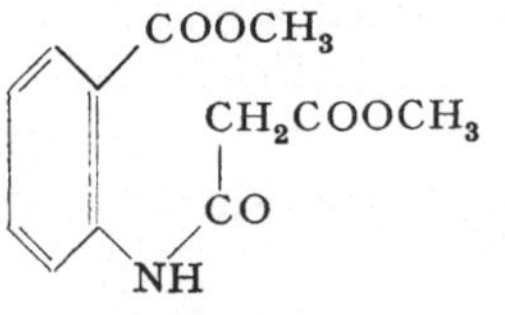

(LII.)

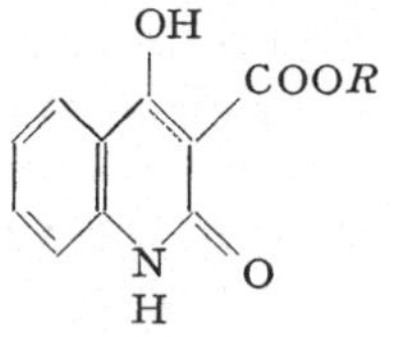

(LI.)
4-Hydroxy-2-quinolone-3-carboxylic acid ester.

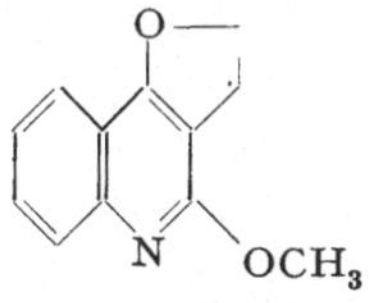

(LIII.)

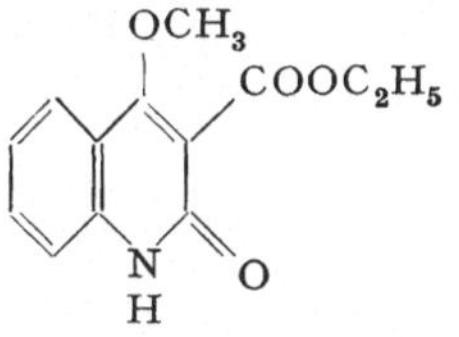

(LIV.)

undoubtedly the same as that obtained by Asahina et al. But this substance proved to be the methyl ester of 4-hydroxy-2-quinolone-3-carboxylic acid (LI; $R = CH_3$) which had been synthesized unequivocally by cyclizing with sodium methoxide in dry methanol the monoamide (LII) prepared from methyl anthranilate and dimethyl malonate.

Since, therefore, the acid synthesized by Asahina, Ohta and Inubuse (*14*) is not (XLIX), their conclusion concerning the structures of dictamnic acid and dictamnine is not necessarily correct and the angular structure for dictamnine (LIII) must be taken into account. However, Arndt, Ergener and Kutlu (*2*) have shown that 4-hydroxy-2-quinolone gives with diazomethane 4-methoxy-2-quinolone, and for this reason the product Brown, Hobbs, Hughes and Ritchie obtained by methylating the ethyl ester of 4-hydroxy-2-quinolone-3-carboxylic acid (LI; $R = C_2H_5$) should be the 4-methoxy compound (LIV)*. Alkaline hydrolysis of (LIV) yields an acid, $C_{11}H_9O_4N$, m. p. 260°, identical with dictamnic acid. Moreover, methylation of dictamnic acid and of the ester (LI; $R = CH_3$) gives the same product, viz. the methyl ester of dictamnic acid. Consequently, dictamnic acid *is* (L) as originally proposed by Asahina and his collaborators, and dictamnine is the linear 4-methoxyfuro-(2,3–b)quinoline (XLV, p. 318).

Earlier, Asahina and Inubuse (*5*) attempted to establish the structure of dictamnine synthetically, but their synthesis is not conclusive as the

---

* This deduction is no doubt sound, but it should be kept in mind that the reaction between quinolones and diazomethane varies according to the nature of the quinolone; compare, for example, skimmianinic and kokusagininic acids (p. 324).

product could be either (XLVI) or (LVII). Starting with 4-hydroxy-2-quinolone-3-aldehyde [*nor*dictamnal (LV)], the α-pyrone (LVI) is prepared by condensation with cyanoacetic acid, hydrolysis and cyclization with sulphuric acid. The bromo derivative obtained after decarboxylation is then converted by alkali to the furan carboxylic acid which, after methylation can be decarboxylated to (LVII) not identical with *iso*dictamnine. A recent attempt by OHTA and MORI (*85*) to synthesize dictamnal from (LV) is likewise indecisive.

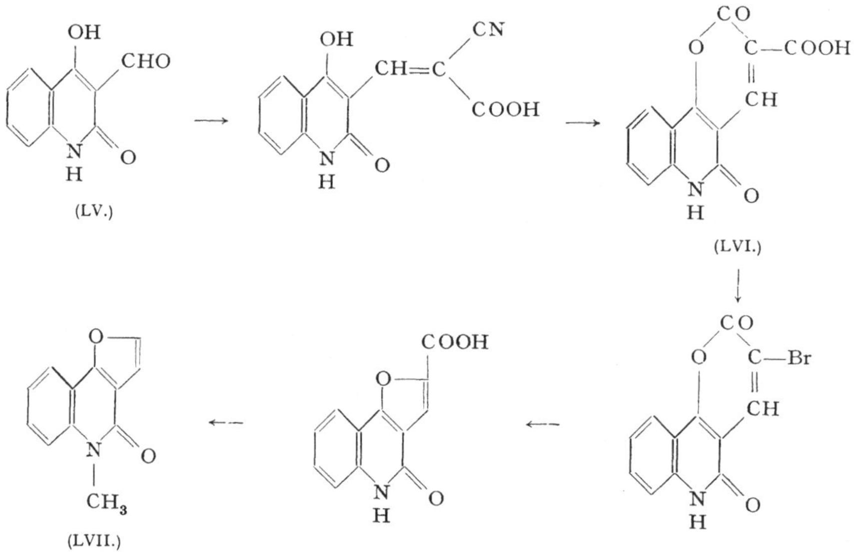

(LV.)   (LVI.)

(LVII.)

The first *synthesis of linear furoquinolines* appears to be that of HAQ, KAPUR and RÂY (*55*) who condensed the lactone (LVIII) prepared from β-3,4-dimethoxybenzoylpropionic acid with *o*-nitrobenzaldehyde, 6-nitro-piperonal and 6-nitroveratraldehyde. Reduction of the arylidene derivatives with zinc dust and acetic acid gave a series of veratryl-furo-quinolines (LIX).

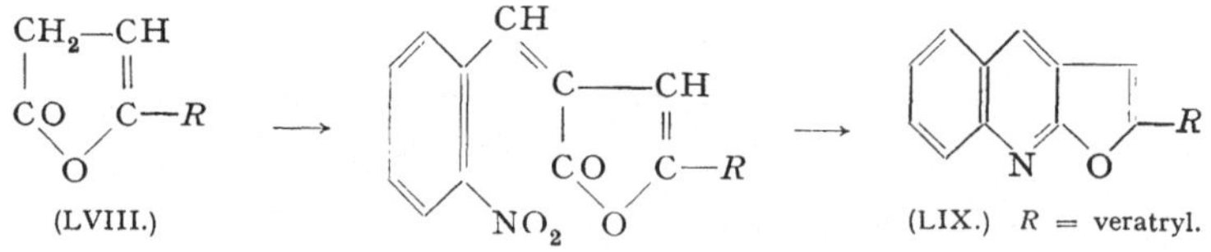

(LVIII.)   (LIX.)   R = veratryl.

However, furo-(2,3–b)quinoline (LXVI) itself has been synthesized by KING, LATHAM and PARTRIDGE (*66*) as follows. Saponification of the ester (LX) obtained by condensing *o*-nitrobenzyl bromide with α-carbeth-oxy-γ-butyrolactone, followed by decarboxylation, gives compound

(LXI). By hydrogenation over Raney nickel (LXI) is converted to 3-ω-hydroxyethyl-dihydrocarbostyril (LXII) and thence with phosphorus pentachloride to the chloroquinoline (LXIII). Hydrolysis with concentrated hydrochloric acid effects conversion to a mixture of the dihydrofuroquinoline (LXV) and 3-ω-chloroethyl-2-quinolone (LXIV). The latter can be cyclized by pyridine/aqueous caustic soda to (LXV) which undergoes dehydrogenation with palladium/charcoal in Dowtherm to the desired furoquinoline (LXVI).

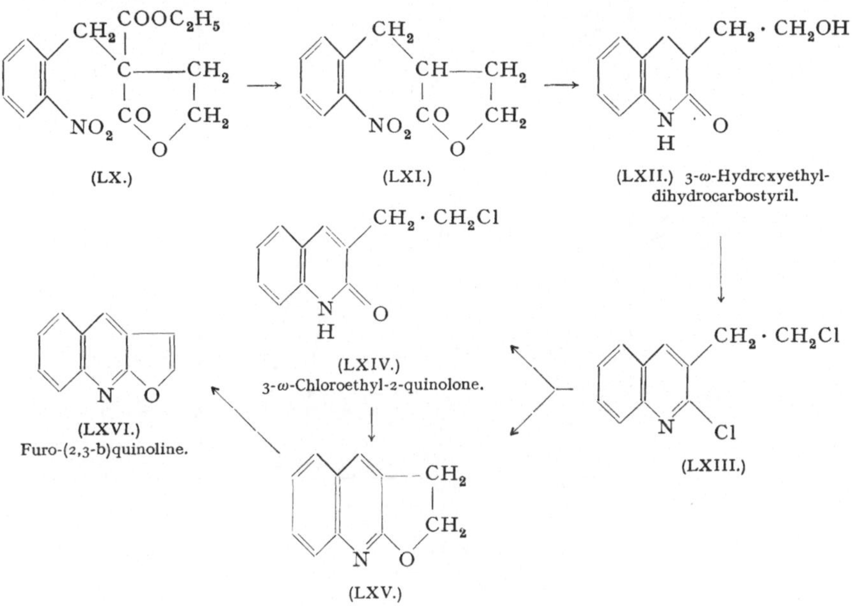

(LX.)

(LXI.)

(LXII.) 3-ω-Hydroxyethyl-dihydrocarbostyril.

(LXIV.) 3-ω-Chloroethyl-2-quinolone.

(LXVI.) Furo-(2,3-b)quinoline.

(LXIII.)

(LXV.)

Subsequently Cooke and Haynes (34) synthesized dihydrodictamnine (LXIX) as follows. The ethyl ester of 4-hydroxy-2-quinolone-3-acetic acid (LXVII), obtained by condensing ethyl anthranilate with succinic ester, was methylated with diazomethane to the 4-methoxy compound. Reduction with lithium aluminium hydride to 4-methoxy-3-ω-hydroxyethyl-2-quinolone (LXVIII) and cyclization with polyphosphoric

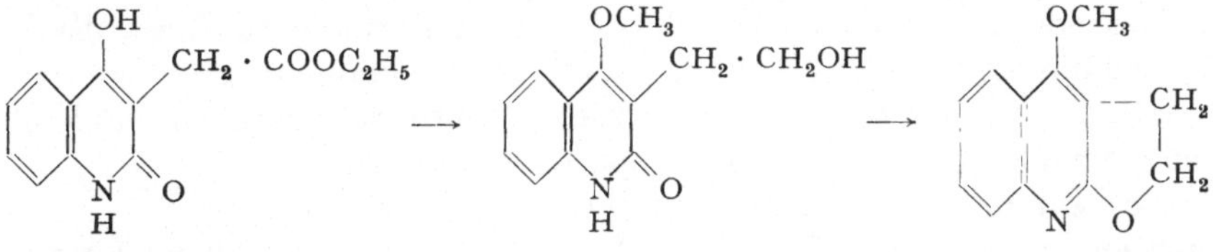

(LXVII.) 4-Hydroxy-2-quinolone-3-acetic acid ethyl ester.

(LXVIII.)

(LXIX.) Dihydro-dictamnine.

acid yields dihydrodictamnine identical with that prepared from the natural alkaloid. Dehydrogenation to dictamnine has not yet been effected.

The main *structural features* of most of the furoquinoline alkaloids have been determined by oxidation of the double bond in the furan ring and conversion of the resulting acid to the corresponding 4-hydroxy-2-quinolone.

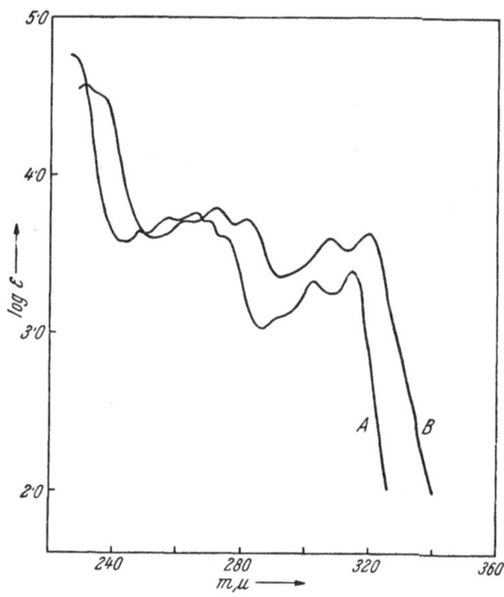

Fig. 1. Molecular extinction curves in ethanol: *A*, 2,4-Dimethoxyquinoline; and *B*, Dihydrodictamnine (unpublished).

But increasing use is being made of hydrogenolysis which has the advantage that it requires less material. When dictamnine is hydrogenated over palladium (*33, 83*) dihydrodictamnine (LXIX) is formed, the ultraviolet spectrum of which closely resembles that of 2,4-dimethoxyquinoline (*91*) (*Fig. 1*). With a platinum oxide catalyst, however, hydrogenolysis occurs with the formation of 3-ethyl-4-methoxy-2-quinolone (LXX) (*81*). The identity of (LXX) with a specimen prepared by methylating synthetic 3-ethyl-4-hydroxy-2-quinolone with diazomethane has been established by Chta and Mori (*85a*) thus confirming the linearity of the dictamnine molecule. Hydrogenolysis of *nor*dictamnine gives 3-ethyl-4-hydroxy-2-quinolone identical with a synthetic specimen, and of *iso*dictamnine, 1-methyl-3-ethyl-4-hydroxy-2-quinolone.

An example of the use of this procedure is furnished by the determination by Cooke and Haynes (*33*) of the structure of *evolitrine* (LXXI) from *Evodia littoralis*.

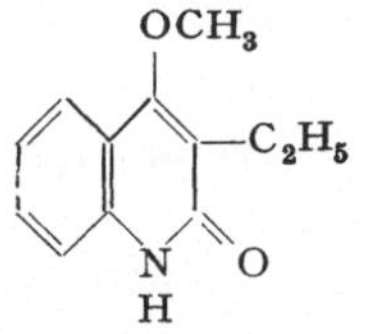

(LXX.) 3-Ethyl-4-methoxy-2-quinolone.

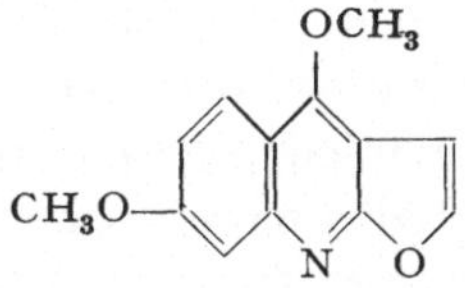

(LXXI.) Evolitrine.

Hydrogenolysis of *iso*evolitrine gives 1-methyl-3-ethyl-4-hydroxy-7-methoxy-2-quinolone (LXXII) identical with a specimen prepared by condensing *m*-anisidine and diethyl ethylmalonate to the monoanilide (LXXIII), cyclizing to the 4-hydroxy-2-quinolone, methylating with dimethyl sulphate and finally demethylating the 4-methoxyl group with hydrochloric acid. (LXXII) was also prepared from the 2-nitro-4-methoxybenzoyl derivative of diethyl ethylmalonate (LXXIV) thus eliminating the possibility that cyclization in the first synthesis had given rise to the 5-methoxy isomer. The alternative angular furoquinoline structure for evolitrine is not excluded.

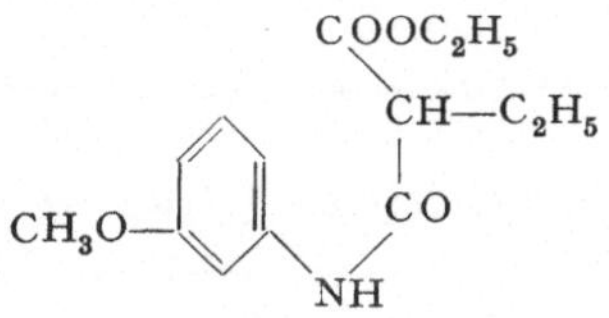

(LXXIII.)

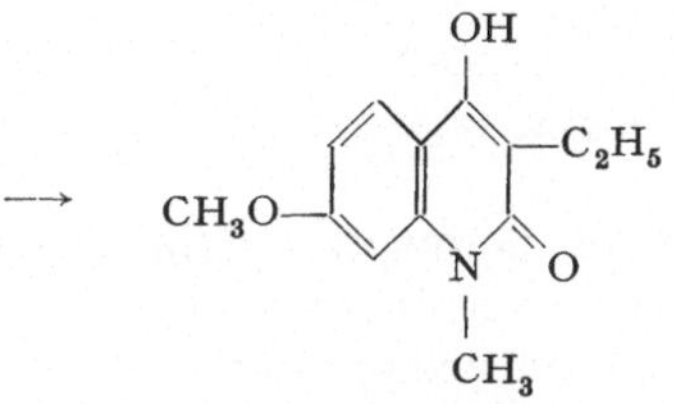

(LXXII.) 1-Methyl-3-ethyl-4-hydroxy-7-methoxy-2-quinolone.

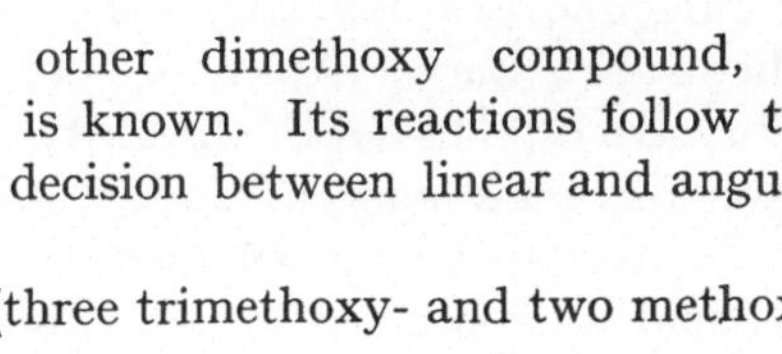

(LXXIV.) Diethyl 2-nitro-4-methoxybenzoyl-ethylmalonate.

Fagarine. One other dimethoxy compound, 4,8-dimethoxyfuroquinoline, fagarine, is known. Its reactions follow the familiar pattern (*19, 41*) and again, decision between linear and angular structures is not yet possible.

Five trialkoxy- (three trimethoxy- and two methoxymethylenedioxy-) furoquinolines are known.

Skimmianine, 4,7,8-trimethoxyfuroquinoline (LXXV), the most widespread member of the group (*86*) resembles dictamnine closely in its reactions (*6, 11, 81–84*). A linear structure has been assigned to it by ASAHINA and INUBUSE (*6*) on the basis of the similarity of the ultraviolet spectrum to that of dictamnine, but more conclusive evidence has recently been put forward. The arguments for the linear structure for both natural dictamnine (*27*) and synthetic dihydrodictamnine (*34*) rest on the preferential 4-methylation of 4-hydroxy-2-quinolone (*2*). Applying the same considerations to skimmianine, BROWN (*25*) finds that methylation of the methyl ester of synthetic 7,8-dimethoxy-4-hydroxy-2-quinolone-3-carboxylic acid with diazomethane gives a product identical with the methyl ester of skimmianinic acid (LXXVI), so confirming the correctness of the linear structure. Likewise, OHTA and MORI (*85a*) find that methylation of synthetic 7,8-dimethoxy-3-ethyl-4-hydroxy-2-quinolone with diazomethane gives a product identical with that resulting from hydrogenolysis of skimmianine.

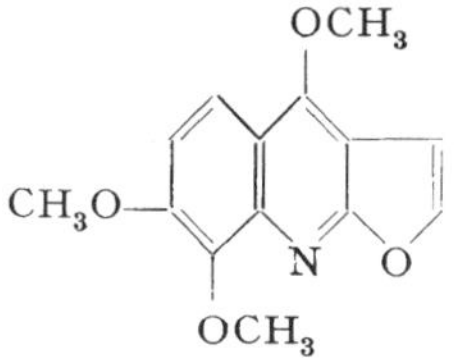

(LXXV.) Skimmianine.

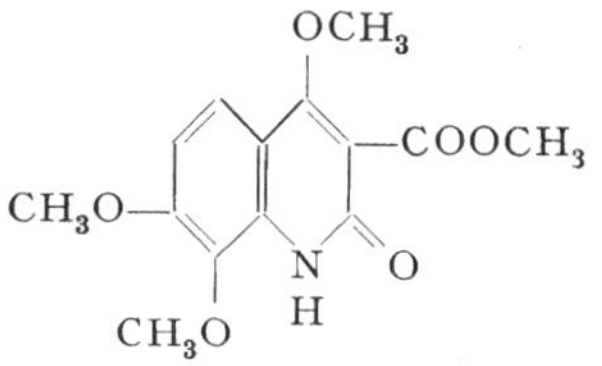

(LXXVI.) Skimmianinic acid methyl ester.

There is, however, no direct evidence of linearity for *kokusagine* (4-methoxy-7,8-methylenedioxy-furoquinoline) (*111*), *maculine* (4-methoxy-6,7-methylenedioxy-furoquinoline) (*26*) or *maculosidine* (4,6,8-trimethoxy-furoquinoline) (*26*). The structure of the latter requires corroboration.

Kokusaginine. The structure of kokusaginine (4,6,7-trimethoxy-furoquinoline) (*86, 110, 110a*) rests on its degradation to 6,7-dimethoxy-4-hydroxy-2-quinolone (*1*) and, by way of hydrogenolysis, to 6,7-dimethoxy-3-ethyl-4-hydroxy-2-quinolone (*79, 110a*) identical with a synthetic specimen. BROWN (*25*) has attempted to establish the linearity of kokusaginine in the same way as for skimmianine, but methylation of kokusagininic acid with diazomethane proceeds further than expected, giving methyl 2,4,6,7-tetramethoxy-quinoline-3-carboxylate. The same result is obtained with the synthetic acid. Earlier, ANET, GILHAM, GOW, HUGHES and RITCHIE (*1*) claimed to have established the linearity of the kokusaginine ring system on the grounds that kokusagininic acid is not identical with a synthetic acid believed to be 4-hydroxy-2,6,7-trimethoxyquinoline-3-carboxylic acid. This acid was obtained by reductive cyclization of the appropriate *o*-nitrobenzoylmalonic ester but BROWN (*25*) states that there is little doubt that it was, in fact, methyl 6,7-dimethoxy-4-hydroxy-2-quinolone-3-carboxylate. However, the point is established by the properties of the hydrogenolysis product of acronidine, a derivative of kokusaginine, and is discussed below.

Acronycidine. Evidence of a different kind is available to establish the linear structure for one of the two known tetraalkoxy-furoquinolines, acronycidine (LXXVII; 4,5,7,8-tetramethoxy-furoquinoline) (69). When the hydrochloride of *iso*acronycidine (LXXVIII) is heated at its melting point, one methoxyl group is demethylated giving *nor-iso*acronycidine (LXXIX) which is insoluble in caustic soda, but forms a monoacetyl derivative and can be remethylated to *iso*acronycidine. It is clearly comparable with the *nor* compounds formed from the acridone alkaloids, and its formation establishes that the pyridone carbonyl group of *iso*-acronycidine is so situated that it can form a hydrogen bond with the oxygen atom of the hydroxyl group produced by demethylation; in other words it must be located *peri* to a methoxyl group in the benzenoid ring, a requirement which is compatible only with a linear structure.

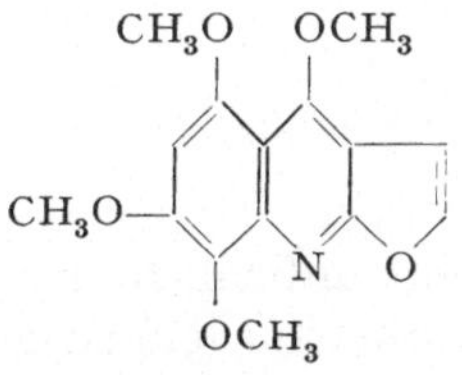

(LXXVII.) Acronycidine. 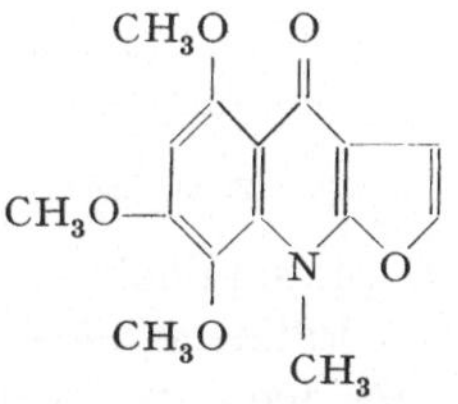   (LXXVIII.) *iso*Acronycidine. 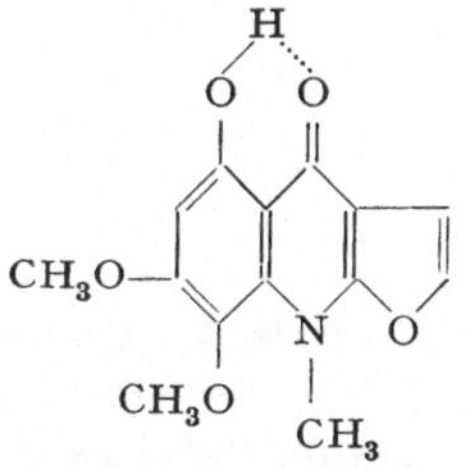   (LXXIX.) *nor-iso*Acronycidine.

The 4-hydroxy-2-quinolone from acronycidine was not easily accessible synthetically and recourse was had to further degradation. It was found that the action of acids on the 3-nitroso derivative of the hydroxy-

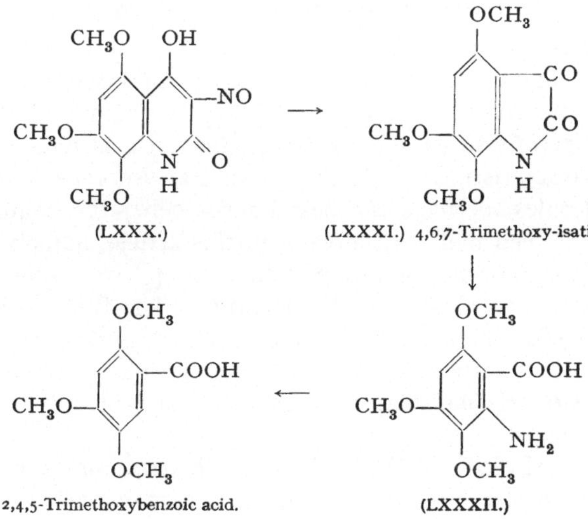

(LXXX.)   (LXXXI.) 4,6,7-Trimethoxy-isatin.

2,4,5-Trimethoxybenzoic acid.   (LXXXII.)

quinolone (LXXX) brings about a BECKMANN-type rearrangement with the formation of the trimethoxy-isatin (LXXXI) which is oxidized by alkaline hydrogen peroxide to the trimethoxy-anthranilic acid (LXXXII). Deamination of (LXXXII) gives 2,4,5-trimethoxybenzoic acid.

The most characteristic property of acronycidine and its derivatives is the ease with which they undergo oxidative demethylation with either nitric or nitrous acids to 1,4-quinones, resembling in this respect certain of the acridine alkaloids.

Flindersiamine (LXXXIII) (*1*) contains two methoxyl groups and one methylenedioxy group and is degraded to 6,7-methylenedioxy-8-methoxy-4-hydroxy-2-quinolone. Again, the angular structure is not excluded.

(LXXXIII.) Flindersiamine.

The *biogenetic pattern* of the furoquinolines is more difficult to unravel than that of the co-occurring acridine alkaloids. ROBINSON (*92*) suggests that the furan ring may be a residue of the tryptophan side-chain by way of (XVIII, p. 310) or some similar intermediate, but he also regards as possible that it may arise by degradation of an aromatic ring of an acridone or even that it may result from condensation of 4-hydroxy-2-quinolone with a suitable intermediate.

Furoquinoline *iso*Pentane Ethers.

One characteristic of members ot the family Rutaceae is their capacity to synthesize highly oxygenated (hydroxy-, methoxy- or methylene-dioxy-) aromatic compounds as exemplified by certain of the acridine alkaloids, by acronycidine and by flavonoids such as meliternatin (*22*). Still more characteristic is the ability to add an isoprene unit to a wide range of molecules, perhaps the best known being the coumarins (*39*). Examples have been met with already in this article, namely flindersine and acronycine, and in the furoquinoline group two types have been reported, viz. O- and C-isoprenoid derivatives. The first are represented by two isoprenoid ethers, evoxine and evolatine. Several alkaloids of unknown structure described in the literature may well be of the same type, for example, *haploperine* isolated by YUNUSOV and SIDYAKIN (*117*) from *Haplophyllum* spp.

Evoxine, $C_{18}H_{21}O_6N$ (LXXXIV) from *Evodia xanthoxyloides* (*43*) resembles the simpler furoquinolines in its conversion to *iso*evoxine and to *nor*evoxine. Hydrolysis with methanolic potash gives a mixture of *nor*evoxine and a phenol (LXXXV) which may also be obtained by fusion with potash. Methylation of (LXXXV) yields skimmianine, thus establishing the main features of the molecule, and ethylation a *homo*skimmianine which is degraded by oxidation, hydrolysis and decarboxylation to 7-ethoxy-8-methoxy-4-hydroxy-2-quinolone (LXXXVI). The identity of (LXXXVI) with a specimen synthesized from 4-ethoxy-3-methoxy-2-nitrobenzoic acid establishes the structure of the phenol (LXXXV).

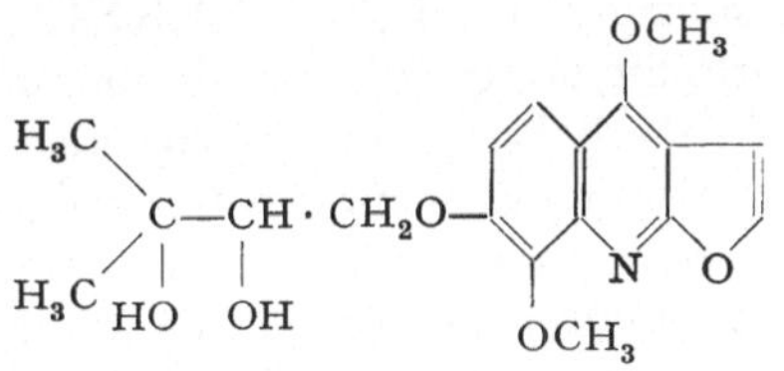

(LXXXIV.) Evoxine. 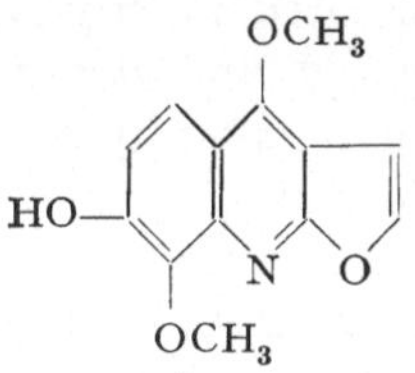 (LXXXV.) 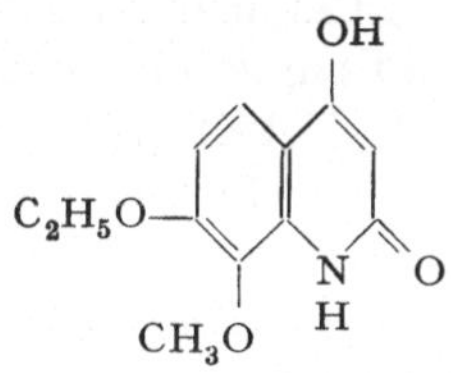 (LXXXVI) 7-Ethoxy-8-methoxy-4-hydroxy-2-quinolone.

The two remaining oxygen atoms are present in the side-chain as hydroxyl groups since acetylation of the alkaloid gives a diacetyl derivative. Evoxine consumes one mole of periodic acid with the formation of acetone and an optically inactive aldehyde, $C_{15}H_{15}O_5N$, which is converted to (LXXXV) by oxidation with alkaline hydrogen peroxide. Evoxine is therefore a 1,2-glycol of structure (LXXXIV). Treatment of evoxine with acids gives, depending on the conditions, varying amounts of (LXXXV) and a substance, $C_{18}H_{19}O_5N$, which had originally been obtained from the plant extract and had been given the name *evoxoidine*. Evoxoidine (LXXXVII) which contains a carbonyl group and evidently arises by a pinacol-pinacolone rearrangement, has been synthesized by reacting the sodium salt of (LXXXV) with 1-chloro-3-methyl-2-butanone. It is thought to be an artifact arising through the use of acid in working up the crude alkaloids.

Evolatine from *Evodia alata* (*48*) is isomeric with evoxine. Its reactions resemble those of evoxine, except that it can be degraded to kokusaginine instead of to skimmianine. The structure (LXXXVIII), assuming linearity of the kokusaginine molecule, has been established in a similar manner.

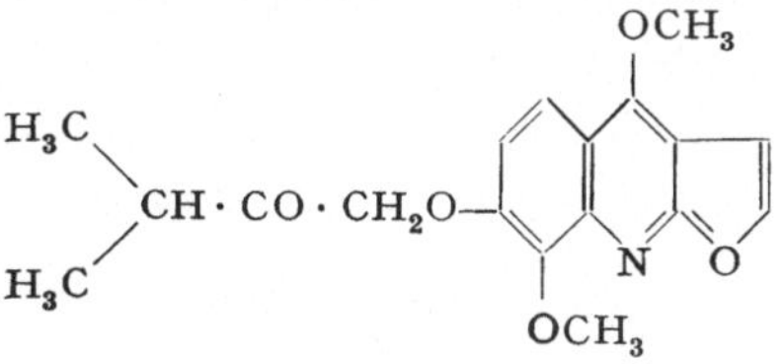

(LXXXVII.) Evoxoidine. 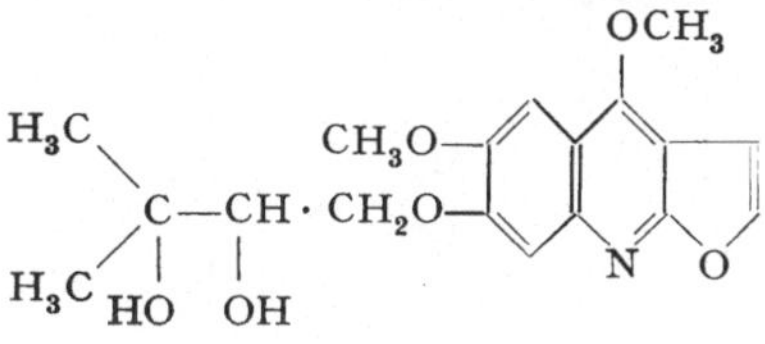 (LXXXVIII.) Evolatine.

Dimethyl-pyranofuroquinolines.

Medicosmine. From *Medicosma cunninghamii*, LAMBERTON and PRICE (*70*) isolated an alkaloid, medicosmine, $C_{17}H_{15}O_3N$, which can be isomerized to *iso*medicosmine and demethylated to *nor*medicosmine. Oxidation with permanganate gives α-hydroxy*iso*butyric acid suggesting the presence of a dimethylpyran ring, and this is confirmed by the formation of both acetone and acetaldehyde on vigorous alkaline hydrolysis. The other product from this hydrolysis is a phenol which has not been isolated as such but methylated directly to a substance $C_{13}H_{11}O_3N$ (LXXXIX) containing one methoxyl and one methylimino group. Oxidation of (LXXXIX) gives 6-methoxy-4-hydroxy-1-methyl-2-quinolone, the identity of which is established by synthesis. Hydrogenation of medicosmine gives two products, a tetrahydro derivative and the dihydropyrano-ethylquinolone (XC a or XC b) arising by hydrogenolysis of the furan ring. Like carbostyril and unlike tetrahydro-medicosmine, (XC) is only sparingly soluble in hot 1% hydrochloric acid; unlike carbostyril it is insoluble in hot aqueous caustic soda. This is in accord with the properties of 3-alkylcarbostyrils, 3-ethyl-4-methylcarbostyril, for example, being insoluble even in hot 30% caustic soda. On the other hand, 2-methyl-3-ethyl-4-hydroxyquinoline dissolves readily in 5% caustic soda. Consequently the hydrogenolysis product must be (XC a or XC b) and not either of the corresponding 2-methoxy-4-quinolones. It follows that the furoquinoline system in medicosmine [and in (LXXXIX)] must be linearly arranged and this is in accord with the ease of solution of *nor*medicosmine in alkali. The alkaloid is, therefore, either (XCI a) or (XCI b).

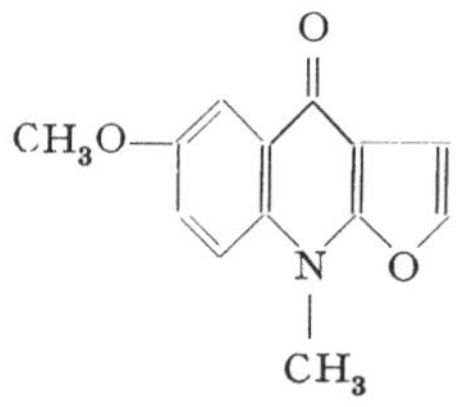

(LXXXIX.)

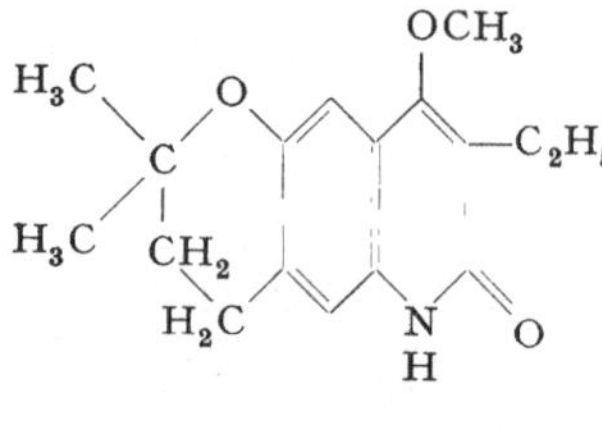

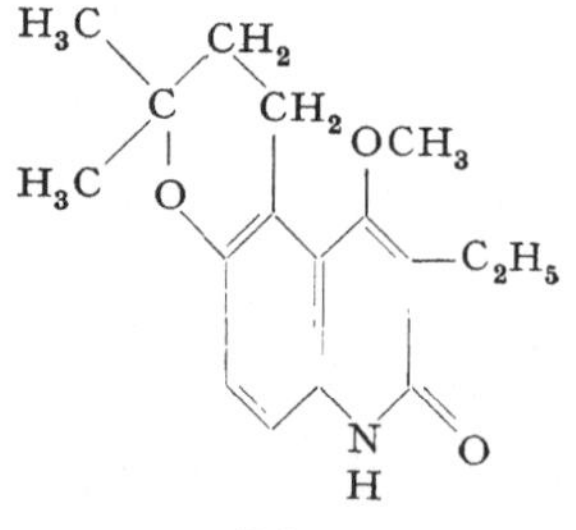

(XC a.)   Medicosmine.   (XC b.)

(XCI a.) (XCI b.)

Acronidine, $C_{18}H_{17}O_4N$, from *Acronychia baueri* (*71*) closely resembles medicosmine in its behaviour. It gives α-hydroxy*iso*butyric acid on oxidation while *nor*ac:onidine is degraded by alkaline hydrolysis to acetone, acetaldehyde and a phenol which can be methylated to *iso*-kokusaginine. Treatment with hydrogen over Raney nickel again results in competitive hydrogenation and hydrogenolysis, the products being tetrahydro-acronidine and a dihydropyrano-ethylquinolone. The properties of the latter resemble those of (XC a or XC b); it is insoluble in hot aqueous caustic soda, whereas *nor*acronidine is easily soluble in aqueous caustic soda. It follows, therefore, as in the case of medicosmine, that the furoquinoline system in acronidine is linear and that acronidine is either (XCII a or XCII b). Since *nor*acronidine couples with diazonium salts the formula (XCII a) with a free 8-position is preferred.

It should be noted that, while acetone is formed by alkaline hydrolysis of the majority of 2,2-dimethylpyrans, *nor*acronidine and *nor*medicosmine provide the only instances in which acetaldehyde has also been isolated. In general, all five carbon atoms of the isoprene unit are split off, but the conditions of hydrolysis are severe and the fate of carbon atoms 3 and 4 has been disregarded. This point is discussed by Lamberton and Price (*71*).

(XCII a.) Acronidine. (XCII b.)

An important consequence of establishing the linearity of the furoquinoline system in acronidine is that it establishes at the same time the linearity of both the kokusaginine and evolatine ring systems. It can thus be stated with reasonable assurance that eight of the fourteen furo-quinoline alkaloids are furo-(2,3-b)quinolines. The considerable amount of data accumulated by DEULOFEU (*40*), PRICE and SHELTON (*91*) and others concerning the ultraviolet spectra of the furoquinoline alkaloids and their derivatives support this conclusion and make it probable that at least the majority of the remainder also have the linear arrangement of the three rings. Not only are the spectra of the simpler members (with the

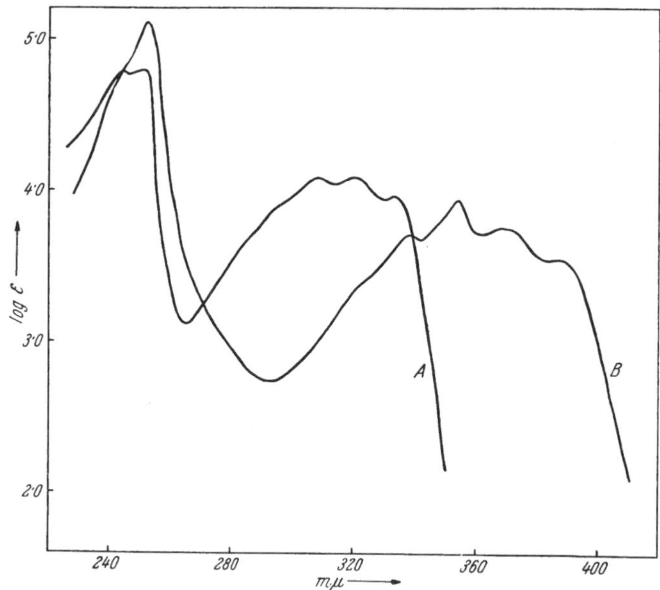

Fig. 2. Molecular extinction curves in ethanol: *A*, Kokusaginine; and *B*, 5-Ethoxyacridine (unpublished).

exception of that of maculosidine) very similar to one another, but they resemble the spectra of anthracene or acridine, rather than phenanthrene, derivatives (cf. *Fig. 2*).

VI. Quinazoline Alkaloids.

Although no more than a half-dozen quinazoline alkaloids have been described, the interest attaching to them is greater, because of their distribution, than for example, to the acridines or furoquinolines. While the two latter groups are confined to the Rutaceae, the quinazolines are found not only in that family (evodiamine, rutaecarpine, arborine) but also in the Acanthaceae (vasicine), Zygophyllaceae (vasicine), Saxi-fragaceae (febrifugine, *iso*febrifugine) and Palmaceae (vasicine).

Arborine, $C_{16}H_{14}ON_2$, has been isolated from *Glycosmis arborea* (*30*) and from *G. pentaphylla* (*31*)*. It is a weak monoacid tertiary base giving on hydrogenation a dihydro derivative. Alkaline hydrolysis of arborine yields N-methylanthranilic acid, phenylacetic acid, ammonia, a small amount of a substance $C_{16}H_{13}O_2N$ and, under certain conditions, some N-methylanthranilamide. Dihydro-arborine is stable to alkali but undergoes acid hydrolysis with the formation of phenylacetaldehyde together with N-methylanthranilic acid and its amide. On this evidence CHAKRAVARTI, CHAKRAVARTI and CHAKRAVARTI (*30*) put forward the benzylquinazolone structure (XCIII), but the formation of benzaldehyde — either by ozonolysis or periodic acid oxidation — as reported by CHATTERJEE and GHOSH MAJUMDAR (*31*), leads to (XCIV) as the preferred formulation. Arborine has been synthesized by heating the amide (XCV) obtained from N-methylanthranilamide and phenylacetylchloride for half-an hour at 170–190° (*30*), or by condensing N-methylanthranilamide directly with phenylacetic acid in the presence of phosphorus pentoxide (*31*).

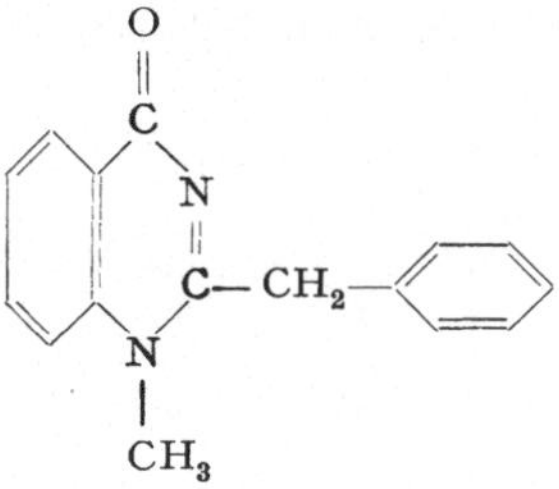

(XCIII.) 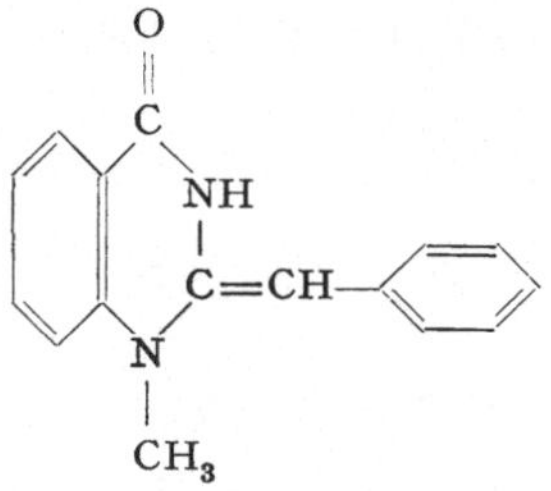   (XCIV.) Arborine.

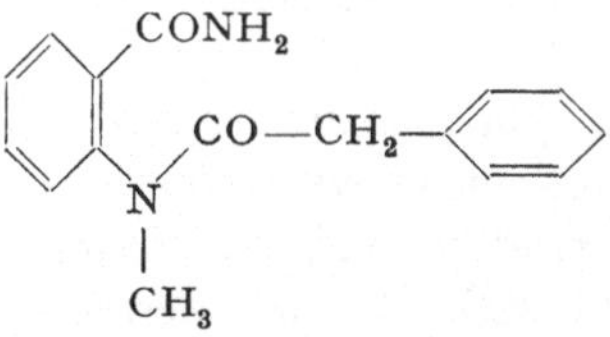

(XCV.) 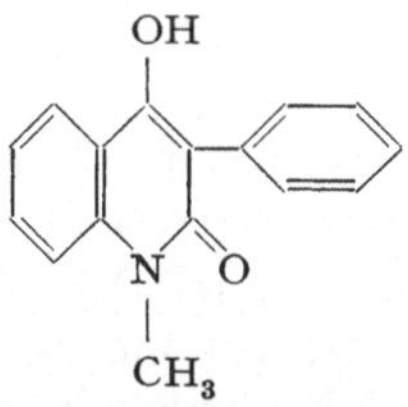   (XCVI.)

The substance $C_{16}H_{13}O_2N$, formed as a minor product during the alkaline hydrolysis of arborine, has been shown to be (XCVI) by synthesis from ethyl phenylmalonate and N-methylaniline. It evidently arises by way of (XCV).

---

* There appears to be some uncertainty as whether the two species are involved or whether both investigations were carried out with *G. arborea*. Only skimmianine and kokusaginine have been isolated from a northern Australian specimen of *G. pentaphylla* (*77*). The name "arborine" would appear to have precedence over "glycosin" used by CHATTERJEE and GHOSH MAJUMDAR (*31*).

The most numerous and widespread group of alkaloids are those containing the benzyl*iso*quinoline structuie or various modifications of it. It is generally accepted that the members of this group are built up from two units derivable from a phenylalanine, usually dihydroxy-phenylalanine (*92*). Though not believed to be the actual intermediates, phenylethylamine and phenylacetaldehyde are the forms in which the two phenylalanine units are usually represented in illustrating the architecture of the benzyl*iso*quinolines.

Phenylalanine, of course, is regarded as the precursor of numerous natural products, non-nitrogenous as well as nitrogenous, and the benzyl-*iso*quinolines are by no means the only alkaloids originating from it. Simple *iso*quinolines, for example, such as salsoline, evidently result from the combination of a phenylalanine (phenylethylamine) unit with an acetate unit or its equivalent. It is readily seen in the case of arborine that this alkaloid could arise from the union of an anthranilic acid unit with a phenylethylamine unit, as shown below, and it is significant that in the same family we should find the alkaloid (XIX) which could result from the union of an anthranilic acid unit with a phenylacetaldehyde unit.

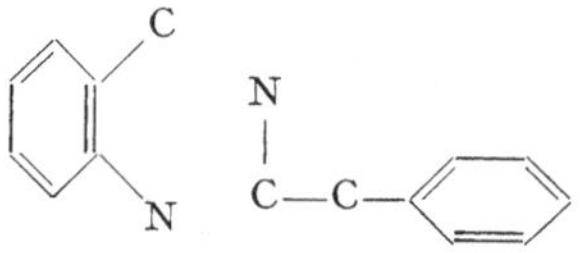

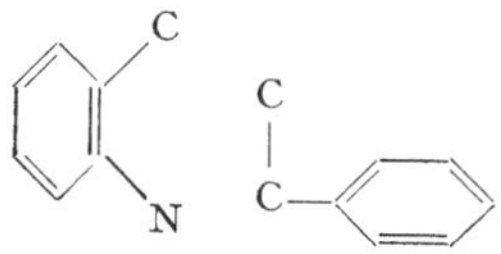

Vasicine. The best known quinazoline alkaloid is vasicine $C_{11}H_{12}ON_2$, which was first isolated in 1888 from *Adhatoda vasica* (Acanthaceae), later from *Peganum harmala* (Zygophyllaceae) and has recently been reported to occur in ester combination in the bark of *Daemonorops draca* (Palmaceae; "Sangre de Drago") (*87*). Vasicine is a monoacid base containing an alcoholic hydroxyl group. Treatment with phosphorus oxychloride converts it to chlorodesoxyvasicine which on reduction with zinc and acid gives desoxyvasicine $C_{11}H_{12}N_2$. Vasicine is readily oxidized, giving, with acid permanganate, 4-quinazolone (*50*) and with alkaline permanganate, an acid $C_{10}H_8O_3N_2$ which is hydrolysed by alkali to anthranilic acid and glycine and is decarboxylated to 3-methyl-4-quinazolone (*104*). The acid, therefore, has the structure (XCVII) and this has been confirmed

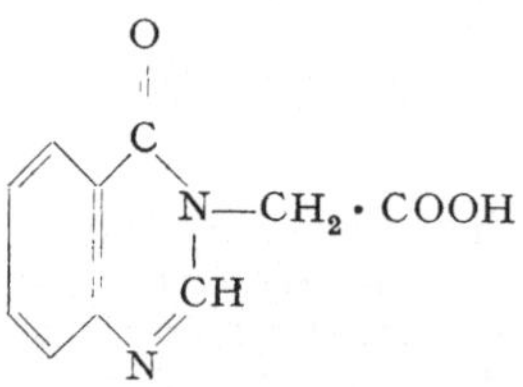

(XCVII.) 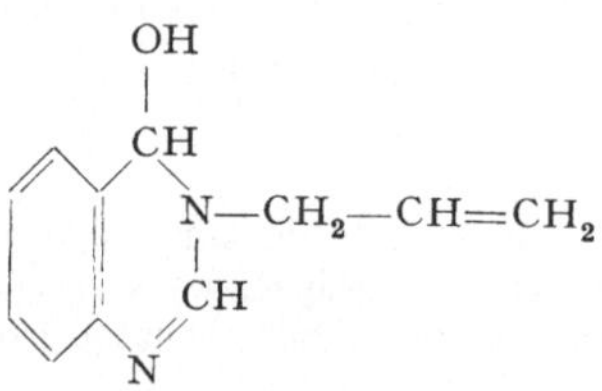   (XCVIII)

by synthesis (*101*). The possibility of bicyclic structures such as (XCVIII), for the alkaloid can be eliminated on various grounds leaving a choice between the linear and angular tricyclic formulations. A decision in

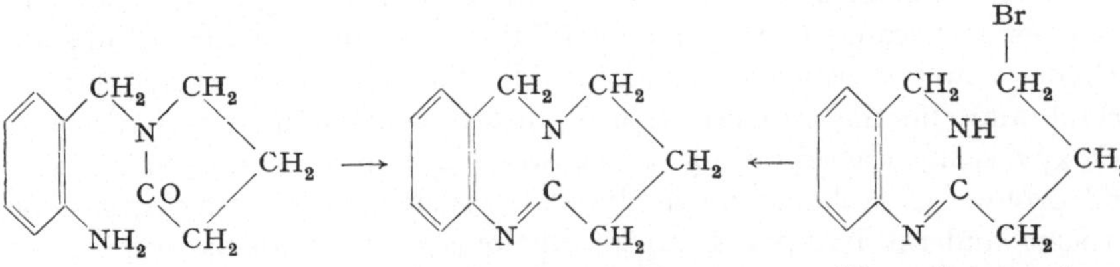

(C.)   (XCIX.) Desoxyvasicine.   (CI.) 2-γ-Bromopropyl-3,4-dihydroquinazoline.

favour of the linear arrangement was reached by the synthesis of desoxyvasicine (XCIX) by cyclizing N-*o*-aminobenzylpyrrolidone (C) with phosphorus oxychloride (*102*) and by cyclizing 2-γ-bromopropyl-3,4-dihydroquinazoline (CI) (*54*). The position of the hydroxyl group in vasicine (CII) has likewise been established by synthesis (*103, 108*), for example, by the condensation of *o*-aminobenzylamine with α-hydroxy-γ-butyro-lactone.

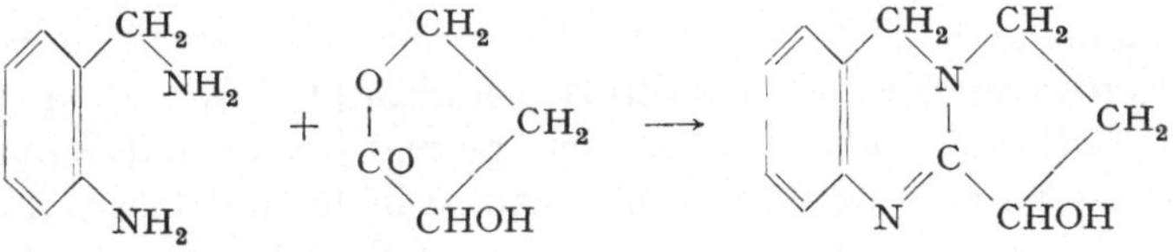

(CII.) Vasicine.

The isolation of vasicine by acid hydrolysis of the water-insoluble red pigment, $C_{46}H_{46}O_4N_4$, of "Dragon's Blood" (from *Daemonorops* spp.) is reported by PALLARES (*88*). The second hydrolytic product, an acid $C_{24}H_{26}O_4$, is converted by ozonolysis to phenyl *n*-propyl ketone and dihydroxytartaric acid. Ozonolysis of the ethyl ester of the acid gives the same ketone and diethyl diketosuccinate. The pigment is therefore regarded as the diester of vasicine with $\alpha\beta$-diphenyl-$\alpha\beta$-di-*n*-propylfulginic acid (CIII).

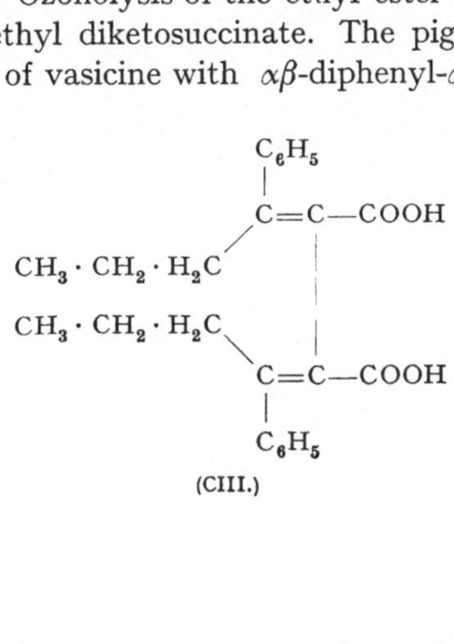

(CIII.)

SCHÖPF and OECHLER (*95*) suggest that vasicine may be produced in the plant from *o*-aminobenzaldehyde and *γ*-amino-*α*-hydroxybutyraldehyde followed by isomerization of the intermediate quaternary base to the 3,4-dihydroquinazoline. The practicability of this suggestion has not been tested because of the unavailability of *γ*-amino-*α*-hydroxybutyraldehyde; however, using *γ*-aminobutyraldehyde, in the form of its diethylacetal, and effecting isomerization by means of palladium and hydrogen, desoxyvasicine was obtained. ROBINSON (*92*) considers anthranilic acid and proline as probable progenitors of vasicine. The co-occurrence of vasicine and the harmine group of alkaloids in *Peganum harmala* is of interest in view of the close metabolic relation between anthranilic acid and tryptophan.

Febrifugine and iso*Febrifugine.* Because of its reputed antimalarial properties, the root of *Dichroa febrifuga* (Saxifragaceae) received considerable attention during the war years and the alkaloids febrifugine and *iso*febrifugine, $C_{16}H_{19}O_3N$, have been isolated from it. CHOU, FU and KAO (*32*) also isolated 4-quinazolone but this has not been reported by other workers and may have been an artifact. Febrifugine, which is apparently largely responsible for the antimalarial activity of *D. febrifuga*, has also been isolated from *Hydrangea* (hort.).

Febrifugine and *iso*febrifugine, both of which are optically active, are readily interconvertible and their ultraviolet absorption spectra are almost identical. Both are oxidized by alkaline permanganate to 4-quinazolone and both give anthranilic acid, formic acid, a little ammonia and an oily base on alkaline hydrolysis. Confirmation of the quinazolone structure is supplied by the ultraviolet spectra which resemble closely the spectra

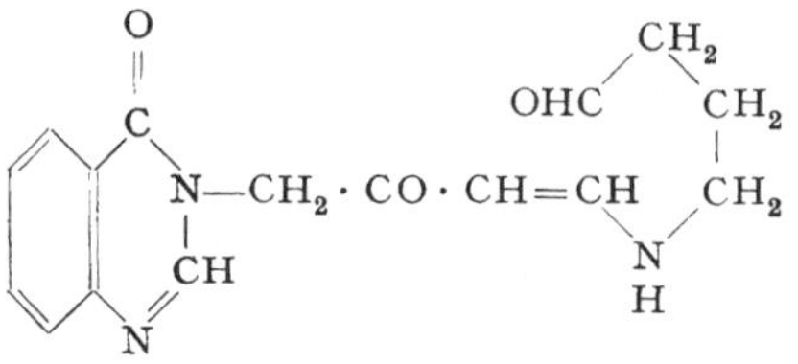

(CIV.)

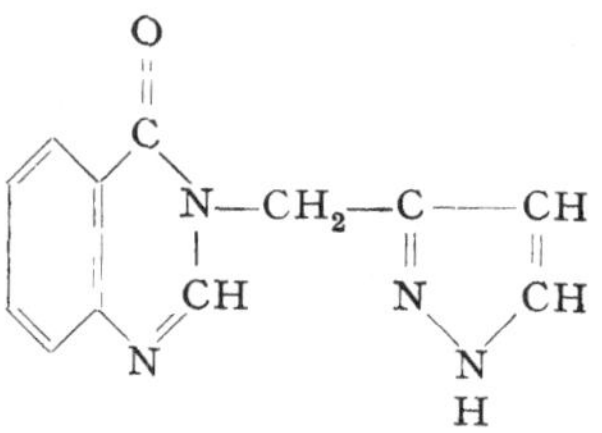

(CV.)
3-(3-Pyrazolylmethyl)-4-quinazolone.

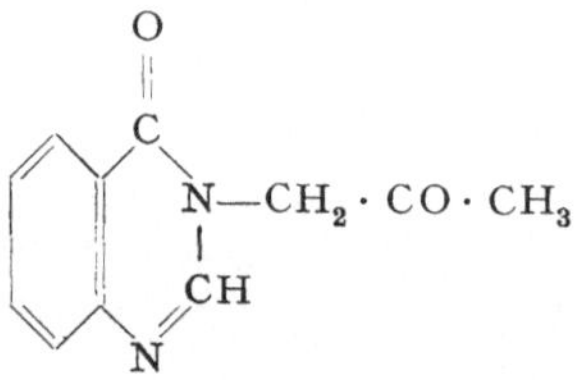

(CVI.) 3-(*β*-Ketopropyl)-4-quinazolone.

of 4-quinazolones, but particularly the spectrum of 3-allyl-4-quinazolone [KOEPFLI, MEAD and BROCKMAN (*68*)].

Both alkaloids contain an imino group as shown by the formation of a nitroso derivative, and both form non-basic, non-acidic bisbenzenesulphonyl derivatives. Febrifugine evidently contains a carbonyl group since it forms an oxime and a semicarbazone, whereas *iso*febrifugine does not. Both give dihydro derivatives on hydrogenation; and hydrolysis of dihydrofebrifugine to anthranilic acid and an unidentified C_8N_2 compound suggests that the 4-quinazolone nucleus may be substituted at the 3-position only. Periodate oxidation of both febrifugine and *iso*febrifugine gives the same optically inactive product, $C_{16}H_{17}O_3N_3$, for which KOEPFLI, BROCKMAN and MOFFAT (*67*) propose structure (CIV). Heating with semicarbazide converts (CIV) to 3-(3-pyrazolylmethyl)-4-quinazolone (CV). This observation coupled with the formation of the C_8N_2 compound by hydrolysis of dihydrofebrifugine led KOEPFLI et al. to suggest that the alkaloids are diastereoisomeric hemiketals derived from (CX) and differing only in configuration about the hemiketal carbon atom.

HUTCHINGS, GORDON, ABLONDI, WOLF and WILLIAMS (*63*) working with the alkaloid from *Hydrangea*, prior to its identification as febrifugine, showed the absence of C-methyl groups and isolated double bonds and, from the preparation of a monoacetyl derivative, inferred the presence of a hydroxyl group. Oxidation with neutral permanganate to 3-carboxymethyl-4-quinazolone (XCVII) (cf. p. 332) and the formation of 3-(β-ketopropyl)-4-quinazolone (CVI) by heating with zinc dust supported the earlier proposals. Proof was afforded by synthesis. First, BAKER, SCHAUB, McEVOY and WILLIAMS (*17*), by condensing 4-quinazolone

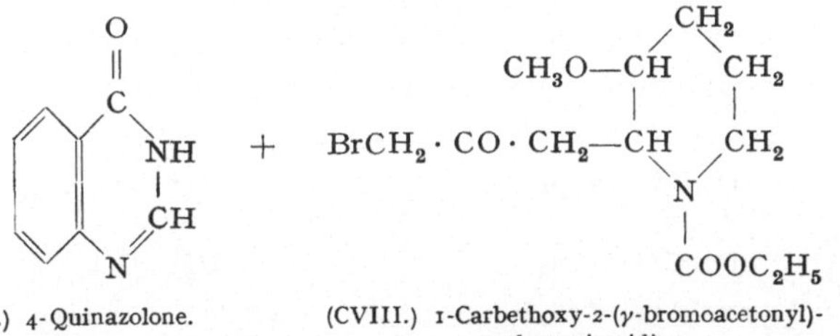

(CVII.) 4-Quinazolone.

(CVIII.) 1-Carbethoxy-2-(γ-bromoacetonyl)-3-methoxypiperidine.

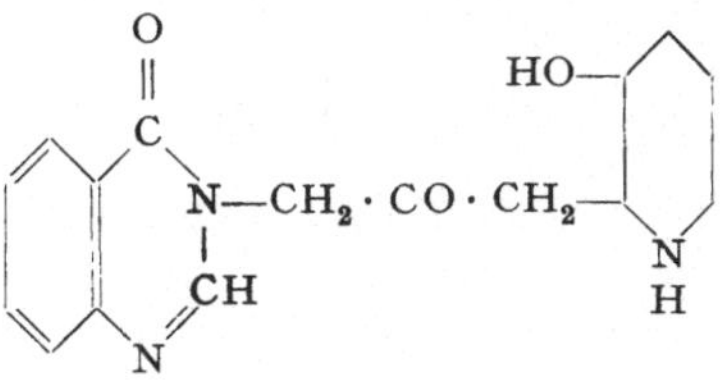

(CX.) Febrifugine.

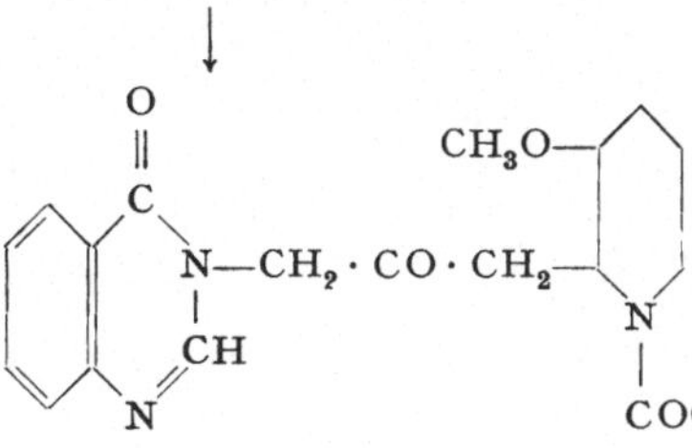

(CIX.)

(CVII) with 1-carbethoxy-2-(γ-bromoacetonyl)-3-methoxypiperidine (CVIII), synthesized (CIX) and hydrolysed it in two stages to (CX). This they were able to convert by periodate oxidation to the substance $C_{16}H_{17}O_3N_3$ identical with that previously obtained by periodate oxidation of febrifugine or *iso*febrifugine. Hence, their synthetic product (CX) which shows half the activity of febrifugine in antimalarial tests is racemic febrifugine. A shorter synthesis of the *DL*-alkaloid with improved yields has since been published by BAKER and McEVOY (*15*).

Finally BAKER, McEVOY, SCHAUB, JOSEPH and WILLIAMS (*16*) synthesized the optically active form of febrifugine and established its relation to *iso*febrifugine. Having shown that hydrogenation of N-benzoyl-β-furyl-β-alanine (CXI) to the tetrahydrofuryl compound gives almost exclusively one of the two possible racemates, (CXI) was resolved and the (—)-isomer hydrogenated. Hydrolysis of the resulting (—)-β-benzamido-β-tetrahydrofuryl-propionic acid (CXII) with hydrobromic acid gives the bromolactone (CXIII) which is recyclized to (—)-3-hydroxy-piperidine-2-acetic acid lactone (CXIV) by means of triethylamine. Benzoylation, followed by hydrolysis of the lactone ring and methylation gives (CXV). After replacing the benzoyl protecting group by carbethoxyl, the synthesis follows the same course as for the racemic alkaloid. The optically active product (CX) is identical with the natural alkaloid and the mother liquors from the last stage of the synthesis were found to contain *iso*febrifugine. The infrared spectrum of febrifugine is in agreement with the open chain ketonic structure (CX) and shows a C=O band at 5.79 μ which is absent in the spectrum of *iso*febrifugine. Similarly, the spectrum of *iso*febrifugine contains strong bands ar 9.05 and 9.48 μ, not present in the febrifugine spectrum, and assigned to a five-membered cyclic ketal ring (*16*). This evidence invalidates the suggestion (*67*) that the alkaloids are diastereoisomeric hemiketals. BAKER et al. (*16*) have

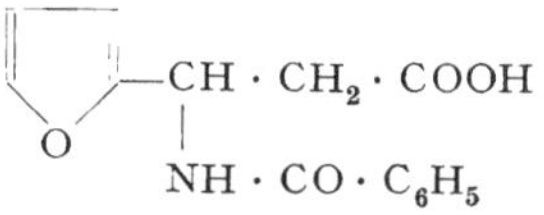

(CXI.) N-Benzoyl-β-furyl-β-alanine.

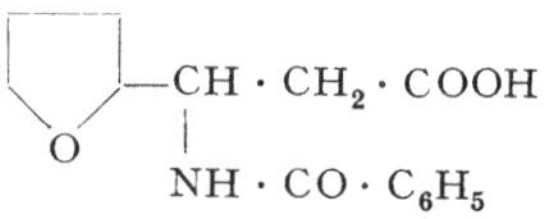

(CXII.) β-Benzamido-β-tetrahydrofuryl-propionic acid.

(CVIII.)

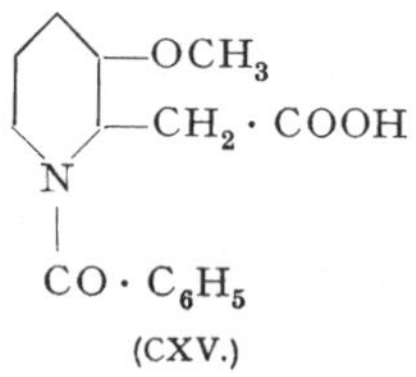

(CXV.)

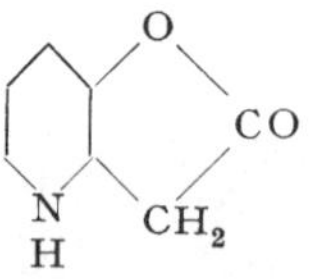

(CXIV.) 3-Hydroxypiperidine-2-acetic acid lactone.

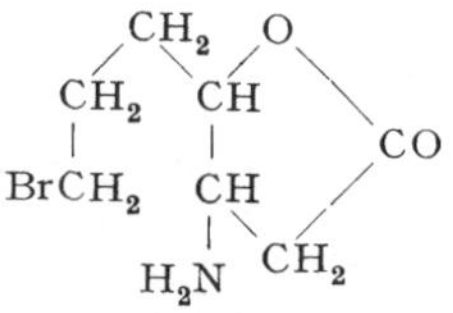

(CXIII.)

pointed out that the hemiketal structure (CXVI) may be stabilized by hydrogen bonding between the oxygen of the hydroxyl group and the quinazolone oxygen atom, but in view of the ease of interconversion of the two, the inability of *iso*febrifugine to form ketonic derivatives is still surprising.

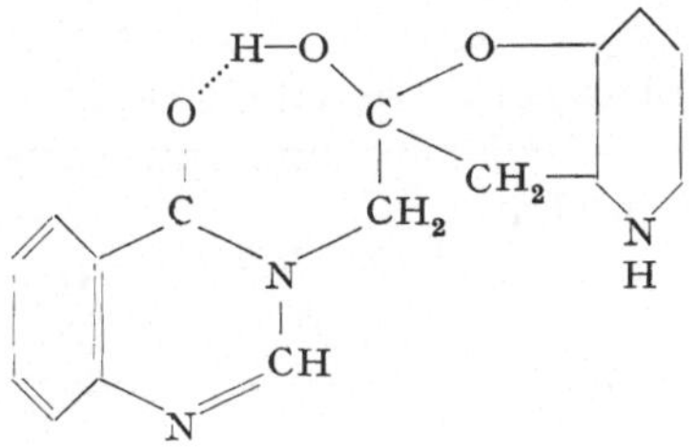

(CXVI.)

Evodiamine and Rutaecarpine. In view of the close biosynthetic relationship between anthranilic acid and tryptophan it is of interest to find, in two alkaloids of *Evodia rutaecarpa*, examples of the combination of an anthranilic acid unit with an almost intact tryptophan unit. Three alkaloids have been isolated from the fruits of *E. rutaecarpa* (*75*) but only evodiamine and rutaecarpine have been studied chemically. The main features of the structure of evodiamine (CXVII) are revealed by alkaline hydrolysis which degrades the alkaloid to N-methyl-anthranilic acid and dihydro-*nor*harman (CXVIII) (*3, 8, 12*).

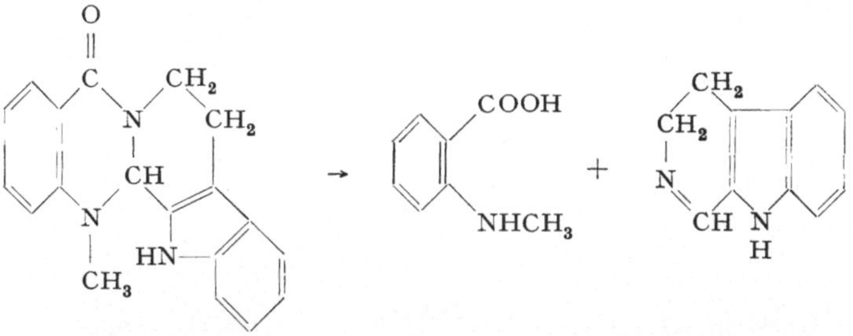

(CXVII.) Evodiamine.

(CXVIII.) Dihydro-*nor*harman.

Boiling with alcoholic hydrochloric acid leads to the addition of a molecule of water with the formation of optically inactive "evodiamine hydrate", from which optically inactive evodiamine is obtained by treatment with acetic anhydride (*8, 10*). "Evodiamine hydrate" is decomposed by alkali to N-methylanthranilic acid, carbon dioxide and a base $C_{10}H_{12}N_2$ which was at first thought, incorrectly, to be 2-β-aminoethylindole. This led to the proposal of structures for the alkaloids which were not derivable from tryptamine. For this reason they were questioned by

KERMACK, PERKIN and ROBINSON (*65*) and the base $C_{10}H_{12}N_2$ was subsequently shown by ASAHINA (*3*) to be tryptamine. The action of amylalcoholic potash on rutaecarpine, $C_{18}H_{13}ON_3$, gives anthranilic acid and an acid, $C_{11}H_{12}O_2N_2$, which when boiled with dilute hydrochloric acid is readily decarboxylated to tryptamine (*4, 10*). Evodiamine can be converted to rutaecarpine by heating the dry hydrochloride of "evodiamine hydrate" (*12*). On the basis of these data, structure (CXVII) and (CXIX) have been allocated to evodiamine and rutaecarpine respectively, and (CXX) to "evodiamine hydrate".

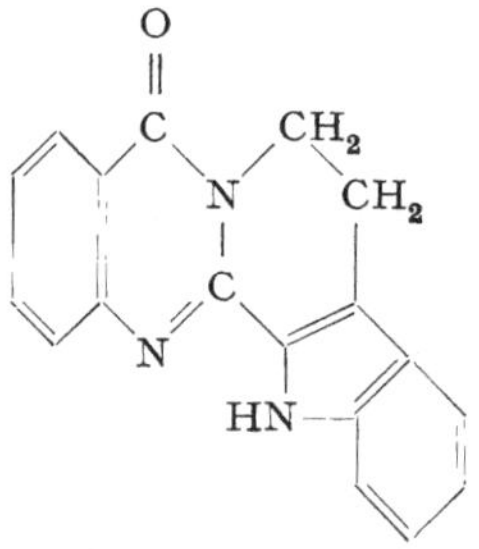

(CXIX.) Rutaecarpine. 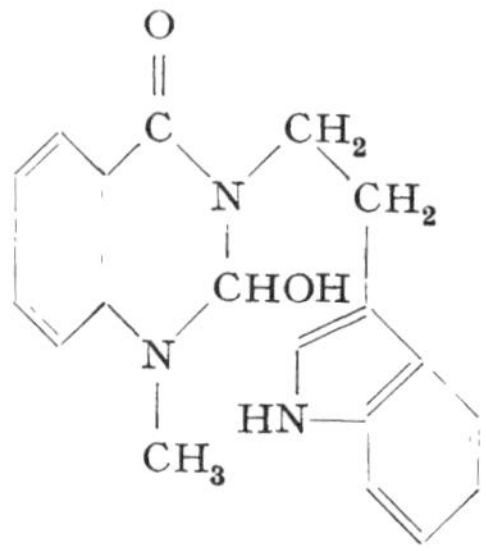   (CXX.) "Evodiamine hydrate".

In confirmation, several syntheses of rutaecarpine have been effected, viz. reduction of *o*-nitrobenzoyl-tryptamine-2-carboxylic acid to the amino compound (CXXI) followed by cyclization with phosphorus oxychloride (*7*), or, more directly, by condensation of 3-keto-3,4,5,6-tetrahydro-β-carboline with either methyl anthranilate (*9*) or isatinic anhydride (*80*).

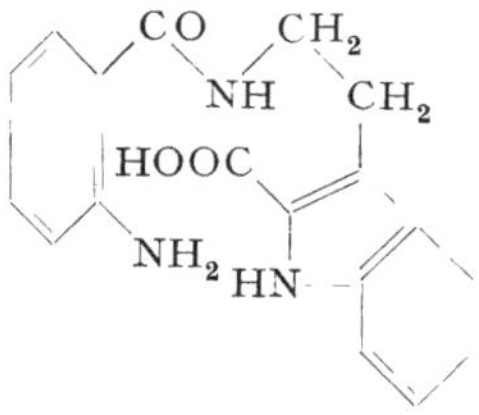

(CXXI.)

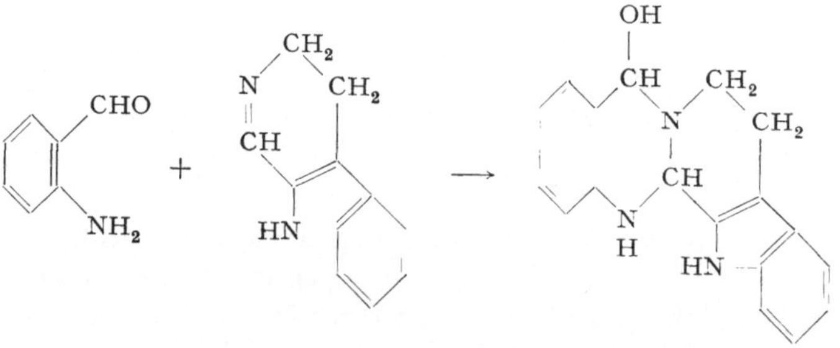

(CXXII.)

Evodiamine has been synthesized by condensing N-methylisatoic anhydride with tryptamine and heating the resulting N-methylanthranoyltryptamine with ethyl orthoformate (*13*).

Finally, Schöpf and Steuer (*96*) condensed *o*-amino-benzaldehyde with dihydro*nor*harman under "physiological conditions" and obtained compound (CXXII) in 75% yield at pH 5. Oxidation with ferricyanide at pH 7 gives rutaecarpine in 70% yield.

VII. Quindoline Alkaloids.

Cryptolepine. One further example of the combination of an anthranilic acid unit with an intact indole nucleus is provided by the unusual purple alkaloid cryptolepine. Cryptolepine, from the roots of *Cryptolepis sanguinolenta* and *C. triangularis*, has the molecular formula $C_{16}H_{12}N_2$ and contains one N-methyl group [Gellert, Raymond-Hamet and Schlittler (*49*)]. It rapidly takes up one mole of hydrogen, the colour of the solution changing from violet to yellow, but the dihydro derivative has not been isolated. The violet colour is regenerated as soon as the solution comes in contact with air. Hydrogenation of the acetate, however, gives tetrahydro-cryptolepine and of the hydrochloride, octahydro-cryptolepine. Selenium dehydrogenation results in the formation of a yellow base, $C_{15}H_{10}N_2$, the methiodide of which is identical with cryptolepine

(CXXIII.) Quindoline.

(CXXIV.) N-Methyl-quindoline (anhydronium form).

(CXXV.) N-Methyl-quindoline (pseudobase form).

hydroiodide, while the tetrahydro derivative of the C_{15}-base gives with methyl iodide tetrahydro-cryptolepine. Clearly, the result of "dehydrogenation" is the removal of the N-methyl group and conversion of a quaternary nitrogen atom to tertiary. When heated with soda lime cryptolepine is converted to a yellow isomeric base, containing one carbon-methyl group, the ultraviolet spectrum of which is almost identical with that of the $C_{15}H_{10}N_2$ base. Cryptolepine evidently contains a very stable ring system with four condensed rings, which, in view of the ultraviolet spectra are probably linear. Comparison shows the $C_{15}H_{10}N_2$ base to be identical with quindoline (CXXIII), while cryptolepine itself is identical with synthetic N-methyl-quindoline, which, on the evidence of ultraviolet spectra of the alkaloid, its salts and derivatives, is best represented in the anhydronium (CXXIV) rather than the pseudobase (CXXV) form.

References.

1. ANET, F. A. L., P. T. GILHAM, P. GOW, G. K. HUGHES and E. RITCHIE: The Chemical Constituents of Australian *Flindersia* Species. III. The Alkaloids of *Flindersia collina* BAIL. Austral. J. Sci. Research **A 5**, 412 (1952).

2. ARNDT, F., L. ERGENER und O. KUTLU: Die Konstitution des 4-Oxy-carbostyrils und seiner Methyl-Derivate. Chem. Ber. **86**, 951 (1953).

3. ASAHINA, Y.: Constitution of Evodiamine and Rutaecarpine. J. pharmac. Soc. Japan, No. **503**, 1 (1924).

4. ASAHINA, Y. and A. FUJITA: Constitution of Rutaecarpine. J. pharmac. Soc. Japan No. **476**, 863 (1921).

5. ASAHINA, Y. und M. INUBUSE: Über die Synthese einer isomeren Verbindung des Dictamnins. Ber. dtsch. chem. Ges. **65**, 61 (1932).

6. — — Über Skimmianin. Ber. dtsch. chem. Ges. **63**, 2052 (1930).

7. ASAHINA, Y., T. IRIE and T. OHTA: Synthesis of Rutaecarpine. J. pharmac. Soc. Japan, No. **543**, 51 (1927).

8. ASAHINA, Y. and K. KASHIWAKI: Chemical Constituents of the Fruits of *Evodia rutaecarpa*. J. pharmac. Soc. Japan, No. **405**, 1293 (1915).

9. ASAHINA, Y., R. H. F. MANSKE and R. ROBINSON: A Synthesis of Rutaecarpine. J. Chem. Soc. (London) **1927**, 1708.

10. ASAHINA, Y. and S. MAYEDA: Evodiamine and Rutaecarpine, Alkaloids of *Evodia rutaecarpa*. J. pharmac. Soc. Japan, No. **416**, 871 (1916).

11. ASAHINA, Y. und S. NAKANISHI: Über Dimethoxy-2,4-dioxy-chinolin. Ber. dtsch. chem. Ges. **63**, 2057 (1930).

12. ASAHINA, Y. and T. OHTA: Transition of Evodiamine into Rutaecarpine. J. pharmac. Soc. Japan, No. **530**, 293 (1926).

13. — — Eine Synthese des Evodiamins. Ber. dtsch. chem. Ges. **61**, 319 (1928).

14. ASAHINA, Y., T. OHTA und M. INUBUSE: Über das Alkaloid von *Skimmia repens*, NAKAI. Ber. dtsch. chem. Ges. **63**, 2045 (1930).

15. BAKER, B. R. and F. J. McEVOY: An Antimalarial Alkaloid from Hydrangea. XXIII. Synthesis by the Pyridine Approach. J. Organ. Chem. (USA) **20**, 136 (1955).

16. BAKER, B. R., F. J. McEVOY, R. E. SCHAUB, J. P. JOSEPH and J. H. WILLIAMS: An Antimalarial Alkaloid from Hydrangea. XXI. Synthesis and Structure of Febrifugine and *iso*Febrifugine. J. Organ. Chem. (USA) **18**, 178 (1953).

17. BAKER, B. R., R. E. SCHAUB, F. J. McEVOY and J. H. WILLIAMS: An Antimalarial Alkaloid from Hydrangea. XII. Synthesis of 3-[β-Keto-γ-(3-hydroxy-2-piperidyl)propyl]-4-quinazolone, the Alkaloid. J. Organ. Chem. (USA) **17**, 132 (1952).

18. BEATTIE, F. S.: Abnormal Biochemical Products of the Rue Anemone. Amer. chem. J. (Johns Hopkins) **40**, 415 (1908).

19. BERINZAGHI, B., A. MURUZABAL, R. LABRIOLA and V. DEULOFEU: Studies on Argentine Plants. VII. The Structure of γ-Fagarine. J. Organ. Chem. (USA) **10**, 181 (1945).

20. BORSCHE, W. und M. WAGNER-ROEMMICH: Über vielkernige kondensierte Systeme mit heterocyclischen Ringen. IX. Liebigs Ann. Chem. **544**, 287 (1940).

21. BRACKEN, A., A. POCKER and H. RAISTRICK: Studies in the Biochemistry of Micro-organisms. Cyclopenin, a Nitrogen-containing Metabolic Product of *Penicillium cyclopium* WESTLING. Biochemic. J. **57**, 587 (1954).

22. BRIGGS, L. H. and R. H. LOCKER: Flavonols from the Bark of *Melicope ternata*. Part I. The Isolation of four new Flavonols, Meliternatin, Meliternin, Ternatin and Wharangin. J. Chem. Soc. (London) **1949**, 2157.

23. BROWN, R. D., L. J. DRUMMOND, F. N. LAHEY and W. C. THOMAS: Alkaloids of the Australian Rutaceae: *Acronychia baueri*. II. Some Reactions of the Alkaloid Acronycine. Austral. J. Sci. Research **A 2**, 622 (1949).

24. BROWN, R. D. and F. N. LAHEY: The Ultraviolet Absorption Spectra of the Acridone Alkaloids. I. Compounds containing the Acridone Nucleus. Austral. J. Sci. Research **A 3**, 593 (1950).

25. BROWN, R. F. C.: The Chemical Constituents of Australian *Flindersia* Species. VII. The Synthesis of Derivatives of Kokusagininic and Skimmianinic Acids. Austral. J. Chem. **8**, 121 (1955).

26. BROWN, R. F. C., P. T. GILHAM, G. K. HUGHES and E. RITCHIE: The Chemical Constituents of Australian *Flindersia* Species. V. The Constituents of *F. maculosa* LINDL. Austral. J. Chem. **7**, 181 (1954).

27. BROWN, R. F. C., J. J. HOBBS, G. K. HUGHES and E. RITCHIE: The Chemical Constituents of Australian *Flindersia* Species. VI. The Structure and Chemistry of Flindersine. Austral. J. Chem. **7**, 348 (1954).

28. BROWN, R. F. C., G. K. HUGHES and E. RITCHIE: A Synthesis of Flindersine. Chem. and Ind. **1955**, 1385.

29. CANNON, J. R., G. K. HUGHES, K. G. NEILL and E. RITCHIE: Alkaloids of the Australian Rutaceae: *Evodia xanthoxyloides* F. MUELL. III. Structures of the Coloured Alkaloids, Evoxanthidine, Xanthevodine and Xanthoxoline. Austral. J. Sci. Research **A 5**, 406 (1952).

30. CHAKRAVARTI, (Mrs.) D., R. N. CHAKRAVARTI and S. C. CHAKRAVARTI: Alkaloids of *Glycosmis arborea*. Part I. Isolation of Arborine and Arborinine: The Structure of Arborine. J. Chem. Soc. (London) **1953**, 3337.

31. CHATTERJEE, A. and S. GHOSH MAJUMDAR: Constitution and Synthesis of Glycosin, the New Alkaloid of *Glycosmis pentaphylla* RETZ. DC. J. Amer. Chem. Soc. **75**, 4365 (1953).

32. CHOU, T. Q., F. Y. FU and Y. S. KAO: Antimalarial Constituents of Chinese Drug, Ch'ang Shan, *Dichroa febrifuga* LOUR. J. Amer. Chem. Soc. **70**, 1765 (1948).

33. COOKE, R. G. and H. F. HAYNES: The Alkaloids of *Evodia littoralis* ENDL. Austral. J. Chem. **7**, 273 (1954).

34. — — Synthesis of Dihydrodictamnine. Personal communication from R. G. COOKE.

35. CORNFORTH, J. W. and A. T. JAMES: Structure of a Naturally Occurring Antagonist of Dihydrostreptomycin. Biochemic. J. **63**, 124 (1956).

36. CROW, W. D. and J. R. PRICE: Alkaloids of the Australian Rutaceae: *Melicope fareana*. V. The Structure of the Alkaloids. Austral. J. Sci. Research **A 2**, 282 (1949).

37. CUNNINGHAM, K. G. and G. G. FREEMAN: The Isolation and Some Chemical Properties of Viridicatin, a Metabolic Product of *Penicillium viridicatum* WESTLING. Biochemic. J. **53**, 328 (1953).

38. DALGLIESH, C. E.: Biological Degradation of Tryptophan. Quart. Rev. Chem. Soc. (London) **5**, 227 (1951).

39. DEAN, F. M.: Naturally Occurring Coumarins. Fortschr. Chem. organ. Naturstoffe **9**, 225 (1952).

40. DEULOFEU, V. and D. BASSI: Espectros de Absorcion Ultravioleta de la Skimianina, Fagarina (γ-Fagarina) y Derivados. Anal. Asoc. Quim. Argentina **40**, 249 (1952).

41. DEULOFEU, V., R. LABRIOLA and J. DE LANGHE: Studies on Argentine Plants. V. Identification and Characterization of some Alkaloids in *Fagara Coco* (GILL) ENGL. J. Amer. Chem. Soc. **64**, 2326 (1942).

42. DRUMMOND, L. J. and F. N. LAHEY: Alkaloids of the Australian Rutaceae: *Acronychia baueri*. III. The Structure of Acronycine. Austral. J. Sci. Research **A 2**, 630 (1949).

43. EASTWOOD, F. W., G. K. HUGHES and E. RITCHIE: Alkaloids of the Australian Rutaceae: *Evodia xanthoxyloides* F. MUELL. IV. The Structures of Evoxine and Evoxoidine. Austral. J. Chem. **7**, 87 (1954).

44. EDWARD, J. T.: Heterocyclic Compounds. Annu. Rep. Chem. Soc. (London) **51**, 240 (1954).

45. ELDERFIELD, R. C., W. J. GENSLER, T. H. BEMBRY, C. B. KREMER, J. D. HEAD, F. BRODY and R. FROHARDT: Synthesis of 2-Phenyl-4-chloroquinolines. J. Amer. Chem. Soc. **68**, 1272 (1946).

46. ELIASBERG, J. und P. FRIEDLÄNDER: Über einige Condensationen des o-Amidobenzaldehyds. Ber. dtsch. chem. Ges. **25**, 1752 (1892).

47. EWINS, A. J.: The Constitution and Synthesis of Damascenine, the Alkaloid of *Nigella damascena*. J. Chem. Soc. (London) **101**, 544 (1912).

48. GELL, R. J., G. K. HUGHES and E. RITCHIE: Alkaloids of *Evodia alata* F. MUELL. Austral. J. Chem. **8**, 114 (1955).

49. GELLERT, E., RAYMOND-HAMET und E. SCHLITTLER: Die Konstitution des Alkaloids Cryptolepin. Helv. Chim. Acta **34**, 642 (1951).

50. GHOSE, T. P., S. KRISHNA, K. S. NARANG and J. N. RÂY: Vasicine. J. Chem. Soc. (London) **1932**, 2740.

51. GOODSON, J. A.: The Alkaloids of the Seeds of *Delphinium elatum*. J. Chem. Soc. (London) **1943**, 139.

52. GRESHOFF, M.: Recherches sur l'Echinopsine, nouvel alcaloïde crystallisé. Rec. trav. chim. Pays-Bas **19**, 360 (1900).

53. GUENTHER, E.: The Essential Oils. New York: Van Nostrand. 1948.

54. HANFORD, W. E. and R. ADAMS: The Structure of Vasicine. II. Synthesis of Desoxyvasicine. J. Amer. Chem. Soc. **57**, 921 (1935).

55. HAQ, M. A., M. L. KAPUR and J. N. RÂY: Quinoline Derivatives. Part I. Furanoquinolines. J. Chem. Soc. (London) **1933**, 1087.

56. HAYS, E. E., I. C. WELLS, P. A. KATZMAN, C. K. CAIN, F. A. JACOBS, S. A. THAYER, E. A. DOISY, W. L. GABY, E. C. ROBERTS, R. D. MUIR, C. J. CARROLL, L. R. JONES and N. J. WADE: Antibiotic Substances Produced by *Pseudomonas aeruginosa*. J. Biol. Chem. **159**, 725 (1945).

57. HENRY, T. A.: The Plant Alkaloids, p. 751, 4th ed. London: J. and A. Churchill Ltd. 1949.

58. Hughes, G. K., N. K. Matheson, A. T. Norman and E. Ritchie: The Demethylation of Methoxyacridones. Austral. J. Sci. Research A 5, 206 (1952).

59. Hughes, G. K. and K. G. Neill: Alkaloids of the Australian Rutaceae: *Evodia xanthoxyloides*. I. Evoxanthine. Austral. J. Sci. Research A 2, 429 (1949).

60. Hughes, G. K., K. G. Neill and E. Ritchie: The Synthesis of Melicopicine and Some Trimethoxy-10-methylacridones. Austral. J. Sci. Research A 3, 497 (1950).

61. — — — Alkaloids of the Australian Rutaceae: *Evodia xanthoxyloides* F. Muell. II. Isolation of the Alkaloids from the Leaves. Austral. J. Sci. Research A 5, 401 (1952).

62. Hughes, G. K. and E. Ritchie: Experiments on the Synthesis of the Acridone Alkaloids. Austral. J. Sci. Research A 4, 423 (1951).

63. Hutchings, B. L., S. Gordon, F. Ablondi, C. F. Wolf and J. H. Williams: An Antimalarial Alkaloid from Hydrangea. III. Degradation. J. Organ. Chem. (USA) 17, 19 (1952).

64. Johnstone, R., J. R. Price and Sir A. R. Todd: unpublished.

65. Kermack, W. O., W. H. Perkin and R. Robinson: Harmine and Harmaline. Part V. The Synthesis of *nor*Harman. J. Chem. Soc. (London) 119, 1602 (1921).

66. King, F. E., K. G. Latham and M. W. Partridge: Synthesis of Furano (2′,3′-2,3)quinoline. Personal communication from Dr. F. E. King; K. G. Latham, Ph. D. Thesis, University of Nottingham, 1953.

67. Koepfli, J. B., J. A. Brockman, Jr. and J. Moffat: The Structure of Febrifugine and *iso*Febrifugine. J. Amer. Chem. Soc. 72, 3323 (1950).

68. Koepfli, J. B., J. F. Mead and J. A. Brockman, Jr.: Alkaloids of *Dichroa febrifuga*. I. Isolation and Degradative Studies. J. Amer. Chem. Soc. 71, 1048 (1949).

69. Lahey, F. N., J. A. Lamberton and J. R. Price: Alkaloids of the Australian Rutaceae. The Structure and Reactions of Acronycidine. Austral. J. Sci. Research A 3, 155 (1950).

70. Lamberton, J. A. and J. R. Price: Alkaloids of the Australian Rutaceae: *Medicosma cunninghamii* Hook F. Austral. J. Chem. 6, 173 (1953).

71. — — Alkaloids of the Australian Rutaceae: *Acronychia baueri* Schott. IV. Alkaloids Present in the Leaves. Austral. J. Chem. 6, 66 (1953).

72. Leete, E., L. Marion and I. D. Spenser: The Biogenesis of Alkaloids. XIV. A Study of the Biosynthesis of Damascenine and Trigonelline. Canad. J. Chem. 33, 405 (1955).

73. Lightbown, J. W.: An Antagonist of Dihydrostreptomycin and Streptomycin Produced by *Pseudomonas pyocyanea*. Nature (London) 166, 356 (1950).

74. — An Antagonist of Streptomycin and Dihydrostreptomycin produced by *Pseudomonas aeruginosa*. J. Gen. Microbiol. 11, 477 (1954).

75. Marion, L.: The Indole Alkaloids. In: R. H. F. Manske and H. L. Holmes, The Alkaloids, Vol. II, p. 369. New York: Academic Press. 1952.

76. Matthes, H. und E. Schreiber: Über hautreizende Hölzer. Ber. dtsch. pharm. Ges. 24, 385 (1914).

77. McKenzie, A. W. and J. R. Price: Alkaloids of the Australian Rutaceae: *Glycosmis pentaphylla* (Retz.) Correa. Austral. J. Sci. Research A 5, 579 (1952).

78. Musajo, L., C. A. Benassi and A. Parpajola: Isolation of Kynurenine and 3-Hydroxykynurenine from Human Pathological Urine. Nature (London) 175, 855 (1955).

79. OHTA, T.: personal communication; cf. *110a*.

80. — Addendum on the Synthesis of Rutaecarpine. J. pharmac. Soc. Formosa **51**, 2 (1938).

81. — Furoquinolines. I. Catalytic Reduction of Dictamnine and Skimmianine. J. pharmac. Soc. Japan **73**, 63 (1953).

82. OHTA, T. and T. MIYAZAKI: Furoquinolines. II. Catalytic Reduction of Skimmianine. Pharmac. Bull. (Japan) **1**, 184 (1953).

83. OHTA, T., T. MIYAZAKI and Y. MORI: Furoquinolines. III. Hydrogenation of Skimmianine with PdO and Raney Nickel Catalysts. J. pharmac. Soc. Japan **74**, 708 (1954).

84. — — — Furoquinolines. IV. Addenda on the Skimmia Bases. Annu. Rep. Tokyo Coll. Pharm. **4**, 7 (1954).

85. OHTA, T. and Y. MORI: Furoquinolines. V. An Attempted Synthesis of Dictamnal. Annu. Rep. Tokyo Coll. Pharm. **4**, 13 (1954).

85a. — — Furoquinolines. VI. Establishment of the Linear Tricyclic Structures for Dictamnine and Skimmianine. Pharmac Bull. (Japan) **3**, 396 (1955).

86. OPENSHAW, H. T.: Quinoline Alkaloids, other than those of Cinchona. In: R. H. F. MANSKE and H. L. HOLMES, The Alkaloids. Vol. III, p. 65. New York: Academic Press. 1953.

87. — The Quinazoline Alkaloids. In: R. H. F. MANSKE and H. L. HOLMES, The Alkaloids. Vol. III, p. 101. New York: Academic Press. 1953.

88. PALLARES, E. S.: The Structure of the Water-Insoluble Pigment of the Bark of "Sangre de Drago". Arch. Biochemistry **10**, 235 (1946).

89. PRICE, J. R.: Acridine Alkaloids. In: R. H. F. MANSKE and H. L. HOLMES, The Alkaloids, Vol. II, p. 353. New York: Academic Press. 1952.

90. — Alkaloids of the Australian Rutaceae: *Melicope fareana*. IV. Some Reactions of 1-Methyl-4-quinolone-3-carboxylic acid, a Degradation Product of the Alkaloids. Austral. J. Sci. Research **A 2**, 272 (1949).

91. PRICE, J. R. and J. P. SHELTON: The Ultra-violet Absorption Spectra of Furoquinoline Alkaloids. Unpublished.

92. ROBINSON, R.: The Structural Relations of Natural Products. Oxford: Univ. Press. 1955.

93. SCHLÄGER, J. und W. LEEB: Konstitution und Synthese von Cusparein. Monatsh. Chem. **81**, 714 (1950).

94. SCHÖPF, C. und G. LEHMANN: Über die Alkaloide der Angosturarinde: Die Synthese des Chinaldins und α-*n*-Amylchinolins unter physiologischen Bedingungen. Liebigs Ann. Chem **497**, 7 (1932).

95. SCHÖPF, C. und F. OECHLER: Zur Frage der Biogenese des Vasicins (Peganins). Die Synthese des Desoxyvasicins unter physiologischen Bedingungen. Liebigs Ann. Chem. **523**, 1 (1936).

96. SCHÖPF, C. und H. STEUER: Zur Frage der Biogenese des Rutaecarpins und Evodiamins. Die Synthese des Rutaecarpins unter zellmoglichen Bedingungen. Liebigs Ann. Chem. **558**, 124 (1947).

97. SCHÖPF, C. und K. THIERFELDER: Die Aldolkondensation zwischen Aldehyden und β-Ketosäuren und ihre Bedeutung für die Biogenese einiger Naturstoffe. Liebigs Ann. Chem. **518**, 127 (1935).

98. SPÄTH, E. und O. BRUNNER: Über die Angostura-Alkaloide. I. Synthese des Cusparins. Ber. dtsch. chem. Ges. **57**, 1243 (1924).

99. SPÄTH, E. und H. EBERSTALLER: Über die Angostura-Alkaloide. II. Synthese des Galipins. Ber. dtsch. chem. Ges. **57**, 1687 (1924).

100. SPÄTH, E. und A. KOLBE: Über das Echinopsin. Monatsh. Chem. **43**, 469 (1922).

101. Spâth, E. und F. Kuffner: Über Peganin und Vasicin. Ber. dtsch. chem. Ges. **67,** 1494 (1934).

102. Spâth, E., F. Kuffner und N. Platzer: Die Konstitution von Peganin (Vasicin). Ber. dtsch. chem. Ges. **68,** 497 (1935).

103. — — — Synthese und Konstitution des Peganins (Vasicins). Ber. dtsch. chem. Ges. **68,** 699 (1935).

104. Spâth, E. und E. Nikawitz: Die Konstitution des Peganins. Ber. dtsch. chem. Ges. **67,** 45 (1934).

105. Spâth, E. und G. Papaioanou: Über Phenolbasen der Angosturarinde: Synthese des Galipolins. Monatsh. Chem. **52,** 129 (1929).

106. Spâth, E. und J. Pikl: 2-*n*-Amyl-4-methoxychinolin, ein basischer Bestandteil der Angostura-Rinde (IV. Mitt. über Angostura-Alkaloide). Ber. dtsch. chem. Ges. **62,** 2244 (1929).

107. — — Über neue Basen der Angosturarinde: Chinolin, 2-Methyl-chinolin, 2-*n*-Amyl-chinolin und 1-Methyl-2-keto-1,2-dihydro-chinolin. Monatsh. Chem. **55,** 352 (1930).

108. Spâth, E. und N. Platzer: Eine neue Synthese von Pegen-(9) und von Peganin (IX. Mitt. über Peganin). Ber. dtsch. chem. Ges. **69,** 255 (1936).

109. Teas, H. J. and E. G. Anderson: Accumulation of Anthranilic Acid by a Mutant of Maize. Proc. Nat. Acad. Sci. (USA) **37,** 645 (1951).

110. Terasaka, M.: Alkaloids of the Root Bark of *Orixa japonica* Thunb. III. J. pharmac. Soc. Japan **53,** 1046 (1933).

110 a. Terasaka, M., K. Narahashi and T. Ohta: Alkaloids of the Root-bark of *Orixa japonica* Thunb. VI. On Kokusaginine. J. pharmac. Soc. Japan **75,** 1040 (1955).

111. Terasaka, M., T. Ohta and K. Narahashi: Alkaloids of the Root Bark of *Orixa japonica* Thunb. V. On the Structure of Kokusagine. Pharmac. Bull. (Japan) **2,** 159 (1954).

112. Thoms, H.: Die chemischen Inhaltsstoffe der Rutaceen. VII. Über den weißen Diptam, *Dictamnus albus* L. Ber. dtsch. pharm. Ges. **33,** 68 (1923).

113. Wehmer, C.: Die Pflanzenstoffe. Bd. I. Jena: G. Fischer. 1929.

114. Wells, I. C.: Antibiotic Substances produced by *Pseudomonas aeruginosa.* Syntheses of Pyo I b, Pyo I c and Pyo III. J. Biol. Chem. **196,** 331 (1952).

115. Wells, I. C., W. H. Elliott, S. A. Thayer and E. A. Doisy: Ozonization of some Antibiotic Substances produced by *Pseudomonas aeruginosa.* J. Biol. Chem. **196,** 321 (1952).

116. Wiss, O.: Die oxydative Spaltung der 3-Oxyanthranilsäure. Z. Naturforsch. **9** b, 740 (1954).

117. Yunusov, S. and G. P. Sidyakin: Alkaloids of *Haplophyllum perforatum, H. pedicellatum, H. dubium, H. bucharicum* and *H. versicolor.* I. Zhur. Obschei Khimii **22,** 1055 (1952).

(Received, November 28, 1955.)

Recent Developments in the Chemistry and Pharmacology of Rauwolfia Alkaloids.

By Asima Chatterjee and Satyesh C. Pakrashi, Calcutta, India, and G. Werner, São Paulo, Brazil.

The authors of the First Part are, A. Chatterjee and S. C. Pakrashi, and the author of the Second Part is G. Werner.

With 18 Figures.

Contents.

Acknowledgement. The authors express their grateful thanks to Dr. D. Chatterjee, Superintendent, Indian Botanic Gardens, Shibpur, India, for the list of the
Rauwolfia species and the plates of *R. canescens* and *R. serpentina.*

First Part:

Chemistry of the Rauwolfia Alkaloids.

I. Introduction.

Rauwolfia represents an important genus of the botanical family
Apocynaceae which holds a unique position in the vegetable kingdom,
since it produces both cardiac glycosides and alkaloids of high therapeutic
efficiency (*74, 208, 247, 351, 397, 398*).

The discovery of the genus *Rauwolfia* dates back to the 16th century. Leonhart
Rauwolf (*Leonardus Rauvolfius*), a German physician and botanist, made an
expedition to Asia and Africa to explore the field of medicinal plants. He published
an account of this excursion in 1582, which created considerable interest. A few
years later, when a new genus was added to the family *Apocynaceae*, it was called
Rauwolfia in his honor. The correct name of this new genus would be Rauvolfia
according to the International Rules of Botanical Nomenclature (*232, 348*).

Up till now, 130 species of *Rauwolfia* (*106, 147–151, 159, 268, 275–277,
323, 366, 376*) have been described and are found in moist tropical and semi-
tropical regions of both hemispheres. To their list published recently (*43a*)
the following species should be added:

R. concolor M. Pichon (Madagascar); *R. confertiflora* M. Pichon (Madagascar);
R. degeneri Sherff (Hawaiian Islands); *R. densiflora* Warb* (Thailand, Indochina,
Phillipines); *R. forbesii* Sherff (Hawaiian Islands); *R. fruticosa* Burck (Indochina,
Malaya, Phillipines); *R. gracilis* Koord. et Val. (Malaya); *R. helleri* Sherff
(Hawaiian Islands); *R. indosinensis* M. Pichon (Indochina); *R. linearisepala*
Guillaumin (North Caledonia); *R. macrantha* K. Schum ex Ule (Tropical America);
R. mauiensis Sherff (Hawaiian Islands); *R. media* M. Pichon (Madagascar);
R. mollis S. Moore (Tropical America); *R. molokaiensis* Sherff (Hawaiian Islands);
R. nana E. A. Bruce (N. Rhodesia); *R. obversa* (Miq.) Koord. (Tropical Asia, Java);
R. remotiflora Degan, and Sherff (Hawaiian Islands); *R. rhonhofiae* Markgraf

---

    * According to the rules of Botanical Nomenclature, a new name should be
given to this species to avoid confusion with *R. densiflora* (Thw.) Benth. ex Hook. f.

(Tropical America); *R. sarapiquensis* R. E. Woodson (Costa Rica); *R. spectabilis* (Miq.) Boerl. (Indochina, Malaya, Phillipines); *R. tetraphylla* L. (Tropical America); *R. viridis* (Muell Arg.) Guillaumin (N. Caledonia); and *R. volkensii* (K. Schum) Stapf (Tropical Africa).

Table 1. Rauwolfia Species Investigated Chemically till 1955.

Rauwolfia canescens L.
R. serpentina Benth
R. obscura K. Schum
R. vomitoria Afzel
R. heterophylla Willd ex Roem and Schult
R. hirsuta Jacq.
R. densiflora Warb
R. perakensis King and Gamble
R. indecora R. E. Woodson
R. micrantha Hook

R. tetraphylla L.
R. sellowii Muell Arg.
R. semperflorens Muell Arg. Schlechter
R. caffra Sond.
R. natalensis Sond.
R. mombasiana Stapf
R. grandiflora Mart ex A. Dc.
R. cumminsii Stapf
R. verticillata Lour. H. Bn.
R. beddomei Hook. f.
R degeneri Sherff

Chemical investigation of *Rauwolfia* plants shows the presence of therapeutically active alkaloids (distributed in stems, roots, leaves, and fruits).

In a pertinent review published in 1953 in the present *Series* by Chatterjee (*43a*) the literature on *Rauwolfia* was covered up to 1952. Since then the increasing medicinal use of *Rauwolfia* bases as sedative and hypotensive drugs has aroused a great interest in these plants. Thus, a voluminous literature has accumulated during the past three years.

II. The Alkaloids of *Rauwolfia canescens* L.

Rauwolfia canescens L. (*Fig. 1*) grows abundantly in India, Australia (*374*) and the East Indies. Its occurrence along the Atlantic coast and in some other regions of Colombia has been reported by Vergara

Fig. 1. *Rauwolfia canescens* L.

(*372*). The presence of alkaloids in this species was first observed by Greshoff (*131*) in 1890. A detailed investigation into its chemical and pharmacological aspects was initiated by Chatterjee (née Mookerjee) in 1941. It was demonstrated that the plant contains a physiologically active alkaloid, rauwolscine, distributed all over the roots, stems and leaves. The percentage of the alkaloid is highest in the leaves. The "Alkaloid saturation point" in the plant is attained in October-December as indicated from the study of the seasonal variations of the alkaloids present in the tissues. From the same species growing in Madras (India) (*359a, 360*) as well as from a commercial sample (*334, 135*) eighteen more alkaloids have been isolated and appear in *Table 2*.

Table 2. The Alkaloids of Rauwolfia canescens L.

Name and References	Source	Composition	M. p. (decomp.)	Optical Rotation $[\alpha]_D$
Rauwolscine (*39–44, 56–65*)	leaves, root, stem	$C_{21}H_{26}O_3N_2$	232°	— 40° (alcohol)
Yohimbine (*135*)	root	$C_{21}H_{26}O_3N_2$	234–36°	— 83° (pyridine)
β-Yohimbine (*157*)	root	$C_{21}H_{26}O_3N_2$	246–49°	— 48° (pyridine)
ψ-Yohimbine (*359*)	root	$C_{21}H_{26}O_3N_2$	265–78°	+ 27° ± 20° (pyridine)
Reserpine (*214*)	root	$C_{33}H_{40}O_9N_2$	254°	— 121.8° (chlorof.)
ψ-Reserpine (*215a*)	root	$C_{32}H_{38}O_9N_2$	257–58°	— 65° ± 2° (chl.)
Canescine (*215, 359*) = = Recanescine (*257*) = = Deserpidine (*334*)	root	$C_{32}H_{38}O_8N_2$	228–32°	— 137° ± 1° (chloroform)
Ajmalicine (*204*)	root	$C_{21}H_{24}O_3N_2$	250–51°	— 60° (chloroform)
Reserpinine (*204*)	root	$C_{22}H_{26}O_4N_2$	233–34°	— 120° (chloroform)
Isoreserpinine (*360*)	leaves	$C_{22}H_{26}O_4N_2$	225–26°	— 5° (pyridine)
Aricine (*360*)	leaves	$C_{22}H_{26}O_4N_2$	186–87°	— 69° (pyridine)
Reserpiline (*360*)	leaves	$C_{23}H_{28}O_5N_4$	205–7°	— 38.4° (ethanol)
Isoreserpiline (*360*)	leaves	$C_{23}H_{28}O_5N_2$	211–12°	— 82° (pyridine)
Ajmaline (*204*)	root	$C_{20}H_{26}O_2N_2$	158–60°	+ 119° (chloroform)
Sarpagine (*21, 204*) = = Raupine (*25, 26*)	root	$C_{19}H_{22}O_2N_2$	340–44°	+ 52° (pyridine)
Raunescine (*22*)	root	$C_{22}H_{28}O_3N_2$	228–29°	+ 57° (alcohol)
Serpentine (*135*)	root	$C_{21}H_{20}O_3N_2$	157–58°	
Raunescine (*160*)		$C_{31}H_{36}O_8N_2$	160–70°	— 74° (chloroform)
Isoraunescine (*160*)		$C_{31}H_{36}O_8N_2$	241–42.5°	— 70° (chloroform)
A new β-carboline alkaloid (*359a*)	root	$C_{14}H_{16}N_2$	190–92°	— 132° (pyridine)

Rauwolscine, $C_{21}H_{26}O_3N_2$, the only base in *R. canescens* L. occurring in the Bengal variety, has been isolated from the cold alcoholic percolate of the leaves. The crude base was purified through its hydrochloride and subsequent chromatography of the free base. It represents a tetra-hydro-β-carboline derivative as indicated by its color reactions (*97, 315*).

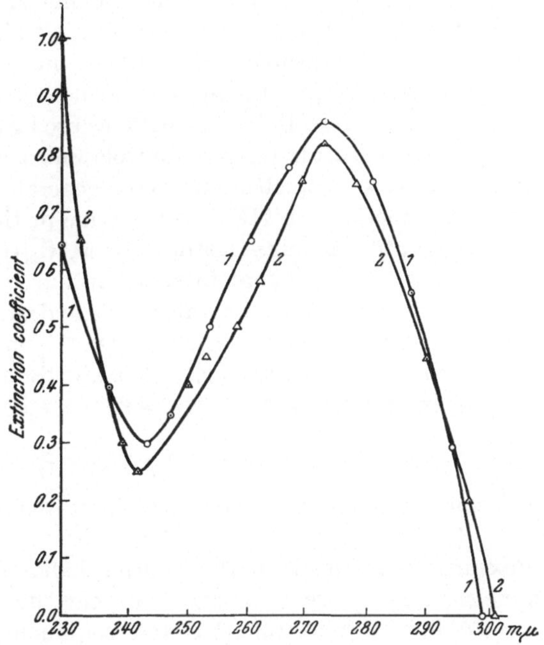

Fig. 2. Ultraviolet spectra, in water, of the hydrochlorides of yohimbine (1) and rauwolscine (2), according to CHATTERJEE (*40*). [From: J. Indian Chem. Soc. *18*, 485 (1941).]

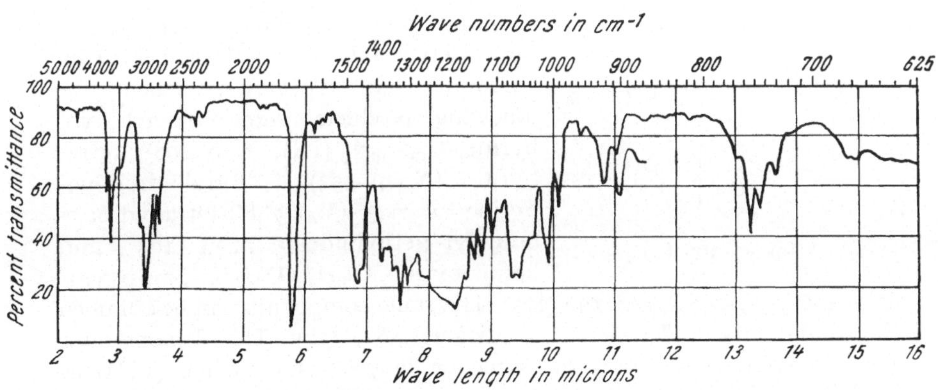

Fig. 3. Infrared spectrum of rauwolscine in chloroform and Nujol, according to CHATTERJEE and PAKRASHI (*59*). [From: J. Indian Chem. Soc. *31*, 29 (1954).]

Rauwolscine is a monoacidic tertiary base which readily forms well crystalline salts with mineral and organic acids. It contains an indole chromophoric group (*280, 324, 385*) and shows characteristic spectral maxima at 232 mμ, 283 and 292; for the hydrochloride cf. p. 351.

The infrared spectrum as given in *Fig. 3* shows bands at 2.8 μ, 2.87 μ, 5.82 μ, 6.0–6.6 μ and 13.3 μ indicating, respectively, —OH, >NH, and —COOCH$_3$ groups, an α,β-disubstituted indole nucleus, and an o-disubstituted benzene ring (*283*). Chemical evidence of the presence of a carbomethoxyl group in rauwolscine has been secured by saponifying the base to a methoxyl-free acid, termed rauwolscinic acid (I, p. 353) (*39, 40*). The latter is amphoteric in character and regenerates rauwolscine upon esterification with methanolic HCl. Rauwolscine methiodide yields with alkali the methoxyl-free betaine isorauwolscine, C$_{21}$H$_{26}$O$_3$N$_2$. The hydroxyl group in rauwolscine like that in strophanthin (*110*) has been reported to be resistant towards acid-catalyzed dehydrating and oxidizing agents, even to chromic acid (*57, 233, 273*). Upon acetylation, however, it forms an O-acyl derivative (*59, 118, 222*), the characteristic band of which appears at 8.1 μ.

The Structure of Rauwolscine (cf. Chart 1).

On thermal decomposition rauwolscine yields β-ethylindole (II) and harman (III, p. 353) (*41*). When fused with KOH the base decomposes (*41, 43*) giving a mixture from which harman, isophthalic acid (IV) and indole-2-carboxylic acid (V) have been isolated. Zinc dust distillation of rauwolscine breaks it up into skatole (VI) and isoquinoline (VII) (*42*).

The isolation of these products led Chatterjee (*42*) to suggest the pentacyclic structure (VIII) for the basic ring system of rauwolscine.

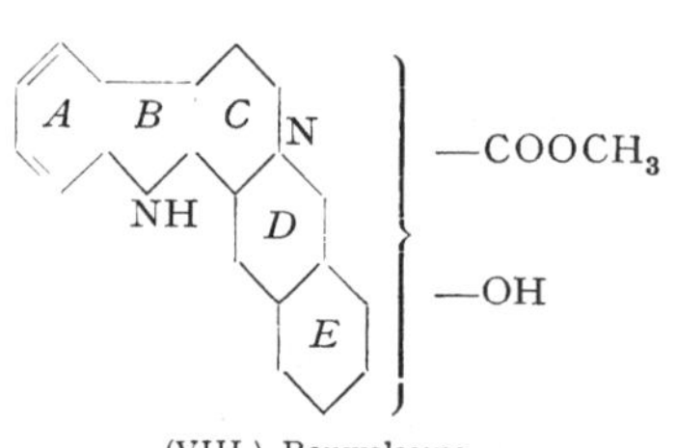

(VIII.) Rauwolscine.

Final proof of this system in the rauwolscine molecule has been gained by selenium dehydrogenation, when the following products were obtained: yobyrine, C$_{19}$H$_{16}$N$_2$ (IX), m. p. 208°, tetrabyrine (X, p. 353) (= tetrahydroisoyobyrine or 2,3'-(5',6',7',8')-tetrahydroisoquinolyl-3-ethylindole), m. p. 165°, and ketoyobyrine, C$_{20}$H$_{16}$ON$_2$ (XI), m. p. 320° (*61, 58, 60, 56, 57, 78, 192, 193, 333*). The same compounds can be obtained from yohimbine under similar conditions (*382, 177*). This demonstrates that rauwolscine possesses the same pentacyclic ring system (VIII) as yohimbine (XII) (*7, 8, 146, 355, 79, 80, 97, 191, 203, 199, 239, 240, 265, 298, 192, 384, 299, 300, 341*). The formation of ketoyobyrine from rauwolscine determines the C$_{(16)}$-position of the carbomethoxyl (*78, 193, 392, 333*). Additional evidence of this position was gained by the isolation of methyl yobyrine (XIII) from the selenium dehydrogenation products of

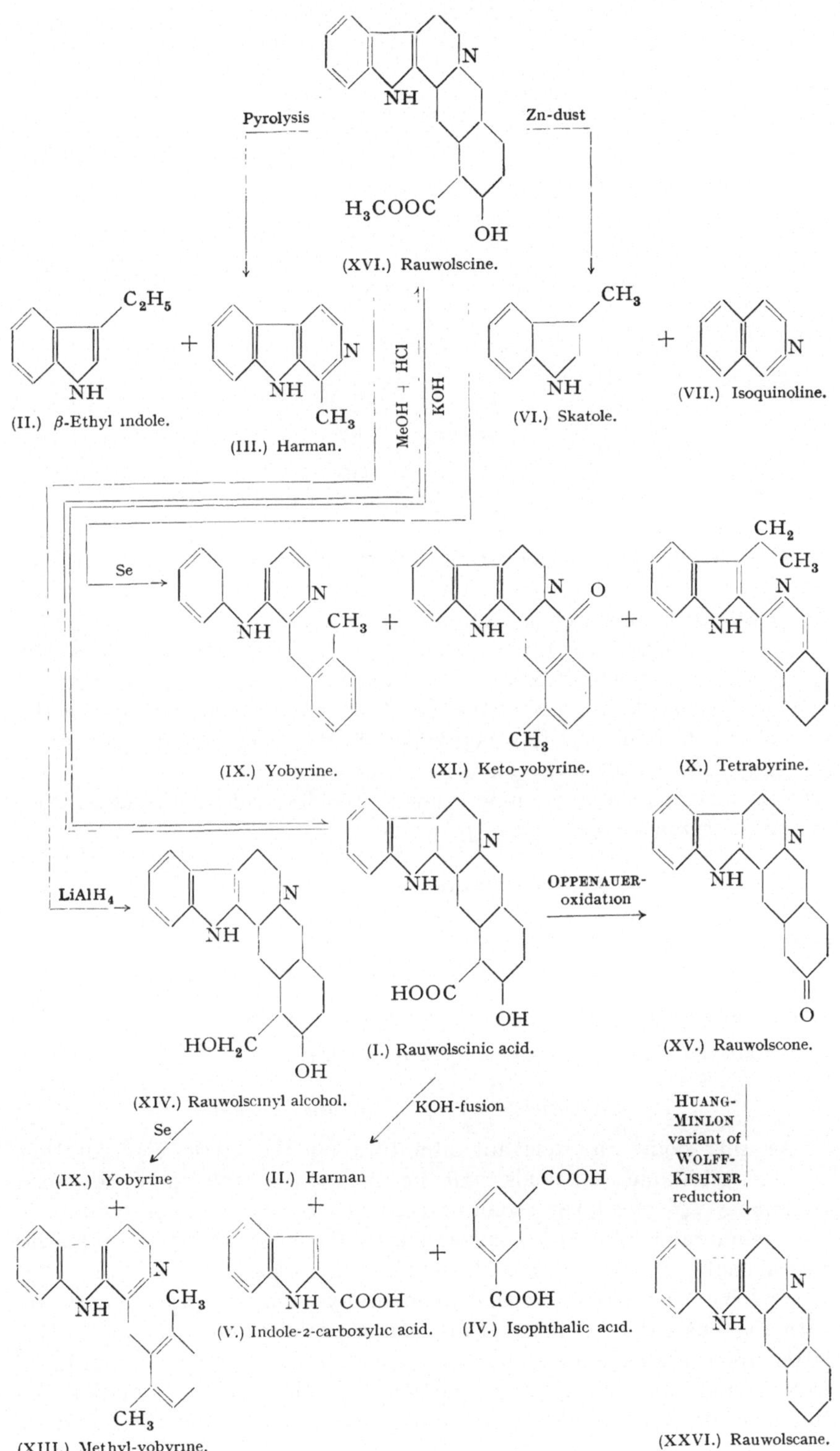

Chart 1. Conversions of Rauwolscine.

rauwolscinyl alcohol (*203, 58*) (XIV, *Chart 1*), prepared from rauwolscine by lithium aluminum hydride reduction (*55*).

(XII.) Yohimbine.

The nature and position of the hydroxyl group in rauwolscine was clarified by OPPENAUER oxidation according to WITKOP (*384*). During this oxidation rauwolscone (XV, Chart 1), $C_{19}H_{22}N_2O$, m. p. 230—232° (dec.), a carbomethoxy-free ketone appears (*58*) which secures the $C_{(17)}$-position for the hydroxyl group. Based on the experiments mentioned and some other data CHATTERJEE and PAKRASHI (*63*) have proposed the rauwolscine formula (XVI, p. 353) which includes five asymmetric centers (at 3, 15, 16, 17 and 20). They have also shown that rauwolscine is a stereoisomer of yohimbine, nine isomers of which had been isolated from various natural sources (*Table 3*).

Table 3. Yohimbine and its Isomers.

Yohimbine (*146, 355*)	α-Yohimbine (= Corynanthidine)
Corynanthine (*113–116, 263, 264, 340*)	(*373, 176*)
β-Yohimbine (*142, 138*)	Serpine (*48, 52*)
ψ-Yohimbine (*180, 181, 200, 137*)	Isorauhimbine (*228, 155*)
Allo-yohimbine (*140, 142, 137 a*)	3-Epi-α-yohimbine (*4, 3*)

Mitraphylline (*236*) is not a yohimbine isomer (*236 a*).

The Stereochemistry of Yohimbine Alkaloids.

At this point the relevant literature on the steric configuration of the yohimbine alkaloids will be mentioned, the more so since **yohimbine** (*146*, cf. *227*) and some of its congeners, viz. β- and ψ-yohimbines occur simultaneously with rauwolscine in *R. canescens* (*157, 359 a*). The spatial configuration of yohimbine was first investigated by WITKOP (*386*). He observed that yohimbic acid (demethyl-yohimbine), upon treatment with thallous oxide at 300°, yields chanodesoxy-yohimbol (XVII). The latter upon hydrogenation, followed by HOFMANN degradation, produces D-N-methyl-*trans*-decahydro-isoquinoline (XVIII). The isolation of this *trans*-derivative induced WITKOP to suggest that the rings D/E in

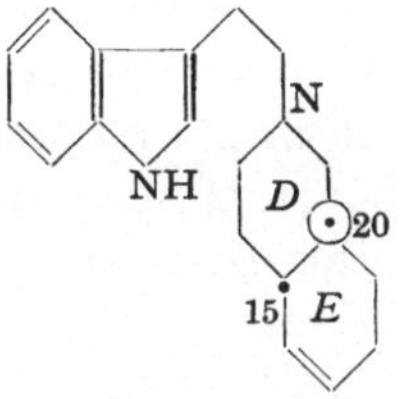

(XVII.) Chanodesoxy-yohimbol.

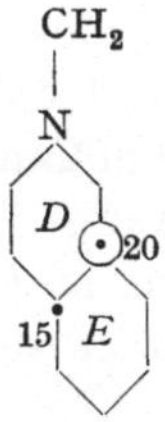

(XVIII.) *D*-N-Methyl-*trans*-decahydro-isoquinoline.

yohimbine are *trans*-locked. This has since been confirmed by KLYNE (*218*) and by VAN TAMELEN and SHAMMA (*365*) by the following synthesis of *DL*-yohimbane (XIX), starting from tryptamine (XX):

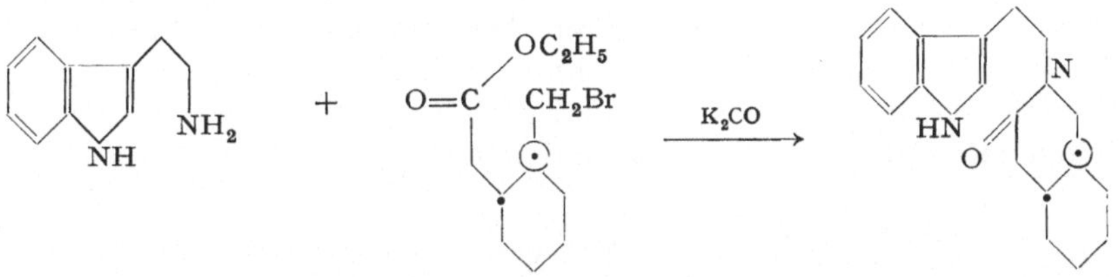

(XX.) Tryptamine.

(XXI.) Ethyl *DL*-trans-2-bromomethyl hexahydrophenyl-acetate.

(XXII.) Lactam of *DL*-trans-N(β-3′-indolyl-ethyl)-2-aminomethyl-hexahydrophenylacetic acid.

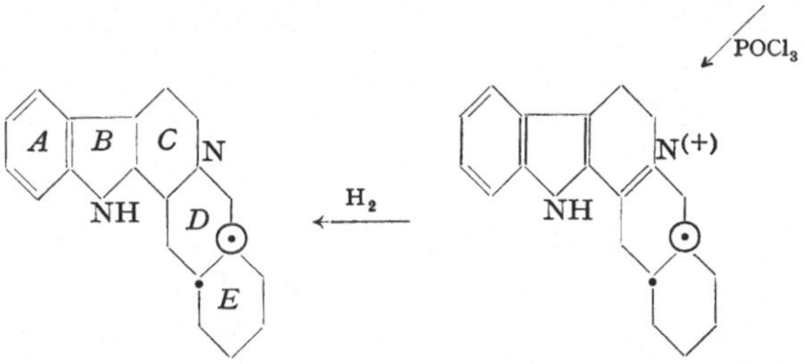

(XIX.) *DL*-Yohimbane.

(XXIII.) *DL*-Δ^3-Dehydroyohimbane.

COOKSON (*81*) is of the opinion that the orientation of the carbomethoxyl and hydroxyl groups in yohimbine is equatorial and axial, respectively. From an examination of models he tentatively suggests that in yohimbine the hydrogen atoms at $C_{(3)}$, $C_{(15)}$ and $C_{(20)}$ are *syn* and *anti*, respectively.

In the yohimbine alkaloids series OPPENAUER oxidation of the bases and their subsequent reduction to oxygen-free compounds (according to WOLFF-KISHNER) have been found valuable in determining steric configurations, especially at ring junctures.

ψ-**Yohimbine.** This yohimbine isomer occurs in the stem bark of *Corynanthe yohimbé* (*200*) and also in the *R. canescens* roots (*359*). It forms a crystalline hydrochloride, m. p. 258–260° (dec.). Like yohimbine, this alkaloid yields a brownish-violet coloration with glacial acetic acid + $FeCl_3$ and conc. sulfuric acid. The ultraviolet spectrum shows maxima at 226 mμ and 280 mμ, with a small peak at 291 mμ (in ethanol). JANOT, GOUTAREL, LE HIR, AMIN and PRELOG (*183*) suggest that the hydrogen atoms at $C_{(3)}$ and $C_{(15)}$ are *anti* in this base, the rings D/E having *trans* juncture. Their conclusion is based on the observations *a*) and *b*):

a) ψ-Yohimbine yields upon OPPENAUER oxidation ψ-yohimbone (a stereoisomer of yohimbone), that is a carbomethoxyl-free ketone, $C_{19}H_{22}N_2O$ (*183*). ψ-Yohimbone when treated according to the HUANG-MINLON variant of the WOLFF-KISHNER reduction method, is converted into yohimbane (XIX). *b*) ψ-Yohimbine and yohimbine when dehydrogenated with lead tetraacetate lose their asymmetric center at $C_{(3)}$ and yield the same tetradehydro-yohimbine (XXIV) which by catalytic hydrogenation can be reconverted into yohimbine (XII). In other words, ψ-yohimbine is the $C_{(3)}$-epimer of yohimbine.

ψ-Yohimbine. $\xrightarrow{\text{Pb(OAc)}_4}$ [structure: N$^{(+)}$ / N$_{(-)}$ / H_3COOC / OH] $\underset{\xrightarrow{\text{H}_2}}{\xleftarrow{\text{Pb(OAc)}_4}}$ (XII.) Yohimbine (p. 354).

(XXIV.) Tetradehydro-yohimbine.

Corynanthine (*113, 116, 263, 264*) has been proved to be the $C_{(16)}$-epimer of yohimbine by studies of the saponification of the base followed by OPPENAUER oxidation and WOLFF-KISHNER reduction (*183*).

β-Yohimbine, $C_{21}H_{26}O_3N_2$, first isolated by HEINEMANN (*142*) from the stem bark of *Corynanthe yohimbé* (Fam. *Rubiaceae*), has now been obtained from the root of *R. canescens* (Madras species) [HOFMANN (*157*)] and also from *Amsonia elliptica* ROEM et SCHULT (*205a*) whose absorption spectra and KELLER reaction are identical with those of β-yohimbine. The stereochemistry of β-yohimbine was studied by LE HIR and GOUTAREL (*224*) who suggest that it represents the $C_{(17)}$-epimer of yohimbine.

Serpine, $C_{21}H_{26}O_3N_2$, recently isolated from the root of *R. serpentina* (cf. p. 368), is interpreted by CHATTERJEE and BOSE (*47, 48, 52*) as the $C_{(17)}$-epimer of corynanthine.

Alloyohimbine (*137a, 140, 184*) *ex Corynanthe yohimbé* has all-*syn*-configuration at $C_{(3)}$, $C_{(15)}$ and $C_{(20)}$, the rings C/D having *cis* juncture (*229, 223, 225, 226*). This conclusion is based on the following observations:

a) The alkaloid when oxidized by the Oppenauer procedure gave a ketone, termed alloyohimbone, which upon Wolff-Kishner reduction was converted into the oxygen-free base alloyohimbane, a yohimbane isomer. *b*) When treated with lead tetraacetate, alloyohimbine gave tetradehydro-alloyohimbane from which alloyohimbane could be recovered upon catalytic reduction. *c*) Sempervirine (XXV) (*278, 393, 126, 173, 391*), was reduced catalytically (pH > 10) to racemic alloyohimbane which, upon resolution with *L*-tartaric acid, afforded *L*-alloyohimbane (*225, 226*). *d*) Upon distillation with soda lime, the base was converted into alloyohimbone (*140*).

(XXV.) Sempervirine.

Further evidence of the all-*syn*-configuration of alloyohimbine at the carbon atoms 3, 15 and 20 has become available by the stereospecific synthesis of rac. alloyohimbane by Stork and Hill (*361*) who started from *cis*-cyclohexane-1,2-diacetic acid and followed essentially the same route as in the synthesis of yohimbane.

Le Hir, Janot and Goutarel (*229*) presume that the carbomethoxyl function in alloyohimbine has equatorial orientation, that of the hydroxyl group being axial. Chatterjee and Pakrashi (*63*) consider the latter as equatorial.

α-Yohimbine (*373, 176*) was claimed to be the $C_{(16)}$-epimer of alloyohimbine (*229*) but it is the $C_{(17)}$-epimer of the latter (*63*).

3-Epi-α-yohimbine (XXX) constitutes one of the alkaloidal ingredients of *R. serpentina* roots (*4, 3*). It has the same configuration as α-yohimbine except for $C_{(3)}$ which is *anti* to $C_{(15)}$ or $C_{(20)}$ with respect to the hydrogens (*3*).

The stereochemistry of iso-rauhimbine (*228, 155*) has not yet been clarified. It can only be inferred that the carbomethoxyl function in rauhimbine is equatorial (*228*).

The Stereochemistry of Rauwolscine (cf. Table 4).

Rauwolscine yields, upon Oppenauer oxidation, a carbomethoxyl-free ketone, rauwolscone (*58, 63*) (p. 353). It has the same composition as alloyohimbone to which rauwolscone resembles in its physical and chemical properties. Rauwolscone when subjected to the Huang-Minlon variant of the Wolff-Kishner reduction yields the oxygen-free base rauwolscane (XXVI, p. 353) identical with alloyohimbane which compound upon refluxing with acetic anhydride produces 3-epi-rauwolscane in small yields. Chatterjee and Pakrashi (*63*) concluded from these observations that the rings *C/D* are *cis* locked in rauwolscane, the hydrogen atoms at $C_{(3)}$ and $C_{(15)}$ being *syn*; the same relation also holds for the parent base, rauwolscine. From saponification experiments these authors have derived the equatorial nature of the carbomethoxyl group in rauwolscine and its acetate. On the basis of infrared readings the hydroxyl group appears to be axial; however, pertinent chemical evidence is still lacking. A molecular model of rauwolscine (XXVII) (LEYBOLD type) has been further examined (*44, 62, 63*), assuming "half-chair" conformation for the piperidine ring C (XXVII) (*12, 349*) and chair conformation for the rings D and E (*10, 11, 13–15, 320*), rings C and D representing an octahydro-pyridocoline system (*81*). CHATTERJEE et al. (*44, 63*) found that in this model, with $C_{(3)}$, $C_{(15)}$ and $C_{(20)}$ in *syn-syn* arrangement, the equatorial carbomethoxyl at $C_{(16)}$ and the axial hydrogen at $C_{(14)}$ should cause considerable hindrance to the acetylation of the indole-NH group; this was borne out by the experiment (*59, 64*). However, such hindrance is not discernible when the hydrogen atoms at $C_{(3)}$, $C_{(15)}$ and $C_{(20)}$ are *anti-syn* or *syn-anti* as in 3-epi-α-yohimbine and

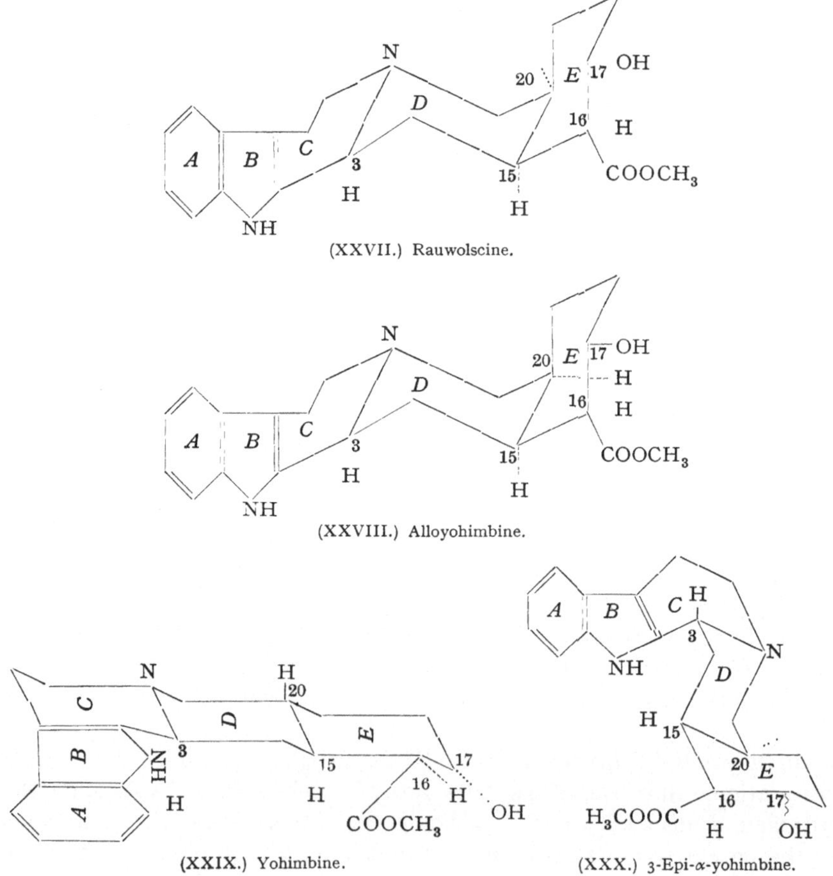

(XXVII.) Rauwolscine.

(XXVIII.) Alloyohimbine.

(XXIX.) Yohimbine.   (XXX.) 3-Epi-α-yohimbine.

yohimbine. From the stereochemical studies and conformational analysis of rauwolscine these investigators conclude that (XXVII) is a $C_{(17)}$-epimer of alloyohimbine and is identical with α-yohimbine (44). This was subsequently confirmed by the identity of the X-ray diffraction patterns.

For a crystallographic and X-ray study of *rauwolscane* cf. RAY (317, 318).

Table 4. Steric Configuration of Rauwolscine, Yohimbine and Isomers.

Isomers $C_{21}H_{26}O_3N_2$	M. p.	Configuration at $C_{(3)}$, $C_{(15)}$ and $C_{(20)}$	Conformation at $C_{(3)}$, $C_{(15)}$ and $C_{(20)}$	Conformation at $C_{(16)}$ (carbomethoxyl group)	Conformation at $C_{(17)}$ (hydroxyl group)
Yohimbine	234–35°	syn-anti	trans-trans	equatorial	axial
ψ-Yohimbine Yohimbene	268°	anti-anti	trans-trans	equatorial	axial
β-Yohimbine	235–36°	syn-anti	trans-trans	equatorial	equatorial
Corynanthine	230–31°	syn-anti	trans-trans	axial	axial
Alloyohimbine α-Yohimbine Rauwolscine	135–40°	syn-syn	trans-cis	equatorial	equatorial
Corynanthidine	232°	syn-syn	trans-cis	equatorial	axial
Serpine	213–14°	syn-anti	trans-trans	axial	equatorial
3-Epi-α-yohimbine	240°	anti-syn	trans-cis	equatorial	axial
Iso-rauhimbine	225–26°	—	—	equatorial	—

Reserpine. This ester alkaloid (p. 367) was first isolated from *R. serpentina* by MÜLLER, SCHLITTLER and BEIN (252). The hypotensive and hypnotic effects of *Rauwolfia* have been mainly attributed to it. It also occurs in *R. canescens* (214).

Deserpidine (334) [**Canescine** (215), **Recanescine** (257)]. The occurrence of this alkaloid, $C_{32}H_{38}O_8N_2$, in *R. canescens* has been recently announced almost simultaneously by STOLL and HOFMANN (359), SCHLITTLER and his group (334), KLOHS and his associates (215) and NEUSS and his collaborators (257). MACPHILLAMY, DORFMAN, HUEBNER, SCHLITTLER and ST. ANDRÉ (234) have observed that it is a minor alkaloid of many *Rauwolfia* species. Pure deserpidine has been prepared from the crude reserpine fraction of *R. canescens* as well as from its mother liquor by chromatography. KLOHS et al. (215) have purified their sample by counter current distribution. NEUSS, BOAZ and FORBES have noticed that on a formamide-pretreated paperchromatogram deserpidine has a higher R_f value (ca. 0.65) compared to reserpine (ca. 0.4). The X-ray powder diffraction pattern of this compound is identical with that of reserpine, but the refraction indices are different.

The ultraviolet spectrum of deserpidine (*Fig. 4*), has been identified with the computed spectrum of yohimbine and 3,4,5-trimethoxybenzoate (in a mole per mole ratio). The maxima [216 mμ (log ε 4.78); 271 mμ

(log ε 4.25); and 289 mμ (log ε 4.06)] are indicative of the presence of 2,3-disubstituted indole and 3,4,5-trimethoxybenzoate chromophores.

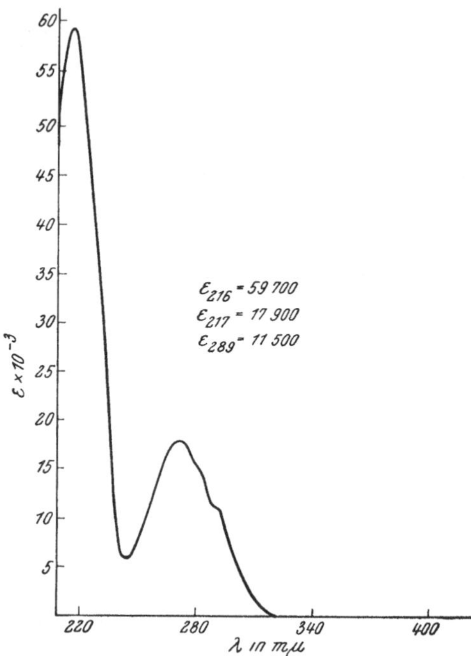

Fig. 4. Ultraviolet spectrum of deserpidine (canescine) in methanol (*231*). [From: Physical Data of Indole and Dihydroindole Alkaloids. Lilly Research Laboratories, Indianapolis, USA.]

The I. R. spectrum exhibits the characteristic >NH band at 2.95 μ, the carbonyl bands at 5.81 μ and 5.88 μ and aromatic absorption at 6.33, 6.67, 8.67 and 9.70 μ (*Fig. 5*). The spectrum, however, lacks the peak absorption at 6.13–6.2 μ which arises due to the polarisation of an indole nucleus by a methoxyl group at the 6-position (*257*).

Deserpidine, containing five methoxyl groups, yields upon basic hydrolysis two acidic components, viz. 3,4,5-trimethoxybenzoic acid and deserpidic acid, $C_{21}H_{26}O_4N_2$, the infrared spectrum of which is typical of a zwitterionic amino acid. With diazomethane this compound forms methyl deserpidate having an α,β-disubstituted indole chromophore. This methyl ester when treated with 3,4,5-trimethoxybenzoylchloride in pyridine restores deserpidine. When heated with selenium, it yields yobyrine (IX, p. 353) as a major product, thus showing the presence of a yohimbane-like ring system. Deserpidine

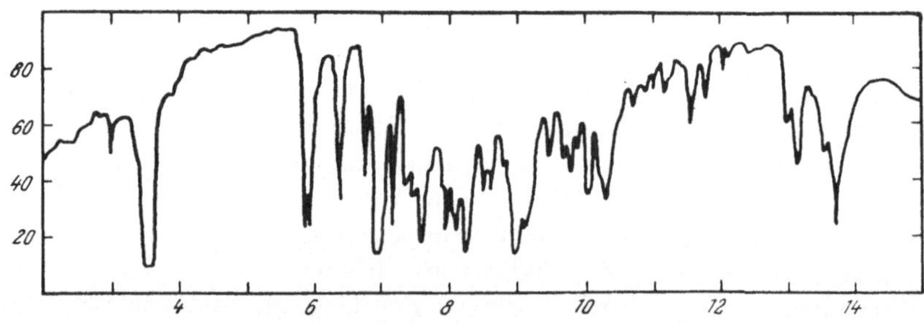

Fig 5. Infrared spectrum of deserpidine (canescine) in Nujol, according to Klohs et al. (*215*). [From: J. Amer. Chem. Soc. 77, 4084 (1955).]

yields, upon reductive cleavage with lithium aluminium hydride, deserpidinediol, $C_{21}H_{28}O_3N_2$ (*235*), which on selenium dehydrogenation affords methyl-yobyrine (XIII, p. 353). This is a direct proof of the $C_{(16)}$-position of the $COOCH_3$ group (XXXI).

The formation of a γ-lactone from deserpidic acid, on treatment with acetic anhydride in pyridine (*234, 235*), suggests that the carbomethoxyl and trimethyl-galloyl moiety occupy the positions 16 and 18 (XXXI). The methoxyl is assumed to be located at $C_{(17)}$ as confirmed by the conversion of deserpidine into α-yohimbine.

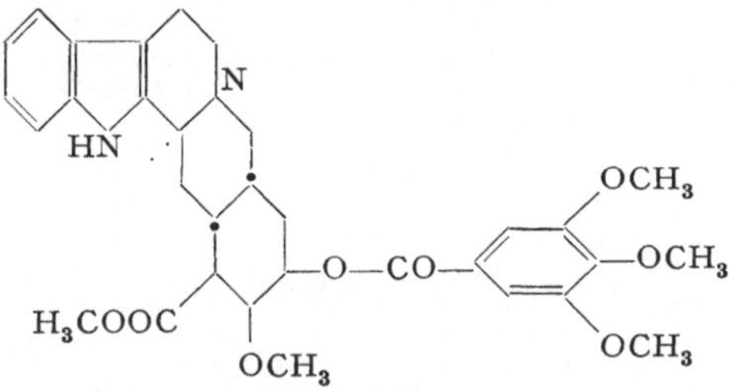

(XXXI.) Deserpidine (canescine; 11-desmethoxy-reserpine).

The Stereochemistry of Deserpidine.

Recently MacPhillamy et al. (*234, 235*) have shown that methyl deserpidate tosylate (XXXII) and NaI or LiBr afford the corresponding 18-halogen compound. The latter, when dehalogenated with zinc and acetic acid is converted into the 18-desoxy derivative (XXXIII). This compound, on demethylation with HBr and reesterification, yields rauwolscine [= α-yohimbine (*63*)]. Rauwolscinyl alcohol (XIV, p. 353) (*58*) which is a reduction product of rauwolscine is also obtained from methyl deserpidate tosylate (XXXII) by direct reduction to deserpidinol, $C_{21}H_{28}N_2O_2$ (XXXIV) followed by demethylation.

The degradation of deserpidine to rauwolscine permits certain conclusions in regard to its configuration at ring junctures, viz. in (XXXI) at least the rings *D/E* must be *cis* locked. MacPhillamy and his associates have shown that the hydrogens at $C_{(3)}$ and $C_{(15)}$ are *anti* in deserpidine which therefore possesses 3-epi-allo-configuration at $C_{(3)}$, $C_{(15)}$ and $C_{(20)}$. These conclusions are based on the observations (*a*)–(*b*):

a) Hydrobromic acid or strong alkali (*235*), applied at 200° in diethylene glycol, brings about epimerisation at the asymmetric center $C_{(3)}$ of deserpidinol (XXXIV) or 18-desoxy-deserpidate (XXXIII). (*b*) Deserpidinediol (XXXV) upon dehydrogenation with lead tetraacetate yields the tetradehydro base (XXXVI) which regenerates the stable 3-isodeserpidine-diol (XXXVII).

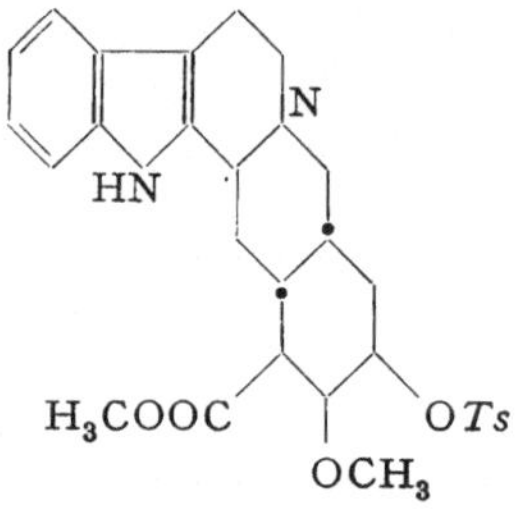

(XXXII.) Methyl deserpidate tosylate.

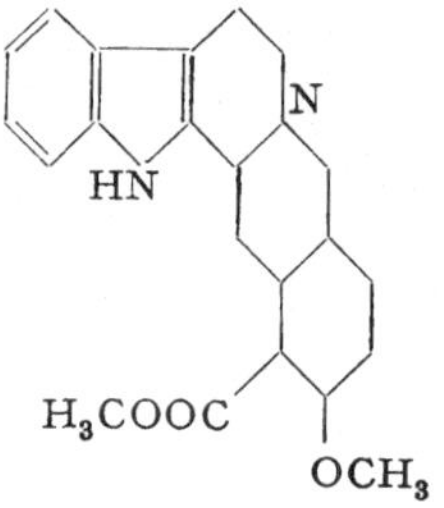

(XXXIII.) 18-Desoxy-deserpidate.

LiAlH$_4$

1. HBr
2. CH$_2$N$_2$

α-Yohimbine.

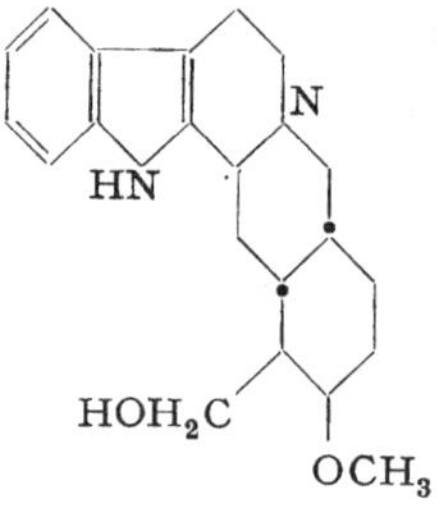

(XXXIV.) Deserpidinol. $\longrightarrow$ (XIV., p. 353.)

HBr

(XXXI.) Deserpidine.

LiAlH$_4$

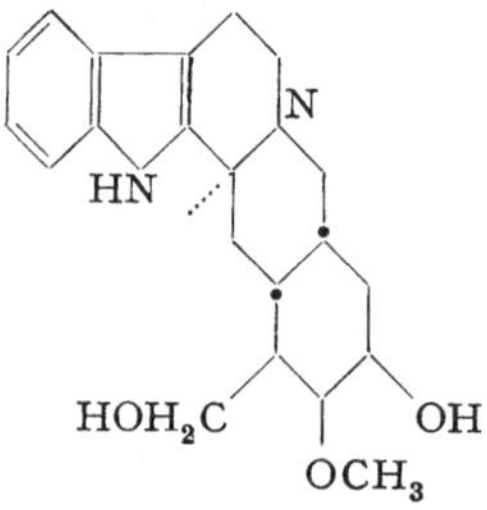

Pb(OAc)$_4$

(XXXV.) Deserpidinediol.

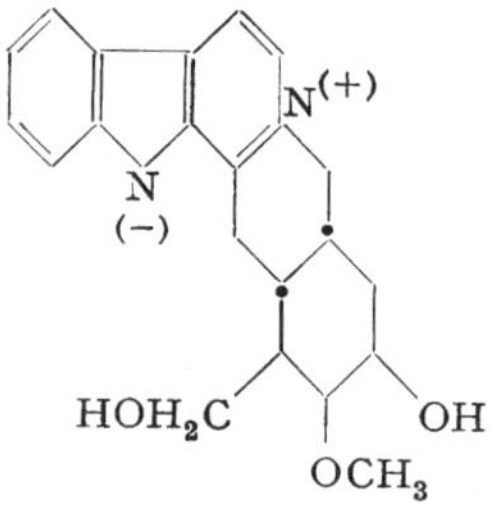

(XXXVI.) Tetradehydro-deserpidinediol.

NaBH$_4$

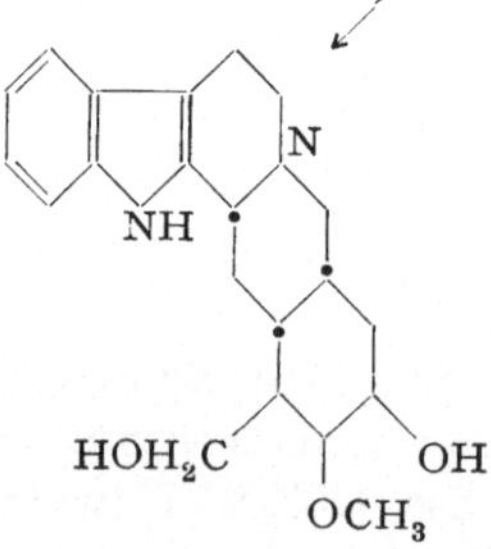

(XXXVII.) 3-Iso-deserpidinediol.

These findings make it clear that deserpidine is a derivative of 3-epialloyohimbane [nomenclature of STORK and HILL (*361*)]. Lactone formation in deserpidic acid further indicates the *cis* arrangement at the $C_{(16)}$-carbomethoxyl group with respect to the $C_{(18)}$-function. From these data and from the close steric analogy of deserpidine with reserpine (XXXI, p. 361) appeared to be the most plausible configuration of deserpidine. Recently, however, the spatial formulation of reserpine has been revised (*363*); accordingly, the absolute configuration of deserpidine is postulated as given in (XXXVIII).

(XXXVIII.) Configuration of deserpidine.

Aricine, $C_{22}H_{26}O_4N_2$, was first isolated by PELLETIER and CORIOL (*262*) in 1829 from *Cinchona pelletierina* WEDDELL (Fam. *Rubiaceae*) and has now been found to occur in *R. canescens* (*360*) and other species, such as *R. heterophylla* (*154*) and *R. sellowii* (*259*). MOISSAN and LANDRIN (*245*) suggested the formula $C_{23}H_{26}O_4N_2$. From the studies of color reactions, pyrolysis with soda lime, zinc dust distillation, U. V. spectrum, RAYMOND-HAMET (*302*) concluded that aricine is an indole alkaloid. He further observed that the ultraviolet spectrum of this alkaloid is very similar to that of corynantheine (*55, 186, 178, 179, 195, 197, 202*), mayumbine (*185*) and δ-yohimbine (*128, 188*). JANOT and his associates (*125, 174*) have revised the formula to $C_{22}H_{26}N_2O_4$. The spectrum of the base (containing two methoxyls) has the characteristic maxima of a normal αβ-disubstituted indole derivative at 225, 280 and 292 mμ with an inflection at 250 mμ, indicative of an ester carbonyl conjugated with an unsaturated ether grouping (XXXIX):

$$H_3COOC—C=C—O—C—$$

(XXXIX.) β-Alkoxy-acrylic ester group.

The infrared spectrum of aricine (*Fig. 6*) exhibits a single sharp band at 3.0 μ (attributed to >NH) and also a characteristic doublet at 5.9 and 6.1 μ [conjugated ester-enolether function (XXXIX)]. Upon

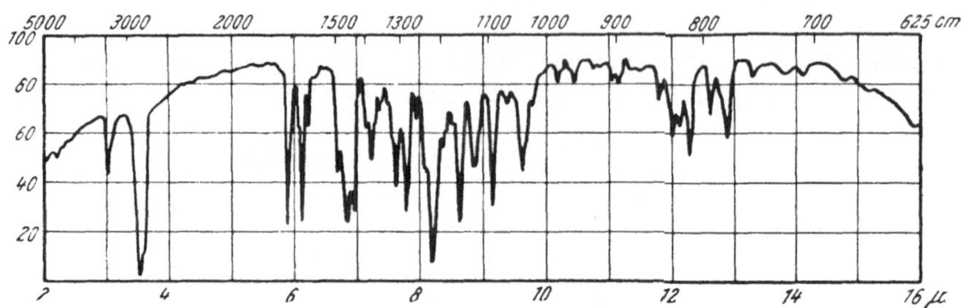

Fig. 6. Infrared spectrum of aricine in Nujol, according to Stoll et al. (*360*).
[From: Helv. Chim. Acta *38*, 270 (1955).]

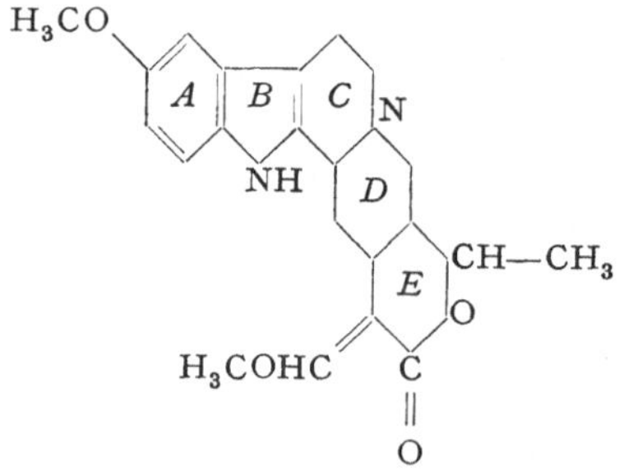

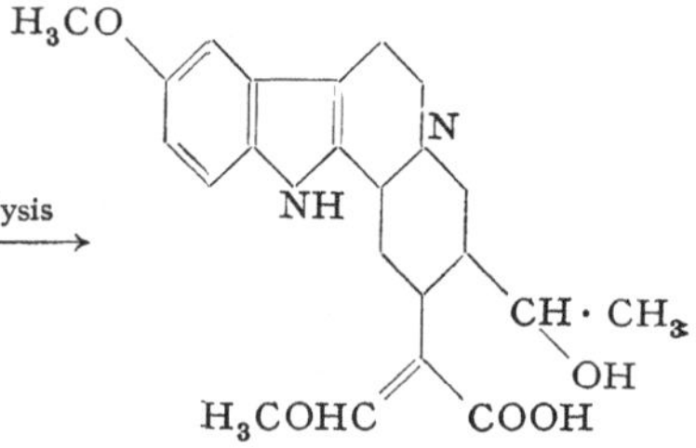

(XL.) Aricine.

Hydrolysis

(XLI.) Aricinic acid.

↓ Se

↓ HCl

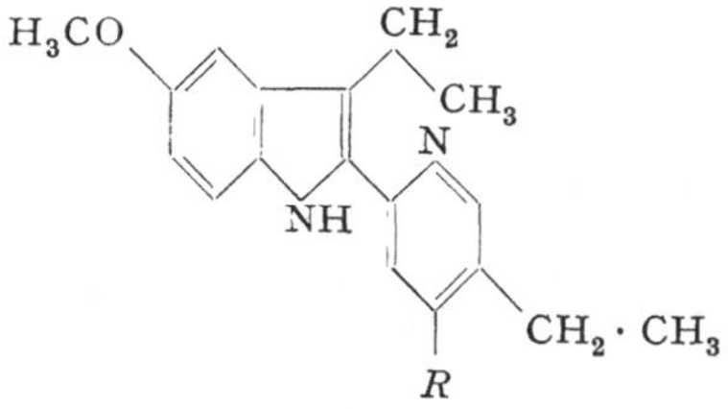

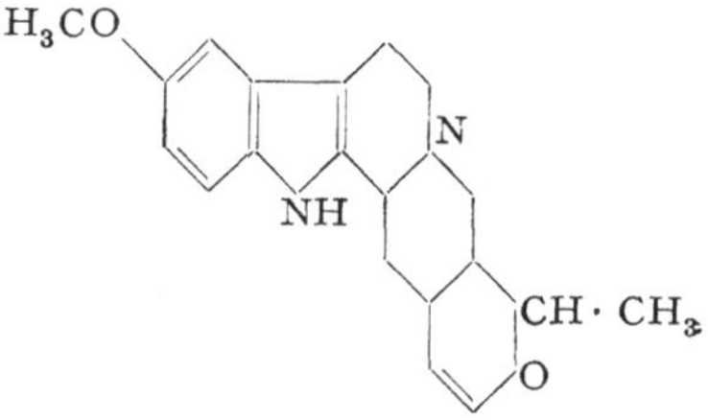

XLIIIa.) Methoxy-alstyrine ($R = CH_2 \cdot CH_3$).
(XLIIIb.) Methoxy-desethyl-alstyrine ($R = H$).

(XLII.)

H_2O_2 (oxid. degradation) and hydrolysis

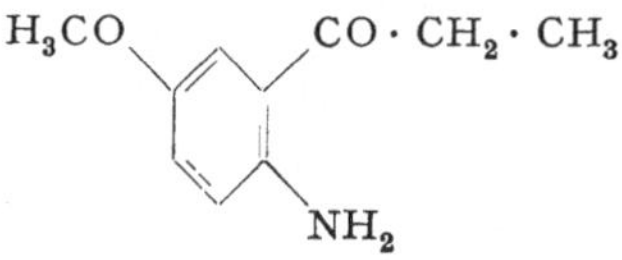

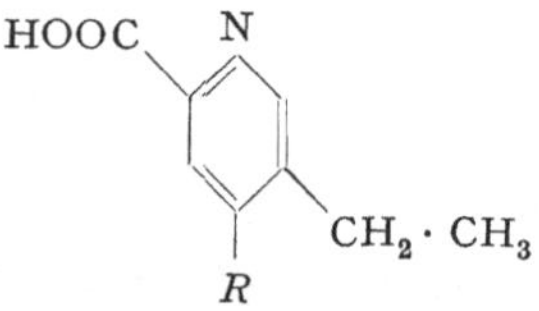

(XLIV.) 2-Amino-5-methoxy-propiophenone. (XLV.) 3-Ethylpyridine-6-carboxylic acid ($R = H$).

(XLVI.) 3,4-Diethyl-pyridine-6-carboxylic acid ($R = C_2H_5$).

Chart 2. Degradation of Aricine.

hydrolysis with methanolic alkali the base (XL) gives aricinic acid, $C_{22}H_{28}O_5N_2$ (XLI), which on warming with hydrochloric acid yields a compound, $C_{20}H_{24}O_2N_2$ (XLII) having only one methoxyl group (*Chart 2*).

Selenium dehydrogenation of aricine affords mainly a methoxy-desethyl-alstyrine, $C_{19}H_{20-22}N_2O$ (XLIIIb) (maxima, 229, 263–264, 312–313, 388 mμ) whose spectrum is similar to that of alstyrine or coryline (XLIII, p. 397) and methoxy-alstyrine (XLIIIa) (cf. p. 364). The dehydrogenation products mentioned, when oxidized with hydrogen peroxide, liberate on hydrolysis 2-amino-5-methoxy-propiophenone (XLIV), 3-ethylpyridine-6-carboxylic acid (XLV), and 3,4-diethyl-pyridine-6-carboxylic acid (XLVI). On the basis of these results, GOUTAREL, JANOT, LE HIR, CORRODI and PRELOG (*125*) interpreted aricine as a δ-lactone (XL).

STOLL, HOFMANN and BRUNNER (*360*) find that aricinic acid, the hydrolysis product of aricine, has the composition $C_{21}H_{24}O_4N_2$ (XLVII), and not $C_{22}H_{28}O_5N$ proposed earlier by GOUTAREL et al. (*125*). They have also observed that the acid when esterified with diazomethane reforms aricine that is therefore an ester and not a δ-lactone. In the light of these new observations the structure of the base (XL) had to be altered to (XLVIII).

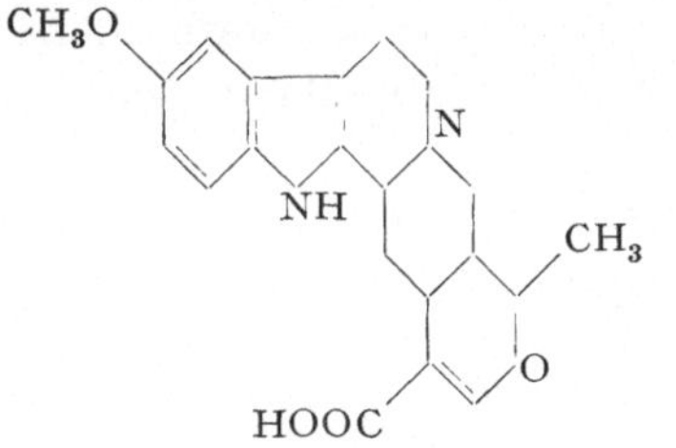

(XLVII.) Aricinic acid.

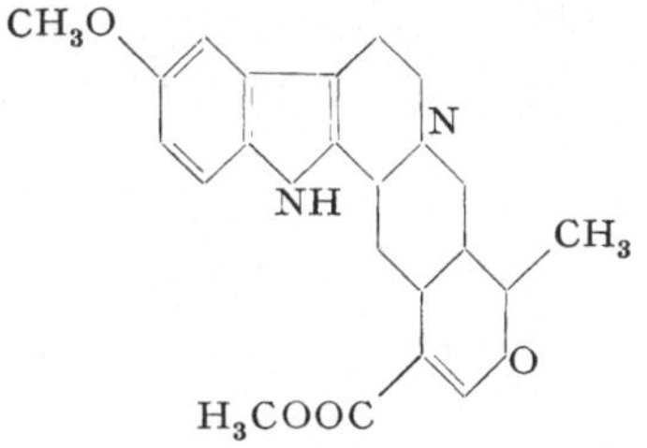

(XLVIII.) Aricine.

Isoreserpinine, $C_{22}H_{26}O_4N_2$, m. p. 225 to 226°, contains two methoxyls and a side-chain methyl group. It gives a blue coloration with KELLER's reagent and possesses a 6-methoxy indole and an αβ-alkoxy-acrylic ester group as ultra-violet chromophores. The I. R. spectrum is identical with that of reserpinine (p. 392), except for the fingerprint region.

(XLIX.) Isoreserpinine.

According to STOLL et al. (*360*) (XLIX) is a stereoisomer of reserpinine.

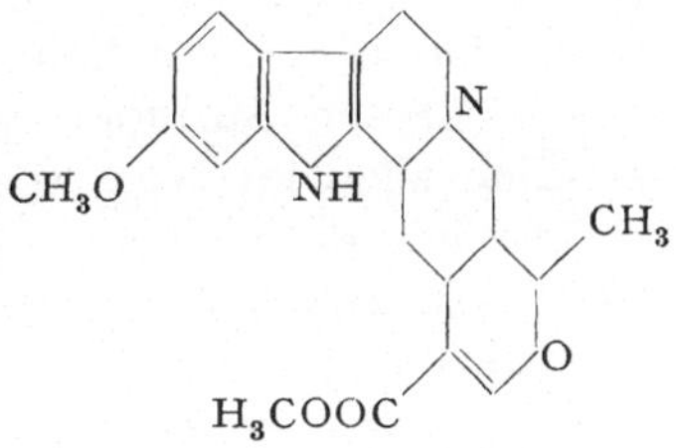

Reserpiline and Isoreserpiline. The presence of reserpiline, $C_{23}H_{28}O_5N_2$, one of the constituents of *R. canescens* (*360*) was first observed in *R.*

serpentina roots by Klohs, Draper, Keller and Malesh (*211*). Iso-reserpiline which accompanies reserpiline in *R. canescens* has been suggested to be a stereoisomer (*360*). Both bases are tetrahydro-β-carboline derivatives with methoxyl groups at $C_{(10)}$ and $C_{(11)}$, the ring *E* being heterocyclic.

Ajmaline, Ajmalicine, Reserpinine, and Sarpagine. These weak bases have been isolated from the root of *R. canescens* by Keck (*204*) as well as by Bhattacharji, Dhar and Dhar (*21*).

Raunescine and Isoraunescine. These new isomeric ester-alkaloids, $C_{31}H_{36}O_8N_2$, both yielding 3, 4, 5-trimethoxybenzoic acid on hydrolysis have been isolated from *R. canescens* (*160*). They appear to be very similar in structure to canescine (XXXI, p. 361) and probably carry a hydroxyl group at $C_{(17)}$.

According to an unpublished work of Bhattacharji, Dhar and Dhar (*22*) "raunescine" isolated from *R. canescens* has the composition $C_{22}H_{28}O_3N_2$; its ultra-violet and infrared bands are in conformity with a yohimbine-like structure.

Serpentine. Haack, Popelak, Spingler and Kaiser (*135*) found in *R. canescens* this strong base, isolated earlier by Siddiqui and Siddiqui from *R. serpentina* roots (*351, 352*).

ψ-**Reserpine** (*215 a*), $C_{32}H_{38}O_9N_2$, m. p. 257–58°, a derivative of either 3-epi-allo-yohimbane or of allo-yohimbane, has been shown to be the 3,4,5-trimethoxybenzoylester of methyl-17-nor-reserpate or -3-iso-reserpate.

III. The Alkaloids of *Rauwolfia serpentina*.

Rauwolfia serpentina Benth ex Kurz (*Fig. 7*) is a low shrub known in India as "Chottachanda" or "Sarpagandha" (*207*). It has long, tapering, crooked roots that contain the active principles, mostly alkaloids. The therapeutic application of this drug (*72, 74, 133, 132*) has been known in India for several hundred years.

Siddiqui and Siddiqui (*351, 352, 354*) succeeded in isolating five crystalline alkaloids from the roots (from Bihar) and classified them into two groups, viz. the ajmaline group comprising three colorless, weak bases, ajmaline, ajmalinine and ajmalicine; and the serpentine group, including two yellow, strong bases, serpentine and serpentinine (*Table 5*). From roots originating from Dehra Dun, Siddiqui (*350*) obtained ajmaline, isoajmaline, neoajmaline, serpentinine and two uncharacterised alkaloids. At the same time van Itallie and Steen-hauer (*171, 172*) reported that *R. serpentina* roots contain three alkaloids, two of which were identical with ajmaline and serpentine. Later Bose (*27*) observed the presence of a new hypotensive alkaloid rauwolfinine in the same material; and MÜLLER, SCHLITTLER and BEIN announced the isolation of reserpine (*252*), the sedative principle of the drug, in pure, crystalline form. Some summaries of the work in this field have appeared in 1954—1956 (*330, 345, 266, 267, 222 a, 323 a*).

Fig. 7. *Rauwolfia serpentina* BENTH.

Chemical investigations have secured the presence of twenty-three crystallizable alkaloids in the roots (*Table 5*), besides some sterols (*33, 121*), an oxymethyl-anthraquinone derivative, fumaric acid, oleic acid, glucose, sucrose, an unsaturated alcohol, calcium oxalate and an unidentified fluorescent substance (*395*).

Table 5. Alkaloids obtained from the Roots of *Rauwolfia serpentina*.

Alkaloid	Composition	M. p.	$[\alpha]_D$
Ajmaline (*1, 46, 49, 50, 51, 111, 251, 310, 351, 352, 354*)	$C_{20}H_{26}O_2N_2$	158–60°	+ 128° (chloroform)
= Rauwolfine (*171*) ...		160°	+ 131.1°
= Neoajmaline (*1, 350*).		205–7°	
Isoajmaline (*350, 1*) ...... = Isorauwolfine (*172*) ..	$C_{20}H_{26}O_2N_2$	263–5°	+ 52° (methanol)
Ajmalinine (*352*)	$C_{20}H_{26}O_3N_2$	180–1°	— 97° (chloroform)
Ajmalicine (*352*) .... = δ-Yohimbine (*156, 212, 377*)	$C_{21}H_{24}O_3N_2$	250–2° 252°	— 45° (pyridine)
= Raubasin (*135, 274*) .	$C_{21}H_{24}O_3N_2$	247–8°	— 61° (pyridine)
= Alkaloid "F" (*256*) ..	$C_{21}H_{24}O_3N_2$	253–4°	— 47.7° (pyridine)
Reserpine (*48, 103, 104, 117, 164, 213, 252, 172, 254, 397, 255, 356, 357*)	$C_{33}H_{40}O_9N_2$	264–5°	— 118° (chloroform)
Rescinnamine (*209, 210*) ..	$C_{35}H_{42}O_9N_2$	238–9°	— 97° (chloroform)
= "Reserpinine" (*136, 134*)	$C_{35}H_{42}O_9N_2$	224–26°	— 98° (chloroform)
Sarpagine (*358, 367*)	$C_{19}H_{22}O_2N_2$	320°	+ 54° (pyridine)
= Raupine (*25, 26*)	$C_{20}H_{26}O_3N_2$	325°	+ 63° (dil. acetic acid)
Rauhimbine (*155, 156*) ... = Corynanthine (*113, 116*)	$C_{21}H_{26}O_3N_2$	218–25° 218–25°	— 82° (pyridine) — 82° (pyridine)
Isorauhimbine (*228, 155, 156*)	$C_{21}H_{26}O_3N_2$	225–8°	— 104° (pyridine)
Reserpinine (*329, 377*)	$C_{22}H_{26}O_4N_2$	240–41°	— 125° (chloroform)
= Alkaloid "C" (*156*) ..		240°	— 127° (pyridine)
= Raubasinine (*274*) ...		228°	— 123° (chloroform)
= Alkaloid "A" (*256*) ..		243–44°	— 131° (pyridine)
Reserpiline (*211*)	$C_{23}H_{28}O_5N_2$	205–7°	— 38° (ethanol)
Yohimbine (*5, 156*)	$C_{21}H_{26}O_3N_2$	235–7°	+ 105° (pyridine)
Reserpic acid methyl ester (*156*)	$C_{23}H_{30}O_5N_2$	244–5°	— 106° (pyridine)
Serpine (*47, 48, 52*)	$C_{21}H_{26}O_3N_2$	213° (dec.)	+ 70.1° (pyridine)
Serpinine (*31, 31a*)	$C_{20}H_{24}ON_2$	sublimed 315–17° (dec.)	
Rauwolfinine (*27–30a*) ...	$C_{19}H_{24}O_2N_2$	235–6° (dec.)	— 34.7° (alcohol)

Alkaloid	Composition	M. p.	$[\alpha]_D$
3-Epi-α-yohimbine = 3-Epi-rauwolscine (*4, 3*)	$C_{21}H_{26}O_3N_2$	125–8° or 181–3° (dimorph.)	— 91° (pyridine)
Serpentine (*6, 331, 352, 377*)	$C_{21}H_{20}O_3N_2$	157–8°	+ 292° (methanol)
Serpentinine (*328, 352*) . . .	$C_{21}H_{20-22}O_3N_2$	163–5°	+ 52° (methanol)
Thebaine (*156*)	$C_{19}H_{21}O_3N$	195°	— 279° (pyridine)
Papaverine (*156*)	$C_{20}H_{21}O_4N$	147°	
Chandrine (*281*)	$C_{25}H_{30}O_8N_2$	230–1°	
Rauwolscine (*152*) see Table 2, p. 350			

Methods of Isolation. A variety of solvents may be employed to obtain a total alkaloid extract from the finely ground root. Ethanol and ethylene chloride are frequently used. The techniques of chromatography and counter-current distribution can then be applied for the isolation of individual alkaloids. A schematic representation of a chromatographic elution sequence follows.

<table>
<tr><td>10% methanol
in chloroform</td><td>Sarpagine
Serpentine
Serpentinine
Ajmaline
Yohimbine and isomers
Rescinnamine
Reserpine
Py-tetrahydroserpentine</td></tr>
<tr><td align="center">↑

benzene</td><td>Reserpinine</td></tr>
</table>

Ajmaline (*351, 352, 354*), $C_{20}H_{26}O_2N_2$, 3 H_2O, m. p. 158–160° (dec.), the most abundant alkaloid of *R. serpentina*, has been identified with rauwolfine by van Itallie and Steenhauer (*171, 172*). Siddiqui and Siddiqui's assumption that this alkaloid is a betaine was disproved by Robinson et al. (*251*) who established its nature as a monoacidic, ditertiary base that shows strychnidine-like color reactions. (Earlier they believed that the base contained an isolated double bond and a semiacetal group.) Upon soda lime and zinc dust distillation, ajmaline yields Ind-N-methyl-

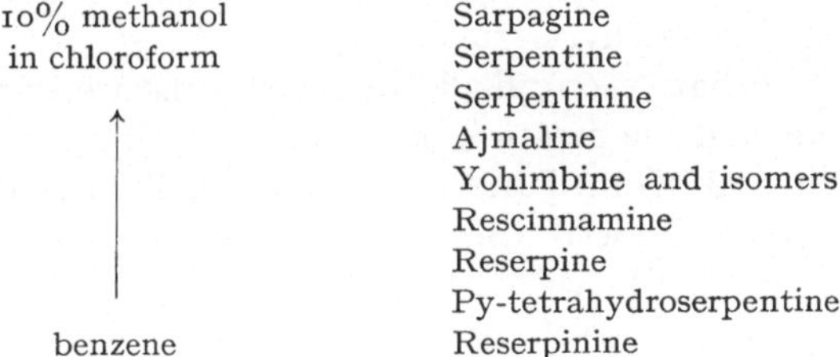

(L.) Ind-N-methyl-harman.　　　　　　　　　(LI.) Carbazole.

harman (L). ROBINSON et al. took the alternative structures (LII) and (LIII) into consideration and gave preference to (LII), since the dihydro-indole moiety required by (LIII) had not yet been found in nature.

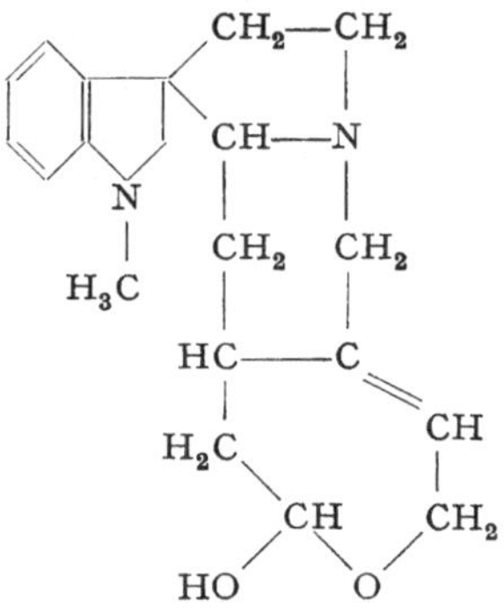

(LII.)

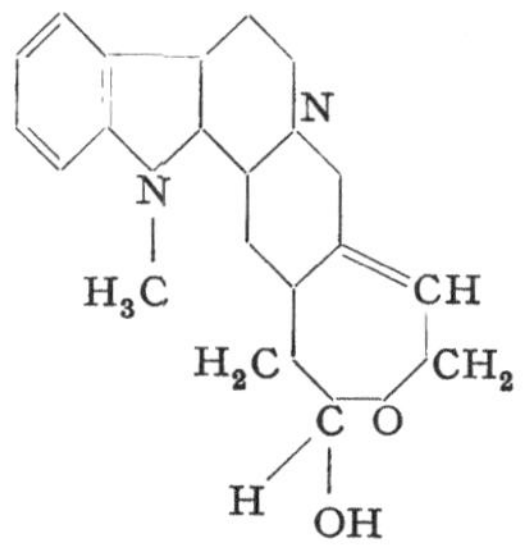

(LIII.)

According to CHATTERJEE and BOSE (*46, 49*) either of these structures appeared to be improbable. They found that, on fusion with potassium hydroxide, ajmaline yields indole-2-carboxylic acid, while on heating with selenium, Ind-N-methyl-harman appears. These observations, the failure to isolate carbazole on alkali fusion, the nature of the dehydrogenation products, and the presence of a terminal methyl group (revealed by the spectrum and KUHN-ROTH analysis), finally the dihydroindole color reactions have induced CHATTERJEE and BOSE to suggest the ajmaline structure (LIV), with reservations regarding ring E. The double bond appeared to be diquaternary as that in rubremetine (*198*).

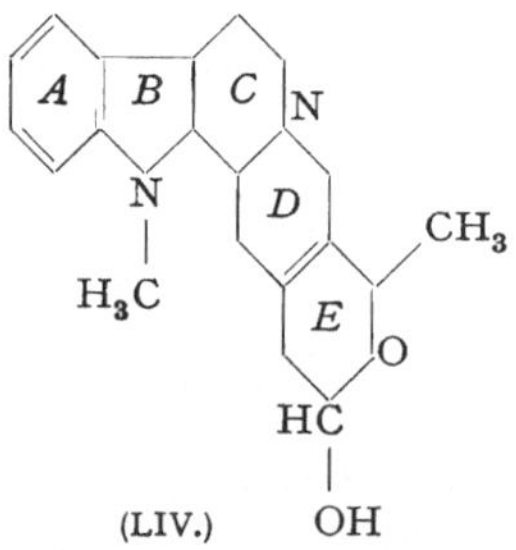

(LIV.)

The Indian workers further suggested that ring D is involved in a weak linkage which is responsible for the facile cleavage of ajmaline to give Ind-N-methyl-harman. For the infrared spectrum of ajmaline cf. *Fig. 8*.

Ajmaline has been further investigated by ANET, CHAKRAVARTI, ROBINSON and SCHLITTLER (*1*). These authors have revised their previous interpretation of the structure since some of their earlier observations (*251*) could not be confirmed. They propose that ajmaline contains two hydroxyl groups, capable of acetylation. Furthermore, the arguments (*a*)–(*c*) are offered in favor of the presence of the reducing group $>N$—CHOH.

a) Ajmaline oxime hydrochloride, on acetylation and subsequent treatment with methanol and some KOH, affords anhydroajmaline oxime, which on acid hydrolysis is converted into an imino-acid. This resists lactamization. However, anhydroajmaline oxime, when reduced

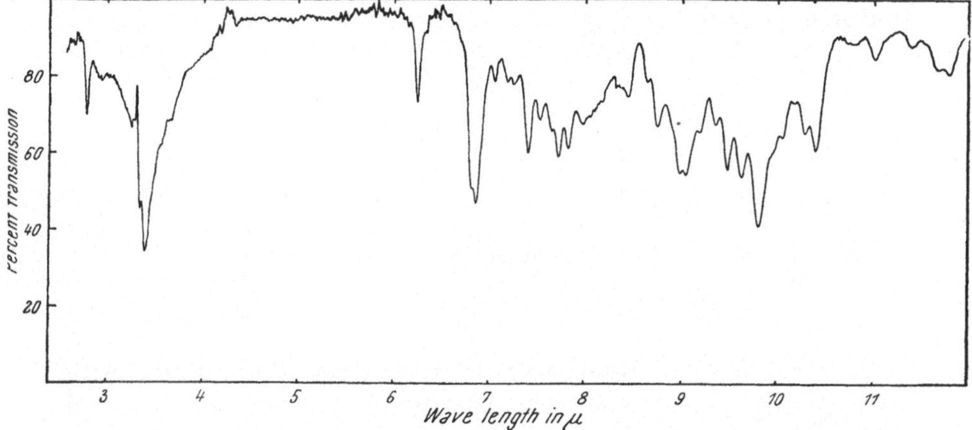

Fig. 8. Infrared spectrum of ajmaline in chloroform (*231*). [From: Physical Data of Indole and Dihydroindole Alkaloids. Lilly Research Laboratories, Indianapolis, USA.]

with LiAlH$_4$, is reconverted into ajmaline, while the latter is unaffected by the reagent. Since anhydroajmaline oxime is a secondary base, the group $>N$. CHOH rather than —O—CHOH is indicated in ajmaline.

b) Ajmaline when submitted to the HUANG-MINLON variant of the WOLFF-KISHNER reduction, forms a secondary base, deoxy-dihydro-ajmaline C$_{20}$H$_{28}$ON$_2$. Thus, $>N$. CHOH is converted into $>N$—CH$_3$ as evidenced by an increased value found in KUHN-ROTH estimations.

c) Ajmaline methiodide gives on distillation an N-methyl-ajmaline containing a conjugated carbonyl group. The latter can be converted into monodeoxy-ajmaline, C$_{19}$H$_{24}$ON(:N . CH$_2$), m. p. 311°.

ROBINSON and his associates suggest that ajmaline contains a latent aldehyde group in the form of a carbinolamine group of an "unusual" kind. This function confers such properties to the base which are not associated with ordinary carbinolamine structure. Thus, the compound is not a strong base; it is not reducible with zinc dust and HCl; and it forms an O-acetyl derivative without suffering fission.

24*

When treated with chromic acid, deoxy-dihydroajmaline affords ethyl-methyl-ketone indicating the presence of either $> N \cdot CHOH$ or

$$> N \cdot \underset{\underset{\displaystyle CH_2-\underset{|}{CH}-CH_3}{|}}{CHOH} \qquad -CH-C_2H_5$$

$> N \cdot CHOH$ in the ajmaline molecule. The latter formulation is the preferred one since it explains the isomerism of ajmaline and iso-ajmaline (see below). Considering these findings Robinson et al. (*1*) proposed the structure (LV). [The biogenetic origin of the ethyl group seems to be analogous to that in emetine (*311*).]

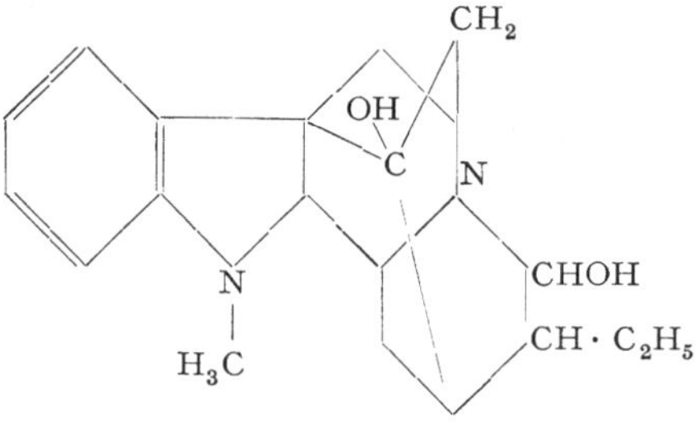

CH$_2$

OH

C   N

CHOH

CH · C$_2$H$_5$

H$_3$C

(LV.)

N

CHOH

CH · C$_2$H$_5$

H$_3$C

—C

—C—

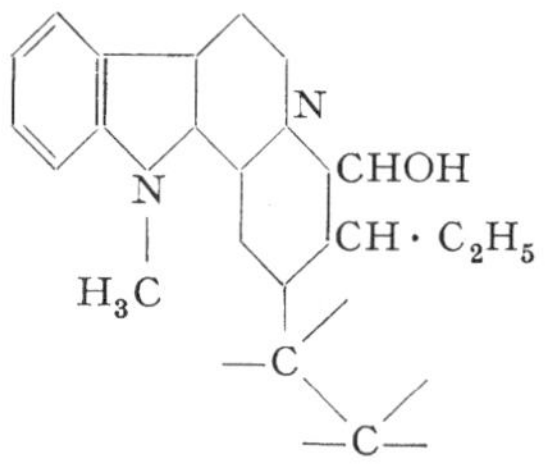

(LVI.)

In some respects Robinson's interpretations now coincide with those given by Chatterjee and Bose (presence of only one C-methyl group and of a weak linkage in ring *D*).

Recently, the ajmaline formula (LV) was modified to the skeletal arrangement (LVI) by Robinson (*111, 310*). Ajmaline (LVII), on treatment with Raney-nickel in boiling xylene, loses the elements of CO and yields a secondary base, decarbono-ajmaline, $C_{19}H_{26}ON_2$ (LVIII).

By potassium borohydride ajmaline is reduced to dihydro-ajmaline, $C_{20}H_{28}N_2O_2$, that is converted by HBr into deoxy-ajmaline, $C_{20}H_{26}ON_2$. Dihydro-deoxyajmaline (obtained from ajmaline by Wolff-Kishner reduction) when dehydrogenated with palladized charcoal gives three or four carboline derivatives, one of which appears to be Ind-N-methyl-alstyrine (LIX) (*1*), although the possibility of the presence of two methyl groups and one ethyl in the pyridine ring is not unlikely. A similar behavior of deoxyajmaline was observed.

According to Anet et al. (*1*) the second hydroxyl group in ajmaline appears to be tertiary as concluded from its resistance to dehydration with thionyl chloride and from the stability towards chromic acid of deoxy-octahydroajmaline, $C_{20}H_{31}ON_2$, prepared from dihydrodeoxy-ajmaline by catalytic hydrogenation.

Consequently, Robinson and his collaborators favored the formula (LX) which is also stereochemically satisfactory.

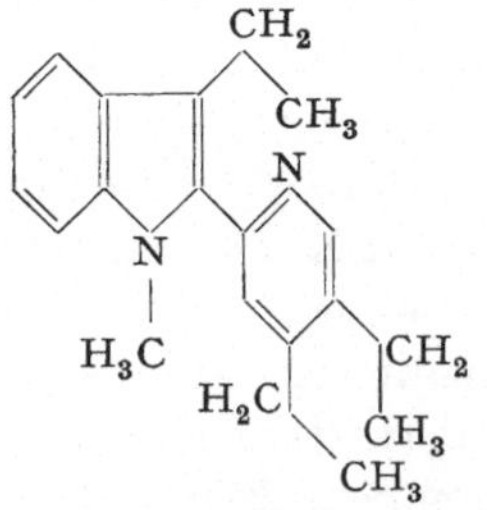

(LIX.) Ind-N-methyl-alstyrine.

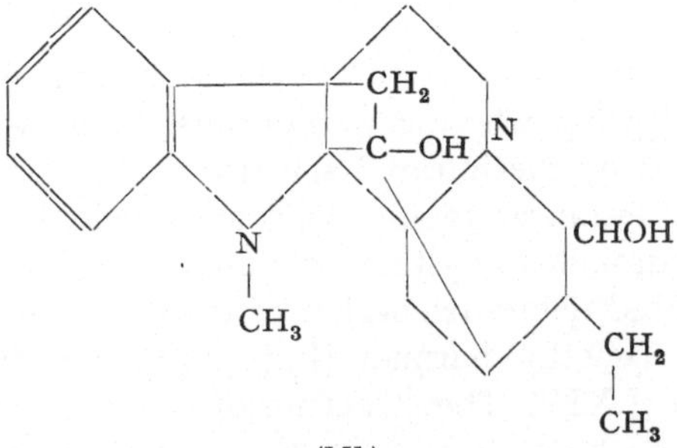

(LX.)

Independently, CHATTERJEE and BOSE (*50, 51*) also reported some of the observations just mentioned but interpreted them differently. According to these authors, ajmaline does not show a carbinolamine band in the infrared region (*387*). Furthermore, ajmaline when reduced either with sodium in liquid NH_3 or with sodium borohydride (*32*) in methanol yields the same compound, viz. ajmalindiol, $C_{20}H_{28}N_2(OH)_2$, while the molecule acquires an active hydrogen. The diol does not respond to the iodoform test and when dehydrated with *p*-toluene-sulfonyl-chloride in pyridine, it is converted into dehydro-ajmaline, $C_{20}H_{26}ON_2$. The latter contains two less active H-atoms than ajmalindiol, and yields with selenium at high temperature mainly Ind-N-methyl-harman (L, p. 369). Under the influence of warm sulfuric acid ajmaline develops an indole chromophore. Hence, CHATTERJEE and BOSE proposed the ajmaline formula (LXI).

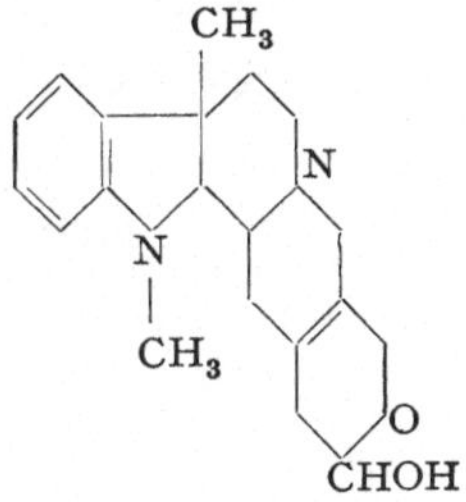

(LXI.)

Recently, as the result of an important study, WOODWARD (*388a*) has established the complete ajmaline structure (LXI a), based on biogenetic considerations and (among others) on the following chemical arguments:

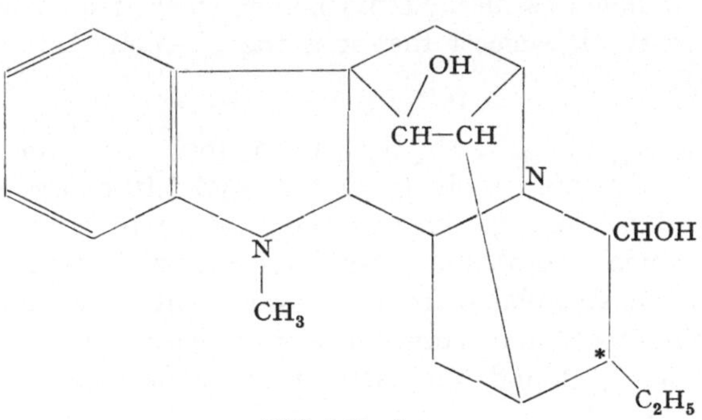

(LXI a.) Ajmaline.

(a) Oxidation of deoxy-ajmaline (*50, 51, 310*), $C_{20}H_{26}ON_2$, with potassium tert.-butylate in benzophenone results in the formation of a ketone, $C_{20}H_{24}ON_2$, whose carbonyl must be present in a five-membered ring as shown by the infrared spectrum. (b) Deoxy-ajmaline when treated with lead tetraacetate develops an aldehyde group (LXI b), whereby the dihydroindole system is converted into the indole system proper. (c) Mild dehydrogenation of deoxy-dihydro-ajmaline (*1*) affords [besides Ind-N-methyl-harman (L, p. 369)], a harman derivative (LXI c) and a base (LXII). The structure of the latter two compounds has since been proved by synthesis (R. B. Woodward, private communication).

(LXI b.)

(LXI c.)

(LXII.)

Isoajmaline obtained from ajmaline by heating or warm alkali is also a natural product. Chemically and spectroscopically it is similar to ajmaline. Upon Huang-Minlon reduction it forms a deoxy-dihydro-derivative that liberates methylethyl ketone when treated with chromic acid. Anet et al. (*1*) suggest that it is the C*-epimer of ajmaline (cf. LXI a, p. 373).

Ajmalinine, $C_{20}H_{26}O_3N_2$, $1^{1}/_{2} H_2O$, m. p. 180°, is another weak base present in *R. serpentina* roots (*305, 351, 352*). It carries a methoxyl group and suffers cleavage at high temperature to give apoajmaline, $C_{13}H_{17}O_3N_2$, which is probably a betaine. Bader, Dickel, Lucas and Schlittler (*4*) have isolated from the same source a yohimbine isomer termed "Alkaloid 3078" and claimed its identity with ajmalinine. However, Huebner et al. (*3*) found the latter to be a new base, viz. 3-epi-α-yohimbine (cf. p. 357).

Ajmalicine [δ-Yohimbine (*128*), Raubasin, Alkaloid "F"]. This alkaloid whose presence in *R. serpentina* was first observed by SIDDIQUI and SIDDIQUI (*351*) has since been isolated from the same source by several American and European workers (*135, 156, 212, 256, 274, 377*) and described under different names, almost simultaneously. KLOHS, DRAPER,

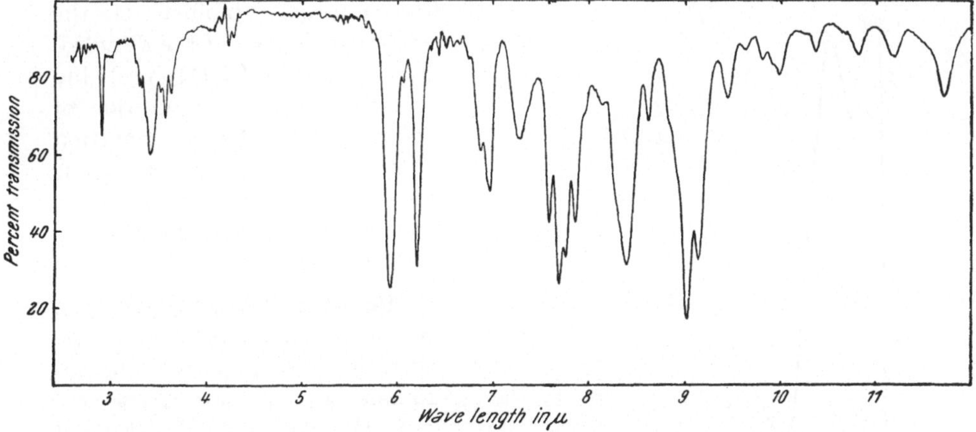

Fig. 9. Infrared spectrum of ajmalicine (Py-tetrahydroserpentine) in chloroform (*231*). [From: Physical Data of Indole and Dihydroindole Alkaloids. Lilly Research Laboratories, Indianapolis, USA.]

KELLER, MALESCH and PETRACEK (*212*) designate this base as "Py-tetrahydro-serpentine" while NEUSS, BOAZ and FORBES (*256*) use the term "Alkaloid F". The name "δ-Yohimbine" was proposed by WEISEN-

(LXIII.) Ajmalicine (δ-yohimbine).

BORN et al. (*377*) and "Raubasin" by POPELAK and his collaborators (*135, 274*). The ultraviolet and infrared (*Fig. 9*) spectra of the base demonstrate the presence of an αβ-alkoxyacrylic ester system (XXXIX, p. 363), and the structure is (LXIII).

The Stereochemistry of Ajmalicine

has been clarified recently by CHATTERJEE and TALAPATRA (66). Ajmalicine when dehydrogenated with lead tetraacetate yields serpentine (p. 366), the latter being reconverted upon catalytic hydrogenation into the original base. This reversible reaction studied by Weisenborn et al. (*377*) and by Chatterjee et al. (*66*) settles the configuration at $C_{(3)}$–$C_{(15)}$ as *syn* (with respect to H). The latter investigators have also observed that ajmalicine (δ-yohimbine) does not react with 2,4-dinitrophenyl-hydrazine in contrast to tetrahydro-alstonine (p. 400). This indicates the unusual stability of the enolether linkage of the dihydro-pyran ring *E* in (LXIII), probably due to the *cis* configuration of the rings *D/E*. Axial orientation of the methyl group at $C_{(19)}$ is also supposed to make some contribution towards the stability of the oxygen bridge.

Reserpine (*Chart 3*, p. 378-379), has acquired importance in the treatment of hypertension, insomnia and mental cases (*1a, 208, 397, 398*). The isolation of this alkaloid, $C_{33}H_{40}O_9N_4$, m. p. 264–265° (dec.), was first described by Müller, Schlitt-ler and Bein (*252*) in 1952 and then, independently, by two other research groups (*213, 254*).

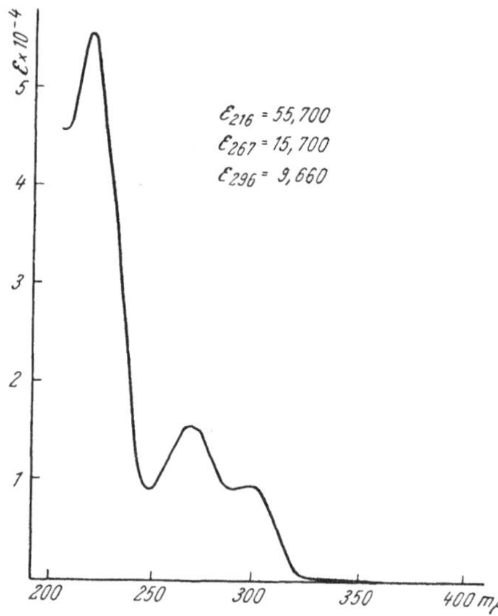

Fig. 10. Ultraviolet spectrum of reserpine in methanol (*231*). [From: Physical Data of Indole and Dihydroindole Alkaloids. Lilly Research Laboratories, Indianapolis, USA.]

Djerassi et al. (*101*) announced the presence of this base in *R. heterophylla*. It also occur in *R. vomitoria* Afzel (*271*), *R. densiflora* Benth and Hook (*65*), *R. perakensis* King and Gamble (*65*), *R. canescens* L. (*214*), *R. micrantha* Hook (*282, 322*), *R. tetraphylla* L. (*100*), *R. beddomei* Hook (*33a*), *R. cumminsii* Stapf (*341a*), *R. natalensis* Sond (*341b*), and *R. sellowii* Muell' (*259*) as well as in a different genus, viz. Australian *Alstonia constricta* (*85a, 86*). Steenhauer (*356*) claims that the alkaloid "B" isolated by her in collaboration with van Itallie (*171*) is identical with reserpine.

The ultraviolet spectrum of this base (*Fig. 10*), shows maxima at 215 mµ (log ε = 4.79); 267 mµ (log ε = 4.23); 295 mµ (log ε = 4.07); shoulder at 225 mµ. The main contributor to the 298 mµ band is a 6-methoxy-indole chromophore. In the infrared spectrum (*Fig. 11*) in chloroform solution the following bands appear: 2.87 µ (>NH); 6.33 and 6.67 µ (two carbonyls); 8.67 and 9.70 µ (aromatic); and 6.2 µ (6-methoxy-indole group).

Based on this spectral evidence and the positive Adamkiewicz color test the reserpine molecule is supposed to include a 11-methoxy-tetra-

hydro-β-carboline moiety. Chemical proof of this feature was given by
SCHLITTLER (*103, 104*) and NEUSS (*254, 255*). Reserpine, a monoacidic
tertiary base, contains six methoxyl groups, yields upon saponification (*104,
117, 213, 254, 255*) trimethylgallic acid (LXIV, p. 378) and reserpic acid,
$C_{22}H_{27}N_2O_4(OH)$ (LXV, p. 378) [= reserpinolic acid (*213*)]. This
demonstrates that reserpine is a diester containing a carbomethoxyl

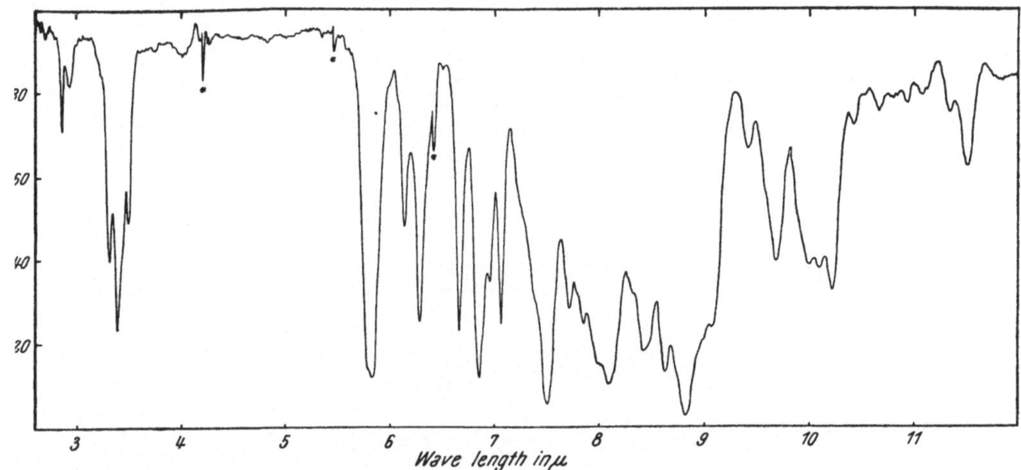

Fig. 11. Infrared spectrum of reserpine in chloroform (*231*). [From: Physical Data of Indole and Dihydroindole Alkaloids. Lilly Research Laboratories, Indianapolis, USA.]

group and a hydroxyl esterified with 3,4,5-trimethoxybenzoic acid.
Accordingly, when treated with lithium aluminum hydride, the base
yields 3,4,5-trimethoxybenzylalcohol (LXVI) and a diol termed reserpic
alcohol (*254*) [reserpine-diol (*235*)], $C_{22}H_{30}O_4N_2$ (LXVII). The demethylation
product of reserpic alcohol reduces FEHLING solution, ammoniacal silver
nitrate in the cold and develops a blue color with the FOLIN-CIOCALTEU
reagent (*112*).

The following conversions of reserpic acid have revealed valuable
features of the reserpine structure (LXVIII) (*Chart 3*, next page).

a) Esterification of reserpic acid with methanol and HCl yields
methyl reserpate (LXVa). The latter reforms reserpine on acylation
with trimethylgalloyl-chloride in pyridine. *b*) When oxidized with
alkaline permanganate reserpic acid is cleaved to give 4-methoxy-N-
oxalyl-anthranilic acid (LXIX) (*103, 104*). *c*) When reserpic acid is
heated with selenium, one of the products is 7-hydroxy-yobyrine (LXX),
characterized by its methyl ether, as synthesized from harmine [HUEBNER
et al. (*164*)]. *d*) When fused with alkali reserpic acid decomposes into
5-hydroxy-isophthalic acid (LXXI) (*103, 104*). *e*) Reserpinol (LXXII),

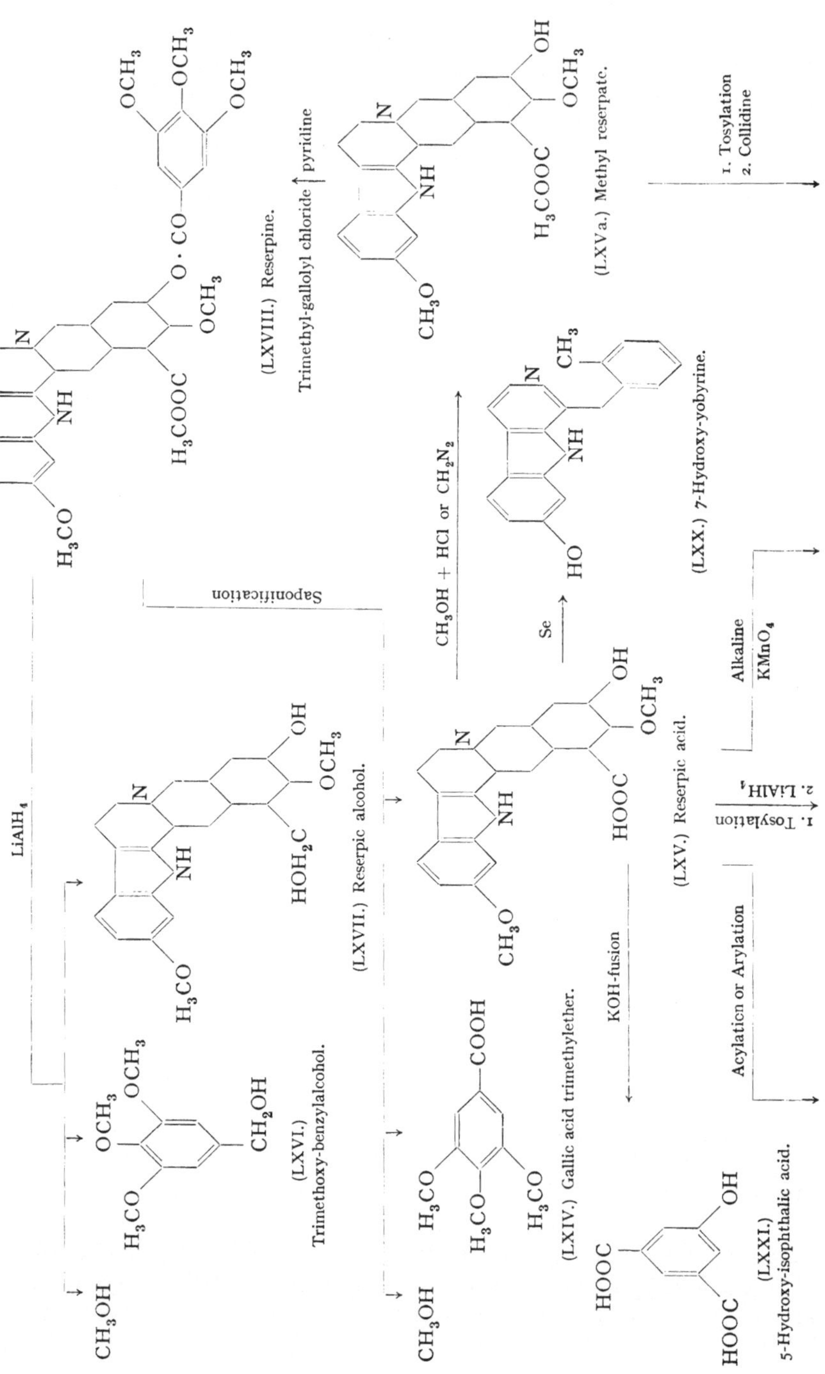

(LXVIII.) Reserpine.

Trimethyl-gallolyl chloride / pyridine

(LXVa.) Methyl reserpate.

1. Tosylation
2. Collidine

Saponification

LiAlH$_4$

(LXVII.) Reserpic alcohol.

(LXVI.) Trimethoxy-benzylalcohol.

CH$_3$OH + HCl or CH$_2$N$_2$

Se

(LXX.) 7-Hydroxy-yobyrine.

(LXV.) Reserpic acid.

Alkaline KMnO$_4$

1. Tosylation
2. LiAlH$_4$

Acylation or Arylation

KOH-fusion

(LXIV.) Gallic acid trimethylether.

CH$_3$OH

(LXXI.) 5-Hydroxy-isophthalic acid.

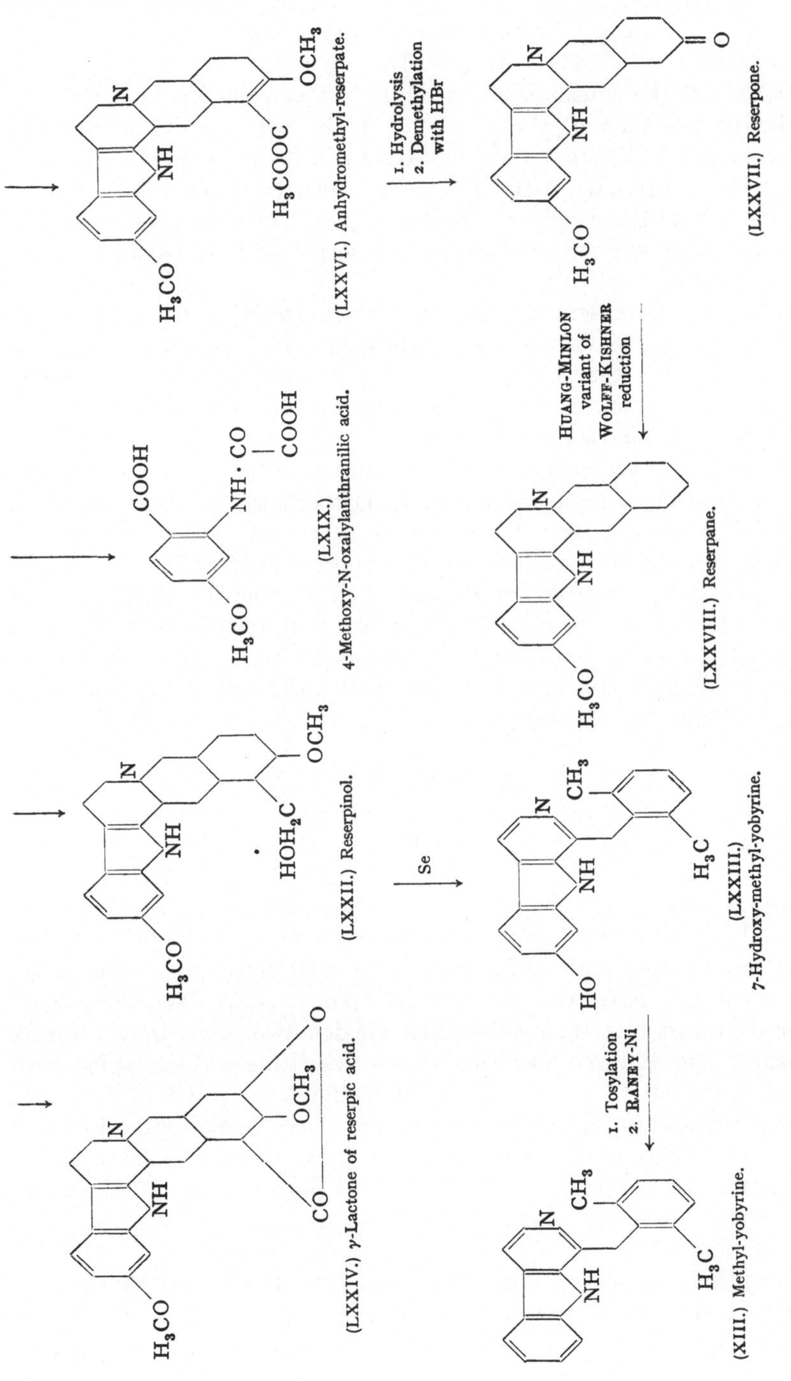

Chart 3. Some Conversions of Reserpine.

a reduction product of tosyl-reserpic acid, yields with selenium 7-hydroxy-methyl-yobyrine (LXXIII) (*164*). The structure of the latter was established by the synthesis of its methyl ether from harmine and 2,6-dimethyl-cyclohexanone (*164*; cf. *203*). *f*) The acid (LXV) readily forms a γ-lactone (LXXIV) (*103*). *g*) The tosylate of methyl reserpate (LXXV, p. 386) when detosylated with collidine is converted into methyl anhydro-reserpate (LXXVI) containing a conjugated carbonyl chromophore (*103*). *h*) The characteristic infrared bands of reserpic acid are located at 1613 and 1709 cm^{-1}.

Methyl anhydroreserpate (LXXVI) when treated with acid suffers demethylation and simultaneous decarboxylation to the ketone reserpone (LXXVII) (*103*); the latter is further reduced to reserpane (*166*). Reserpane is also obtained when 11-methoxy-alloyohimbane is refluxed with acetic acid for two days (*235*).

The above conversions *a*)–*c*) prove the presence of an 11-methoxy-yohimbane nucleus in reserpine (LXVIII) (11-methoxy-epi-alloyohimbane). The earlier postulate that (for biogenetical reasons) the carbomethoxyl group occupies the 16-position (*103*) is supported by *d*) and *e*). Conversion *d*) also locates the hydroxyl group at $C_{(18)}$ in reserpic acid (LXV); this is confirmed by *f*). The methoxyl function has been tentatively assigned to $C_{(17)}$ (*164*). A conclusive evidence for the location of the carbomethoxyl, methoxyl and hydroxyl groups in ring *E* (LXVIII) at the carbon atoms 16–18 is offered by reaction *g*).

The Stereochemistry of Reserpine.

The complexity of the reserpine molecule with six asymmetric centers offers an interesting stereochemical problem, studied by several workers (*98, 99, 161, 166, 167, 234, 235, 364, 363, 379*). The *cis* fusion of rings *D*/*E* has been suggested by MAC PHILLAMY et al. (*234, 235*) and confirmed by SCHLITTLER and his coworkers (*166*). They observed that methyl anhydroreserpate (LXXVI), upon hydrolysis and subsequent decarboxylation, gave rise to two reserpones (LXXVII) that when further degraded (via their thioketals) yielded two, oppositely rotating reserpanes. The infrared spectrum of (—)-reserpane was identified with that of synthetic *DL*-11-methoxy-alloyohimbane (LXXVIII) (*166*). Racemic reserpane [HUEBNER (*161*)] has also been synthesized by SCHLITTLER et al. (*166*) by catalytic hydrogenation (in a basic medium) of 11-methoxy-sempervirine (LXXIX), the latter being obtained by condensing lithium harmine (LXXX) with isopropoxymethylene-cyclo-hexanone (LXXXI).

DIASSI, WEISENBORN, DYLION and WINTERSTEINER (*98*) have drawn the same conclusion concerning the ring juncture *D*/*E*. They found that the collidine detosylation product of methyl reserpate tosylate (LXXV)

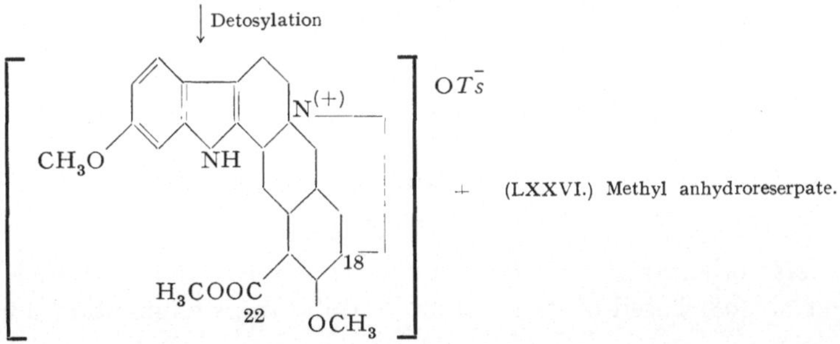

(LXXIX.) 11-Methoxy-sempervirine.

(LXXX.) Lithium harmine.

(LXXXI.) Isopropoxymethylene-cyclohexanone.

1. Catalytic hydrogenation
2. Glacial acetic acid

(LXXVIII.) DL-Reserpane.

is a mixture of methyl anhydro-reserpate (LXXVI) and the tosylate of the quaternary base (LXXXII), an isomeride of methyl reserpate tosylate (LXXV). This quaternization also occurs when methyl reserpate tosylate is refluxed in dimethyl-formamide. The formation of a $N_{(4)}$–$C_{(18)}$ bond (LXXXII) in this process requires an allo (cis) ring juncture at D/E. Such juncture in reserpine has been unambiguously proved by VAN TAMELEN and his associates (364) by a stereospecific synthesis of DL-reserpane (LXXVIII) from 6-methoxy-tryptamine and ethyl DL-cis-2-bromomethyl-cyclohexane acetate, following essentially the same procedure as in the preparation of racemic yohimbane (XIX, p. 355) (365).

(LXXV.) Methyl reserpate tosylate.

Detosylation

(LXXVI.) Methyl anhydroreserpate.

(LXXXII.) Tosylate of a quaternary base.

Several derivatives related to reserpine having appropriate substituents of the yohimbane nucleus at $C_{(16)}$ and $C_{(18)}$ have been prepared by WEISENBORN and APPLEGATE ($376\,a$) who also described the total synthesis of 16-carbomethoxy-18-hydroxy-$\Delta^{15(20)}$-yohimbene.

MAC PHILLAMY et al. (234, 235) have also reported that reserpine and its derivatives undergo acid-or base-catalyzed epimerization at $C_{(3)}$ (LXVIII). This is demonstrated by the lead tetraacetate oxidation of

reserpinediol (LXXXIII) to tetradehydro-reserpinediol (LXXXIV) and its
subsequent sodium borohydride reduction to 3-iso-reserpinediol (LXXXV).
These observations (*235*) coupled with the allo juncture at D/E in reserpine
postulate 3-epi-allo configuration at $C_{(3)-(15)}$ and $C_{(15)-(20)}$. With regard
to the spatial arrangement of the substituents in ring E of (LXVIII,
p. 378), Mac Phillamy et al. were of the opinion that the carbomethoxy,
methoxy and trimethylgalloyloxy groups are in all-*cis* configuration (*234*).
This conclusion was based on the easy γ-lactonization of reserpic acid (LXV).
Equatorial and axial orientation is assigned, respectively, to the carbomethoxyl and methoxyl groups (LXXXVI).

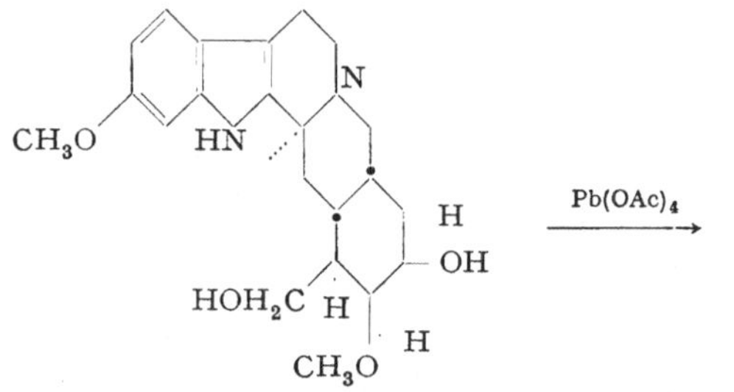

(LXXXIII.) Reserpinediol.

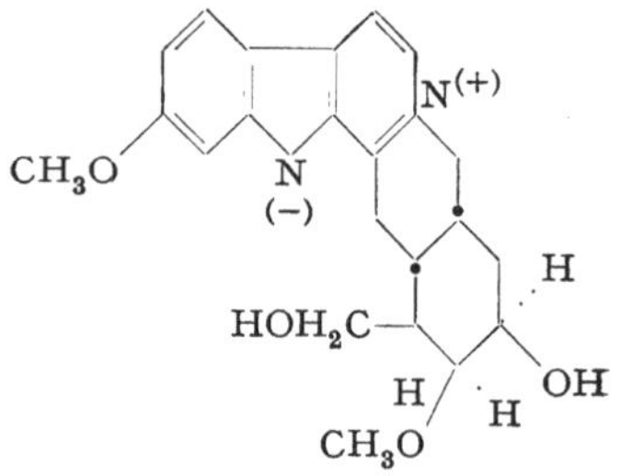

(LXXXIV.) Tetradehydro-reserpinediol.

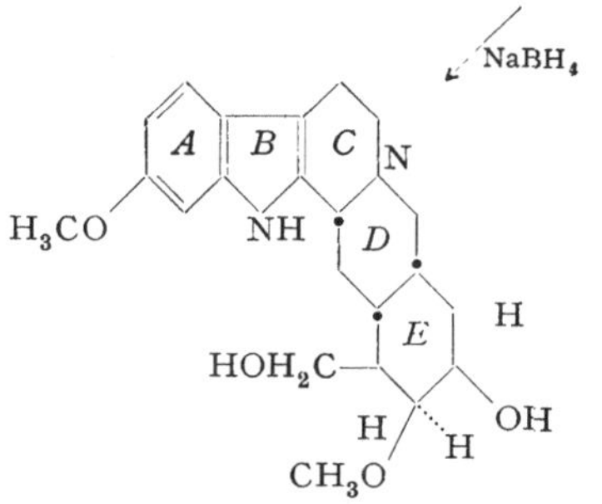

(LXXXV.) 3-Isoreserpinediol.

The *cis* orientation of the methoxyl group was proposed by
Diassi et al. (*98*) based on the assumption of a *trans* elimination during

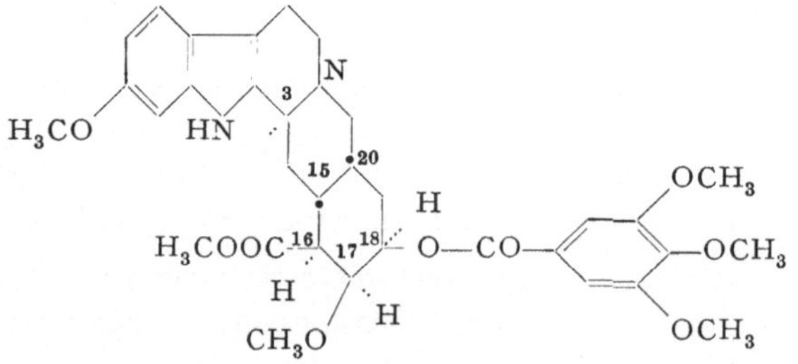

(LXXXVI.)

the formation of methyl anhydroreserpate (LXXVI) from methyl reserpate tosylate (LXXV).

DIASSI, WEISENBORN, DYLION and WINTERSTEINER (98) believed earlier that the quaternary salt (LXXXII) was formed by the tertiary nitrogen atom $N_{(4)}$ with the displacement of the tosylate-ion resulting in an inversion at $C_{(18)}$. This requires the $C_{(18)}$-acyloxy and $C_{(16)}$-carbomethoxy functions to be *cis* oriented with respect to the hydrogens at

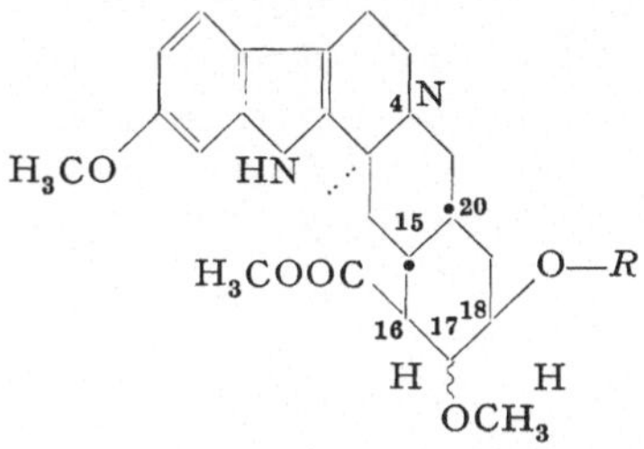

(LXXXVII.) (R = Trimethylgalloyl.)

$C_{(15)}$ and $C_{(20)}$. The application of HUDSON's lactone rule (98) to reserpic acid lactone (103) according to KLYNE (216, 217, 243, 244) induced DIASSI et al. to assign β-orientation to the $C_{(18)}$ function and to other substituents mentioned above, as given in (LXXXVII). This has been modified in the light of some new findings reported by the same group (99), by VAN TAMELEN and HANCE (363), as well as by HUEBNER and WENKERT (167). A β-orientation of the $C_{(3)}$-hydrogen, $C_{(16)}$-carbomethoxyl and $C_{(18)}$-acyloxy groups was demonstrated in studies

Ts Cl in pyridine

(LXXXVIII.) 3-Iso-reserpinol.

(LXXXIX.)
Quaternary base from 3-iso-reserpinol.

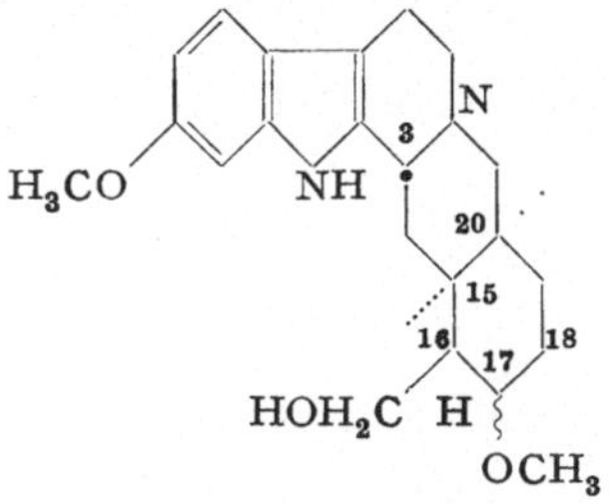

(XC.) Configuration of reserpinol.

of the intramolecular $N_{(4)}$-quaternization in 3-iso-reserpinol (*167*), (LXXXVIII) → (LXXXIX). A similar phenomenon has also been observed in the reserpinol molecule [Huebner, MacPhillamy, St. André and Schlittler (*165*)], indicating configuration (XC).

The $C_{(3)}$-epimerization in reserpine gave further information on the configurations of reserpine and 3-isoreserpine. It was shown (*165, 166*) that reserpine and its derivatives can be completely epimerized to the allo form by acid treatment, while epimerization of isoreserpine and derivatives to compounds of the reserpine series takes place with mercuric acetate and perchloric acid in acetic acid at 60° [Weisenborn and Diassi (*376 b*)]. However, reserpane (11-methoxy-epi-alloyohimbane) and 3-isoreserpane (11-methoxy-3-alloyohimbane) that have no substituents at $C_{(16)}$, $C_{(17)}$ and $C_{(18)}$, are approximately of equal stability (*165, 166*). Furthermore, Wenkert and Liu (*379*) have shown that the equilibrium ratio, 3-epi-alloyohimbane/alloyohimbane, is 3.6 : 1. This also indicates β-orientation of the $C_{(3)}$-hydrogen in reserpine. On the basis of a conformational analysis, the structure of iso-reserpine

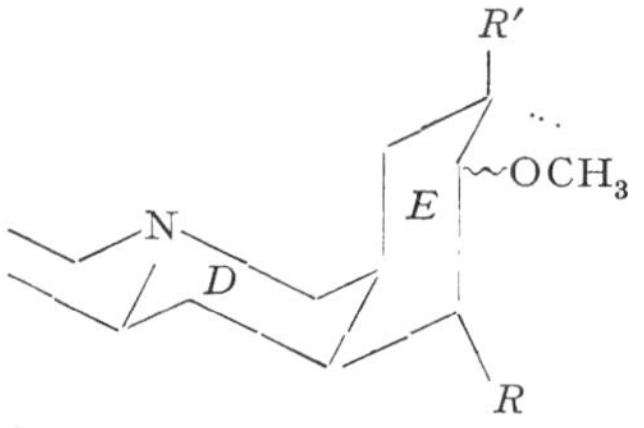

(XCI.) $R = COOCH_3$; $R' = OCO \cdot C_6H_2(OCH_3)_3$. 3-Isoreserpine.
(XCII.) $R = COOH$; $R' = OH$. 3-Isoreserpic acid.

(3-epi-reserpine) at equilibrium can be given as (XCI). The structure (XCI) accounts for the greater stability of the iso compound and also explains the failure of 3-isoreserpic acid (XCII) to lactonize when treated with acetic anhydride-pyridine, in contrast to reserpic acid hydrochloride.

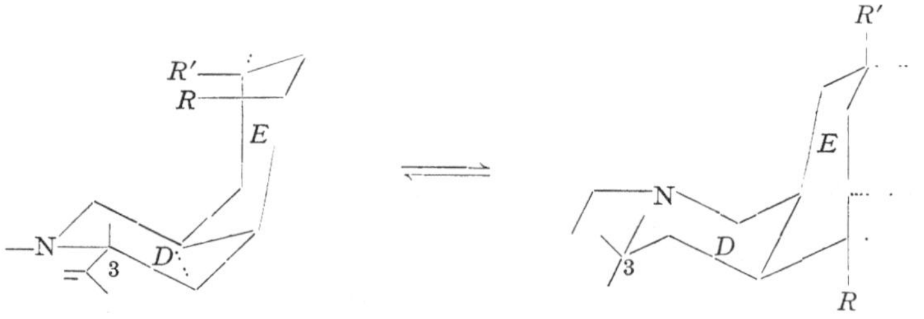

(XCIII.)

(XCIV.)

(XCIIIa.) $R = COOCH_3$; $R' = O \cdot CO \cdot C_6H_2(OCH_3)_3$. (XCIVa.) $R = COOCH_3$; $R' = O \cdot CO \cdot C_6H_2(OCH_3)_3$.
(XCIIIb.) $R—R' = —C—O—$.   (XCIVb.) $R = CH_2OH$; $R' = OH$.
                    O

According to SCHLITTLER et al. (*165*) reserpine may be regarded as an equilibrium mixture of (XCIIIa) and (XCIVa), based on the *trans-anti-cis* and *cis-anti-cis* ring skeletons, the latter being energetically the preferred one. The predominance of (XCIVa) is indicated by the behavior of reserpic acid lactone (XCIIIb) (*trans-anti-cis* conformation) and reserpinediol (XCIVb) in catalytic dehydrogenation over Pd, in the presence of maleic acid as a hydrogen acceptor.

Recently, DIASSI, WEISENBORN, DYLION and WINTERSTEINER (*99*) have reported that the $C_{(16)}$-carbomethoxyl group and the 18-acyloxy function in reserpine are *trans* to the $C_{(15)}$ and $C_{(20)}$ hydrogens and not *cis* as previously postulated (*98*). This is evidenced by the observation that reserpinol, upon treatment with *p*-toluenesulfonylchloride in pyridine, forms a quaternary salt, $C_{29}H_{36}N_2O_5S$ (XCV), that shows bands at 8.56, 8.95, 9.7 and 9.94 μ characteristic for the tosylate ion, while no bands are located in the 3.8–4.0 μ region. Hence, the hydroxymethylene group

(XC.) Reserpinol. $\xrightarrow[\text{pyridine}]{Ts \text{ Cl in}}$

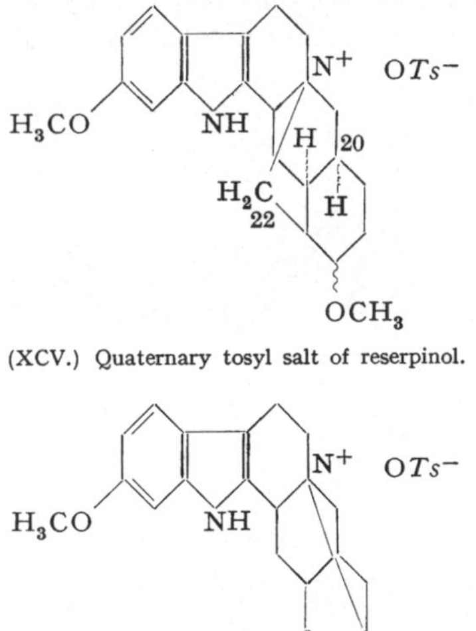

(XCV.) Quaternary tosyl salt of reserpinol.

(LXXV.) Methyl reserpate tosylate. $\xrightarrow[\text{pyridine}]{Ts \text{ Cl in}}$

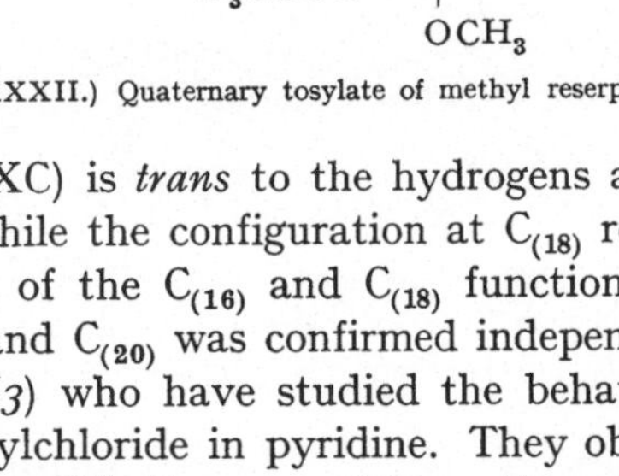

(LXXXII.) Quaternary tosylate of methyl reserpate.

at $C_{(16)}$, i. e. —COOCH$_3$ at $C_{(16)}$ (XC) is *trans* to the hydrogens at $C_{(15)}$ and $C_{(20)}$, (LXXV → LXXXII), while the configuration at $C_{(18)}$ remains unchanged. The *trans* orientation of the $C_{(16)}$ and $C_{(18)}$ functions with respect to the hydrogens at $C_{(15)}$ and $C_{(20)}$ was confirmed independently by VAN TAMELEN and HANCE (*363*) who have studied the behavior of reserpinol towards *p*-toluenesulfonylchloride in pyridine. They obtained the same quaternary salt (XCV) as did DIASSI and his colleagues (*99*) [maxima, 222 mμ (log ε 4.70), 269 mμ (log ε 3.80), and 294 mμ (log ε 3.93)].

According to van Tamelen and Hance (*363*), since (XCV) possesses a bridged bicyclic system defined by the atoms 4 and 15–22, it can arise from an $N_{(4)}$-attack on $C_{(22)}$ of the unisolated O-tosylate of reserpinol (XC) only, if $C_{(21)}$ and the carbomethoxyl function are present in an arrangement *cis* to the ring *E* (LXVIII). It was also postulated (*99, 363*) that the easy epimerization, i. e. the less stable configuration at $C_{(3)}$, implies that $C_{(2)}$ is linked through the axial bond to $C_{(3)}$. Hence the $C_{(3)}$ hydrogen should be equatorial and *trans* to the hydrogens at $C_{(15)}$ and $C_{(20)}$.

It has also been suggested (*363*) that the stereochemical course of elimination of the tosyloxy group from methyl reserpate tosylate gives methyl anhydroreserpate (LXXVI). Its formation and the simultaneous internal quaternization of (LXXV) to (LXXXII) are explicable by the participation of a $C_{(17)}$ methoxyl, in *trans* position to the groups at $C_{(16)}$ and $C_{(18)}$. First an oxonium salt (XCVI) is produced from (LXXV) which reacts in two different ways giving rise to (LXXVI) and (LXXXII). (LXXVI) arises from (XCVI) by loss of a proton, while (LXXXII) is formed by a second displacement at $C_{(18)}$ as shown below:

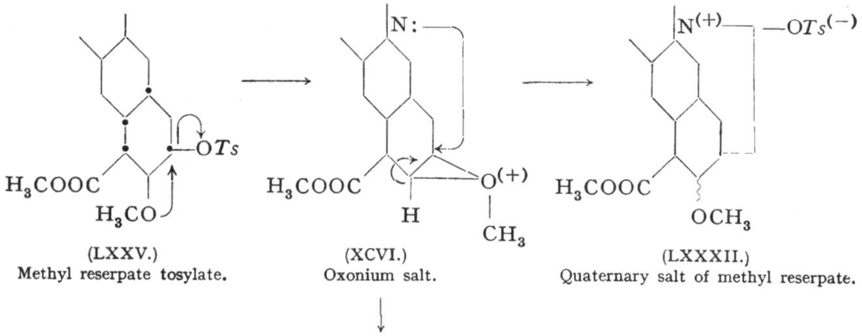

(LXXV.)
Methyl reserpate tosylate.

(XCVI.)
Oxonium salt.

(LXXXII.)
Quaternary salt of methyl reserpate.

(LXXVI.) Methyl anhydroreserpate.

From the rotational evidence bearing on the relative configurations at $C_{(3)}$ and $C_{(16)}$ it follows that the $C_{(16)}$-hydrogen is *trans* to that at $C_{(3)}$ (XCVIa) (*363*).

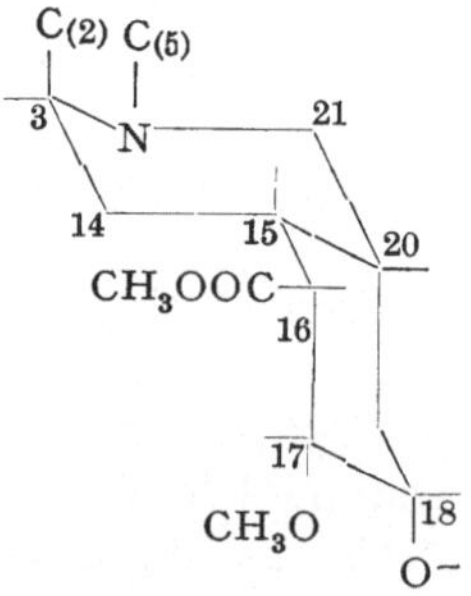

(XCVIa.) Stereochemical formulation of reserpine

Schlittler and his colleagues (*165, 166*) believe that the $C_{(17)}$-methoxyl should preferably have α-orientation, since the β-configuration would sterically hinder the formation of a quaternary salt (LXXXII) from (LXXV).

Considering these data, the best stereochemical formulations for reserpine and 3-isoreserpine are, (XCVII) and (XCVIII); cf. also formula (XCVIII p; p. 389).

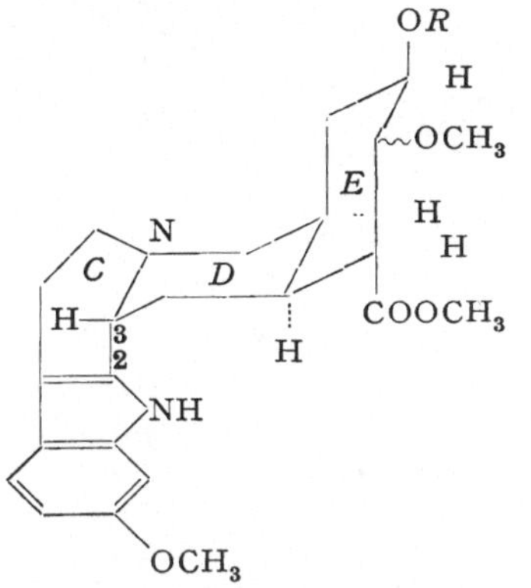

(XCVII.) Configuration of reserpine.

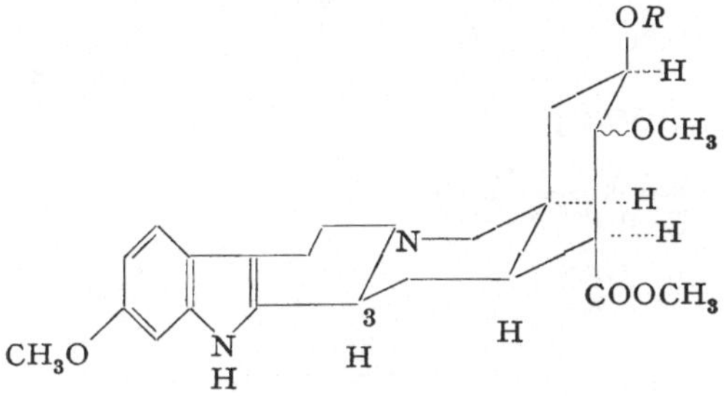

(XCVIII.) Configuration of 3-isoreserpine.

Total Synthesis of Reserpine.

Quite recently, Woodward, Bader, Bickel, Frey and Kierstead (*388 b*) have announced a brilliant synthesis of reserpine (*Chart 4*, p. 388).

Structure-Action Relation in Reserpine.

It is interesting to note that the sedative properties of reserpine are lost when it is hydrolyzed to methyl reserpate, trimethylgallic acid and methyl alcohol. It appears therefore that the trimethoxybenzoyl group potentiates the biological activity. On the basis of a study of the negati-α tranquilizing effects of a reserpine analogue of rauwolscine, 3-epiveyohimbine and methyl reserpate, Chatterjee and Talapatra (*67, 68*) believe that both the additive effect of the three substituents in ring E (LXVIII) in reserpine and the trimethylgalloyloxy group at $C_{(18)}$ are responsible for the hypnotic action of the drug. The same observation was made by Huebner, Lucas, MacPhillamy and Troxell (*163*). Stoll and Hofmann (*359*) have shown that the methoxyl group in position 11 is not required for the pharmacodynamic effect of reserpine. Huebner (*162*) finds that N-methyl-reserpine acts as a reserpine antagonist. The biological properties of reserpine appear to be largely dependent on steric factors as indicated by the inactivity of 3-isoreserpine (*235*).

Rescinnamine (*209, 210*) ("Reserpinine"; *136*). A comparative study of pure reserpine and the "Oleoresin fraction" (ex *R. serpentina*) has shown that the latter was physiologically more active than could be explained by the reserpine content alone. Hence, other potent components must be present in this fraction. In fact, Klohs, Draper and Keller (*209, 210*) as well as Haack, Popelak, Spingler and Kaiser (*136*) have

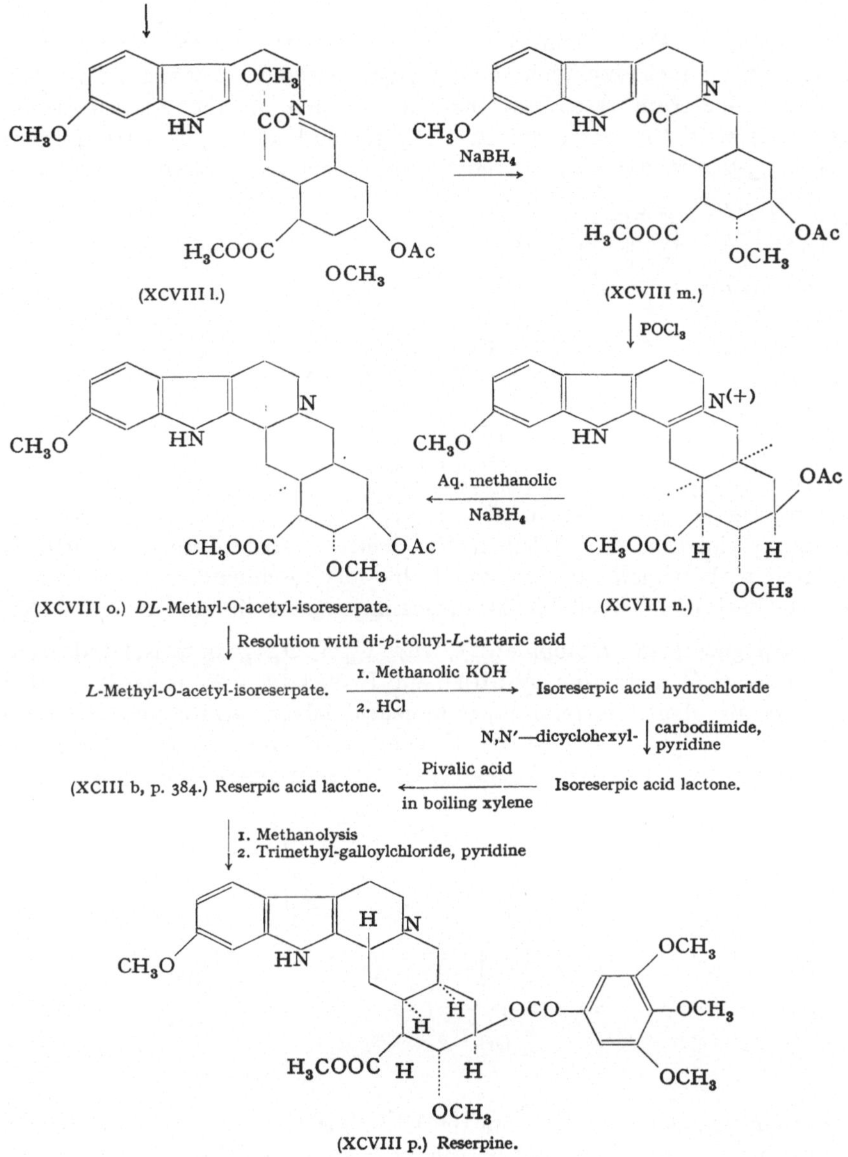
CH$_3$O
HN
OCH$_3$
CON
H$_3$COOC
OAc
OCH$_3$
(XCVIII l.)
NaBH$_4$
CH$_3$O
HN
OC
N
H$_3$COOC
OAc
OCH$_3$
(XCVIII m.)
POCl$_3$
CH$_3$O
HN
N(+)
OAc
CH$_3$OOC
H
H
OCH$_3$
(XCVIII n.)
Aq. methanolic
NaBH$_4$
CH$_3$O
HN
N
CH$_3$OOC
OAc
OCH$_3$
(XCVIII o.) *DL*-Methyl-O-acetyl-isoreserpate.
Resolution with di-*p*-toluyl-*L*-tartaric acid
L-Methyl-O-acetyl-isoreserpate.
1. Methanolic KOH
2. HCl
Isoreserpic acid hydrochloride
N,N'—dicyclohexyl- carbodiimide, pyridine
Isoreserpic acid lactone.
Pivalic acid
in boiling xylene
(XCIII b, p. 384.) Reserpic acid lactone.
1. Methanolysis
2. Trimethyl-galloylchloride, pyridine
CH$_3$O
HN
H
N
H
H
OCO
OCH$_3$
OCH$_3$
OCH$_3$
H$_3$COOC
H
H
OCH$_3$
(XCVIII p.) Reserpine.

Chart 4. Total Synthesis of Reserpine [WOODWARD et al. (*388 b*)]. For a simplification cf. (*388 c*)

announced the presence of a sedative base, rescinnamine, isolated in the pure state. Rescinnamine, $C_{35}H_{42}O_9N_2$, contains six methoxy groups. Its ultraviolet spectrum is similar to that of reserpine (*Fig. 10*, p. 376);

maxima at 229 mμ (log $\varepsilon = 4.73$) and 302 mμ (log $\varepsilon = 4.39$), the latter arising out of the combined effects of two chromophores, viz., $\alpha\beta$-disubstituted-6-methoxyindole and 3,4,5-trimethoxy-cinnamic acid. The infrared spectrum resembles that of reserpine in the 2.5–7 μ region (*Fig. 11*) with the exception of a sharp band at 6.19 μ, attributed to the 3,4,5-trimethoxy-cinnamic acid unit. Alcoholic potassium hydroxide

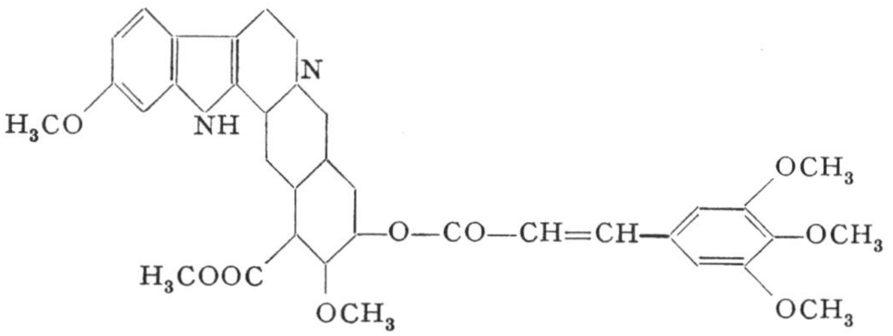

(XCIX.) Rescinnamine.

hydrolyzes rescinnamine to give reserpic acid (LXV, p. 378) and tri-methoxy-cinnamic acid. When the methyl ester of reserpic acid is treated with trimethoxy-cinnamoyl chloride, rescinnamine is recovered. On the basis of the available data rescinnamine has the structure (XCIX).

Sarpagine (*358*) **(Raupine;** *25, 26*), $C_{19}H_{22}O_2N_2$, first isolated from the roots of *R. serpentina* by Stoll and Hofmann (*358*), is easily soluble in aqueous alkali; it reduces ammoniacal silver nitrate and Fehling solution. It gives a positive Keller test and forms a diacetate. According to Raymond-Hamet (*304*), it contains a 5-methoxyindole

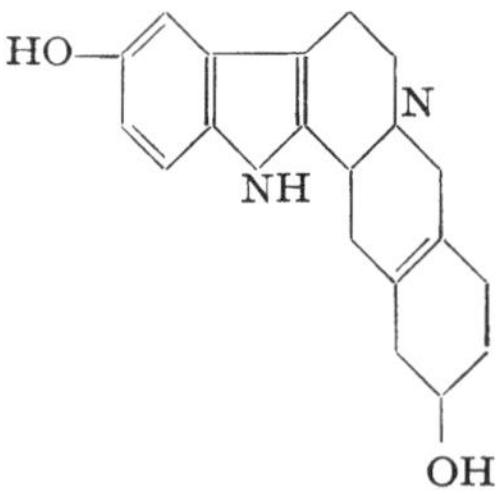

(C.) Sarpagine   or 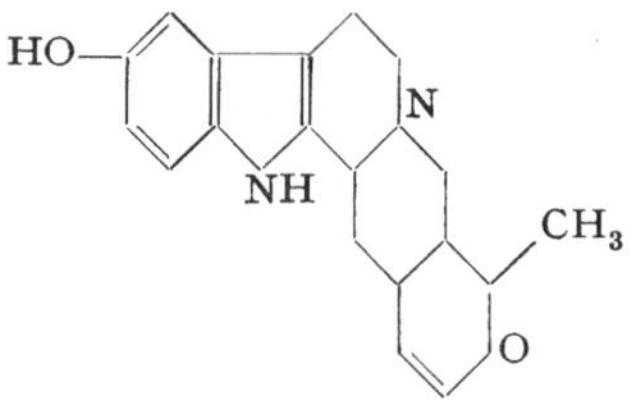   (CI.) Sarpagine.

chromophore as evidenced from the U. V. spectral data. Two alternative sarpagine structures, (C) and (CI), have been proposed by Thomas (*367*). "Raupine" was identified with sarpagine by Bodendorf and Eder (*25, 26*).

Rauhimbine (Corynanthine). The presence of this alkaloid in *R. serpentina* roots was reported by Hofmann (*155*). It is identical with corynanthine (p. 356), the $C_{(16)}$-epimer of yohimbine (*156*).

Isorauhimbine (*228*, *155*, *156*). This isomer of rauhimbine occurs together with the latter. Its structure was partially clarified by HOF-

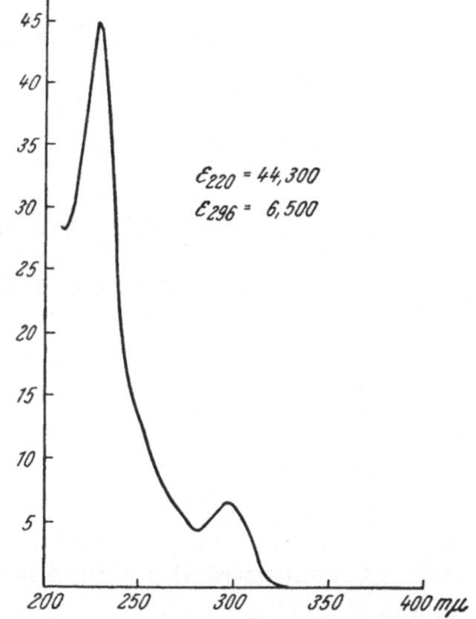

(CII.) Dehydro-ketoyobyrine. (CIII.) Isorauhimbine.

MANN (*155*) and was shown to be an isomer of yohimbine, $C_{21}H_{26}O_3N_2$, containing a hydroxyl group and a carbomethoxyl group. It has two active hydrogen atoms, forms a diacetyl derivative and shows the U. V. spectrum of yohimbine. Infrared readings have confirmed the presence of —OH (2.8 μ), >NH (3.1 μ), —COOCH$_3$ (5.8 μ), and an o-disubstituted benzene (13.4 μ). The position of the carbomethoxyl and the hydroxyl group and the stereochemistry of isorauhimbine have been investigated by LE HIR, GOUTAREL, JANOT and HOFMANN (*228*). When heated with selenium, at high temperature, the alkaloid gives rise to yobyrine (IX, p. 353), "tetrabyrine" (X, p. 353) and dehydro-ketoyobyrine (CII). The formation of yobyrine and tetrabyrine secures the yohimbine structure and that of dehydro-keto-yobyrine (CII) locates the carbomethoxyl group at $C_{(16)}$ (CIII). The position of the hydroxyl (assumed to be tertiary) and the spatial configuration are unclarified.

Fig. 12. Ultraviolet spectrum of reserpinine in methanol (*231*). [From: Physical Data of Indole and Dihydroindole Alkaloids. Lilly Research Laboratories, Indianapolis, USA.]

Reserpinine (*329*, *377*, *187*) [**Alkaloid "C"** (*156*); **Raubasinin** (*274*); **Alkaloid "A"** (*256*)]. Reserpinine, $C_{22}H_{26}O_4N_2$, carries a side methyl and two methoxyl groups (one in COOCH$_3$) and it shows the presence of one active hydrogen. The ultraviolet spectrum (*Fig. 12*) demonstrates the presence of a 6-methoxyindole chromophore (maxima at 229, 298 mμ). The infrared data indicate a β-alkoxyacrylic ester system (*Fig. 13*). The best available structure is given in (CIV).

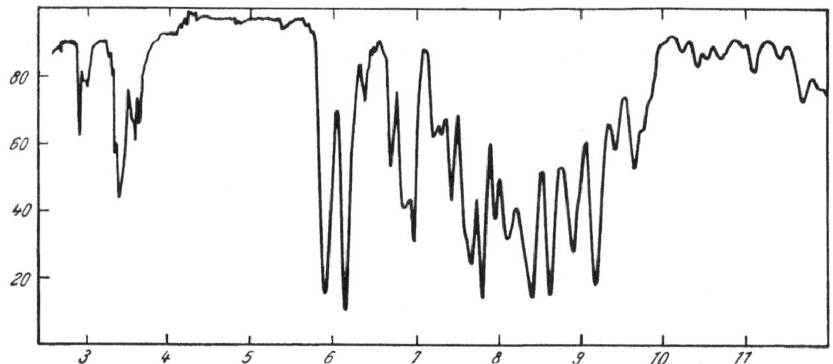

Fig. 13. Infrared spectrum of reserpinine in chloroform (*231*). [From: Physical Data of Indole and Dihydro-indole Alkaloids. Lilly Research Laboratories, Indianapolis, USA.]

(CIV.) Reserpinine.

Reserpiline. The presence of this base in *R. serpentina* roots, was observed by Klohs and his coworkers (*211*). The amorphous base forms a number of crystalline salts, analyses of which favor the formula $C_{23}H_{28}O_5N_2$. This tetrahydro-β-carboline derivative carries three methoxyl groups (one carbomethoxyl), and a terminal methyl group. The infrared spectrum includes a band at $3.0\,\mu$ ($>$NH), others at 5.99 and $6.2\,\mu$ (ester conjugated with an enol ether) (*314*). The lithium aluminum hydride reduction product shows a maximum at $298\,m\mu$ characteristic for a 2,3-disubstituted-5,6-dimethoxyindole.

(CV.) Reserpiline.

Serpine (*47, 48, 52*) was found in the roots of *R. serpentina* Benth of Cochin variety. Serpine, $C_{21}H_{26}O_3N_2$, is a tetrahydro-β-carboline alkaloid and an isomer of yohimbine. Both the ultraviolet and the infrared spectra of the base are very similar to those of yohimbine. On selenium dehydrogenation serpine decomposes into yobyrine (IX), "tetrabyrine" (X) and ketoyobyrine (XI, p. 353). OPPENAUER oxidation converts serpine into yohimbone which proves the *syn-anti* configuration at $C_{(3)}$–$C_{(15)}$ and $C_{(15)}$–$C_{(20)}$. The axial nature of the carbomethoxyl group has been postulated on the basis of some saponification data. Possibly, serpine is the $C_{(17)}$-epimer of corynanthine or the $C_{(16)}$-epimer of β-yohimbine.

Serpinine (*31*) was obtained from the mother liquors of rauwolfinine [BOSE (*31*)]. Serpinine, $C_{20}H_{24}ON_2$, is free from methoxyl, methylenedioxy and phenolic groups. It feebly reduces ammoniacal silver nitrate. It is an indoline derivative as evidenced from its chemical properties, color reactions and spectrum (maxima at 250 and 293 mμ). Bands in the I. R. region: 6.25 μ, 6.84 μ, 8.96 μ, and 13.55 μ. BOSE suggested (*31*) that serpinine is an indoline alkaloid, structurally analogous with ajmaline. It has now been found by BOSE (*31 a*) that serpinine contains a hydroxyl group and not an oxygen bridge as postulated earlier. Serpinine is probably identical with tetraphyllicine and has the structure (CXIX, p. 403) (*31 a*).

Yohimbine. SCHLITTLER, BADER and DICKEL (*5*) have isolated yohimbine from an amorphous alkaloid fraction (of *Rauwolfia* roots); it was freed from reserpine by chromatography.

3-Epi-α-yohimbine (*4, 3*) is a new stereoisomer of yohimbine. The presence of this base ("Alkaloid 3078") in the accumulated mother liquors of ajmaline was announced by BADER, DICKEL, LUCAS and SCHLITTLER (*4*). It shows two crystal forms, m. p. 125–128° and 181–183°. Infrared readings showed the presence of the following bands (with an inflection at 3527 cm.$^{-1}$ (—OH); 3365 cm.$^{-1}$ (>NH); 1720 cm.$^{-1}$ (CH$_3$OOC—); and 738 cm.$^{-1}$ (*o*-disubstituted benzene ring); maxima in the ultraviolet: 226, 282 and 290 mμ. 3-Epi-α-yohimbine yields on

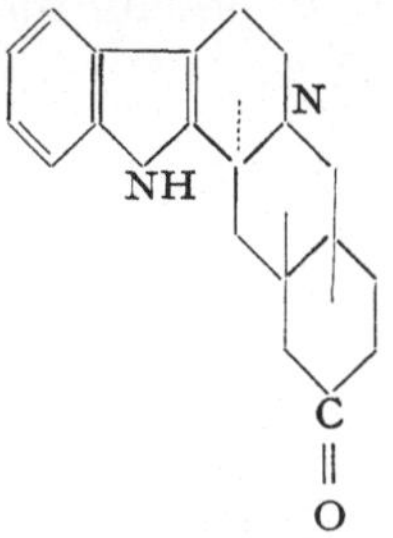

CVI.) 3-Epi-alloyohimbone
(not abs. configuration).

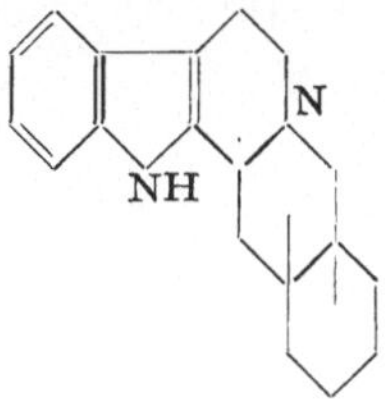

(CVII.) 3-Epi-alloyohimbane
(not abs. configuration).

saponification the corresponding methoxyl-free acid which with diazo-methane regenerates the original base. The formation of tetrabyrine (X) and yobyrine (IX, p. 353) by selenium dehydrogenation of 3-epi-α-yohimbine proves the presence of the yohimbine skeleton. The isolation of methyl-yobyrine (XIII, p. 353) from the dehydrogenation products of 3-epi-α-yohimbylalcohol (a reduction product of the ester-alkaloid), locates the carbomethoxyl group at $C_{(16)}$. The position of the secondary hydroxyl group at $C_{(17)}$ has been secured by the formation of 3-epi-

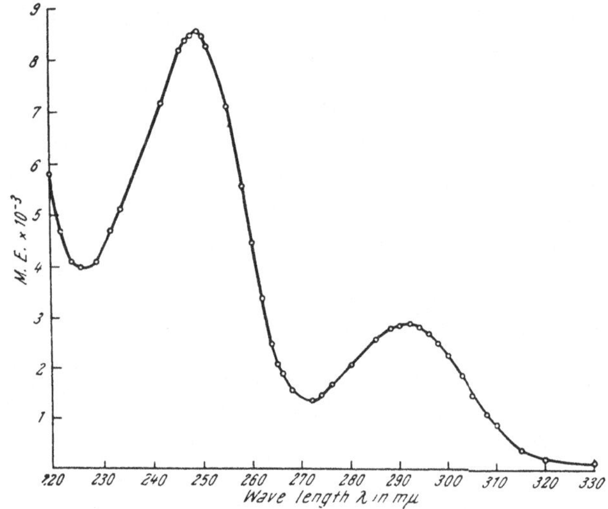

Fig. 14. Molecular extinction curve of rauwolfinine in alcohol, according to Bose (29). [From: J. Indian Chem. Soc. 31, 311 (1954).]

alloyohimbone (CVI) during Oppenauer oxidation, the first dextrototary ketone prepared in the yohimbine series. The reduction of this ketone by the Huang-Minlon method yielded two isomeric oxygen-free bases, viz. alloyohimbane (XXVI, p. 353) and 3-epi-alloyohimbane (CVII). The ethylene-mercaptol of 3-epi-alloyohimbone gave on desulfurization with

(CVIII.) 3-Epi-α-yohimbine.

Raney-nickel 3-epi-alloyohimbane only. Hence, 3-epi-α-yohimbine differs from α-yohimbine only in the configuration at $C_{(3)}$. This has been confirmed by its conversion to α-yohimbine by lead-tetraacetate oxidation of the base and subsequent reduction of the py-tetradehydro-α-yohimbine with sodium borohydride. Wenkert and Liu (*379*) as well as Huebner, Schlittler et al. (*165*) have assigned the absolute configuration (CVIII) to 3-epi-α-yohimbine at $C_{(3)}$, $C_{(15)}$ and $C_{(20)}$.

Rauwolfinine. Bose (*27* to *30*) observed that *R. serpentina*, growing in North-Western India, yielded no ajmaline but a new base, rauwolfinine, m. p. 235 to 236°, (dec.). Rauwolfinine, $C_{19}H_{24}O_2N_2$, does not contain methoxyl or methylenedioxy groups. It is a monoacidic base, with a $>N . CH_3$, and a $\longrightarrow C . CH_3$ function. The ultraviolet spectrum (*Fig. 14*) indicates indoline character (*27, 28*), supported by chemical observations.

The I. R. spectrum (*Fig. 15*) shows the following bands: 9.0 μ (ether linkage), 7.24 μ ($\longrightarrow C.CH_3$), and 2.8 μ (hydroxyl).

Rauwolfinine is not a carbinolamine base as shown by its behavior towards LiAlH$_4$ and the negative result of the Huang-Minlon reduction. Rauwolfinine is not very stable towards acids. On warming with 2 N-H$_2$SO$_4$ it suffers dehydration and develops an indole chromo-

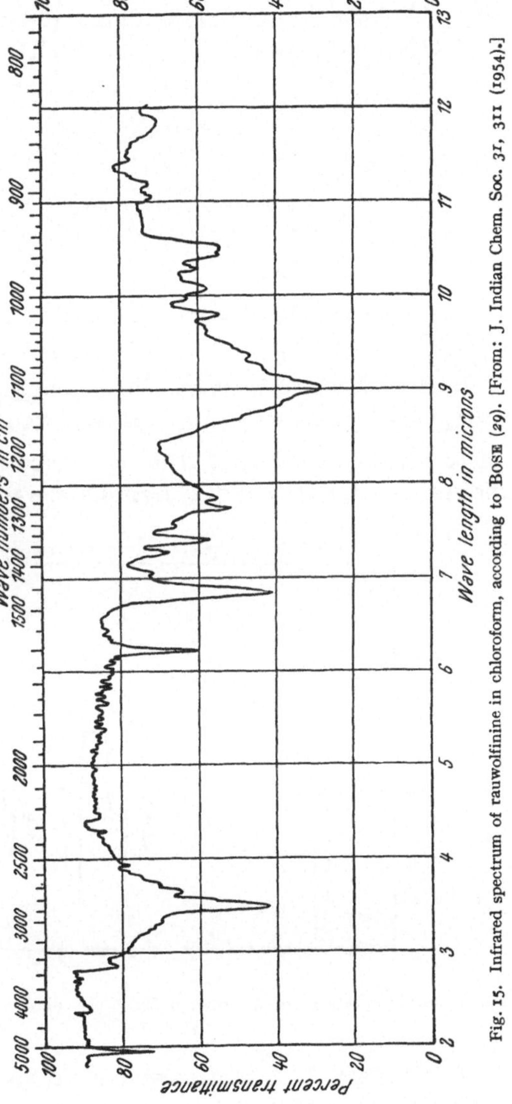

Fig. 15. Infrared spectrum of rauwolfinine in chloroform, according to Bose (29). [From: J. Indian Chem. Soc. *31*, 311 (1954).]

phore. Zinc dust distillation (*30*) or selenium dehydrogenation yielded Ind-N-methyl-harman (L, p. 369). Upon fusion with potassium hydroxide (*30*), the alkaloid undergoes facile cleavage to give indole-2-carboxylic acid, Ind-N-methyl-harman and *n*-butyric acid (cf. *206*). Recently, Bose (*30a*) proposed a structure for rauwolfinine containing the same ring system as in (CXIX a, p. 404) and the group (CIX).

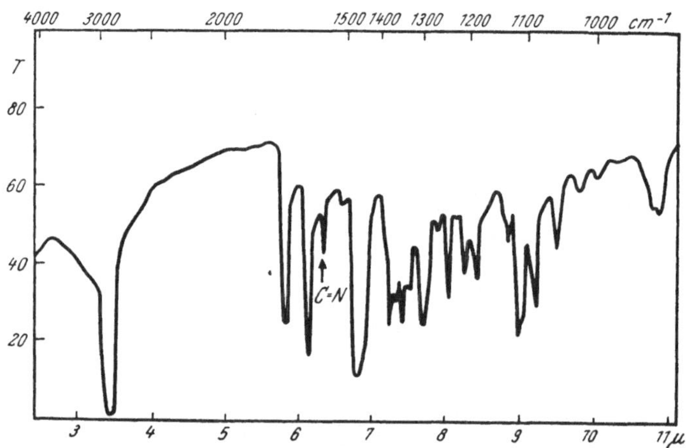

Thebaine and Papaverine. According to Hofmann (*156*), thebaine, $C_{19}H_{21}O_3N$, and papaverine, $C_{20}H_{21}O_4N$ occur in *R. serpentina* roots.

Reserpic Acid Methylester. The presence of this ester (cf. LXVa, p. 378), $C_{23}H_{30}O_5N_2$, in *R. serpentina* was reported by Hofmann (*156*).

Serpentine (cf. *Chart 5*). The occurrence of serpentine, which belongs to the group of strong bases of *R. serpentina*, was first reported by Siddiqui and

Fig. 16. Infrared spectrum of serpentine in Nujol, according to Bader and Schwarz (*6*). [From: Helv. Chim. Acta *35*, 1594 (1952).]

Siddiqui (*352*). It is also present in *R. hirsuta* (*241, 372*) and *R. heterophylla* (*102, 154, 182*), *R. canescens* (*135*), furthermore, in *Lochnera rosea* L. (fam. Apocynaceae) (*245a*). According to Schlittler and Schwarz (*331*) its correct formula is, $C_{21}H_{22}O_3N_2$ (instead of $C_{20}H_{20}O_3N_2$). They confirmed the presence of a carbomethoxy group and showed that the remaining

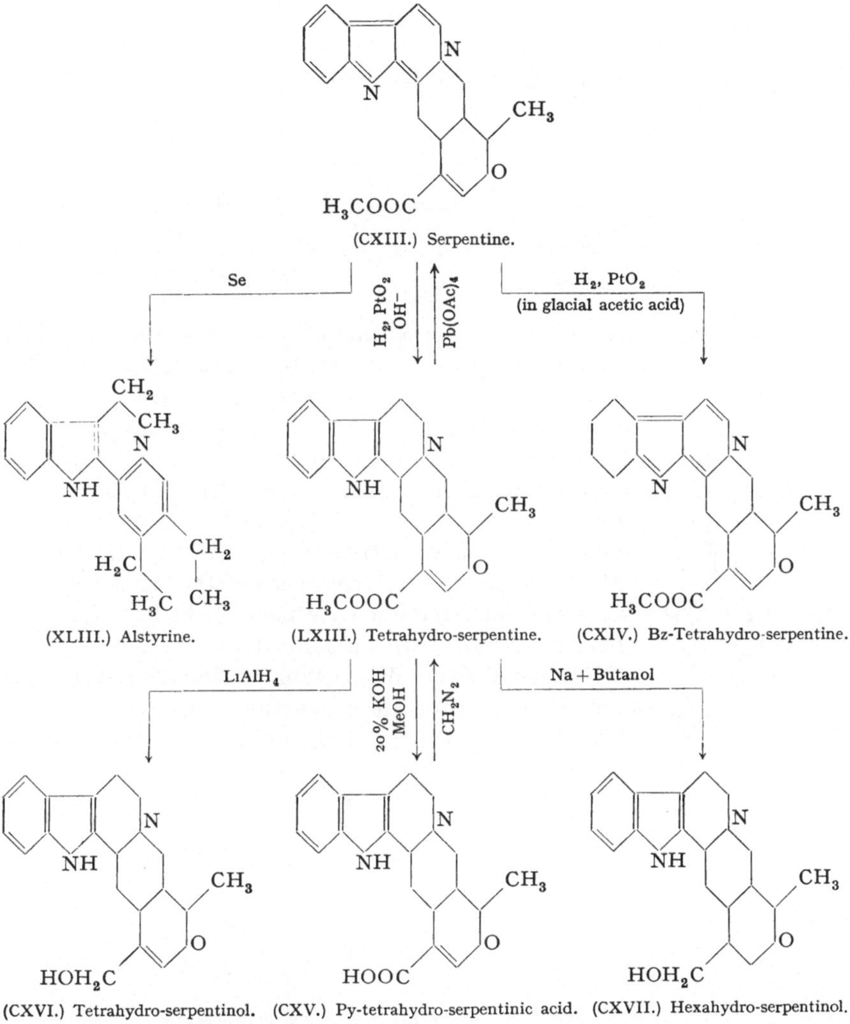

(CXIII.) Serpentine.

Se

H_2, PtO_2
OH^-

$Pb(OAc)_4$

H_2, PtO_2
(in glacial acetic acid)

(XLIII.) Alstyrine. (LXIII.) Tetrahydro-serpentine. (CXIV.) Bz-Tetrahydro-serpentine.

LiAlH$_4$

20% KOH
MeOH

CH_2N_2

Na + Butanol

(CXVI.) Tetrahydro-serpentinol. (CXV.) Py-tetrahydro-serpentinic acid. (CXVII.) Hexahydro-serpentinol.

Chart 5. Some Conversions of Serpentine.

oxygen-atom was not present in a free hydroxyl group but probably as an ether linkage (indicated by the I. R. spectrum; *Fig. 16*).

Serpentine is a β-carboline alkaloid (*331*) and a monoacidic base containing easily reducible double bonds. From its I. R. spectrum the indole >NH band is missing; however, it appears in the spectrum of the py-tetra-hydro derivative. Consequently, serpentine is an anhydro-nium base, like sempervirine (*393, 173, 391*) or tetra-

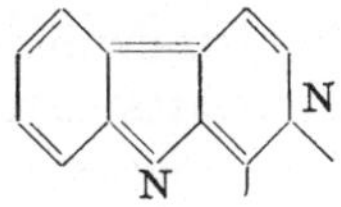

(CX.)

dehydro-yohimbine (*342*), and contains the chromophoric group (CX) (*342*). During salt formation or catalytic reduction, this chromophore is supposed to tautomerize to the structure (CXa) which undergoes conjugation to (CX).

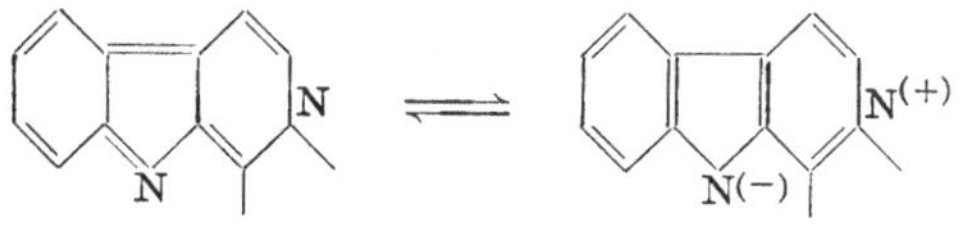

(CX.)   (CXa.)

Selenium dehydrogenation of serpentine afforded alstyrine, (XLIII, p. 364). The structure of alstyrine that can be prepared in a similar manner from corynantheine (*195, 202*), alstonine (*230*) or vincain (*66*), had been clarified by Karrer and Enslin (*196*). Schlittler and Schwarz (*331*) suggested a formulation of serpentine as (CXI). Later Bader and Schwarz (*6*) reported that the serpentine molecule contained a side methyl group, the logical place of which is in ring *E* considering the formation of alstyrine on selenium treatment of the base (p. 397). This necessitated the modification of the structure (CXI) to (CXII). Weisenborn, Moore and Diassi (*377*) have reinvestigated serpentine in connection with their studies on δ-yohimbine (LXIII, p. 375) (*142, 128, 188*). The latter when

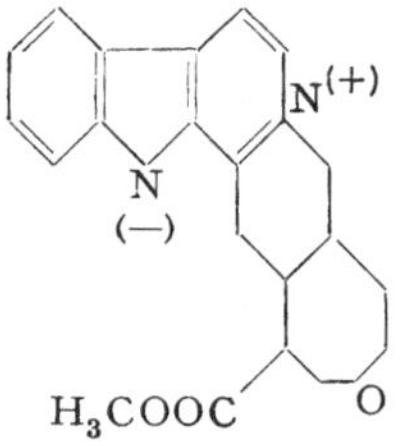

(CXI.)

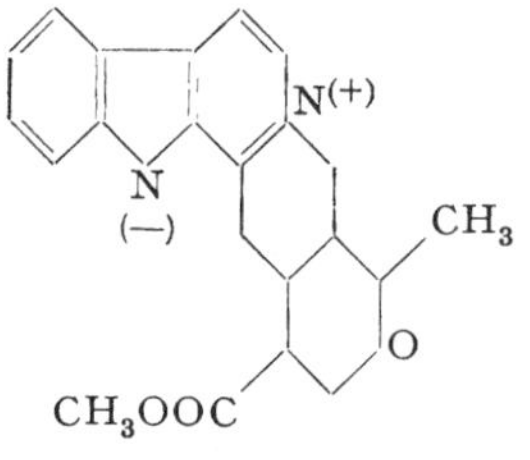

(CXII.)

dehydrogenated with lead tetraacetate yielded serpentine that is reconverted to δ-yohimbine by catalytic hydrogenation.

They have also observed that the serpentine spectrum shows maxima at 225 and 292 mμ, with a strong absorption in the 250 mμ region, indicating a $CH_3OOC—C=C—O—$ system that is also present in δ-yohimbine. This was found simultaneously by Klohs and his associates (*212*) in an elegant series of investigations. These authors have succeeded in identifying Siddiqui's ajmalicine (p. 375) with py-tetrahydro-serpentine [Bader and Schwarz (*6*)] by catalytic hydrogenation of serpentine. The analysis of py-tetrahydro-serpentine showed 2 H-atoms less than in the formula $C_{21}H_{26}O_3N_2$, earlier proposed by SCHLITTLER and SCHWARZ (*331*); this indicated an additional center of unsaturation. Spectral studies of py-tetrahydro-serpentine revealed that this double bond must be present in a $CH_3COOC-C=C-O-$ system. In the light of these new observations the formula is, $C_{21}H_{20}O_3N_2$ and the structure (CXII) of serpentine has been modified to (CXIII) [KLOHS et al. (*212*); WEISENBORN et al. (*377*)]. For the conversions of serpentine cf. *Chart 5*, p. 397.

(CXIII.) Serpentine.

Serpentinine. In a study of this second strong base of *R. serpentina,* SCHLITTLER, HUBER, BADER and ZAHND (*328*) modified the earlier formula, $C_{20}H_{20}O_5N_2$, proposed by SIDDIQUI and SIDDIQUI (*352, 354*) to $C_{21}H_{20-22}O_3N_2$. This alkaloid also occurs in *R. tetraphylla* L. as reported by DJERASSI and FISHMAN (*100*).

The intense yellow color of the free base and the pale yellow color of its salts as well as the ultraviolet spectrum showing the characteristic shift of the maxima in alkaline solution (*328*) offer an unambiguous proof of the anhydronium and quaternary indole basic character of serpentinine. Like serpentine (*212, 377*) serpentinine contains a β-alkoxyacrylic ester grouping as indicated by strong bands at $5.83\,\mu$ and $6.16\,\mu$.

Upon dehydrogenation with selenium, serpentinine yields alstyrine (LXIII, p. 397), and affords indole-2-carboxylic acid, and pyridine-3-4b-indole-1(2)-one (CXIIIa) (*190*) upon caustic potash fusion. No definite structure has been proposed for serpentinine.

(CXIIIa.)

IV. The Alkaloids of *Rauwolfia vomitoria* and *R. obscura.*

These plants of African origin possess hypotensive properties. The root of *R. obscura* contains the yellow base alstonine, $C_{21}H_{20}O_3N_2$ (*332, 144, 145*), the principal alkaloid of *Alstonia constricta* MUELL (*346, 347*). From *R. vomitoria* a number of alkaloids have been isolated by French and Swiss investigators (*271, 129, 332*). PARIS (*260*) first reported the isolation of ajmaline, iso-ajmaline, ajmalicine, ajmalinine and serpentine.

Almost 10 years later Schlittler, Schwarz and Bader (*332*) confirmed the presence of ajmaline. They have also obtained alstonine, $C_{21}H_{20}O_3N_2$. Recently, Goutarel, Le Hir, Poisson and Janot (*129, 271*) have observed the presence of reserpine in the plant besides three other new alkaloids, *raumitorine*, $C_{22}H_{26}O_4N_2$, *seredine*, $C_{23}H_{30}O_5N_2$ (*129, 271, 272*) and *rauvomitine* (*270 a*); the latter was also obtained by Haack et al. (*134 a*). Kidd has isolated *rescinnamine* from the same source (*205*).

Alstonine, $C_{21}H_{20}O_3N_2$, cannot be recrystallized without decomposition but it forms well crystalline, yellow or orange salts whose aqueous solutions show blue fluorescence. Alstonine is a monoacidic, tertiary base containing a carbomethoxyl group. It also contains two easily reducible double bonds and, upon catalytic reduction, affords a tetrahydro derivative which on basic hydrolysis yields the methoxyl-free tetrahydro-alstoninic acid, $C_{20}H_{22}O_3N_2$. By esterification the tetrahydro base is recovered. The positive Adamkiewicz reaction of tetrahydro-alstonine and the isolation of N-oxalylanthranilic acid from the base indicate a β-carboline structure. This was confirmed by Leonard and Elderfield (*230*) who obtained harman by alkali fusion and zinc dust distillation. Sharp (*346, 347*) observed that by selenium dehydrogenation of alstonine dimethyldiselenide and an oxygen-free base, alstyrine (cf. XLIII, p. 397) are obtained.

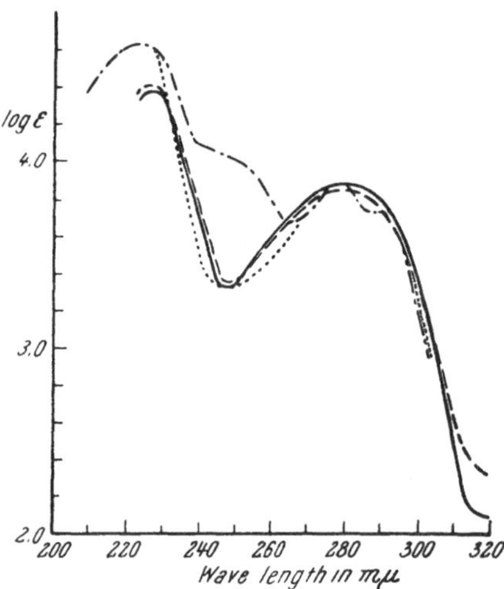

Fig. 17. Ultraviolet spectra; ———— tetrahydro-alstonol; ————— desoxy-hexahydro-alstonol; ————— tetrahydro-alstonine; and ————— hexahydro-alstonol, according to Elderfield and Gray (*109*). [From: J. Organ. Chem. (USA) *16*, 506 (1951).]

The structure of alstyrine was clarified by Karrer and Enslin (*196*) who conducted oxidative degradation experiments with subsequent hydrolysis, and by a recent synthesis [Lee and Swan (*222 b*)]. Elderfield and Gray (*109*) have shown that tetrahydro-alstonine carries a $\rightarrow$C.CH_3 group and, when treated with sodium in boiling butanol, it affords hexahydro-alstonol, $C_{20}H_{26}N_2O_2$; however, with lithium aluminum hydride the tetrahydro derivative is obtained. The ultraviolet spectra (*Fig. 17*) of tetrahydro-alstonol, hexahydro-alstonol, and tetrahydro-alstonine reveal the presence of a double bond, resistant to saturation, in conjugation with the ester group.

The shift of the ester bands in alstonine and tetrahydro-alstonine from 1736 cm.$^{-1}$ (as discernible in yohimbine) to 1710–1720 cm.$^{-1}$ confirms the presence of the conjugated ester system (XXXIX, p. 363) in these bases. The peak at 1668 cm.$^{-1}$ exhibited by tetrahydro-alstonol is indicative of a cyclic double bond. During ozonolysis no formaldehyde is liberated. This eliminates the possibility of the double bond being exocyclic. Tetrahydro-alstonol is unusually acid-labile in contrast to the hexahydro derivative. ELDERFIELD and GRAY proposed the tentative structure (CXIVa) for alstonine.

(CXIVa.) Alstonine.   (CXIVb.)

SCHLITTLER, SCHWARZ and BADER (*332*) have suggested the same nuclear structure for alstonine. From a careful analysis of infrared and ultraviolet data (maxima at 250, 310 and 370 mμ) of py-tetrahydro-alstonine, SCHLITTLER and his associates concluded the existence of an enol-ether conjugated ester group (CH$_3$OOC—C=C—O—) in ring E which locates the double bond at $\Delta^{16,\,17}$. The dihydropyran structure of ring E (CXIVb) has since been confirmed by BADER (*2*) considering its acid lability and the formation of 2,4-dinitrophenyl-hydrazone with simultaneous fission at the oxygen bridge.

Raumitorine, C$_{22}$H$_{26}$O$_4$N$_2$ (*ex. R. vomitoria*) (*129, 272*) is a weak base containing a side methyl and two methoxyl groups, one of these being associated with the chromophore, CH$_3$OOC—C=C—O— (also confirmed by U. V. and I. R. data which are identical with those for aricine). Raumitorine is a β-tetrahydro-carboline alkaloid with a methoxy group at C$_{(10)}$. According to JANOT, GOUTAREL, LE HIR and POISSON, raumitorine is a 10-methoxy-δ-yohimbine or an isomer of aricine (XLVIII, p. 365).

Seredine, C$_{23}$H$_{30}$O$_5$N$_2$ (*ex R. vomitoria*) (*129, 272*), is also a β-tetrahydro-carboline base with two active hydrogens and two methoxyl groups. JANOT et al. interpret it as a stereoisomer of methyl reserpate (LXVa, p. 378).

Rauvomitine, C$_{30}$H$_{34}$O$_5$N$_2$. HAACK et al. (*134a*) and JANOT et al. (*270a*) have recently detected this alkaloid in *R. vomitoria* roots. Ajmaline, rescinnamine and reserpine were also isolated (*270 a*). Upon hydrolysis rauvomitine yielded the base anhydro-ajmaline and, furthermore, trimethoxybenzoic acid (*270 a*); hence it is a trimethoxybenzoic ester of

tetraphyllicine (*102 a*). This has been established independently by
Djerassi, Gorman, Pakrashi and Woodward (*102 a*).

V. The Alkaloids of *Rauwolfia heterophylla*
(*101, 102, 168, 154, 261, 306*).

R. heterophylla Roem and Schult grows widely in Mexico, Guatemala
("Chalchupa"), Costa Rica and Colombia ("Pinique-pinique"). According
to Dr. R. E. Woodson the plant is identical with *R. hirsuta* Jacq. Earlier
investigation of this species [Deger (*96*)] revealed the presence of two
amorphous products, chalchupins A and B, to which the formulas
$C_{14}H_{21}N_3O_{12}$ and $C_{15}H_{24}N_6O_{11}$ were assigned. Djerassi, Gorman,
Nussbaum and Reynoso (*101*) have reexamined the alkaloidal components
of this plant. · A fraction yielded reserpine and *L*-narcotine which was
shown later (*102*) to be a contaminant.

Djerassi et al. also observed (*102*) that a crude alkaloid fraction
when subjected to a countercurrent distribution gave ajmaline and
serpentine. Janot, Goutarel and Le Hir (*182*), have confirmed the
presence of serpentine. Ishidate, Okada and Saito (*168*) have recently
isolated three more alkaloids from the weakly basic fraction, viz. sarpagine
(raupine), yohimbine and δ-yohimbine.

Hochstein, Murai and Boegemann (*154*) have obtained (besides
reserpine) seven alkaloids from the same source, viz. yohimbine,
rauwolscine, ajmalicine, heterophylline (aricine), ajmaline, serpentine,
and sarpagine (raupine). At least four more alkaloids are present in
the roots (*154*). Leaf extracts contain little or no reserpine and their
alkaloidal constituents are different from those found in root extracts.

VI. The Alkaloids of Further *Rauwolfia* Species.

R. hirsuta Jacq. which is claimed to be identical with *R. heterophylla*
Roem and Schult (*154*) is also identical with *R. canescens* L. (*241, 242,
372*). *R. hirsuta* (known as Pinique-pinique, Cruceto or Pepa de culebra
in various parts of South America) is used in Colombia as a medicine
against snake bites. This Colombian species (alkaloid content 1.0% in
the roots) contains reserpine, rauwolscine, alstonine and a minor base,
probably sarpagine [Uribe Vergara (*372*)].

R. densiflora Benth and Hook is abundant in Southern India and
grows widely on the Malabar coast. Chatterjee and Talapatra (*65*)
have isolated reserpine and ajmaline from the roots.

R. perakensis King and Gamble is a Malayan species. Chromatographic resolution (*65*) of its non-basic resin fraction affords γ-sitosterol
and reserpine besides an indoline base which appears to be rauwolfinine (*27*).

R. indecora R. E. WOODSON according to ISHIDATE, OKADA and SAITO (*169*) produces reserpine, ajmaline and sarpagine in the roots.

R. micrantha HOOK is a species commonly growing in South Travancore, India (*395, 396*) and is used as a substitute for *R. serpentina*. RAO and RAO (*282*) have isolated the four following crystalline bases from the "oleoresin fraction", viz. Base A, presumably reserpine, the presence of which has also been recorded by SHAVEL et al. (cf. *322*); Base B, m. p. 247–248°; Base C, m. p. 157–159°; and Base D, apparently identical with ajmalicine.

R. tetraphylla L. Four main basic components have been isolated (*100*) from *R. tetraphylla*, a tree common in the West Indies: tetraphyllin, tetraphyllicine, reserpine, and serpentinine. Ajmaline and ψ-yohimbine have recently been isolated (*100 a*).

Tetraphyllin appears to be a new base (m. p. 220–223°; dec.) whose ultraviolet spectrum is identical with that of reserpinine (p. 391). The infrared spectrum shows bands at 5.92 and 6.17 μ, and an inflection at 6.1 μ typical of the system, $H_3COOC—C=C—O—$. Further resolution of this spectrum in the region, 6.0–6.5 μ shows bands at 5.96, 6.14 and 6.22 μ and indicates the presence of a methoxylated benzene nucleus. On saponification, tetraphyllin yields an amorphous acid that is reconverted into the original base when treated with diazomethane. DJERASSI and FISHMAN (*100*) consider tetraphyllin (CXVIII) as a stereoisomer of reserpinine (CIV, p. 392).

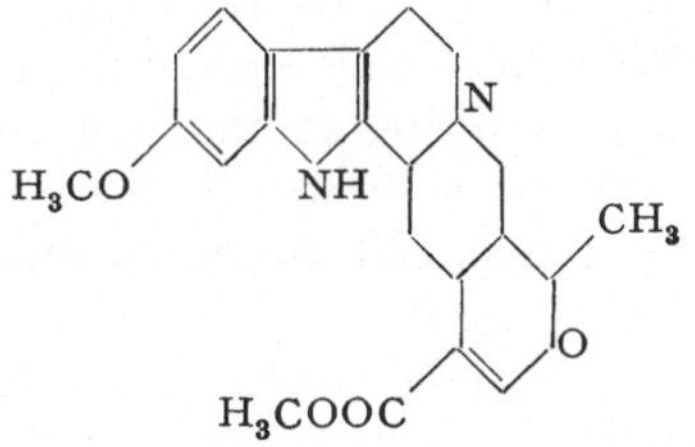

(CXVIII.) Tetraphyllin.

Tetraphyllicine, $C_{20}H_{26}ON_2$, was obtained by chromatographing the crude extract (*100*). It contains a terminal methyl and a $>N . CH_3$ group. Its ultraviolet spectrum is identical with that of ajmaline.

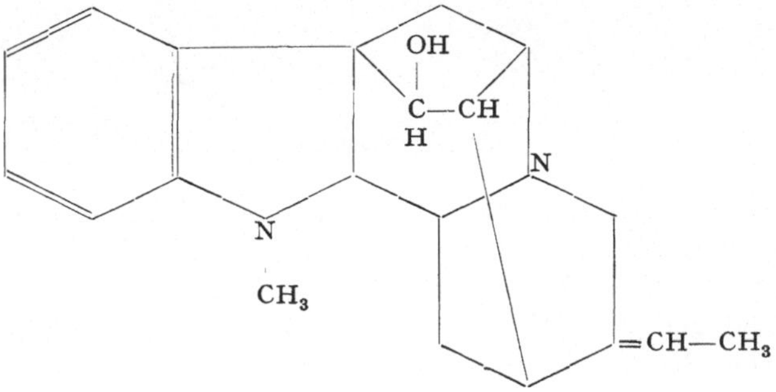

(CXIX.) Tetraphyllicine.

The empirical formula and the structure (CXIX) of tetraphyllicine have been secured by Djerassi, Gorman, Pakrashi and Woodward (*102 a*) on the basis of extensive analytical data and degradation experiments. Dihydro-tetraphyllicine, obtained by catalytic hydrogenation proved to be identical with desoxy-ajmaline. A 55% yield of acetaldehyde upon ozonolysis has located the double bond.

R. sellowii Muell. Argov. The histochemistry of the alkaloids present in this Brazilian tree was studied by Neubern and Wasicky (*253*). The isolation of ajmaline, ajmalinine, serpentine and two unidentified crystalline alkaloids from this plant has been reported. Recently, Pakrashi, Djerassi, Wasicky and Neuss (*259*) as well as Hochstein (*153*) have reinvestigated this plant. Pakrashi et al. have recorded the presence of seven alkaloids, viz. ajmaline, aricine (major bases), reserpine, ajmalicine, tetraphyllicine, py-tetrahydroalstonine (first occurrence), and ajmalidine. *R. sellowii* is the richest source for ajmaline (1.2%) (*259*). For the physiological action (*343, 344*) the reserpine content is responsible.

Ajmalidine, $C_{20}H_{24}N_2O_2$, is a new base, m. p. 241–242°. Its color reaction with nitric acid and its ultraviolet spectrum are practically the same as those of ajmaline. The I. R. spectrum is also identical with that of ajmaline except for a strong band at $5.77\,\mu$, ascribed to a five-membered ring ketone. Ajmalidine is a much weaker base than ajmaline. A similar shift in pK' has been ascribed in the dihydroindole series to the conversion of an α-aminoketone to the corresponding α-amino-alcohol (*220*).

According to Djerassi, Gorman, Pakrashi and Woodward (*102 a*) ajmalidine is best represented by structure (CXIX a).

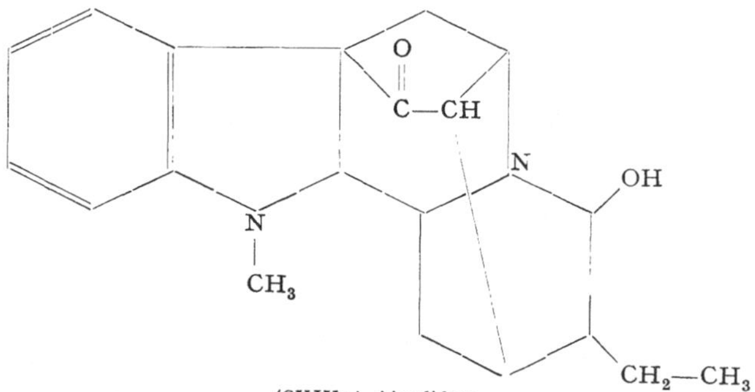

(CXIXa.) Ajmalidine.

R. semperflorens Schlechter. From the bark of this species Schlittler and Furlenmeier (*325*) have obtained an indoline base, semperflorine, m. p. 295°, besides a new indole alkaloid. Semperflorine carries a side methyl and an $>$N—CH$_3$ group. It forms a crystalline hydrochloride. The ultraviolet spectrum (*Fig. 18*) reveals indoline character, in conformity with the color reactions observed. The infrared spectrum of semperflorine hydrochloride shows a band at 3.08 μ assignable to a >NH function rather than to —OH. The hydrochloride band appears at 4.02 μ. An intense band at 9.28 μ suggests the presence of an ether linkage. The spectral evidence excludes the possibility of the presence of an ester or carbonyl function in the molecule. No structure has yet been assigned to this alkaloid.

R. caffra SOND. From this species, the bark of which is reputed for its therapeutic value, KOEPFLI (*219*) obtained three new and physiologically active alkaloids; Base A, m. p. 294–295°, forming hairlike colorless needles; Base B, crystalline but yet uncharacterized; and *rauwolfine*, $C_{20}H_{26}O_3N_2 \cdot 2^1/_2 H_2O$; m. p. 235 to 238°, as the major alkaloid, representing a quaternary ammonium base type. It forms crystalline salts and is free from alkoxyl groups. The base is insoluble in most solvents and shows characteristic color reactions. With sulfuric acid it produces a brilliant yellow solution, the color of which gradually disappears; on addition of a drop of conc. nitric acid a brilliant indigoblue color appears changing rapidly to purple and golden brown.

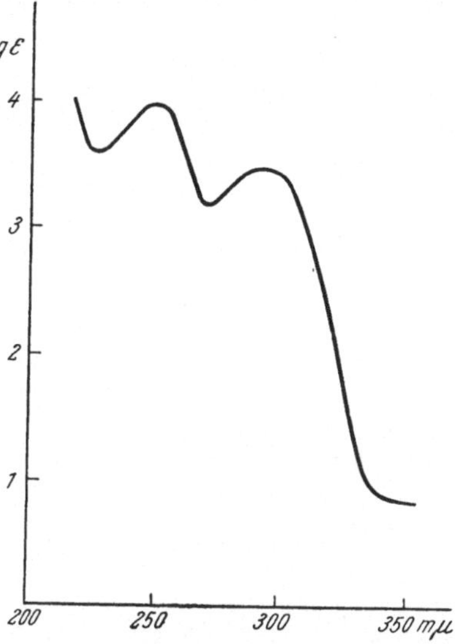

Fig. 18. Ultraviolet spectrum of semperflorine, according to SCHLITTLER and FURLENMEIER (*325*). [From: Helv. Chim. Acta *36*, 996 (1953.)]

R. natalensis SOND (Africa) contains three active alkaloids (*309*), among them reserpine and ajmaline (*341 b*). – *R. mombasiana* STAPF. has distinct physiological effects (*301*). – *R. grandiflora* MART. contains reserpine and other alkaloids (*245 a*). – *R. cumminsii* STAPF. From the root bark reserpine has been isolated (*341 a*). – *R. verticillata*. The bark contains δ-yohimbine (*1 b*). – *R. beddomei* HOOK. The occurrence of sarpagaine in the roots has been reported (*33 a*). — *R. degeneri* SHERFF. DJERASSI et al. (*102 b*) have isolated serpentinine, ajmaline, tetraphyllin and tetraphyllicine from this species which resembles in its alkaloid composition *R. tetraphylla* (p. 403).

VII. On the Biogenesis of the Rauwolfia Alkaloids.

The literature reveals the versatility of *Rauwolfia* spp. to produce physiologically active alkaloids. According to *Table 6*, the *Rauwolfia* alkaloids can be divided into four main groups.

Table 6 shows that the yohimbane and hetero-yohimbane systems constitute the basic structure of the majority of *R*. bases. The yohimbane

Table 6. Classification

Indole Bases

Yohimbane group	Basic structure	Heteroyohimbane group	Basic structure

Yohimbine
β-Yohimbine
ψ-Yohimbine
Rauhimbine
  (= Corynanthine)
Isorauhimbine
Rauwolscine
  (= α-Yohimbine)
3-Epi-rauwolscine
Serpine
Reserpine
Rescinnamine
Methyl ester of reserpic acid
Canescine (= Deserpidine)
Sarpagine (= Raupine)
		Ajmalicine
		Reserpinine
		Isoreserpinine
		Reserpiline
		Isoreserpiline
		Aricine
		Raumitorine
		Seredine
		Tetraphyllin
		Serpentine
		Alstonine
		Serpentinine } Anhydronium bases	

alkaloids are probably precursors of the hetero-yohimbane compounds, the latter originating from yohimbanes by bio-oxidation and simultaneous fission at $C_{(18)}$. The anhydronium compounds of the serpentine type might be produced from hetero-yohimbanes by further oxidation of ring C.

Karrer (*194*) in his article on "Curare Alkaloids" has discussed Woodward's hypothesis (*388*) concerning the biogenesis of strychnos alkaloids, especially the genetical relationship between yohimbine and strychnine alkaloids. If in the hypothetical substance (CXX) a bond between the carbon atoms 2 and 3 is formed, a compound of the β-carboline or yohimbine type (CXXI) appears. If in (CXX) the carbon atoms 3 and 7 on one side and the carbons 2 and 16 on the other, are joined, a compound of the strychnine type (CXXII) would arise.

Mavacurine (*23*), a curare alkaloid (*Loganiaceae*) (*194, 201*) can also be related structurally to (CXXI).

Assuming the validity of this hypothesis it can be postulated that (CXX), the precursor of both yohimbine and strychnine alkaloids, is biosynthesized in the same manner in plants belonging to the families *Apocynaceae* (*43 a, 189, 304, 312, 326, 330, 377*), *Loganiaceae* (*127, 201, 278, 23, 327, 393*) and *Rubiaceae* (*173, 355, 303*).

Rauwolfia Alkaloids.

Dihydroindole bases	Basic structure	Isoquinoline group
Ajmaline Serpinine Tetraphyllicine Ajmalidine Rauvomitine	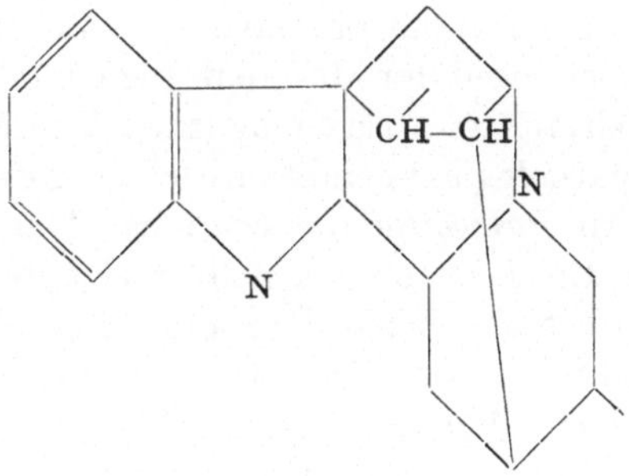	Thebaine Papaverine
Rauwolfinine Semperflorine	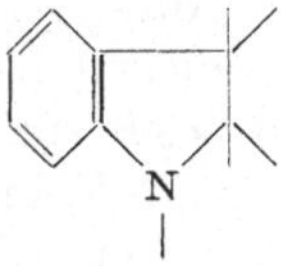	

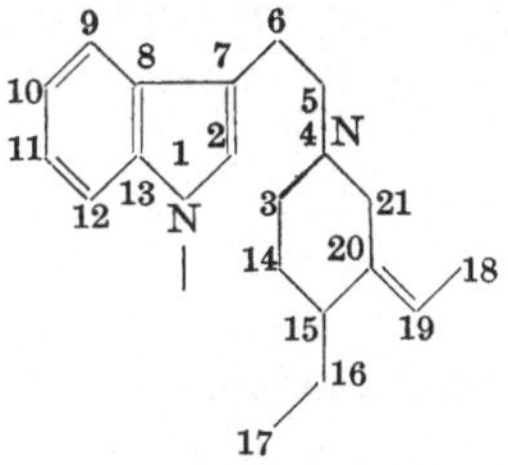

(CXX.)

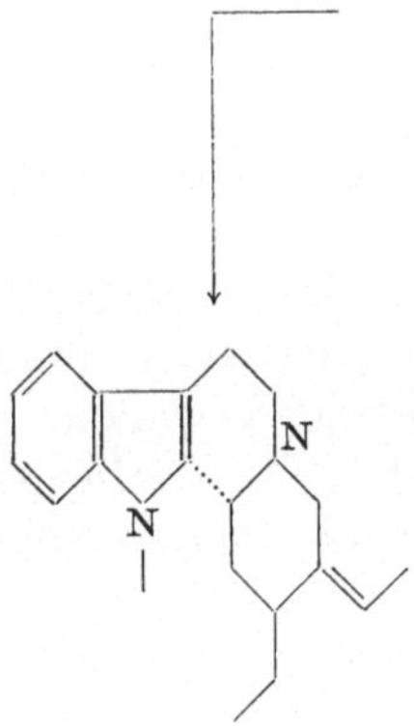

(CXXI.)

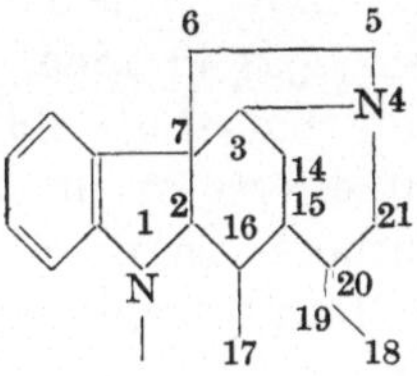

(CXXII.)

The reader is referred to WOODWARD's recent survey article (*388 a*) on the biogenesis of alkaloids and some other natural products.

Second Part:

Pharmacology of the Rauwolfia Alkaloids.

VIII. Historical Introduction.

Preparations of roots or leaves of various *Rauwolfia* species have apparently been used for different medicinal purposes by Indian and Malayan physicians for a very long time. There was, however, an amazing diversity of pathological conditions which were reported to respond to a treatment with *Rauwolfia* preparations: The root of *R. serpentina* was used, both in the form of internally and externally applied infusions or decoctions, in cases of snake bite and stings of poisonous insects. It has also been used in cases of diarrhoea and dysentery, partly in combination with some other indigeneous drugs. The juice of leaves of *R. serpentina*, according to various old reports, had a high reputation as a remedy for opacity of the cornea. Although no mention can be traced in the old literature on Indian medicinal plants regarding a sedative or hypnotic action of *Rauwolfia*, it seems to have been well known for these properties to poor people (*94*, *95*); in fact, it was common practice in some parts of India (e. g. Bihar) to use *R. serpentina* for keeping children quiet.

The first important report about the therapeutic application of *R. serpentina* in hypertensive disease that can be traced appeared in 1918 (*207*) and since the plant has gained increasing popularity with physicians and patients in India, not only for its blood pressure lowering effect in hypertension but also for its sedative action. It was not before 1933, however, that the first attempt was made by Chopra, Gupta and Mukherjee (*76*) to elucidate the pharmacological basis of this therapeutic effect, at the same time giving a considerable impetus to a still more extensive medical application of active plant products. Their paper became important for further developments for several reasons. It was shown that the biological activity of *R. serpentina* preparations could be attributed to an alkaloidal fraction, an observation which had been although based on less conclusive evidence, already pointed out by Greshoff (*131*) in 1890. Moreover, Chopra and co-workers (*76*) were already at that stage able to characterize, at least in principle, all the important effects of the *R. serpentina* alkaloids. It now appears that these authors were investigating a mixture of several alkaloids, although they thought to have been working with Siddiqui's ajmaline only (*351*).

The contribution of Chopra et al. (*74*, *75*, *76*) can be considered as the beginning of the present era of stepwise elucidation of the manifold pharmacological effects of the alkaloids of various *Rauwolfia* species that will be discussed below.

IX. Pharmacological Effects of *Rauwolfia serpentina* Alkaloids.

1. Alkaloid Mixtures.

As pointed out before, the investigations of CHOPRA et al. (*76*) outlined the entire spectrum of pharmacological activities of *R. serpentina* and their results can be considered as typical for a mixture of several alkaloids of this species. The material used was obtained from powdered root by means of a conventional extraction procedure for alkaloids. A similar preparation of alkaloids from the powdered root of the "Dehra Dun" variety was used some twenty years later by another group of workers (*90, 91, 285, 284*) likewise in the Pharmacology Department of the School of Tropical Medicine, Calcutta. The results of these two groups of investigators show good agreement in many points and will be described together.

The long lasting depression of arterial blood pressure caused by intravenous injection of either alcoholic solutions of the alkaloid bases or of aqueous solutions of the hydrochlorides is mainly due to the following three causes: a) a peripheral vasodilatory effect, noted in isolated perfused limbs (*75*); b) a peripheral adrenergic blocking activity (*90*); and c) a reduction of the central vasomotor tone and reflex activity (*75, 91, 285, 284*).

The transient diminution of the cardiac output after intravenous injection of high doses of alkaloids (*75*) which was associated with marked electrocardiographic changes (*90*) did not appear to contribute to the long lasting reduction of arterial pressure. In fact the drop of pressure could regularly be obtained with such doses that did not reduce the efficiency of the heart as far as could be ascertained through simultaneous measurements of the venous pressure (*90*).

Some additional comments can be made regarding the adrenergic blocking activity of the alkaloid mixtures under review (*90*): It developed without appreciable delay following intravenous injection, simultaneously with the drop of blood pressure, and led to a reduction of the pressor effect of adrenaline as well as of electric stimulation of the splanchnic nerve to about the same extent. The antagonism of the alkaloid mixture against the pressor effect of adrenaline is of a competitive type and resembles therefore the adrenolytic activity of substances belonging to the imidazoline or benzodioxane series (*258*). It is interesting to note that the intensities of the hypotensive and of the anti-adrenaline effect take a different time course during the recovery period after drug administration, the former outlasting the latter considerably. This can be taken as evidence for the participation of other factors in the hypotensive response, in addition to the adrenergic blocking activity. The adrenaline antagonism of the alkaloid mixture did not concern inhibitory actions of adrenaline on smooth muscle organs, nor did it, over a considerable dosage range, alter the postive inotropic effect of adrenaline on the heart or the stimulating effect of adrenaline on the nictitating membrane. This is in conformity with the characteristic activity pattern of most adrenergic blocking drugs (*258*).

Several arguments could be presented in favor of a central depressant effect on vasomotor tone and reflex activity as a contributing factor to the hypotensive action of the alkaloid mixture. Chopra et al. (*75*) noted that decerebrated animals responded with much less reduction of the arterial pressure than animals with an intact central nervous system. On intracisternal injection in monkeys, the alkaloids were found to be ten to twenty times more potent as regards reduction of arterial pressure than on intravenous injection (*91, 380*). On intracisternal application the alkaloids caused a suppression of pressor reflexes originating from the carotid sinus pressor receptors also much more effectively than on intravenous application (*380*). In the dosage range used, a suppression of vasomotor reflexes originating from chemoreceptors (through stimulation with local cyanide injection) was, however, not observed (*285*). There was sufficient evidence available to show conclusively that neither afferent nor efferent activity of the vagus nerves was involved in the hypotensive response obtained on systemic or central application (*75, 91*). Some support for the assumption of a central depressant action of the alkaloid mixture on the vasomotor tone and reflex activity was obtained from experiments with metrazole and picrotoxine, both substances being known to increase the reflex excitability of the vasomotor center for stimuli originating from the carotid sinus (*36, 105*). It was, in fact, possible in some experiments to counteract the hypotensive and vasomotor reflex blocking action of the *Rauwolfia* alkaloids by means of these drugs, though doses close to the convulsive level had to be administered (*91*).

In the studies of Chopra et al. (*75*), this alleged depression of the vasomotor center was associated with a general depressant effect on the central nervous system. Thus, after injection of sublethal doses frogs fell quickly into deep narcosis; drowsiness and somnolence was noted after intracisternal injection in cats or monkeys and was accompanied by reduction of convulsive activity of strychnine; and antipyretic action was observed in rabbits, and high doses caused death through central respiratory failure.

Except for the last-mentioned effect, central depressant actions other than described for the vasomotor center were not noted with alkaloid mixtures used by later investigators (*91, 380*). Two reasons can be advanced to explain this discrepancy: a) there might have been differences in the isolation procedure of the alkaloids with the result that the alkaloids with hypnotic action were not contained in the mixture applied by later investigators; b) the roots used by the two groups of workers differed in their alkaloidal composition. As a matter of fact such differences have been noted in relation to different areas from which the plants were collected (*20, 350*).

In any case, it is obvious that the central depressant effect of *Rauwolfia* alkaloids on the vasomotor tone and reflex activity is not necessarily and invariably associated with a sedative or hypnotic action, a conclusion which is well supported by more recent results obtained with pure alkaloids from *R. serpentina*.

This discrepancy illustrates the necessity of using pure and well defined alkaloids for pharmacological investigations, and most of the recent work was done with pure alkaloids. However, studies with crude drug preparations are not fully in vain since, for economical reasons, many of the preparations used in medicine are only partially purified, thereby making some sort of standardization imperative. Before, however, turning to the unsatisfactory state of the standardization problem in the field of *Rauwolfia* preparations, two observations with an alkaloid mixture should be mentioned which undoubtedly bear significant relation to the therapeutic application of *R. serpentina* in hypertension, although no decision can be made as to which of its individual pharmacological activities or alkaloids this effect could be ascribed. By means of an angiographic technique (*279*) which permitted direct visualisation of the vascular system of the kidney, it was observed that an alkaloid mixture of *R. serpentina* produced marked dilatation of the renal vessels in all instances of a reduced blood supply to the renal cortex, induced by such manipulations as haemorrhage or asphyxia. In other words, the alkaloid mixture was effective in increasing the blood supply of the renal cortex during a state of reflexly induced vasoconstriction (i. e. shunt of the renal cortex) (*37*). A significant effect was also obtained by feeding a diet to rats that contained 5% powdered *R. serpentina* root (*362*). Such rats exhibited a sustained reduction of blood pressure though they were in a state of "metacorticoid hypertension" (*130*), i. e. a self-sustained, generalized cardiovascular disease attained in rats about two months after implantation of a desoxycorticosterone-acetate pellet and simultaneous substitution of saline for drinking water.

The standardization of crude *R. serpentina* preparations can be approached from different angles. Pharmacognostic identification of roots by means of histological (*369, 395*) and histochemical (*253*) criteria have been worked out. For the pharmacologist and drug controller, however, the important aspect consists in means to express and standardize the biological potency. It has been reliably established that no other active constituents but alkaloids are present in *R. serpentina* (*345*), thereby reducing the task to the evaluation of the alkaloid fraction. The claim of earlier investigators (*133*) that a resinous non-alkaloidal fraction exerted a hypnotic activity is, in view of more recent conclusive evidence, no longer tenable. Evidently, a complete characterization of crude preparations can only be performed by comparing each of the individual actions (i. e. adrenergic blocking activity, hypotensive and vasomotor reflex blocking activity, hypnotic activity) with the corresponding effect of a standard sample, since the relative potency of a preparation with respect to each individual type of action can vary independently, due to variations in the alkaloid composition. In other words, a crude preparation would require at least three indices for complete characterization. Evidently, such a procedure would be too laborious, but it is equally obvious that preparations cannot be compared simply on the basis of their alkaloid content, as suggested by the British Pharmaceutical Codex 1949 or the Indian Pharmaceutical Codex 1953. It should also be pointed out that the extraction method as recommended in both pharmaceutical codices (*107*) does not permit the estimation of the reserpine group because of their low basicity. The method of HOLT and COSTELLO (*158*) would appear far superior since it allows the resolution of the total alkaloids into groups for which the pattern of biological activity can be predicted.

A simple bioassay procedure for a mixture of *Rauwolfia* alkaloids could be based on their adrenergic blocking or vasomotor reflex blocking activity, in conformity with similar methods currently in use for other drugs. For instance, the estimation of a dose of alkaloids required to reduce a certain pressor response caused by adrenaline or by temporary occlusion of the carotid arteries to 50% of its original value would serve this purpose. Another suggestion was to assess the blood pressure lowering action of such preparations in cats made hypertensive by injection of hypertensine (*319, 237*). A straightforward method for the quantitative estimation of the hypotensive activity of *Rauwolfia* preparations is being used by CRONHEIM et al. (*82—85*). It consists of feeding the preparation to dogs for five subsequent days, after which the blood pressure is measured under pentobarbital anesthesia. This method permits evaluation of a dose response relation. Pretreatment of dogs over a specified period with alkaloidal extracts permits, furthermore, the observation and quantitative evaluation of such effects in the anesthetized animal as reduced heart rates or depression of carotid sinus reflexes (*124*). The most obvious approach to estimate the hypnotic activity consists in measuring the sleeping time or the potentiation of the barbiturate sleeping time by *Rauwolfia* alkaloids. A most specific action appears in the "mouse-ptosis" test of RUBIN and BURKE (*321*) or the *Rauwolfia*-induced emesis in pigeons (*108*). In both cases the responses to drug administration are suitable for statistical treatment. In both tests the responses are associated with the presence of, what is now called, "reserpine-like" activity in the alkaloid mixtures. In un-anesthetised dogs, however, an alkaloidal extract was found to elevate the emetic threshold to Veriloid (*123*).

2. Individual Alkaloids.

For the systematic description of the pharmacological effects of pure alkaloids *ex R. serpentina* a chemical classification recently given by SCHLITTLER, SCHNEIDER and PLUMMER (*330*) will be followed. According to these authors three groups of alkaloids can be distinguished, viz. the tertiary indoline alkaloids, the quaternary anhydronium bases, and the tertiary indole bases.

a) Tertiary Indoline Alkaloids.

In this section the alkaloids ajmaline, isoajmaline, neoajmaline and rauwolfinine will be discussed.

Some conflicting reports exist in the literature regarding the action of ajmaline on the blood pressure. RAYMOND-HAMET (*294*) found that ajmaline lowered the blood pressure in dogs in doses of 5 mg/kg; this effect was accompanied by inhibition of the intestinal movements. One year later CHOPRA and CHAKRAVARTI (*73*) stated that ajmaline acted predominantly as a stimulant on the vasomotor center, thereby increasing the blood pressure; only in spinal animals was a hypotensive action noted, due to a peripheral vasodilatation. These results were confirmed in a later paper from the same laboratory (*72*), though more recently evidence was again put forward in favor of a weak hypotensive action of ajmaline. On intravenous injection of doses ranging from 1 to 3 mg/kg the depressor effect was found to be associated with a weak adrenergic blocking activity. The alkaloid caused fall of pressure also on intracisternal injection in monkeys with a simultaneous depression of carotid sinus reflexes (*53*). Other vasomotor reflexes, such as the pressor response elicited by electric stimulation of the sciatic nerve were also supressed by ajmaline after intravenous injection. Moreover, it was noted in the course of these studies that the sensitivity of the experimental animals for the vasomotor reflex blocking activity of *ajmaline* could be markedly increased by suitable transection of the central nervous system; the animals were most sensitive when parts of the rhinencephalon remained intact after ablation of the neocortex. Pressor effects due to direct electrical stimulation of hypothalamic pressor areas and the "sham-rage" of diencephalic animals were, however, not suppressed specifically, i. e. more than could be accounted for by the peripheral adrenergic blocking action (*92*). Hypotensive effects of ajmaline were also noted by Bein and Gross (*18*), though only after the administration of higher doses and not accompanied by adrenaline antagonism. Raymond-Hamet (*288*), on the other hand, had previously reported on a paralysing action of ajmaline with respect to the vasoconstrictor activity of adrenaline.

One possible reason for the discrepancy between the results of Chopra et al. (*72, 73*) and the other workers (*53, 92*) is the non-identity of the respective alkaloid samples used under the designation of "ajmaline". An additional reason to question the identity of the preparations is that Raymond-Hamet (*293*) described for his product inhibition of the intestinal motility whereas Chopra et al. (*72, 73*) drew attention to the intestinal stimulating effect of their sample. The substance used by Chatterjee et al. (*53*) and Das Gupta et al. (*92*) which showed hypotensive activity, fulfilled all criteria for identification with ajmaline and it would therefore appear justified to ascribe finally a hypotensive activity to ajmaline. This activity is, however, in any case comparatively weak and does not contribute much to the hypotensive activity of an alkaloid mixture.

The rauwolfine "A" of van Itallie and Steenhauer is apparently identical with ajmaline; it was found to interfere with the impulse conduction in the bundle of Hiss in the hearts of warm and cold blooded animals (*141*).

Pharmacological studies with *iso-* and *neo-ajmaline*, obtained by Siddiqui (*350*) from the Dehra Dun variety of *R. serpentina* were carried out by Bhatia and Kapur (*20*). Both alkaloids were found to act predominantly on the central nervous system as depressants, although an initial and transient excitatory effect was regularly observed. The blood pressure was lowered in intact, decerebrated and spinal animals. It must be pointed out, however, that the existence of neo-ajmaline is doubtful (*330*), whereas iso-ajmaline has been recognized as a stereoisomer of ajmaline (*1*).

In a comparative study, rauwolfinine on intravenous injection was found to be equipotent with ajmaline. The antagonism against the pressor effect of adrenaline was less marked in the case of rauwolfinine, since the adrenaline dose had only to be doubled after 3 to 5 mg/kg of rauwolfinine in order to obtain the same pressor response as before (after 2 to 3 mg/kg of ajmaline a three-to five-fold increase of the adrenaline test dose was required to obtain equipressor responses). On the other hand, rauwolfinine was, when injected intracisternally, about twice as potent as ajmaline in respect to its hypotensive activity (*53*). Rauwolfinine did in no instance show any hypnotic or sedative action (*54*).

b) Quaternary Anhydronium Bases.

In this group serpentine and serpentinine will be discussed.

All investigators who have been working with *serpentine* agree on its hypotensive activity (*73, 72, 18, 53*) which can be observed in experimental animals both before and after destruction of the vasomotor center (*73, 72*). It decreases the blood pressure on intracisternal injection of doses which amount only to $^1/_5$ to $^1/_{10}$ of the equipotent dose required on intravenous injection (*53*). However, no other evidence for a central action could be obtained. The pressor effect of adrenaline was reduced to about the same extent as after equal doses of ajmaline (*53*). Serpentine decreased the intestinal motility concomitant with a reduction of tone, though only after an initial phase of stimulation (*292*). Serpentine nitrate in a concentration of 10^{-5} g/ml lowered the sensitivity of the isolated guinea pig intestine for histamine to about 50% (*313*).

A similar discrepancy as in the case of ajmaline exists also regarding *serpentinine*. Whereas RAYMOND-HAMET reported vasodilatation and hypotensive activity, associated in higher doses with an adrenergic blocking action particularly with respect to the renal vessels (*286, 295*), CHOPRA et al. (*73, 72*) insisted on a blood pressure rising action in animals with intact vasomotor center. BEIN and GROSS (*18*) found a weak hypotensive action but no evidence for an adrenergic or ganglionic blocking effect.

c) Tertiary Indole Bases.

Two types of such bases are distinguished by SCHLITTLER et al. (*330*), one being related to tetrahydro-alstonine, the other consisting of alkaloids with yohimbine-like structures. To the former group belong ajmalicine and reserpinine.

SIDDIQUI's *ajmalicine* (*351*) is apparently identical with py-tetrahydro-serpentine (*212*). An alkaloid isolated from *R. serpentina* by HAACK, POPELAK, SPINGLER and KAISER was first named "raubasine" and later also identified as py-tetrahydro-serpentine (*135*); the same alkaloid had previously been designated as alkaloid "II" (*274*). It showed in doses of 1 to 2 mg/kg in cats a marked adrenergic blocking activity, leading to a reversal of the adrenaline pressor effect and to a reduction of pressor reflexes originating from the carotid sinus (*221*). A quantitative comparison of the adrenergic blocking activity was carried out on the guinea pig's seminal vesicles. Ajmalicine was 100 times more potent than serpentine and only about 10 times less active than ergotamine (*274*). No sedative action of subtoxic doses was noted (*274*).

Reserpinine (*329*) was found to lack any specific hypotensive or central-sedative activity.

To the second group of tertiary indole bases belongs *yohimbine* that was isolated from *R. serpentina* by BADER, DICKEL and SCHLITTLER (*5*). Its pharmacological actions were reviewed by NICKERSON (*258*) and will not be discussed here. *Rauhimbine* as well as *iso-rauhimbine* (*155*) are apparently yohimbine isomers; rauhimbine was subsequently identified with *corynanthine* from *Pseudocinchona africana* (*156*). This alkaloid was investigated for its pharmacological activities as early as 1935 by ROTHLIN and RAYMOND-HAMET (*316*), and as an adrenergic blocking substance it was found to be more potent and less toxic than either yohimbine or pseudo-corynanthine.

Rauwolscine (*39*) was recently identified with α-*yohimbine* (*44*); although it belongs, from the chemical point of view, to the group of tertiary indole bases, it will be discussed, in accordance with its occurrence in nature, in Chapter X, p. 421.

The first publication of MÜLLER, SCHLITTLER and BEIN (*252*) reporting briefly on the isolation of *reserpine* and its remarkable pharmacological actions, was soon followed by a considerable number of papers, with the result that a fairly satisfactory outline of the pattern of activity of this alkaloid is now available. The experimental studies of TRAPOLD, PLUMMER and YONKMAN (*368*) resulted in the demonstration of a significant depression of blood pressure and respiration in dogs under barbital anesthesia. The maximum intensity of the action was reached only after a considerable latency period subsequent to the intravenous injection, due either to a metabolic conversion of the alkaloid into an active substance or to slow penetration to the site of action. The fall of pressure was predominantly caused by a decrease of peripheral resistance, the cardiac output remaining essentially unaltered (*19*). Adrenergic or ganglionic blocking as well as parasympathicomimetic effects were excluded by BEIN (*16*). This author also showed that stimulation of peripheral vagal receptors as observed with veratrine, for example, was not involved in the hypotensive response. By this process of elimination of peripheral actions TRAPOLD et al. (*368*) and BEIN (*16*) were led to the conclusion that the reserpine-induced hypotension is primarily of central origin. The earlier observation that reserpine did abolish the pressor response caused by direct stimulation of the vasomotor center through increase of the intracranial pressure (*394*), was further interpreted as an indication for interference by reserpine with the impulse transmission through or from a higher center, involved in sympathetic reflex activity (*368*).

Since the fall of blood pressure was associated with bradycardia, depression of vasomotor reflexes, relaxation of the nictitating membrane, myosis, increased peristalsis and lowering of the body temperature, a status was produced which closely resembled the behavioral pattern noted by W. R. Hess during electric stimulation of certain diencephalic areas. Additional observations which bear relation to a possible correspondence between the drug effect and the establishment of a generalized "trophotropic" reaction pattern will be presented later.

Before turning to a more detailed description of the nature of the central action, the following additional findings concerning the circulatory response to reserpine will be discussed. Haemodynamic studies in man confirmed the primary role of the decrease of peripheral resistance, associated with reduction of renal plasma flow, following intravenous injection. Electrocardiographic changes were not noticed, though signs of coronary insufficiency were occasionally accentuated (246, 307). No consistent effect on renal haemodynamics was observed in the course of prolonged oral administration (246). Although a possible contribution of peripheral actions on blood vessels had been excluded as a cause of the hypotensive effect of reserpine (16, 19), the problem was taken up by Tripod and Meier (370, 371). They extended the previously reported observation that reserpine exerted a very marked inhibition on the constricting effect of BaCl$_2$ in isolated perfused hind limbs, without causing any vasodilatation itself. In this respect reserpine closely resembles hydrazinophthalazine. On the coronary vessels of the isolated mammalian heart, on the other hand, reserpine regularly produced primary dilatation which was inhibited by an antihistaminic (pyribenzamine), an adrenolytic (regitine) agent, and by coronary vasoconstrictors such as barium chloride or pitressin.

The suggestion that the hypotensive effect of reserpine originates (383) from a central site of action appears highly plausible in view of other interesting actions which this alkaloid exerts on the central nervous system (336). The most obvious effect of reserpine when administered to unanesthetised animals was the production of a state of quietude and sedation (16, 270). In dogs or monkeys, for example, it was most impressive to see how the animals lost the active interest in their surroundings to such an extent that the term "insulation from environment" was considered suitable to describe the condition. This state of quiescence persisted over considerable periods of time after administration of single doses; some reduction of spontaneous activity could usually still be noted on the following day after the drug administration. Even in monkeys, however, which responded with most dramatic changes of behavior after doses of 0.5 to 1.0 mg/kg, reserpine did not cause unconsciousness or abolition of co-ordinated movements. Both avoidance behavior and food-rewarded responses were found to be severely depressed by reserpine (378). Lesions of the amygdaloid nucleus did not alter the characteristic behavioral changes. No significant electroencephalographic changes could be noted during the state of quietude (270), nor was any effect on the electroencephalogram of epileptic monkeys observable (*77, 89*). Even with increase of the reserpine dose to a multiple of the minimum effective dose no anesthesia could be brought about nor was there a sleep pattern in the electroencephalogram; the animals feel only asleep from which state they could readily be aroused. Consequently, the term "tranquilizing action" was introduced to characterize this type of drug action on the central nervous system.

In any case, reserpine cannot be considered a hypnotic drug when its effects on the cerebral electrical activity are compared with those of barbiturates. Reserpine causes in the rabbit, in striking contrast to barbiturates, an alert electroencephalographic pattern, apparently as a result of a stimulation of the mesodiencephalic activating system (*308*).

Considerable efforts have been made to define the nature of the central action of reserpine more precisely as an entity, differentiated from the effects of sedatives like bromide or barbiturates (*238, 370, 89*). In fact, in most of these tests reserpine behaved differently from either bromide or barbiturates. Though it showed definite antagonism against such psychomotor stimulant drugs as caffeine, morphine and cocaine (but not pervitine), it did not counteract central convulsants like strychnine, nicotine, picrotoxine or metrazole. It was inactive against electroshock seizures but afforded significant protection against audiogenic seizures. Rather surprising was the discovery that reserpine even facilitated the onset of maximum tonic extensure seizures in mice receiving metrazole or caffeine (but not strychnine). It was furthermore found to lower the convulsive threshold of mice to electric stimuli; moreover, reserpine antagonised the anticonvulsive activity of dilantin in a competitive manner (*71*). The facilitory effect of reserpine on electric seizures in dilantin-treated animals was, however, eliminated by barbital (*69*). The obvious interpretation was that reserpine exerted its facilitory action at the same site where dilantin interferred with seizure spread. Barbital, on the other hand, known to raise the threshold for convulsive discharges, was only slightly affected in its anticonvulsive activity by reserpine (*70*). Of great interest was also the demonstration of facilitory influences of reserpine on synaptic transmission in the spinal cord (*339*). Studies of the interaction between reserpine and other drugs known to affect the central nervous system revealed a marked antagonism to the analgesic action of morphine (*19, 335*).

A significant contribution to the elucidation of the central actions of reserpine was recently made by Schneider (*336, 337*). Cats, made over-excitable by means of suitable transections of the central nervous system ("sham-rage preparations") lost their responsiveness to outside stimuli within $1/_2$ to 1 hour after the intravenous injection of 0.5–1.0 mg/kg reserpine. It appeared that only the reflex component of the sham-rage behavior was abolished by reserpine in so far as, under the influence of the drug, outside stimuli did not reach the sympathetic center, though it was in a state of hyperactivity consequent to the decortication. The operated animals did not develop myosis or bradycardia after reserpine administration as the inact animals regularly do. Another important point for the characterization of the central effects of reserpine was the demonstration that the blood pressure rise elicited by diencephalic stimulation was not decreased after reserpine in anesthetised animals, whereas the carotid occlusion reflex was suppressed (*339*).

The experimental evidence regarding the nature of the hypotensive activity of reserpine was recently summarized and further extended by BEIN (*17*) who pointed out that the pressor reflex, originating from the carotid sinus and certain reflex respiratory changes, was blocked by reserpine in doses as low as 10–20 μg/kg (intravenously, in cats). After transection of the brain stem at the level of the inferior colliculus, the same respiratory reflexes were not inhibited by reserpine in doses up to 3 mg/kg (intravenously), whereas the pressor reflex could still be inhibited, though higher doses were required.

Reserpine inhibited the augmentation of respiration which normally results from breathing of gas mixtures rich in CO_2. This effect was not due to a diminished sensitivity of the carotid sinus chemoreceptors and the original reactivity could be restored by coramine. Reserpine did not, on the other hand, alter the effects of direct electrical stimulation of medullary substrates which modify the inspiratory or expiratory phase of respiration, respectively.

The high specificity and distinct localization of the central actions of reserpine is thereby well illustrated.

In certain points some similarity exists between the central effects of reserpine and those of chlorpromazine, particularly with respect to the behavior changes as best observed in monkeys (*88*). Chlorpromazine was also most effective in suppressing the sham-rage of diencephalic animals (*91*); however, it appeared to block not only the reflex excitability of the sympathetic centers but also their direct electrical excitability (*93*). Further differences between the central actions of reserpine and chlorpromazine consisted in the presence of considerable electroencephalographic changes (*88*) and in the lack of an antagonism against morphine in case of chlorpromazine (*335*). The main differences between the central activity of reserpine and chlorpromazine (the latter taken as an example for a substance which comes comparatively close to the overall pattern of actions of reserpine) consists apparently of two points: a) the integrity of the cerebral cortex is essential for the establishment of the complete pattern of central actions of reserpine, whereas in the case of chlorpromazine the drug sensitivity of the hypothalamus increases after disconnection of the cortex; b) the existence of facilitory effects of reserpine, not demonstrable under comparative conditions for chlorpromazine.

Considering all the experimental facts obtained for reserpine, Schneider et al. (*339*) arrived at the conclusion that reserpine is likely to cause an increase of cortical inhibition on hypothalamic structures. Although this concept still requires further corroboration, it serves best at the present time to characterize succinctly the uniqueness of the central effects of reserpine. It will be interesting to observe in more details differences in the respective clinical effects of the two drugs mentioned, in connection with their application in psychiatric disorders (cf. *375*).

There cannot be any doubt that the action pattern of reserpine is largely determined by its central effects. However, one more recent observation points also to a peripheral action with some possible significance in view of the generalized alteration of the balance between the activity of sympathetic and parasympathetic nervous system. Both in the vagally innervated gastric fistula pouch and in the Heidenhain pouch, reserpine evoked equal flow of hydrochloric acid; the response was inhibited by antrenyl and hexamethonium in both cases (*9*). The authors conclude that the cause for this action of reserpine must be ascribed to a stimulating effect on parasympathetic ganglia.

It is not surprising that the central action of reserpine is, at least to some extent, associated with certain effects on endocrine glands, possibly taking place by interference with the hypothalamic control of the hypophysis. Only the following preliminary pertinent results are available (*120*): Reserpine interfered with the oestrous cycle of rats, caused mild stimulation of the adrenal cortex and may stimulate the release of antidiuretic hormone. The hypertension associated with chronic cortisone treatment was prevented by reserpine. The depletion of ascorbic acid from the adrenal glands took a different time course than the hypotensive effect and might, therefore, be interpreted as a specific response, not linked with an unspecific stress reaction [*83*, cf. also *250*). Gaunt et al. (*119*) investigated particularly the effect of prolonged treatment with reserpine in metacorticoid hypertension and succeeded in demonstrating a significant protective action.

Investigations in man (*383*) have essentially confirmed and extended the conclusions reached from animal experimentation, in particular concerning the site of the principal action. Further clinical literature will, however, not be included in the present review.

The neuro-biochemical mechanism of the reserpine action on the central nervous system is possibly linked with the drug's property to liberate bound serotonin from tissues, including nervous tissue (*269*).

Another alkaloid of *R. serpentina* with similar chemical constitution and pharmacological activity was isolated by KLOHS, DRAPER and KELLER (*209*) and HAACK, POPELAK, SPINGLER and KAISER (*136*). The latter authors named this alkaloid first reserpinine, but since this name was meanwhile given to another alkaloid by SCHLITTLER et al. (*329*), they accepted the term used by KLOHS et al. (*209*), i. e. *rescinnamine* (*134*). The pharmacological properties of this alkaloid were found to be very similar to those of reserpine (*82*) in so far as it produced hypotension, sedation, bradycardia and the characteristic response of "eye-lid ptosis" in mice (*85*). Only some minor, mainly quantitative, differences were noted in the course of a detailed comparison between the two alkaloids (*84,85*). The activity of reserpine and rescinnamine did not run parallel if tested by different methods in different species; for instance, rescinnamine was comparatively more potent in dogs than in other animals.

d) Other Alkaloids, Not Classified Chemically.

In this group, an alkaloid RS 51 (isolated by Gluconate Ltd., India) must be mentioned; its pharmacological actions were recently described by ISWARIAH, SUBRAMANIAM and GURUSWAMI (*170*). The hypotensive activity of this alkaloid which had previously been used clinically (*38*) was, according to these authors, mainly based on peripheral vasodilation, caused partly by adrenergic blockade, partly by histamine release; in fact, anti-histaminics prevented the fall of blood pressure. No evidence for central actions has yet been obtained.

X. Pharmacological Effects of *Rauwolfia canescens* Alkaloids.

The first alkaloid isolated from leaves of *R. canescens* was named rauwolscine (*39*) and was recently identified with α-*yohimbine* (*44*). Later, the alkaloids deserpidine (*334*) and canescine (*359*) were detected in this plant and, in addition, serpentine and yohimbine have been isolated (*135*).

1. Rauwolscine.

An adrenergic blocking activity was demonstrated for rauwolscine by MUKHERJEE (*248, 249*). Subsequent investigations confirmed and extended this observation (*54*). On isolated perfused limbs a vasodilator effect was noticed with doses as low as 50 μg., whereas a significant antiadrenaline effect appeared only after at least the double of this dose. In search for a central vasomotor depressant effect, rauwolscine was found to lower the blood pressure and inhibit carotid sinus reflexes in monkeys, if injected intracisternally in doses of 0.1–0.2 mg (per animal), whereas 0.2–0.5 mg/kg were required to produce a similar response following intravenous injection (*54*). No other evidence could, however, be obtained for a central action. The direct electrical excitability of hypothalamic pressor areas was not specifically influenced, since whatever reduction of the pressor response was noted after higher doses, was fully accounted for by the peripheral adrenergic blocking action. It has been by no means definitely decided whether such apparent absence of central actions in short term experiments is conclusive. The authors mentioned rather favored the idea that, due to a slow passage across the blood-brain barrier, considerable time had elapsed until central actions became manifest (*54*).

In haemodynamic studies, a dramatic fall of peripheral resistance could be produced by rauwolscine, leaving at the same time the cardiac output practically unaltered (*87*). The nature of the changes of the pulse wave (i. e. abolition of reflected waves in the arterial tree) was essentially the same upon intravenous, intraperitoneal or intracisternal injection of the drug, though a smaller dose was required in the latter instance (*87*).

Rauwolscine represents one more example of a *Rauwolfia* alkaloid that produces ascorbic acid depletion in the adrenal glands of rats (*250*).

2. Deserpidine (Canescine).

Comparative pharmacological studies with these alkaloids led to the conclusion that no significant differences existed between them either in quantitative or in qualitative respect (*338*). Indeed, recently these two compounds have been proved to be identical.

The circulatory effects of canescine resemble those of reserpine to such an extent that neither qualitative nor even quantitative differences could be detected (*35*). Like reserpine, canescine exhibits also a tranquilizing and myotic activity (*35*).

XI. The Pharmacological Action of Further *Rauwolfia* Species.

1. R. caffra.

The alkaloid rauwolfine (of Koepfli; not identical with the rauwolfine of van Itallie and Steenhauer), showed definite curare-like activity in frogs. This is not surprising in view of Koepfli's suggestion that the alkaloid belongs to the quaternary ammonium type. To what extent the hypotensive action observed in cats and dogs has any significance, is difficult to decide on the basis of the existing evidence (*219*). Sympathicolytic actions were reported by Raymond-Hamet (*287*).

2. R. heterophylla.

Working with an aqueous extract from *R. heterophylla*, Raymond-Hamet (*289*) observed an antagonistic effect against the blood pressure rise caused by adrenaline. The carotid sinus pressor reflex was not reduced by the same dose, nor was any effect on the cardiac action of electric stimulation of the vagus nerve noted. On the intestine the same extract exerted an inhibitory effect (*289*), thereby deviating from the pattern of activity observed for an extract of *R. vomitoria*. The sensitivity of the heart for acetylcholine remained unaltered under the influence of the extract (*292*). Since this work with crude extracts had been undertaken some pure alkaloids (reserpine, serpentine and narcotine) have been isolated from *R. heterophylla*.

3. R. vomitoria.

The effects of an aqueous extract of *R. vomitoria* (*290*) on the circulation were found to resemble those of *R. heterophylla* (*291*); a difference was observed only with respect to the action on the intestine which in the case of *R. vomitoria* was stimulated (*296*). A comparison with SIDDIQUI's alkaloids *ex R. serpentina* showed that neither of these had a similar pattern of activity as found for the total alkaloids of *R. vomitoria* (*297*).

4. R. hirsuta.

The total alkaloids were investigated by MEZEY and URIBE (*242*) who found a persistent fall of blood pressure, bradycardia and respiratory depression. Inhibition of the adrenaline pressor effect was also noted. The alkaloids produced a decrease of motility on smooth muscle organs.

5. R. sellowii *Muell-Arg.*

A number of alkaloids originally isolated from *R. serpentina* have also been found in this species which occurs in Brazil. Various fractions of extracts obtained from roots, barks and leaves of *R. sellowii* produce hypotension (*343, 344*). Central effects, apparently similar to the pattern observed with reserpine, were also described (*34*).

XII. Concluding Remarks.

The pharmacological studies of the *Rauwolfia* alkaloids have resulted in the discovery of a most interesting type of action on the central nervous system which is now usually referred to as "reserpine-like" or "tranquilising". The characteristic features of this action were outlined in Section 2 (p. 413). It is interesting to note that the discovery of this pattern of activity coincided to some extent with the stepwise elucidation of the action of chlorpromazine on the central nervous system: A period of considerable progress in the understanding of drug actions on the central nervous system was thereby initiated which is likely to advance to a great extent and in the near future our theoretical knowledge and the practical application in the field of "neuropharmacology". The most significant contribution so far obtained with reserpine-like alkaloids was the demonstration of an integrated and balanced readjustment of functions of the autonomic nervous system and alertness, similar to what can be obtained with electric stimulation of suitable diencephalic foci (W. R. Hess); however, the mechanisms are distinctly different.

The precise site of action of reserpine-like alkaloids as well as other *Rauwolfia* alkaloids with an alleged central action on vasomotor tone and reflex activity (though not associated with a tranquilising effect), is still rather uncertain. A comparison with other hypotensive drugs which affect in some manner the activity of structures of the central nervous system (e. g. ergot alkaloids) reveals distinct differences. Furthermore, there are several structures of the central nervous system known to exert controlling influences on vasomotor tone and/or reflex activity (cf. *381*). The clarification of these effects is made more difficult by the existence of a considerable latency period, apparently required for both the reserpine-like and other alkaloids (e. g. rauwolscine), to establish whatever action on the central nervous system they might have. Whether this is only a question of penetrating the blood-brain barrier or whether an active compound is slowly formed in the organism from the injected material cannot yet be decided, although immediate effects of comparatively low doses can be obtained with some alkaloids on intracisternal injection, favoring thereby the first assumption.

The *peripheral* actions of *Rauwolfia* alkaloids are in comparison with their central effects much less interesting from the pharmacologist's point of view and also with respect to their practical application. Except for the reserpine-like alkaloids, they follow in general the common pattern of adrenergic blocking and vasodilator effects.

References.

1. Anet, F. A. L., D. Chakravarti (née Mukherjee), R. Robinson and E. Schlittler: Ajmaline. I. J. Chem. Soc. (London) **1954**, 1242.
1a. Anonymous: Pharmacology of Reserpine. Lancet **268**, 548 (1955).
1b. Arthur, H. R.: δ-Yohimbine from the Bark of *Rauwolfia verticillata*. Chem. and Ind. **1956**, 85.
2. Bader, F. E.: Die Lage der Doppelbindung in Ring E des Alstonins. Helv. Chim. Acta **36**, 215 (1953).
3. Bader, F. E., D. F. Dickel, C. F. Huebner, R. A. Lucas and E. Schlittler: Rauwolfia Alkaloids. XVII. 3-Epi-α-yohimbine. J. Amer. Chem. Soc. **77**, 3547 (1955).
4. Bader, F. E., D. F. Dickel, R. A. Lucas and E. Schlittler: Rauwolfia Alkaloids. XI. Isolation of an Isomer of Yohimbine. Experientia **10**, 298 (1954).
5. Bader, F. E., D. F. Dickel and E. Schlittler: Rauwolfia Alkaloids. IX. Isolation of Yohimbine from *Rauwolfia serpentina* Benth. J. Amer. Chem. Soc. **76**, 1695 (1954).

6. BADER, F. E. und H. SCHWARZ: Zur Konstitution des Serpentins. Helv. Chim. Acta 35, 1594 (1952).

7. BARGER, G. and E. FIELD: Yohimbine (Quebrachine). II. *apo*-Yohimbine and Deoxy-yohimbine. J. Chem. Soc. (London) 123, 1038 (1923).

8. BARGER, G. und C. SCHOLZ: Über Yohimbin. Helv. Chim. Acta 16, 1343 (1933).

9. BARRETT, W. E., A. J. PLUMMER, A. E. EARL and B. ROGIE: Effect of Reserpine on Gastric Secretion of the Dog. J. Pharmacol. exp. Therapeut. 113, 3 (1955).

10. BARTON, D. H. R.: The Stereochemistry of *cyclo*Hexane Derivatives. J. Chem. Soc. (London) 1953, 1027.

11. — The Conformation of the Steroid Nucleus. Experientia 6, 316 (1950).

12. BARTON, D. H. R., R. C. COOKSON, W. KLYNE and C. W. SHOPPEE: The Conformation of *cyclo*Hexene. Chem. and Ind. 1954, 21.

13. BARTON, D. H. R. and J. S. FAWCETT: The Nature of Anhydrocevine. Chem. and Ind. 1953, 615.

14. BARTON, D. H. R., O. HASSEL, K. S. PITZER and V. PRELOG: Nomenclature of *cyclo*Hexane Bonds. Nature (London) 172, 1096 (1953).

15. BARTON, D. H. R. and W. J. ROSENFELDER: The Stereochemistry of Steroids. IV. The Concept of Equatorial and Polar Bonds. J. Chem. Soc. (London) 1951, 1048.

16. BEIN, H. J.: Zur Pharmakologie des Reserpins, eines neuen Alkaloids aus *Rauwolfia serpentina* BENTH. Experientia 9, 107 (1953).

17. — Significance of Selected Central Mechanisms for the Analysis of the Action of Reserpine. Ann. New York Acad. Sci. 61, 4 (1955).

18. BEIN, H. J. und F. GROSS: Pharmakologische Untersuchungen über verschiedene Alkaloide (Serpasil, Ajmalin und Serpentin) aus *Rauwolfia serpentina*. Verh. dtsch. Ges. Krebsforsch. 19, 277 (1954).

19. BEIN, H. J., F. GROSS, J. TRIPOD und R. MEIER: Experimentelle Untersuchungen über Serpasil (Reserpin), ein neues, sehr wirksames Rauwolfiaalkaloid mit neuartiger zentraler Wirkung. Schweiz. med. Wschr. 83, 1007 (1953).

20. BHATIA, B. B. and R. D. KAPUR: Pharmacological Action of Alkaloids of *Rauwolfia serpentina* BENTH. I. Neo-ajmaline and Iso-ajmaline. Indian J. med. Res. 32, 177 (1944).

21. BHATTACHARJI, S., M. M. DHAR and M. L. DHAR: Isolation of Sarpagine from *Rauwolfia canescens* LINN. J. Sci. Industr. Research (India) 14 B, 310 (1955).

22. — — — Raunescine—a New Alkaloid from *Rauwolfia canescens* LINN. Read at the Symposium on *Rauwolfia* of the Council of Scientific and Industrial Research (India), Calcutta, October 1955.

23. BICKEL, H., E. GIESBRECHT, J. KEBRLE, H. SCHMID und P. KARRER: Zur Kenntnis des Fluorocurins. Helv. Chim. Acta 37, 553 (1954).

24. BLUMENTHAL, A., C. H. EUGSTER und P. KARRER: Trennung von Corynanthein und Dihydro-corynanthein. Helv. Chim. Acta 37, 787 (1954).

25. BODENDORF, K. und H. EDER: Raupin, ein neues Alkaloid aus *Rauwolfia serpentina*. Naturwiss. 40, 342 (1953).

26. — — Über Raupin. Chem. Ber. 87, 818 (1954).

27. BOSE, S.: Rauwolfinine, the New Alkaloid of the Root of *Rauwolfia serpentina* BENTH. Science and Culture (India) 18, 98 (1952).

28. — Rauwolfinine: A New Alkaloid of *Rauwolfia serpentina* BENTH. I. J. Indian Chem. Soc. 31, 47 (1954).

29. — Rauwolfinine: A New Alkaloid of *Rauwolfia serpentina* BENTH. II. Studies on the Ultraviolet and Infrared Absorption Spectra. J. Indian Chem. Soc. 31, 311 (1954).

30. Bose, S.: Rauwolfinine: A New Alkaloid of *Rauwolfia serpentina* Benth. III. Studies on the Degradation of Rauwolfinine. J. Indian Chem. Soc. **31,** 691 (1954).

30a. — Rauwolfinine: A New Alkaloid of *Rauwolfia serpentina* Benth. IV. Studies on the Constitution of Rauwolfinine. J. Indian Chem. Soc. (1956) (in press).

31. — Serpinine, a Minor Alkaloid of *Rauwolfia serpentina* Benth. Naturwiss. **42,** 71 (1955).

31a. — On the Constitution of Serpinine, a Minor Alkaloid of *Rauwolfia serpentina* Benth. J. Indian Chem. Soc. (1956) (in press).

32. — Reduction of Carbinolamine Bases with Sodium Borohydride. J. Indian Chem. Soc. **32,** 450 (1955).

33. — Isolation of γ-Sitosterol from the Resin of *Rauwolfia serpentina* Benth. Science and Culture (India) **21,** 43 (1955).

33a. Bose, S., S. K. Talapatra and A. Chatterjee: The Alkaloids of *Rauwolfia beddomei* Hook. f. I. J. Indian Chem. Soc. (1956) (in press).

33b. Brodie, B. B., A. Pletscher and P. A. Shore: Evidence that Serotonin has a Role in Brain Function. Science (Washington) **122,** 968 (1955).

34. Campos, J. S., R. A. Seba and O. F. Pinto: Algumas observações sôbre o emprêgo da *Rauwolfia sellowii* em psiquiatria. Bol. Inst. Vital (Brazil) **5,** 199 (1954).

35. Cerletti, A., H. Konzett und M. Taeschler: Canescin, ein mit Reserpin wirkungsgleiches Alkaloid aus *Rauwolfia canescens*. Experientia **11,** 98 (1955).

36. Chakravarty, M.: A Quantitative Comparison of Different Analeptics. J. Pharmacol. exp. Therapeut. **67,** 153 (1939).

37. Chakravarty, N. K., S. P. Basu and G. Werner: Vasodilatation in the Kidney Following the Injection of *Rauwolfia serpentina*. Bull. Calcutta School Trop. Med. **2,** 1 (1954).

38. Chakravarty, N. K., M. N. Rai Chaudhuri and R. N. Chaudhuri: *Rauwolfia serpentina* in Essential Hypertension. Indian med. Gaz. **86,** 348 (1951).

39. Chatterjee (née Mookerjee), A.: The Alkaloids of *Rauwolfia canescens* Linn. I. J. Indian Chem. Soc. **18,** 33 (1941).

40. — The Alkaloids of *Rauwolfia canescens* Linn. II. J. Indian Chem. Soc. **18,** 485 (1941).

41. — The Alkaloid of *Rauwolfia canescens* Linn. III. Some Degradation Products of Rauwolscine. J. Indian Chem. Soc. **20,** 11 (1943).

42. — The Alkaloid of *Rauwolfia canescens* Linn. IV. On the Constitution of Rauwolscine. J. Indian Chem. Soc. **23,** 6 (1946).

43. — On the Alkaloid of *Rauwolfia canescens*. V. J. Indian Chem. Soc. **28,** 29 (1951).

43a. — Rauwolfia Alkaloids. Fortschr. Chem. organ. Naturstoffe **10,** 390 (1953).

44. Chatterjee, A., A. K. Bose and S. Pakrashi: Conformation of Rauwolscine, Alloyohimbine and their Congeners. Chem. and Ind. **1954,** 491.

45. Chatterjee, A. and S. Bose: A New Alkaloid from the Root of *Rauwolfia serpentina* Benth. Science and Culture (India) **17,** 139 (1951).

46. — — The Constitution of Ajmaline. Experientia **9,** 254 (1953).

47. — — Serpine — A New Isomeride of Yohimbine Isolated from *Rauwolfia serpentina* Benth. Experientia **10,** 246 (1954).

48. — — Isolation of Serpine, a New Isomer of Yohimbine from the Root of *Rauwolfia serpentina* Benth. Science and Culture (India) **19,** 512 (1954).

49. — — The Constitution of Ajmaline. I. J. Indian Chem. Soc. **31,** 17 (1954).

50. — — Constitution of Ajmaline, the Alkaloid of *Rauwolfia serpentina* Benth. Proc. Indian Sci. Congr. (42nd) Part III, p. 137 (1955).

51. CHATTERJEE, A. and S. BOSE: On the Constitution of Ajmaline, the Alkaloid of *Rauwolfia serpentina* BENTH. Science and Culture (India) **20**, 606 (1955).

52. — — The Constitution and Stereochemistry of Serpine. Science and Culture (India) **21**, 110 (1955).

53. CHATTERJEE, A., S. BOSE, S. R. DAS GUPTA, G. K. ROY and G. WERNER: The Chemistry and Pharmacology of Ajmaline, Rauwolfinine and Serpentine. Bull. Nat. Inst. Sci. (India) **4**, 31 (1955).

54. CHATTERJEE, A., S. R. DAS GUPTA and G. WERNER: Central and Peripheral Actions of Two Yohimbine Isomers. Indian J. med. Res. **42**, 613 (1954).

55. CHATTERJEE, A. und P. KARRER: Untersuchung über Corynanthein. II. Helv. Chim. Acta **33**, 802 (1950).

56. CHATTERJEE, A. and S. PAKRASHI: Yobyrine, the Selenium Dehydrogenation Product of Rauwolscine, the Alkaloid of *Rauwolfia canescens* LINN. Proc. Indian Sci. Congr. (40th) Part III, p. 355 (1953).

57. — — Yobyrine, the Selenium Dehydrogenation Product of Rauwolscine, the Alkaloid of *Rauwolfia canescens* LINN. Science and Culture (India) **18**, 443 (1953).

58. — — On the Constitution of Rauwolscine, the Alkaloid of *Rauwolfia canescens* LINN. Science and Culture (India) **19**, 109 (1953).

59. — — The Alkaloids of *Rauwolfia canescens* LINN. VII. Studies on the Infra-red Spectra of Rauwolscine and its Acetyl Derivative. J. Indian Chem. Soc. **31**, 29 (1954).

60. — — The Alkaloids of *Rauwolfia canescens* LINN. VIII. Selenium Dioxide Oxidation of Yobyrine — the Selenium Dehydrogenation Product of Rauwolscine. J. Indian Chem. Soc. **31**, 31 (1954).

61. — — The Alkaloids of *Rauwolfia canescens* LINN. VI. Yobyrine — the Selenium Dehydrogenation Product of Rauwolscine. J. Indian Chem. Soc. **31**, 25 (1954).

62. — — On the Stereochemistry of Rauwolscine, the Alkaloid of *Rauwolfia canescens* LINN. Proc. Indian Sci. Congr. (41st) Part III, p. 74 (1954).

63. — — On the Stereochemistry of Rauwolscine, the Alkaloid of *Rauwolfia canescens* LINN. Naturwiss. **41**, 215 (1954).

64. — — Studies on the Steric Hindrance to N-Acylation of Indole-NH Group in Rauwolscine, the Alkaloid of *Rauwolfia canescens* LINN. Proc. Indian Sci. Congr. (42nd) Part III, p. 136 (1955).

65. CHATTERJEE, A. and S. K. TALAPATRA: Alkaloids of the Roots of *Rauwolfia densiflora* BENTH. and HOOK, *Rauwolfia perakensis* KING and GAMBLE, *Rauwolfia canescens* LINN. and *Rauwolfia serpentina* BENTH. Naturwiss. **42**, 182 (1955).

66. — — Constitution of Vincain, a β-Carboline Alkaloid Isolated from *Vinca rosea* LINN. Science and Culture (India) **20**, 568 (1955).

67. — — Synthesis of Reserpine Analogue from Rauwolscine, the Alkaloid of *Rauwolfia canescens* LINN. J. Sci. Industr. Res. (India) **14 C**, 237 (1955).

68. — — Structure-Action Relation in Reserpine. Proc. Indian Sci. Congress (43rd) Part III, p. 101 (1956).

69. CHEN, G.: Antagonism Studies on Reserpine, Dilantin and Barbital by Electrically-induced Hind-leg Extensor Response in Mice. J. Pharmacol. exp. Therapeut. **113**, 10 (1955).

70. CHEN, G. and C. R. ENSOR: Antagonism Studies on Reserpine and Certain CNS Depressants. Proc. Soc. exp. Biol. Med. **87**, 602 (1954).

71. CHEN, G., C. R. ENSOR and B. BOHNER: A Facilitation Action of Reserpine on the Central Nervous System. Proc. Soc. exp. Biol. Med. **86**, 507 (1954).

72. Chopra, R. N., B. C. Bose, J. C. Gupta and I. C. Chopra: Alkaloids of *Rauwolfia serpentina*, a Comparative Study of their Pharmacological Action and their Role in Experimental Hypertension. Indian J. med. Res. **30**, 319 (1942).

73. Chopra, R. N. and M. Chakravarti: A Preliminary Note on the Pharmacological Action of the Alkaloids of *Rauwolfia serpentina*. Indian J. med. Res. **29**, 763 (1941).

74. Chopra, R. N., N. N. Das and S. N. Mukherjee: The Action of Ajmaline on Nerve Impulses. Indian J. med. Res. **24**, 1125 (1937).

75. Chopra, R. N., J. C. Gupta, B. C. Bose and I. C. Chopra: Hypnotic Effect of *Rauwolfia serpentina*: the Principle Underlying this Action, its Probable Nature. Indian J. med. Res. **31**, 71 (1943).

76. Chopra, R. N., J. C. Gupta and B. Mukherjee: The Pharmacological Action of an Alkaloid Obtained from *Rauwolfia serpentina* Benth. Indian J. med. Res. **21**, 261 (1933).

77. Chusid, J. G., L. M. Kopeloff and N. Kopeloff: Reserpine (Serpasil) Effects on Epileptic Monkeys. Proc. Soc. exp. Biol. Med. **88**, 276 (1955).

78. Clemo, G. R. and G. A. Swan: Structure of Ketoyobyrine. Nature (London) **162**, 693 (1948).

79. — — The Constitution of Yohimbine. I. J. Chem. Soc. (London) **1946**, 617.

80. — — The Constitution of Yohimbine. II. J. Chem. Soc. (London) **1949**, 487.

81. Cookson, R. C.: The Stereochemistry of Alkaloids. Chem. and Ind. **1953**, 337.

82. Cronheim, G., W. Brown, J. Cawthorne, M. I. Toekes and J. Ungari: Pharmacological Studies with Rescinnamine, a New Alkaloid Isolated from *Rauwolfia serpentina*. Proc. Soc. exp. Biol. Med. **86**, 120 (1954).

83. Cronheim, G. and S. Koster: The Effect of Rauwiloid, an Alkaloidal Extract of *Rauwolfia serpentina*, and of some *Rauwolfia* Alkaloids on Adrenal Ascorbic Acid. J. Pharmacol. exp. Therapeut. **113**, 12 (1955).

84. Cronheim, G., C. Stipp and W. Brown: Hypotensive Agents from *Rauwolfia serpentina*: Reserpine and Other Alkaloids. J. Pharmacol. exp. Therapeut. **110**, 13 (1954).

85. Cronheim, G. and M. I. Toekes: Comparison of Some Pharmacological Properties of Rescinnamine and Reserpine, two Alkaloids Isolated from *Rauwolfia serpentina*. J. Pharmacol. exp. Therapeut. **113**, 13 (1955).

85a. Crow, W. D. and Y. M. Greet: Occurrence of Reserpine in *Alstonia constricta* F. Muell. Austral. J. Chem. **8**, 461 (1955).

86. Curtis, R. G., G. J. Handley and T. C. Somers: Reserpine from *Alstonia constricta* F. Muell. Chem. and Ind. **1955**, 1598.

87. Das, N. N., S. R. Das Gupta, K. L. Mukherjee and G. Werner: Hemodynamic Effects of Rauwolscine. Indian J. med. Res. **43**, 101 (1955).

88. Das, N. N., S. R. Das Gupta and G. Werner: Changes of Behaviour and Electroencephalogram in Rhesus Monkeys Caused by Chlorpromazine. Arch. int. pharmacodyn. thérap. **99**, 451 (1954).

89. Das Gupta, S. R., K. L. Mukherjee and G. Werner: The Activity of some Central Depressant Drugs in Acute Decorticate and Diencephalic Preparations. Arch. int. pharmacodyn. thérap. **97**, 149 (1954).

90. Das Gupta, S. R., G. K. Roy, P. K. Roy and G. Werner: The Sympathicolytic Activity of *Rauwolfia serpentina* Benth. Indian J. med. Sci. **7**, 229 (1953).

91. Das Gupta, S. R., G. K. Roy and G. Werner: Central Action of *Rauwolfia serpentina*. Indian J. med. Sci. **7**, 597 (1953).

92. Das Gupta, S. R. and G. Werner: Inhibition of Vasomotor Reflexes by Ajmaline. Indian J. med. Res. **42**, 393 (1954).

93. DAS GUPTA, S. R. and G. WERNER: Inhibition of Hypothalmic, Medullary and Reflex Vasomotor Responses by Chlorpromazine. Brit. J. Pharmacol. **9**, 389 (1954).

94. DE, M. N. and T. CHATTERJEE: Combined Digitalis and Rauwolfia Poisoning in a Human Subject. Indian Med. Gaz. **76**, 724 (1941).

95. DEB, A. K.: Role of *Rauwolfia serpentina* in Treatment of Mental Disorders. Indian med. Record **63**, 359 (1943).

96. DEGER, E. C.: Allgemeine Untersuchung einer von den Indianern gegen Schlangengift und Malaria verwandten Pflanze. Arch. Pharmaz. Ber. dtsch. pharmaz. Ges. **275**, 496 (1937).

97. DEWAR, M. J. S. and F. E. KING: Constitution of Yohimbine. Nature (London) **148**, 25 (1941).

98. DIASSI, P. A., F. L. WEISENBORN, C. M. DYLION and O. WINTERSTEINER: The Stereochemistry of Reserpine. J. Amer. Chem. Soc. **77**, 2028 (1955).

99. — — — — On the Stereochemistry of Reserpine. J. Amer. Chem. Soc. **77**, 4687 (1955).

100. DJERASSI, C. and J. FISHMAN: Tetraphyllin and Tetraphyllicine, two New Alkaloids from *Rauwolfia tetraphylla* L. Chem. and Ind. **1955**, 627.

100 a. DJERASSI, C., S. C. PAKRASHI and M. GORMAN: Unpublished work.

101. DJERASSI, C., M. GORMAN, A. L. NUSSBAUM and J. REYNOSO: Alkaloid Studies. II. Isolation of Reserpine and Narcotine from *Rauwolfia heterophylla* ROEM. and SCHULT. J. Amer. Chem. Soc. **75**, 5446 (1953).

102. — — — — Alkaloid Studies. IV. The Isolation of Reserpine, Serpentine and Ajmaline from *Rauwolfia heterophylla* ROEM. and SCHULT. J. Amer. Chem. Soc. **76**, 4463 (1954).

102a. DJERASSI, C., M. GORMAN, S. C. PAKRASHI and R. B. WOODWARD: The Structures of Tetraphyllicine, Ajmalidine and Rauvomitine. J. Amer. Chem. Soc. **78**, 1259 (1956).

102b. DJERASSI, C., J. P. KUTNEY and P. J. SCHEUER: Private communication.

103. DORFMAN, L., A. FURLENMEIER, C. F. HUEBNER, R. LUCAS, H. B. MACPHIL-LAMY, J. M. MÜLLER, E. SCHLITTLER, R. SCHWYZER und A. F. ST. ANDRÉ: Die Konstitution des Reserpins. 8. Mitt. über Rauwolfia-Alkaloide. Helv. Chim. Acta **37**, 59 (1954).

104. DORFMAN, L., C. F. HUEBNER, H. B. MACPHILLAMY, E. SCHLITTLER and A. F. ST. ANDRÉ: On the Constitution of Reserpine from *Rauwolfia serpentina* BENTH. Experientia **9**, 368 (1953).

105. DOUGLAS, W. W., I. R. INNES and H. W. KOSTERLITZ: The Vasomotor Responses due to Electrical Stimulation of the Sinus and Vagus Nerves of the Cat and their Modification by Large Doses of Pentobarbital (Nembutal). J. Physiol. **111**, 215 (1950).

106. DURAND, TH. and B. D. JACKSON: Index Kewensis. Oxford: Clarendon Press. Suppl. I, 358 (1886—1895).

107. DUTT, A., J. C. GUPTA, S. GHOSH and B. S. KAHALI: *Rauwolfia serpentina* BENTH. Comparative Chemical Investigation of Some of the Constituents of the Drug Obtained from Different Sources and Isolation of the Active Resin. Indian J. Pharm. **9**, 54 (1947).

108. EARL, A. E.: Reserpine-induced Emesis in Pigeons: a Possible Assay. J. Pharmacol. exp. Therapeut. **113**, 17 (1955).

109. ELDERFIELD, R. C. and A. P. GRAY: Alstonia Alkaloids. III. Further Investigation of Alstonine. Reduction, Ozonolysis, and Spectrographic Studies. The Structure of Alstonine. J. Organ. Chem. (USA) **16**, 506 (1951).

110. Elderfield, R. C. and W. A. Jacobs: Strophanthin. XXIX. The Dehydrogenation of Strophanthidin. Science (Washington) **79,** 279 (1934).

111. Finch, F. C., J. D. Hobson, Sir R. Robinson and E. Schlittler: The Structure of Ajmaline. Chem. and Ind. **1955,** 653.

112. Folin, O. and V. Ciocalteu: On Tyrosine and Tryptophane Determinations in Proteins. J. Biol. Chem. **73,** 629 (1927).

113. Fourneau, E.: Sur un nouvel alcaloïde retiré de l'écorce du *Pseudocinchona africana* (Rubiacées). C. R. hebd. Séances Acad. Sci. **148,** 1770 (1909).

114. — Préparation de l'alcaloïde cristallisé du *Pseudocinchona africana* A. Chev. Bull. sci. pharmacol. **17,** 190 (1910).

115. Fourneau, E. et G. Benoit: Sur l'acide corynanthique. Bull. soc. chim. France **12,** 934 (1945).

116. Fourneau, E. et Fiore: Sur l'isomérie entre la base du pseudocinchona (corynanthine) et la yohimbine. Bull. soc. chim. France [4] **9,** 1037 (1911).

117. Furlenmeier, A., R. Lucas, H. B. MacPhillamy, J. M. Müller and E. Schlittler: On the Constitution of Reserpine. Experientia **9,** 331 (1953).

118. Fürst, A., H. H. Kuhn, R. Scotoni, Jr. und Hs. H. Günthard: Infrarotspektren epimerer Alkohole und Acetoxy-Verbindungen. Helv. Chim. Acta **35,** 951 (1952).

119. Gaunt, R., N. Antonchak, G. J. Miller and A. A. Renzi: Effect of Reserpine (Serpasil) and Hydralazine (Apresoline) on Experimental Steroid Hypertension. Amer. J. Physiol. **183,** 63 (1955).

120. Gaunt, R., A. A. Renzi, N. Antonchak, G. J. Miller and M. Gilman: Endocrine Aspects of the Pharmacology of Reserpine. Ann. New York Acad. Sci. **59,** 22 (1954).

121. Ghosh, B. P. and R. K. Basu: A Sterol (Serposterol) for Use in Mental Cases from *Rauwolfia serpentina.* Naturwiss. **42,** 130 (1955).

122. Goodson, J. A.: Echitamine in *Alstonia* Barks. J. Chem. Soc. (London) **1932,** 2626.

123. Gourzis, J. T.: Influence of Rauwiloid, an Alkaloidal Extract of *Rauwolfia serpentina* on Veratrum-induced Emesis in Dogs. J. Pharmacol. exp. Therapeut. **113,** 24 (1955).

124. Gourzis, J. T., R. Sonnenschein and R. Barden: Alterations in Cardiovascular Responses of the Dog Following Rauwiloid, an Alkaloidal Extract of *Rauwolfia serpentina.* J. Pharmacol. exp. Therapeut. **110,** 21 (1954).

125. Goutarel, R., M. M. Janot, A. Le Hir, H. Corrodi und V. Prelog: Über die Konstitution des Aricins. Helv. Chim. Acta **37,** 1805 (1954).

126. Goutarel, R., M. M. Janot und V. Prelog: Über die Konstitution des Sempervirins. Experientia **4,** 24 (1948).

127. Goutarel, R., M. M. Janot, V. Prelog, R. P. A. Sneeden und W. I. Taylor: Über das Gelsemin. Helv. Chim. Acta **34,** 1139 (1951).

128. Goutarel, R. et A. Le Hir: Sur la δ-yohimbine. Bull. soc. chim. France **18,** 909 (1951).

129. Goutarel, R., A. Le Hir, J. Poisson et M. M. Janot: Raumitorine et sérédine. Bull. soc. chim. France **1954,** 1481.

130. Green, D. M., F. J. Saunders, N. Wahlgren and R. L. Craig: Self-Sustaining Post-DCA Hypertensive Cardiovascular Disease. Amer. J. Physiol. **170,** 94 (1952).

131. Greshoff, M.: Mittheilungen aus dem chemisch-pharmakologischen Laboratorium des Botanischen Gartens zu Buitenzorg (Java). Ber. dtsch. chem. Ges. **23,** 3537 (1890).

132. GUPTA, J. C., S. GHOSH, A. T. DUTTA and B. S. KAHALI: A Note on the Hypnotic Principle of *Rauwolfia serpentina*. J. Amer. Pharmaceut. Assoc., Sci. Ed. 36, 416 (1947).

133. GUPTA, J. C., B. S. KAHALI and A. DUTTA: The Hypnotic Effect of a Resin Fraction Isolated from the Root of *Rauwolfia serpentina* Obtained from Dehra Dun. Indian J. med. Res. 32, 183 (1944).

134. HAACK, E., A. POPELAK und H. SPINGLER: Rauwolfia-Alkaloide Reserpinin und Rescinnamin. Naturwiss. 42, 47 (1955).

134a. — — — Rauvomitin, ein neues Alkaloid aus *Rauwolfia vomitoria* AFZ. Naturwiss. 42, 627 (1955).

135. HAACK, E., A. POPELAK, H. SPINGLER und F. KAISER: Isolierung von Yohimbin und Serpentin aus *Rauwolfia canescens* LINN. Raubasin identisch mit Py-Tetrahydroserpentin. Naturwiss. 41, 479 (1954).

136. — — — — Reserpinin, ein neues Alkaloid aus *Rauwolfia serpentina* BENTH. Naturwiss. 41, 214 (1954).

137. HAHN, G. und W. BRANDENBERG: Über Yohimbehe-Alkaloide. I. Mitt. Yohimben, ein neues Yohimbehe-Alkaloid. Ber. dtsch. chem. Ges. 59, 2189 (1926).

137a. — — Über Yohimbehe-Alkaloide. II. Mitt. Zwei weitere Nebenalkaloide des Yohimbins. Ber. dtsch. chem. Ges. 60, 669 (1927).

138. HAHN, G. und F. JUST: Über die Existenz des Iso-yohimbins und die Identität von Yohimbin und Quebrachin (Entgegnung auf K. WARNAT: Über Yohimbin und Quebrachin). Ber. dtsch. chem. Ges. 65, 714 (1932).

139. HAHN, G. und W. SCHUCH: Über Yohimbehe-Alkaloide. V. Mitt. Die Acetylierung des Iso-yohimbins. Ber. dtsch. chem. Ges. 62, 2953 (1929) — VI. Mitt. Zwei weitere Nebenalkaloide des Yohimbins. Ber. dtsch. chem. Ges. 63, 1638 (1930).

140. HAHN, G. und W. STENNER: Über Yohimbehe-Alkaloide. IV. Mitt. Ber. dtsch. chem. Ges. 61, 278 (1928).

141. HARTOG, J.: Die Wirkung von Rauwolfin auf das Herz. Arch. int. pharmacodyn. thérap. 51, 10 (1935).

142. HEINEMANN, H.: Über Yohimbehe-Alkaloide. Ber. dtsch. chem. Ges. 67, 15 (1934).

143. HENRY, T. A.: The Plant Alkaloids, 4th ed. Philadelphia and Toronto: Blakiston Co. 1949.

144. HESSE, O.: Über die Alkaloide der Ditarinde. Liebigs Ann. Chem. 203, 144 (1880).

145. — Beitrag zur Kenntnis der australischen Alstoniarinde. Liebigs Ann. Chem. 205, 360 (1880).

146. — Studien über argentinische Quebracho-Droguen. Liebigs Ann. Chem. 211, 249 (1882).

147. HILL, A. W.: Index Kewensis. Oxford: Clarendon Press. Suppl. VI, 172 (1916—1920).

148. — Index Kewensis. Suppl. VII, 205 (1921—1925).

149. — Index Kewensis. Suppl. VIII, 202 (1926—1930).

150. — Index Kewensis. Suppl. IX, 232 (1931—1935).

151. HILL, A. W. and E. J. SALISBURY: Index Kewensis. Suppl. X, 191 (1936 to 1940).

152. HOCHSTEIN, F. A.: Private communication.

153. — Alkaloids of *Rauwolfia sellowii*. J. Amer. Chem. Soc. 77, 5744 (1955).

154. HOCHSTEIN, F. A., K. MURAI and W. H. BOEGEMANN: Alkaloids of *Rauwolfia heterophylla*. J. Amer. Chem. Soc. 77, 3551 (1955).

155. Hofmann, A.: Rauhimbin und Isorauhimbin, zwei neue Alkaloide aus *Rauwolfia serpentina* Benth. 2. Mitt. über Rauwolfia-Alkaloide. Helv. Chim. Acta **37**, 314 (1954).

156. — Die Isolierung weiterer Alkaloide aus *Rauwolfia serpentina* Benth. 3. Mitt. über Rauwolfia-Alkaloide. Helv. Chim. Acta **37**, 849 (1954).

157. — β-Yohimbin aus den Wurzeln von *Rauwolfia canescens* L. 6. Mitt. über Rauwolfia-Alkaloide. Helv. Chim. Acta **38**, 536 (1955).

158. Holt, W. L. and C. H. Costello: A Preliminary Chemical Investigation of *Rauwolfia serpentina*. J. Amer. Pharmaceut. Assoc., Sci. Ed. **43**, 144 (1954).

159. Hooker, J. D. and B. D. Jackson: Index Kewensis. Oxford: Clarendon Press. **2**, 692 (1895).

160. Hosansky, N. and E. Smith: Raunescine and Isoraunescine from *Rauwolfia canescens* L. J. Amer. Pharmaceut. Assoc., Sci. Ed. **44**, 639 (1955).

161. Huebner, C. F.: Rauwolfia Alkaloids. XXIII. The Synthesis of *dl*-Reserpane. Chem. and Ind. **1955**, 1186.

162. — Derivatives of Reserpine. Communication on the Rauwolfia Alkaloids. XIII. J. Amer. Chem. Soc. **76**, 5792 (1954).

163. Huebner, C. F., R. Lucas, H. B. MacPhillamy and H. A. Troxell: Rauwolfia Alkaloids. XIV. Derivatives of Yohimbé Alkaloids. J. Amer. Chem. Soc. **77**, 469 (1955).

164. Huebner, C. F., H. B. MacPhillamy, A. F. St. André and E. Schlittler: Rauwolfia Alkaloids. XV. The Constitution of Reserpic Acid: Position of Substituents in Ring E. J. Amer. Chem. Soc. **77**, 472 (1955).

165. Huebner, C. F., H. B. MacPhillamy, E. Schlittler and A. F. St. André: The Stereochemistry of Reserpine and Deserpidine. Experientia **11**, 303 (1955).

166. Huebner, C. F., A. F. St. André, E. Schlittler and A. Uffer: Rauwolfia Alkaloids. XX. 11-Methoxyalloyohimbane from Reserpine. J. Amer. Chem. Soc. **77**, 5725 (1955).

167. Huebner, C. F. and E. Wenkert: Rauwolfia Alkaloids. XXII. Further Observations on the Stereochemistry of Reserpine. J. Amer. Chem. Soc. **77**, 4180 (1955).

168. Ishidate, M., M. Okada and K. Saito: Isolation of Alkaloids from *Rauwolfia* spp. Isolation of Sarpagine, Yohimbine and δ-Yohimbine from the Roots of *Rauwolfia heterophylla* Roem et Schult. Bull. Soc. pharm. (Japan) **3**, 319 (1955).

169. — — — Isolation of Reserpine, Ajmaline and Sarpagine from the Roots of *Rauwolfia indecora* R. E. Woodson. Bull. Soc. pharm. (Japan) **3**, 320 (1955).

170. Iswariah, V., R. Subramaniam and M. N. Guruswami: Indian J. med. Sci. **8**, 254 (1954).

171. Itallie, L. van and A. J. Steenhauer: *Rauwolfia serpentina* Benth. Arch. Pharmaz. **270**, 313 (1932).

172. — — Über *Rauwolfia serpentina* Benth. Pharmac. Weekbl. **69**, 334 (1932).

173. Janot, M. M. et A. Berton: Étude comparée des spectres d'absorption, dans l'ultraviolet, de solutions de gelsémine, de strychnine, de sempervirine et de cinchonamine. C. R. hebd. Séances Acad. Sci. **216**, 564 (1943).

174. Janot, M. M. et R. Goutarel: Sur l'aricine. C. R. hebd. Séances Acad. Sci. **229**, 724 (1949).

175. — — Sur la corynanthine. Bull. soc. chim. France **16**, 659 (1949).

176. — — Sur la corynanthidine, quatrième alcaloïde cristallisé isolé des écorces de *Pseudocinchona africana* Aug. Chev. — Son identité avec l'α-yohimbine de Lillig. Bull. soc. chim. France **13**, 535 (1946).

177. JANOT, M. M. et R. GOUTAREL: Sur la constitution de la cétoyobyrine. Ann. pharm. franç. **6**, 254 (1948).

178. — — Spectres d'absorption, dans l'ultraviolet, de la corynanthéine et de quelques-uns de ses dérivés. Ann. pharm. franç. **7**, 552 (1949).

179. — — Sur la corynanthéine. Bull. soc. chim. France **18**, 588 (1951).

180. JANOT, M. M., R. GOUTAREL et M. AMIN: Sur la pseudo-yohimbine. C. R. hebd. Séances Acad. Sci. **230**, 2041 (1950).

181. — — — La pseudo-yohimbine est identique au yohimbène. C. R. hebd. Séances Acad. Sci. **231**, 582 (1950).

182. JANOT, M. M., R. GOUTAREL et A. LE HIR: Isolement de la serpentine des racines de *Rauwolfia heterophylla* ROEM et SCHULT. C. R. hebd. Séances Acad. Sci. **238**, 720 (1954).

183. JANOT, M. M., R. GOUTAREL, A. LE HIR, M. AMIN et V. PRELOG: Stéréochimie de la pseudo-yohimbine, de la yohimbine et de la corynanthine. Bull. soc. chim. France **19**, 1085 (1952).

184. JANOT, M. M., R. GOUTAREL, A. LE HIR, G. TSATSAS und V. PRELOG: Konfiguration des Kohlenstoff-Atoms 3 von Dihydrocorynanthean, Corynantheidan, Yohimban und Allo-yohimban. Helv. Chim. Acta **38**, 1073 (1955).

185. JANOT, M. M., R. GOUTAREL et J. MASSONNEAU: Structure de la mayumbine. C. R. hebd. Séances Acad. Sci. **234**, 850 (1952).

186. JANOT, M. M., R. GOUTAREL und V. PRELOG: Über die Konstitution des Corynantheins. Helv. Chim. Acta **34**, 1207 (1951).

187. JANOT, M. M. et J. LE MEN: Alcaloïdes de la grande pervenche (*Vinca major* L.). Présence de réserpinine ou méthoxy 11-δ-yohimbine. C. R. hebd. Séances Acad. Sci. **238**, 2550 (1954).

188. — — Présence de la δ-yohimbine dans *Lochnera lancea* BOJ. (ex A. DC) K. SCHUM ou *Vinca lancea* BOJ. (ex A. DC). C. R. hebd. Séances Acad. Sci. **239**, 1311 (1954).

189. JOBST, J. und O. HESSE: Über die Ditarinde. Liebigs Ann. Chem. **178**, 49 (1875).

190. JOHNSON, J. R., A. A. LARSEN, A. D. HOLLEY and K. GERZON: Gliotoxin. VII. Synthesis of Pyrazinoindolones and Pyridindolones. J. Amer. Chem. Soc. **69**, 2364 (1947).

191. JOST, J.: Über das Ringgerüst des Yohimbins. IV. Yohimban. Helv. Chim. Acta **32**, 1297 (1949).

192. JULIAN, P. L., W. J. KARPEL, A. MAGNANI and E. W. MEYER: Studies in the Indole Series. X. Yohimbine, Part 2. The Synthesis of Yobyrine, Yobyrone and "Tetrahydroyobyrine". J. Amer. Chem. Soc. **70**, 180 (1948).

193. — — — — The Synthesis of Ketoyobyrine. J. Amer. Chem. Soc. **70**, 2834 (1948).

194. KARRER, P.: Curare Alkaloids. Nature (London) **176**, 277 (1955).

195. KARRER, P. und P. ENSLIN: Untersuchungen über Corynanthein. Helv. Chim. Acta **32**, 1390 (1949).

196. — — Die Konstitution des „Alstyrins". Helv. Chim. Acta **33**, 100 (1950).

197. KARRER, P. und ST. MAINONI: Ein weiterer Beitrag zur Konstitutionsaufklärung des Corynantheins. Helv. Chim. Acta **36**, 127 (1953).

198. KARRER, P. und O. RÜTTNER: Zur Kenntnis des Dehydroemetins. Helv. Chim. Acta **33**, 291 (1950).

199. KARRER, P. und R. SAEMANN: Über zwei neue Derivate des Yohimbins (16-Methyl-yohimbol und 16-Methyl-yohimban). Helv. Chim. Acta **35**, 1932 (1952).

200. KARRER, P. und H. SALOMON: Über zwei neue Alkaloide aus der Yohimbérinde. Helv. Chim. Acta **9**, 1059 (1926).

201. Karrer, P. und H. Schmid: Neuere Arbeiten uber Curare, insbesondere Calebassen-Curare und Alkaloide aus Strychnos-Rinden. Angew. Chem. **67**, 361 (1955).

202. Karrer, P., R. Schwyzer und A. Flam: Die Konstitution des Corynantheins und Dihydro-corynantheins. Helv. Chim. Acta **35**, 851 (1952).

203. Karrer, P., R. Schwyzer, A. Flam und R. Saemann: Dehydrierung von Yohimbylalkohol mit Selen und Palladiumkohle. Helv. Chim. Acta **35**, 865 (1952).

204. Keck, J.: Über die Isolierung weiterer Inhaltsstoffe aus den Wurzeln von *Rauwolfia canescens* L. Naturwiss. **42**, 391 (1955).

205. Kidd, D. A. A.: Alkaloids of *Rauwolfia* Species. The Isolation of Rescinnamine from *Rauwolfia vomitoria* Afz. Chem. and Ind. **1955**, 1481.

205a. Kimoto, S. and M. Okamoto: Studies on the Alkaloids of *Amsonia elliptica* Roem et Schult. III. Identity of Amsonine with β-Yohimbine. Pharmac. Bull. (Japan) **3**, 392 (1955).

206. Kirchner, J. G., A. N. Prater and A. J. Haagen-Smit: Separation of Acids by Chromatographic Adsorption of Their p-Phenylphenacyl Esters. Ind. Eng. Chem., Analyt. Ed. **18**, 31 (1946).

207. Kirtikar, R. K. and B. D. Basu: Indian Medicinal Plants, Vol. II, p. 777. Bahadurganj, Allahabad (India). 1918 (in English).

208. Kline, N. S.: Use of *Rauwolfia serpentina* Benth in Neuropsychiatric Conditions. Ann. New York Acad. Sci. **59**, 107 (1954).

209. Klohs, M. W., M. D. Draper and F. Keller: Alkaloids of *Rauwolfia serpentina* Benth. III. Rescinnamine, a New Hypotensive and Sedative Principle. J. Amer. Chem. Soc. **76**, 2843 (1954).

210. — — — Alkaloids of *Rauwolfia serpentina* Benth. V. Rescinnamine. J. Amer. Chem. Soc. **77**, 2241 (1955).

211. Klohs, M. W., M. D. Draper, F. Keller and W. Malesh: Alkaloids of the *Rauwolfia serpentina* Benth. IV. Reserpiline. Chem. and Ind. **1954**, 1264.

212. Klohs, M. W., M. D. Draper, F. Keller, W. Malesh and F. J. Petracek: Alkaloids of *Rauwolfia serpentina* Benth. II. The Isolation of Naturally Occurring Py-tetrahydroserpentine (Ajmalicine) and a Contribution Towards its Structure. J. Amer. Chem. Soc. **76**, 1332 (1954).

213. Klohs, M. W., M. D. Draper, F. Keller and F. J. Petracek: Alkaloids of *Rauwolfia serpentina* Benth. I. The Characterization of Reserpine and its Hydrolysis Products. J. Amer. Chem. Soc. **75**, 4867 (1953).

214. — — — — The Isolation of Reserpine from *Rauwolfia canescens* Linn. J. Amer. Chem. Soc. **76**, 1381 (1954).

215. Klohs, M. W., F. Keller, R. E. Williams and G. W. Kusserow: Alkaloids of *Rauwolfia canescens* Linn. II. The Isolation and Structure of Canescine. J. Amer. Chem. Soc. **77**, 4084 (1955).

215a. — — — — Alkaloids of *Rauwolfia canescens* Linn. III. Pseudo-reserpine. Chem. and Ind. **1956**, 187.

216. Klyne, W.: The Molecular Rotations of Polycyclic Compounds. I. General Principles and the Correlation of the Triterpenoids with the Steroids. J. Chem. Soc. (London) **1952**, 2916.

217. — The Application of Hudson's Lactone Rule to the Molecular Rotations of Polycyclic Compounds. Chem. and Ind. **1954**, 1198.

218. — Stereochemical Correlation of the Yohimbine Alkaloids with the Steroids. Chem. and Ind. **1953**, 1032.

219. Koepfli, J. B.: Chemical Investigation of *Rauwolfia caffra*. I. Rauwolfine. J. Amer. Chem. Soc. **54**, 2412 (1932).

220. KORNFELD, E. C.: Private communication.

221. KRONEBERG, G. und J. D. ACHELIS: Adrenolytische und sympathicolytische Wirkungen von zwei neuen Rauwolfia-Alkaloiden, Raupin und Raubasin, am Blutdruck und an der Nickhaut der Katze. Arzneimitt.-Forsch. **4**, 270 (1954).

222. KUNZ, A. and C. S. HUDSON: Relations between Rotatory Power and Structure in the Sugar Group. XV. Conversion of Lactose to another Disaccharide, Neolactose. The Chloro-hepta-acetate and two Octa-acetates of Neolactose. J. Amer. Chem. Soc. **48**, 1978 (1926).

222a. LEE, C. P.: Rauwolfia Alkaloids. Chem. Quart. Chinese Chem. Soc. Formosa **1955**, 349.

222b. LEE, T. B. and G. A. SWAN: The Constitution of Yohimbine and Related Alkaloids. Part IX. Synthesis of 2-(4:5-Diethyl-2-pyridyl)-3-ethylindole (Alstyrine or Coryline) and two Related Compounds. J. Chem. Soc. (London) **1956**, 771.

223. LE HIR, A.: α-Yohimbine et allo-yohimbine. C. R. hebd. Séances Acad. Sci. **234**, 2613 (1952).

224. LE HIR, A. et R. GOUTAREL: Stéréoïsomères de la yohimbine. II: β-yohimbine. Bull. soc. chim. France **20**, 1023 (1953).

225. LE HIR, A., R. GOUTAREL et M. M. JANOT: Hydrogénation de la sempervirine. Obtention des trois alloyohimbanes: racémique, dextrogyre et lévogyre. Bull. soc. chim. France **19**, 1091 (1952).

226. — — — Passage de la sempervirine à l'alloyohimbane. C. R. hebd. Séances Acad. Sci. **235**, 63 (1952).

227. — — — Extraction et séparation de la yohimbine et de ses stéréoïsomères. Ann. pharm. franç. **11**, 546 (1953).

228. LE HIR, A., R. GOUTAREL, M. M. JANOT et A. HOFMANN: Sur la constitution de l'isorauhimbine. Helv. Chim. Acta **37**, 2161 (1954).

229. LE HIR, A., M. M. JANOT et R. GOUTAREL: Stéréoïsomères de la yohimbine. III: Allo-yohimbine et α-yohimbine. Bull. soc. chim. France **20**, 1027 (1953).

230. LEONARD, N. J. and R. C. ELDERFIELD: Alstonia Alkaloids. I. Degradation of Alstonine to β-Carboline Bases and the Reduction of Tetrahydroalstonine with Sodium and Butyl Alcohol. J. Organ. Chem. (USA) **7**, 556 (1942).

231. Lilly Research Laboratories: Physical Data of Indole and Dihydroindole Alkaloids. Indianapolis. 1955.

232. LINNAEUS, C.: Species Plantarum, Vol. I, p. 208. Holmiae, Impensis Laurentri Salvii. 1753.

233. LUKES, R. M., G. I. POOS, R. E. BEYLER, W. F. JOHNS and L. H. SARETT: Approaches to the Total Synthesis of Adrenal Steroids. VI. 2,4b-Dimethyl-7-ethylenedioxy-1,2,3,4,4aα,4b,5,6,7,8,10,10aβ-dodecahydrophenanthrene-4β-ol-1-one and Related Compounds. J. Amer. Chem. Soc. **75**, 1707 (1953).

234. MACPHILLAMY, H. B., L. DORFMAN, C. F. HUEBNER, E. SCHLITTLER and A. F. ST. ANDRÉ: Rauwolfia Alkaloids. XVIII. On the Constitution of Deserpidine and Reserpine. J. Amer. Chem. Soc. **77**, 1071 (1955).

235. MACPHILLAMY, H. B., C. F. HUEBNER, E. SCHLITTLER, A. F. ST. ANDRÉ and P. R. ULSHAFER: Rauwolfia Alkaloids. XIX. The Constitution of Deserpidine and Reserpine. J. Amer. Chem. Soc. **77**, 4335 (1955).

236. MARION, L.: The Indole Alkaloids. In: MANSKE, R. H. F. and H. L. HOLMES: The Alkaloids, Vol. II, pp. 406ff. New York: Academic Press. 1952.

236a. — Private communication.

237. MEIER, R. and H. J. BEIN: Private communication; cf. Ref. *319*.

238. Meier, R., H. J. Bein, F. Gross, J. Tripod et H. Tuchmann-Duplessis: Effets de la réserpine, un nouvel alcaloïde de la *Rauwolfia serpentina* Benth. sur le système nerveux central de l'animal. C. R. hebd. Séances Acad. Sci. **238**, 961 (1954).

239. Mendlik, F. und J. P. Wibaut: Die Dehydrierung des Yohimbins (Vorl. Mitt.). Rec. trav. chim. Pays-Bas **48**, 191 (1929).

240. — — Über Yohimbin (2. Mitt.). Rec. trav. chim. Pays-Bas **50**, 91 (1931).

241. Mezey, K. and B. Uribe: *Rauwolfia hirsuta*, una Planta Medicinal Colombiana. Anal. Socied. Biol. Bogota **6**, 127 (1954).

242. — — *Rauwolfia hirsuta* — a Colombian Medicinal Plant. Arch. int. pharmacodyn. thérap. **98**, 273 (1954).

243. Mills, J. A.: Correlation between Monocyclic and Polycyclic Unsaturated Compounds from Molecular Rotation Differences. J. Chem. Soc. (London) **1952**, 4976.

244. — Correlation of Hydroxysteroids with Simpler Alcohols. Chem. and Ind. **1953**, 218.

245. Moissan, H. et E. Landrin: Recherches sur la préparation et sur les propriétés de l'aricine. Bull. soc. chim. Paris [3] **4**, 257 (1890).

245a. Mors, W. B., P. Zaltzman, J. J. Beereboom, S. C. Pakrashi and C. Djerassi: Alkaloids of two Brazilian Apocynaceae: *Rauwolfia grandiflora* Mart. and *Lochnera* (*Vinca*)*rosea* (L.) Reichb. *var. alba* (Sweet) Hubbd. Chem. and Ind. **1956**, 173.

246. Moyer, J. H.: Cardiovascular and Renal Hemodynamic Response to Reserpine (Serpasil), and Clinical Results of Using this Agent for the Treatment of Hypertension. Ann. New York Acad. Sci. **59**, 82 (1954).

247. Mukerji, B., B. K. Ghosh and L. B. Siddons: Search for an Antimalarial Drug in the Indigenous Materia Medica. Indian med. Gaz. **77**, 723 (1942).

248. Mukherjee, J. N.: Studies on the Pharmacology of Rauwolscine, the Alkaloid of *Rauwolfia canescens* L. Science and Culture (India) **18**, 6 (1952).

249. — Studies on the Alkaloids from *Rauwolfia canescens* L. Thesis, Calcutta Univ. 1955.

250. Mukherjee, K. L. and G. Werner: Depletion of Ascorbic Acid from the Adrenal Glands through Rauwolscine. Science and Culture (India) **20**, 144 (1954).

251. Mukherji, (Miss) D., R. Robinson and E. Schlittler: Chemistry of Ajmaline, Rauwolfine of van Itallie and Steenhauer. Experientia **5**, 215 (1949).

252. Müller, J. M., E. Schlittler und H. J. Bein: Reserpin, der sedative Wirkstoff aus *Rauwolfia serpentina* Benth. Experientia **8**, 338 (1952).

253. Neubern de Toledo, Th. A. und R. Wasicky: Der Alkaloidgehalt der brasilianischen *Rauwolfia sellowii*, einige Reaktionen und die histochemische Charakterisierung der Alkaloide. Sci. pharmaceut. **22**, 217 (1954).

254. Neuss, N., H. E. Boaz and J. W. Forbes: Structure of Reserpine. J. Amer. Chem. Soc. **75**, 4870 (1953).

255. — — — *Rauwolfia serpentina* Alkaloids. I. Structure of Reserpine. J. Amer. Chem. Soc. **76**, 2463 (1954).

256. — — — Rauwolfia Alkaloids. II. Isolation and Characterization of Two New Alkaloids from *Rauwolfia serpentina* Benth. J. Amer. Chem. Soc. **76**, 3234 (1954).

257. — — — Rauwolfia Alkaloids. III. Recanescine, a New Sedative Principle of *Rauwolfia canescens* Linn. J. Amer. Chem. Soc. **77**, 4087 (1955).

258. Nickerson, M.: The Pharmacology of Adrenergic Blockade. Pharmacol. Rev. **1**, 27 (1949).

259. PAKRASHI, S. C., C. DJERASSI, R. WASICKY and N. NEUSS: Alkaloid Studies. IX. Rauwolfia Alkaloids. IV. Isolation of Reserpine and Other Alkaloids from *R. sellowii* MUELL. ARGOV. J. Amer. Chem. Soc. **77**, 6687 (1955).

260. PARIS, R.: Sur une Apocynacée africaine, le *Rauwolfia vomitoria*. Ann. pharm. franç. **1**, 138 (1943).

261. PARIS, R. et R. MENDOZA: Sur une Apocynacée de Colombia: pinique-pinique (*Rauwolfia heterophylla* ROEM et SCHULT). Bull. sci. pharmacol. **48**, 146 (1941).

262. PELLETIER, J. et CORIOL: Sur une nouvelle base salifiable organique. J. pharmac. sci. access. **15**, 565 (1829).

263. PERROT, É.: Sur une écorce médicinale nouvelle de la Côte d'Ivoire et son alcaloïde. C. R. hebd. séances Acad. Sci. **148**, 1465 (1909).

264. — Sur le *Pseudocinchona africana* A. CHEV. Bull. sci. pharmacol. **17**, 187 (1910).

265. PERROT, É. et RAYMOND-HAMET: Sur une nouvelle méthode de tirage des écorces de Yohimbé et sur son application a la détermination de la teneur en alcaloïdes totaux et en yohimbine de seize écorces récoltées au Cameroun. Bull. sci. pharmacol. **39**, 593 (1932).

266. PHILLIPS, D. D. and M. S. CHADHA: The Alkaloids of *Rauwolfia serpentina* BENTH. Chem. and Ind. **1955**, 414.

267. — — The Alkaloids of *Rauwolfia serpentina* BENTH. J. Amer. Pharmaceut. Assoc., Sci. Ed. **44**, 553 (1955).

268. PICHON, M.: Classification des Apocynacées II, genre "Rauvolfia". Bull. soc. bot. France **94**, 31 (1947).

269. PLETSCHER, A., P. A. SHORE and B. B. BRODIE: Serotonin Release as a Possible Mechanism of Reserpine Action. Science (Washington) **122**, 374 (1955).

270. PLUMMER, A. J., A. E. EARL, J. A. SCHNEIDER, J. TRAPOLD and W. BARRETT: Pharmacology of *Rauwolfia* Alkaloids, including Reserpine. Ann. New York Acad. Sci. **59**, 8 (1954).

270a. POISSON, J., R. GOUTAREL et M. M. JANOT: Présence dans les racines du *Rauwolfia vomitoria* AFZ. de l'ester triméthoxybenzoïque d'un alcaloïde du type de l'ajmaline. C. R. hebd. Séances Acad. Sci. **241**, 1840 (1955).

271. POISSON, J., A. LE HIR, R. GOUTAREL et M. M. JANOT: Isolement de la réserpine des racines de *Rauwolfia vomitoria* AFZ. C. R. hebd. Séances Acad. Sci. **238**, 1607 (1954).

272. — — — — La raumitorine et la sérédine, deux nouveaux alcaloïdes, isolés des racines de *Rauwolfia vomitoria* AFZ. C. R. hebd. Séances Acad. Sci. **239**, 302 (1954).

273. POOS, G. I., G. E. ARTH, R. E. BEYLER and L. H. SARETT: Approaches to the Total Synthesis of Adrenal Steroids. V. 4b-Methyl-7-ethylenedioxy-1,2,3,4,4aα,4b,5,6,7,8,10,10aβ-dodecahydrophenanthrene-4β-ol-1-one and Related Tricyclic Derivatives. J. Amer. Chem. Soc. **75**, 422 (1953).

274. POPELAK, A., H. SPINGLER und F. KAISER: Neue Alkaloide aus *Rauwolfia serpentina*. Naturwiss. **40**, 625 (1953).

275. PRAIN, D.: Index Kewensis. Oxford: Clarendon Press. Suppl. III, 149 (1901 to 1905).

276. — Index Kewensis. Suppl. IV, 198 (1906—1910).

277. — Index Kewensis. Suppl. V, 214 (1911—1915).

278. PRELOG, V.: Über das Sempervirin. Helv. Chim. Acta **31**, 588 (1948).

279. PRICHARD, M. M. L. and P. M. DANIEL: Some Features of the Vascular Arrangement of Kidney and the Liver. Ciba Found. Symp. on Visceral Circulation p. 60 (1953).

280. PRUCKNER, F. und B. WITKOP: Die Konstitution einiger Derivate der Harman-reihe im Lichte ihrer UV.-Spektren. Liebigs Ann. Chem. **554**, 127 (1943).

281. Rakshit, B.: Isolation of a New Alkaloid "Chandrine" from *Rauwolfia serpentina*. Indian Pharmacist **9**, 226 (1954).

282. Rao, D. S. and S. B. Rao: A Note on the Alkaloids of *R. micrantha* Hook F. J. Amer. Pharmaceut. Assoc., Sci. Ed. **44**, 253 (1955).

283. Rasmussen, R. S. and R. R. Brattain: Infrared Spectra of Some Carboxylic Acid Derivatives. J. Amer. Chem. Soc. **71**, 1073 (1949).

284. Ray, G. K., S. R. Das Gupta and G. Werner: The Central Action of some Hypotensive Drugs. Indian J. Physiol. all. Sci. **8**, 1 (1954).

285. Ray, G. K., P. K. Roy, S. R. Das Gupta and G. Werner: Action of *Rauwolfia serpentina* on Vasomotor Reflexes. Arch. exp. Pathol. Pharmakol. **219**, 310 (1953).

286. Raymond-Hamet: Influence de la serpentinine sur les effects de l'adrénaline, de l'occlusion carotidienne et de la faradisation du pneumogastrique. C. R. hebd. Séances Acad. Sci. **211**, 414 (1940).

287. — Sur un nouveau sympathicolytique vrai: la Rauvolfine de Koepfli. C. R. hebd. Séances Acad. Sci. **201**, 1050 (1935).

288. — Sur un nouveau paralysant électif des vasoconstricteurs adrénalino-sensibles, l'ajmalinine, alcaloïde cristallisé de l',,*Ophioxylum serpentinum*" Willd. Bull. sci. pharmacol. **43**, 364 (1936).

289. — Sur une curieuse propriété physiologique de l'extrait aqueux de *Rauwolfia heterophylla* Roem et Sch. C. R. hebd. Séances Acad. Sci. **209**, 384 (1939).

290. — Le *Rauwolfia vomitoria*, possède-t-il réellement les vertus thérapeutiques que lui attribuent les guérisseurs indigènes ? Bull. acad. méd. (Paris) **122**, 9 (1939).

291. — Sur les effects intestinaux directs et indirects de l'extrait de *Rauwolfia heterophylla* Roem et Sch. C. R. hebd. Séances Acad. Sci. **209**, 599 (1939).

292. — Effets de l'extrait de *Rauwolfia heterophylla* Roem et Sch. sur le pneumogastrique cardiaque. C. R. Séances Soc. Biol. **132**, 213 (1939).

293. — Sur quelques propriétés physiologiques de la serpentine, alcaloïde cristallisé du *Rauwolfia serpentina* Bentham. C. R. Séances Soc. Biol. **134**, 94 (1940).

294. — Effets intestinaux de l'ajmaline pure. C. R. Séances Soc. Biol. **134**, 369 (1940).

295. — Effets tenseurs et vasculaires d'un des alcaloïdes colorés du *Rauwolfia serpentina* Benth: la serpentinine. C. R. hebd. Séances Acad. Sci. **223**, 927 (1940).

296. — Effets intestinaux des principes actifs hydrosolubles du *Rauwolfia vomitoria* Afzelius et l'influence de ces principes sur l'action intestino-inhibitrice de l'adrénaline. C. R. Séances Soc. Biol. **135**, 627 (1941).

297. — Sur quelques propriétés physiologiques des alcaloïdes totaux des racines de *Rauwolfia vomitoria* Afzelius. C. R. Séances Soc. Biol. **138**, 40 (1944).

298. — Sur l'identité de la yohimbine et de la québrachine. C. R. hebd. Séances Acad. Sci. **187**, 142 (1928).

299. — Sur les manifestations initiales de l'action sympathicolytique de la yohimbine. C. R. hebd. Séances Acad. Sci. **198**, 977 (1934).

300. — Production d'un isomère de la corynanthine par l'estérification méthylique de son produit de saponification alcaline. C. R. hebd. Séances Acad. Sci. **199**, 1658 (1934).

301. — Sur un nouveau faux Iboga pharmacologiquement actif. C. R. hebd. Séances Acad. Sci. **210**, 789 (1940).

302. — Sur un alcaloïde des Quinquinas n'appartenant pas au type quinolyl — quinuclidique. C. R. hebd. Séances Acad. Sci. **221**, 307 (1945).

303. — Sur un nouvel alcaloïde cristallisé extrait d'une Rubiacée gabonaise et appartenant à un type chimique inédit. C. R. hebd. Séances Acad. Sci. **232**, 2354 (1951).

304. RAYMOND-HAMET: Essai de classification des alcaloïdes des *Rauwolfia* (Apocynacées). C. R. hebd. Séances Acad. Sci. **237**, 1435 (1953).

305. — Sur un nouveau paralysant électif des vasoconstricteurs adrénalinosensibles, l'ajmalinine, alcaloïde cristallisé de l'„*Ophioxylum serpentinum*" WILLD. Bull. acad. méd. Paris **115**, 452 (1936).

306. — Sur quelques propriétés pharmacologiques d'une drogue alexitère et antimalarique du Guatemala: le chalchupa. C. R. Séances Soc. Biol. **129**, 462 (1938).

307. REUBI, F., P. MÜLLER et P. STUCKI: Effets circulatoires de la réserpine (serpasil). Helv. Med. Acta **21**, 493 (1954).

308. RINALDI, F. and H. E. HIMWICH: A Comparison of Effects of Reserpine and some Barbiturates on the Electrical Activity of Cortical and Subcortical Structures of the Brain of Rabbits. Ann. New York Acad. Sci. **61**, 27 (1955).

309. RINDL, M. and P. W. G. GROENEWOUD: The Chemistry of *Rauwolfia natalensis*. Trans. Roy. Soc. South Africa **21**, 55 (1932).

310. ROBINSON, R.: Further Observations on the Chemistry of Ajmaline. Chem. and Ind. **1955**, 285.

311. — Structure and Biogenesis of Emetine. Nature (London) **162**, 524 (1948).

312. ROBINSON, R. and A. F. THOMAS: The Alkaloids of *Picralima nitida* STAPF, TH. and H. DURAND. I. The Structure of Akuammigine. J. Chem. Soc. (London) **1954**, 3479.

313. ROHDEN, A. v.: Vergleichende Untersuchungen über die Wirkung der indischen Droge *Rauwolfia serpentina* und deren Alkaloid-Fraktionen Serpentin und SIV an Darm, Uterus und Samenblase des Meerschweinchens. Dissert., Univ. Giessen, 1953.

314. ROSENKRANTZ, H. und M. GUT: Deutung einer Infrarot-Absorptionsbande in der Gegend von 6 μ, verursacht durch eine —O—C=CH-Gruppe. Helv. Chim. Acta **36**, 1000 (1953).

315. ROSSI, L., A. DEL BOCA and R. LOBO: Analytical Studies of Yohimbine. IV. A New Reaction for the Recognition of Yohimbine. Anal. farm. bioquím. (Buenos Aires) **3**, 51 (1932).

316. ROTHLIN, E. et RAYMOND-HAMET: Sur la toxicité et l'activité adrénalinolytique de la pseudocorynanthine comparées avec celles de la corynanthine et de la yohimbine. Arch. int. pharmacodyn. thérap. **50**, 241 (1935).

317. RAY, L.: Cell Dimension and Symmetry Determination of Rauwolscane Crystal. Science and Culture (India) **20**, 507 (1955).

318. — The Unit-cell Dimensions and the Space-group of Rauwolscane. Acta Crystallogr. (manuscript submitted).

319. ROY, P. K., G. K. RAY and B. MUKHERJEE: Unpublished; cf. E. STEINEGGER: Bulletin Galenica **17**, 50 (1954).

320. ROY CHOUDHURI, D. K. and A. K. BOSE: Conformational Analysis. A New Tool. Science and Culture (India) **19**, 135 (1953).

321. RUBIN, B. and J. C. BURKE: Mouse Ptosis Bioassay of *Rauwolfia serpentina* for Reserpine-like Activity. Federat. Proc. (Amer. Soc. exp. Biol.) **13**, 400 (1954).

322. SAKAL, E. H. and E. J. MERRILL: Ultraviolet Spectrophotometric Determination of Reserpine. J. Amer. Pharmaceut. Assoc., Sci. Ed. **43**, 709 (1954).

323. SALISBURY, E. J.: Index Kewensis. Oxford: Clarendon Press. Suppl. XI, (1941—1945).

323 a. SAXTON, J. E.: The Indole Alkaloids Excluding Harmine and Strychnine. Quart. Rev. Chem. Soc. (London) **10**, 108 (1956).

324. Schlemmer, F. und H. Schmitt: Spektrographische Untersuchungen über Mutterkorn. I. Die wirksamen Inhaltsstoffe des Mutterkorns. Arch. Pharmaz. Ber. dtsch. pharmaz. Ges. **270**, 15 (1932).

325. Schlittler, E. und A. Furlenmeier: Über Alkaloide aus *Rauwolfia semperflorens* Schlechter (Mitt. 5 über Rauwolfia-Alkaloide). Helv. Chim. Acta **36**, 996 (1953).

326. — — Vincamin, ein Alkaloid aus *Vinca minor* L. (Apocynaceae). Helv. Chim. Acta **36**, 2017 (1953).

327. Schlittler, E. und J. Hohl: Über die Alkaloide aus *Strychnos melinoniana* Baillon. Helv. Chim. Acta **35**, 29 (1952).

328. Schlittler, E., H. U. Huber, F. E. Bader und H. Zahnd: Über das Alkaloid Serpentinin aus *Rauwolfia serpentina* Benth. Mitt. XII über Rauwolfia-alkaloide. Helv. Chim. Acta **37**, 1912 (1954).

329. Schlittler, E., H. Saner und J. M. Müller: Reserpinin, ein neues Alkaloid aus *Rauwolfia serpentina* Benth. Experientia **10**, 133 (1954).

330. Schlittler, E., J. A. Schneider und A. J. Plummer: Über Rauwolfia-alkaloide. Angew. Chem. **66**, 386 (1954).

331. Schlittler, E. und H. Schwarz: Über das Alkaloid Serpentin aus *Rauwolfia serpentina* Benth. Helv. Chim. Acta **33**, 1463 (1950).

332. Schlittler, E., H. Schwarz und F. Bader: Isolierung von Alstonin aus afrikanischen Rauwolfia-Arten. Helv. Chim. Acta **35**, 271 (1952).

333. Schlittler, E. und P. Speitel: Über das Ringgerüst des Yohimbins. II. Keto-yobyrin. Helv. Chim. Acta **31**, 1199 (1948).

334. Schlittler, E., P. R. Ulshafer, M. L. Pandow, R. M. Hunt and L. Dorfman: Rauwolfia Alkaloids. XVI. Deserpidine, a New Alkaloid from *Rauwolfia canescens*. Experientia **11**, 64 (1955).

335. Schneider, J. A.: Reserpine Antagonism of Morphine Analgesia in Mice. Proc. Soc. exp. Biol. Med. **87**, 614 (1954).

336. — Further Studies on the Central Action of Reserpine (Serpasil). Amer. J. Physiol. **179**, 670 (1954).

337. — Further Characterization of the Central Effects of Reserpine (Serpasil). Amer. J. Physiol. **181**, 64 (1955).

338. Schneider, J. A., A. J. Plummer, A. E. Earl, W. E. Barrett, R. Reinhart and R. C. Dibble: Pharmacological Studies with Deserpidine, a New Alkaloid from *Rauwolfia canescens*. J. Pharmacol. exp. Therapeut. **114**, 10 (1955).

339. Schneider, J. A., A. J. Plummer, A. E. Earl and R. Gaunt: Neuropharmacological Aspects of Reserpine. Ann. New York Acad. Sci. **61**, 17 (1955).

340. Scholz, C. R.: La constitution de la corynanthine. C. R. hebd. Séances Acad. Sci. **200**, 1624 (1935).

341. Schomer, A.: Bestimmung des Yohimbins in der Yohimbérinde. Pharmaz. Zentralhalle **62**, 169 (1921); **63**, 385 (1922).

341a. Schüler, B. O. G. and F. L. Warren: Reserpine from *Rauwolfia cumminsii* Stapf. Chem. and Ind. **1955**, 1593.

341b. — — Rauwolfia Alkaloids. Part I. Reserpine and Ajmaline from *Rauwolfia natalensis* Sond. (*R. caffra*). J. Chem. Soc. (London) **1956**, 215.

342. Schwarz, H. und E. Schlittler: Über die Konstitution natürlicher und synthetischer quaternärer β-Carboline, ihre Absorptionsspektren und die Konstitutionsermittlung ihrer Hydrierungsprodukte. Helv. Chim. Acta **34**, 629 (1951).

343. SEBA, R. A., J. S. CAMPOS and J. G. KUHLMANN: Introdução ao Estudo químico e farmacologico da *Rauwolfia sellowii*. Bol. Inst. Vital (Brazil) **5**, 175 (1954).

344. — — — Pharmacological Aspects of *Rauwolfia sellowii*. Its probable Indications in the Treatment of Arterial Hypertension. Rev. quím. farm. (Rio de Janeiro) **19**, 229 (1954).

345. SHARMA, V. N., J. D. KOHLI and B. MUKERJI: Chemistry and Pharmacology of *Rauwolfia*. J. Sci. Industr. Res. (India) **13 A**, 261 (1954).

346. SHARP, T. M.: The Alkaloids of *Alstonia* Barks. I. *A. constricta* F. MUELL. J. Chem. Soc. (London) **1934**, 287.

347. — The Alkaloids of *Alstonia* Barks. III. Alstonine. J. Chem. Soc. (London) **1938**, 1353.

348. SHERFF, E. E.: A Preliminary Study of Hawaiian Species of the Genus *Rauvolfia*. Field Museum of Natural History, Chicago (Botan. Ser.) **23**, 321 (1947).

349. SHOPPEE, C. W.: The Boat Conformation of *cyclo*Hexane. Chem. and Ind. **1952**, 86.

350. SIDDIQUI, S.: A Note on the Alkaloids of *Rauwolfia serpentina* BENTH. J. Indian Chem. Soc. **16**, 421 (1939).

351. SIDDIQUI, S. and R. H. SIDDIQUI: Chemical Examination of the Roots of *Rauwolfia serpentina* BENTH. J. Indian Chem. Soc. **8**, 667 (1931).

352. — — The Alkaloids of *Rauwolfia serpentina* BENTH. I. J. Indian Chem. Soc. **9**, 539 (1932).

353. — — The Alkaloids of *Holarrhena Antidysenterica*. III. Studies in the Action of BrCN on Conessine and its *N*-Demethylation to *iso*Conessimine and Conimine. J. Indian Chem. Soc. **11**, 787 (1934).

354. — — The Alkaloids of *Rauwolfia serpentina* BENTH. II. Studies in the Ajmaline Series. J. Indian Chem. Soc. **12**, 37 (1935).

355. SPIEGEL, L.: Untersuchungen einiger neuer Droguen. Chem.-Ztg. **20**, 970 (1896).

356. STEENHAUER, A. J.: On Reserpine, an Alkaloid from *Rauwolfia serpentina* BENTH. Pharmac. Weekbl. **89**, 161 (1954).

357. — Reserpine, an Alkaloid from *Rauwolfia serpentina*. II. Pharmac. Weekbl. **89**, 617 (1954).

358. STOLL, A. und A. HOFMANN: Sarpagin, ein neues Alkaloid aus *Rauwolfia serpentina* BENTH. Helv. Chim. Acta **36**, 1143 (1953).

359. — — Canescine and Pseudoyohimbine from the Roots of *Rauwolfia canescens* L. J. Amer. Chem. Soc. **77**, 820 (1955).

359a. — — Alkaloids from the Leaves and Roots of *Rauwolfia canescens*. Soc. Biol. Chem. (India), Souvenir **1955**, 248.

360. STOLL, A., A. HOFMANN und R. BRUNNER: Alkaloide aus den Blättern von *Rauwolfia canescens* L. 4. Mitt. über Rauwolfia-Alkaloide. Helv. Chim. Acta **38**, 270 (1955).

361. STORK, G. and R. K. HILL: The Stereospecific Synthesis of *dl*-Alloyohimbane and *dl*-3-Epialloyohimbane. J. Amer. Chem. Soc. **76**, 949 (1954).

362. STURTEVANT, F. M.: Response of Metacoroticoid Hypertension to Bistrium, Apresoline, Veriloid and Serpentina. Proc. Soc. exp. Biol. Med. **84**, 101 (1953).

363. TAMELEN, E. E. VAN and P. D. HANCE: The Stereochemical Formulation of Reserpine. J. Amer. Chem. Soc. **77**, 4692 (1955).

364. TAMELEN, E. E. VAN, P. D. HANCE, K. V. SIEBRASSE and P. E. ALDRICH: The D/E *cis* Ring Juncture of Reserpine. J. Amer. Chem. Soc. **77**, 3930 (1955).

365. TAMELEN, E. E. VAN and M. SHAMMA: The Stereospecific Synthesis of *dl*-Yohimbane. J. Amer. Chem. Soc. **76**, 951 (1954).

366. Thiselton-Dyer, W. T.: Index Kewensis., Suppl. III, p. 156. Oxford: Clarendon Press. 1901–1905.

367. Thomas, A. F.: The Constitution of Sarpagine. Chem. and Ind. **1954**, 488.

368. Trapold, J. H., A. J. Plummer and F. F. Yonkman: Cardiovascular and Respiratory Effects of Serpasil, a New Crystalline Alkaloid from *Rauwolfia serpentina*, in the Dog. J. Pharmacol. exp. Therapeut. **110**, 205 (1954).

369. Trease, G. E. and W. C. Evans: *Rauwolfia serpentina* and Other Species. Pharmac. J. **172**, 351 (1954).

370. Tripod, J., H. J. Bein and R. Meier: Characterization of Central Effects of Serpasil (Reserpine, a New Alkaloid of *Rauwolfia serpentina*) and of their Antagonistic Reactions. Arch. int. pharmacodyn. thérap. **96**, 406 (1954).

371. Tripod, J. et R. Meier: Détermination et classification pharmacodynamique de l'action vasculaire périphérique de l'aprésoline, du néprésol et du serpasil. Arch. int. pharmacodyn. thérap. **99**, 104 (1954).

372. Uribe Vergara, B.: Extraction of Reserpine and Other Alkaloids from Colombian *Rauwolfia hirsuta*. J. Amer. Chem. Soc. **77**, 1864 (1955).

373. Warnat, K.: Über Yohimbin und Quebrachin. (Bemerkung zu der Abhandlung von G. Hahn und W. Schuch: Über die Identität von α-Yohimbin mit Iso-yohimbin.) Ber. dtsch. chem. Ges. **64**, 1408 (1931).

374. Webb, L. G.: Guide to the Medicinal and Poisonous Plants of Queensland. Council for Sci. Ind. Res. Australia, Bull. No. 232, p. 19 (1948).

375. Weber, E.: Ein Rauwolfia-Alkaloid in der Psychiatrie — seine Wirkungs-ähnlichkeit mit Chloropromazin. Schweiz. med. Wschr. **84**, 968 (1954).

376. Wehmer, C.: Die Pflanzenstoffe, Bd. II, S. 990. Jena: G. Fischer. 1931.

376 a. Weisenborn, F. L. and H. E. Applegate: Synthesis of Compounds Related to Reserpine. Conversion of an Intermediate to 16-Methylyohimbane. J. Amer. Chem. Soc. **78**, 2021 (1956).

376 b. Weisenborn, F. L. and P. A. Diassi: The Reaction of Rauwolfia Alkaloids with Mercuric Acetate. Conversion of 3-Isoreserpine to Reserpine. J. Amer. Chem. Soc. **78**, 2022 (1956).

377. Weisenborn, F. L., M. Moore and P. A. Diassi: Isolation of δ-Yohimbine and a New Related Alkaloid from *Rauwolfia serpentina* Benth. Correlation of δ-Yohimbine with Serpentine. Chem. and Ind. **1954**, 375.

378. Weiskrantz, L. and W. A. Wilson, Jr.: The Effects of Reserpine on Emotional Behavior of Normal and Brain-Operated Monkeys. Ann. New York Acad. Sci. **61**, 36 (1955).

379. Wenkert, E. and L. H. Liu: The Constitution of the Alloyohimbanes. Experientia **11**, 302 (1955).

380. Werner, G.: Zur Wirkung von *Rauwolfia serpentina*. Arzneim.-Forschg. **4**, 40 (1954).

381. — The Central Control of Blood Pressure. Indian med. Gaz. **88**, No. 3, p. 111 (1953).

382. Wibaut, J. P. und A. J. P. van Gastel: Über Yohimbin (3. Mitt.). Rec. trav. chim. Pays-Bas **54**, 85 (1935).

383. Winsor, T.: Human Pharmacology of Reserpine. Ann. New York Acad. Sci. **59**, 61 (1954).

384. Witkop, B.: Zur Konstitution des Yohimbins und seiner Abbauprodukte. Liebigs Ann. Chem. **554**, 83 (1943).

385. — Über den Abbau von Indol-Verbindungen durch Ozon. Liebigs Ann. Chem. **556**, 103 (1944).

386. — On the Stereochemistry of Yohimbine. J. Amer. Chem. Soc. **71**, 2559 (1949).

387. WITKOP, B.: Studies on Carboline Anhydronium Bases. J. Amer. Chem. Soc. **75**, 3361 (1953).

388. WOODWARD, R. B.: Biogenesis of the Strychnos Alkaloids. Nature (London) **162**, 155 (1948).

388a. — Neuere Entwicklungen in der Chemie der Naturstoffe. Angew. Chem. **68**, 13 (1956).

388b. WOODWARD, R. B., F. E. BADER, H. BICKEL, A. J. FREY and R. W. KIERSTEAD: The Total Synthesis of Reserpine. J. Amer. Chem. Soc. **78**, 2023, (1956).

388 c. — — — — — A Simplified Route to a Key Intermediate in the Total Synthesis of Reserpine. J. Amer. Chem. Soc. **78**, 2657 (1956).

389. WOODWARD, R. B., W. J. BREHM and A. L. NELSON: The Structure of Strychnine. J. Amer. Chem. Soc. **69**, 2250 (1947).

390. WOODWARD, R. B., M. P. CAVA, W. D. OLLIS, A. HUNGER, H. U. DAENIKER and K. SCHENKER: The Total Synthesis of Strychnine. J. Amer. Chem. Soc. **76**, 4749 (1954).

391. WOODWARD, R. B. and W. M. McLAMORE: The Synthesis of Sempervirine Methochloride. J. Amer. Chem. Soc. **71**, 379 (1949).

392. WOODWARD, R. B. and B. WITKOP: The Structure of Ketoyobyrine. J. Amer. Chem. Soc. **70**, 2409 (1948).

393. — — The Structure of Sempervirine. J. Amer. Chem. Soc. **71**, 379 (1949).

394. YESINICK, L. and E. GELLHORN: Studies on Increased Intracranial Pressure and its Effect During Anoxia and Hypoglycemia. Amer. J. Physiol. **128**, 185 (1939).

395. YOUNGKEN, H. W.: A Pharmacognostical Study of *Rauwolfia*. J. Amer. Pharmaceut. Assoc., Sci. Ed. **43**, 70 (1954).

396. — Malabar Rauwolfia, *Rauwolfia micrantha* HOOK. J. Amer. Pharmaceut. Assoc., Sci. Ed. **43**, 141 (1954).

397. Conference on Reserpine (Serpasil) and Other Alkaloids of *Rauwolfia serpentina*: Chemistry, Pharmacology and Clinical Applications. Ann. New York Acad. Sci. **59**, No. 1 (1954).

398. Conference on Reserpine in the Treatment of Neuropsychiatric, Neurological, and Related Clinical Problems. Ann. New York Acad. Sci. **61**, No. 1 (1955).

(Received, December 31, 1955.)

Synthese von Peptiden.

Von **W. Grassmann** und **E. Wünsch**, Regensburg.

Inhaltsübersicht.

        α) Der 2-Nitrophenoxy-acetyl-rest 462. — β) Der (2-Nitro-4-carbomethoxyphenyl-)-glycyl-rest 463. — γ) Der „Chloracetyl-(2-aminophenyl-)-glycyl-rest 463.

Einleitung.

Seitdem das wesentliche Bauprinzip der Eiweißstoffe in der amidartigen Verknüpfung von Aminosäuren erkannt war, ist die Peptidsynthese eine der wichtigsten Aufgaben der Eiweißforschung gewesen. Synthetische Polypeptide sind erstmals von Curtius (*114*) hergestellt worden. Die von ihm aufgefundene Selbstkondensation freier Aminosäureester konnte ihrer Natur nach nur zu Gemischen von undefinierter Kettenlänge führen. Hingegen zeigen seine, unter Verwendung von Hippurylchlorid und Hippurylazid durchgeführten Synthesen von Glycin-Polypeptiden bereits die wesentlichen Merkmale der zum Aufbau definierter Peptide üblichen Schritte: Die Verwendung von Schutzgruppen zur Abdeckung der Aminogruppe und von energiereichen und reaktionsfähigen Derivaten der Carboxylgruppe. Aber die von Curtius eingeführte Benzoylgruppe kann nicht unter Bedingungen abgespalten werden, unter denen die Peptidbindung widerstandsfähig ist; die Herstellung freier Peptide ist daher Curtius auf dem von ihm eingeschlagenen Wege nicht gelungen.

Dieser Erfolg blieb Emil Fischer (*144*) und seiner Schule vorbehalten: Die halbseitige Öffnung der Diketopiperazine, die Verwendung von α-Halogen-acylchloriden und ihre nachfolgende Aminierung und die Verwendung der Säurechloride der Aminosäuren sind die wesentlichen Schritte der peptidsynthetischen Verfahren Fischers. Aber die Methoden Fischers haben sich, namentlich im Hinblick auf die Synthese langkettiger und unter Beteiligung polyfunktioneller Aminosäuren aufgebauter Peptide, als weniger allgemein anwendbar erwiesen, als man dies zunächst erhofft hatte. So folgte dem von Fischer erarbeiteten und von seinem Schüler Abderhalden bis an die Grenze der damaligen experimentellen Möglichkeit weitergeführten methodischen Fortschritten eine Periode des Stillstandes. Das eigentliche Zeitalter der systematischen Peptidsynthese begann erst, als Bergmann und Zervas (*56*) mit der Einführung des Carbobenzoxyrests die lange angestrebte, leicht wieder rückgängig zu machende Maskierungsmöglichkeit der Aminogruppe verwirklichten.

Im Laufe des letzten Jahrzehnts hat, angeregt durch wichtige Erkenntnisse über die Funktion der Eiweißkörper und getragen von entscheidenden Fortschritten der eiweiß-analytischen Methodik, eine imponierende Neuentwicklung auf dem Gesamtgebiet der Eiweißforschung eingesetzt. Dieser Entwicklung laufen bedeutende Fortschritte der Peptidsynthese parallel. Sie sind gekennzeichnet durch die Auffindung zahlreicher neuer Schutzgruppen, die Verwendung energiereicher Ester, Thioester und Anhydride, sowie durch weitere neuere Synthesemethoden, die praktisch auf eine Aktivierung der Aminogruppe hinauslaufen. Auch die Theorie der zur Knüpfung der Peptidbindung führenden Reaktionen beginnt sich abzuzeichnen.

Eine zusammenfassende detaillierte Darstellung der Peptidsynthese hat FRUTON (*175*) 1949 gegeben. Die stürmische Entwicklung der letzten Jahre auf diesem Gebiet hat dazu geführt, daß nicht nur diese Darstellung überholt ist, sondern in wesentlichen Teilen auch die kürzeren neueren Übersichtsreferate, die von TH. WIELAND (*452, 453*), SHAPIRO (*382*) sowie FROMAGEOT und JUTISZ (*174*) gegeben wurden.

Die folgende Darstellung soll diese Lücke schließen. Sie beschränkt sich auf die Wege der rein chemischen Synthese und wird das wichtige Problem der enzymatischen und biologischen Eiweiß- und Peptidsynthese nur gelegentlich streifen. Eine erschöpfende Behandlung dieses Fragenkomplexes wäre nur im Zusammenhang mit den Gesamtproblemen der Peptidase- und Proteinasewirkung möglich.

I. Theoretische Grundlagen der Peptidsynthese.

Die zur Synthese der Säureamid- und damit auch der Peptidbindung führenden Reaktionen sind bimolekulare nucleophile Substitutionen (S_{N_2}); ihr wesentlicher Teilvorgang ist die Anlagerung des freien Elektronenpaares des Stickstoffs an das positivierte Kohlenstoffatom der Carboxylgruppe [vgl. z. B. SCHWYZER (*379*)].

$$R'-\ddot{N}H_2 + \underset{R}{\overset{X}{C}}{=}O \rightleftharpoons \left[R'-\underset{H}{\overset{H}{N}} : \underset{R}{\overset{X}{C}}\overset{(+)}{-}\overline{O}|^{(-)} \right] \rightleftharpoons \left[R'-\underset{H}{\overset{H_{(+)}}{N}}-\underset{R}{\overset{X}{C}}-\overline{O}|^{(-)} \right] \rightleftharpoons$$

(I.)

$$\rightleftharpoons R'-\underset{H}{\ddot{N}}-\underset{R}{C}{=}O + HX$$

Die *Aminokomponente* reagiert daher nur in der Form der freien Base, nicht als Ammonsalz; und die Reaktion der freien Basen ihrerseits ist —

abgesehen von sterischen Einflüssen (*25*, *446*) — erschwert, wenn ihr freies Elektronenpaar durch Mesomerie beansprucht wird, wie dies z. B. bei aromatischen Aminen der Fall ist. Demzufolge wird man erwarten dürfen, daß die Bereitschaft der freien Base zur aminolytischen Substitution etwa mit der Basenstärke parallel geht. Sie ist am größten beim Benzylamin, dann folgen Aminosäureester und primäre aliphatische Amine, während gewisse sekundäre Amine, wie z. B. Piperidin und schließlich das schwächer basische Anilin, am Ende der Reihe stehen (*373*):

$$H_2N\text{---}Ar \;<\; H_2N\text{---}Alk \;<\; H_2N\text{---}CH_2\text{---}Ar$$

Ar = aromatischer Rest, Alk = aliphatischer Rest.

Die Umsetzungsgeschwindigkeit kann nach Betts und Hammett (*70*), sowie Schwyzer (*375*) durch Basenkatalyse (Alkoholat, tertiäres Amin) erheblich gesteigert werden. Nach der Hypothese dieser Autoren spielt die durch intermolekulare Protonenverschiebung entstehende Imino-Anionform als nucleophiles Reagens eine bedeutende Rolle für den Reaktionsablauf.

$$2\,R'\text{---}NH_2 \;\rightleftharpoons\; R'\text{---}NH^{(-)} + R'\text{---}NH_3^{(+)}$$

Bei Umsetzungen von Estern in flüssigem NH_3 oder Aminen ist interessanterweise von Audrieth und Mitarb. (*134*, *28*, *189*) auch ein katalytischer Einfluß von Säuren, und zwar besonders der Essigsäure, beobachtet worden, ein Befund, der von Schwyzer (*375*) bestätigt wird [vgl. auch Shatenshtein (*383*), Jung, Miller und Day (*251*); vgl. dagegen Betts und Hammett (*70*)].

In der *Carbonylkomponente* hängt die positive Polarisierung von der Natur der Reste X und R ab; alle Änderungen dieser beiden Substituenten, welche die Positivierung des Carbonyl-Kohlenstoffatoms fördern, werden die Neigung zur Aminolyse begünstigen. Nach dem Einfluß des *Restes R* lassen sich die Carbonsäuren und Derivate etwa in der folgenden Reihe fallender Reaktionsfähigkeit ordnen (*469*, *373*, *420*):

α-Halogenacyl > Formyl > N-Acylaminosäuren > Acetyl > langkettige und verzweigte aliphatische sowie aromatische Säuren.

Die hohe Reaktionsfähigkeit der halogen-substituierten Säuren ist dabei dem induktiven Effekt der elektronen-anziehenden Halogene, der Unterschied zwischen Formyl- und Acetylverbindung dem Fehlen bzw. Vorhandensein der Hyperkonjugation zuzuschreiben. Die geringe Reaktionsfähigkeit der letzten Gruppe anderseits dürfte, abgesehen von sterischen Hinderungen (*420*), zurückzuführen sein auf die elektronen-abstoßende Wirkung großer und verzweigter Alkylgruppen, die der Positivierung des Carbonyl-C-Atoms entgegenwirkt, im Falle der aromatischen Carbonsäuren auch auf die Resonanzstabilisierung des aro-

matischen Systems, in die die Elektronen des Carbonyls mit einbezogen
werden.

Die angeführte Reihenfolge, die besonders auch für das Verhalten
gemischter Säureanhydride wichtig ist (s. unten), geht im großen und
ganzen der Dissoziation der zugrunde liegenden Säuren parallel. Dies ist
zu erwarten; denn Elektronenmangel am Carboxyl-C ist gleichbedeutend
mit verstärkter Tendenz zur Abstoßung des Protons von der Hydroxylgruppe und zum Übergang in die anionische Form. Auch die Bedeutung
der *Funktionsgruppen* X für die positive Polarisierung des Carbonyl-
Kohlenstoffs und damit für die Neigung zur Amidbildung kann unter
ähnlichen Gesichtspunkten verstanden werden.

Carbonsäuren (X = OH) und Thiolcarbonsäuren (X = SH) werden
der Reaktion zugänglich sein in Form der undissozierten Säure,

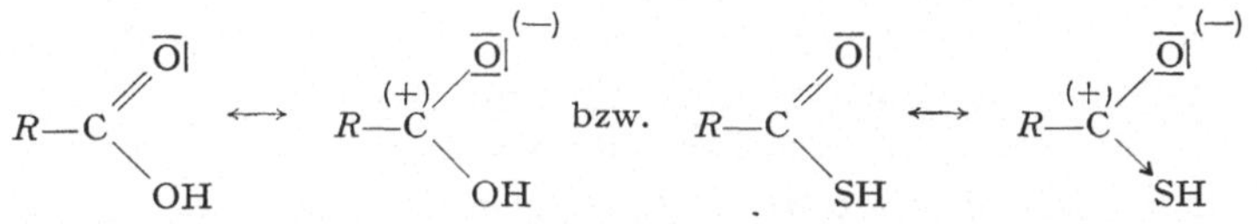

$$R-C\underset{OH}{\overset{\bar{O}}{{<}}} \longleftrightarrow R-\overset{(+)}{C}\underset{OH}{\overset{\bar{O}|^{(-)}}{{<}}} \quad \text{bzw.} \quad R-C\underset{SH}{\overset{\bar{O}}{{<}}} \longleftrightarrow R-\overset{(+)}{C}\underset{SH}{\overset{\bar{O}|^{(-)}}{{<}}}$$

nicht (oder im Fall der Thiolcarbonsäuren nur schwer) dagegen in Form
des Säureanions, in dem wegen der Mesomerie $A \rightleftharpoons B$ eine Grenzform (C)
mit positiviertem Kohlenstoffatom kaum gebildet wird.

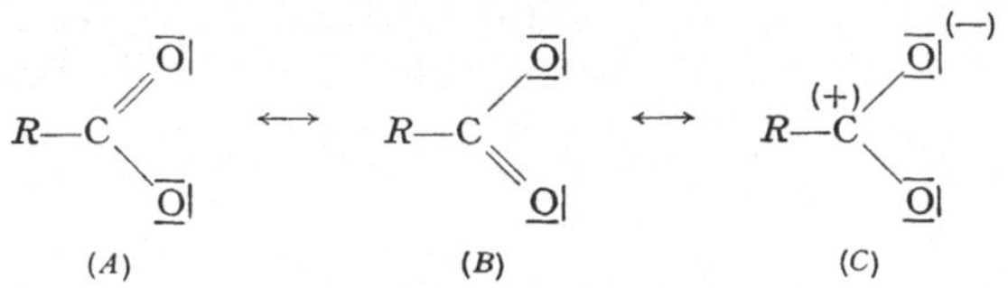

$$R-C\underset{\bar{O}|}{\overset{\bar{O}}{{<}}} \longleftrightarrow R-C\underset{\bar{O}|}{\overset{\bar{O}|}{{<}}} \longleftrightarrow R-\overset{(+)}{C}\underset{\bar{O}|}{\overset{\bar{O}|^{(-)}}{{<}}}$$

$$(A) \qquad\qquad (B) \qquad\qquad (C)$$

Damit lassen sich zunächst die Verhältnisse für die Umsetzungen der
Carbonsäuren und Thiolcarbonsäuren mit Aminen in wäßrigen oder
wasserähnlichen Medien überschauen: Durch Erhöhung der Wasser-
stoffionen-konzentration wird die Dissoziation der Carboxylgruppe zurückgedrängt, gleichzeitig aber die Aminkomponente in die Ammoniumform übergeführt; die COOH-Gruppe, nicht aber die Aminkomponente,
hat einen der Reaktion günstigen Zustand. Bei Erhöhung des p_H-Wertes
erfolgt das Umgekehrte: Das Amin ist reaktionsbereit, nicht aber die nun
zum Anion ionisierte Carboxylgruppe. Bei aliphatischen Carbonsäuren und
aliphatischen Aminen und erst recht bei den in mittleren p_H-Bereichen
ausschließlich als Zwitterionen vorliegenden Aminosäuren gibt es im
allgemeinen in wäßriger Lösung keinen p_H-Bereich, in dem undissoziierte
Carboxylgruppe und freies Amin in größerer Konzentration coexistent
sind. Bei größerem Abstand zwischen Amino- und Carboxylgruppe, also
bei längeren Peptiden, rücken, verglichen mit den Verhältnissen bei

Aminosäuren, die p_K-Werte der Carboxylgruppen zu höheren, die der Aminogruppen zu niedrigeren p_H-Werten. Die an sich sehr ungünstigen energetischen Voraussetzungen für eine direkte Kondensation zwischen Amino- und Carboxylgruppe werden dadurch etwas verbessert (*452*).

Im Falle der reaktionsfähigen Thiolsäuren sollte die Reaktionsgeschwindigkeit in wäßriger Lösung dem Produkt aus den Konzentrationen der undissoziierten Thiolsäure [—SH] und des freien Amins [—NH$_2$] proportional sein, und der p_H-Wert maximaler Umsetzungsgeschwindigkeit sollte dem Mittel der p_K-Werte von beiden entsprechen. Daß diese Beziehung annähernd gilt, ist von Schwyzer (*373*) gezeigt worden: $p_H{}^{max} = {}^1\!/_2\,(p_{K_{SH}} + p_{K_{NH_2}})$.

Bei Umsetzungen von Thiolcarbonsäuren und den meisten anderen Carbonsäure-derivaten in wäßriger Lösung ist zu berücksichtigen, daß die Aminolyse mit der Hydrolyse zu konkurrieren hat und daß die tatsächlichen Ausbeuten an Amid das Ergebnis dieser beiden Konkurrenzreaktionen sind. Da im wäßrigen Medium das molekulare Verhältnis Wasser : Amin sehr stark nach der Seite des Wassers gelegen ist, muß die Geschwindigkeitskonstante der Aminolyse bedeutend höher sein als die der Hydrolyse, wenn beträchtliche Ausbeuten erhalten werden sollen. Für den Fall der Thiolcarbonsäuren und ihrer Ester liegt das Verhältnis der beiden Geschwindigkeitskonstanten beispielsweise in der Größenordnung von 400 (*373*).

Die Frage, durch welche substitutionschemischen oder katalytischen Einflüsse die Aminolyse *spezifisch*, also ohne gleichzeitige Beschleunigung der Hydrolyse, begünstigt werden kann, bedarf weiterer Untersuchung. Im Falle der Umsetzung von Thiolcarbonsäureestern mit Aminen kommt den Ionen des Ag, Cu, Hg und Pb eine solche spezifische Wirkung zu, nach einer hypothetischen Vorstellung Schwyzers (*377*) wahrscheinlich deswegen, weil diese positiven Ionen Elektronen vom Schwefel und damit indirekt auch vom Carbonyl-Kohlenstoff abziehen und zugleich durch ihre komplexe Bindung an das Amin dieses in die räumliche Nähe des positivierten C-Atoms bringen.

$$
\begin{array}{c}
\overset{\displaystyle O}{\underset{\displaystyle }{\|}} \\
R\!-\!C\!\rightarrow\!S\!-\!R'' \\
\underset{\displaystyle H}{}\quad \downarrow \\
R'\!-\!N\,\cdots\!\rightarrow\!Ag \\
H
\end{array}
\quad\longrightarrow\quad
\begin{array}{c}
O \\
\| \\
R\!-\!C \\
| \\
R'\!-\!NH
\end{array}
\;+\;
\begin{array}{c}
Ag\!-\!S\!-\!R'' \\
H^+
\end{array}
$$

Auf diesem Wege ist von Schwyzer die Synthese von Benzoylanilid und Hippursäure aus Benzoyl-pantethein und Anilin bzw. Glykokoll in wäßriger Lösung unter Bedingungen durchgeführt worden, unter denen sie ohne Schwermetallzusatz nicht gelingt. Die Bedeutung solcher Reaktionen als Aminoacylase-Modelle wird allerdings eingeschränkt durch die

Tatsache, daß für die Reaktion nicht katalytische, sondern stöchiometrische Schwermetallmengen benötigt werden. Dies ist zu erwarten, da das Schwermetall an der entstehenden Sulfhydrylverbindung gebunden bleibt. Wird an Stelle von Schwermetallionen das gleichfalls elektrophile Jod verwendet, so entsteht kein Amid, sondern es erfolgt rasch Hydrolyse der COS-Bindung. Auch diese Reaktion gelingt aber nicht mit katalytischen, sondern nur mit stöchiometrischen Mengen, weil Jod mit der gebildeten SH-Verbindung zum Disulfid weiterreagiert.

Ersetzt man in der Formel (I) (S. 447) das Carboxylhydroxyl (X) durch Reste von höherer Elektronenaffinität ($-OR''$, $-SH$, $-SR''$, $-N_3$, $-Cl$), so gelangt man zu reaktionsfähigen und energiereichen Säurederivaten, von denen die meisten schon lange als Acylierungsmittel von Bedeutung sind. Die gesteigerte Elektronen-affinität dieser Reste begünstigt die Elektronen-verarmung am Carbonylkohlenstoff und zugleich die nachfolgende Abspaltung von X in anionischer Form. Weiterhin wird durch die Einführung dieser Reste (mit Ausnahme der SH-Gruppe) die Ausbildung der der Amidbildung nicht oder kaum zugänglichen Carboxylat-anionen verhindert, so daß es nun keine Schwierigkeiten macht, sowohl in wäßrigen wie in nicht-wäßrigen Medien Bedingungen zu wählen, unter denen die Aminkomponente als freie Base vorliegt. Bei Umsetzung in wäßrigen Medien aber bleibt natürlich auch hier die Hydrolyse als Konkurrenzreaktion zu beachten.

Da die Reste $-SH$, $-SR''$, $-N_3$, $-Cl$ (und im weiteren Sinne auch $-OR''$) Säurereste sind, kann man die entsprechenden substituierten Carboxylderivate auch als gemischte Anhydride auffassen und sie den in neuerer Zeit zu Peptidsynthesen vielfach herangezogenen gemischten Säureanhydriden mit Phosphor-, Phosphorig-, Schwefel-, Schwefliger- und Kohlensäure als Anhydridpartner an die Seite stellen. Daß die Säurestärke der zugrunde liegenden freien Säuren HX und die Fähigkeit, in Bindung an die Acylgruppe $R-CO-X$ die Positivierung des Carbonylkohlenstoffs zu begünstigen, parallel gehen sollten, ergibt sich daraus, daß die Fähigkeit, die Protonen festzuhalten und die Fähigkeit, Elektronen vom Carbonylkohlenstoff abzuziehen, offenbar gegenläufig sein müssen. Dementsprechend sind in der Aminolyse Phenylester und Enolester reaktionsfähiger als gewöhnliche Alkylester, und Thioester reaktionsfähiger als die entsprechenden Sauerstoffester.

Schließlich läßt sich die Reaktionsfähigkeit von Carbonsäureestern und Thiolcarbonsäureestern, denen die sehr schwachen Säuren Alkohol bzw. Thioalkohol zugrunde liegen, durch Einführung negativer Substituenten in den Rest R'' (Substituenten mit I^--Wirkung) beträchtlich steigern. Der negative Substituent vermindert durch einen zusätzlichen Induktionseffekt die Elektronendichte am Carbonyl-C-Atom vor und während des nucleophilen Angriffs, wodurch die Reaktionsgeschwindigkeit

erhöht wird. Als besonders geeignet hat sich der Cyanmethylester
($R'' = $ —CH$_2$CN) erwiesen; etwas weniger reaktionsfähig und zum Teil
auch aus anderen Gründen weniger geeignet sind z. B. Carbäthoxy-
($R'' = $ —CH$_2$COOC$_2$H$_5$), Dicarbäthoxy- [$R'' = $ —CH$_2$(COOC$_2$H$_5$)$_2$] und
Acetomethylester ($R'' = $ —CH$_2$COCH$_3$) sowie p-Nitrobenzylester. Auch
die Ester des Cholins sind wegen der I$^-$-Wirkung der quarternären
Ammoniumgruppe hierher zu zählen (*379, 376, 375*).

In der Reihe der Thiolcarbonsäureester sind von Schwyzer (*373*)
auch Carboxymethyl- ($R'' = $ —CH$_2$COOH) und die β-Carboxyäthyl-
ester ($R'' = $ —CH$_2$CH$_2$COOH) als Ester von hoher Reaktionsfähigkeit
verwendet worden. Dabei ist zu beachten, daß nur die undissoziierte
Carboxylgruppe dieser Thioestersäuren als reaktions-erleichternder
I$^-$-Substituent anzusprechen ist, nicht dagegen deren Anion; für die
p$_H$-Abhängigkeit ihrer Umsetzung mit Aminen in wäßriger Lösung ergeben sich damit ähnliche Verhältnisse wie sie oben für die Thiolcarbonsäuren geschildert worden sind.

Durch die Begrenzung der Polarisierbarkeit des Sauerstoffs und wohl
auch des Schwefels ist allerdings der Erfolg des I$^-$-Effekts beschränkt.
Die Einführung *mehrerer* negativer Substituenten in die Methylgruppe
der Ester führt nicht zu einer Beschleunigung der Aminolyse, sondern
zu einer Änderung des Reaktionsablaufs; unter Regenerierung der Carbonsäure wird die OH-Gruppe des Alkohols gegen den Aminrest ausgetauscht
(*100*), z. B.:

$$R-C\overset{O}{\underset{O-CH\underset{CH_2NO_2}{\diagdown}}{\diagdown}}CCl_3 \ + \ H_2N-R' \ \longrightarrow \ R-C\overset{O}{\underset{OH}{\diagdown}} \ + \ H_2N-CH\underset{CH_2NO_2}{\overset{CCl_3}{\diagdown}}$$

Eine Aktivierung des Acylrests kann schließlich durch Bindung an
die saure Iminogruppe des *Imidazols* und verwandter Verbindungen
[Purinderivate und dgl.; vgl. Gerngross (*184*), Bergmann (*54*), Wieland
(*465*)] erreicht werden. Der Elektronenmangel des Imidazol-Stickstoffs,
dessen Elektronenpaar durch Einbeziehung in das heterocyclisch-aro-
matische System weggenommen ist, führt hier zur Positivierung des
Carbonyl-C-Atoms.

$$R-C\overset{O}{\underset{O-CH-OR'''}{\diagdown}}R''$$

Den aktivierten Estern werden häufig, aber zu Unrecht,
als eine weitere Gruppe von Verbindungen mit aktiven
Acylgruppen die Verbindungen der allgemeinen neben-
stehenden Formel also z. B. die „Methoxymethylester"
(*379*) oder die „Ester" des α-Oxytetrahydropyrans (*248*)
zugezählt. Der Partner des Säurerests ist in diesem Falle kein Alkohol, sondern
ein in der Halbacetalform vorliegender Aldehyd. Es handelt sich also um *Acyl-
halbacetale*; ihr Reaktionsverhalten steht dem der aktivierten Ester nahe.

Als letzte und zugleich als eine der für die neuere Peptidsynthese wichtigsten Gruppen von Verbindungen mit aktivierten Acylgruppen seien die *symmetrischen* und die *unsymmetrischen* (oder „gemischten") *Säureanhydride* angeführt. Die Verwendung von Säureanhydriden zu Acylierungen ist lange bekannt. Aber Anhydride von Acylaminosäuren sind erstmals fast gleichzeitig von WIELAND (*461*), VAUGHAN (*416*) und BOISSONNAS (*75*) hergestellt und für die Peptidsynthese verwendet worden. Eine große Bedeutung haben dabei insbesondere auch die gemischten Säureanhydride mit anorganischen Säuren (S. 499) gewonnen. Von den nicht N-acylierten Aminosäuren sind bisher nur in einigen wenigen Fällen gemischte Anhydride mit Phosphorsäure beschrieben (*42, 254*); im allgemeinen dürften diese Anhydride wegen der gleichzeitigen Anwesenheit einer ungeschützten Aminogruppe und einer aktivierten Carboxylgruppe sehr wenig beständig sein.

In den Säureanhydriden sind die C-Atome beider Carbonylgruppen durch die elektronen-anziehende Wirkung sowohl der Carbonyl-sauerstoffe wie des Brückensauerstoffs positiviert und der Aminolyse deswegen leicht zugänglich. Wegen der Resonanz zwischen den beiden symmetrischen Grenzformen

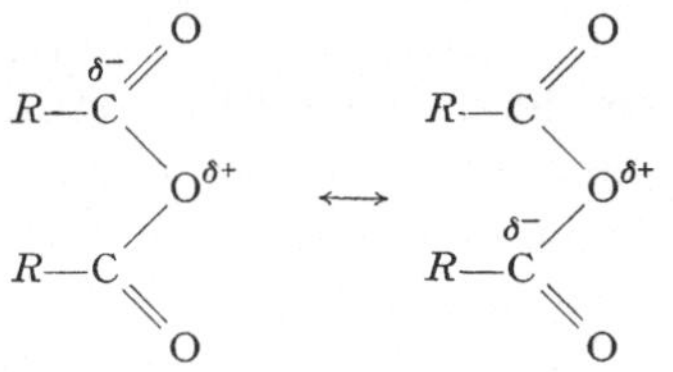

$$
R\text{—}C\underset{\underset{O}{\diagdown}}{\overset{\overset{O}{\diagup^{\delta-}}}{}}\ O^{\delta+}\ \diagup R\text{—}C \quad\longleftrightarrow\quad R\text{—}C\ O^{\delta+}\ R\text{—}C
$$

sind dabei energetisch die symmetrischen Anhydride gegenüber den unsymmetrischen begünstigt, so daß bei gemischten Anhydriden mit einer Disproportionierung zu rechnen ist:

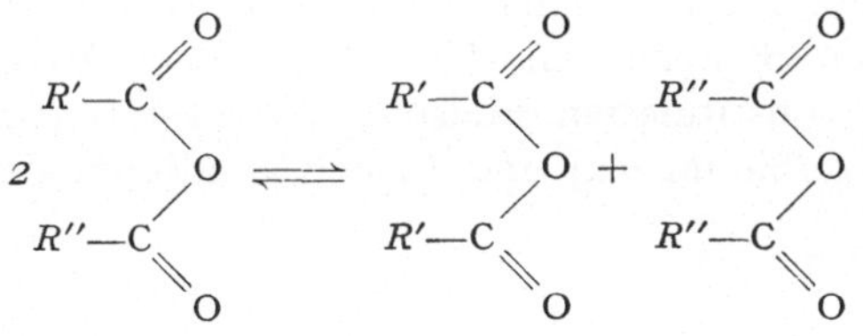

$$
2\ \ \begin{matrix} R'\text{—}C\diagup^{O}\\ \diagdown O \diagup \\ R''\text{—}C\diagdown_{O}\end{matrix} \ \rightleftharpoons\ \begin{matrix} R'\text{—}C\diagup^{O}\\ \diagdown O \diagup \\ R'\text{—}C\diagdown_{O}\end{matrix} \ +\ \begin{matrix} R''\text{—}C\diagup^{O}\\ \diagdown O \diagup \\ R''\text{—}C\diagdown_{O}\end{matrix}
$$

Bei der Aminolyse symmetrischer Anhydride wird bekanntlich einer der beiden Säurereste als Amid, der andere als freie Säure erhalten und man gewinnt, unabhängig davon, welche der beiden mesomeren Grenzformen reagiert hat, immer 50% der vorhandenen Säure als Amid. Bei der Verwendung unsymmetrischer Säureanhydride einer Acylaminosäure und einer (anorganischen oder organischen) „Hilfssäure" wird man diese und die Bedingungen so wählen müssen, daß die Disproportionierung weitgehend vermieden und die Aminolyse an der Aminosäure-Hälfte und nicht an der Hilfssäure verläuft (vgl. S. 504).

Es besteht hier die bereits oben geschilderte Reihenfolge der Reaktionsbereitschaft verschiedener Säurereste. Wieland und seine Schule (469) haben festgestellt, daß in jedem Falle diejenige Acylhälfte bevorzugt dem Angriff nucleophiler Reagenzien unterliegt, deren Esterderivate z. B. leichter der Hydrolyse anheimfallen. Im ganzen gesehen besteht auch hier die schon oben diskutierte Parallelität mit den Dissoziationskonstanten der freien Säuren. Bei der Aminolyse gemischter Anhydride aus Acylaminosäure und anorganischer Säure, wie z. B. Phosphor- oder Schwefelsäure, aber tritt der Stickstoff immer an den organischen Säurerest, während der Brückensauerstoff grundsätzlich am anorganischen Rest verbleibt; dies entspricht vollkommen den Verhältnissen bei der Hydrolyse anorganischer Säureester.

Den geschilderten Verfahren der Peptidsynthese, denen eine Aktivierung der Carboxylgruppe zugrunde liegt, kann man im gewissen Sinne eine Gruppe in neuerer Zeit aufgefundenen Peptidsynthesen gegenüberstellen, deren Wesen in einer *„Aktivierung" der Aminogruppe* gesehen wird. In Wahrheit handelt es sich dabei nicht um eine wirkliche Aktivierung der Aminogruppe, deren freies Elektronenpaar zu einer nucleophilen Reaktion keiner besonderen Aktivierung mehr bedarf, sondern um Säureamid- bzw. -imidverbindungen, deren ungesättigtes Säure-zentralatom sich an die Carboxylgruppierung anzulagern vermag. Zum Beispiel setzen sich Isocyanate (N-Carbonylderivate von primären Aminen) in völliger Analogie zum Keten mit Carbonsäuren zu den thermodynamisch labilen gemischten Anhydriden — in diesem Falle von N-Carbonsäuren (II) — um. Diese in einigen Fällen gefaßten Zwischenprodukte (IIb) (328) enthalten den Stickstoff und das positivierte Carbonyl-C-Atom in räumlicher günstiger Lage und zerfallen daher besonders leicht unter CO_2-Abspaltung zu den stabileren Säureamiden (III). In völliger Analogie erfolgt die Umsetzung von substituierten Phosphat-, Phosphit- (einschließlich der „Phosphorazo"-verbindungen) und Arsenit-amide mit Carbonsäuren.

$$R\text{—COOH} + O{=}C{=}N\text{—}R'$$

$$
\left[
\begin{array}{c}
R\text{—C}\overset{\displaystyle O}{\diagdown}\ :N\text{—}R' \\
O\text{—C} \\
OH
\end{array}
\right.
\quad \rightleftharpoons \quad
\left.
\begin{array}{c}
R\text{—C}^{(+)}\quad \overset{H}{:N\text{—}R'} \\
O^{(-)}\text{—C} \\
O
\end{array}
\right]
\quad \xrightarrow{-CO_2} \quad
R\text{—C}\overset{\displaystyle O}{\diagup}\diagdown NH\text{—}R'
$$

(II a.)　　　　　　　(II b.)　　　　　　　(III.)

Die Knüpfung der Amidbindung als solche ist vom rein präparativen Standpunkt weitgehend gelöst. Die Übertragung auf das Peptidgebiet ist jedoch mit einer erheblichen Schwierigkeit verbunden. Sie resultiert daraus, daß die miteinander zu verbindenden Komponenten gleichzeitig Amino- und Carboxylgruppen im Molekül enthalten, und als Zwitterionen vorliegen. Jeder Versuch einer „Aktivierung" dieser Systeme — wenn sie in der Zwitterionenform überhaupt möglich wäre — würde sofort zu einer unkontrollierbaren Polykondensation ausarten. Um die Voraussetzung für eine gelenkte „systematische", nucleophile Addition zu schaffen, ist es daher notwendig, die zweite Funktion des Aminosäuremoleküls zu blockieren. Man erreicht aber damit *mehr* als die Verhinderung einer unerwünschten Reaktion: erst durch die Blockierung der Carboxylgruppe wird die Aminogruppe statt in der Ammoniumform als freie Base und erst durch die Blockierung der Amino- die Carboxylgruppe als freie Carbonsäure verfügbar und für die oben besprochenen Aktivierungsreaktionen zugänglich. Erst in zweiter Linie sind die hierfür herangezogenen Derivate „Schutzgruppen". Selbstverständlich werden nur solche „Blockierungs"- bzw. „Schutz"-gruppen der Carboxyl- wie Aminogruppierung für die Peptidsynthese brauchbar sein, die nachher leicht, ohne jeglichen Angriff auf die gebildete Peptidbindung, wieder entfernt werden können.

II. Methodische Voraussetzungen der Peptidsynthese.

A. Leicht abspaltbare α-Amino-Schutzgruppen.

1. Die „Acyl-blockierung".

a) Carbamidsäureester (Urethane).

α) *Der „Carbobenzoxy-rest".* Der Anwendungsbereich der von FISCHER und von CURTIUS gegebenen Methoden zur Synthese der Peptidbindung blieb durch das Fehlen von leicht abspaltbaren α-Aminoschutzgruppen — auch der Carbäthoxyrest hatte den langjährigen Bemühungen FISCHERS getrotzt (S. 499) — doch sehr beschränkt. Wie erwähnt, leiteten BERGMANN und ZERVAS (*56*) einen neuen Abschnitt der Peptidchemie ein. Die verhältnismäßig leichte Abspaltbarkeit von O- und N-Benzylgruppen durch katalytische Hydrierung (*347, 172, 51*) veranlaßte die BERGMANN-Schule, nach ihren Mißerfolgen mit den Acetylderivaten (*51, 48, 55, 68*) die Darstellung von „Carbobenzoxy-aminosäuren" und ihre Verwendung in der Peptidsynthese zu versuchen.

Das hierzu benötigte „Carbobenzoxychlorid" (Chlorkohlensäure-benzylester) gewinnt man leicht aus Phosgen und Benzylalkohol in Toluol (*56, 97*) bzw. neuerdings (*132*) besonders rein und bei 0° jahrelang haltbar, direkt aus den Komponenten bei —20° bis —30°; dieses kuppelt leicht mit Aminosäuren bzw. deren Estern zu den gewünschten substituierten Urethanen.

$$C_6H_5CH_2O\text{---}COCl + H_2NCHCOOR'' \xrightarrow[\text{---HCl}]{} Cbo\text{---}NHCHCOOR'' \longrightarrow$$

(mit R' über den CH-Gruppen; $\underbrace{C_6H_5CH_2O\text{---}COCl}_{Cbo\text{---}}$)

$$\xrightarrow[\text{---}CO_2]{H_2(Pd)} C_6H_5CH_3 + H_2NCHCOOR''$$

R' = Aminosäure-(AS)-Seitenkette. R'' = H oder Alkyl.

Bei den optisch-aktiven Ausgangsmaterialien tritt hierbei keine Racemisierung ein. Durch einfache katalytische Hydrierung wird die Carbobenzoxy-verbindung in Toluol, Kohlendioxyd und Aminoderivat gespalten (s. oben).

Dieses Verfahrens haben sich bis in die heutige Zeit zahlreiche Autoren bedient. Nachdem es gelungen war, zusätzliche Abspaltungs-reaktionen für den Carbobenzoxyrest erfolgreich zu verwerten, hat sich der Anwendungsbereich noch umfangreicher gestaltet.

So empfehlen zur Entfernung dieser Schutzgruppe: White (*451*) verdunnte Salzsäure bei 60°, du Vigneaud und Mitarb. (*393, 293*) Natrium in flüssigem Ammoniak, Harington und Mead (*217*) Phosphoniumjodid in Eisessig bei 40° (*83*), Waldschmidt-Leitz und Kühn (*434*) Jodwasserstoff in Eisessig, Anderson (*19*), Ben-Ishai (*39*), sowie Boissonnas (*77*) Bromwasserstoff in Eisessig bzw. Albertson und McKay (*15*) in Nitromethan, Boissonnas und Preitner (*77*) schließlich Chlor-wasserstoff in Eisessig. Die letztgenannten Autoren geben in ihrer Arbeit eine Zu-sammenstellung über die Abspaltbarkeit von Carbo-alkoxy-, Carbo-aryloxy-, Tosyl-, Formyl- und Phthalylresten nach den genannten Methoden an.

Im Gegensatz zur hydrierenden Spaltung ermöglichen diese Ent-acylierungsmethoden die Heranziehung von Carbobenzoxy-verbindungen S-haltiger Aminosäuren zur Peptidsynthese (*37, 191*). (Zur Darstellung von Methioninpeptiden unter Anwendung dieser Verfahren s. auch S. 490.)

Der Abspaltung des Carbobenzoxyrestes mittels Halogenwasser-stoffen in Eisessig liegt nachstehendes Schema einer durch Protonen katalysierten Solvolyse zugrunde. Besonders dieses Verfahren der Urethanspaltung hat sich gut bewährt: Die Benzyl-thioäther (z. B. S-Benzylcystein), primäre Carbonsäureamid- und die Estergruppierung bleiben hierbei unangegriffen (*76, 40*). Entgegen der Mitteilung von Ben-Ishai können wir die vollkommene Beständigkeit der Benzylester nicht bestätigen (*207*).

$$C_6H_5CH_2\text{---}O\text{---}CO\text{---}NH\text{---}R' \xrightarrow{H^{(+)}} C_6H_5CH_2\overset{(+)}{\underset{H}{\text{---}O}}\text{---}CO\text{---}NH\text{---}R'$$

$$\xrightarrow[\text{---}CO_2]{H^{(+)}} C_6H_5\overset{(+)}{CH_2} + H_3\overset{(+)}{N}\text{---}R'$$

$$C_6H_5CH_2{}^{(+)} + Br^{(-)} \longrightarrow C_6H_5CH_2Br$$

R' = Aminosaure- oder Peptidrest.

Als bislang zutage getretene Nachteile der Carbobenzoxygruppierung muß ihre nicht vollkommene Alkalibeständigkeit genannt werden. So entstehen aus den Carbobenzoxy-peptidestern bei ihrer Verseifung als Nebenprodukte Harnstoff- bzw. Hydantoinderivate (*359, 437, 207*); vgl. auch S. 492—493.

$$C_6H_5CH_2OCO—NHCHCO—NHCHCO—NH\ldots\ldots COOR''$$

$$R' \qquad R'$$

$$\downarrow NaOH$$

$$\left[\begin{array}{c} R' \\ HC—CO \\ \quad\quad\diagdown \\ \quad\quad\quad N—CHCO—NH\ldots\ldots COOH \\ \quad\quad\diagup \\ HN—CO \quad R' \end{array}\right] + C_6H_5CH_2OH$$

$$H^{(+)}\uparrow \downarrow OH^{(-)}$$

$$HOOCCHNH—CO—NHCHCO—NH\ldots\ldots COOH$$

$$R' \qquad\qquad R'$$

R' = Aminosäure-Seitenkette. R'' = Alkyl.

Neuerdings ist man zur Verseifung der Ester mit verdünnter Salzsäure (*18*), insbesondere nach vorgehender Entfernung des Carbobenzoxyrestes mit Halogenwasserstoffen in Eisessig, übergegangen (*418*).

Weiterhin beschränkt die verhältnismäßig große Unbeständigkeit der Carbobenzoxy-aminosäurechloride, die sich leicht zu N-Carbonsäureanhydriden zersetzen, die Anwendung der ausgezeichneten „FISCHERSCHEN Säurechloridmethode" oft weitgehend. Anderseits aber stellt diese Reaktion einen der bedeutendsten Wege zur Gewinnung der Oxazoliddione dar.

β) Modifizierte „Carbobenzoxy-reste". In den letzten Jahren wurde durch Einführung von *p*-substituierten Carbobenzoxy-resten die Kristallisationsfreudigkeit verschiedener Aminosäurederivate und Peptidzwischenprodukte erheblich erhöht und damit deren Handhabung sowie die Herstellung von Peptiden verbessert. In Analogie zum „normalen" Carbobenzoxy-rest ist auch die *p*-Brom- von CHANNIG und Mitarb. (*98*) bzw. die *p*-Nitro-verbindung von CARPENTER und GISH (*96, 187*) leicht wieder abspaltbar.

γ) Weitere leicht spaltbare Urethane. STEVENS und WATANABE (*404*) konnten zeigen, daß das Allylchlorocarbonat mit Aminosäuren sehr gut kristallisierte Derivate liefert. Die Carballyloxy-schutzgruppe ist durch Phosphoniumjodid oder Bromwasserstoff (*77*) in Eisessig quantitativ abspaltbar, mit Hilfe der katalytischen Hydrierung bzw. der Einwirkung von Natrium in flüssigem Ammoniak jedoch nur zu 70—80%. Ein Teil des Carballyloxy-restes wird hierbei zur hydrogenolytisch-unangreifbaren N-Carbo-propyloxy-gruppierung reduziert. Boissonnas und Preitner konnten ferner N-Carbo-phenoxy- und N-Carbo-p-tolyloxy-aminoverbindungen mit Natrium in flüssigem Ammoniak bzw. durch katalytische Hydrierung quantitativ spalten und somit für die Peptidsynthese nutzbar machen (77).

b) Thio-urethane.

Ehrensvärd (126) hat den Phenyl-thiocarbonyl-rest (PTC) als Aminoschutzgruppe vorgeschlagen. Seine besonderen Vorzüge sollten die relative Säurebeständigkeit bzw. eine leichte Abspaltung mittels Bleiacetat in 70%igem Alkohol oder Bleihydroxyd $+$ 0,1 n-Natronlauge sein. Hofmann und Mitarb. (289) konnten jedoch zeigen, daß bei der Abspaltung des PTC-restes unter genannten Bedingungen keinesfalls freie Peptide oder Peptidester erhalten werden. Zwar wird unter dem

$$
\begin{array}{c}
\overset{O}{\underset{\|}{}}\quad\overset{R'}{\underset{|}{}}\\
C_6H_5S-C-NH-CH + Pb/2^{(++)}\\
|\\
HN-CO\\
|\\
R'-CH\\
|\\
COOR''
\end{array}
\;\xrightarrow[-C_6H_5SPb/2]{}\;
\left[
\begin{array}{c}
\overset{O}{\underset{\|}{}}\;\overset{H}{\underset{|}{}}\;\overset{R'}{\underset{|}{}}\\
^{(+)}C-N-CH\\
|\\
HN-CO\\
|\\
R'-CO\\
|\\
COOR''
\end{array}
\right]\;\longrightarrow
$$

(IV.)

$$
\xrightarrow[-H^{(+)}]{}
\left[
\begin{array}{c}
R'\\
|\\
O=C=N-CH\\
|\\
HN-CO\\
|\\
R'-CH\\
|\\
COOR''
\end{array}
\right]
\;\xrightarrow[\text{Ringschluß}]{}\;
\begin{array}{c}
\quad NH-CH-R'\\
O=C\diagup\qquad\quad |\\
\diagdown\; N-C=O\\
\quad\;|\\
R'-CH\\
|\\
COOR''
\end{array}
$$

(V.)   (VI.)

R' = Aminosäure-Seitenkette. R'' = H oder Alkyl.

Einfluß von Pb^{++}-Ionen Bleithiophenolat gebildet, doch das dabei entstehende N-Carbonyl-kation (IV) lagert sich spontan in das Isocyanat (V) um, das nun seinerseits durch innermolekularen Ringschluß Hydantoin-essigsäure (-ester) (VI) ergibt. Diese Reaktion wurde neuerdings von Wessely und seiner Schule (444) erfolgreich beim Peptidabbau angewandt. Noguchi und Hayakawa (318) nutzen die enorme Alkaliempfindlichkeit der PTC-Aminosäuren aus, um durch Erhitzen in Dioxan in Gegenwart von Pyridin Polypeptide zu erhalten.

KOLLONITSCH und Mitarb. (275) ist es vor kurzem gelungen, die Thiourethangruppierung in Richtung freies Amin zu spalten und der Peptidsynthese nutzbar zu machen: Phenylthiocarbonyl-peptidester (VII)

$$C_6H_5S—CO—NHCHCO—NHCHCOOR''$$

(VII.)

Perbenzoesaure $\downarrow -10°$

$$C_6H_5—\overset{O}{\underset{O}{S}}—CO—NHCHCO—NHCHCOOR \xrightarrow[-CO_2]{pH\ 4—5} H_2NCHCO—NHCHCOOR''$$

(VIII.)   (IX.)

$$+ C_6H_5—SO_2H$$

R' = Aminosäure-Seitenkette. R'' = Alkyl.

lassen sich durch oxydative Behandlung mit Wasserstoffperoxyd in Eisessig, Ozon oder am besten mit organischen Persäuren und folgende milde saure Hydrolyse des Oxydationsprodukts (VIII) in guter Ausbeute in die freien Peptidester überführen.

Weiter konnte KOLLONITSCH mit Einführung besonders der Benzyl-thiocarbonyl-gruppierung (BTC) die sehr störende Alkali-empfindlichkeit der PTC-Verbindungen bannen. Der aus Benzylmercaptan und Phosgen in Gegenwart katalytischer Mengen Aluminiumchlorid leicht erhält-liche Chlorkohlensäureester (X) kuppelt mit Alkalisalzen von Amino-säuren in wäßrig-bicarbonatalkalischer Lösung. Bei optisch-aktiven Aminosäuren tritt hierbei keine Racemisierung ein. Die erhaltenen Benzylthiocarbonyl-aminosäuren (XI) sind im Gegensatz zu den PTC-Derivaten der Peptidknüpfung mittels der WIELAND-BOISSONNAS-VAUGHANschen Anhydrid- sowie der SHEEHANschen Carbodiimid-methode (SS. 510, 513) zugänglich.

$$C_6H_5CH_2SH + COCl_2 \xrightarrow[-HCl]{AlCl_3} C_6H_5CH_2S—COCl$$

(X.)

$$C_6H_5CH_2S—COCl + H_2NCHCOOH \xrightarrow[(MgO)]{NaHCO_3} C_6H_5CH_2S—CO—NHCHCOOH$$

(XI.)

R' = Aminosäure-Seitenkette.

Erwähnenswert ist ferner die Beständigkeit der Benzylthio-urethane gegenüber Bromwasserstoff-Eisessig (275).

c) Der „Formyl-rest".

Über die Verwendung des Formyl-rests zur Maskierung der α-Aminogruppe berichtet Waley (*435*). N-Formyl-aminosäuren lassen sich nach der „Anhydrid-methode" mit Aminosäureestern (hier speziell Benzylestern) kuppeln; die erhaltenen N-Formyl-derivate spalten leicht die schützende Gruppe beim Behandeln mit 0,5 n-alkoholischer (-benzylalkoholischer) Salzsäure ab. Waley gelang so die Darstellung von L-Tyrosyl-ε-carbobenzoxy-L-lysinbenzylester-hydrochlorid.

$$O{=}CH{-}NHCHCO{-}O{-}COOR'' + H_2NCHCOOR''' \longrightarrow$$
$$\overset{\displaystyle |}{R'} \qquad\qquad\qquad\qquad \overset{\displaystyle |}{R'}$$

$$\longrightarrow O{=}CH{-}NHCHCO{-}NHCHCOOR''' \xrightarrow[-HCOOH]{HCl} H_2NCHCO{-}NHCHCOOR'''$$
$$\overset{\displaystyle |}{R'} \quad\ \overset{\displaystyle |}{R'} \qquad\qquad\qquad\qquad \overset{\displaystyle |}{R'} \quad\ \overset{\displaystyle |}{R'}$$

$$R' = \text{Aminosäure-Seitenkette.} \quad R'' = \text{Alkyl.} \quad R''' = \text{Benzyl.}$$

Hillmann (*228*) hatte auf schlechte Ausbeuten bei der Synthese von Formylpeptiden aus Formyl-aminosäuren und Aminosäuren (Natriumsalze) mittels der „Anhydridmethode" hingewiesen. Dies bestätigen neuerdings King und Mitarb. (*269*), die fanden, daß das gemischte Anhydrid aus N-Formyl-glycin und Chlorkohlensäureester mit p-Aminobenzoesäure-natrium in wäßriger Lösung knapp 50% des gesuchten Derivats liefert. Als sie den entsprechenden Ester als Kupplungskomponente wählten, erhielten sie jedoch den N-Carbäthoxy-aminobenzoesäureester. Zur Begründung dieses Reaktionsverlaufs wird eine völlig abnormale Bildung des Kations $R''OOC^+$ an Stelle von $OHC{-}NHCH_2CO^{(+)}$ angenommen (*269*).

d) Der „Trifluoracetyl-rest".

Weygand und Mitarb. (*448a*) haben durch Einwirkung von Trifluoressigsäure-anhydrid „*TFA*-Aminosäuren" (XII) in sehr guter Ausbeute synthetisiert. Diese sind sehr beständige und zum Teil im Vakuum sublimierbare Körper, die aber rasch der Spaltung mit verdünnter Natronlauge (*448*), Bariumhydroxyd-lösung (*411*) oder selbst wäßrigem Ammoniak bei p_H 10 unterliegen (*358*). Damit stellt diese Schutzgruppe die einzige alkalisch leicht spaltbare Amino-Blockierung dar, ohne daß unter diesen Bedingungen die Gefahr einer Hydrolyse der Peptidbindung besteht. Interessanterweise wird die sonst relativ säurebeständige Amidbindung von alkoholischer Salzsäure unter Regenerierung der freien Aminogruppe angegriffen (*450*), eine Tatsache, die mit Erfolg zur Darstellung von Peptidester-hydrochloriden angewandt werden kann.

Die TFA-Aminosäuren lassen sich mit Thionylchlorid in Benzol glatt in die zur Peptidknüpfung geeigneten Säurechloride (XIII) überführen (*448*), die nach Weygand und Reiher (*450*) mittels Dicyclohexylaminazid in die TFA-Aminosäureazide (XIV), nach Calvin und Mitarb. (*358*) mittels Thiophenol in die Thioester (XV) umwandelbar sind. In

$$\underset{R'}{H_2NCHCO}-\underset{R'}{NHCHCOOH}$$

$\uparrow$ NaOH

$$\underset{R'}{CF_3CO}-\underset{R'}{NHCHCO}-\underset{R'}{NHCHCOOH} \quad \xleftarrow[H^{(+)}]{\underset{R'}{H_2NCHCOONa}} \quad \underset{R'}{CF_3CO}-NHCHCO-SC_6H_5$$

(XV.)

Thiophenol

$$\underset{R'}{CF_3CO}-NHCHCOOH \quad \xrightarrow{+\ SOCl_2} \quad \underset{R'}{CF_3CO}-NHCHCOCl \quad \longrightarrow \quad \underset{R'}{H_2NCHCOOR''}$$

(XII.) (XIII.)

Dicyclohexyl-amin-azid $\downarrow$

$$\underset{R'}{CF_3CO}-NHCHCON_3$$

(XIV.)

$$(CF_3CO)_2O \quad P \nearrow \underset{R'}{NCHCOOR''} \searrow \underset{R'}{NHCHCOOR''}$$

$$\underset{R'}{CF_3CO}-NHCHCO-\underset{R'}{NHCHCOOR''}$$

$$\underset{R'}{CF_3CO}-NHCHCO \diagdown$$
$$\qquad\qquad O$$
$$CF_3CO \diagup$$

(XVI.)

$+ \underset{R'}{H_2NCHCOOR}$

HCl-Alkohol

$$\underset{R'}{H_2NCHCO}-\underset{R'}{NHCHCOOR''}$$
,HCl

$H_2NCHCOOBz$ (mit R')

$$\left[\begin{array}{c} \underset{R'}{CF_3CO}-NHCHCO \diagdown \\ \qquad\qquad O + O \\ \underset{R'}{CF_3CO}-NHCHCO \diagup \end{array} \begin{array}{c} \diagup COCF_3 \\ \\ \diagdown COCF_3 \end{array} \right]$$

(XVII.)

$$\longrightarrow \quad \underset{R'}{CF_3CO}-NHCHCOOH +$$
$$+\ \underset{R'}{H_2NCHCOOH} +$$
$$+CF_3COOBz + CF_3COOH$$

R' = Aminosäure-Seitenkette. R'' = Alkyl. Bz = Benzyl.
(Betr. Phosphorazokörper s. die Fußnote auf S. 533.)

Formelübersicht 1.

besonders gelagerten Fällen der Peptidsynthese können diese energiereichen Derivate — trotz ihrer umständlichen Darstellungsweise — von erheblicher Bedeutung sein (*Formelübersicht 1*).

Darüber hinaus sind für die Synthese von Peptiden aus TFA-Aminosäuren brauchbar: 1. die Phosphorazomethode (*195, 450*) und 2. die Bildung von gemischten Anhydriden mit Trifluoressigsäure (XVI, S. 461) und deren Umsetzung mit Aminosäureestern.

Als empfindliche Nachteile wurden hierbei beobachtet: die vollständig eintretende Racemisierung bei der unsymmetrischen Anhydridbildung (Ausnahme: TFA-*L*-Prolin) und eine anormale Rückbildung der Ausgangs-TFA-Aminosäure bei der Kupplung mit Aminosäurebenzylestern. In diesem Falle scheint als Ursache eine plötzlich ausgelöste Disproportionierung des gemischten in die beiden symmetrischen Anhydride (XVII) maßgebend zu sein (*449, 447, 450*).

Versuche einer Peptidknupfung mit Hilfe der WIELAND-BOISSONNAS-VAUGHANschen Anhydridmethode waren unbefriedigend (*450*). Die aminolytische Aufspaltung eines TFA-Alanin-benzoesäureanhydrids durch Anilin z. B. verlief in beiden möglichen Richtungen; zudem erwies sich die Trennung der beiden Reaktionsprodukte (Benzanilid und TFA-Alanin-anilid) als undurchführbar.

Über die peptid-synthetische Verwendung symmetrischer TFA-Aminosäureanhydride (s. S. 536), der aus TFA-Aminosäuren und Aldehyden darstellbaren N-TFA-Oxazolidone (s. S. 524).

Eine zusätzliche Darstellungsmethode für TFA-Aminosäuren beschreiben neuerdings CALVIN und Mitarb. (*358*); als acylierendes Agens verwenden sie Trifluoressigsäure-thioäthylester (*224*) in wäßriger Lösung bei p_H 8—9. Unter diesen Bedingungen gelingt es, u. a. Tyrosin, Tryptophan, Asparagin und Arginin glatt in ihre N_α-TFA-Verbindungen zu überführen. Als wichtig ist ferner die geglückte direkte ω-Acylierung der freien Aminosäuren Lysin und Ornithin zu verzeichnen (vgl. auch S. 479).

e) Die „Lactam-Schutzgruppen".

Schon PLÖCHL (*332*) hatte eine rasche Lactam-cyclisierung bei *o*-Aminophenyl-glycin, THATE (*412*) bzw. JACOBS und HEIDELBERGER (*249*) eine solche von *o*-Aminophenoxy-essigsäure beobachtet, die neuerdings HOLLEY und HOLLEY in einer Reihe von Arbeiten für den Auf- und Abbau von Peptiden verwertet haben (*234, 235, 236*). Als Aminoschutzgruppen werden von diesen Autoren empfohlen:

α) *Der 2-Nitrophenoxy-acetyl-rest.* (2-Nitrophenoxy)-acetyl-aminosäuren lassen sich nach der Azid- oder „Anhydrid"-methode mit Aminosäureestern umsetzen; nach Reduktion der *o*-Nitro- zur *o*-Amino-gruppe wird die cyclisierende Entacylierung durch Erwärmen mit Wasser auf 100° unter Freiwerden des gewünschten Peptids erreicht.

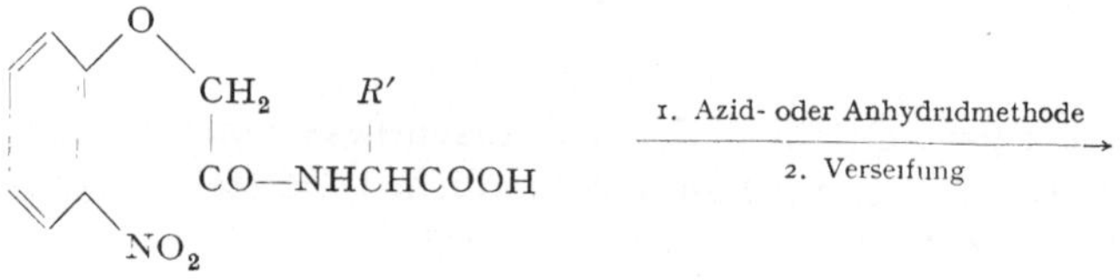

$$\begin{array}{c}\text{CH}_2 \quad R' \\ | \qquad | \\ \text{CO—NHCHCOOH}\end{array}$$

1. Azid- oder Anhydridmethode
$\longrightarrow$
2. Verseifung

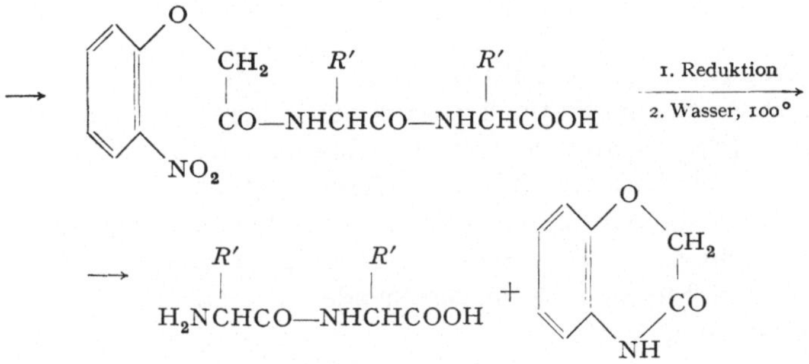

$$\longrightarrow \quad \text{(2-Nitro-phenyl-glycyl-Struktur)} \quad CO-NHCHCO-NHCHCOOH \qquad \text{1. Reduktion} \atop \text{2. Wasser, 100°}$$

$$\longrightarrow \quad H_2NCHCO-NHCHCOOH \quad + \quad \text{(Lactam-Struktur)}$$

$R' =$ Aminosäure-Seitenkette.

β) Der (2-Nitro-4-carbomethoxy-phenyl-)-glycyl-rest. Diese Gruppierung, die mittels 2-Nitro-4-carbomethoxy-phenyl nach SANGER eingeführt werden kann, dient praktisch jedoch nur dem systematischen Abbau von Peptiden.

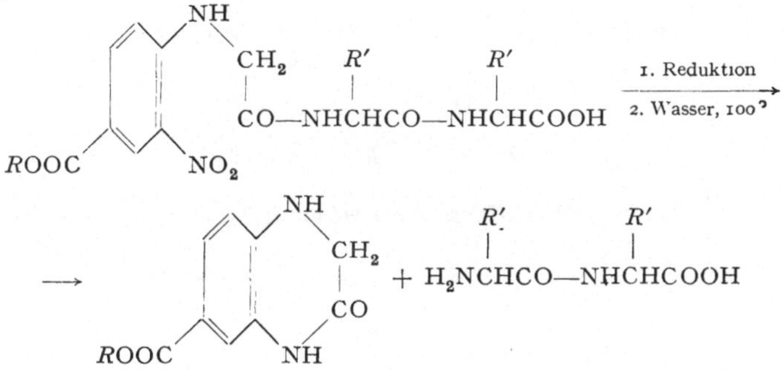

$$ROOC,\ NO_2\text{-}\text{phenyl}\quad CH_2\text{-}CO-NHCHCO-NHCHCOOH \qquad \text{1. Reduktion} \atop \text{2. Wasser, 100°}$$

$$\longrightarrow \quad \text{(Lactam-Struktur, ROOC)} \quad + \quad H_2NCHCO-NHCHCOOH$$

$R' =$ Aminosäure-Seitenkette. $R =$ Methyl.

γ) Der „Chloracetyl- (2-amino-phenyl-)-glycyl-rest". Chloracetylpeptide, nach dem FISCHERschen Verfahren des öfteren hergestellt (S. 501), lassen sich vom Chloracetylrest befreien, indem man sie mit *o*-Phenylendiamin zum *o*-Aminophenyl-glycylpeptid umsetzt und durch anschließendes Erhitzen mit Wasser die „Lactam-cyclisierung" auslöst:

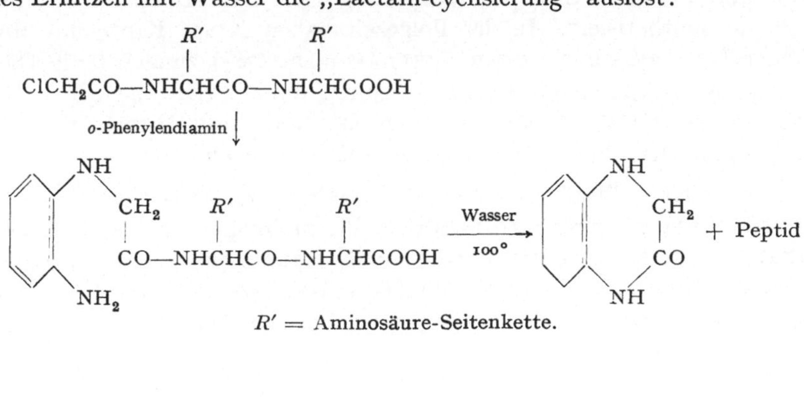

$$ClCH_2CO-NHCHCO-NHCHCOOH$$

o-Phenylendiamin $\downarrow$

$$\text{(2-Amino-phenyl-glycyl-Struktur)}\quad CO-NHCHCO-NHCHCOOH \qquad \xrightarrow[100°]{\text{Wasser}} \qquad \text{(Lactam-Struktur)} \quad + \text{ Peptid}$$

$R' =$ Aminosäure-Seitenkette.

f) Der „Phthalyl-rest".

Die Aufspaltung der Phthalamido-gruppierung mit Hydrazinhydrat zum freien Amin und Phthalhydrazid wurde bereits von Radenhausen beobachtet (*341*), später von Ing und Manske (*246*) näher untersucht. Jedoch erst 1948 berichten Kidd und King (*267*) über eine neue Methode der Synthese von Glutamylpeptiden unter Verwendung des Phthalylrestes als leicht abspaltbare Schutzgruppe. Etwas später veröffentlichten Sheehan und Frank (*385*) nachstehenden allgemeingültigen Weg der Peptiddarstellung:

$$
\begin{array}{l}
\text{Phthalsäureanhydrid} + H_2NCHCOOH \;(R') \longrightarrow \text{Phthalyl-}N{-}CHCOOH\;(R') \longrightarrow \\[4pt]
\xrightarrow[\text{oder } PCl_5]{SOCl_2} \text{Phthalyl-}N{-}CHCOCl\;(R') \xrightarrow{\;H_2NCHCOOR''\;(R')\;} \\[4pt]
\longrightarrow \text{Phthalyl-}N{-}CHCO{-}NHCHCOOR''\;(R',\,R') \xrightarrow[\text{2. } HCl,\,Na_2CO_3 \text{ oder Essigsäure}]{\text{1. } H_2N{-}NH_2,\,H_2O} \\[4pt]
\longrightarrow \text{Phthalhydrazid (NH-NH)} + H_2NCHCO{-}NHCHCOOR''\;(R',\,R')
\end{array}
$$

R' = Aminosäure-Seitenkette. R'' = H bzw. Na.

Nach diesem Schema haben Sheehan und Frank (*385*) sowie Grassmann und Schulte-Uebbing (*202*) einige Di- und Tripeptide in guter Ausbeute synthetisiert. In der Folgezeit haben King, Kidd und Mitarbeiter (*272, 271, 273*), sowie Sheehan u. a. (*384*) zunächst die Darstellung der Ausgangs-phthalylaminosäuren unter Erhaltung der vollen optischen Aktivität geklärt [s. auch Sheehan und Mitarb. (*386a*)]. Die Vorteile der Phthalylmethode gegenüber den anderen liegen in:

1. hoher Kristallisationsneigung der Ausgangs- und Zwischenprodukte 2. glatter Entacylierung auch S-haltiger Verbindungen mittels Hydrazinhydrat; 3. besserer Beständigkeit und Handhabung der Phthalylaminosäurechloride; 4. höchster Widerstandsfähigkeit gegenüber den zur Abspaltung der anderen Schutzgruppen gebräuchlichen Reagenzien (*77*).

Unter Verwendung des Carbobenzoxyhydrazid-rests als Carboxyl-maskierung wurde auch die CURTIUSsche Azidmethode zur Synthese von Phthalylpeptiden ermöglicht (*231*). In guter Ausbeute haben solche Peptide ferner BOISSONNAS und Mitarb. (*75, 369*), WIELAND u. a. (*466*), sowie KING u. a. (*269*) nach dem Anhydridverfahren hergestellt. SCHUMANN und BOISSONNAS (*369*) berichten ferner über die geglückte Verseifung von Phthalyl-peptidestern zu den Phthalyl-peptiden mittels verdünnter Natronlauge; demgegenüber weisen SHEEHAN (*384*), sowie HANSON und ILLHARDT (*216*) auf eine erhebliche Alkaliempfindlichkeit des Phthalylrests hin. Besser bewährt sich nach SHEEHAN und Mitarb. (*386 a*) die Verseifung mit Aceton-Salzsäure. KING, KIDD und Mitarb. (*268, 269*) zeigen anderseits die Spaltung von Phthalylpeptidestern mit Hydrazinhydrat zu den entsprechenden freien Estern auf:

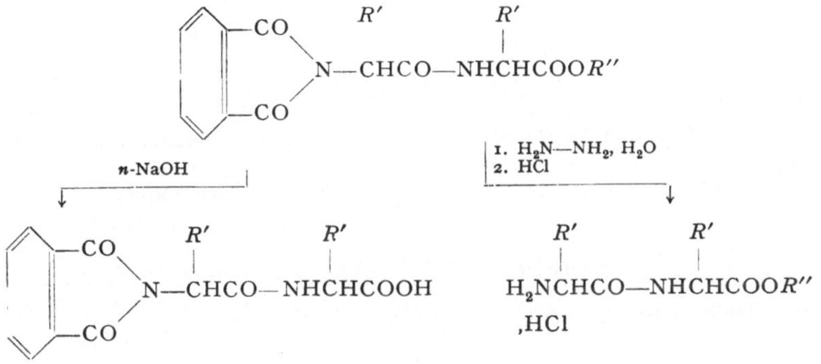

$R' = $ Aminosäure-Seitenkette. $R'' = $ Alkyl.

Den Phthalylrest mittels Phenylhydrazin als weniger agressives Reagenz zu entfernen, wird neuerdings von SCHUMANN und BOISSONNAS (*367, 368*) empfohlen.

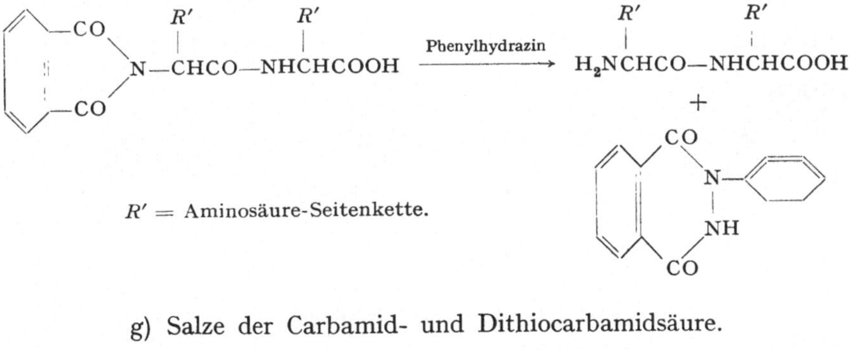

$R' = $ Aminosäure-Seitenkette.

g) Salze der Carbamid- und Dithiocarbamidsäure.

Die systematische Peptidsynthese mittels N-Carbonsäureanhydriden von Aminosäuren bzw. ihrer Dithio-analoga (*33, 107*) basiert auf einer Vereinigung von Aminoschutzgruppe und eines energiereichen Anhydrids in einem cyclischen System. Nach der ersten Stufe der Aminolyse sind, wie aus dem folgenden Schema ersichtlich, als α-Aminoschutzgruppe die Ammonsalze der N-Carbonsäure (bzw. Dithio-analogon) mit tertiären Basen erkennbar, die bei der anschließenden Behandlung — Erwärmen auf Zimmertemperatur oder Umsatz mit alkoholischer Salzsäure — unter Freiwerden des Peptidesters zersetzt werden.

$$
\begin{array}{l}
\overset{\displaystyle H_2C-CO}{\underset{\displaystyle HN-CO}{\Big\rangle}} O + H_2NCH\overset{R'}{|}COOR'' \longrightarrow CH_2CO-NH\overset{R'}{|}CHCOOR'' \longrightarrow \\[2mm]
\qquad\qquad\qquad\qquad\qquad\qquad\qquad\qquad\quad NH-COOH,\ N(R''')_3
\end{array}
$$

$$
\xrightarrow[\ -CO_2\ ]{\ Erwarmen\ } CH_2CO-NH\overset{R'}{|}CHCOOR'' + N(R''')_3
$$
$$
\qquad\qquad\qquad\quad NH_2
$$

$$
\overset{\displaystyle H_2C-CO}{\underset{\displaystyle HN-CS}{\Big\rangle}} S + H_2NCH\overset{R'}{|}COOR'' \longrightarrow CH_2CO-NH\overset{R'}{|}CHCOOR'' \longrightarrow
$$
$$
\qquad\qquad\qquad\qquad\qquad\qquad\qquad\quad NH-CSSH,\ N(R''')_3
$$

$$
\xrightarrow{\ HCl\ } HCl,\ H_2NCH_2CO-NH\overset{R'}{|}CHCOOR'' + CS_2 + (R''')_3N,HCl
$$

R' = Aminosäure-Seitenkette. R'' = Ester-alkyl. R''' = tert. Basen-alkyl.

h) Der Pyrrolidon-ring.

Eine bereits von der Natur vorgezeichnete Blockierungsmöglichkeit der α-Aminogruppe ist der Glutaminsäure eigentümlich. Unter Wasserabspaltung — die z. B. schon beim Kochen der wäßrigen Lösung vor sich

$$
\begin{array}{l}
CH_2CH_2COOH \\
| \\
CH-NH_2 \\
| \\
COOH
\end{array}
\longrightarrow
\begin{array}{l}
CH_2\!-\!\!-CH_2 \\
|\qquad\ | \\
CO\quad CH-COOH \\
\ \diagdown\ \diagup \\
\quad NH
\end{array}
\longrightarrow Saurechlorid\ oder\ \text{-}azid \longrightarrow
$$

$$
\xrightarrow{\ H_2NCH\overset{R'}{|}C'OOR''\ }
\begin{array}{l}
CH_2\!-\!\!-CH_2 \\
|\qquad\ | \\
CO\quad CH-CO-NH\overset{R'}{|}CHCOOR'' \\
\ \diagdown\ \diagup \\
\quad NH
\end{array}
\xrightarrow{\ alkohol.\ HCl\ }
$$

$$
\longrightarrow
\begin{array}{l}
CH_2CH_2COOR'' \\
| \\
CH\,CO-NH\,CHCOOR'' \\
|\qquad\qquad\quad | \\
HCl,\,NH_2\qquad R'
\end{array}
$$

R' = Aminosäure-Seitenkette. R'' = Alkyl.

geht — tritt Lactam-Ringschluß zwischen α-Amino- und γ-Carboxyl-gruppe ein; die Pyrrolidoncarbonsäure läßt sich nunmehr mittels der Säurechlorid-, -azid- oder dergleichen Methode leicht in einen Pyr-rolidonyl-aminosäure- bzw. -peptidester überführen (*340*). Anschließende Ringöffnung durch Behandeln mit alkoholischer Salzsäure in der Wärme ergibt den gewünschten α-Glutamyl-(-γ-ester)-peptidester in Form des Hydrochlorids. Boothe und Mitarb. (*23*) machen hiervon bei der Darstellung von α-Glutamylglutaminsäure-triäthyl-, und dem nächst höheren homologen -tetraäthylester Gebrauch.

Über Tosyl-pyrrolidoncarbonsäure vgl. S. 493.

j) Der p-Toluolsulfonyl-(,,Tosyl''-)-rest.

Die Entdeckung Fischers (*148*), daß p-Tosyl-aminosäuren durch Reduktion mit Jodwasserstoff und Phosphoniumjodid das Ausgangs-produkt unverändert zurückgeben, machte sich Schönheimer (*362*) zu-nutze. Er konnte zeigen, daß unter den Bedingungen der Entfernung des Tosylrests die Peptidbindung nicht angegriffen wird. Damit war eine geeignete Schutzgruppe für die Säurechlorid- und -azidmethode gefunden.

$$H_3CC_6H_4SO_2{-}NHCHCON_3 + H_2NCHCOOR'' \longrightarrow$$
$$\underset{R'}{|} \qquad\qquad \underset{R'}{|}$$

$$\longrightarrow H_3CC_6H_4SO_2{-}NHCHCO{-}NHCHCOOR'' \xrightarrow[\text{HJ} + \text{PH}_4\text{J}]{\text{Verseifen}}$$
$$\underset{R'}{|} \qquad\quad \underset{R'}{|}$$

$$\longrightarrow H_2NCHCO{-}NHCHCOOH + H_3CC_6H_4SH$$
$$\underset{R'}{|} \qquad\quad \underset{R'}{|}$$

R' = Aminosäure-Seitenkette. R'' = Alkyl.

Du Vigneaud und Behrens (*426*) zeigten, daß sich der Tosyl-, analog dem Carbobenzoxy-rest, reduktiv mit Natrium in flüssigem Ammoniak abspalten läßt. Auf Grund dieser Tatsache sieht neuerdings die Toluol-sulfonyl-schutzgruppe einer erfolgreichen Verwendung entgegen (*477, 130, 407, 262, 431, 333*), vor allem bei der Darstellung solcher Peptide, bei denen mehrere (zusätzliche) funktionelle Gruppen zeitlich verschieden geschützt werden sollen. Interessant ist eine Feststellung von Hillmann (*228*), wonach Tosyl-glycyl-(*TSG*)-aminosäuren und -peptide beim Er-wärmen mit 4—5 n-alkoholischer Salzsäure unter Abspaltung des TSG-restes Aminosäure- bzw. Peptidester-hydrochloride in 60—80%iger Aus-beute ergeben. In der gleichen Arbeit wird auf die Nichtverwendbarkeit der ,,Anhydrid-methoden'' bei N-Tosyl-aminosäuren hingewiesen, da der stärker acide Wasserstoff der Sulfonamid-gruppe zu Nebenreaktionen Anlaß gibt (*228*).

k) Phosphatamide.

GOLDSCHMIDT und Mitarb. (*196, 194*) konnten zeigen, daß sich substituierte Phosphatamide nur unter sehr energischen Bedingungen, z. B. in der Schmelze über 150°, der Umsetzung mit Carbonsäuren zum Säureamid unterwerfen lassen (vgl. S. 529). ZERVAS (*479 a*) hat diese Reaktionsträgheit der substituierten Dialkyl-phosphorsäureamide als α-Aminoschutzgruppe ausgenutzt. N-Di-(*p*-nitrobenzyl)-phosphataminosäuren (XVIII) z. B. lassen sich mit Hilfe der FISCHERschen Säurechloridmethode mit Aminosäuren zu den substituierten Peptiden (XIX) kuppeln. Die Abspaltung des Phosphatrests gelingt durch katalytisch erregten Wasserstoff in Gegenwart von Wasserstoffionen, da die P—N-Bindung des zuerst erhaltenen Phosphorsäure-monoamids (XX) bei p_H 4 sehr leicht hydrolysiert wird.

$$(O_2NC_6H_4CH_2O)_2{-}\overset{\displaystyle O}{P}{-}NH\overset{\displaystyle R'}{CH}COCl + H_2N\overset{\displaystyle R'}{CH}COOH \xrightarrow{-HCl}$$

(XVIII.)

$$\longrightarrow (O_2NC_6H_4CH_2O)_2{-}\overset{\displaystyle O}{P}{-}NH\overset{\displaystyle R'}{CH}CO{-}NH\overset{\displaystyle R'}{CH}COOH \xrightarrow{H_2/Pd}$$

(XIX.)

$$\longrightarrow (HO)_2\overset{\displaystyle O}{P}{-}NH\overset{\displaystyle R'}{CH}CO{-}NH\overset{\displaystyle R'}{CH}COOH \xrightarrow{H^+} H_2N\overset{\displaystyle R'}{CH}CO{-}NH\overset{\displaystyle R'}{CH}COOH$$

(XX.) $\qquad\qquad\qquad + H_3PO_4$

R' = Aminosäure-Seitenkette.

2. Die „Alkyl-blockierung".

a) Mono- und Dibenzyl-aminosäuren.

Die Verwendung von N-Benzyl-*DL*-asparaginsäure (*169*) zur Synthese von Asparagyl-peptiden beschreiben erstmals LIWSCHITZ und ZILKHA (*292*). Über ein gemischtes Anhydrid mit Chlorkohlensäure (XXI) einerseits bzw. ein daraus entstehendes „innermolekulares Anhydrid" (XXII) anderseits erfolgt Peptidknüpfung mit Aminosäureestern. Nach Verseifung läßt sich die N-Benzylgruppe mit $PdCl_2$-C-Katalysator bei 60—70° und geringem Druck hydrogenolytisch abspalten. Gemäß nachstehendem Schema werden die Synthesen von α-Asparagyl-peptiden aus (XXI), der β-Isomeren aus (XXII) beschrieben:

$$\begin{array}{ccc}
CH_2{-}COOH & CH_2{-}COOH & CH_2{-}COOH \\
| & | & | \\
CH{-}COOH \xrightarrow{COCl_2} & CH{-}CO{-}O{-}COCl \xrightarrow{H_2NR} & CH{-}CO{-}NH{-}R \\
| & | & | \\
NH{-}Bz & NH{-}Bz & NH{-}Bz
\end{array}$$

(XXI.)

$$\downarrow \begin{array}{l} {-}\,CO_2 \\ {-}\,HCl \end{array} \qquad\qquad \downarrow \begin{array}{l} \text{1. Verseifung} \\ \text{2. } H_2/Pd{-}C \end{array}$$

$$
\begin{array}{ccc}
\begin{matrix}
\mathrm{CH_2CO-NH}-R \\
| \\
\mathrm{CH-COOH} \\
| \\
\mathrm{NH}-Bz
\end{matrix}
& \longleftarrow \quad
\begin{matrix}
\mathrm{CH_2-CO} \\
\quad\quad\quad\diagdown \\
| \quad\quad\quad\quad \mathrm{O} \\
\quad\quad\quad\diagup \\
\mathrm{CH-CO} \\
| \\
\mathrm{NH}-Bz
\end{matrix}
&
\begin{matrix}
\mathrm{CH_2-COOH} \\
| \\
\mathrm{CHCO-NHCHCOOH} \\
| \\
R'
\end{matrix}
\end{array}
$$

(XXII.)

1. Verseifung
2. $\mathrm{H_2/Pd-C}$

$$
\begin{matrix}
\mathrm{CH_2CO-NHCHCOOH} \\
\vdots \quad\quad\quad\quad \vdots \\
\mathrm{CHCOOH} \quad R' \\
| \\
\mathrm{NH_2}
\end{matrix}
$$

$$R = \mathrm{-CHCOOR''}. \quad R' = \text{Aminosäure-Seitenkette.}$$
$$R''' = \text{Alkyl.} \quad Bz = \mathrm{C_6H_5CH_2}.$$

Unabhängig von den letztgenannten Autoren haben VELLUZ und Mitarb. (*422*) die Verwendung des N-Benzyl- bzw. N,N-Dibenzylrests als α-Aminoschutzgruppe mit Erfolg bearbeitet. N,N-Dibenzyl- bzw. N-Benzylaminosäuren* lassen sich mittels Phosphorpentachlorid in Benzol in die Säurechlorid-hydrochloride überführen, die schließlich mit Aminosäureestern in wäßrigem Dioxan bei pH 7—8 unter Bildung von Dibenzyl-(Benzyl)-peptidester-hydrochloriden reagieren. [Dieser Reaktionsverlauf entspricht praktisch der Blockierung der α-Aminogruppe durch Salzbildung (S. 471); doch wird durch den zusätzlichen N-Benzylrest eine erhebliche Stabilität gegenüber den α-Aminosäurechlorid-hydrochloriden erreicht.] Die Verseifung zu den N-benzylierten Peptiden erfordert allerdings Erhitzen mit 20%iger methanolischer Kalilauge (vgl. dazu die alkalische Hydrolyse von N-Trityl-aminosäureestern, S. 470). Dagegen gelingt die Abspaltung der Aminoschutzgruppe glatt und in hervorragender Ausbeute mit katalytisch erregtem Wasserstoff (Pd-schwarz) in Eisessig bei 80°. Bei N,N-Dibenzylaminosäuren war auch eine Peptidknüpfung mittels der WIELAND-BOISSONNAS-VAUGHANschen Anhydridmethode mit sehr gutem Erfolg möglich (*422 b*). Nach diesen beiden Verfahren haben VELLUZ und Mitarb. (*422, 422 b*) mehrere Di- und Tripeptide synthetisiert.

b) N-Trityl-aminosäuren.

Die Darstellung von Trityl-aminosäuren wird erstmals von HELFERICH (*227*), später von HILLMANN und Mitarb. (*229*) be-

---

* Diese werden zum Beispiel durch Umsetzung von Aminosäureestern mit Benzaldehyd bei Gegenwart von wasserfreiem Natriumsulfat, folgende Hydrierung der SCHIFFschen Base und anschließende alkalische Verseifung der N-Alkylester erhalten. Weitere Darstellungsmöglichkeiten für Benzylderivate vgl. (*422*), für N,N-Dibenzylaminosäuren ANATOL und Mitarb. (*17*), VELLUZ und Mitarb. (*422 a*).

schrieben. Die Umsetzung von Tritylchlorid mit zwei Äquivalenten Aminosäureester in Benzol verläuft nahezu quantitativ, während die folgende alkalische Verseifung zu den freien Säuren nur teilweise befriedigend ist.

$$(C_6H_5)_3CCl + 2\ H_2NCHCOOR'' \longrightarrow$$
$$R'$$

$$\longrightarrow (C_6H_5)_3C{-}NHCHCOOR'' \xrightarrow[\text{2. Weinsäure}]{\text{1. 30\%-ige NaOH/Propylenglykol}} (C_6H_5)_3C{-}NHCHCOOH$$
$$R' \qquad\qquad\qquad\qquad\qquad\qquad\qquad\qquad\qquad R'$$

$$R' = \text{Aminosäure-Seitenkette.} \quad R'' = \text{Alkyl.}$$

Eine Vereinfachung der Trityl-aminosäureester-darstellung gibt neuerdings ZERVAS (*480*) an; Aminosäureester-hydrochloride reagieren glatt mit Tritylchlorid in Gegenwart von Triäthylamin. Die Schwierigkeit der folgenden Verseifung nimmt, wie ZERVAS weiter zeigen konnte, vom Glycin- zum Phenylalanin-derivat enorm zu, mit steigender Peptidkettenlänge dagegen ab. Für eine Hydrazidbildung gelten analoge Verhältnisse. ZERVAS umgeht diese Klippe durch direkte Tritylierung von Aminosäuren in Isopropanol-Wasser bei 20° in Gegenwart von Diäthylamin. Die Verwendung von Trityl-aminosäuren (XXIII) zur Peptidsynthese

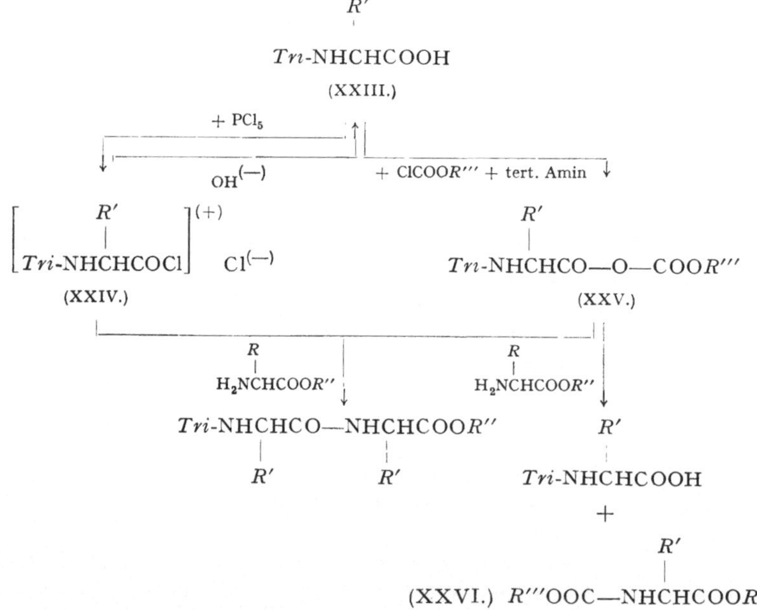

Tri = Trityl. *R'* = Aminosäure-Seitenkette. *R''* = Esteralkyl der Aminosäure-Komponente. *R'''* = Esteralkyl der Kohlensäure.

gelingt nach Zervas (*480*) in guter Ausbeute über Tritylaminosäure-chlorid-hydrochloride (XXIV), die sich in indifferenten Lösungsmitteln mittels Phosphorpentachlorid herstellen lassen, sowie über die gemischten Kohlensäureanhydride (XXV) nach Wieland-Boissonnas-Vaughan. Zervas konnte zeigen, daß die Umsetzung mit Aminosäureestern nach letzterem Verfahren wegen Hinderung durch den Tritylrest nur beim Glycinderivat erfolgreich ist [vgl. auch Hillmann (*229*)], während beim Phenylalanin-analogen Carbäthoxy-aminosäureester (XXVI) entstehen. Wiederum führt eine Verlängerung der Peptidkette (schon ab Trityl-dipeptid) eindeutig zur Bevorzugung der gewünschten Amidbildung.

Die Abspaltung des Tritylrests läßt sich unter schonenden Bedingungen durch Säuren sowie auch mittels katalytisch erregtem Wasserstoff leicht bewerkstelligen; für eine einwandfreie Synthese von Peptidthio-ester-hydrochloriden hat diese Möglichkeit große Bedeutung (*229*).

c) Die „Schiffschen Basen".

Aus Arbeiten der Bergmann-Schule (*45, 53*) sind seit langem Erd-alkali- bzw. Alkaloidsalze von Alkyliden- bzw. Aryliden-verbindungen der Aminosäuren bekannt. Erstmals haben Wieland und Mitarb. (*463*) Kaliumsalze der Benzyliden-derivate rein hergestellt und diese auch zu Umsetzungen herangezogen, die in unmittelbarer Verwandtschaft zu Peptidknüpfungs-methoden stehen. Mit Chlorkohlensäureestern gelang die Bildung eines gemischten Anhydrids, das mit Thiophenol zum S-Phenylester weiterreagierte. Nach Ansäuern wurde sofort das Thio-ester-hydrochlorid erhalten.

$$C_6H_5CH{=}N{-}CH_2COOK + ClCOOR'' \xrightarrow{\;-\;KCl\;} C_6H_5CH{=}N{-}CH_2COO{-}COOR'' \longrightarrow$$

$$\xrightarrow{HSC_6H_5} C_6H_5CH{=}N{-}CH_2CO{-}S{-}C_6H_5 \xrightarrow{\;HCl\;} H_2NCH_2CO{-}S{-}C_6H_5 + C_6H_5CHO$$
$$,HCl$$

$$R'' = \text{Alkyl.}$$

3. Die „Ammonsalzbildung".

Diese Salzbildung an der α-Aminogruppe ermöglicht die Umsetzung der von Fischer (*139*) 1905 aufgefundenen Aminosäurechlorid-hydro-chloride mit einer freien „Aminokomponente", z. B. mit einem Amino-säure- oder Peptidester in indifferenten Lösungsmitteln, nicht jedoch mit Salzen von Aminosäuren in wäßriger Lösung. Im letzteren Falle würde als erstes die Blockierung der NH_2- als $NH_3{}^+$-Gruppierung aufgehoben; auf der anderen Seite scheint die Zersetzungstendenz der Aminosäure-chloride in wäßrig-alkalischer Lösung doch auch so enorm zu sein, daß in diesem Falle eine systematische Säureamidbildung unmöglich wird.

Vgl. auch S. 469, über die Verwendung der Ammonsalze von N-Benzyl-, N,N-Dibenzyl- bzw. N-Trityl-aminosäurechloriden zur Peptidsynthese.

B. Die nachträgliche Einführung der α-Aminogruppe.

1. α-Halogen-acyl-verbindungen.

In Ermangelung von leicht wieder abspaltbaren α-Aminoschutzgruppen hatten Fischer und Otto (*159*) die Halogenfettsäuren in die Peptidsynthese einbezogen. Die im Enderfolg stets erhaltenen α-Halogenacyl-aminosäuren bzw. -peptide tauschen beim Behandeln mit konz. Ammoniak ihr Halogen gegen die Aminogruppe aus, geben also die gewünschten Peptide.

Leider ist dieser Austauschvorgang von Nebenreaktionen in mehr oder weniger umfassendem Maße begleitet. Neben den α-Amino- können sich disubstituierte Imino-, α-Oxy- und (unter Halogenwasserstoff-abspaltung) α,β-Dehydroderivate bilden, die stets, und das ist das besonders Unerfreuliche, ninhydrin-negative Substanzen darstellen, so daß ihre Anwesenheit durch die lange Zeit üblichen chromatographischen Untersuchungsmethoden nicht festgestellt werden konnte. So berichtet u. a. bereits die Fischersche Schule über „Ausnahmefälle" bei der Aminierung von α-*D*-Brom-isocapronyl-*L*-prolin (*160*) und -N-phenylglycin (*154*) bzw. α-Brom-β-phenylpropionyl-glycin (*150*), wobei α-Oxyacyl-aminosäureamide bzw. bei letzterem ein Zimtsäurederivat entstanden. Der Heranziehung von flüssigem an Stelle von wäßrigem Ammoniak zur Aminierung dürfte zur Vermeidung einer Hydrolyse und zur Unterdrückung der Bildung von sekundären Aminen der Vorzug zu geben sein. (Bezüglich der Herstellung optisch-aktiver Peptide vgl. S. 501.)

2. α-Azido-acyl-verbindungen.

Die Verwendung von α-Azido- an Stelle der α-Halogenfettsäuren (*395, 173, 69*) ist vom präparativen Standpunkt weitaus günstiger, da durch katalytische Hydrierung die Überführung in die α-Aminogruppe glatt gelingt. Die soeben genannten Nebenprodukte scheiden hier aus.

$$\overset{R'}{\underset{|}{N_3CHCO}}-\overset{R'}{\underset{|}{NHCHCOOH}} \quad \xrightarrow[-N_2]{H_2/Pd} \quad \overset{R'}{\underset{|}{H_2NCHCO}}-\overset{R'}{\underset{|}{NHCHCOOH}}$$

$R' =$ Aminosäure-Seitenkette.

Wegen der umständlichen Herstellung der α-Azido- aus den entsprechenden α-Halogen-verbindungen blieb diese Methode bislang auf die Synthese von Aminosäuren beschränkt; nur vereinzelt findet sie Eingang in peptidsynthetische Arbeiten (*239*). Neuerdings berichtet eine amerikanische Arbeitsgruppe (*315*) über die Darstellung von O-Glycyl-serin aus O-Azidoacetyl-serin.

3. α-Keto-acyl-verbindungen.

Fußend auf der Aminosäuresynthese von Hartung (*222*) haben Shemin und Herbst (*391*) α-Keto-acyl-aminosäuren in Form ihrer Oxime durch katalytische Hydrierung in freie Peptide übergeführt:

$$R'-\underset{\underset{\textstyle NOH}{\|}}{C}-CO-NH\underset{\underset{\textstyle R'}{|}}{CH}COOH \quad \xrightarrow{H_2/Pd} \quad H_2NCH\underset{\underset{\textstyle R'}{|}}{CO}-NH\underset{\underset{\textstyle R'}{|}}{CH}COOH$$

$$R' = \text{Aminosäure-Seitenkette.}$$

HARTUNG und Mitarb. (*438*, *439*) schlagen einen etwas anderen Weg ein. Die von ihnen bevorzugten α-(O-Benzyl-oximino)-acyl-derivate (XXVII) lassen sich gleichfalls katalytisch hydrieren, wobei jedoch die Gegenwart von Wasserstoffionen hauptsächlich zur Bildung von Diketopiperazinen (XXVIII) neben nur sehr wenig Peptid (XXIX) Anlaß gibt. Die Ausführung der Hydrierung in ammoniakalischem Medium hingegen (*223*) liefert das gewünschte freie Peptid (XXIX).

$$R'-C=NO-CH_2C_6H_5$$
$$|$$
$$CO-NHCHCOOH$$
$$|$$
$$R'$$

(XXVII.)

$$\xleftarrow[\text{H}^{(+)}]{H_2/Pd-C} \qquad \xrightarrow[\text{NH}_4\text{OH}]{H_2/Pd-C}$$

$$R'-CH-NH-CO \qquad\qquad H_2NCHCO-NHCHCOOH$$
$$|\qquad\qquad| \qquad\qquad\qquad\qquad | \qquad\qquad |$$
$$CO-NH-CH-R' \qquad\qquad\qquad R' \qquad\qquad R'$$

(XXVIII.) $\qquad\qquad\qquad\qquad$ (XXIX.)

$$R' = \text{Aminosäure-Seitenkette.}$$

4. α,β-Ungesättigte Acyl-verbindungen.

FISCHER und KOENIGS (*156*) gelang die Synthese von α- bzw. β-Asparagyl-peptiden, als sie den aus Fumarsäurechlorid und Glycinester erhaltenen Fumaryl-bis-glycinester nach Verseifung mit konz. Ammoniak bei 100° behandelten. Über eine durchsichtigere Synthese gleicher Art berichten

$$\begin{matrix} CH-CO \\ \| \qquad\quad \diagdown \\ \qquad\qquad O \;+\; H_2NCHCOOR'' \\ \| \qquad\quad \diagup \qquad\qquad | \\ CH-CO \qquad\qquad\quad R' \end{matrix} \longrightarrow \begin{matrix} CH-CO-NHCHCOOR'' \\ \| \qquad\qquad\qquad | \\ CH-COOH \qquad R' \end{matrix} \longrightarrow$$

(XXX.)

$$\xrightarrow{C_6H_5CH_2NH_2} \begin{matrix} R' \\ | \\ CH_2CO-NHCHCOOR'' \\ | \\ CH-COOH \\ | \\ NH-CH_2C_6H_5 \end{matrix} \xrightarrow[\text{2. } H_2/Pd]{\text{1. Verseifung}} \text{(XXXI.)}$$

$$R' = \text{Aminosäure-Seitenkette.} \quad R'' = \text{Alkyl.}$$

neuerdings Liwschitz und Zilkha (*292*). Maleinsäureanhydrid setzt sich mit Aminosäureestern zu einem Fumarsäure-halbamid (XXX) um; aus diesem wird durch Anlagerung von Benzylamin, anschließende Verseifung und Abhydrierung des N-Benzylrestes ein Peptid der *DL*-Asparaginsäure (XXXI) erhalten.

C. Leicht abspaltbare α-Carbonsäure-Schutzgruppen.

1. Ester und Alkalisalze.

Das Problem eines Schutzes der Carboxylgruppe im engeren Sinn, also ihre Blockierung zu dem Zwecke, sie einer unerwünschten Reaktion zu entziehen, tritt sichtbar nur in Erscheinung bei der Synthese von Peptiden der Aminodicarbonsäuren. In den Halbestern dieser Säuren, die auf verschiedenem Wege herstellbar sind (vgl. SS. 491, 526), kann die *freie* Carboxylgruppe durch Umsetzung mit Isocyanat oder Phosphorazo-aminosäureestern oder durch Umwandlung in das Säurechlorid und folgende Kupplung zum Peptid verknüpft werden, während die veresterte Carboxylgruppe sich nicht an der Reaktion beteiligt. Umgekehrt ist die *veresterte* Carboxylgruppe auf dem Wege über Hydrazid und Azid oder, falls es sich um Ester genügend hoher Reaktionsfähigkeit handelt, auch direkt der Peptidverknüpfung zugänglich, während die freie Carboxyl-gruppe an diesen Reaktionen nicht teilnimmt. Gerade das Beispiel der Veresterung macht aber die in Kap. I erwähnte doppelte Funktion solcher „Schutzgruppen" besonders deutlich. Denn die wesentliche und all-gemeine Bedeutung der Veresterung — insbesondere der α-Carboxyl-gruppe — für die Peptidsynthese liegt in der Rückwirkung, die sie auf die Reaktionsfähigkeit der Aminogruppe ausübt. Erst nach der Aufhebung der Zwitterionenform durch die Veresterung der Carboxylgruppe ist die freie Aminogruppe für die nucleophile Addition verfügbar.

Die Verwendung der *Aminosäureester* zur Peptiddarstellung geht daher auf die Anfänge der synthetischen Arbeiten auf diesem Gebiet zurück.

Als Herstellungsvariationen für die Gewinnung der Ester-hydrochloride der Aminosäuren und der Peptide, aus denen die freien Ester nach verschiedenen lange bekannten Verfahren gewonnen werden können, seien die Veresterung mit HCl-Alkohol (*409*) oder mit Thionylchlorid-Alkohol (*86*) bzw. der Umweg über Carbobenz-oxy-aminosäureester [Veresterung der N-acylierten Säuren mit Diazoalkanen oder mit HCl-Alkohol (*204*)] oder über die N-Trityl-aminosäureester-hydrochloride (*480*) mit folgender Abspaltung der α-Aminoschutzgruppe oder schließlich die Ring-öffnung von N-Carbonsäureanhydriden mit alkoholischer HCl (*65, 82*) angeführt. Über Nebenreaktionen bei der Veresterung von Peptiden, insbesondere von solchen der Oxyaminosäuren, vgl. (*242*).

Die mit der Aufhebung der Zwitterionenstruktur verbundene erhöhte Reaktionsbereitschaft der Aminogruppe ist, mindestens in gewissem Grade, allerdings auch bei den *Alkalisalzen der Aminosäuren* gegeben und man kann daher diese ebenso wie die freien Aminosäure- bzw. Peptidester mit den verschiedensten reaktionsfähigen Derivaten acylierter Aminosäuren, z. B. den FISCHERschen Acylaminosäurechloriden, den Phosphor-, Schwefelsäure-, Alkylkohlensäureanhydriden, Thiophenylestern usw. unter Peptidknüpfung im wäßrigen Medium zur Umsetzung bringen. Dabei haben aber doch die Ester die größeren Vorzüge. Erlauben doch die Acylpeptidester (XXXII) stets eine saubere Abtrennung von den Ausgangsmaterialien auf Grund ihres neutralen Charakters in Verbindung mit der Löslichkeit in organischen Lösungsmitteln, speziell Essigester.

$$
\begin{array}{c}
R' \\
| \\
\text{CH—CO—X} \\
| \\
\text{NH—}Ac
\end{array}
\quad
\xrightarrow{\;\;H_2NCHCOOR''\;\;}
\quad
\begin{array}{c}
R' \qquad\qquad R' \\
| \qquad\qquad | \\
Ac\text{—NHCHCO—NHCHCOO}R'' + \text{H}X \\
\text{(XXXII.)}
\end{array}
$$

R' = Aminosäure-Seitenkette. R'' = Alkyl. Ac = Acyl. X = Anhydrid-komponente.

Vorsichtige Verseifung der reinen Acylpeptidester (XXXII) liefert das Acylpeptid, das nach Abspaltung der Aminoschutzgruppe das gesuchte Peptid in hervorragender Reinheit ergibt.

Da in letzter Zeit festgestellt wurde, daß bestimmte End-aminosäureester in Peptiden sich schwer (vor allem alkalisch) verseifen lassen (besonders die höheren Acyl-peptidester), und daß bei einem Überschuß an Alkali die Gefahr eines Angriffs an der α-Aminoschutzgruppe (speziell dem meistbenutzten Carbobenzoxyrest) gegeben ist (359, 207), geht man in steigendem Maße zu Peptidsynthesen mit Hilfe von *Benzylestern* über (65, 207, 218, 397, 215, 165, 82, 84, 359, 352, 437, 354, 419). Diese durch katalytische Hydrierung, aber auch mittels Natrium in flüssigem Ammoniak (346) leicht spaltbaren Ester sind teilweise durch die übliche Veresterung mit HCl-Benzylalkohol schon von ABDERHALDEN und SUZUKI (11) erhalten worden [s. auch SACHS und BRAND (353), DU VIGNEAUD und Mitarb. (345)]; weiters wurden dieselben bereitet: von FISCHER (141) aus Aminosäurechlorid-hydrochloriden, von MILLER und WAELSCH (304) bzw. ERLANGER und HALL (129) unter Verwendung von Benzolsulfo- bzw. Polyphosphorsäure als Veresterungsmittel, von BRAND und Mitarb. (127) durch Alkoholyse von N-Carbonsäureanhydriden bei HCl-Gegenwart und schließlich von BEN-ISHAI und BERGER (40) auf dem Weg Carbobenzoxy-aminosäure, -benzylester [s. auch BAER und MAURUKAS (31)], Aminosäure-benzylester-hydrobromid. α- bzw. γ-Benzylester acylierter Glutaminsäure lassen sich über deren inneres Anhydrid (66, 337) oder durch partielle Verseifung der entsprechenden Di-benzylester (215) gewinnen; das gleiche gilt für die Benzylester der N-Acyl-asparaginsäure (165). Für die direkte Darstellung des freien Glutaminsäure-α- bzw. -γ-benzylesters aus der Aminosäure

sind lediglich die von Sachs und Brand (*352*), Hanby und Mitarb.
(*215*), sowie Blout und Karlson (*73 a*) beschriebenen Methoden brauchbar.

2. N'-Phenylhydrazide.

Im Zusammenhang mit enzymatischen Peptidsynthesen aus Carbobenzoxy-aminosäuren und N-Aminoacyl-N'-phenylhydrazid konnten
Waldschmidt-Leitz und Kühn (*434*) zeigen, daß das nach Entfernen
des Carbobenzoxy-rests erhaltene N-Dipeptidyl-N'-phenylhydrazid
(XXXIII) durch Erwärmen mit Cuprisalzlösung unter Regenerierung
der Carboxylgruppe zum Peptid gespalten werden kann.

$$Cbo-NHCH_2CO-NHCH_2CO-NHNH-C_6H_5$$

$$\downarrow HJ/Eisessig$$

(XXXIII.)

$$HJ,H_2NCH_2CO-NHCH_2CO-NHNH-C_6H_5$$

$$\downarrow \begin{array}{l} -HJ \\ Cupriacetat,\ 95° \end{array}$$

$$H_2NCH_2CO-NHCH_2COOH$$

Cbo = Carbobenzoxy.

Als Varianten sollen sich auch Mono- und Dinitrophenylhydrazide verwenden
lassen (*434*).

3. N'-Carbobenzoxy-hydrazide.

Diese erstmals von Hofmann und Mitarb. (*231*) eingeführte
Carboxyl-schutzgruppe ist nur für die Darstellung von Acylpeptid-
hydraziden und N-Peptidyl-N'-carbobenzoxy-hydraziden brauchbar, nicht
aber für die Darstellung freier Peptide, da es nicht gelingt, den Hydrazid-
rest ohne Gefährdung der Peptidbindung abzuspalten. Doch wird
die Heranziehung dieser Zwischenprodukte für die Synthese höherer
Peptide sicher nicht erfolglos sein. Die Synthese von N-Aminoacyl-
N'-carbobenzoxy-hydraziden (XXXIV) gelingt über die Mercapto-

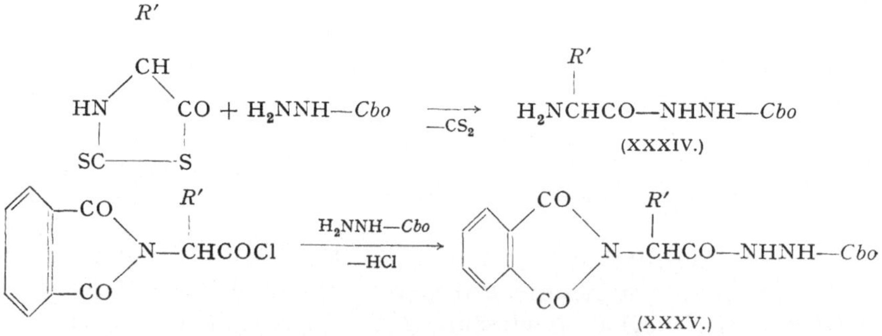

R' = Aminosäure-Seitenkette.   *Cbo* = Carbobenzoxy.

thiazolone, von N-Phthalylaminoacyl-derivaten (XXXV) aus Phthalyl-aminosäurechloriden.

Die Variationsmöglichkeiten bei der Verwendung gerade dieser Derivate (XXXV) sind leicht abzusehen, wenn man bedenkt, daß sowohl der Phthalyl- wie der Carbobenzoxyrest selektiv abspaltbar sind (*Formelübersicht 2*).

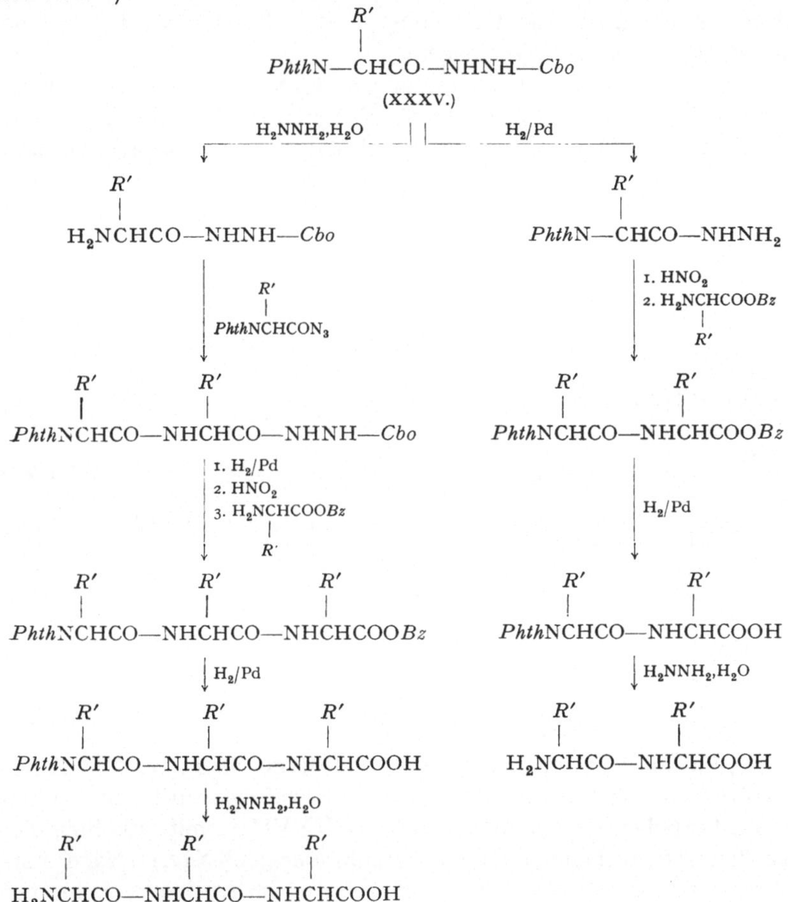

$R' =$ Aminosäure-Seitenkette. $Bz =$ Benzyl. $Cbo =$ Carbobenzoxy. $Phth =$ Phthalyl.

Formelübersicht 2.

D. „Mehrfunktionelle" Aminosäuren und ihre Einbeziehung in die Synthese.

1. Die ω-Aminogruppe.

Diese zusätzliche funktionelle Gruppe der Diaminosäuren — vor allem des Lysins und Ornithins — ist leicht und einwandfrei durch Überführung in ω-Acylamide zu blockieren, wobei als leicht abspaltbare Reste hauptsächlich Carbobenzoxy bzw. Tosyl gewählt wurden. Während die Verwendung von α,ω-Di-acyl-diaminosäuren als „Kopf"-Aminosäuren bei den Synthesen von Peptiden denen der Monoaminosäuren völlig analog ist, müssen dort, wo es gilt, Lysin, Ornithin und dgl. als mittel- bzw. endständig einzuführen, die Mono-ω-acylderivate bzw. deren Ester als Ausgangssubstanzen zur Verfügung stehen. Hierfür wurden bislang die drei Wege a) bis c) eingeschlagen.

a) $N_{\alpha,\,\omega}$-Dicarbobenzoxy-verbindungen (XXXVI) werden mittels PCl_5 oder PBr_3 in die Säurehalogenide übergeführt, die beim Erwärmen in indifferenten Lösungsmitteln unter Abspaltung von Benzylhalogenid

$$
\begin{array}{c}
HN\!-\!COOCH_2C_6H_5 \\
| \\
(CH_2)_n \\
| \\
CH \\
H_2N \diagup \ \diagdown COOH \\
(XXXVIII.)
\end{array}
$$

$\uparrow n\text{-HCl}$

$$
\begin{array}{ccccc}
HN\!-\!COOCH_2C_6H_5 & & HN\!-\!COOCH_2C_6H_5 & & HN\!-\!COOCH_2C_6H_5 \\
| & & | & & | \\
(CH_2)_n & \xrightarrow[-C_6H_5CH_2Cl]{PCl_5,\ 50^\circ} & (CH_2)_n & \xrightarrow[HCl]{alkohol.} & (CH_2)_n \\
| & & | & & | \\
CH & & CH & & CH \\
HN \diagup \ \diagdown COOH & & HN \diagup \ \diagdown CO & & ClH,\ H_2N \diagup \ \diagdown COOR'' \\
| & & | \quad\quad | & & \\
CO & & OC\!-\!-\!-\!O & & \\
| & & & & \\
OCH_2C_6H_5 & & & & \\
(XXXVI.) & & (XXXVII) & & (XXXIX.)
\end{array}
$$

$$R'' = \text{Alkyl.}$$

in die N_α-Carbonsäureanhydride (XXXVII) (vgl. S. 515) der N_ω-Acyl-aminosäuren übergehen. Die Aufspaltung mit wäßriger HCl führt zu den freien N_ω-Carbobenzoxy-diaminosäuren (XXXVIII), mit alkoholischer HCl zu den entsprechenden Esterhydrochloriden (XXXIX). Nach erstmals von Bergmann und Mitarb. (65) für ε-Carbobenzoxy-lysin und das zugehörige Methylester-hydrochlorid gegebenen Vorschriften [vgl. auch Brand und Mitarb (82); Waley und Watson (437)] wurden später auch die entsprechenden Ornithin-derivate synthetisiert (409, 261).

b) Kupfer-Komplexsalze (XL) des Lysins und Ornithins sind lange bekannt. Kurtz (279) hat darauf eine Isolierungsmethode für die erstere Aminosäure begründet, indem er den Kupfer-komplex in schwach alkalischem Medium mit Benzoylchlorid zum ε-Benzoyl-lysin-kupfer kuppelte. Neuberger und Sanger (313) haben erstmals ε-Carbobenz-

oxy-lysin-kupfer bzw. durch Zerlegen mit Schwefelwasserstoff die freie
ε-Acyl-verbindung hergestellt [vgl. SCHLÖGL und FABITSCHOWITZ (*359*)].

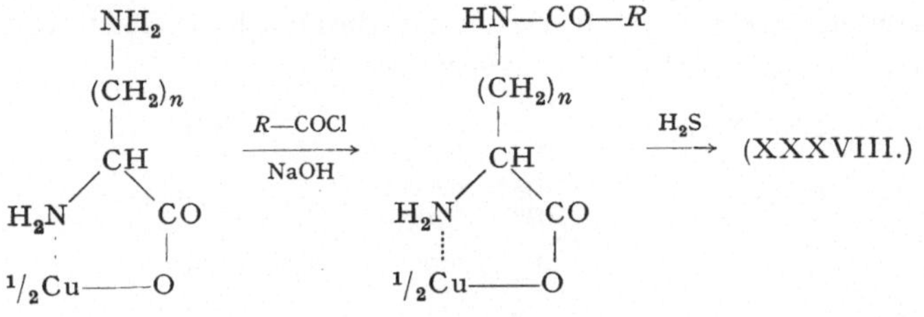

$$R = \text{Acyl-Seitenkette.}$$

SYNGE (*409*) sowie HARRIS und WORK (*221*) synthetisierten nach
diesem Verfahren δ-Carbobenzoxy-*L*-ornithin; BRAND und Mitarb.
(*130*) δ-Tosyl-*L*-ornithin, und STEVENS und WATANABE (*404*) ε-Carballyl-
oxy-*L*-lysin. Damit zeigt sich eine allgemeine Anwendbarkeit im Gegen-
satz zu Weg a), der auf Carboalkoxy-verbindungen beschränkt ist. Nach
SYNGE (*409*) gelingt die Veresterung von ω-Carbobenzoxy-aminosäuren
glatt mit alkoholischer HCl bei Zimmertemperatur.

c) Als dritten und neuesten Weg empfehlen SCHALLENBERG und
CALVIN (*358*) die Umsetzung von freiem Lysin bzw. Ornithin mit Tri-
fluoressigsäure-thioestern (XLI) zum ω-Acyl-derivat (XLII).

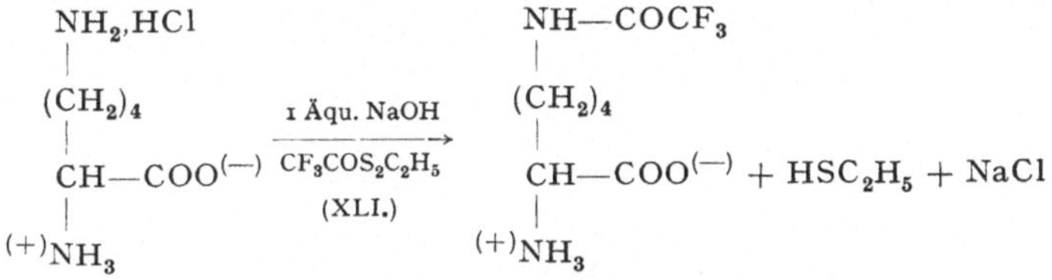

2. Die ω-Guanidogruppe.

Der natürliche Vertreter ist das Arginin, dessen Einbeziehung in die
Peptidsynthese ein sehr wichtiges, aber schwieriges Problem darstellt.

a) Die einzige bislang brauchbare Blockierungsmöglichkeit der
ω-Guanidogruppe ist durch ihre Nitrierung gegeben. N_ω-Nitroarginin
wurde erstmals von KOSSEL und KENNAWAY (*276*) synthetisiert; BERG-
MANN und Mitarb. (*64*) haben später gezeigt, daß diese NO_2-Gruppe
durch katalytische Hydrierung ohne Veränderung des Arginin-moleküls
abgespalten werden kann [vgl. auch VAN ORDEN und SMITH (*320*)].
Doch brachte die Verwendung dieser Substanz zur Synthese von Arginin-
peptiden nur teilweise Erfolg (Einsatz als End-aminosäure), da die Her-

stellung des Säurechlorids bzw. -hydrazids von N_α-Carbobenzoxy-N_ω-nitroarginin — wie später auch die von ω-acylierten Argininderivaten (*188*) — mißlang. Mit Heranziehung der „gemischten Anhydride" der Alkyl-kohlensäure konnte in der Folge diese Schwierigkeit überwunden werden (*232, 232a, 320*) (*Formelübersicht 3*).

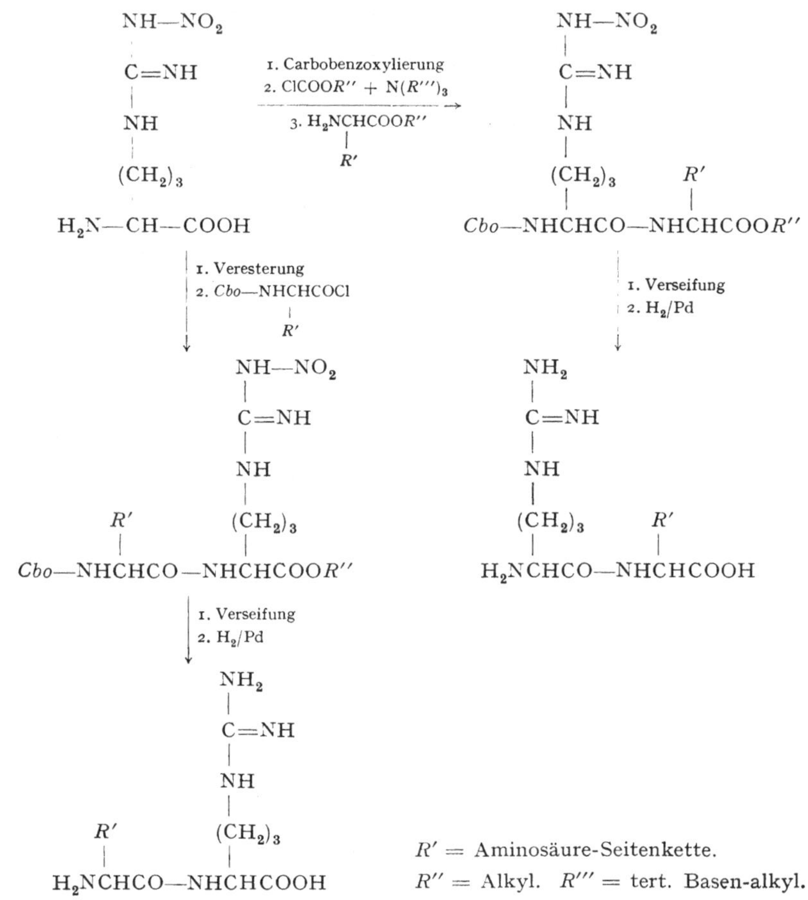

Formelübersicht 3.

b) Arginin mit völlig ungeschützter Guanido-gruppe wurde bereits von Bergmann (*48, 68*) als „End"-aminosäure-komponente zum Umsatz mit „Azlactonen" herangezogen, wobei allerdings im allgemeinen Diastereomere auftreten. Neuerdings veröffentlichen Gish und Carpenter (*187, 188*), sowie Anderson (*18*) Synthesen von Arginylpeptiden: N_α-carbobenzoxylierte Argininderivate wurden nach der Säurechlorid- bzw. der Pyrophosphit-methode mit Aminosäureestern gekuppelt.

c) Als dritter Weg kann der nachträgliche Aufbau der ω-Guanido-gruppe bei N_α-acylierten, aber N_ω-freien Ornithin-peptiden gegangen werden; ihm liegen langbekannte Reaktionen der Aminogruppe mit Cyanamid (*38, 366*), Guanidin (*342*), am günstigsten aber mit S-Methyl-isothioharnstoff bzw. O-Methyl-isoharnstoff (*252, 210, 370, 414*) zugrunde.

$$
\begin{array}{ccc}
\begin{array}{c} \text{NH}_2,\ \text{H}X \\ | \\ (\text{CH}_2)_3 \\ | \\ Ac\!-\!\text{NHCHCOOH} \end{array}
&
\xrightarrow[-\text{CH}_3\text{OH}]{\ \text{H}_3\text{CO}\!-\!\text{C}\diagup^{\text{NH}}_{\diagdown\text{NH}_2}\ }
&
\begin{array}{c} \text{NH} \\ || \\ \text{NH}\!-\!\text{C}\!-\!\text{NH}_2,\ \text{H}X \\ | \\ (\text{CH}_2)_3 \\ | \\ Ac\!-\!\text{NHCHCOOH} \end{array}
\end{array}
$$

Ac = Acyl. X = Cl, NO_3 und dergleichen.

3. Die heterocyclischen Ringsysteme.

a) Imidazole.

Die Einbeziehung des Histidins in die Synthese von Peptiden der Zusammensetzung „X-His" bot keine Schwierigkeiten; mit Hilfe der Säurechlorid- bzw. -azidmethode konnten schon von FISCHER und CONE (*151*), von ABDERHALDEN und GOHDES (*5*) und vor allem von DU VIGNEAUD und seiner Schule (*243, 244, 428, 393*) zahlreiche Peptide erhalten werden. Von einigen unwesentlichen Versuchen abgesehen (*8, 163, 55*), blieb jedoch die Darstellung von Peptiden der Formel „His-X" unbearbeitet, obgleich DU VIGNEAUD und BEHRENS (*426*) mit der Darstellung von N(Im)-Benzyl-histidin (XLIII) durch Umsetzung von Histidin mit Natrium in flüssigem Ammoniak + Benzylchlorid eine reversible Maskierungs-möglichkeit der NH-Imidazol-gruppierung aufgezeigt hatten. Diese wird neuerdings von KATCHALSKI und PATCHORNIK (*258*) für die einwandfreie Darstellung von Poly-histidin (XLV) über das N(Im)-Benzylhistidin-N-carbonsäureanhydrid (XLIV) herangezogen [vgl. auch OVERELL und PETROW (*321*)]:

(XLIII.)

(XLIV.)

$$\xrightarrow[-CO_2]{\text{Base}} \left[\begin{array}{c} \text{CH} \quad\quad \text{CH—CH}_2\text{CHCO—..} \\ \text{N} \diagdown\diagup \text{N} \quad\quad \text{NH—..} \\ \text{CH} \quad\quad \text{CH}_2 \\ \\ \text{C}_6\text{H}_5 \end{array} \right] \xrightarrow{\text{Na/NH}_3} \quad \text{(XLV.)}$$

Erst 1954 gelang Holley und Sondheimer (*236*) und etwas später R. F. Fischer und Whetstone (*164*) die Isolierung von Carbobenzoxy-histidin-azid aus carbonat-alkalischer Lösung und damit der Einsatz des Histidins als „Kopf"-aminosäure. Mit dem N(Im)-Trityl-histidin-methyl-ester-hydrochlorid hat Zervas (*479, 480*) in letzter Zeit ein weiteres, zur eindeutigen Peptidsynthese brauchbares Imidazolring-geschütztes Histidinderivat geschaffen.

b) Indole.

Die Synthese von Peptiden des β-Indolyl-alanins (Tryptophan) wurde, von einer Veröffentlichung der Abderhaldenschen Schule (*6*) abgesehen, erst in neuerer Zeit mit Arbeiten von Fruton (*177*) und Smith (*396*) begonnen; sie gelang mit Hilfe der Säurechloridmethode unter Verwendung der Carbobenzoxy-schutzgruppe. Weitere wichtige Erkenntnisse stammen von Katchalski u. a. (*324*) mit der Darstellung bzw. Polymerisation von Tryptophan-N-carbonsäureanhydrid, sowie von Vaughan und Eichler (*419*), denen die Synthese von Tryptophan-peptiden mittels der „Anhydridmethode" gelang. In keinem Falle wurden durch das Indolringsystem bedingte Nebenreaktionen beobachtet [vgl. auch Wieland und Hörlein (*460*)].

4. Die alkoholische Hydroxylgruppe.

Die Zahl der aliphatischen Oxy-aminosäuren ist stetig angewachsen; neben dem biologisch wichtigen Serin, Threonin, Oxyprolin und Oxylysin konnten weitere dieser trifunktionellen Aminosäuren als spezifische Bausteine gewisser Eiweißkörper bzw. von Alkaloiden aufgefunden werden. Die Synthese von Oxyaminosäure-peptiden blieb ähnlich der Histidinanaloga neben einigen unwesentlichen Versuchen über die Diketopiperazine (*145, 49, 3*) praktisch den letzten 15 Jahren vorbehalten; 1942 konnten Fruton (*176*) und später Smith und Bergmann (*397*) zeigen, daß auch Carbobenzoxy-serin bzw. -oxyprolin in ihre Azide überführbar sind. Diese kuppeln mit Aminosäureestern zu den entsprechenden Carbobenzoxy-peptidestern (XLVI), sofern durch Einhaltung niedriger Temperaturen jegliche Nebenreaktionen als Resultat einer vorausgehenden Umlagerung zum Isocyanat (XLVII) weitgehend zurückgedrängt werden. Nach diesem Verfahren sind in der Folgezeit eine

Anzahl Serin- (*477, 220, 405, 164, 30, 14, 230 a*) bzw. Oxyprolin-peptide (*314*) mit guten Ausbeuten synthetisiert worden:

$$\begin{array}{ccc}
\overset{\displaystyle CH_2OH}{\underset{\displaystyle Cbo\text{—}NHCHCON_3}{|}} & \xrightarrow{25^\circ} & \left[\overset{\displaystyle CH_2OH}{\underset{\displaystyle Cbo\text{—}NHCH\text{—}N=C=O}{|}}\right]
\end{array}$$

(XLVII.)

$$4^\circ \downarrow \overset{R'}{\underset{H_2NCHCOOR''}{|}}$$

$$\overset{CH_2OH \quad R'}{\underset{Cbo\text{—}NHCHCO\text{—}NHCHCOOR''}{|\qquad\quad|}}$$

(XLVI.)

$$\overset{CH_2\text{—}O}{\underset{Cbo\text{—}NHCH\text{—}NH}{|}} \!\!\!\diagdown_{CO}\!\!\!\diagup$$

R' = Aminosäure-Seitenkette. R'' = Alkyl.

Neuerdings berichten WIELAND und Mitarb. (*462*), sowie SCHWYZER (*371*) über die Synthese von substituierten Oxysäure-amiden, wobei sie die gemischten Anhydride der Oxysäure mit Alkylkohlensäure herstellten und im gewünschten Sinne aminolytisch umsetzten. Die Übertragung dieser Methodik auf N-acylierte Oxyaminosäuren wurde von GRASSMANN und WÜNSCH (*204*) bei der Darstellung von Oxyprolyl-glycin mit ausgezeichnetem Erfolg angewandt.

Die Darstellung von Serinpeptiden der Formel „*X*-Ser" nach dem FISCHERSchen Säurechlorid-verfahren beschreiben in letzter Zeit BOTWINIK und Mitarb. (*79, 80*); z. B. kuppelt Phthalyl-(oder Tosyl)-glycylchlorid mit Serinester in Dioxan in Gegenwart von Diäthylanilin zum N-Phthalyl-glycylserinester (*80*), während der Einsatz von zwei Äquivalenten Säurechlorid zum N,O-Bis-acylderivat führt.

Eine Möglichkeit der Maskierung aliphatischer Hydroxylgruppen bietet die Veresterung zu O-acylierten Oxyaminosäuren (*44, 50, 408, 356*). Doch haben bislang nur FRANKEL und HALMANN (*167*), sowie SHEEHAN und Mitarb. (*386 a*) bei der Synthese von Poly-O-acetyl- und Poly-O-carbobenzoxy-serin bzw. N-Phthalyl-O-acetyl-serylpeptiden hiervon Gebrauch gemacht.

Als eine zweite Möglichkeit diskutiert FRUTON (*175*) die Verwendung von O-Benzyläthern. GRASSMANN und WÜNSCH (*205*) haben diesen Vorschlag aufgegriffen und die peptid-synthetische Verwendung von O-Benzyl-serin erprobt. Der Benzyläther ist leicht und rasch durch katalytische Hydrierung in Serin und Toluol spaltbar:

$$\overset{CH_2O\text{—}CH_2C_6H_5}{\underset{R\text{—}NHCHCOOR''}{|}} \xrightarrow{H_2/Pd} \overset{CH_2OH}{\underset{R\text{—}NHCHCOOR''}{|}} + C_6H_5CH_3$$

R = H oder Acyl. R'' = H oder Alkyl.

Nach Woods und Kramer (*475*), sowie Parham und Anderson (*323*) addiert sich Dihydropyran (XLVIII) leicht an alkoholische Hydroxylgruppen; die entstandenen alkalibeständigen Acetale können mit Säuren unter Regenerierung des Alkohols wieder gespalten werden. Schwyzer und Mitarb. (*248*) haben daraus eine zusätzliche O-Blockierung der aliphatischen bzw. aromatischen Oxyaminosäuren abgeleitet, indem sie N-Carbobenzoxy-oxyaminosäureester (XLIX) mit Dihydropyran zum Acetal (L) umsetzen. Nach alkalischer Verseifung der Estergruppe wird die N,O-blockierte Oxyaminosäure (LI) z. B. in der Form des aktivierten Esters (s. S. 495) mit Aminosäureestern zum Peptidderivat (LII) gekuppelt. Ein kleiner Nachteil liegt in der Isolierung und weiteren Umsetzung von (LI), da die Säure wegen des Auftretens eines zusätzlichen asymmetrischen C-Atoms am Tetrahydropyranring als Isomerengemisch vorliegt.

$$
\begin{array}{ccc}
\text{CH}_2\text{OH} & & \\
| & & \\
Cbo\text{—NHCHCOO}R'' + \text{(XLVIII)} & \longrightarrow & \text{CH}_2\text{—O—(L)} \quad \xrightarrow{n\text{-NaOH}}
\end{array}
$$

(XLIX.)   (XLVIII.)   (L.)

$$
Cbo\text{—NHCHCOOH} \quad \xrightarrow[\substack{\text{H}_2\text{NCHCOO}R'' \\ R'}]{\substack{\text{ClCH}_2\text{CN} \\ \text{tert. Amin}}} \quad Cbo\text{—NHCHCO—NHCHCOO}R''
$$

(LI.)   (LII.) R'

Cbo = Carbobenzoxy. R' = Aminosäure-Seitenkette. R'' = Alkyl.

Je nach Wunsch kann das Dipeptid-derivat (LII) durch selektive Abspaltung der drei Schutzgruppen als Ausgangsmaterial für weitere Umsetzungen zu höheren Peptiden dienen.

Ein natürliches Esterpeptid des Serins liegt im Antibioticum „Azaserin" vor (*182*). Auf Grund neuerer Ergebnisse kann ferner auch mit der Existenz von „O-Serinpeptiden" bei natürlichen Eiweißkörpern gerechnet werden (*199*). Die Synthese dieser Verbindungsklasse wurde bereits in Angriff genommen (*309, 315*).

5. Die phenolische Hydroxylgruppe.

Von den in der Natur vorkommenden Aminosäuren dieser Klasse (Tyrosin, 3,4-Dioxyphenylalanin, Jodgorgosäure und Thyroxin) sind Peptide des Tyrosins schon seit längerer Zeit Gegenstand synthetischer Arbeiten. Wie Histidin und die aliphatischen Oxy-aminosäuren läßt sich auch das Tyrosin mit ungeschützter phenolischer Hydroxylgruppe als carboxyl-endständige Komponente (Natriumsalz bzw. Ester) mit α-Halogenacyl-halogeniden (*147*), mit Carbobenzoxy-aminosäurechloriden (*9, 1*), -aziden (*164*) bzw. -anhydriden (*67, 420*) glatt zur Vorstufe des gewünschten Peptids kuppeln. Auch die Umsetzung von N-Acyl-tyrosin-aziden (*46, 219, 164*) und -gemischten Anhydriden nach der Pyrophosphit-Standardmethode (*19, 431*) führt ohne Nebenreaktionen zum Ziel. Die Übertragung der FISCHERschen Säurechlorid-methode auf N-Acyl-tyrosin (vorwiegend N-Carbobenzoxy-verbindungen) verlangte jedoch eine Schutzgruppe für das phenolische Hydroxyl; nachdem schon ABDERHALDEN und BAHN (*2*) die O-carbobenzoxylierte Verbindung dafür

R' = Aminosäure-Seitenkette. R'' = Alkyl.

Formelübersicht 4.

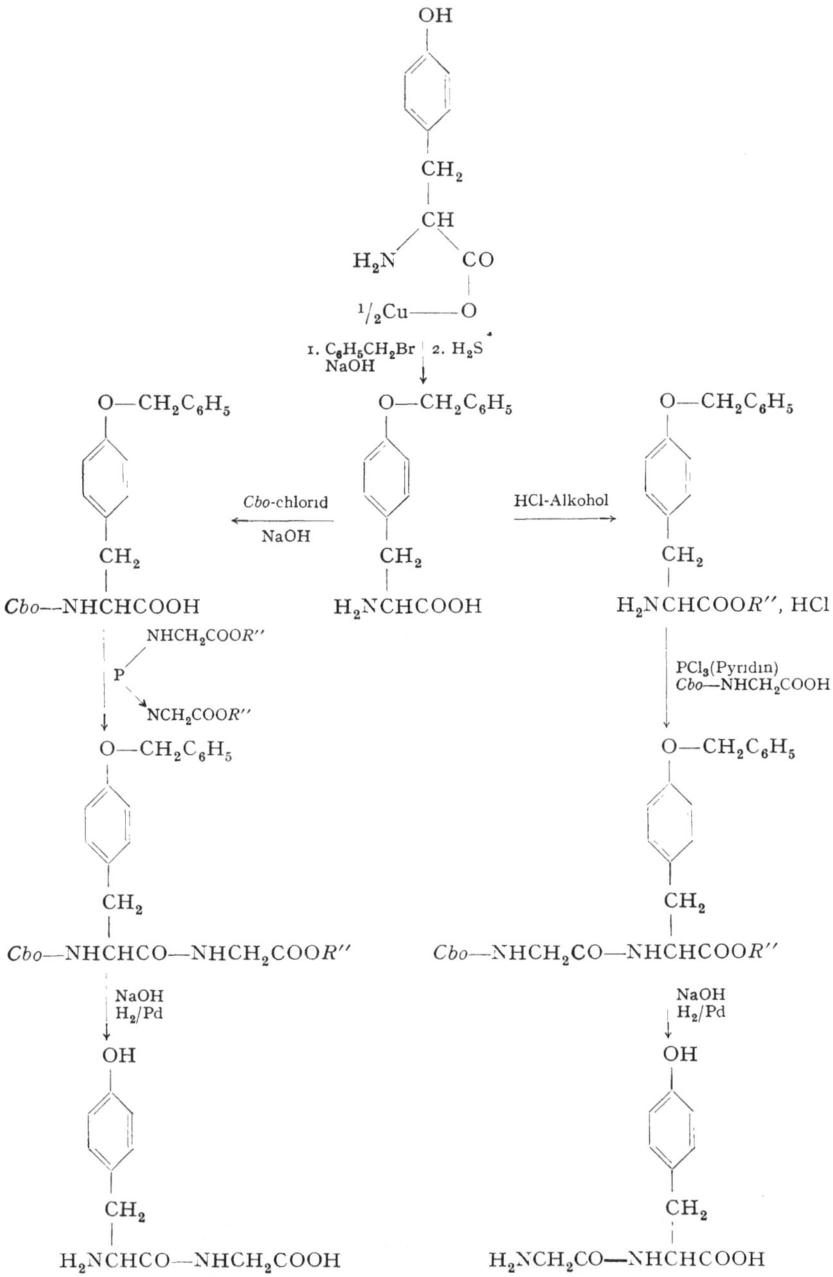

OH

CH$_2$

CH

H$_2$N   CO

$^1/_2$Cu——O

1. C$_6$H$_5$CH$_2$Br 2. H$_2$S
NaOH

O—CH$_2$C$_6$H$_5$ O—CH$_2$C$_6$H$_5$ O—CH$_2$C$_6$H$_5$

Cbo-chlorid   HCl-Alkohol
NaOH

CH$_2$ CH$_2$ CH$_2$

Cbo—NHCHCOOH   H$_2$NCHCOOH   H$_2$NCHCOO*R''*, HCl

NHCH$_2$COO*R''*

P

NCH$_2$COO*R''*

PCl$_3$(Pyridin)
Cbo—NHCH$_2$COOH

O—CH$_2$C$_6$H$_5$ O—CH$_2$C$_6$H$_5$

CH$_2$ CH$_2$

Cbo—NHCHCO—NHCH$_2$COO*R''*   *Cbo*—NHCH$_2$CO—NHCHCOO*R''*

NaOH   NaOH
H$_2$/Pd   H$_2$/Pd

OH   OH

CH$_2$ CH$_2$

H$_2$NCHCO—NHCH$_2$COOH   H$_2$NCH$_2$CO—NHCHCOOH

Cbo = Carbobenzoxy. *R''* = Alkyl.

Formelübersicht 5.

herangezogen hatten, nehmen Bergmann und Mitarb. (*67*) das O-Acetyl-, Harington und Pitt Rivers (*219*) das O-Benzoyl-derivat zu Hilfe, deren Spaltung nach erfolgter Umsetzung mittels alkalischer Hydrolyse ohne Angriff auf die Peptidbindung gelingt (*Formelübersicht 4*).

Von Barkdoll und Ross (*37*) wird nach dieser Methode die Synthese von Tri-tyrosin, von Bailey (*33*) diejenige einer Anzahl von Di- und Tripeptiden des Tyrosins über das N-Carbonsäureanhydrid (LIIa) beschrieben [vgl. auch Katchalski und Sela (*259*), Schwyzer u. a. (*247*), Overell und Petrow (*321*); die letzteren konnten zeigen, daß die O-Carbobenzoxylierung auch direkt über den Kupferkomplex gelingt]. Grassmann und Mitarb. (*206*) haben erstmals den Benzyläther des *L*-Tyrosins auf einfache Weise hergestellt und mit sehr gutem Erfolg zu peptid-synthetischen Arbeiten (Phosphorazo-methode) herangezogen. Wie im Falle des Serins ist der O-Benzylrest in der Folge leicht mit katalytisch erregtem Wasserstoff oder Natrium in flüssigem Ammoniak [vgl. auch du Vigneaud (*424*)] unter Regenerierung der freien phenolischen Hydroxylgruppe abspaltbar (*Formelübersicht 5*).

In völliger Analogie zum Serin (s. oben) addiert sich Dihydropyran auch an die phenolische Hydroxylgruppe; Schwyzer und Mitarb. (*247, 248*) machen von dieser Blockierungsmöglichkeit bei der Synthese von Tyrosinpeptiden Gebrauch.

6. *Die Sulfhydrylgruppe.*

Die große physiologische Bedeutung des Cysteins führte schon frühzeitig — vor allem seit der Entdeckung des Glutathions durch Hopkins (*238*) — zum Bedürfnis nach der Synthese von Peptiden dieser Aminosäure. Die neuen Arbeiten über Coenzym A, Penicillin, Oxytocin-Vasopressin und verwandte Substanzen, die als entscheidenden Faktor Cystein (Cystin) bzw. Cysteamin im Molekül enthalten, haben die Richtigkeit dieser Forderung bestätigt. Nach einigen ersten Schritten der Fischerschen Schule (*162a, 153*), sowie von Abderhalden (*7, 12*) glückte Bergmann und Zervas (*56*) mittels des Carbobenzoxy-verfahrens der entscheidende Vorstoß. Die Wege a) und b), die beide auf einer eindeutigen Blockierung der Sulfhydrylgruppe basieren, erwiesen sich besonders für Peptide der Formel „Cys-*X*" gangbar.

a) Das „Cystin-verfahren".

Nach Bergmann und Zervas (*56*) läßt sich Cystin in die N,N'-Dicarbobenzoxy-verbindung (LIII) überführen, diese mit Phosphorpentachlorid in das zugehörige Bis-säurechlorid (LIV). Die übliche „Fischer-Synthese" führt zu den gewünschten Dicarbobenzoxy-cystin-peptiden (bzw. -estern). Die Abspaltung der Schutzgruppe, die hier durch Hydrierung nicht möglich ist, gelang dann White (*451*), Harington und Mead (*217*), sowie Loring und du Vigneaud (*293*) auf den auf S. 456 angeführten Wegen. Bei der Abspaltung mit Phosphoniumjodid oder Natrium in flüssigem Ammoniak tritt gleichzeitig reduktive Öffnung der Disulfid-brücke auf, die bei nichtreduktiver Entfernung der Schutzgruppe in einem gesonderten Schritt mit Natrium in flüssigem Ammoniak (*425*) bzw. mit Zink und Säure (*329*) erfolgen muß. Auf der anderen Seite ist die Überführung der Cystein- in die Cystin-peptide mittels Luftoxydation leicht wieder zu bewerkstelligen (*293, 219, 211, 431*). Neuerdings machen

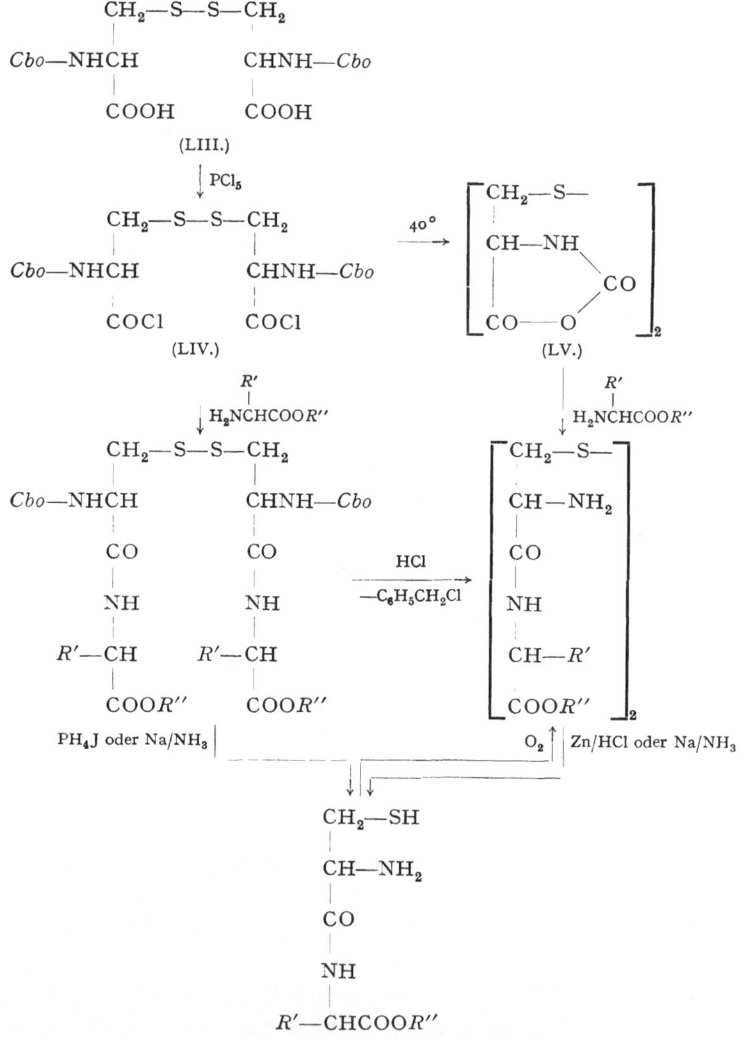

R' = Aminosäure-Seitenkette. R'' = H bzw. Alkyl.

Formelübersicht 6.

jedoch sowohl HOLLY und Mitarb. (*237*), als auch GOLDSCHMIDT und JUTZ (*191*) darauf aufmerksam, daß die direkte Carbobenzoxylierung des Cystins in alkalischem Medium keine reinen Produkte liefert. BAILEY (*33*) gelangte über das Carbobenzoxy-cystinylchlorid (LIV) zum Bis-N-Carbonsäure-anhydrid (LV), das seinerseits für die Synthese von Cystinpeptiden Verwendung finden kann (*Formelübersicht 6*).

b) Das „S-Benzyläther-verfahren".

DU VIGNEAUD (*425*) hat mit der Darstellung des S-Benzyläthers des Cysteins und der Feststellung, daß diese Blockierung der Sulfhydrylgruppe mittels Natrium in flüssigem Ammoniak ebenso leicht wieder rückgängig gemacht werden kann, den bahnbrechenden Hinweis für alle synthetischen Arbeiten auf dem Cystein-Cystin-gebiet gegeben (*293*):

$$
\begin{array}{ccccc}
CH_2\text{—}SH & & CH_2\text{—}SNa & & \left[CH_2\text{—}S\text{—}\right. \\
| & \xrightarrow{Na/NH_3} & | & \xleftarrow{\quad} & | \\
CH\text{—}NH_2 & \xleftarrow{\;H^{(+)}} & CH\text{—}NH_2 & {}_{Na/NH_3} & CH\text{—}NH_2 \\
| & & | & & | \\
COOH & & COONa & & \left.COOH\right]_2
\end{array}
$$

$$
\begin{array}{c}
{}^{Na/NH_3}_{-C_6H_5CH_3} \uparrow \;\; \downarrow {}^{C_6H_5CH_2Cl}_{H^{(+)}} \\
CH_2\text{—}S\text{—}CH_2C_6H_5 \\
| \\
CH\text{—}NH_2 \\
| \\
COOH
\end{array}
$$

Zahlreiche Veröffentlichungen, die vor allem die Synthesen von Glutathion und neuerdings von Oxytocin bzw. Vasopressin behandeln, bedienen sich mit bestem Erfolg des S-Benzyl-cysteins in Form der Carbobenzoxy-verbindung bzw. des Esters. Als besonders vorteilhaft erwies sich dabei, daß mittels Halogenwasserstoffen in indifferenten Lösungsmitteln eine selektive Abspaltung des Carbobenzoxyrests möglich ist (*218, 430, 429, 452, 303, 219, 274, 225, 37, 191, 345, 431, 427*):

$$
\begin{array}{c}
CH_2\text{—}S\text{—}CH_2C_6H_5 \\
| \\
\textit{Cbo}\text{—}NHCHCO\text{—}NHCHCOO\textit{R}'' \\
| \\
R'
\end{array}
$$

Alkal. Verseifung ← | → HCl-Alkohol

$$
\begin{array}{cc}
CH_2\text{—}S\text{—}CH_2C_6H_5 & \qquad CH_2\text{—}S\text{—}CH_2C_6H_5 \\
| & \qquad | \\
\textit{Cbo}\text{—}NHCHCO\text{—}NHCHCOOH & \qquad H_2NCHCO\text{—}NHCHCOO\textit{R}'', HCl \\
| & \qquad | \\
R' & \qquad R'
\end{array}
$$

Cbo = Carbobenzoxy. *R'* = Aminosäure-Seitenkette. *R''* = Alkyl.

Die Darstellung der N-Phthalyl-verbindung des S-Benzylcysteins ist neuerdings Balenović und Mitarb. *(34)* gelungen; sie wurde jedoch noch nicht zu Peptidsynthesen herangezogen.

c) S-Aminoacyl-derivate.

Einen neuen, äußerst interessanten Weg zur Synthese von Cysteinpeptiden der Formel „*X*-Cys" haben Wieland und Mitarb. *(458)* aufgezeigt. Unter bestimmten Reaktionsbedingungen setzen sich S-Aminoacyl-derivate des Cysteins (die als nicht isolierbare Zwischenprodukte einer Umesterung Thiophenyl-: S-Cysteinester entstehen), spontan zu N-Aminoacyl-cystein um (vgl. auch S. 496).

$$
\begin{array}{ccc}
& R' & \\
& | & \\
C_6H_5\text{—}S\text{—}CO\text{—}CHNH_2 & & \\
+ & & \\
\text{SH} \quad NH_2 & & \\
| \quad\quad | & & \\
CH_2\text{—}CHCOOH & &
\end{array}
\longrightarrow
\begin{array}{c}
R' \\
| \\
OC\text{—}CHNH_2 \\
| \\
S \quad NH_2 \\
| \quad\quad | \\
CH_2\text{—}CHCOOH
\end{array}
\longrightarrow
\begin{array}{c}
R' \\
| \\
OC\text{—}CHNH_2 \\
| \\
SH \quad NH \\
| \quad\quad | \\
CH_2\text{—}CHCOOH
\end{array}
$$

R' = Aminosäure-Seitenkette.

7. Die „Thioäther".

Neben Methionin und Lanthionin als natürliche Vertreter dieser Aminosäureklasse ist der synthetische S-Benzyläther des Cysteins zu beachten. Im allgemeinen gelten somit die auf S. 489 gegebenen Hinweise. In einigen Fällen der Synthese von Methionin-peptiden war es sogar möglich, den Carbobenzoxy-rest mit katalytisch erregtem Wasserstoff zu entfernen *(125)*. Wählt man eine Abspaltung der Aminoschutzgruppe mittels Natrium in flüssigem Ammoniak, so ist für eine nachträgliche Methylierung von hierbei entstehendem Homocystein-derivat zu sorgen *(125, 403, 86 a, 230 a)*, während die Halogenwasserstoff-solvolyse in Eisessig [Dekker *(124)*] zu Austauschreaktionen Anlaß gibt. So werden Methioninderivate (LVI) in S-Benzyl-homocystein-derivate (LVII) übergeführt:

$$
\begin{array}{c}
C_7H_7OCO\text{—}NHCHCO\text{—}R \\
| \\
CH_2 \\
CH_2 \\
| \\
S \\
| \\
CH_3 \\
\text{(LVI.)}
\end{array}
\xrightarrow[\text{Eisessig}]{HX}
\left[
\begin{array}{c}
HX,\ NH_2CHCO\text{—}R \\
| \\
CH_2 \\
CH_2 \\
| \\
S \\
H_7C_7 \diagup \diagdown CH_3
\end{array}
\right]^{(+)} X^{(-)}
\xrightarrow{-CH_3X}
\begin{array}{c}
HX,NH_2CHCO\text{—}R \\
| \\
CH_2 \\
CH_2 \\
| \\
S \\
| \\
C_7H_7 \\
\text{(LVII.)}
\end{array}
$$

Cbo = Carbobenzoxy. R = OH, OR'' oder —NHCHCOOH(R''). R'' = Alkyl.

R'

8. Die ω-Carboxylgruppe.

Peptide der Amino-dicarbonsäuren, insbesondere der in natürlichen Eiweißkörpern stark vertretenen Glutamin- und Asparaginsäure, sind mehrfach, vor allem seit der Einführung der Carbobenzoxy-methode, synthetisiert worden. Die Heranziehung von „inneren" Anhydriden von Derivaten der beiden Aminosäuren — in gewissem Sinne doch eine Blockierung der ω-Carboxylgruppe — ist auf SS. 524—526 diskutiert. Eine weitere Maskierungsmöglichkeit ist mit Hilfe des Pyrrolidonrings für die Glutaminsäure gegeben (S. 466). Zwei weitere spezifische Verfahren a) und b) wurden beschrieben:

a) Synthesen mit ungeschützter ω-Carboxylgruppe.

Nach BERGMANN und Mitarb. (*61*) gelingt die Überführung von N-Acyl-glutamyl-aminosäure-monoestern bzw. -peptid-monoestern in die Azide und die weitere Umsetzung dieser zu höheren Acyl-peptidestern, ohne daß die hierbei ungeschützt verbleibende ω-Carboxylgruppe zu Nebenreaktionen Anlaß gibt (*231*). HEGEDÜS (*225*) sowie LE QUESNE und YOUNG (*338*) beschreiben die Darstellung von Carbobenzoxy-glutamyl-monoaziden und deren Verwendung zur Synthese von α- und γ-Glutamyl-peptiden. Jedoch erheben neuerdings SACHS und BRAND (*355*) berechtigte Einwände, da sie bei der Nachbearbeitung dieses Verfahrens jeweils beide Derivate isolieren konnten. Eindeutig dürfte dagegen die Synthese über N-Acyl-glutaminsäure-α- bzw. -γ-thiophenylester verlaufen (*470, 355 a*).

b) Synthesen mit veresterter ω-Carboxylgruppe.

ω-Halbester der Asparagin- bzw. Glutaminsäure sind lange bekannt (*330, 214, 58*), seit 1933 auch das N-Carbobenzoxy-derivat des Glutamin-säure-γ-esters (*10*). Die Darstellung gelingt durch direkte Carbobenzoxy-lierung des freien γ-Esters in schwach alkalischem Medium (*10, 317, 225, 215, 337*) [für den α-Ester siehe SACHS und BRAND (*353*)], ferner durch Alkoholyse der N-Acylaminodicarbonsäure-anhydride (hierbei werden α- und γ-Ester gebildet; vgl. S. 526) und schließlich durch partielle Ver-seifung N-acylierter Diester (*215*). Die so erhaltenen N-Acyl-amino-dicarbonsäure-monoester lassen sich nun nach beliebigen Methoden mit Aminosäuren bzw. Peptiden (Natriumsalze bzw. Ester) kuppeln. In der Folge ist die Verseifung der ω-Estergruppe durchzuführen; sie ist quantitativ nur mit überschüssiger Lauge zu erreichen, was bei der Alkali-empfindlichkeit des Carbobenzoxy-restes wenig vorteilhaft erscheint. Die Heranziehung von ω-Benzylestern (*215, 352*) ist daher wünschenswert (vgl. auch SS. 525—526).

Für die Synthese von γ-Glutamyl- bzw. β-Asparagyl-peptiden über die α-Ester der beiden Aminosäuren gelten analoge Verhältnisse; vgl. auch S. 475.

9. Die primäre Carbonsäureamid-gruppe.

Wie bekannt, beteiligen sich Glutamin bzw. Asparagin in weitgehendem Maße am Aufbau natürlicher Eiweißkörper. Neuerdings wurde Glycinamid als Baustein der Hormone Oxytocin und Vasopressin aufgefunden; Peptide mit endständiger primärer α-Carbonsäureamid-gruppe haben Bedeutung für enzymatische Arbeiten erlangt (*46, 47, 230*).

a) Synthesen mit ungeschützter —$CONH_2$-gruppe.

Schon frühzeitig gelang Fischer und Koenigs (*156, 157*), sowie Thierfelder und v. Cramm (*413*) die Kupplung von Asparagin bzw. Glutamin mit Halogen-acylchloriden nach Schotten-Baumann; die Aminierung der α-Halogenacylderivate lieferte die gesuchten Peptide. Erst neuerdings haben sich du Vigneaud und Mitarb. (*262, 431, 427*) wieder des Säurechlorid-verfahrens zur Darstellung der Glutaminyl-asparaginyl-gruppierung bedient, wobei sie Magnesiumoxyd in Wasser als HCl-bindendes Agens benutzten. Gleichzeitig synthetisierten Sondheimer und Holley (*399*) Glutaminylpeptide nach der Curtiusschen Azid-methode, nachdem ihnen die Darstellung von Carbobenzoxy-glutaminylhydrazid bzw. -azid aus dem entsprechenden Methylester gelungen war (*398*) [vgl. auch Boissonnas und Mitarb. (*76*)].

b) Nachträglicher Aufbau der —$CONH_2$-gruppe.

Melville (*299*), später Harington und Mead (*218*), sowie Fruton und Bergmann (*179*) führten Carbobenzoxy-glutamylderivate in die γ-Säurechloride über und diese in die γ-Säureamide. Allgemeiner anwendbar erwies sich aber die Aminolyse von Estern mittels alkoholischem Ammoniak, die vor allem die Bergmann-Schule (*46, 47, 230, 178*) studierte. Sie wurde von Miller und Waelsch (*305*) zur Synthese von Glutaminylpeptiden, von Ressler und du Vigneaud (*345*) zur

$$
\begin{array}{ccc}
R' & & R' \\
| & & | \\
\end{array}
$$

$$Cbo\!-\!NHCHCO\!-\!NHCHCOOR''$$

$$\downarrow NH_3 \qquad\qquad NH_3 \downarrow$$

$$
\begin{array}{ccc}
R' & R' & R' \\
| & | & | \\
\end{array}
$$

$$Cbo\!-\!NHCHCO\!-\!NHCHCONH_2 \qquad\qquad CH\!-\!CO$$

$$\downarrow H_2/Pd$$

on the right (LIX.):

$$
\begin{array}{l}
R' \\
| \\
CH\!-\!CO \\
| \qquad\quad \diagdown \\
| \qquad\qquad N\!-\!CHCONH_2 \\
| \qquad\quad \diagup \\
NH\!-\!CO
\end{array}
$$

$$
\begin{array}{cc}
R' & R' \\
| & | \\
\end{array}
$$

$$H_2NCHCN\!-\!NHCHCONH_2$$

(LVIII.)   (LIX.)

$R' = $ Aminosäure-Seitenkette. $R'' = $ **Alkyl.**

Darstellung von Peptidamiden (LVIII) benutzt. Nach BERGMANN und FRUTON (*178*,) nimmt jedoch diese Reaktion bei Carbobenzoxy-phenylalanyl- bzw. -leucylglycinestern einen abnormalen Verlauf: Unter Freiwerden von Benzylalkohol bilden sich die Amide der substituierten Hydantoin-essigsäuren [LIX, s. auch (*125*)].

Für den Aufbau von Glutaminyl-peptiden hat die in den letzten Jahren von DU VIGNEAUD und seiner Schule (*407*), sowie gleichzeitig von RUDINGER und Mitarb. (*349, 350*) aufgefundene Ringöffnung von N-Tosyl-pyroglutamylaminosäuren bzw. -peptiden (LX) einen sehr erheblichen Fortschritt gebracht:

$$
\begin{array}{ccc}
\mathrm{CH_2CH_2COOH} & \xrightarrow{\mathrm{PCl_5}} & OC\!\!\diagup^{\textstyle CH_2}\!\!\diagdown CH_2 \longrightarrow \\
| & & | \qquad\quad | \\
Tos\!-\!\mathrm{NHCHCOOH} & & Tos\!-\!\mathrm{N}\!-\!\!-\!\!-\!\mathrm{CH}\!-\!\mathrm{COCl}
\end{array}
$$

$$
\begin{array}{c}
R' \\
| \\
\mathrm{H_2NCHCOOH} \\
\xrightarrow{\hspace{2cm}}
\end{array}
\quad
OC\!\!\diagup^{\textstyle CH_2}\!\!\diagdown CH_2 \qquad
\begin{array}{c} R' \\ | \end{array}
\longrightarrow
\begin{array}{c} \mathrm{CH_2CH_2CONH_2} \\ | \end{array}
$$

$$
Tos\!-\!\mathrm{N}\!-\!\!-\!\!-\!\mathrm{CHCO}\!-\!\mathrm{NHCHCOOH} \qquad Tos\!-\!\mathrm{NHCHCO}\!-\!\mathrm{NHCHCOOH}
$$

$$(LX.) \hspace{3cm} R'$$

Tos = p-Toluolsulfonyl. R' = Aminosäure-Seitenkette.

10. Aminozucker und Phosphorsäureester.

Auf Grund vierfacher Erfahrungen [vgl. z. B. GRASSMANN und KÜHN (*200*)] darf mit dem Auftreten von Aminozuckern als Bausteine natürlicher Eiweißkörper gerechnet werden. Die Einbeziehung dieser Körperklasse in die Peptidsynthese — von BERGMANN und ZERVAS (*57*) mit der Darstellung von Glycyl- bzw. Alanyl-glucosamin begonnen [s. auch (*69*)] — wird notwendig.

Analoge Gesichtspunkte ergeben sich für die Phosphorsäureester der Oxyaminosäuren; aus Casein-hydrolysaten (Trypsin-abbau) wurden z. B. Phosphoserin (*290*), Phosphothreonin (*423*), Phospho-seryl-glutaminsäure (*285*) und höhere Peptid-phosphorsäureester (*334*) isoliert. Mit der Synthese von Phospho-peptiden hat man sich erst verhältnismäßig spät befaßt. POSTERNAK und GRAFL (*335*) beschreiben erstmals Phospho-tyrosin-peptide; die Einführung des Phosphorsäurerestes gelang mittels Phosphoroxychlorid. Interesse verdienen ferner die Ergebnisse von WAGNER-JAUREGG (*331*): durch Wasserstoffionen hervorgerufene N : O-Acylwanderung ermöglicht die einwandfreie Darstellung von Oxyaminosäure-O-phosphorsäureestern. Erst in letzter Zeit veröffentlichten BAER und MAURUKAS (*31*), sowie WILSON und Mitarb. (*472*) ein allgemein anwendbares Phosphorylierungsverfahren für Oxyaminosäuren, das auch auf Peptide übertragbar ist:

$$\begin{array}{c}CH_2OH \\ Cbo—NHCHCOOR'' \end{array} \xrightarrow{ClP(OC_6H_5)_2} \begin{array}{c}CH_2O—P(OC_6H_5)_2 \\ \;\;\;\;\;\;| \;\;\;\;\; O \\ Cbo—NHCHCOOR'' \end{array} \xrightarrow[\text{Eisessig}]{HBr}$$

$$\xrightarrow{} \begin{array}{c}CH_2O—P(OC_6H_5)_2 \\ |\;\;\;\;\;\; O \\ H_2NCHCOOR'' \end{array} \xrightarrow{NaOH} \begin{array}{c}CH_2O—P(OH)_2 \\ |\;\;\;\;\; O \\ H_2NCHCOOH \end{array}$$

$R'' =$ Alkyl. $Cbo =$ Carbobenzoxy.

III. Methoden der Peptidknüpfung.

E. Esterkondensationen.

1. Diketopiperazine und ihre Aufspaltung.

Fußend auf einer Arbeit von ABENIUS und WIDMAN (*13*) gelang FISCHER und FOURNEAU (*152*) 1901 durch saure Hydrolyse von Diketopiperazin zum ersten Male die künstliche Herstellung eines Peptides:

$$\begin{array}{ccc} & CH_2—CO & \\ HN & & NH \\ & CO—CH_2 & \end{array} \longrightarrow H_2NCH_2CO—NHCH_2COOH$$

Diese Methode versagte bei den symmetrischen „Anhydriden" des Alanins und Leucins (*135*); erst die vorsichtige Hydrolyse mit verdünntem Alkali bei Raumtemperatur führte zum Erfolg (*139*). In der Folge wurde jedoch festgestellt, daß hierbei ganz oder teilweise Racemisierung eintrat (*155*, *161*). LEVENE (*286*) und BERGMANN (*63*) zeigten später, daß die optische Aktivität der Diketopiperazine bzw. der Peptide vor allem im alkalischen Medium verlorengeht, viel weniger dagegen im sauren (*209*, *52*). Gemischte (asymmetrische) Diketopiperazine ergeben Mischungen der beiden möglichen Dipeptide, die meist kaum trennbar sind (*161*) [vgl. aber BERGMANN und TIETZMAN (*52*)].

Neuerdings lassen sich die „Aminosäure-anhydride" auch direkt aus den Aminosäuren durch Wasserabspaltung bei höherer Temperatur erhalten (*357*, *363*). Wie MEGGY (*298*) am Beispiel des Glycins zeigen konnte, verläuft diese Piperazin-dion-2,5-darstellung bei 170—180° unter gleichzeitiger Bildung von höheren Peptiden und sogar polymeren Produkten, worauf bereits MAILLARD (*296*) hingedeutet hatte.

2. Freie lineare Esterkondensation.

Schon 1883 hatte CURTIUS (*113*, *116*) die spontane Umwandlung von Glycinmethylester zur „Biuretbase" (Tetraglycinmethylester) beobachtet. Triglycinmethylester wiederum unterliegt beim Erwärmen größtenteils einer Dimerisation zum Hexaglycinester (*142*); als Nebenprodukte entstehen Polyglycinester mit einer Kettenlänge bis zu 100

Glycinresten (*322*). Mit analogem Erfolg verlief die Reaktion bei Leucyl-glycyl-glycin- und Alanyl-glycyl-glycinmethylester (*143*). FRANKEL und KATCHALSKI (*168*), sowie PACSU und WILSON (*322, 471*) haben dann das Studium dieser Polykondensationen auf breiter Grundlage untersucht [vgl. auch KATCHALSKI (*255*)]. Neuerdings befaßten sich SCHRAMM und Mitarb. (*364, 365*), BROCKMANN und MUSSO (*88, 89*), YOUNG und Mitarb. (*344*), sowie RYDON und SMITH (*351*) mit den zur Polykondensation von Aminosäure- bzw. Peptidestern führenden Bedingungen.

3. Cyclische Esterkondensation.

Neuerdings machen BROCKMANN u. a. (*90*) auf die Bildung cyclischer Peptide als Nebenprodukte bei der Polykondensation von Tripeptid-methylestern (in Methanol in Gegenwart von Ammoniak bzw. Piperidin) aufmerksam. Es gelang ihnen die Isolierung gut kristallisierter Verbindungen, in denen weder freie Amino- noch Carboxylgruppen nachweisbar waren, deren Hydrolysat jedoch alle Aminosäuren des eingesetzten Peptidesters lieferte [s. auch SCHRAMM und THUMM (*365*)]. Selbstverständlich ist hierher auch die Bildung der symmetrischen bzw. unsymmetrischen Diketopiperazine aus Aminosäure- bzw. Dipeptidestern zu zählen (vgl. S. 494).

4. Systematische Esterkondensation.

a) Energiereiche „O-Ester".

Beim Zusammenschmelzen von Benzoylglycinester mit Glycin fand CURTIUS (*114*) im Reaktionsgemisch Benzoylhexaglycin. Die damit anbahnte Erkenntnis, daß Ester acylierter Aminosäuren energiereiche Verbindungen sind, erlaubte E. FISCHER (*135*) die Synthese eines Carbäthoxy-tripeptidesters, als er Carbäthoxy-glycyl-glycinmethylester mit Leucinester unter Alkoholabspaltung kondensierte. Es ist ein Verdienst SCHWYZERS (*379, 376, 375, 378*) sich dieser fast vergessenen Peptid-knüpfung angenommen zu haben. Er hat hierbei auf die schon im Zusammenhang mit den Arbeiten über Thioester gemachten Erfahrungen (s. unten) zurückgegriffen und versuchte, durch Einführung negativer Substituenten in die Alkohol-komponente die Reaktionsfähigkeit der Carbonsäure-methylester gegenüber dem nucleophilen Angriff zu steigern. Als solche aktivierende Gruppen erwiesen sich die Cyan-, Carboäthoxy-, Dicarbäthoxy-, Aceto-, Nitrophenyl- und Diäthylamino-äthyl-gruppe (in fallender Stärke; abnehmender I^--Effekt); vgl. S. 451.

Als weitaus am geeignetsten behaupteten sich jedoch nur die Ester des Glykolsäurenitrils (LXI). Die Vorteile dieser „Cyanmethylester-methode" seien wie folgt dargelegt: 1. Leichte Zugänglichkeit der aktivierten Ester durch direkten Umsatz von Acylaminosäuren bzw. -peptiden mit Chloracetonitril in Gegenwart tertiärer Basen. 2. Gute Stabilität und Lagerfähigkeit; ausgezeichnete Kristallisationsfreudigkeit und leichte Reinigung. 3. Die aminolytische Aufspaltung durch Aminosäureester gelingt rasch und in guter Ausbeute, am besten in Essigester oder Tetrahydrofuran $(+ {}^1/_{10}$—${}^1/_{20}$ Mol Eisessig als Katalysator) und bei genügend hoher Konzentration der Reaktionspartner. 4. Darstellung und Umsetzung der Cyanmethylester verlaufen unter Erhalt der optischen Reinheit, sofern man hohe Temperaturen vermeidet [s. dazu ISELIN, FEURER und SCHWYZER (247)].

$$\underset{R'}{Cbo—NHCHCOOH} + ClCH_2CN \quad \xrightarrow[\text{Amin}]{\text{tert.}} \quad \underset{R'}{Cbo—NHCHCOOCH_2CN} \quad \longrightarrow$$

(LXI.)

$$\xrightarrow{\underset{R'}{H_2NCHCOOR''}} \quad \underset{R'}{Cbo—NHCHCO}—\underset{R'}{NHCHCOOR''} + HOCH_2CN$$

(LXII.)

Cbo = Carbobenzoxy. R' = Aminosäure-Seitenkette. R'' = Alkyl.

Ein Nachteil dieses Verfahrens ist die erforderliche Anwendung eines Überschusses der Endkomponente, da nur so die hohen Ausbeuten erzielbar sind (375). Eine Störung des Reaktionsverlaufs dagegen durch eine mögliche Kondensation des als Nebenprodukt entstehenden Glykolsäurenitrils (Cyanmethylalkohol) mit der Aminokomponente wurde nicht beobachtet (379).

BODÁNSZKY (74) veröffentlichte eine kurze Mitteilung über die Synthese von Peptiden mit Hilfe von Nitrophenylestern. In Dioxan lassen sich Acylaminosäure-nitrophenylester mit Aminosäureestern in ausgezeichneten Ausbeuten zu Acylpeptidestern kondensieren. Dem Autor gelang nach dieser Methode die Darstellung von S-Benzyl-cysteinyl-prolyl-leucyl-glycin-benzyl-ester-hydrochlorid sowie einiger Phthalyl-peptidamide (74a).

b) Energiereiche „S-Ester".

In die Arbeiten auf dem peptidsynthetischen Gebiet wurden von der WIELAND-Schule auch die Thiophenylester von Acyl-aminosäuren einbezogen (464). Ihre Darstellung gelingt über die Säurechloride oder besser über die gemischten Anhydride von Kohlensäureestern durch Umsetzen mit Thiophenol. Die S-Phenylester acylierter Amino-thiosäuren (LXIII) zeigen eine beträchtliche Hydrolysebeständigkeit. Dagegen ist die aminolytische Aufspaltung relativ viel rascher erzielbar, obgleich es hier gegenüber den vorher besprochenen gemischten Anhydriden schärferer Reaktionsbedingungen bedarf (längere Dauer und höhere Temperatur). Dieses

Verhalten läßt sogar das Arbeiten in alkoholischem Medium zu. Schwermetallkationen, vor allem Silber-ionen fördern spezifisch die Aminolyse (*372*); vgl. S. 450).

$$Ac-NHCHCO-SC_6H_5 \quad (LXIII.) \xrightarrow{H_2NCHCOOR''} \left[\begin{array}{c} R' \quad O^{(-)} \\ | \quad | \\ Ac-NHCH-C-SC_6H_5 \\ | \\ {}^{(+)}NH_2 \\ | \\ R'-CHCOOR'' \end{array} \right] \longrightarrow$$

$$\longrightarrow Ac-NHCHCO-NHCHCOOR'' + HSC_6H_5$$

R' = Aminosäure-Seitenkette. R'' = Alkyl oder Na.

Die Ausbeute an Carbobenzoxy-di- und -tripeptiden nach dieser Methode, die im Enderfolg auf eine Aminolyse der COS-Bindung analog den Mercapto-thiazolonen (*72*) bzw. Thiazoliddionen (*27, 265*) hinausläuft (vgl. S. 521) ist eine sehr gute.

WIELAND und Mitarb. gelang auch die Darstellung der freien Aminosäure- bzw. Di- und Tripeptid-thiophenylester (*463, 457, 455*), die sich durch Erwärmen oder unter dem Einfluß von tertiären Basen (z. B. in Aceton) polykondensieren lassen. Schließlich konnten aus den energiereichen Thiophenylestern durch Umacylierung S-Acyl-derivate aliphatischer Mercaptane synthetisiert werden (*458*). Auch diese sind wiederum in der Lage, den Acyl-(bzw. Aminoacyl-)-rest zu übertragen, besonders schnell auf in „idealer Nachbarschaft" befindliche Aminogruppen des gleichen Moleküls (vgl. S. 490).

Fast gleichzeitig haben sich auch SCHWYZER und Mitarb. (*373, 377*) mit aliphatischen Thioestern und ihrer Aminolyse befaßt. Sie konnten zeigen, daß negative Substituenten der Alkoholkomponente — analog zu den geschilderten Ergebnissen der aktivierten O-Ester — den nucleophilen Angriff auf die Thiocarbonsäureester-gruppierung erleichtern. Besonders erfolgreich verliefen die Versuche mit den Carboxy-thiomethylestern (LXVI), die leicht und in guter Ausbeute durch Umsatz der gemischten Carbonsäure-alkylkohlensäure-anhydride (LXIV) mit Thioglykolsäure (LXV) entstehen (*377, 374*). Diese Thioester (LXVI) stellen

$$R-COOH \xrightarrow[\text{tert. Amin}]{ClCOOR''} R-CO-O-COOR'' \xrightarrow{HSCH_2COOH} R-CO-SCH_2COOH \longrightarrow$$
$$\qquad\qquad (LXIV.) \qquad\qquad\qquad (LXVI.)$$

$$\xrightarrow[\text{pH 6, Ag}^{(+)}]{H_2NCH_2COO^{(-)}} R-CO-NHCH_2COO^{(-)} + HSCH_2COOH$$
$$\qquad\qquad\qquad\qquad (LXV.)$$

R = Carbonsäure-substituent. R'' = Alkyl.

durchaus stabile und teilweise aus Wasser umkristallisierbare Verbindungen dar, die, am zweckmäßigsten in nicht zu verdünnter Lösung, mit Aminosäure-natriumsalzen unter Amidbildung reagieren. Besonders günstig verlief die Umsetzung bei p_H 6 und in Gegenwart von Silberionen.

Freie Aminosäure- bzw. -peptid-thioglykolester, die besondere Bedeutung für Polykondensationen besitzen, sind in Form ihrer Hydrobromide leicht durch Spaltung der entsprechenden Carbobenzoxy-derivate mit Bromwasserstoff-Eisessig zugänglich (*374*).

Im Zusammenhang mit dem biochemischen Prozeß der Acyl-übertragung durch das „Coenzym A" (z. B. S-Acetyl-Coenzym A, als „aktivierte Essigsäure" bezeichnet) (*294*), wird eine biologische Synthese von Peptiden über Aminoacyl-thioester in den Bereich des Möglichen gerückt (*295*).

F. O-Acyl-halbacetale.

Im Rahmen ihrer Arbeiten über aktivierte Ester haben Schwyzer und Mitarb. (*379, 248*) unter anderem „Methoxy-methyl"- (LXVII) und „2-Oxy-tetrahydropyranyl"-ester (LXXI) hergestellt und auf ihre peptidsynthetische Verwendung geprüft. So reagiert zwar (LXVII) rasch mit Amino-verbindungen (LXVIII) unter Amidbildung; doch reagiert auch der als Nebenprodukt anfallende Methoxy-methylalkohol (LXIX) — bzw. dessen Zerfallsprodukte Formaldehyd und Methanol — spontan mit der nucleophilen Komponente (LXVIII) zur Schiffschen Base (LXX), so daß mindestens zwei Äquivalente von (LXVIII) eingesetzt werden müssen, um eine quantitative Umsetzung von (LXVII) zu erreichen (*379*).

$$R{-}COOH + ClCH_2OCH_2 \xrightarrow{\;N(R'')_3\;} R{-}COOCH_2OCH_3 \longrightarrow$$
$$\text{(LXVII.)}$$
$$\xrightarrow{\;H_2N{-}R'\;\text{(LXVIII.)}\;} R{-}CO{-}NH{-}R' + HOCH_2OCH_3 \;\text{(LXIX.)}$$
$$HOCH_2OCH_3 \xrightarrow{\;H_2NR'\;\text{(LXVIII.)}\;} CH_2 = N{-}R' + CH_3OH + H_2O$$
$$\text{(LXIX.)} \qquad\qquad \text{(LXX.)}$$

R = Carbonsäure-substituent. R' = Amin-substituent. R'' = Alkyl.

Die Darstellung der „Oxytetrahydropyranyl-ester" (LXXI) gelingt leichter als diejenige von (LXVII) durch Anlagerung von Dihydropyran an freie Carboxylgruppen. Die Reaktionsfähigkeit dieser Derivate liegt etwas unterhalb der erwähnten Cyanmethylester.

$$R{-}COOH + \text{(Dihydropyran)} \longrightarrow R{-}COO{-}\text{(Tetrahydropyranyl)}$$
$$\text{(LXXI.)}$$

R = Carbonsäure-substituent.

Eine interessante Entdeckung machte ARENS (*24*); er fand, daß Acylaminosäuren und Aminosäureester in indifferenten Lösungsmitteln durch Zugabe von Methoxy-acetylen (LXXII) zum Acylpeptidester (LXXIV) kondensiert werden:

$$R'$$
$$Ac\text{—NHCHCOOH} + HC\equiv C\text{—OCH}_3 \longrightarrow Ac\text{—NHCHCO—O—C—OCH}_3 \longrightarrow$$
$$\text{(LXXII.)} \qquad\qquad \text{(LXXIII.)}$$

$$Ac\text{—NHCHCO—NHCHCOO}R'' + HO\text{—C—OCH}_3 \longrightarrow CH_3COOCH_3$$
$$\text{(LXXIV.)} \qquad \text{(LXXV.)} \qquad \text{(LXXVI.)}$$

$$Ac = \text{Acyl.} \quad R' = \text{Aminosäure-Seitenkette.} \quad R'' = \text{Alkyl.}$$

Wie ersichtlich, lagert sich die Carboxylkomponente an die Äthinbindung von (LXXII) an unter Bildung eines O-Acylhalbketals (LXXIII) des Methylenketons (Ketens); (LXXIII) reagiert mit dem nucleophilen Aminosäureester unter Bildung des Acylpeptidesters (LXXIV) und des Halbacetals des Ketens (LXXV), das sich spontan in Essigester (LXXVI) umlagert.

Eine Acyl-übertragung durch Halbacetale (-ketale) könnte auch in vivo z. B. als biologische Peptidsynthese erfolgen, wobei in erster Linie an die Halbacetal-Hydroxylgruppen der Zucker zu denken ist.

G. Gemischte Anhydride aus Acylaminosäure und anorganischen bzw. organischen Säuren.

*1. Die „*FISCHER*sche Säurechlorid-methode".*

Wie bekannt, versuchte FISCHER (*149*) 1903 zum ersten Male Peptide durch systematischen Aufbau zu synthetisieren. Durch Behandeln von N-Carbäthoxy-aminosäuren bzw. -peptiden mit Thionylchlorid konnten die entsprechenden Säurechloride erhalten werden, die mit Aminosäuren nach SCHOTTEN-BAUMANN oder mit zwei Äquivalenten Aminosäureestern in Chloroform der Äther kuppelten (*136*). Im letzteren Falle mußte der erhaltene Ester verseift werden. Mit Wiederholung der Operation war die Möglichkeit der Kettenverlängerung gegeben. Die Herstellung der freien Peptide scheiterte jedoch an dem Fehlen einer geeigneten Abspaltungsreaktion der N-Carbäthoxygruppe unter gelinden Bedingungen. Bei längerer alkalischer Hydrolyse erhielt FISCHER (*135*) ein Produkt, das er als die N-Carbonsäure des gesuchten Peptids ansprach. Es blieb WESSELY (*441, 442*) vorbehalten, das „FISCHERsche Verseifungsprodukt" als Carbonyl-bisglycin (Harnstoff-N,N'-diessigsäure) (LXXVII) zu identifizieren.

$$
\begin{array}{ccc}
\begin{array}{l}
\mathrm{NH{-}CH_2} \\
\;\;|\quad\;\;| \\
\mathrm{CO}\;\;\mathrm{CO} \\
\;\;|\quad\;\;| \\
\mathrm{C_2H_5O}\;\;\mathrm{NH} \\
\quad\;\;| \\
\mathrm{CH_2} \\
\quad\;\;| \\
\mathrm{COO}R''
\end{array}
&
\xrightarrow{\;\text{NaOH}\;}
&
\left[
\begin{array}{c}
\mathrm{NH{-}\!-\!-CH_2} \\
|\qquad\qquad| \\
\mathrm{OC}\quad\;\mathrm{CO} \\
\;\;\diagdown\;\;\diagup \\
\mathrm{N} \\
| \\
\mathrm{CH_2} \\
| \\
\mathrm{COOH}
\end{array}
\right]
\longrightarrow
\begin{array}{l}
\mathrm{NH{-}CH_2COOH} \\
\quad| \\
\mathrm{CO} \\
\quad| \\
\mathrm{NH} \\
\quad| \\
\mathrm{CH_2} \\
\quad| \\
\mathrm{COOH}
\end{array}
\end{array}
$$

$R'' =$ Alkyl.   (LXXVII.)

Mit der Einführung leicht abspaltbarer Schutzgruppen hat diese Methode der Verwendung von Säurechloriden eine neue Blütezeit erlebt (*384, 432*), besonders bei der Synthese von Tryptophan- (*396*) und Arginin-peptiden (*188*).

$$
Cbo\mathrm{{-}NHCHCOCl} + \mathrm{H_2NCHCOO}R''
\xrightarrow{\;-\text{HCl}\;}
Cbo\mathrm{{-}NHCHCO{-}NHCHCOO}R''
$$

(mit Seitenketten R')

$R' =$ Aminosäure-Seitenkette. $R'' =$ H oder Alkyl. $Cbo =$ Carbobenzoxy.

Pacsu und Wilson (*322*) erheben Bedenken gegen eine Verwendung dieser Derivate zur Synthese höherer Peptide und weisen darauf hin, daß bei der Darstellung von Acyl-peptidchloriden auch eine Einwirkung auf die Peptidbindung gemäß:

$$
\mathrm{{-}CO{-}NH{-}} \;\rightleftarrows\; \underset{\mathrm{OH}}{\mathrm{{-}C{=}N{-}}} \;\longrightarrow\; \underset{\mathrm{Cl}}{\mathrm{{-}C{=}N{-}}}
$$

in Betracht zu ziehen ist [vgl. aber Frankel u. a. (*171*), sowie Wieland (*463*)].

Der nächste erfolgreiche Schritt zur Darstellung der freien Peptide gelang wiederum Fischer und seiner Schule (*159*). α-Halogenfettsäurechloride kuppeln leicht mit Aminosäure- bzw. Peptidestern in organischen Lösungsmitteln zu den entsprechenden α-Halogenacyl-derivaten, wobei der freiwerdende Chlorwasserstoff durch ein Äquivalent des Esters oder einer tertiären Base gebunden wird. Nach vorsichtiger Verseifung liefert die erhaltene α-Halogenacyl-aminosäure bzw. das -peptid (die auch direkt nach Schotten-Baumann aus α-Halogenfettsäurechlorid und Aminosäure bzw. Peptid in alkalischem Medium erhalten werden können) beim Behandeln mit konzentriertem, besser mit flüssigem Ammoniak das jeweils höhere Peptid:

$$
\mathrm{Br{-}CHCOCl} + \mathrm{H_2NCHCOO}R'' \longrightarrow \mathrm{Br{-}CHCO{-}NHCHCOO}R'' \longrightarrow
$$
$$
\xrightarrow{\;\text{NH}_3\;} \mathrm{H_2NCHCO{-}NHCHCOO}R''
$$

$R' =$ Aminosäure-Seitenkette. $R'' =$ Alkyl, Na bzw. H.

Durch Heranziehung entsprechender α-Halogenfettsäuren konnten in der Folgezeit neben Glycyl- auch *DL*-Alanyl (*137*), *DL*-Leucyl- (*139*), *DL*-Phenylalanyl- (*138*) und *DL*-Prolyl-peptide (*162*, *9*) synthetisiert werden. Einen erweiterten Anwendungsbereich fand diese Methode durch:

1. Überführung der gewonnenen α-Halogenacyl-aminosäuren bzw. -peptide in die entsprechenden Säurechloride [vgl. aber Pacsu und Wilson (*322*)] und erneute „Fischer-Synthese". Erst in der letzten Stufe werden dann durch den Ersatz des α-Halogenatoms durch die NH_2-Gruppe die höheren Peptide selbst erhalten.

2. Verwendung der Säurechloride optisch-aktiver α-halogenierter Säuren (*140*). Letztere konnten mittels Racematspaltung bzw. aus optisch-aktiven Aminosäuren durch Behandeln mit Nitrosylbromid (*433*) erhalten werden. Allerdings mußte bei der folgenden Aminierung eine Waldensche Umkehrung in Kauf genommen werden (*143*, *312*). Die volle Erhaltung der optischen Aktivität wird jedoch neuerdings angezweifelt (*363*).

Bekanntlich krönte Fischer sein Werk mit der Synthese eines Oktadeca-peptids (*146*); Abderhalden und Fodor (*4*) gelang, ein Nonadeca-peptid aufzubauen.

Zur Darstellung von Glycyl-peptiden ist diese Synthese bis heute wertvoll geblieben (*343*, *212*), obgleich beim Umsatz der α-Halogenkörper mit konz., wäßrigem Ammoniak die Bildung z. B. sekundärer Amine in Betracht zu ziehen ist. Ferner wird über Ausnahmefälle bei der Aminierung von α-*D*-Brom-isocapronyl-*L*-prolin und -N-phenylglycin (*160*, *154*), sowie von α-Brom-β-phenyl-propionyl-glycin (*150*) berichtet, wobei α-Hydroxyacyl-aminosäureamide bzw. bei letzterem unter Bromwasserstoff-abspaltung ein Zimtsäurederivat entstehen (vgl. S. 472).

Die von Fischer vorgenommene Heranziehung der nicht substituierten α-Aminosäurechloride hat trotz einiger anfänglicher Erfolge (*139*, *141*, *158*, *6*) in der Folgezeit nicht befriedigt. Neuerdings empfehlen Levine (*287*), sowie Frankel und Mitarb. (*171*) verbesserte Darstellungsvorschriften für Aminosäure- bzw. Peptidchloride [vgl. auch Wieland u. a. (*457*)]. Frankel gelang deren Polymerisation durch mehrstündiges Erhitzen im Hochvakuum auf 180°.

Auch der Vorschlag, an Stelle der α-Halogen- die α-Azidoacylchloride (*69*, *173*) zu benutzen, deren Azidogruppe schließlich durch katalytische Hydrierung gegen die Aminogruppe ausgetauscht wird, hat wenig Anklang gefunden (vgl. auch S. 472).

2. *Die „Curtiussche Azid-methode".*

Die umfangreichen Arbeiten von Curtius zeigten u. a. die Verwendbarkeit der Säureazide für die „Schotten-Baumann-Reaktion". In Übertragung auf die Peptidsynthese gelang Curtius und seinen Mitarb. die Darstellung benzoylierter Di- bis Hexapeptide (*115, 122, 120, 117—119*) in guter Ausbeute. Bei der Verwendung von Aminosäureestern als Kupplungskomponente erwies es sich als glücklich, daß der entstehende Stickstoffwasserstoff gasförmig aus dem Reaktionsgemisch austrat; daher war im Gegensatz zur Säurechloridmethode kein zweites Äquivalent Aminosäureester bzw. tertiäres Amin mehr erforderlich. Mit der Auffindung leicht abspaltbarer Amino-schutzgruppen trat die Azid-methode einen Siegeszug an (*62*). Sie stellt auch heute eine der bedeutendsten Methoden zur Knüpfung der Peptidbindung, vor allem bei der Synthese höherer Peptide (*83*) dar, da es mit ihrer Hilfe möglich ist, auch Aminodicarbonsäuren (*61, 225, 338*) und Oxyaminosäuren (*176, 397*), sowie primäre Aminosäureamide (Asparagin, Glutamin, Glycinamid) (*399, 76*) ohne Verwendung besonderer Schutzgruppen einzubeziehen. Eine Racemisierung, selbst in geringem Maße, wurde dabei nicht beobachtet.

$$
\begin{array}{c}
\overset{\displaystyle R'}{\underset{\displaystyle |}{}} \\
Cbo\text{—}NHCHCO\text{—}NHNH_2 \xrightarrow{\ HNO_2\ } Cbo\text{—}NHCHCON_3 \longrightarrow
\end{array}
$$

$$
\xrightarrow[\ \ \ H_2NCHCOOR''\ \ \ ]{} Cbo\text{—}NHCHCO\text{—}NHCHCOOR'' + HN_3
$$

$$R' = \text{Aminosäure-Seitenkette.} \quad R'' = \text{Alkyl.}$$

In Erweiterung der Methode gelang Magee und Hofmann (*295a*) die Darstellung hochpolymerer Glycinpeptide, als sie Triglycin-hydrazid-hydrochlorid mit Nitrit und dann mit Alkali behandelten.

$$
\underset{,HCl}{H_2NCH_2CO\text{—}(NHCH_2CO)_2\text{—}NHNH_2} \xrightarrow{\ HNO_2\ } \underset{,HCl}{H_2NCH_2CO\text{—}(NHCH_2CO)_2N_3} \longrightarrow
$$

$$
\xrightarrow{\ Alkali\ } \ldots\text{—}CO\text{—}(NHCH_2CO\text{—}NHCH_2CO\text{—}NHCH_2CO)_x\text{—}\ldots
$$

Bei der Umsetzung mit salpetriger Säure erhält somit die Azidbildung vor der möglichen Desaminierung den Vorzug; beim Alkalischmachen kann dann die Azid- mit der freien Aminogruppe in Reaktion treten. Wird dagegen mit Bicarbonat bei 0—4° gearbeitet, so erfolgt nach Sheehan und Richardson (*390*) Ringschluß zum makrocyclischen Peptid (*387, 36*) [vgl. auch Rydon und Smith (*351*)].

Über eine weitere Verwendung der Azidmethode zur Synthese von cyclischen Polypeptiden berichten Fruton und Mitarb. (*473*). Carbobenzoxy-tripeptidazide werden mit katalytisch-erregtem Wasserstoff unter Einhaltung des „Verdünnungsprinzips" behandelt, wobei Cyclisierung eintritt [vgl. auch Boissonnas und Schumann (*78*); S. 512].

$$Cbo—NHCHCO—NHCH_2CO—NHCH_2CON_3 \xrightarrow{H_2/Pd} \begin{array}{c} CH_2C_6H_5 \\ | \\ NH—CH—CO—NH \\ | \qquad\qquad\qquad \diagdown \\ | \qquad\qquad\qquad\quad CH_2 \\ | \qquad\qquad\qquad \diagup \\ CO—CH_2—NH—CO \end{array}$$

$$\underset{CH_2C_6H_5}{|}$$

$$Cbo = \text{Carbobenzoxy.}$$

Auch bei der Azidmethode sind einige Nebenreaktionen festgestellt worden. So findet gelegentlich eine Umlagerung des Azids zum Isocyanat statt, das dann mit Aminogruppen unter Harnstoffbildung (*59*) bzw. mit benachbarten Hydroxylgruppen unter cyclischer Urethanbildung (LXXVIII) reagiert (*176, 405*).

$$Ac—NHCHCON_3 \atop \underset{R}{\overset{|}{\underset{|}{CHOH}}} \longrightarrow \left[Ac—NHCH—N=C=O \atop \underset{R}{\overset{|}{\underset{|}{CHOH}}} \right] \longrightarrow \begin{array}{c} \qquad\quad NH \\ \qquad \diagup \quad \diagdown \\ Ac—NH—CH \qquad CO \\ | \qquad\qquad | \\ CH————O \\ | \\ R \end{array}$$

(LXXVIII.)

$$R = \text{H oder } CH_3.$$

Ferner berichten PRELOG und WIELAND (*336*) über die Isolierung von Dicarbobenzoxy-lysinamid bei der Verwendung von Dicarbobenzoxy-lysinazid zu Synthesen [vgl. auch BRENNER und BURCKHARDT (*85*)].

3. Anhydride der Phosphorsäure.

CHANTRENNE (*99*) sowie SHEEHAN und FRANK (*386*) veröffentlichten Arbeiten über energiereiche Anhydride aus Carbobenzoxy- bzw. Phthalyl-aminosäuren und substituierten Phosphorsäuren (Phenyl- bzw. Dibenzyl-ester), die bei pH 7,4 und 37° mit Aminosäuren unter Peptidbildung reagieren.

$$Cbo—NHCH—\overset{\displaystyle R'}{\overset{|}{C}} \begin{array}{c} \diagup O \\ \diagdown \\ O—P=O \\ \diagdown \\ OC_6H_5 \end{array} O^{(-)} \xrightarrow{H_2NCHCOO^{(-)} \atop \overset{|}{R'}}$$

$$\xrightarrow[\text{pH 7,4}]{37°} Cbo—NHCHCO—NHCHCOO^{(-)} + HO—P\underset{OC_6H_5}{\overset{O^{(-)}}{\diagup\diagdown}}O$$
$$\qquad\qquad\qquad\quad \underset{R'}{|} \qquad\quad \underset{R'}{|}$$

$$Cbo = \text{Carbobenzoxy.} \quad R' = \text{Aminosäure-Seitenkette.}$$

Diese Reaktionsfolge schlägt eine Brücke zur biologischen Peptidsynthese, für die als wahrscheinliche Zwischenstufen Acyl-phosphate angenommen werden (*291*, dort S. 152) [vgl. AVISON (*29*)].

Die präparative Anwendung der Synthese ist durch die umständliche Darstellung der gemischten Phosphorsäureanhydride aus dem Säurechlorid einer Acylaminosäure und dem Ag-salz der substituierten Phosphorsäure sehr eingeschränkt. Bemerkenswert ist ferner die Umlagerung der unsymmetrischen Anhydride in symmetrische beim Stehen (*386, 456*).

Neuerdings haben Bentler und Netter (*42*) freie Aminophosphorsäureanhydride auf folgendem Wege aufgebaut:

$$\alpha\text{-Azido-fettsäurechlorid} + \text{Dibenzylphosphorsäure-Ag}$$
$$\downarrow - AgCl$$
$$\alpha\text{-Azido-fettsäure-(dibenzyl)-phosphorsäureanhydrid}$$
$$\downarrow H_2$$
$$\alpha\text{-Aminosäure-(dibenzyl)-phosphorsäureanhydrid}$$
$$\downarrow H_2$$
$$\alpha\text{-Aminosäure-phosphorsäureanhydrid}$$

Die Umkehr der Chantrenne-Sheehanschen Methodik ist vor kurzem Katchalsky und Paecht (*254*) geglückt: Umsetzung der Silbersalze von Carbobenzoxy-aminosäuren mit Dibenzyl-chlorophosphat (LXXIX) ergab in glatter Reaktion die gemischten substituierten Anhydride (LXXX), die beim Behandeln mit gasförmigem Bromwasserstoff in Tetrachlorkohlenstoff in die freien Aminosäure-phosphorsäureanhydride (LXXXI) übergehen. Diese polymerisieren beim Aufbewahren in wäßriger Lösung. Im Falle der ω-Anhydride der Asparagin- bzw. Glutaminsäure ließ sich die aminolytische Aufspaltung der gemischten Anhydride mit konz. Ammoniak zum ω-Amid (LXXXII) mit guter Ausbeute bewerkstelligen.

$$\begin{array}{ccc}
\begin{matrix} CH_2COOAg \\ | \\ Cbo\text{—}NHCHCOOC_7H_7 \end{matrix}
& \xrightarrow[\text{(LXXIX)}]{\overset{O}{Cl\text{—}P(OC_7H_7)_2}}
& \begin{matrix} CH_2COO\text{—}\overset{O}{P}(OC_7H_7)_2 \\ | \\ Cbo\text{—}NHCHCOOC_7H_7 \\ \text{(LXXX.)} \end{matrix} \longrightarrow
\end{array}$$

$$\begin{array}{ccc}
\xrightarrow[\substack{-4\,C_7H_7Br \\ -CO_2}]{4\,HBr}
& \begin{matrix} CH_2COO\text{—}\overset{O}{P}(OH)_2 \\ | \\ H_2NCHCOOH \\ \text{(LXXXI.)} \end{matrix}
& \xrightarrow{NH_3} \begin{matrix} CH_2CONH_2 + H_3PO_4 \\ | \\ H_2NCHCOOH \\ \text{(LXXXII.)} \end{matrix}
\end{array}$$

$$Cbo = \text{Carbobenzoxy.}$$

4. Anhydride der Phosphorigsäure.

Anderson und Mitarb. (*21, 22*) stießen auf das bereits von Cook u. a. (*109*) beschriebene Diäthylchlorophosphit als Anhydridbildner; wegen der

leichteren Darstellung und größeren Beständigkeit wurde dann das
o-Phenylen-analogon vorgezogen. Nach diesem Schema wurden einige
Peptide synthetisiert (Racemisierung wurde bei Dipeptiden nicht beob-
achtet):

$$Cbo\text{—NHCHCOOH} + \text{Cl—P}\overset{OR''}{\underset{OR''}{\diagdown}} \xrightarrow{\text{— HCl}} Cbo\text{—NHCHCOO—P}\overset{OR''}{\underset{OR''}{\diagdown}} \longrightarrow$$

$$\xrightarrow{H_2NCHCOOR''} Cbo\text{—NHCHCO—NHCHCOO}R'' + \text{HOP}(OR'')_2$$

$$Cbo = \text{Carbobenzoxy.} \quad R' = \text{Aminosäure-Seitenkette.} \quad R'' = \text{Alkyl.}$$

Wie bei den gemischten Phosphorsäureanhydriden, verläuft auch hier
eine Umlagerung im Sinne: 2 Mol unsymmetrisches $\rightarrow$ je 1 Mol sym-
metrisches Anhydrid,

$$2\ R\text{—COO—P}(OR'')_2 \longrightarrow (R\text{—CO})_2O + (R''O)_2P\text{—O—P}(OR'')_2$$

$$R = \text{Carbonsäure-substituent.} \quad R'' = \text{Alkyl.}$$

also u. a. zu einem tetraalkyliertem Pyrophosphit (LXXXIII). Dieses
bildet im Gegensatz zu dem entsprechenden Phosphorsäurederivat
erneut mit Carbonsäuren ein gemischtes Anhydrid (LXXXIV) (19), mit
Aminogruppen zur Knüpfung der Peptidbindung energiereiche Phosphitamide (s. auch S. 530).

$$Cbo\text{—NHCHCOOH} + \overset{R''O}{\underset{R''O}{\diagup}}\text{P—O—P}\overset{OR''}{\underset{OR''}{\diagdown}} \longrightarrow$$

$$\text{(LXXXIII.)}$$

$$\longrightarrow Cbo\text{—NHCHCOO—P}\overset{OR''}{\underset{OR''}{\diagdown}} + \text{HO—P}\overset{OR''}{\underset{OR''}{\diagdown}}$$

$$\text{(LXXXIV.)}$$

$$Cbo = \text{Carbobenzoxy.} \quad R' = \text{Aminosäure-Seitenkette.} \quad R'' = \text{Alkyl.}$$

Dieses Verhalten führte zur Ausarbeitung einer ,,Standard-methode'' mit Tetra-
äthyl-pyrophosphit als Anhydrid- und Phosphitamid-bildner (vgl. SS. 530—531).

5. Anhydride der Arsenigsäure.

Wie VAUGHAN (415) fand, liefert das Diäthyl-chloro-arsenit (297)
gleich seinem Phosphoranalogon (s. oben) mit Acylaminosäuren gemischte
Anhydride, die ohne vorherige Isolierung mit Aminosäureestern in inerten

Lösungsmitteln Acyl-peptidester in 60—75%iger Ausbeute ergeben. (Das als Nebenprodukt entstehende Diäthylarsenit wird von Wasser zu Arsentrioxyd zersetzt.)

$$Cbo\text{—NHCHCOOH} + Cl\text{—As}\begin{smallmatrix}OR'' \\ \\ OR''\end{smallmatrix} \longrightarrow Cbo\text{—NHCHCOO—As}\begin{smallmatrix}OR'' \\ \\ OR''\end{smallmatrix} \longrightarrow$$

mit R' unter NHCHCOOH bzw. NHCHCOO.

$$\xrightarrow{\text{H}_2\text{NCHCOO}R''} Cbo\text{—NHCHCO—NHCHCOO}R'' + \text{HO—As}\begin{smallmatrix}OR'' \\ \\ OR''\end{smallmatrix}$$

mit R' unter den NHCHCO-Gruppen.

Cbo = Carbobenzoxy. R' = Aminosäure-Seitenkette. R'' = Alkyl.

6. Thiosäuren.

Daß organische Thiosäuren hervorragende Acylierungsreagentien darstellen, erkannte zuerst Pawlewski (326, 327). Neuerdings haben Cronyn und Jiu (110), sowie Sheehan und Johnson (389) die Herstellung von Acylthio-aminosäuren (LXXXV) aus gemischten Anhydriden bzw. Säurechloriden beschrieben.

$$Ac\text{—NHCHCOO—COO}R'' \qquad\qquad Ac\text{—NHCHCOCl}$$

mit R' unter beiden NHCH-Gruppen.

$$\underset{\text{H}_2\text{S} + \text{N}(R)_3}{\xleftarrow{\hspace{1cm}-20°\hspace{1cm}}} \Big\downarrow\ \Big\downarrow\ \underset{\text{NaSH}}{\xrightarrow{\hspace{1cm}}}$$

$$Ac\text{—NHCHCOSH}$$

mit R' darunter.

(LXXXV.)

Ac = Acyl. R' = Aminosäure-Seitenkette. R'' = Alkyl. R = tert. Basen-alkyl.

In Dimethylformamid als Lösungsmittel lassen sich diese Thiosäuren (LXXXV) mit den Natriumsalzen oder Estern von Aminosäuren bzw. Peptiden glatt zu den Peptiden (LXXXVI) umsetzen:

$$Ac\text{—NHCHCOSH} + \text{H}_2\text{NCHCOO}R'' \xrightarrow{-\text{H}_2\text{S}} Ac\text{—NHCHCO—NHCHCOO}R''$$

mit R' unter den NHCH-Gruppen.

(LXXXV.)   (LXXXVI.)

Ac = Acyl. R' = Aminosäure-Seitenkette. R'' = Alkyl.

Freie Amino-thiosäuren, sowie deren Überführung in Carbobenzoxy-derivate beschrieben Wieland und Mitarb. (467, 468, 459).

7. *Anhydride der Schwefel- und Schwefligsäure.*

Leitet man Schwefeltrioxyd in Dimethylformamid unter Kühlung ein, so entsteht ein SO_3-Dimethylformamidkomplex (LXXXVII), der seinerseits mit wasserfreien Salzen im gleichen Lösungsmittel sich zu einem gemischten Anhydrid (LXXXVIII) umsetzt. In wäßriger Phase reagiert das letztere mit den Natriumsalzen von Aminosäuren nach folgendem Schema:

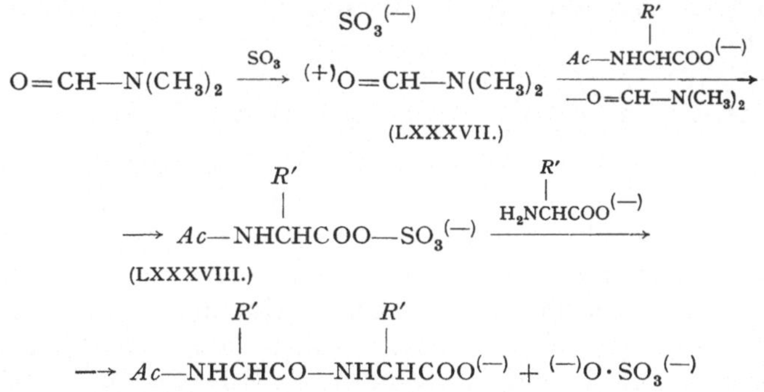

$$O{=}CH{-}N(CH_3)_2 \xrightarrow{SO_3} {}^{(+)}O{=}CH{-}N(CH_3)_2 \xleftrightarrow[\;-O{=}CH{-}N(CH_3)_2\;]{Ac{-}NHCHCOO^{(-)} \; (R')} $$

$$SO_3^{(-)}$$

(LXXXVII.)

$$\longrightarrow Ac{-}NHCHCOO{-}SO_3^{(-)} \xrightarrow{H_2NCHCOO^{(-)} \; (R')}$$

(LXXXVIII.)

$$\longrightarrow Ac{-}NHCHCO{-}NHCHCOO^{(-)} + {}^{(-)}O\cdot SO_3^{(-)}$$

$R' =$ Aminosäure-Seitenkette.

Tosyl-*DL*-alanyl-glycin, Carbobenzoxy-glycyl-phenylalanin (*DL* und *L*) und Carbobenzoxy-glycyl-phenylalanyl-glycin (*DL* und *L*) wurden u. a. mit guter Ausbeute synthetisiert (*263*).

Schon vorher hatten WIELAND und BERNHARD (*456*) über gemischte Anhydride aus zwei Äquivalenten Acylaminosäure und ein Äquivalent Sulfuryl- bzw. Thionylchlorid berichtet. Bei der Aminolyse wurden jedoch stets nur Ausbeuten unter 50% an Acylpeptiden erzielt. Am Bis-anhydrid (LXXXIX) konnte beobachtet werden, daß die Disproportionierungs-

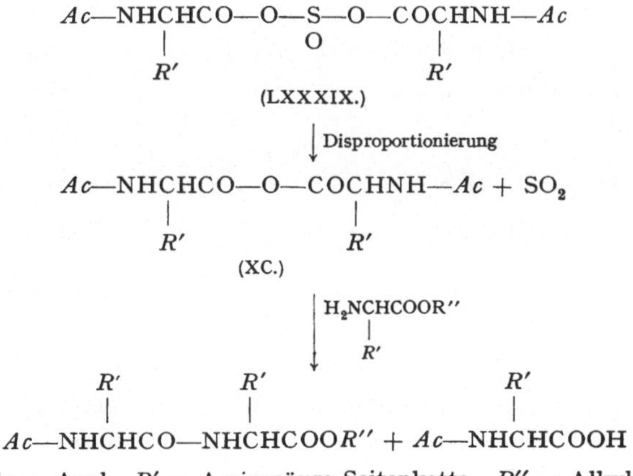

$$Ac{-}NHCHCO{-}O{-}S{-}O{-}COCHNH{-}Ac$$

(LXXXIX.)

$$\downarrow \text{Disproportionierung}$$

$$Ac{-}NHCHCO{-}O{-}COCHNH{-}Ac + SO_2$$

(XC.)

$$\downarrow H_2NCHCOOR''$$

$$Ac{-}NHCHCO{-}NHCHCOOR'' + Ac{-}NHCHCOOH$$

$Ac =$ Acyl. $R' =$ Aminosäure-Seitenkette. $R'' =$ Alkyl.

tendenz solcher Anhydride sehr groß ist. Für die Peptidknüpfung kommt dann nur mehr das symmetrische Acylaminosäure-anhydrid (XC) in Betracht, wodurch zwangsläufig nur Ausbeuten bis 50% möglich werden.

Diese Disproportionierung dürfte mehr oder minder bei allen unsymmetrischen Anhydriden stattfinden, wodurch als Nebenprodukt der aminolytischen Umsetzung kleinere oder größere Mengen Ausgangsmaterial anfallen. Beim Umsatz mit Natriumsalzen von Aminosäuren besteht dann die Gefahr, daß sich Acylaminosäure und Acylpeptid beim Aufarbeiten des Reaktionsgemisches nicht trennen lassen (vgl. auch S. 511).

8. Anhydride aliphatischer und aromatischer Carbonsäuren.

Curtius (*III, 112*) berichtete über eine Reaktion, mit der er praktisch zum Begründer der modernen „Anhydrid-methode" wurde. Bei der Einwirkung von Benzoylchlorid auf Glycin-silber in siedendem Benzol, die zum Zwecke der Benzoylierung der Aminosäure vorgenommen war, fanden sich im Reaktionsprodukt neben Hippursäure auch Benzoyldiglycin und -hexaglycin vor. Curtius schlug damals als Erklärung vor, daß aus Benzoylchlorid und Glycin-silber (XCI) Hippursäure (XCII) entsteht, die dann von einem weiteren Molekül Benzoylchlorid in das Säurechlorid (XCIII) übergeführt wird, welches sich schließlich mit Glycinsilber zum Benzoyldipeptid (XCIV) kondensiert.

$$C_6H_5COCl + H_2NCH_2COOAg \longrightarrow C_6H_5CO-NHCH_2COOH \xrightarrow{C_6H_5COCl}$$

(XCI.) (XCII.)

$$\longrightarrow C_6H_5COOH + C_6H_5CO-NHCH_2COCl \xrightarrow{H_2NCH_2COOAg}$$

(XCIII.)

$$\longrightarrow C_6H_5CO-NHCH_2CO-NHCH_2COOH$$

(XCIV.)

Curtius sprach auch die bereits von Kraut und Hartmann (*277*) geäußerte Vermutung aus, daß die erste Stufe über ein gemischtes Anhydrid (XCV) verläuft:

$$H_2NCH_2COO-COC_6H_5 \qquad\qquad C_6H_5CO-NHCH_2COOH$$

(XCV.) (XCVI.)

Es ist einleuchtend, daß sowohl aus (XCV) wie (XCVI) mit einem weiteren Molekül Benzoylchlorid das gemischte Anhydrid (XCVII) entstehen kann, wobei der gebildete Chlorwasserstoff durch Glycin-silber als Silberchlorid unter Freiwerden eines Glycin-moleküls gebunden wird. Dieses setzt sich mit dem Anhydrid (XCVII) zum Benzoyldipeptid (XCVIII) um:

$$(XCV) \text{ oder } (XCVI) + ClCOC_6H_5 \xrightarrow{-HCl} C_6H_5CO-NHCH_2COO-COC_6H_5$$
(XCVII.)

$$AgOOCCH_2NH_2 \xrightarrow{HCl} H_2NCH_2COOH + AgCl$$

$$C_6H_5CO-NHCH_2COO-COC_6H_5 + H_2NCH_2COOH \longrightarrow$$
(XCVII.)

$$\longrightarrow C_6H_5CO-NHCH_2CO-NHCH_2COOH + C_6H_5COOH$$
(XCVIII.)

Nach dem gleichen Schema kann nun die Synthese weiter über Tri-, Tetra-, Penta- zum Hexaglycin-derivat verlaufen.

Neuerdings haben WIELAND und Mitarb. (*461, 466, 454, 469*) diese Reaktion für die systematische Synthese von Peptiden nutzbar gemacht. Dabei zeigte sich, daß eine Isolierung bzw. Reindarstellung der gemischten Anhydride nicht erforderlich ist. Sie bilden sich besonders glatt aus acylierten Aminosäuren bzw. Peptiden und einem „Fremdsäurechlorid" unter Zusatz von ein Äquivalent tertiärer Base in indifferenten Lösungsmitteln. Die aminolytische Aufspaltung kann nach zwei Wegen hin verlaufen:

$$Ac-NHCHCO-NHCHCOOR'' + C_6H_5COOH$$
$$R' \qquad\qquad R'$$
(C.)

$$\uparrow Weg\ A$$

$$Ac-NHCHCO-O-COC_6H_5 + H_2NCHCOOR''$$
$$R' \qquad\qquad R'$$
(IC.)

$$\downarrow Weg\ B$$

$$Ac-NHCHCOOH + C_6H_5CO-NHCHCOOR''$$
$$R' \qquad\qquad R'$$
(CI.)

(IC.) Gemischtes Anhydrid. — (C.) Acyl-peptid. — (CI.) „Fremd"-acyl-aminosäureester. R' = Aminosäure-Seitenkette. R'' = Na oder Alkyl.

Für die Synthese von Peptiden ist nur der Weg *A*, d. h. Freiwerden der „Fremdsäure" und Knüpfung der Peptidbindung Acylaminosäure-Aminosäure-(ester) brauchbar.

Die Aufspaltungsweise dieser unsymmetrischen Anhydride durch nucleophile Reagenzien wird vom Elektronenmangel der beiden konkurrierenden —C=O-Zentren bestimmt und davon die Geschwindigkeit der Addition der nucleophilen an die elektrophile Gruppierung (vgl. S. 453).

Mit Hilfe einer eleganten Methode hat WIELAND und seine Schule (*454, 469*) die Verwendbarkeit von „Hilfssäuren" als Anhydridpartner festgestellt. Als reaktionsschwächste und damit Weg *B* verwerfende Acylhälften haben sich Benzoesäure, verzweigte aliphatische Carbonsäuren [vgl. Vaughan und Osato (*420*)], des weiteren Trifluoressigsäure (S. 462) und vor allem Kohlensäureester (s. unten) erwiesen. Eine Anzahl von teils optisch-aktiven Peptiden wurde nach diesem Verfahren in meist ausgezeichneter Ausbeute hergestellt (*466, 420, 437, 345*).

9. *Anhydride der Kohlensäure.*

Die Einbeziehung der Anhydride der Kohlensäure bzw. substituierter Kohlensäure mit Acylaminosäuren in die Peptidsynthese brachte einen besonders guten Erfolg. Sie sind bequem und billig herstellbar und liefern die gewünschten Verbindungen in hoher Reinheit und bester Ausbeute. Neben der Azid- ist wohl diese Methode der Peptidknüpfung zur meist angewandten geworden; so berichten u. a. über die Synthese von teilweise höheren Peptiden Boissonnas (*75*), Brand (*354, 355, 130*) und Vaughan (*419*).

a) Bisanhydride.

Wie Sehring fand, bildet auch Phosgen mit Acylaminosäuren in indifferenten Lösungsmitteln in Gegenwart tertiärer Basen „gemischte Anhydride" (CII), deren Tendenz, unter Disproportionierung zu zerfallen, jedoch selbst bei 0° groß ist. Hierbei entstehen die symmetrischen Anhydride der Acylaminosäuren (CIII) (*456*):

$$Ac\text{—}NHCHCO\text{—}O\text{—}CO\text{—}O\text{—}COCHNH\text{—}Ac$$

$$R' \qquad\qquad\qquad R'$$

$$\text{(CII.)}$$

$$\Big\downarrow - CO_2$$

$$Ac\text{—}NHCHCO\text{—}O\text{—}COCHNH\text{—}Ac$$

$$R' \qquad\qquad R'$$

$$\text{(CIII.)}$$

$$R' = \text{Aminosäure-Seitenkette.}$$

Wenngleich diese Tatsache für die Peptidsynthese wenig günstig ist, so stellt doch die leicht und nebenprodukt-freie Bildung symmetrischer Anhydride einen Fortschritt dar.

Über ein gemischtes Anhydrid der Chlorkohlensäure berichten Liwschitz und Zilkha (*292*) (vgl. S. 468).

b) Anhydride der Mono-alkyl-kohlensäure.

Es wurden von Wieland und seiner Schule (*456*) verschiedene Halbesterchloride der Kohlensäure mit positivem Resultat als Anhydrid-

Synthese von Peptiden. | 511

bildner erprobt. Hält man die Temperatur unter 0°, so wird jedwede Disproportionierung des gemischten Anhydrids zurückgedrängt; dies äußert sich u. a. am Ausbleiben einer Kohlendioxyd-Entwicklung. Die aminolytische Aufspaltung durch Aminosäuren verläuft in Richtung Acyl-peptid + Kohlensäurehalbester (CIV), der sofort in Kohlendioxyd und einen Alkohol zerfällt.

Fast gleichzeitig veröffentlichten auch BOISSONNAS (*75*) und VAUGHAN (*416*, *421*) analoge Arbeiten. Einheitlich wird festgestellt, daß längere und vor allem verzweigte Alkylreste als Substituenten der Kohlensäure-acylhälfte die Ausbeute an Acylpeptiden steigern. Der Grund liegt hier in einer Beeinflussung der Elektronenkonfiguration an der Carbonsäure-anhydridgruppierung zugunsten der für eine erfolgreiche Anlagerung der nucleophilen Amino-Verbindungen notwendigen Kationisierung der Acyl-aminosäure-hälfte. Diese Kationisierung äußert sich auch in der Erschwerung der hydrolytischen Spaltung solcher Ester und ihrer Hydrazid-bildung bei der Einwirkung von Hydrazinhydrat.

$$Ac\text{—}NHCHCOOH + Cl\text{—}COOR''' $$

$$\downarrow -HCl$$

$$Ac\text{—}NHCHCO\text{—}O\text{—}COOR''' \longleftrightarrow \left[\, Ac\text{—}NHCH\text{—}C \ldots \right]$$

$$\downarrow \; H_2NCHCOOR''$$

$$Ac\text{—}NHCHCO\text{—}NHCHCOOR'' + HO\text{—}C \xrightarrow{-CO_2} R'''OH$$

(CIV.)

R' = Aminosäure-Seitenkette. R'' = R''' = Alkyl.

Es erwies sich auch vorteilhafter (*75*, *416*, *278*), als Kupplungskomponente Aminosäureester in indifferenten Lösungsmitteln an Stelle der Natriumsalze der Säuren zu verwenden; in einigen Fällen wurde eine nicht abtrennbare Verunreinigung des Acylpeptids mit dem Acyl-ausgangsmaterial nachgewiesen (*359*, *478*).

In weiteren Arbeiten stellten VAUGHAN und Mitarb. (*417*, *418*) die Bedingungen fest, unter denen ganz oder teilweise Racemisierung während der Reaktion erfolgen kann. Vor allem tritt dies bei der Darstellung höherer Peptide aus Acyl-, aber auch Carbobenzoxy-peptiden auf, sofern die das Anhydrid bildende letzte Aminosäure optisch-aktiv ist. Als maßgebender Faktor scheint hierbei das indifferente Lösungsmittel zu fungieren: so wurde z. B. in Dioxan bzw. Tetrahydrofuran im Gegensatz zu Chloroform-Toluol kaum eine Racemisierung beobachtet.

Die Synthese von hydroxyl-haltigen Peptiden mit Hilfe der gemischten Alkylkohlensäureanhydride erwies sich ebenfalls als gangbar, wie Schwyzer (*371*), King (*270*) bzw. Grassmann und Mitarb. (*242, 204*) zeigen konnten. Schwyzer gelang gleichzeitig die Darstellung von höchst säureempfindlichen Amiden des Äthylenimins, als er die entstehende Esterkohlensäure mit einem Äquivalent einer tertiären Base blockierte.

Als Boissonnas und Schumann (*78*) dieses Anhydrid-verfahren auf das freie Tripeptid *D*-Leucyl-glycyl-glycin (CV) übertrugen, erhielten sie eine ninhydrin-negative Substanz, die sie als ein Cyclopeptid (CVI) ansprechen:

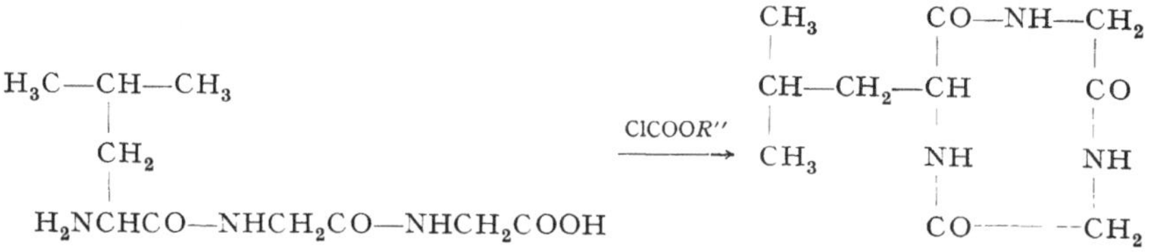

$$
\begin{array}{ccc}
\mathrm{H_3C-CH-CH_3} & & \\
| & & \\
\mathrm{CH_2} & \xrightarrow{\;\mathrm{ClCOOR''}\;} & \\
| & & \\
\mathrm{H_2NCHCO-NHCH_2CO-NHCH_2COOH} & &
\end{array}
$$

(CV.)

$$
\begin{array}{lll}
\mathrm{CH_3} & \mathrm{CO-NH-CH_2} \\
| & | & | \\
\mathrm{CH-CH_2-CH} & & \mathrm{CO} \\
| & | & | \\
\mathrm{CH_3} & \mathrm{NH} & \mathrm{NH} \\
& | & | \\
& \mathrm{CO} \!-\!-\!-\! \mathrm{CH_2}
\end{array}
$$

(CVI.)

$R'' = \text{Alkyl.}$

10. Die N(Im)-Acyl-imidazole.

Aus Untersuchungen von Gerngross ist bekannt, daß N-acylierte Imidazole, wie N-Benzoylimidazol (*184, 185*) oder N,N(Im)-Dibenzoyl-histidinester (*186*), äußerst leicht durch Wasser gespalten werden. Bergmann und Zervas (*54*) haben festgestellt, daß der Imidazolacyl-rest noch rascher der aminolytischen Abspaltung unterliegt; es gelang ihnen u. a. die Darstellung von Benzoyl-diglycin aus N,N(Im)-Di-(benzoyl-glycyl)-histidinmethylester (CVII) und Glycin-Natrium:

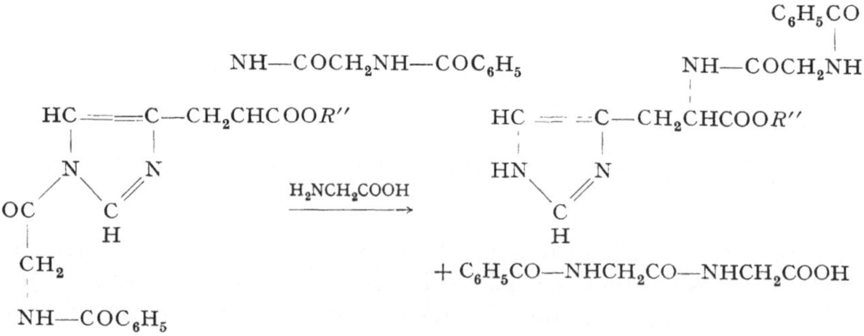

$$
\begin{array}{l}
\mathrm{NH-COCH_2NH-COC_6H_5} \\
| \\
\mathrm{HC} \!=\!=\! \mathrm{C-CH_2CHCOOR''} \\
| \qquad\quad | \\
\;\mathrm{N} \quad\;\; \mathrm{N} \\
/ \quad\;\; \diagdown \quad \diagup \\
\mathrm{OC} \qquad \mathrm{C} \\
| \qquad\quad \mathrm{H} \\
\mathrm{CH_2} \\
| \\
\mathrm{NH-COC_6H_5}
\end{array}
$$

(CVII.)

$$\xrightarrow{\;\mathrm{H_2NCH_2COOH}\;}$$

$$
\begin{array}{l}
\mathrm{C_6H_5CO} \\
| \\
\mathrm{NH-COCH_2NH} \\
| \\
\mathrm{HC} \!=\!=\! \mathrm{C-CH_2CHCOOR''} \\
| \qquad\quad | \\
\mathrm{HN} \quad\;\; \mathrm{N} \\
\diagdown \quad\; \diagup \\
\mathrm{C} \\
\mathrm{H}
\end{array}
$$

$$+\ \mathrm{C_6H_5CO-NHCH_2CO-NHCH_2COOH}$$

$R'' = \text{Alkyl.}$

Da nach Zervas und Bergmann auch Benzoyl-theobromin befähigt ist, seinen Acylrest auf „auffangbereite" Verbindungen, wie Aminosäuren, Zucker usw., zu übertragen, liegt es auf der Hand, daß allen Purinen und damit den Nucleinsäuren diese Fähigkeit zukommen dürfte. Somit steht neben energiereichen Phosphaten, S-Acyl-derivaten und O-Acyl-halbacetalen eine vierte Möglichkeit der biologischen Peptidsynthese offen [s. auch Stadtman und White (401)].

Neuerdings haben Wieland und Schneider (465) das Studium dieser Verbindungsklasse weitergeführt.

11. Das „Carbodiimid-Verfahren".

Die bisher einfachste und eleganteste Methode der Peptidknüpfung wurde von Sheehan und Hess (388) aufgefunden. Acylaminosäuren (Carbobenzoxy- oder Phthalylderivate) kuppeln mit Aminosäureestern in indifferenten Lösungsmitteln und bei Raumtemperatur nach Zugabe von ein Mol N,N'-Dicyclohexyl-carbodiimid (CVIII) (360, 361) in guter Ausbeute zu den gewünschten Acylpeptidestern (CXI). Der als Nebenprodukt anfallende N,N'-Dicyclohexyl-harnstoff (CXII) ist schwer löslich und leicht abzutrennen.

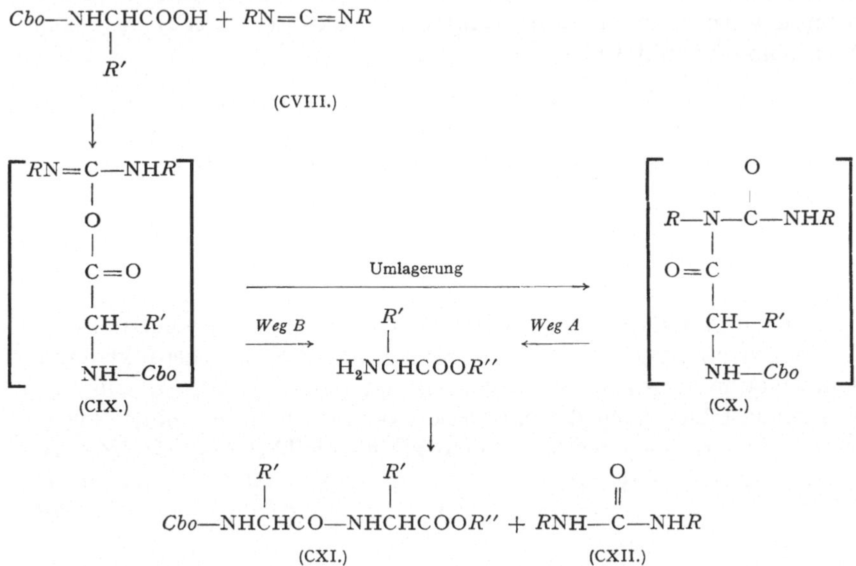

Nach Khorana (266) erfolgt als erste Stufe des Reaktionsverlaufs die Addition der Carboxylkomponente zum O-Acyllactim (CIX), das sich leicht durch Acylwanderung in ein N-Acyl-ureid (CX) umlagert. Nach

Kent (*264*) wird dann (CX) aminolytisch zum Säureamid (CXI) und disubstituierten Harnstoff (CXII) gespalten (Weg *A*)*.

Nach Auffassung der Verfasser dürfte jedoch eine nucleophile Additionsreaktion der Aminoverbindung an das O-Acyl-lactim (CIX), die letztlich zu den gleichen Endprodukten führt (Weg *B*), wahrscheinlicher sein.

Die Vorteile dieser Peptidsynthese werden noch durch folgende Umstände gesteigert (*388*): 1. Feuchtigkeit ist ohne Einfluß auf den Reaktionsablauf; die Umsetzung läßt sich sogar in wasserhaltiger Lösung ermöglichen. 2. Die Einbeziehung von Oxy-aminosäuren gelingt einwandfrei, ohne zusätzliche O-Schutzgruppen [s. auch Sheehan und Mitarb. (*386 a*)]. 3. Racemisierung wurde selbst beim Aufbau von höheren Peptiden aus Carbobenzoxy-dipeptiden mit carboxyl-endständiger, optisch-aktiver Aminosäure nicht beobachtet.

12. Cyclische „innermolekulare" Anhydride.

a) Die Carbonsäure-anhydride (Oxazolid-2,5-dione).

In Fortsetzung der Fischerschen Arbeiten mit Carbäthoxy-aminosäuren fand Leuchs (*282*), daß sich Carbäthoxy-glycylchlorid (CXIII) beim Erwärmen unter Abspaltung von Chloräthyl (CXIV) zu Glycin-N-carbonsäure-anhydrid (CXV) zersetzt:

$$R''O-CO-NHCHCOCl \longrightarrow R''Cl + OC\underset{O}{\overset{HN-CH-R'}{<}}CO$$

$$\text{(CXIII.)} \qquad \text{(CXIV.)} \qquad \text{(CXV.)}$$

$$R' = H. \quad R'' = \text{Äthyl.}$$

Bessere Ausbeuten wurden bei Verwendung der Carbomethoxy-verbindungen erhalten; schließlich erwiesen sich die Carbobenzoxy-aminosäurechloride als günstigstes Ausgangsmaterial (*56*). Auf diesem Wege wurden fast alle, auch die optisch-aktiven Aminosäuren, ohne Racemisierung in ihre N-Carbonsäureanhydride übergeführt (*65, 261, 167, 128, 33, 215*), wobei weitere funktionelle Gruppen geschützt sein müssen. Die leichter erfolgende Cyclisierung von Carboalkoxy-aminosäure-bromiden beschreiben Ben-Ishai und Katchalski (*41*).

Einen zweiten wichtigen Weg zur Darstellung der Oxazoliddione beschritt Fuchs (*181*) durch direkte Einwirkung von Phosgen auf N-Phenyl-

---

* Ein ähnlicher Umsetzungsmechanismus energiereicher N-Diacyl-verbindungen — man denke an N-Acyl-diketopiperazine, sowie an die N(Im)-Acylimidazole — könnte auch in vivo verlaufen.

glycin. Diese direkte „Phosgenierung" hat eine englische Forschergruppe neuerdings auf N-unsubstituierte *DL*- und *L*-Aminosäuren übertragen und in zahlreichen Arbeiten [FARTHING, COLEMAN u. a. (*91, 288, 133, 132, 102, 101*)] die günstigsten Reaktionsbedingungen (speziell in Dioxan und Tetrahydrofuran) ermittelt.

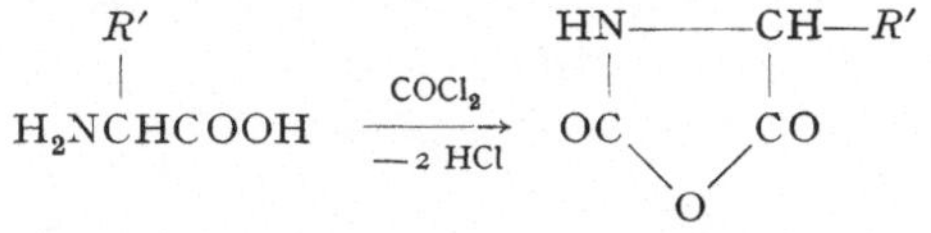

$$R' = \text{Aminosäure-Seitenkette.}$$

BRUCKNER und Mitarb. (*95*), sowie BIRKOFER und KACHEL (*73*) konnten nach diesem Verfahren erstmals die N-Carbonsäureanhydride von β-Aminosäuren gewinnen. Während im Falle des Asparaginsäure-α-äthylesters der direkte Ringschluß gelang, erhielt BIRKOFER aus β-Alanin als erstes Produkt eine N-Chloroformyl-β-aminosäure, die schließlich unter Einwirkung tertiärer Basen das gewünschte Derivat lieferte. Weitere Darstellungsmethoden für Oxazoliddione stammen von CURTIUS (*121*) bzw. HURD und BUESS (*245*), die von der Umlagerung der „Halbazide" bzw. „Halbhydroxamsäuren" substituierter Malonsäuren (CXVI bzw. CXVII) Gebrauch machen:

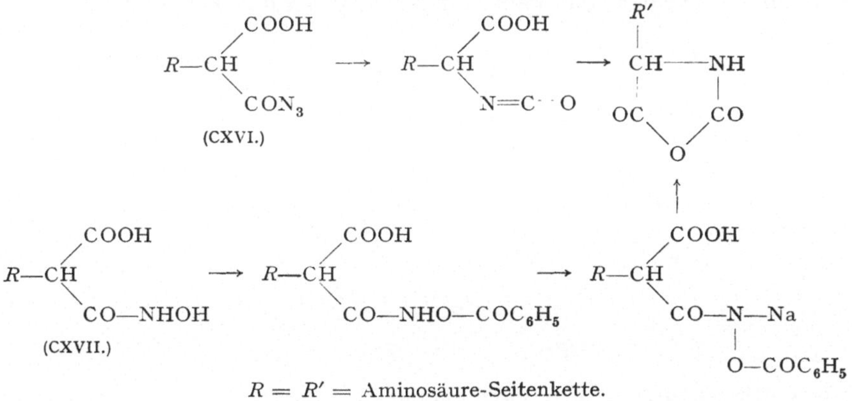

(CXVI.)

(CXVII.)

$$R = R' = \text{Aminosäure-Seitenkette.}$$

Bereits LEUCHS und MANASSE (*284*) hatten die leichte aminolytische Aufspaltbarkeit des N-Carbonsäureanhydrids von N-Phenylglycin (CXVIII) festgestellt. Hierbei entsteht, wie auch FUCHS (*181*) später aufzeigte, sogleich das Amid der substituierten Aminosäure (CXIX), da durch einen Basenüberschuß als Folgereaktion die N-Carboxylgruppe abgespalten wird.

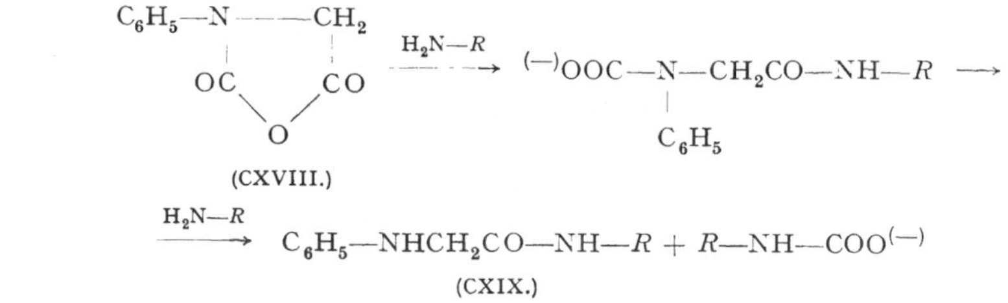

$$C_6H_5-N-\!-\!-CH_2$$
$$OC\diagdown\quad\diagup CO \xrightarrow{\;H_2N-R\;} {}^{(-)}OOC-N-CH_2CO-NH-R \longrightarrow$$
$$\diagdown O \diagup \qquad\qquad\qquad\qquad C_6H_5$$

(CXVIII.)

$$\xrightarrow{\;H_2N-R\;} C_6H_5-NHCH_2CO-NH-R + R-NH-COO^{(-)}$$

(CXIX.)

$$R = \text{Amino-substituent.}$$

Wessely und seine Schule (*440, 394, 445*) haben diese Reaktion am Phenylalanin-N-carbonsäureanhydrid näher untersucht. Dabei gelang ihnen erstmals die Synthese von Peptiden, als sie auf das genannte Oxazoliddion die Natriumsalze von Aminosäuren bzw. Peptiden in wäßriger Lösung einwirken ließen. Die aminolytische Aufspaltung mit Aminosäureestern beschrieben später Hunt und du Vigneaud (*243*). Doch hat erst Bailey (*32, 33*) einen die Nebenreaktionen ausschließenden Weg der Synthese aufgefunden: Die gleichzeitige Einwirkung von Aminosäureestern und tertiären Basen auf die „inneren Anhydride" (CXV) in indifferenten Lösungsmitteln bei tiefen Temperaturen (—10 bis —70°) liefert das carbaminsaure Salz eines Peptidesters (CXX). Bei 25—40° erfolgt thermische Spaltung in tertiäre Base, Kohlendioxyd und Peptidester (CXXI); Behandeln mit zwei Äquivalenten alkoholischer Salzsäure liefert u. a. das Peptidester-hydrochlorid (CXXII).

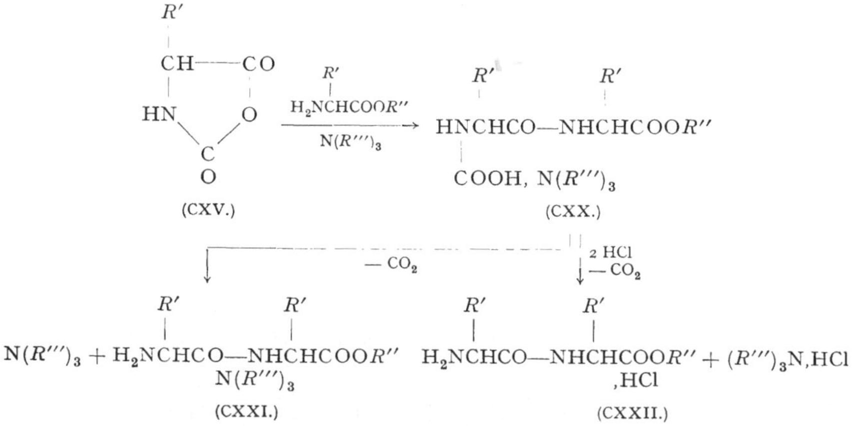

$$R'$$
$$CH-\!-\!-CO \qquad R'$$
$$HN\diagdown\quad\diagup O \xrightarrow[N(R''')_3]{H_2NCHCOOR''} HNCHCO-NHCHCOOR''$$
$$\diagdown C \diagup \qquad\qquad\qquad\qquad COOH,\ N(R''')_3$$
$$O$$

(CXV.) $\qquad\qquad\qquad\qquad\qquad\qquad$ (CXX.)

$$-CO_2 \qquad\qquad \downarrow \begin{array}{l}2\ HCl\\ -CO_2\end{array}$$

$$R' \qquad\quad R' \qquad\qquad\qquad R' \qquad\quad R'$$
$$N(R''')_3 + H_2NCHCO-NHCHCOOR'' \qquad H_2NCHCO-NHCHCOOR'' + (R''')_3N,HCl$$
$$N(R''')_3 \qquad\qquad\qquad\qquad\qquad\qquad ,HCl$$

(CXXI.) $\qquad\qquad\qquad\qquad\qquad\qquad$ (CXXII.)

$$R' = \text{Aminosäure-Seitenkette.} \quad R'' = R''' = \text{Alkyl.}$$

Als Verbesserung dieses Reaktionsweges schlagen Langenbeck und Kresse (*280*) die Verwendung von Tribenzylamin als tertiäre Base vor, wegen der Schwerlöslichkeit des carbaminsauren Salzes. — Außer der aminolytischen hatte LEUCHS auch die hydrolytische Spaltung der N-Carbonsäureanhydride untersucht. Es zeigte sich, daß zur völligen Hydrolyse unzureichende Mengen Wasser unter Kohlendioxyd-Entwicklung ein unlösliches, hochmolekulares Produkt ergeben (283). Seit den Arbeiten von K. H. MEYER u. a. (300) werden so gebildete Stoffe als echte Polypeptide angesehen (16, 91). WOODWARD und SCHRAMM (476) haben Polymere des Leucins und Phenylalanins erhalten, als sie deren Oxazoliddione in gewöhnlichem Benzol aufbewahrten; hierbei leiten Spuren von Wasser mit der Aufspaltung des Ringsystems einiger Anhydrid-moleküle die Polymerisation ein.

$$
\begin{array}{ccccc}
R' & & R' & & R' \\
| & & | & & | \\
CH\!\!-\!\!-\!\!CO & \xrightarrow{\ H_2O\ } & CH\!\!-\!\!COOH & \xrightarrow{\ -CO_2\ } & CH\!\!-\!\!COOH \\
| \quad\ | & & | & & | \\
HN \quad\ O & & NH\!\!-\!\!COOH & & NH_2 \\
\ \ \diagdown\!\!\diagup & & & & \\
\ \ CO & & & & \\
(CXV.) & & & &
\end{array}
$$

$$
\begin{array}{c}
R' \\
| \\
CH\!\!-\!\!COOH \\
| \\
NH_2
\end{array}
$$

$$
\begin{array}{ccc}
R' \qquad\ R' & & R' \qquad\quad\ R' \\
| \qquad\quad\ | & & | \qquad\qquad\ | \\
CHCO\!\!-\!\!NHCHCOOH & \xrightarrow{\ -CO_2\ } & CHCO\!\!-\!\!NHCHCOOH \ \ usw. \\
| & & | \\
NH\!\!-\!\!COOH & & NH_2
\end{array}
$$

$$R' = \text{Aminosäure-Seitenkette.}$$

Die Polymerisation gelingt auch leicht in Gegenwart tert. Basen [(26); betr. Mechanismus vgl. (452—455)].

Als spezielle Polypeptide konnten so Poly-lysin (256, 257, 321), Poly-ornithin und daraus Poly-arginin (260), Poly-glutaminsäure (215, 92, 321, 94) und -glutamin (93), Poly-β-asparaginsäure (95), Poly-N-β-benzyl-asparagin (170), Poly-serin (166), sowie Poly-histidin (258), Poly-prolin (43) und Poly-tyrosin (259, 321) synthetisiert werden. Neuerdings berichten KATCHALSKI und Mitarb. (324) über die Darstellung des Poly-tryptophans, die aus dem N-Carbonsäureanhydrid bei 150° und Hochvakuum unter Kohlendioxyd-Abspaltung gelang. Vernetzung zwischen α-Carboxyl- und Indol-NH-Gruppen trat nicht ein.

Auch die Synthese unregelmäßig gemischter Polypeptide, z. B. von Lysin : Tyrosin (302, 321), Lysin : Glutaminsäure (321) und Glutaminsäure : Tyrosin (321) wird neuerdings beschrieben.

In den letzten Jahren sind zahlreiche Arbeiten über den Mechanismus dieser Polykondensationsreaktion veröffentlicht worden [s. auch KATCHALSKY (254)]. So zeigten WALEY und WATSON (436), daß die Polykondensation von N-Methyl-glycin-N-carbonsäureanhydrid (CXXIII) auch

durch geringe Mengen eines niedermolekularen Poly-sarkosin-diäthylamids
(CXXIV) eingeleitet werden kann:

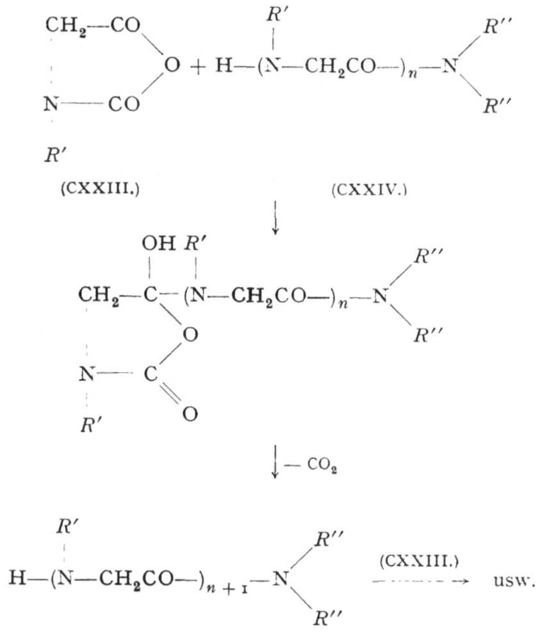

$$
\begin{array}{c}
\text{CH}_2\text{—CO} \\
\diagdown \\
\text{O} + \text{H—(N—CH}_2\text{CO—)}_n\text{—N} \\
\diagup \\
\text{N——CO} \\
\\
R'
\end{array}
\qquad
\begin{array}{c}
R' \\
| \\
\\
\nearrow R'' \\
\searrow R''
\end{array}
$$

(CXXIII.)					(CXXIV.)

$$
\downarrow
$$

$$
\begin{array}{c}
\text{OH } R' \\
| \quad | \\
\text{CH}_2\text{—C—(N—CH}_2\text{CO—)}_n\text{—N} \\
\diagdown \\
\text{O} \\
\\
\text{N——} \quad \text{C} \\
\diagdown \\
R' \qquad \text{O}
\end{array}
\qquad
\begin{array}{c}
\nearrow R'' \\
\searrow R''
\end{array}
$$

$$
\downarrow - CO_2
$$

$$
\begin{array}{c}
R' \\
| \\
\text{H—(N—CH}_2\text{CO—)}_{n+1}\text{—N} \\
\diagup \diagdown \\
R'' \quad R''
\end{array}
\qquad
\begin{array}{c}
\text{(CXXIII.)} \\
\text{------} \rightarrow \quad \text{usw.}
\end{array}
$$

$$R' = \text{Methyl.} \quad R'' = \text{Äthyl.}$$

Analogen Gesetzmäßigkeiten gehorcht nach Heyns und Brockmann
(*227a*) die thermische Zersetzung der Leuchsschen Körper, wobei das
gebildete Kohlendioxyd eindeutig vom N-Carboxyl stammt (Isotopen-
methode):

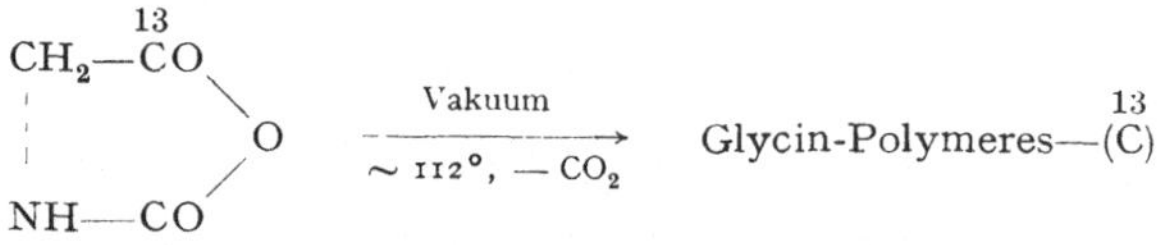

$$
\begin{array}{c}
\overset{13}{\text{CH}_2}\text{—CO} \\
\diagdown \\
\quad\quad \text{O} \\
\diagup \\
\text{NH—CO}
\end{array}
\quad
\xrightarrow[\sim 112°, \; - CO_2]{\text{Vakuum}}
\quad
\text{Glycin-Polymeres—}\overset{13}{\text{(C)}}
$$

Sela und Berger (*380*) isolierten eine substituierte Hydantoinessig-
säure (CXXV) vom Reaktionsgemisch einer thermischen Polymerisation;
ähnliche Ergebnisse erzielten auch Wessely und Mitarb. (*443, 71*). Damit
scheint festzustehen, daß das gemischte innermolekulare N-Carbonsäureanhydrid (CXV), wenn auch nur im geringen Maße, neben dem vor-
herrschenden Polarisationszustand (CXXVI) auch den entgegengesetzten
(CXXVII) ausbilden kann. Mit Weg *B* ist stets ein Kettenabbruch des
Systems verbunden.

$$R' \quad O$$

(CXXVII.) $\longleftrightarrow$ (CXV.) $\longleftrightarrow$ (CXXVI.)

$$\text{CH—C} \overset{O}{\underset{O^{\delta-}}{\Big\langle}} \qquad \text{CH—CO} \overset{}{\underset{O}{\Big\langle}} \qquad \text{CH—C} \overset{O}{\underset{O^{\delta-}}{\Big\langle}}$$

$$\text{NH—C} \overset{\delta+}{\underset{O}{\Big\langle}} \qquad \text{NH—CO} \qquad \text{NH—C} \overset{}{\underset{O}{\Big\langle}}$$

Weg B $\Big\downarrow \; \underset{H_2NCHCOOH}{\overset{R'}{|}}$
$\qquad\qquad\qquad$ *Weg A* $\Big\downarrow \; \underset{H_2NCHCOOH}{\overset{R'}{|}}$

$$\underset{HN—CO—NHCHCOOH}{\overset{R'}{\underset{|}{\overset{|}{CH—COOH}}}}$$
$$\underset{\underset{R'}{|}}{}$$

$$\underset{NH—COOH}{\overset{R' \qquad R'}{CHCO—NHCHCOOH}}$$

$\Big\downarrow H^{(+)}$
$\qquad\qquad\qquad$ $\Big\downarrow \left[\begin{array}{c} -CO_2 \\ +\,(CXXVI.) \end{array} \right]_{n-1}$

$$\underset{NH—CO}{\overset{R'}{CH—CO}} \overset{R'}{\underset{N—CHCOOH}{\Big\rangle}}$$
(CXXV.)

$$\underset{NH—COOH}{\overset{R' \qquad\qquad R'}{CHCO—(NHCHCO)_n—OH}}$$

$$R' = \text{Aminosäure-Seitenkette.}$$

Bamford und Mitarb. (35) konnten die Hydantoin-essigsäurebildung in teilweise erheblichen Prozentsätzen neben der von cyclischen Polypeptiden ebenfalls feststellen, als sie N-Carbonsäureanhydride (CXVa) in N,N-Dialkyl-formamid mit in diesem Lösungsmittel löslichen anorganischen Salzen oder tertiären Basen behandelten. Sie legen folgenden Mechanismus zugrunde:

$$\underset{NH—CO}{\overset{CH_2—CO}{\Big\rangle}} O + Li^{(+)} \longrightarrow \underset{N=\!=\!C—OLi}{\overset{CH_2—CO}{\Big\rangle}} O + H^{(+)}$$

(CXVa.) $\qquad\qquad\qquad\qquad$ (CXXVIII.)

$$(CXV) + (CXXVIII) + H^{(+)} \rightleftharpoons \underset{\underset{NH—CO}{\overset{|}{CH_2—C—OH}}}{\overset{CH_2—CO}{\Big\rangle}} \overset{O}{\underset{N—CO}{\Big\rangle}} + Li \quad \xrightarrow{-CO_2} \quad \underset{\underset{NH_2}{\overset{|}{CH_2—C=O}}}{\overset{CH_2—CO}{\Big\rangle}} \overset{}{\underset{N—CO}{\Big\rangle}} O + Li^{(+)}$$

(CXXIX.)

Das zunächst gebildete N-Aminoacyl-N-carbonsäureanhydrid (CXXIX) reagiert mit weiteren Molekülen des Ausgangsmaterials (CXV) zu N-„Peptidyl"-oxazoliddionen (CXXX), bis zum Kettenabbruch durch innermolekularen Ringschluß, zum Cyclo-polypeptid (CXXXI) unter CO_2-Abspaltung (Weg C); es erfolgt jedoch auch eine innermolekulare Umlagerung zum Hydantoinderivat (CXXXII) (Weg D):

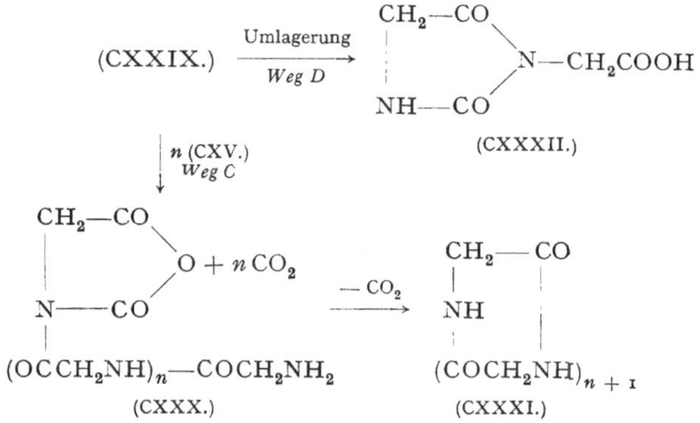

$$
(CXXIX.) \xrightarrow[\textit{Weg D}]{\text{Umlagerung}}
\begin{array}{l}
CH_2\!-\!CO \\
\qquad\qquad\quad N\!-\!CH_2COOH \\
NH\!-\!CO
\end{array}
$$

(CXXXII.)

$$
\Big\downarrow \; \begin{array}{l} n\,(CXV.) \\ \textit{Weg C} \end{array}
$$

$$
\begin{array}{l}
CH_2\!-\!CO \\
\qquad\qquad O + n\,CO_2 \\
N\!-\!CO \\
(OC\,CH_2NH)_n\!-\!COCH_2NH_2
\end{array}
\xrightarrow{-\,CO_2}
\begin{array}{l}
CH_2\!-\!CO \\
NH \\
(COCH_2NH)_{n+1}
\end{array}
$$

(CXXX.)　　　　　　　　(CXXXI.)

Die gleichfalls diskutierte Möglichkeit, daß die Polykondensation der N-Carbonsäureanhydride über freie Radikale verlaufen könnte, wird ausgeschlossen durch einen Versuch von Coleman und Farthing (102), bei dem Acrylester als Lösungsmittel benutzt wurde. Eine Polymerisation des Acrylesters, die im Falle eines Radikal-Mechanismus zu erwarten wäre, trat dabei nicht ein.

Auch die Ringöffnung der Oxazoliddione mittels Alkohol- bzw. Mercaptan-Chlorwasserstoff zu Ester- (65) und Thioester-hydrochloriden (455) ist für die Peptidsynthese eine Bereicherung (vgl. S. 475). In Übertragung auf das Serin als Verbindung mit einer Alkohol-funktion gelang Moore und Mitarb. (309) die Synthese des O-Glycyl-L-Serins und daraus des tumor-hemmenden Antibioticums „Azaserin":

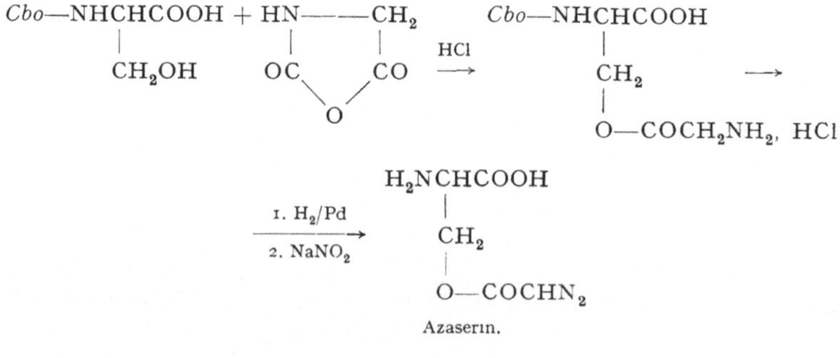

$$
\begin{array}{l}
Cbo\!-\!NHCHCOOH + HN\!-\!\!-\!\!-CH_2 \\
\qquad\;\; CH_2OH \qquad OC\quad CO \\
\qquad\qquad\qquad\qquad\quad\; O
\end{array}
\xrightarrow{HCl}
\begin{array}{l}
Cbo\!-\!NHCHCOOH \\
\qquad\;\; CH_2 \\
\qquad\;\; O\!-\!COCH_2NH_2,\ HCl
\end{array}
\longrightarrow
$$

$$
\xrightarrow[\text{2. } NaNO_2]{\text{1. } H_2/Pd}
\begin{array}{l}
H_2NCHCOOH \\
\quad CH_2 \\
\quad O\!-\!COCHN_2
\end{array}
$$

Azaserin.

Cbo = Carbobenzoxy.

b) Die Mercapto-thiazolone (Thio-thiazolidone).

In einer Reihe von Arbeiten berichten HEILBRON, COOK, LEVY und Mitarb. (*103, 104*) [vgl. auch die Zusammenfassung von HEILBRON (*226*)] über die Darstellung und Eigenschaften von Dithio-analoga der N-Carbonsäureanhydride, der „2-Mercapto-thiazolone" (CXXXIII).

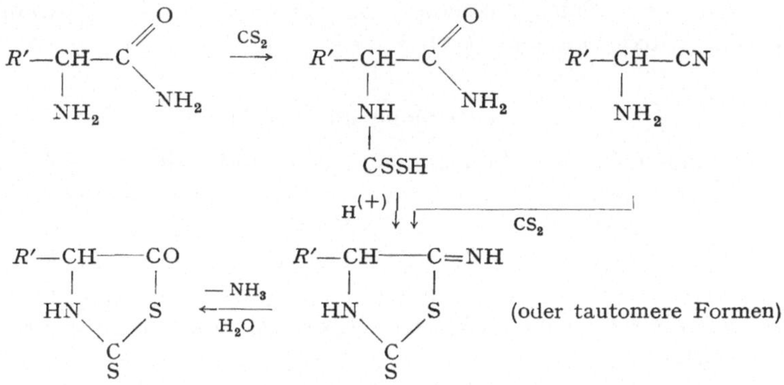

$R' = $ Aminosäure-Seitenkette.

Dieses heterocyclische Ringsystem unterliegt den Oxazoliddionen analogen Spaltungsreaktionen. Durch Einwirkung von primären und sekundären Aminen wird die —CO—S-Bindung unter Amidknüpfung geöffnet (*72, 105, 106, 27, 265*). Mit Aminosäureestern in Gegenwart von Trialkylaminen bilden sich die Ammoniumsalze der Dipeptidester-N-dithiocarbaminsäure (CXXXIV), woraus nach Zugabe von zwei Äquivalenten HCl die Peptidester-hydrochloride (CXXXV) resultieren (*107, 108*).

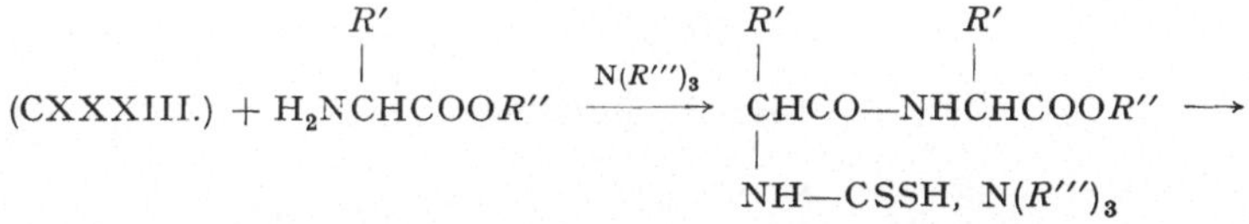

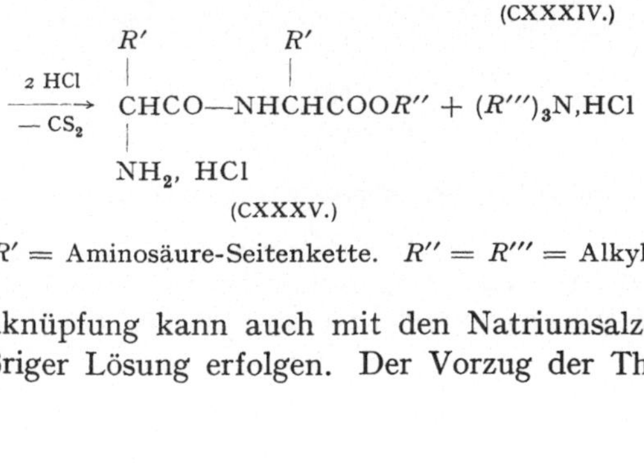

$R' = $ Aminosäure-Seitenkette. $R'' = R''' = $ Alkyl.

Die Peptidknüpfung kann auch mit den Natriumsalzen der Aminosäuren in wäßriger Lösung erfolgen. Der Vorzug der Thio-thiazolidone

gegenüber ihren O-Analoga ist die viel größere Beständigkeit. Anderseits
scheint die synthetische Verwendung auf die Darstellung von Glycyl-
peptiden eingeschränkt werden zu müssen, da substituierte Mercapto-
thiazolone, z. B. das Alanin-derivat, nach erfolgter Kupplung beim Ansäuern unverändert zurückerhalten werden (*108*), ferner optisch wenig
beständig sind und rasch racemisieren (*123*). Hofmann u. a. (*231*) benutzen diese Thio-thiazolidone zur Darstellung der N-Aminoacyl-N'-
carbobenzoxy-hydrazide (vgl. S. 476).

c) Oxazolone und „Azlactone".

Schon frühzeitig berichteten Mohr und Mitarb. (*306—308*) über
eine weitere Klasse zur Knüpfung der Amidbindung energiereicher Körper,
den substituierten Oxazolonen (CXXXVI), die man als innere Anhydride
der Lactimform (CXXXVII) acylierter Aminosäuren ansehen kann. Aus
letzteren gelingt ihre Darstellung durch Wasserabspaltung mittels Essig-
säureanhydrid, wobei Wieland (*452*) als Zwischenstufe ein unsymmetrisches Acetyl-anhydrid (CXXXVIII) annimmt. Auch der Weg über
die Säurechloride (CXXXIX) hat in verschiedenen Variationen zum Erfolg geführt (*253*).

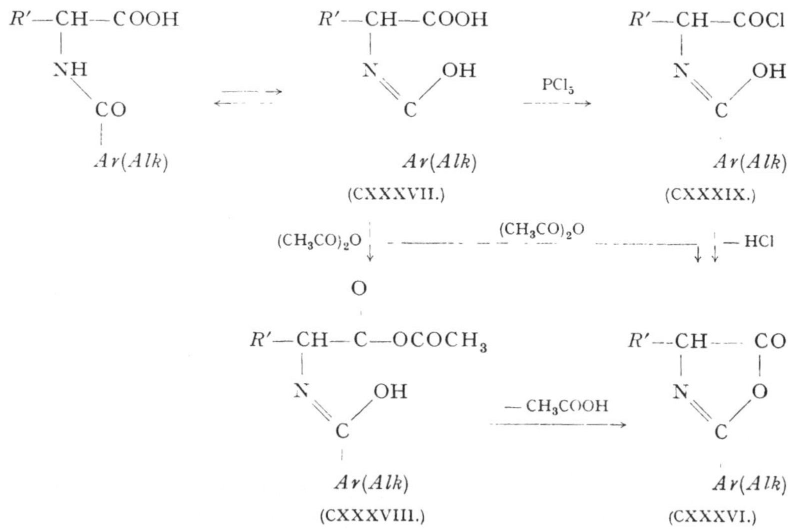

$$R' - \text{CH} - \text{COOH} \qquad R' - \text{CH} - \text{COOH} \qquad R' - \text{CH} - \text{COCl}$$

(CXXXVII.)   (CXXXIX.)

$$R' - \text{CH} - \text{C} - \text{OCOCH}_3 \qquad R' - \text{CH} - \text{CO}$$

(CXXXVIII.)   (CXXXVI.)

R' = Aminosäure-Seitenkette. Ar = Aryl. Alk = Alkyl.

O'Brien und Niemann (*319*) führen diese Oxazolon-bildung auf eine
durch Wasserstoff-ionen katalysierte Cyclisierung der Acylaminosäuren
zurück:

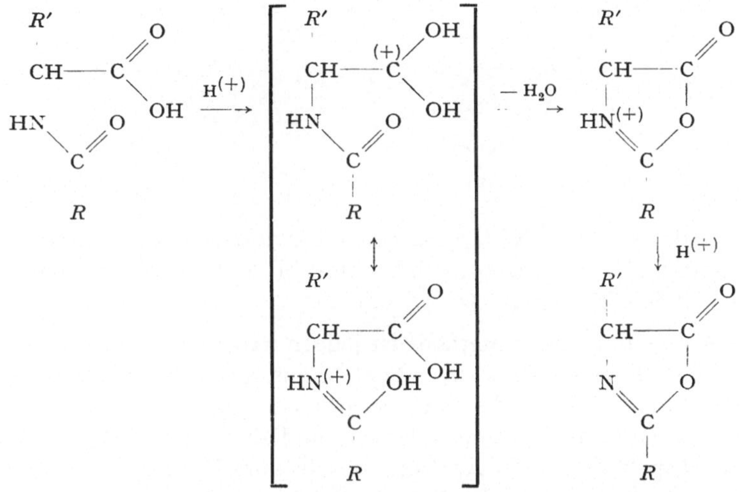

$$R = \text{Aryl oder Alkyl.} \quad R' = \text{Aminosäure-Seitenkette.}$$

(CXL.)

Die substituierten Oxazolone — die MOHRsche Schule befaßte sich mit den von Benzoylaminosäuren abgeleiteten 2-Phenyl-derivaten — setzen sich leicht mit Aminen, Aminosäureestern und auch mit Natriumsalzen von Aminosäuren in wäßriger Lösung zu den entsprechenden Amiden um (*281*):

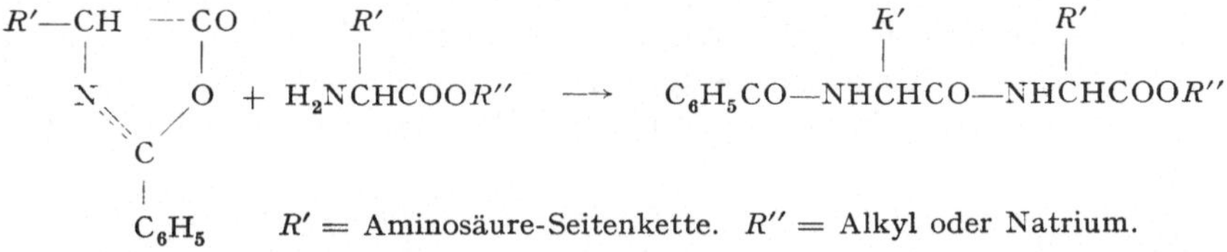

$$R'-CH-CO \quad R' \qquad \qquad R' \qquad R'$$
$$N \cdots O + H_2NCHCOOR'' \longrightarrow C_6H_5CO-NHCHCO-NHCHCOOR''$$
$$C$$
$$C_6H_5 \quad R' = \text{Aminosäure-Seitenkette.} \quad R'' = \text{Alkyl oder Natrium.}$$

BERGMANN und Mitarb. (*51*) bestätigten diese Ergebnisse an Hand der 2-Methyl-oxazolone.

Einer völlig analogen Verwendung unterliegen natürlich die von ERLENMEYER jr. (*131*) entdeckten 4-Alkyliden-2-alkyl(-aryl-)-oxazolone (CXLI), die „Azlactone". Die nach einer „Peptidkupplung" erhaltenen

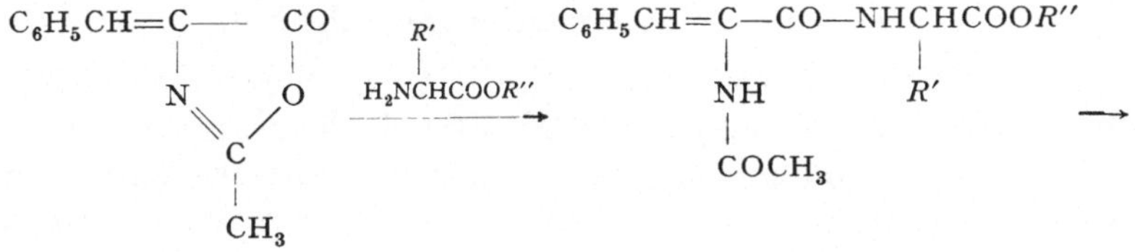

$$C_6H_5CH=C-CO \quad R' \qquad C_6H_5CH=C-CO-NHCHCOOR''$$
$$N \quad O \xrightarrow{H_2NCHCOOR''} NH \quad R' \longrightarrow$$
$$C$$
$$CH_3$$

(CXLI.) (CXLII.)

$$C_6H_5CH_2CHCO{-}NHCHCOOR''$$

H₂/Pt → with NH, R; COCH₃ branch

(CXLIII.) → H₂NCHCO—NHCHCOOH, CH₂, R'; C₆H₅ (CXLIV.)

$R' = $ Aminosäure-Seitenkette. $R'' = $ Alkyl.

Acyl-dehydropeptide (CXLII) lassen sich katalytisch zur gesättigten Verbindung (CXLIII) hydrieren; milde saure Hydrolyse soll zu den freien Peptiden (CXLIV) führen (*51, 48, 68*).

Die Anwendung dieser Methode ist jedoch weitgehend einzuschränken, da 1. die Synthese optisch-aktiver Peptide ausgeschlossen ist, bis auf jene wenigen Fälle, bei denen die Acylpeptid-diastereomeren sich durch fraktionierte Kristallisation trennen lassen; und da 2. die verhältnismäßig schwere Abspaltbarkeit der Acylreste durch saure Hydrolyse zur Bildung von Diketopiperazinen und bei höheren Peptiden sicherlich zur Sprengung der —CO—NH-Bindungen führt.

d) N-Acyl-oxazolidone.

Beim Übertragen der genannten „Azlacton"-synthese auf Trifluoracetyl-aminosäuren erhielten Weygand und Mitarb. (*449, 450*) aus TFA-glycin, Benzaldehyd, Trifluoressigsäureanhydrid und Natrium-trifluoracetat eine Verbindung, die sich schließlich als 5-Oxo-2-phenyl-3-trifluoracetyl-oxazolidin (CXLV) erwies. Dieser Verbindungstyp ließ sich in siedendem Tetrahydrofuran leicht aminolytisch zum Säureamid aufspalten; mit Aminosäure-estern bzw. -amiden konnten in hervorragender Ausbeute die TFA-peptidester (CXLVI) bzw. -amide erhalten werden:

$$H_2NCH_2COOR'' \longrightarrow CF_3CO{-}NHCH_2CO{-}NHCH_2COOR'' + C_6H_5CHO$$

(CXLV.)   (CXLVI.)

$R'' = $ Alkyl.

e) Spezielle innermolekulare Anhydride.

Bergmann und Zervas (*56*) haben mit der Ringöffnung des Carbobenzoxy-glutaminsäureanhydrids (CXLVII) eine klassische Synthese von Glutamyl- und Glutaminylpeptiden begründet. Ihre Wegweiser dürften Pauly und Weir (*325*), sowie Nicolet (*316*) gewesen sein. Nach Bergmann und Zervas wird das Anhydrid (CXLVII) — aus der Acylaminodicarbonsäure beim Behandeln mit Acetanhydrid in der Wärme (*56*), besser bei Zimmertemperatur (*337*), erhalten — leicht aminolytisch aufgespalten. Die Einwirkung von Ammoniak sollte eindeutig das α-Amid (Carbobenzoxy-isoglutamin) (*56*) von Glycinester (*201, 60*), Glycylglycinester (*61*), Glutaminsäurediester (*56*) usw. stets Carbobenzoxy-α-glutamyl-peptide ergeben:

$$CH_2CH_2COOH$$
$$|$$
$$Cbo—NHCHCOOH \xrightarrow{\text{Acetanhydrid}}$$

(CXLVII.)

$$\xrightarrow[H_2N—R]{} Cbo—NHCHCO—NH—R \xrightarrow[H_2/Pd]{} H_2NCHCO—NH—R$$

Cbo = Carbobenzoxy. R = Amin-substituent. Ac = Acetyl.

Völlig analoge Verhältnisse sollten auch für die Asparaginsäure zutreffen, deren Anhydrid (*56, 66, 303*) mit Ammoniak (*56*) und Aminosäureestern (*201, 60*) α-Amide lieferte, mit Ausnahme des Tyrosinesters, der zum β-Peptid kuppelte (*67*). Im Gegensatz dazu beobachtete MELVILLE (*299*) nach NH_3-Aufspaltung des Carbobenzoxy-glutaminsäureanhydrids und folgender hydrogenolytischer Entfernung der Schutzgruppe neben Isoglutamin auch 14% Glutamin. BOOTHE und Mitarb. (*381*) isolierten beide Tripeptid-isomeren beim Umsatz mit α-Glutamylglutaminsäuretriester, und schließlich brachten die Arbeiten von LE QUESNE und YOUNG (*337—339*) bzw. ŠORM und RUDINGER (*400*) die Klarstellung: Die aminolytische Aufspaltung verläuft im Sinne der Bildung der α- und γ-Isomeren bei der Glutaminsäure, von α- und β-Isomeren bei der Asparaginsäure, die sich als Carbobenzoxypeptide, auf Grund ihrer verschiedenen Acidität, mittels Alkali fraktioniert trennen lassen (*250*).

Ähnlich verläuft die Alkoholyse der „inneren Anhydride" der acylierten Glutamin- bzw. Asparaginsäure: α- und γ- bzw. β-Ester entstehen jeweils nebeneinander; ihre Trennung erfolgt (*337, 191*) durch fraktionierte alkalische Extraktion. Die rein erhaltenen Carbobenzoxy-aminosäureester sind das Ausgangsmaterial für die folgenden eindeutigen Peptidsynthesen der Glutamyl- und Asparagylreihe:

1. Überführung der freien Carboxylgruppe in das Säurechlorid (CXLVIII) und folgende FISCHER-Synthese (*218, 274*). HARINGTON und MEAD (*217*) bauten so erstmals das Glutathion auf (*238*). 2. Umsetzung der freien Carboxylgruppe zum Amid mittels der „gemischten Anhydrid-

methode" (*354*) oder mit „energiereichen Aminen" (z. B. Isocyanaten, Phosphorazo-körpern. u. dgl.) (*191*). 3. Überführung der „Halbester" (CXLIX a und b) in die „Halbhydrazide" (CL a und b); diese werden nach Curtius über die Halbazide (CLI a und b) in gewünschtem Sinn umgesetzt (*338, 225, 348*). Neuerdings machen aber Sachs und Brand (*355*) darauf aufmerksam, daß bei der Verwendung von Carbobenzoxy-glutaminsäure-γ-azid zur Synthese von γ-Peptiden infolge innermolekularer Umlagerungen α- und γ-Peptide nebeneinander entstehen (*Formelübersicht 7*).

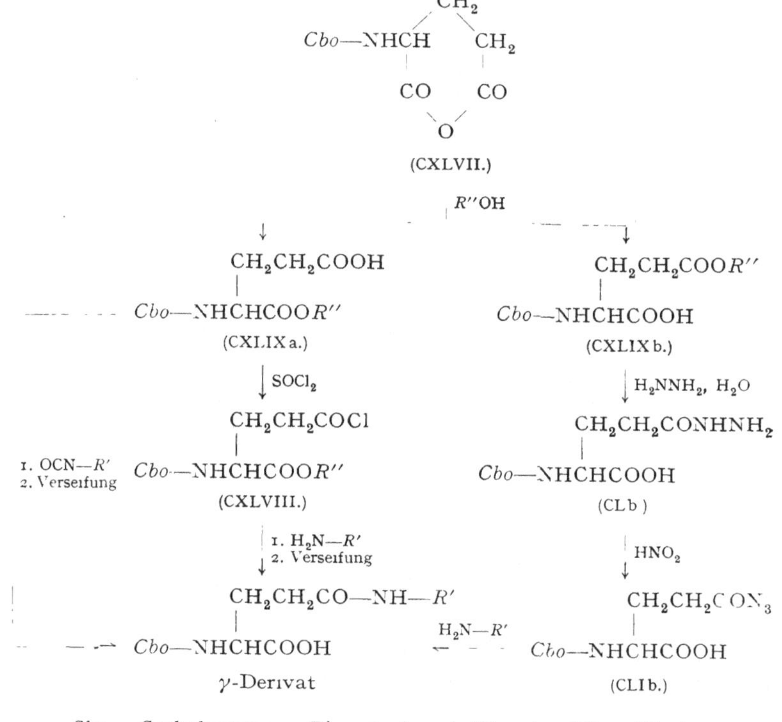

Cbo = Carbobenzoxy. R' = Amin-substituent R'' = Alkyl.

Formelübersicht 7.

Die Verwendung der N-Phthalyl-schutzgruppe bedingt bei der aminolytischen Aufspaltung der Anhydride die bevorzugte Bildung von γ-Glutamyl-amiden und -peptiden bzw. den β-Analoga bei der Asparaginsäure (*272, 268, 269*). Jedoch soll nach Tanenbaum (*410*) bei letzteren das Lösungsmittel eine Rolle spielen. Auch die Alkoholyse ergibt hier beim Asparaginderivat α- und β-Ester nebeneinander (*270*) [vgl. auch N-Tosylasparaginsäure-anhydrid (*81*)].

Interessant ist die Aufspaltung von N-Acyl-glutaminsäureanhydrid mit Thiophenol, die beim Phthalylderivat eindeutig zum γ-S-Phenylester führt, bei der Carbobenzoxyverbindung hingegen von der zugesetzten tertiären Base gelenkt wird — von Pyridin zu 70% in α-, von Triäthylamin zu 80% in γ-Stellung *(453, 470)*.

H. Energiereiche „N-Derivate" der Aminosäureester.

1. N-Carbonyl-aminosäureester (α-Isocyanat-fettsäureester).

Isocyansäureester (CLII) setzen sich mit Carbonsäuren unter Bildung von substituierten Säureamiden (CLIII) um; dabei wird CO_2 frei. Diese Reaktion verläuft über gemischte Anhydride (CLIV), die zuerst von STAUDINGER vermutet wurden *(402)* und später auch isoliert werden konnten *(311, 310, 328)*:

$$R—COOH + O=C=N—R' \longrightarrow R—CO—O—CO—NH—R' \xrightarrow{-CO_2} R—CO—NH—R'$$

(CLII.)   (CLIV.)   (CLIII.)

R = Carbonsäure-substituent. R' = Isocyanat-substituent.

Die Darstellung der erforderlichen Isocyanatfettsäureester durch Einwirkung von Phosgen auf Aminosäureester ist lange bekannt *(183, 240, 241, 309a)*, doch hat erstmals SIEFKEN *(392)* den Carbonyl-glycinester als einfachsten Vertreter dieser Klasse in reiner Form in den Händen gehabt.

GOLDSCHMIDT und WICK *(197)* berichten über die Verwendung dieser energiereichen Ester zu Peptidsynthesen. Sie haben *(198)* eine große Anzahl bislang unbekannter, teils optisch-aktiver Carbonyl-aminosäureester (CLV) hergestellt und ihre Umsetzung mit Carbobenzoxy- bzw. Phthalylaminosäuren (-peptiden) beschrieben. Die Ausbeuten sind hervorragend.

Die Gewinnung von Carbonyl-peptidestern schlug fehl; es entstanden in intramolekularer Reaktion der zuerst gebildeten Isocyanate (CLVI) Hydantoin-derivate (CLVII):

R' = Aminosäure-Seitenkette. R'' = Alkyl.

Die Nachteile dieser sonst erfolgreichen Methode sind: 1. die große Wasserempfindlichkeit; jede Spur Feuchtigkeit bedingt vor der Peptidknüpfung die Bildung von Harnstoffderivaten, deren Abtrennung nicht immer gelingt (*478*). 2. Acyl-oxyaminosäuren geben Urethane (CLVIII), Acylglutamin bzw. -asparagin N-Acyl-harnstoffe (CLIX) (*478*):

$$Ac\text{—}NHCHCOOH \qquad R' \qquad\qquad Ac\text{—}NHCHCOOH \qquad R'$$
$$\overset{|}{CH_2OH} \ + \ O{=}C{=}N\text{—}\overset{|}{CHCOO}R'' \longrightarrow \qquad \overset{|}{CH_2O\text{—}CO\text{—}NHCH}$$
$$\overset{|}{COOR''}$$

(CLV.) $\qquad\qquad\qquad\qquad\qquad$ (CLVIII.)

$$Ac\text{—}NHCHCOOH \qquad\qquad\qquad Ac\text{—}NHCHCOOH$$
$$\overset{|}{CH_2} \qquad\qquad R' \longrightarrow \qquad \overset{|}{CH_2} \qquad\qquad R'$$
$$\overset{|}{CO\text{—}NH_2} + O{=}C{=}N\text{—}CH \qquad\qquad CO\text{—}NH\text{—}CO\text{—}\overset{|}{NHCH}$$
$$\overset{|}{COOR''} \qquad\qquad\qquad\qquad\qquad \overset{|}{COOR''}$$

(CLV.) $\qquad\qquad$ (CLIX.)

$R' =$ Aminosäure-Seitenkette. $\quad R'' =$ Alkyl.

3. Die Einbeziehung von Carbonyl-asparaginsäure- bzw. -glutaminsäureester in die Peptidsynthese (Umsetzung mit Carbobenzoxyaminosäuren) führt zur Bildung von Carbonyl-bis-aminoderivaten der Aminodicarbonsäureester, die sich leicht aus dem Reaktionsgemisch isolieren lassen (*478*). Es dürfte hier, wie schon Naegeli (*310*) beim Phenylisocyanat gezeigt hat, als Konkurrenzreaktion zur Säureamidbildung eine Disproportionierung des gemischten Anhydrid-zwischenproduktes (CLIV) zu einem disubstituierten Harnstoff (CLX) und einem Kohlensäurebisanhydrid (CLXI) verlaufen; letzteres geht unter CO_2-Abspaltung spontan in das symmetrische Anhydrid der Ausgangssäure über (CLXII):

$$2\ R\text{—}CO\text{—}O\text{—}CO\text{—}NH\text{—}R'$$

(CLIV.)

$$\downarrow$$

$$R'\text{—}NH\text{—}CO\text{—}NH\text{—}R' + R\text{—}CO\text{—}O\text{—}CO\text{—}O\text{—}CO\text{—}R$$

(CLX.) $\qquad\qquad\qquad\qquad$ (CLXI.)

$$\downarrow -CO_2$$

(CLXII.) $\quad R\text{—}CO\text{—}O\text{—}CO\text{—}R$

$R =$ Carbonsäure-substituent. $\quad R' =$ Isocyanat-substituent.

Zu diesem Reaktionsverlauf vgl. Fry (*180*).

2. Phosphatamide.

Nach Goldschmidt und Obermeier (*196*) gibt auch die Phosphorsäure einige energiereiche, substituierte Amide, die sich mit Carbonsäuren zu Carbonsäureamiden umsetzen lassen. Interessant ist, daß tri-substituierte Phosphatamide (Monoamiddialkyl-, Diamid-monoalkyl-, Trisamid-derivate) nur unter sehr energischen Bedingungen, etwa in der Schmelze bei 200—220°, aber mit guten Ausbeuten mit Carbonsäuren reagieren. Diese Reaktionsträgheit der Monoamid-diester der Phosphorsäure erlaubt sogar ihre Verwendung als Aminoschutzgruppe (vgl. S. 468). Erfolgversprechender sind die disubstituierten Verbindungen, wie Monoamido-monoester bzw. Diamido-derivate, die sich aus den entsprechenden chlorierten Produkten durch Verseifen leicht herstellen lassen. Für Peptidsynthesen eignen sich wegen der Zugänglichkeit nur die Phosphatdiamide (CLXIII). Allerdings muß in der Folge 50% Verlust an Ausgangsmaterial in Kauf genommen werden, da die Hälfte als Salz der Metaphosphorsäure (CLXIV) zurückerhalten wird.

$$
2 \underset{\underset{\text{COO}R''}{|}}{\overset{\overset{R'}{|}}{\text{CH}}}\!\!-\!\text{NH}_2 \xrightarrow{\text{POCl}_3} \underset{\underset{\text{COO}R''}{|}}{\overset{\overset{R'}{|}}{\text{CH}}}\!\!-\!\text{NH}\!-\!\underset{\underset{\text{Cl}}{|}}{\overset{\overset{\text{O}}{\|}}{\text{P}}}\!\!-\!\text{NH}\!-\!\underset{\underset{\text{COO}R''}{|}}{\overset{\overset{R'}{|}}{\text{CH}}} \xrightarrow{\text{OH}^-} \underset{\underset{\text{COO}R''}{|}}{\overset{\overset{R'}{|}}{\text{CH}}}\!\!-\!\text{NH}\!-\!\underset{\underset{\text{OH}}{|}}{\overset{\overset{\text{O}}{\|}}{\text{P}}}\!\!-\!\text{NHCH}\underset{\text{COO}R''}{\overset{R'}{|}} \longrightarrow
$$

(CLXIII.)

$$
\underset{Ac-\text{NHCHCOOH}}{\overset{R'}{|}} \longrightarrow Ac\!-\!\text{NHCHCO}\!-\!\text{NHCHCOO}R'' + \underset{\underset{\text{COO}R''}{|}}{\overset{\overset{R'}{|}}{\text{CH}}}\!\!-\!\text{NH}_2, \text{HPO}_3
$$

(CLXIV.)

Ac = Acyl. R' = Aminosäure-Seitenkette. R'' = Alkyl.

Die Monoamido-phosphorsäuren (CLXV) sind unbeständig; zwei Moleküle disproportionieren sich wie folgt:

$$
2\ \text{C}_6\text{H}_5\!-\!\text{NH}\!-\!\text{P} \underset{\text{OH}}{\overset{\text{O} \diagup \text{OH}}{\diagdown}} \longrightarrow \text{C}_6\text{H}_5\!-\!\text{NH}\!-\!\underset{\underset{\underset{\text{C}_6\text{H}_5}{|}}{\text{NH}_2}\ \text{OH}}{\overset{\text{O} \diagup \text{OH}}{\diagdown}}\!\!\!\text{P} + {}^{1}\!/_{x}(\text{HPO}_3)_x
$$

(CLXV.)

Versuche der Umsetzung mit Carbonsäuren verliefen negativ, was im Hinblick auf obenstehenden Reaktionsmechanismus der Diamido-phosphorsäuren verständlich ist.

3. *N-Phosphorigsäure-derivate (Phosphitamide, -imide und Phosphorazokörper)*[*].

Die auf S. 504 beschriebenen Dialkyl-chlorophosphite (CLXVI) erwiesen sich nicht nur als Anhydridbildner, sondern auch als Ausgangs-

---

[*] Eine ausgezeichnete Zusammenstellung über energiereiche N-Derivate des drei- und fünfbindigen Phosphors geben GOLDSCHMIDT und KRAUSS (*193*).

$$
\begin{array}{c}
R' \\
Ac\text{—}NHCHCOOH \;+\;
\begin{array}{c}
R''O \\
\diagdown \\
P\text{—}O\text{—}P \\
\diagup \\
R''O
\end{array}
\begin{array}{c}
OR'' \\
\diagup \\
 \\
\diagdown \\
OR''
\end{array}
\;+\; H_2NCHCOOR''' \quad (R')
\end{array}
$$

(CLXX.)

$$
\begin{array}{c}
R' \\
Ac\text{—}NHCHCO\text{—}O\text{—}P
\begin{array}{c}
OR'' \\
\diagup \\
\diagdown \\
OR''
\end{array}
\end{array}
\qquad
\begin{array}{c}
R''O \\
\diagdown \\
P\text{—}NHCHCOOR''' \\
\diagup \\
R''O
\end{array}
\quad (R')
$$

(CLXXII.)   (CLXXI.)

$$
+\; H_2NCHCOOR''' \quad (R')
$$

$$
Ac\text{NHCHCOOH} \quad (R')
$$

$$
\begin{array}{c}
R''O \\
\diagdown \\
P\text{—}O\text{—}P \\
\diagup \\
R''O
\end{array}
\begin{array}{c}
OR'' \\
\diagup \\
\\
\diagdown \\
OR''
\end{array}
\;+\;
\begin{array}{c}
Ac\text{—}NHCHCO \\
\diagdown \\
O \\
\diagup \\
Ac\text{—}NHCHCO
\end{array}
\quad (R')
\qquad
Ac\text{—}NHCHCO\text{—}NHCHCOOR''' \quad (R' \quad R')
$$

(CLXXI.)   (CLXXIII.)

$$
+ \qquad\qquad R'
$$

$$
+\; H_2NCHCOOR''' \quad (R')
$$

$$
\begin{array}{c}
R''O \\
\diagdown \\
P\text{—}OH \\
\diagup \\
R''O
\end{array}
$$

(CLXIX.)

$$
Ac\text{—}NHCHCO\text{—}NHCHCOOR''' \quad (R' \quad R')
$$

(CLXXIII.)

$$
+
$$

$$
Ac\text{—}NHCHCOOH \quad (R')
$$

$$
R' = \text{Aminosäure-Seitenkette.} \quad R'' = \text{Äthyl.} \quad R''' = \text{Alkyl.}
$$

Formelübersicht 8.

material zur Darstellung der „Phosphitamide" (CLXVII), einer weiteren
Form einer aktivierten Aminogruppe, die über einen zur Peptidknüpfung
erforderlichen Energiezustand verfügt (*21, 20*). Sie bilden stabile, teils
destillierbare Öle. Beim Umsatz mit N-Acylaminosäuren z. B. in Diäthylphosphit als Lösungsmittel werden die entsprechenden Amide (CLXVIII)
in ausgezeichneter Ausbeute erhalten; als Nebenprodukt entsteht Dialkyl-
phosphit (CLXIX):

$$R'''\!\!\searrow \atop R''''\!\!\nearrow P\!-\!Cl \;+\; H_2NCHCOOR'' \;\longrightarrow\; R'''\!\!\searrow \atop R''''\!\!\nearrow P\!-\!NHCHCOOR'' \;\longrightarrow$$

(CLXVI.)   (CLXVII.)

$$Ac\!-\!NHCHCOOH \atop \longrightarrow \; Ac\!-\!NHCHCO\!-\!NHCHCOOR'' + HO\!-\!P\!\!{\nearrow R''' \atop \searrow R''''}$$

(CLXVIII.)   (CLXIX.)

$$R' = \text{Aminosäure-Seitenkette.} \quad R'' = \text{Alkyl.}$$

$$R''' = R'''' = \text{Äthoxy; oder } R'''\!-\!R'''' = {-O \atop -O}\!\!>\!C_6H_4(o)$$

Das oben Gesagte gilt gleichfalls für Tetraäthylpyrophosphit (CLXX). Es reagiert einmal mit der Carboxylgruppe zu einem gemischten Anhydrid (CLXXII) (vgl. S. 505), zum zweiten mit einer freien Aminogruppe unter Phosphitamid-bildung (CLXX) (*Formelübersicht 8*). Dies gab ANDERSON und Mitarb. (*19*) die Möglichkeit der Ausarbeitung einer „Standard-methode", da in einer Mischung von Acylaminosäure, Aminosäureester und Tetraäthylpyrophosphit in einem inerten Lösungsmittel (Diäthylphosphit) alle Haupt- und Nebenreaktionen über verschiedene Zwischenstufen zu dem Acylpeptidester (CLXXIII) führen. Diese Methode ist in letzter Zeit erfolgreich zur Synthese von Oxytocin und Vasopressin (*431, 427, 76*) sowie *L*-Arginyl-*L*-leucin (*18*) angewandt worden; für die primären Säureamid- bzw. Guanidogruppen war keine Maskierung erforderlich.

GOLDSCHMIDT und OBERMEIER (*196*) beschreiben die Umsetzung von Aminosäureestern mit Alkyldichlorophosphit (CLXXIV) in Toluol. Je nach stöchiometrischem Verhältnis der beiden Komponenten lassen sich zwei mögliche Zwischenprodukte diskutieren:

Molverhältnis 3 : 1

$$3 \;{R' \atop CHNH_2 \atop COOR''} \;+\; Cl_2POR''' \;\longrightarrow\; {R' \atop CH\!-\!N\!\leftarrow\!P\!-\!OR''' \atop COOR''} \;+\; 2\;{R' \atop CHNH_2, HCl \atop COOR''}$$

(CLXXIV.)   (CLXXV.)

$$R' = \text{Aminosäure-Seitenkette.} \quad R'' = R''' = \text{Alkyl.}$$

Solche N-substituierte Imido-phosphorigsäureester (CLXXV) setzen sich mit Carbonsäuren zu Säureamiden um. Mit Acylaminosäuren entstehen Acyl-peptidester in sehr guter Ausbeute:

34*

$$R'$$
$$|$$
$$(CLXXV) + Ac\text{—}NHCHCOOH \longrightarrow$$

$$R' \qquad\quad R'$$
$$| \qquad\quad\; |$$
$$\longrightarrow Ac\text{—}NHCHCO\text{—}NHCHCOOR'' + {}^1/_x(O{=}P\text{—}OR''')_x$$

$$R' = \text{Aminosäure-Seitenkette.} \quad R'' = R''' = \text{Alkyl.} \quad Ac = \text{Acyl.}$$

Molverhältnis 4 : 1

$$R' \qquad\qquad\qquad\qquad\qquad R' \quad OR''' \; R' \qquad\quad R'$$
$$| \qquad\qquad\qquad\qquad\qquad\quad | \qquad | \qquad |\qquad\qquad\;\;|$$
$$4\,CHNH_2 + Cl_2POR''' \longrightarrow CHNH\text{—}P\text{—}NHCH + 2\,CHNH_2,HCl$$
$$| \qquad\qquad\qquad\qquad\qquad\quad | \qquad\qquad\;\; | \qquad\qquad\; |$$
$$COOR'' \qquad\qquad\qquad\quad COOR'' \qquad COOR'' \; COOR''$$

(CLXXVI.)

$$R' = \text{Aminosäure-Seitenkette.} \quad R'' = R''' = \text{Alkyl.}$$

Es wäre gemäß obigen Reaktionsverlaufs ein substituierter Diamidophosphorigsäureester (CLXXVI) zu erwarten, der mit zwei Molekülen einer Carbonsäure unter Amidbildung reagieren sollte (Weg A, s. unten). Letzeres geschieht auch in der Tat. Da Goldschmidt und Obermeier vergeblich versuchten, den Halbester (CLXXVII) nachzuweisen — stets wurden nur Phosphorigsäure und Phenol aufgefunden —, ferner da das Zwischenprodukt der Umsetzung (Molverhältnis 4 : 1) mit Wasser eine

$$
\begin{array}{ccccc}
& Cl & R' & & R'\\
& / & & & |\\
R'''O\text{—}P & & + \; 2\,H_2NCHCOOR'' & \xrightarrow{-\,2\,HCl} & R'''O\text{—}P\quad NHCHCOOR''\\
& \backslash & & & \backslash\\
& Cl & & & NHCHCOOR''\\
& & & & |\\
& & & & R'
\end{array}
$$

(CLXXVI.)

$$\text{(Weg B)} \quad - R'''OH \qquad\qquad\qquad 2\,R\text{—COOH} \quad \text{(Weg A)}$$

$$R' \qquad\qquad R' \qquad\qquad\qquad\qquad R'$$
$$| \qquad\qquad\qquad\qquad\qquad\qquad\qquad\qquad\; |$$
$$R''OOC\,CHN \leftarrow P\text{—}NHCHCOOR'' \qquad 2\,R\text{—CO—}NHCHCOOR''$$

(CLXXIX.)

$$2\,R\text{—COOH} \qquad\qquad\qquad\qquad +$$

$$R' \qquad\qquad\qquad\qquad\qquad\qquad OH$$
$$| \qquad\qquad\qquad\qquad\qquad\qquad\quad /$$
$$2\,R\text{—CO—}NHCHCOOR'' + {}^1/_x(HPO_2)_x \qquad R'''O\text{—}P$$
$$\qquad\qquad\qquad\qquad\qquad\qquad\qquad\qquad\quad \backslash$$
$$\qquad\qquad\qquad\qquad\qquad\qquad\qquad\qquad\quad OH$$

(CLXXVII)

$$R = \text{Carbonsäure-substituent.} \quad R' = \text{Aminosäure-Seitenkette.} \quad R'' = \text{Alkyl.}$$
$$R''' = \text{Phenyl.}$$

substituierte Diamido-phosphorigsäure (CLXXVIII) ergab, kommen sie zu dem Schluß, daß der zuerst gebildete Diamido-ester (CLXXVI) sofort unter Abspaltung von Phenol in eine „Phosphorazo-verbindung" (CLXXIX) übergeht. Diese reagiert wie üblich mit zwei Molekülen einer Carbonsäure.

Aromatische „*Phosphorazo-verbindungen*" wurden erstmals von MICHAELIS und SCHROETER (*301*) beschrieben, als sie Anilin mit Phosphortrichlorid zu einem „Phosphorazo-chlorid" (CLXXX) umsetzten, das mit einem weiteren Molekül Anilin Phenyl-phosphorazo-anilid (CLXXXI) lieferte:

$$C_6H_5{-}NH_2 + PCl_3 \longrightarrow C_6H_5N \leftarrow P{-}Cl \xrightarrow{\;H_2N{-}C_6H_5\;} C_6H_5{-}N \leftarrow P{-}NH{-}C_6H_5$$

(CLXXX.)   (CLXXXI.)*

GRIMMEL und Mitarb. (*213*) gelang es, CLXXXI leicht rein herzustellen, als sie fünf Mol eines primären Amins mit einem Mol Phosphortrichlorid umsetzten. Als definierte Verbindung konnte u. a. N-Phenyl-phosphorazo-anilid erhalten werden. Auf gleiche Art haben GOLDSCHMIDT und LAUTENSCHLAGER (*190*, *195*) verfahren.

Als Verbesserung führten sie zunächst den Umsatz mit nur zwei Mol Amin, ein Mol Phosphortrichlorid und zur Bindung des freiwerdenden Chlorwasserstoffs zusätzlichen drei Mol eines tertiären Amins aus; bei der Heranziehung von wasserfreiem Pyridin als Lösungsmittel erübrigte sich schließlich auch das tertiäre Amin, ja es erwies sich als möglich, als Ausgangsmaterial die leichter herzustellenden und zu handhabenden Hydrochloride der Aminkomponente anzuwenden:

$$2\,R{-}NH_2 + PCl_3 \xrightarrow[\text{oder } 3\,N(R'')_3]{3\,R{-}NH_2} R{-}N \leftarrow P{-}NH{-}R \xleftarrow[\text{(Pyridin)}]{-\,5\,HCl} 2\,R{-}NH_2 + PCl_3$$

(CLXXXII.)

$R =$ Amin-substituent. $R'' =$ Alkyl.

Die Reaktionsfähigkeit solcher Phosphorazokörper (CLXXXII) ist groß. Wasser wird sofort zu einem „Hydrat" angelagert — von MICHAELIS (*301*) als Anilinsalz der Phenylimido-phosphorigen Säure (CLXXXIV) beschrieben —, das im Falle des Anilin-abkömmlings von GRIMMEL u. a. (*213*) kristallin gefaßt wurde und von GOLDSCHMIDT als Di-anilido-phosphorigsäure (CLXXXIIIa oder b) formuliert wird. Diese Verbindung (CLXXXIII) unterliegt (leicht bei Gegenwart von Mineralsäuren) weiterer hydrolytischer Spaltung zu zwei Molekülen Amin und phosphoriger Säure. Mit organischen Carbonsäuren dagegen erfolgt in

---

* Der Einfachheit halber wird die nebenstehende zunächst von GOLDSCHMIDT und OBERMEIER (*196*) bevorzugte Schreibweise beibehalten, obgleich nach den neuesten Ergebnissen von GOLDSCHMIDT und KRAUSS (*194*) die Phosphorazoverbindung in dimerer Form vorliegt.

$$\begin{array}{l} R{-}NH \\ \qquad\diagdown \\ \qquad\qquad P{-}NH{-}R \\ \qquad\diagup \\ R{-}N \\ \qquad\diagdown \\ \qquad\qquad P = N{-}R \end{array}$$

der Wärme „Amidknüpfung“, ebenfalls unter Abscheidung von phosphoriger Säure, während die ursprünglichen Phosphorazo-verbindungen (CLXXXII) bei analogem Umsatz als Nebenprodukt metaphosphorige Säure liefern:

$$R\text{—N} \leftarrow \text{P—OH}$$
$$,H_2N\text{—}R \xleftarrow{\ H_2O\ } R\text{—N} \leftarrow \text{P—NH—}R \xrightarrow[-\,^1/_x(HPO_2)_x]{\ H_2O\ ,\ 2\,R'\text{—COOH}} 2\,R'\text{—CO—NH—}R$$

(CLXXXIV.)　　　　　　　　　　　(CLXXXII.)

$$\xrightarrow[2\,R'\text{—COOH}]{-\,H_3PO_3}$$

$$R\text{—NH}_2,H_3PO_3 + R\text{—NH}_2 \xleftarrow{\ 2\,H_2O\ }
\begin{array}{c} R\text{—NH} \\ \diagdown \\ P \\ \diagup \\ R\text{—NH} \end{array}
\begin{array}{c} O \\ \diagup\diagup \\ \\ \diagdown \\ H \end{array}
\rightleftharpoons
\begin{array}{c} R\text{—NH} \\ \diagdown \\ P\text{—OH} \\ \diagup \\ R\text{—NH} \end{array}$$

(CLXXXIIIb.)　　　　　　(CLXXXIIIa.)

$R =$ Amin-substituent.　$R' =$ Carbonsäure-substituent.

Goldschmidt und Lautenschlager (*195*) haben diese Reaktion auf die Ausgangsmaterialien der Peptidsynthese übertragen, also auf Aminosäure- bzw. Peptidester und Acylaminosäure bzw. -peptid. Nach folgendem Schema wurden hierbei Acylpeptidester mit hervorragender Ausbeute erhalten (*195, 203, 208*), im Gegensatz zu der Mitteilung von Anderson u. a. (*20*). Goldschmidt und Jutz (*191*) gelang u. a. eine verbesserte Glutathionsynthese.

$$2\ \begin{array}{c} R' \\ | \\ \text{CHNH}_2 \\ | \\ \text{CO—}R'' \end{array}
\xrightarrow[-\,3\,HCl]{\ PCl_3\ }
\begin{array}{c} R' \\ | \\ \text{CHN} \\ | \\ \text{CO—}R'' \end{array} \leftarrow \text{P—NH}\begin{array}{c} R' \\ | \\ \text{CH} \\ | \\ \text{CO—}R'' \end{array} \longrightarrow$$

$$\begin{array}{c} R' \\ | \\ Ac\text{—NHCHCO—}R'' \end{array} \longrightarrow Ac\text{—(NHCHCO)}_n\text{—NHCHCOO}R''' + {}^n/_x\,(HPO_2)_x$$

$R' =$ Aminosäure-Seitenkette.　$R'' =$ —O-Alkyl　oder　—NHCHCOOR'''

$R''' =$ Alkyl.　$Ac =$ Acyl.　　　　　　R'

Als besondere Vorzüge der Phosphorazo-methode seien genannt: 1. Geringer experimenteller Aufwand. 2. Hohe Ausbeuten bei praktisch stöchiometrischen Einsatz der Komponenten. 3. Anwendbarkeit auch auf Peptidester. 4. Direkte Anwendung der Esterhydrochloride. 5. Geringe Feuchtigkeit nach erfolgter Bildung der Phosphorazokörper ist ohne Einfluß auf die Peptidknüpfung. Von großer Bedeutung ist es ferner, daß auch sekundäre Amine mit Phosphortrichlorid „energiereiche“ Reaktionsprodukte (CLXXXV) geben, die sich mit organischen Carbonsäuren zu Säureamiden umsetzen lassen (*196*):

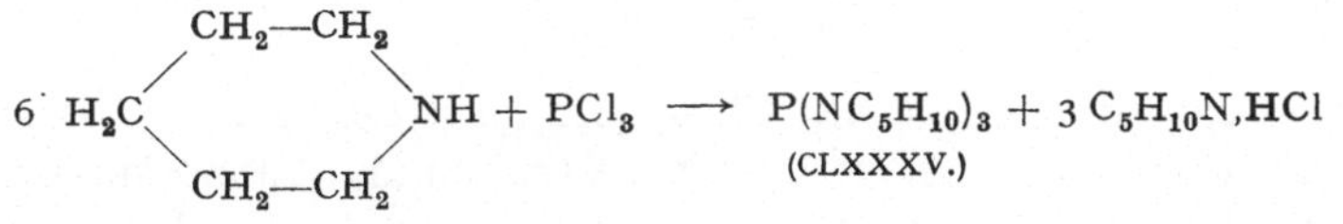

$$6\ H_2C \underset{CH_2-CH_2}{\overset{CH_2-CH_2}{\diagup\diagdown}} NH + PCl_3 \longrightarrow P(NC_5H_{10})_3 + 3\ C_5H_{10}N,HCl$$

(CLXXXV.)

$$(CLXXXV) + 3\ Ac\text{—NHCHCOOH} \longrightarrow 3\ Ac\text{—NHCHCO—NC}_5H_{10} + H_3PO_3$$

(über R')

R' = Aminosäure-Seitenkette.

Als Nachteile der Phosphorazo-methode wurden bekannt: 1. Bei der Umsetzung mit Carbobenzoxypeptiden mit carboxyl-endständiger optisch-aktiver Aminosäure tritt im Gegensatz zu den Feststellungen von GOLDSCHMIDT und Mitarb. (*193, 192*) teilweise Racemisierung ein (*208*). 2. Der Einsatz von Carbobenzoxy-oxyaminosäuren als „Kopfpartner" der Phosphorazo-methode verläuft unbefriedigend. Bei dem neben geringen Mengen an Carbobenzoxy-(oxyaminosäure)-peptidester erhaltenen bicarbonatlöslichen Rohprodukt dürfte es sich (*204*) um ein Gemisch eines Phosphorigsäureester-derivats der acylierten Oxyaminosäure (CLXXXVI) bzw. eines entsprechenden Peptids (CLXXXVIII) handeln, die ihren Ursprung in einer Anlagerung der Phosphorazo-gruppierung an die freie Hydroxylgruppe (CLXXXVI) haben (vgl. dagegen die Ausführungen von GOLDSCHMIDT und OBERMEIER; S. 532).

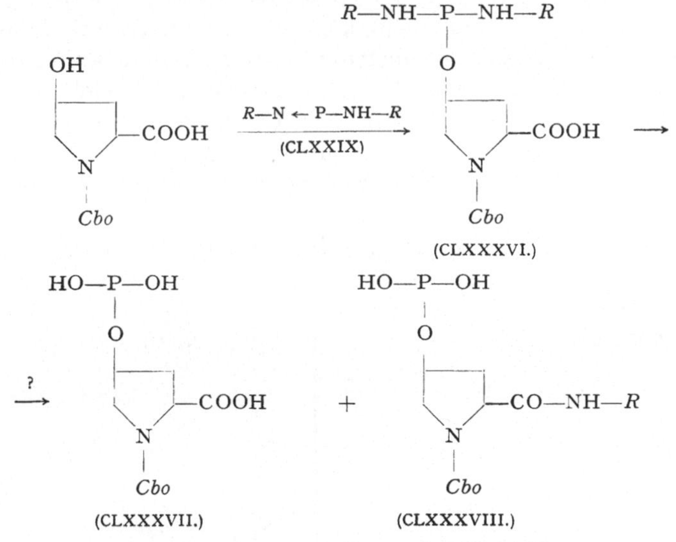

(CLXXXVI.)

(CLXXXVII.)   (CLXXXVIII.)

Cbo = Carbobenzoxy. $R = $ —CH$_2$COOC$_2$H$_5$.

Gleichfalls über eine Synthese mit Hilfe von Phosphortrichlorid berichtet SÜSS (*406*). Ob hierbei Phosphorazo-körper oder Phosphorigsäuretriamide die energetisch wichtigen Zwischenprodukte sind, ist ungeklärt.

4. Arsenitamide.

In völliger Analogie zu den Phosphorigsäure-derivaten vereinigen sich auch Diäthyl-chloro-arsenit (CLXXXIX) und Aminosäureester zu energiereichen Amiden (CXC), wobei der freiwerdende Chlorwasserstoff an tertiäre Basen gebunden wird (415). Die Umsetzung mit Acylamino-säuren erfolgt, wie aus folgendem Schema ersichtlich, zu den in guter Ausbeute erhältlichen Acylpeptidestern (vgl. auch S. 506).

$$R''O\diagdown As—Cl + H_2NCHCOOR''' \longrightarrow R''O\diagdown As—NHCHCOOR''' \xrightarrow{Ac—NHCHCOOH}$$

(CLXXXIX.)   (CXC.)

$$\longrightarrow Ac—NHCHCO—NHCHCOOR''' + R''O\diagdown As—OH \xrightarrow{H_2O} As_2O_3$$

R' = Aminosäure-Seitenkette. R'' = Äthyl. R''' = Alkyl. Ac = Acyl.

J. Spezielle Umlagerungsreaktionen.

1. Symmetrische N-Trifluoracetyl-aminosäureanhydride.

Weygand und Reiher (450) konnten die interessante Feststellung machen, daß sich aus symmetrischen N-TFA-aminosäureanhydriden (CXCI) beim Behandeln mit Dicyclohexylamin in Alkohol TFA-peptide und Trifluoressigsäureester bilden. Sie führen diese Reaktion auf einen induktiven Effekt der CF_3-Gruppe zurück, wodurch der Imidwasserstoff zur Salzbildung mit dem Amin befähigt wird. Nunmehr erfolgt intra-molekulare Umlagerung mit der in räumlicher Nachbarschaft stehenden C=O-gruppe der zweiten Anhydridhälfte zum Ammoniumsalz einer

$$CF_3CO—NHCH_2CO\diagdown_{O}\diagup CO \xrightarrow{HN(C_6H_{11})_2} \left[CF_3CO—NHCH_2CO—N—CH_2COO^- \; CO \; CF_3, H_2\overset{+}{N}(C_6H_{11})_2 \right]$$

(CXCI.)   (CXCII.)

$$\downarrow \text{Alkohol}$$

$$CF_3CO—NHCH_2CO—NHCH_2COO^-, H_2\overset{+}{N}(C_6H_{11})_2 + CF_3COOC_2H_5$$

(CXCIII.)   (CXCIV.)

N,N'-Diacyl-aminosäure (CXCII), das schließlich alkoholytisch zum TFA-dipeptid-dicyclohexylaminsalz (CXCIII) und Trifluoressigester (CXCIV) gespalten wird. (Diese Synthese ist von Racemisierung begleitet.)

2. O-(α-Aminoacyl)-salicylsäuren bzw. -amide.

Ein neues Prinzip zum Aufbau von Peptidketten erblicken BRENNER und Mitarb. (*87*) in ihren Beobachtungen über eigenartige Umlagerungsreaktionen von O-(α-Aminoacyl)-salicylsäuren (CXCVa) bzw. -amiden (CXCVb) in N-Salicylyl-aminosäuren (CXCVIa) bzw. -amide (CXCVIb) unter dem Einfluß von Hydroxylionen:

$$R' \qquad\qquad\qquad OH^- \qquad\qquad OH$$

$$\text{OOCCHNH}_3^+\ X^- \xrightarrow{\ OH^-\ } \qquad\qquad R'$$

$$\text{CO—NH—}R \qquad\qquad \text{CO—NHCHCO—NH—}R$$

(CXCVa.)   (CXCVIb.)

R' = Aminosäure-Seitenkette. R = Amid-substituent.

BRENNER (*87*) gelang so die Darstellung von Salicylylglycyl-phenyl-alanyl-glycinmethylester.

Die Übertragung dieser Methodik auf Serin, Threonin bzw. Cystein an Stelle der Salicylsäure hatte keinen Erfolg [vgl. dazu die intramolekularen Umlagerungsreaktionen von S-Aminoacyl-cystein (S. 490) und O-Acyl-serin (*50, 309*) in N-Acyl-aminosäuren].

Literaturverzeichnis.

1. ABDERHALDEN, E., R. ABDERHALDEN, H. WEIDEL, E. BAERTICH und W. MORNE-WEG: Über das Verhalten von Tetrapeptiden gegenüber Erepsin und Trypsin, an deren Aufbau Glykokoll (2 Mol), *L*-(+)-Alanin (1 Mol) und *L*-(—)-Tyrosin (1 Mol) beteiligt sind. Fermentforsch. **16**, 98 (1938).

2. ABDERHALDEN, E. und A. BAHN: Isolierung von Tyrosyl-seryl-prolyl-tyrosin beim stufenweisen Abbau von Seidenfibroin (*Bombyx mori*). Z. physiol. Chem. (Hoppe-Seyler) **219**, 72 (1933).

3. — — Beitrag zur Synthese von Dipeptiden, in denen Serin die Aminogruppe aufweist. Z. physiol. Chem. (Hoppe-Seyler) **234**, 181 (1935).

4. ABDERHALDEN, E. und A. FODOR: Synthese von hochmolekularen Polypeptiden aus Glykokoll und *L*-Leucin. Ber. dtsch. chem. Ges. **49**, 561 (1916).

5. ABDERHALDEN, E. und W. GOHDES: Über das Verhalten von aus *L*-(+)-Alanin aufgebauten Polypeptiden gegenüber verdünntem Alkali und Fermenten. Fermentforsch. **13**, 52 (1931).

6. ABDERHALDEN, E. und M. KEMPE: Synthese von Polypeptiden. XX. Derivate des Tryptophans. Ber. dtsch. chem. Ges. **40**, 2737 (1907).

7. ABDERHALDEN, E. und W. KÖPPEL: Vergleichende Studien über den fermentativen Abbau von Polypeptiden durch Erepsin und Trypsin-Kinase. Fermentforsch. **9**, 516 (1928).

8. ABDERHALDEN, E. und F. LEINERT: Über das Verhalten von Diketo-piperazinen gegenüber Magensaft. Fermentforsch. **15**, 324 (1937).

9. ABDERHALDEN, E. und H. NIENBURG: Fortgesetzte Studien uber das Verhalten von Prolylpolypeptiden gegenuber dem Erepsin- und dem Trypsin-Kinase-Komplex. Fermentforsch. **13**, 573 (1933).

10. — — Darstellung von *l*-(+)-Glutaminsáure-monoathylester, *l*-(—)-Isoglutamin und *l*-(—)-N-Carbobenzoxy-glutaminsäure-monoäthylester. Z. physiol. Chem. (Hoppe-Seyler) **219**, 155 (1933).

11. ABDERHALDEN, E. und S. SUZUKI: Studien uber den Einfluß der an der Bildung von Aminosäureestern beteiligten Alkoholgruppe auf die Geschwindigkeit der Bildung von 2,5-Dioxo-piperazinen und die Entstehung von Guanidino-verbindungen bei der Einwirkung von Guanidin auf verschiedene Aminosäureester. Z. physiol. Chem. (Hoppe-Seyler) **176**, 101 (1928).

12. ABDERHALDEN, E. und E. WYBERT: Synthese von Polypeptiden, an deren Aufbau die Aminosäuren Glycin, Alanin, Leucin und Cystin beteiligt sind. Ber. dtsch. chem. Ges. **49**, 2449 (1916).

13. ABENIUS, P. W. und O. WIDMAN: Über das Bromazeto-orthotoluid und einige daraus erhaltene Verbindungen. Ber. dtsch. chem. Ges. **21**, 1662 (1888).

14. ADKINS, H., R. M. ROSS, D. C. SCHROEDER, C. L. MAHONEY and W. W. GILBERT: The Synthesis of N-(2-Benzyl-4-Δ^2-oxazolinoyl)-valine. J. Amer. Chem. Soc. **76**, 147 (1954).

15. ALBERTSON, N. F. and F. C. McKAY: Acid-catalyzed Decarbobenzoxylation. J. Amer. Chem Soc. **75**, 5323 (1953).

16. AMBROSE, E. J. and W. E. HANBY: Evidence of Chain Folding in a Synthetic Polypeptide and in Keratin. Nature (London) **163**, 483 (1949).

17. ANATOL, J. et V. TORELLI: Sur l'accès aux N,N-dibenzyl α-amino-acides. Bull. soc. chim. France [5] **21**, 1446 (1954).

18. ANDERSON, G. W.: The Synthesis of an Arginyl Peptide. J. Amer. Chem. Soc. **75**, 6081 (1953).

19. ANDERSON, G. W., J. BLODINGER and A. D. WELCHER: Tetraethyl Pyrophosphite as a Reagent for Peptide Syntheses. J. Amer. Chem. Soc. **74**, 5309 (1952)

20. ANDERSON, G. W., J. BLODINGER, R. W. YOUNG and A. D. WELCHER: The Use of Phosphite Amides in Peptide Syntheses. J. Amer. Chem. Soc. **74**, 5304 (1952).

21. ANDERSON, G. W., A. D. WELCHER and R. W. YOUNG: Diethyl Chlorophosphite as a Reagent for Peptide Syntheses. J. Amer. Chem. Soc. **73**, 501 (1951).

22. ANDERSON, G. W. and R. W. YOUNG: The Use of Diester Chlorophosphites in Peptide Syntheses. Mixed Anhydrides. J. Amer. Chem. Soc. **74**, 5307 (1952).

23. ANGIER, R. B., C. W. WALLER, B. L. HUTCHINGS, J. H. BOOTHE, J. H. MOWAT, J. SEMB and Y. SUBBAROW: Pteroic Acid Derivatives. VI. Unequivocal Syntheses of Some Isomeric Glutamic Acid Peptides. J. Amer. Chem. Soc. **72**, 74 (1950).

24. ARENS, J. F.: The Chemistry of Acetylenic Ethers. XIII. Acetylenic Ethers as Reagents for the Preparation of Amides. Rec. trav. chim. Pays-Bas **74**, 769 (1955).

25. ARNETT, E. McC., J. G. MILLER and A. R. DAY: Effect of Structure on Reacitivity. III. Aminolysis of Esters with Primary Amines. J. Amer. Chem. Soc. **72**, 5635 (1950).

26. ASTBURY, W. T., C. E. DALGLIESH, S. E. DARMON and G. B. B. M. SUTHERLAND: Studies of the Structure of Synthetic Polypeptides. Nature (London) **162**, 596 (1948).

27. AUBERT, P. and E. B. KNOTT: Synthesis of Thiazolid-2:5-dione. Nature (London) **166**, 1039 (1950).

28. AUDRIETH, L. F. and J. KLEINBERG: Acid Catalysis in Liquid Ammonia III. Effect of α-Substituents on the Ammonolysis of Esters. J. Organ. Chem. (USA) **3**, 312 (1938).

29. AVISON, A. W. D.: The Synthesis of Acyl Phosphates in Aqueous Solution. J. Chem. Soc. (London) **1955**, 732.

30. BADDILEY, J. and A. P. MATHIAS: Coenzyme A. IX. The Synthesis of Panthothenoylcysteine, its 4'-Phosphate, and Related Compounds as Possible Precursors of the Coenzyme. J. Chem. Soc. (London) **1954**, 2803.

31. BAER, E. and J. MAURUKAS: Phosphatidyl Serine. J. Biol. Chem. **212**, 25 (1955).

32. BAILEY, J. L.: A New Peptide Synthesis. Nature (London) **164**, 889 (1949).

33. — The Synthesis of Simple Peptides from Anhydro-N-carboxy-amino-acids. J. Chem. Soc. (London) **1950**, 3461.

34. BALENOVIĆ, K. and D. FLEŠ: Synthesis of L-β-Amino-γ-benzyl-thiobutyric Acid (L-β-Amino-S-(benzyl)homocysteine. Amino Acids. VI. J. Organ. Chem. (USA) **17**, 347 (1952).

35. BALLARD, D..G. H., C. H. BAMFORD and F. J. WEYMOUTH: New Observations on the Chemistry of N-Carboxy-alpha-amino-acid Anhydrides. Nature (London) **174**, 173 (1954).

36. BAMFORD, C. H. and F. J. WEYMOUTH: The Structure of Macrocyclic Glycine Peptides. J. Amer. Chem. Soc. **77**, 6368 (1955).

37. BARKDOLL, A. E. and W. F. ROSS: Syntheses of Tyrosyltyrosyltyrosine and Tyrosyltyrosyltyrosyltyrosine. J. Amer. Chem. Soc. **66**, 951 (1944).

38. BAUMANN, E.: Über die Addition von Cyanamid. Liebigs Ann. Chem. **167**, 77 (1873).

39. BEN-ISHAI, D.: The Use of Hydrogen Bromide in Acetic Acid for the Removal of Carbobenzoxy Groups and Benzyl Esters of Peptide Derivatives. J. Organ. Chem. (USA) **19**, 62 (1954).

40. BEN-ISHAI, D. and A. BERGER: Cleavage of N-Carbobenzoxy Groups by Dry Hydrogen Bromide and Hydrogen Chloride. J. Organ. Chem. (USA) **17**, 1564 (1952).

41. BEN-ISHAI, D. and E. KATCHALSKI: Synthesis of N-Carboxy-α-amino Acid Anhydrides from N-Carbalkoxy-α-amino Acids by the Use of Phosphorus Tribromide. J. Amer. Chem. Soc. **74**, 3688 (1952).

42. BENTLER, M. und H. NETTER: Synthese von Aminosäure-Phosphorsäure-Anhydriden. Z. physiol. Chem. (Hoppe-Seyler) **295**, 362 (1953).

43. BERGER, A., J. KURTZ and E. KATCHALSKI: Poly-L-proline. J. Amer. Chem. Soc. **76**, 5552 (1954).

44. BERGMANN, M., E. BRAND und F. WEINMANN: Umlagerungen peptidähnlicher Stoffe. II. Derivate der γ-Amino-β-oxybuttersäure. Z. physiol. Chem. (Hoppe-Seyler) **131**, 1 (1923).

45. BERGMANN, M., H. ENSSLIN und L. ZERVAS: Über die Aldehyd-Verbindungen der Aminosäuren. Ber. dtsch. chem. Ges. **58**, 1034 (1925).

46. BERGMANN, M. and J. S. FRUTON: On Proteolytic Enzymes. XIII. Synthetic Substrates for Chymotrypsin. J. Biol. Chem. **118**, 405 (1937).

47. BERGMANN, M., J. S. FRUTON and H. POLLOK: The Specificity of Trypsin J. Biol. Chem. **127**, 643 (1939).

48. BERGMANN, M. und H. KÖSTER: Synthese argininhaltiger Dipeptide: Isomere Phenylalanyl-arginine und ihre Umwandlung in Phenylalanyl-ornithin. Z. physiol. Chem. (Hoppe-Seyler) **167**, 91 (1926).

49. Bergmann, M. und A. Miekeley: Neue desmotrope Aminosäureanhydride vom Piperazintypus. Zur Kenntnis des Abbaues der Aminosäuren. Serin als Dehydrierungsmittel. Liebigs Ann. Chem. **458**, 40 (1927).

50. — — Umlagerungen peptidähnlicher Stoffe. 3. Mitt. Derivate des *d,l*-Serins. Über neuartige Anhydride des Glycylserins. Z. physiol. Chem. (Hoppe-Seyler) **140**, 128 (1924).

51. Bergmann, M., F. Stern und Ch. Witte: Über neue Verfahren der Synthese von Dipeptiden und Dipeptid-Anhydriden. Liebigs Ann. Chem. **449**, 277 (1926).

52. Bergmann, M. and J. E. Tietzman: Transformations of an Acyl Diketo-piperazine. J. Biol. Chem. **155**, 535 (1944).

53. Bergmann, M. und L. Zervas: Über die Aldehydverbindungen der Amino-säuren und ihre präparative Verwendung. Z. physiol. Chem. (Hoppe-Seyler) **152**, 282 (1926).

54. — — Zur Kenntnis des Histidins. Peptidbildung durch Acylwanderung. 23. Mitt. über Umlagerung peptidähnlicher Stoffe. Z. physiol. Chem. (Hoppe-Seyler) **175**, 145 (1928).

55. — — Notiz über Synthese eines *d,l*-Histidyl-glycin. 24. Mitt. über Umlagerung peptidähnlicher Stoffe. Z. physiol. Chem. (Hoppe-Seyler) **175**, 154 (1928).

56. — — Über ein allgemeines Verfahren der Peptid-Synthese. Ber. dtsch. chem. Ges. **65**, 1192 (1932).

57. — — Über die Synthese von Glucopeptiden des *d*-Glucosamins (N-Glycyl-*d*-glucosamin und N-*d*-Alanyl-*d*-glucosamin). Ber. dtsch. chem. Ges. **65**, 1201 (1932).

58. — — Über Isoglutamin. Z. physiol. Chem. (Hoppe-Seyler) **221**, 51 (1933).

59. — — A Method for the Stepwise Degradation of Polypeptides. J. Biol. Chem. **113**, 341 (1936).

60. Bergmann, M., L. Zervas and J. S. Fruton: On Proteolytic Enzymes. VI. On the Specificity of Papain. J. Biol. Chem. **111**, 225 (1935).

61. — — — On Proteolytic Enzymes. XI. The Specificity of the Enzyme Papain Peptidase. I. J. Biol. Chem. **115**, 593 (1936).

62. Bergmann, M., L. Zervas, J. S. Fruton, F. Schneider and H. Schleich: On Proteolytic Enzymes. V. On the Specificity of Dipeptidase. J. Biol. Chem. **109**, 325 (1935).

63. Bergmann, M., L. Zervas und H. Köster: Autoracemisation argininhaltiger Aminosäure-Anhydride. (Beitrag zur Struktur des Clupeins.) Ber. dtsch. chem. Ges. **62**, 1901 (1929).

64. Bergmann, M., L. Zervas und H. Rinke: Über proteolytische Enzyme. Neues Verfahren zur Synthese von Peptiden des Arginins. Z. physiol. Chem. (Hoppe-Seyler) **224**, 40 (1934).

65. Bergmann, M., L. Zervas and W. F. Ross: On Proteolytic Enzymes. VII. The Synthesis of Peptides of *l*-Lysine and their Behavior with Papain. J. Biol. Chem. **111**, 245 (1935).

66. Bergmann, M., L. Zervas und L. Salzmann: Synthese von *l*-Asparagin und *d*-Glutamin. Ber. dtsch. chem. Ges. **66**, 1288 (1933).

67. Bergmann, M., L. Zervas, L. Salzmann und H. Schleich: Über proteo-lytische Enzyme. Über Dipeptide mit vorwiegend sauren Eigenschaften und ihr fermentatives Verhalten. Z. physiol. Chem. (Hoppe-Seyler) **224**, 17 (1934).

68. Bergmann, M., L. Zervas und V. du Vigneaud: Synthese argininhaltiger Peptide: *d*-Tyroxyl-*d*-arginin und sein Anhydrid. Ber. dtsch. chem. Ges. **62**, 1905 (1929).

69. BERTHO, A. und J. MAIER: Zur Synthese peptidähnlicher Körper aus Amino-zuckern und Aminosäuren. II. Über Dialanyl-N-glucosamin. Liebigs Ann. Chem. **495**, 113 (1932).

70. BETTS, R. L. and L. P. HAMMETT: A Kinetic Study of the Ammonolysis of Phenylacetic Esters in Methanol Solution. J. Amer. Chem. Soc. **59**, 1568 (1937).

71. BILEK, L., J. DERKOSCH, H. MICHL und F. WESSELY: Über die Zersetzung von α-Amino-N-carbonsäureanhydriden mit Pyridin und Pyridinderivaten. (Zur Frage der Bildung von höhermolekularen Cyklopeptiden.) Monatsh. Chem. **84**, 717 (1953).

72. BILLIMORIA, J. D. and A. H. COOK: Studies in the Azole Series. XIX. Reactions with 2-Mercaptothiazol-5-one. J. Chem. Soc. (London) **1949**, 2323.

73. BIRKOFER, L. und H. KACHEL: Synthese eines N-Carboxy-β-aminosäure-anhydrids. Naturwiss. **41**, 576 (1954).

73 a. BLOUT, E. R. and R. H. KARLSON: Polypeptides. III. The Synthesis of High Molecular Weight Poly-γ-benzyl-L-glutamates. J. Amer. Chem. Soc. **78**, 941 (1956).

74. BODÁNSZKY, M.: Synthesis of Peptides by Aminolysis of Nitrophenyl Esters. Nature (London) **175**, 685 (1955).

74 a. BODÁNSZKY, M., M. SZELKE, E. TÖMÖRKÉNY and E. WEISZ: Peptide Synthesis by Aminolysis of Active Esters. Chem. and Ind. **1955**, 1517.

75. BOISSONNAS, R. A.: Une nouvelle méthode de synthèse peptidique. Helv. Chim. Acta **34**, 874 (1951).

76. BOISSONNAS, R. A., ST. GUTTMANN, P. A. JAQUENOUD et J. P. WALLER: Une nouvelle synthèse de l'oxytocine. Helv. Chim. Acta **38**, 1491 (1955).

77. BOISSONNAS, R. A. et G. PREITNER: Étude comparative de la scission de divers groupes de blocage de la fonction α-amino des acides aminés. Helv. Chim. Acta **36**, 875 (1953).

78. BOISSONNAS, R. A. et I. SCHUMANN: Sur la cyclisation des polypeptides. Helv. Chim. Acta **35**, 2229 (1952).

79. BOTWINIK, M. M., Ss. M. AWAJEWA und E. A. MISSTRJUKOW: Synthese von N·O-Peptiden des Serins. Zhur. Obschei Khimii **23** (85) 971 (1953) [Chem. Zbl. **1954**, I, 1696].

80. — — — Synthese von Derivaten der Serinpeptide. Zhur. Obschei Khimii **23** (85), 1716 (1953) [Chem. Zbl. **1955**, I, 563].

81. BOVARNICK, M. R.: Substitution of Heated Asparagine-Glutamate Mixture for Nicotinamide as a Growth Factor for *Bacterium dysenteriae* and other Microörganisms. J. Biol. Chem. **148**, 151 (1943).

82. BRAND, E., B. F. ERLANGER and H. SACHS: Optical Rotation of Peptides. VI. Tetra- and Pentapeptides Containing Alanine and Lysine. J. Amer. Chem. Soc. **74**, 1851 (1952).

83. — — — Optical Rotation of Peptides. V. Alanine Tetra-, Penta- and Hexa-peptides. J. Amer. Chem. Soc. **74**, 1849 (1952).

84. BRAND, E., B. F. ERLANGER, H. SACHS and J. POLATNICK: Optical Rotation of Peptides. II. Glycine and Alanine Tripeptides. J. Amer. Chem. Soc. **73**, 3510 (1951).

85. BRENNER, M. und C. H. BURCKHARDT: Die Adsorption einiger Di- und Tri-peptide an synthetischen organischen Ionenaustauschern. Helv. Chim. Acta **34**, 1070 (1951).

86. BRENNER, M. und W. HUBER: Herstellung von α-Aminosäureestern durch Alkoholyse der Methylester. Helv. Chim. Acta **36**, 1109 (1953).

86 a. Brenner, M. und R. W. Pfister Enzymatische Peptidsynthese. 2. Mitt. Isolierung von enzymatisch gebildeten *L*-Methionyl-*L*-methionin und *L*-Methionyl-*L*-methionyl-*L*-methionin; Vergleich mit synthetischen Produkten. Helv. Chim Acta **34**, 2085 (1951).

87. Brenner, M., J. P. Zimmermann, J. Wehrmüller, P. Quitt und J. Photaki: Eine neue Umlagerungsreaktion und ein neues Prinzip zum Aufbau von Peptidketten. Experientia **11**, 397 (1955).

88. Brockmann, H. und H. Musso: Die papierchromatographische Trennung von Glycin-peptiden. Naturwiss. **38**, 11 (1951).

89. — — Versuche zur Synthese von Polypeptiden durch Kondensation von Aminosäure- und Peptidestern. 1. Mitt. Chem. Ber. **87**, 581 (1954).

90. Brockmann, H., H. Tummes und F. A. v. Metzsch: Zur Bildung zyklischer Polypeptide. Naturwiss. **41**, 37 (1954).

91. Brown, C. J., D. Coleman and A. C. Farthing: Further Studies in Synthetic Polypeptides. Nature (London) **163**, 834 (1949).

92. Bruckner, V., J. Kovács und K. Kovács: Synthese der α-*L*-Polyglutaminsäure. Naturwiss. **39**, 380 (1952).

93. — — — The Structure of Native Poly-*D*-glutamic Acid. IV. The Synthesis of Poly-*L*-glutamine and the Hofmann Degradation thereof. J. Chem. Soc. (London) **1953**, 1512.

94. Bruckner, V., K Kovács, J. Kovács und A. Kótai: Eine vereinfachte Synthese optisch reiner α-Polyglutaminsäure der *L*- und der *D*-Reihe. Experientia **10**, 166 (1954).

95. Bruckner, V., T. Vajda und J. Kovács: Synthese der β-Poly-*DL*- Asparaginsäure. Naturwiss. **41**, 449 (1954).

96. Carpenter, F. H. and D. T. Gish: The Application of *p*-Nitrobenzyl Chloroformate to Peptide Synthesis. J. Amer. Chem. Soc. **74**, 3818 (1952).

97. Carter, H. E., R. L. Frank and H. W. Johnston: Carbobenzoxy Chloride and Derivatives. Organ. Syntheses **23**, 13 (1943).

98. Channig, D. M., P. B. Turner and G. T. Young: Modified Carbobenzyloxy Groups in Peptide Synthesis. Nature (London) **167**, 487 (1951).

99. Chantrenne, H.: Synthèses peptidiques à partir d'un dérivé du glycylphosphate. Biochim. Biophys. Acta **4**, 484 (1950).

100. Chattaway, F. D.: The Action of Ammonia upon Esters. J. Chem. Soc. (London) **1936**, 355.

101. Coleman, D.: Synthetic Polypeptides. III. Initiators for the Co-polymerisation of Oxazolid-2 : 5-diones. J. Chem. Soc. (London) **1950**, 3222.

102. Coleman, D. and A. C. Farthing: Synthetic Polypeptides. II. Properties of Oxazolid-2:5-diones and an Initial Study of the Preparation of Polypeptides therefrom. J. Chem. Soc. (London) **1950**, 3218.

103. Cook, A. H., Sir Ian Heilbron and A. L. Levy: Studies in the Azole Series. II. The Interaction of α-Aminonitriles and Carbon Disulphide. J. Chem. Soc. (London) **1942**, 1598.

104. — — — Studies in the Azole Series. III. The Interaction of Aminoacetonitrile and Carbon Disulphide. J. Chem. Soc. **1948**, 201.

105. Cook, A. H. and A. L. Levy: Studies in the Azole Series. XXV. The Action of Bases on 2-Thio-5-thiazolidone. J. Chem. Soc. (London) **1950**, 637.

106. — — Studies in the Azole Series. XXVI. The Action of Bases on 2-Thio-4-methyl-5-thiazolidone. J. Chem. Soc. (London) **1950**, 642.

107. — — Studies in the Azole Series. XXVII. A New Method of Peptide Synthesis: Glycyl Peptides. J. Chem. Soc. (London) **1950**, 646.

108. Cook, A. H. and A. L. Levy: Studies in the Azole Series. XXVIII. A New Method of Peptide Synthesis: Alanylpeptides. J. Chem. Soc. (London) 1950, 651.

109. Cook, H. G., J. D. Ilett, B. C. Saunders, G. J. Stacey, H. G. Watson, I. G. E. Wilding and S. J. Woodcock: Esters containing Phosphorus. IX. J. Chem. Soc. (London) 1949, 2921.

110. Cronyn, M. W. and J. Jiu: A New Method for the Preparation of Thio Acids and Application to Peptide Chemistry. J. Amer. Chem. Soc. 74, 4726 (1952).

111. Curtius, Th.:. Über die Einwirkung von Chlorbenzoyl auf Glycocollsilber. J. prakt. Chem 24, 239 (1881).

112. — Über einige neue der Hippursäure analog konstituierte synthetisch dargestellte Amidosäuren. J. prakt. Chem. 26, 145 (1882).

113. — Über das Glycocoll. Ber. dtsch. chem. Ges. 16, 753 (1883).

114. — Verkettung von Amidosäuren. I. Abh. J. prakt. Chem. 70, 57 (1904).

115. — Synthetische Versuche mit Hippur-azid. Ber. dtsch. chem. Ges. 35, 3226 (1902).

116. — Über die freiwillige Zersetzung des Glycocollesters. Ber. dtsch. chem. Ges. 37, 1284 (1904).

117. Curtius, Th. und H. Curtius: Verkettung von Amidosäuren. VI. Abh. Über die Bildung von Asparaginsäureketten mit Hippurazid. J. prakt. Chem. 70, 158 (1904).

118. Curtius, Th. und O. Gumlich: Verkettung von Amidosäuren. VII. Abh. Kettenbildung zwischen Hippurazid und β-Amino-α-oxy-propionsäure und β-Aminobuttersäure. J. prakt. Chem. 70, 195 (1904).

119. Curtius, Th. und E. Lambotte: Verkettung von Amidosäuren. IV. Abh. Über die Einwirkung von Hippurazid auf α-Alanin. J. prakt. Chem. 70, 109 (1904).

120. Curtius, Th. und L. Levy: Verkettung von Amidosäuren. III. Abh. Weitere Untersuchungen über die Bildung von Glycylketten mit Hippurazid. J. prakt. Chem. 70, 89 (1904).

121. Curtius, Th. und W. Sieber: Umwandlung von alkylierten Malonsäuren in α-Aminosäuren. II. Mitt. Synthese des β-Phenyl-α-alanins und der α-Amino-n-buttersäure. Ber. dtsch. chem. Ges. 55, 1543 (1922).

122. Curtius, Th. und R. Wüstenfeld. Verkettung von Amidosäuren. II. Abh. Über die Bildung von Glycylketten mit Hippurazid. J. prakt. Chem. 70, 73 (1904).

123. Davis, A. C. and A. L. Levy: Peptide Synthesis from Heterocyclic Intermediates. I. 2-Thio-5-thiazolidone Derivatives of Valine, Leucine, Norleucine, Methionine, *L*-Tyrosine, Glutamine, α-Amino*iso*butyric Acid, and Aminomalonamide. J. Chem. Soc. (London) 1951, 2419.

124. Dekker, C. A. and J. S. Fruton. Preparation of *D*- and *L*-Methionine from *DL*-Methionine by Enzymatic Resolution. J. Biol. Chem. 173, 471 (1948).

125. Dekker, C. A., S. P. Taylor and J. S. Fruton: Synthesis of Peptides of Methionine and their Cleavage by Proteolytic Enzymes. J. Biol. Chem. 180, 155 (1949).

126. Ehrensvärd, G. C. H.: Phenyl-thiocarbonyl Chloride as a Group-protecting Agent. Nature (London) 159, 500 (1947).

127. Erlanger, B. F. and E. Brand: Optical Rotation of Peptides. I. Glycine and Alanine Dipeptides. J. Amer. Chem. Soc. 73, 3508 (1951).

128. — — Optical Rotation of Peptides. III. Lysine Dipeptides. J. Amer. Chem. Soc. 73, 4025 (1951).

129. Erlanger, B. F. and R. M. Hall: Improved Synthesis of Amino Acid Benzyl Esters. J. Amer. Chem. Soc. 76, 5781 (1954).

130. Erlanger, B. F., H. Sachs and E. Brand: The Synthesis of Peptides Related to Gramicidin S. J. Amer. Chem. Soc. **76**, 1806 (1954).

131. Erlenmeyer, E. jun. und E. Früstück: Über Phenyl-α-amidomilchsäure (Phenyl-serin). Liebigs Ann. Chem. **284**, 36 (1895).

132. Farthing, A. C.: Synthetic Polypeptides. I. Synthesis of Oxazolid-2 : 5-diones and a New Reaction of Glycine. J. Chem. Soc. (London) **1950**, 3213.

133. Farthing, A. C. and R. J. W. Reynolds: Anhydro-N-Carboxy-*Dl-β*-Phenylalanine. Nature (London) **165**, 647 (1950).

134. Fellinger, L. L. and L. F. Audrieth: Acid Catalysis in Liquid Ammonia. II. Ammonolysis of Ethyl Benzoate. J. Amer. Chem. Soc. **60**, 579 (1938).

135. Fischer, E.: Über einige Derivate des Glycocolls, Alanins und Leucins. Ber. dtsch. chem. Ges. **35**, 1095 (1902).

136. — Synthese von Derivaten der Polypeptide. Ber. dtsch. chem. Ges. **36**, 2094 (1903).

137. — Synthese von Polypeptiden. I. Ber. dtsch. chem. Ges. **36**, 2982 (1903).

138. — Synthese von Polypeptiden. IV. Derivate des Phenylalanins. Ber. dtsch. chem. Ges. **37**, 3062 (1904).

139. — Synthese von Polypeptiden. IX. Chloride der Aminosäuren und ihrer Acylderivate. Ber. dtsch. chem. Ges. **38**, 605 (1905).

140. — Synthese von Polypeptiden. XI. Liebigs Ann. Chem. **340**, 123 (1905).

141. — Synthese von Polypeptiden. XIII. Chloride der Aminosäuren und Polypeptide und ihre Verwendung zur Synthese. Ber. dtsch. chem. Ges. **38**, 2914 (1905).

142. — Synthese von Polypeptiden. XIV. Ber. dtsch. chem. Ges. **39**, 453 (1906).

143. — Synthese von Polypeptiden. XV. Ber. dtsch. chem. Ges. **39**, 2893 (1906).

144. — Untersuchungen über Aminosäure, Polypeptide und Proteine (1899—1906). Berlin: J. Springer 1906.

145. — Vorkommen von *l*-Serin in der Seide. Ber. dtsch. chem. Ges. **40**, 1501 (1907).

146. — Synthese von Polypeptiden. XVII. Ber. dtsch. chem. Ges. **40**, 1754 (1907).

147. — Synthese von Polypeptiden. XXIII. Ber. dtsch. chem. Ges. **41**, 850 (1908).

148. — Reduktion der Arylsulfamide durch Jodwasserstoff. Ber. dtsch. chem. Ges. **48**, 93 (1915).

149. Fischer, E. und P. Bergell: Über die Derivate einiger Dipeptide und ihr Verhalten gegen Pankreasfermente. Ber. dtsch. chem. Ges. **36**, 2592 (1903).

150. Fischer, E. und P. Blank: Synthese von Polypeptiden. XIX. 1. Derivate des Phenylalanins. Liebigs Ann. Chem. **354**, 1 (1907).

151. Fischer, E. und L. H. Cone: Synthese von Polypeptiden. XXVII. 1. Derivate des Histidins. Liebigs Ann. Chem. **363**, 107 (1908).

152. Fischer, E. und E. Fourneau: Über einige Derivate des Glykokolls. Ber. dtsch. chem Ges. **34**, 2868 (1901).

153. Fischer, E. und O. Gerngross: Synthese von Polypeptiden. XXX. Derivate des *L*-Cystins. Ber. dtsch. chem. Ges. **42**, 1485 (1909).

154. Fischer, E. und W. Gluud: Synthese von Polypeptiden. XXXI. Derivate des Leucins, Alanins und N-Phenylglycins. Liebigs Ann. Chem. **369**, 247 (1909).

155. Fischer, E. und K. Kautzsch: Synthese von Polypeptiden. XII. Alanylalanin und Derivate. Ber. dtsch. chem. Ges. **38**, 2375 (1905).

156. Fischer, E. und E. Koenigs: Synthese von Polypeptiden. VIII. Polypeptide und Amide der Asparaginsäure. Ber. dtsch. chem. Ges. **37**, 4585 (1904).

157. — — Synthese von Polypeptiden. XVIII. Derivate der Asparaginsäure. Ber. dtsch. chem. Ges. **40**, 2048 (1907).

158. Fischer, E. und A. Luniak: Synthese von Polypeptiden. XXXII. Derivate des *l*-Prolins und des Phenyl-alanins. Ber. dtsch. chem. Ges. **42**, 4752 (1909).

159. Fischer, E. und E. Otto: Synthese von Derivaten einiger Dipeptide. Ber. dtsch. chem. Ges. **36**, 2106 (1903).

160. Fischer, E. und G. Reif: Synthese von Polypeptiden. XXVII. 2. Derivate des Prolins. Liebigs Ann. Chem. **363**, 118 (1908).

161. Fischer, E. und W. Schrauth: Synthese von Polypeptiden. XIX. 3. Aufspaltung von Diketopiperazinen und Dipeptide des Tyrosins. Liebigs Ann. Chem. **354**, 21 (1907).

162. Fischer, E. und U. Suzuki: Synthese von Polypeptiden. III. Derivate der α-Pyrrolidincarbonsäure. Ber. dtsch. chem. Ges. **37**, 2842 (1904).

162a. — — Synthese von Polypeptiden. VII. Derivate des Cystins. Ber. dtsch. chem. Ges. **37**, 4575 (1904).

163. — — Synthese von Polypeptiden. X. Polypeptide der Diamino- und Oxyaminosäuren. Ber. dtsch. chem. Ges. **38**, 4173 (1905).

164. Fischer, R. F. and R. R. Whetstone: Peptide Derivatives Containing Two Trifunctional Amino Acids J. Amer. Chem. Soc. **76**, 5076 (1954).

165. Frankel, M. and A. Berger: Synthesis of Polyaspartic Acid. J. Organ. Chem. (USA) **16**, 1513 (1951).

166. Frankel, M., S. Cordova and M. Breuer: Synthesis of Poly-(O-acetyl-*DL*-serine) and of Poly-*DL*-serine. J. Chem. Soc. (London) **1953**, 1991.

167. Frankel, M. and M. Halmann: N-Carboxy-anhydrides of O-Acetyl- and O-Carbobenzyloxy-serine. J. Chem. Soc. (London) **1952**, 2735.

168. Frankel, M. and E. Katchalski: Poly-condensation of α-Amino Acid Esters. I. Poly-condensation of Glycine Esters. J. Amer. Chem. Soc. **64**, 2264 (1942).

169. Frankel, M., Y. Liwschitz and Y. Amiel: Syntheses of *DL*-Aspartic Acid and *DL*-Asparagine *via* Their N-Benzyl Derivatives. J. Amer. Chem. Soc. **75**, 330 (1953).

170. Frankel, M., Y. Liwschitz and A. Zilkha: Synthesis of Poly-Nβ-benzyl-*dl*-asparagine. J. Amer. Chem. Soc. **75**, 3270 (1953).

171. — — — Acyl Chlorides of Amino Acids and Peptides as Monomers for the Preparation of Polymeric Polypeptides. J. Amer. Chem. Soc. **76**, 2814 (1954).

172. Freudenberg, K., W. Dürr und H. v. Hochstetter: Zur Kenntnis der Aceton-Zucker. XIII. Die Hydrolyse einiger Disaccharide, Glucoside und Aceton-Zucker. Ber. dtsch. chem. Ges. **61**, 1735 (1928).

173. Freudenberg, K., H. Eichel und F. Leutert: Synthesen von Abkömmlingen der Aminosäuren. Ber. dtsch. chem. Ges. **65**, 1183 (1932).

174. Fromageot, Cl. and M. Jutisz: Chemistry of Amino Acids, Peptides, and Proteins. Annu. Rev. Biochem. **22**, 629 (1953).

175. Fruton, J. S.: The Synthesis of Peptides. Adv. Protein Chem. **5**, 1 (1949).

176. — Synthesis of Peptides of *l*-Serine. J. Biol. Chem. **146**, 463 (1942).

177. — A Peptide Derivative Related to Gramicidin. J. Amer. Chem. Soc. **70**, 1280 (1948).

178. Fruton, J. S. and M. Bergmann: The Multiple Specificity of Chymotrypsin. J. Biol. Chem. **145**, 253 (1942).

179. Fruton, J. S., M. Bergmann and W. P. Anslow, Jr.: The Specificity of Pepsin J. Biol. Chem. **127**, 627 (1939).

180. Fry, A.: A Tracer Study of the Reaction of Isocyanates with Carboxylic Acids. J. Amer. Chem. Soc. **75**, 2686 (1953).

181. Fuchs, F.: Über N-Carbonsäure-anhydride. Ber. dtsch. chem. Ges. **55**, 2943 (1922).

182. Fusari, S. A., T. H. Haskell, R. P. Frohardt and Q. R. Bartz: Azaserine, a New Tumor-inhibitory Substance. Structural Studies. J. Amer. Chem. Soc. **76**, 2881 (1954).

183. Gattermann, L.: Über Harnstoffchloride und deren synthetische Anwendung. Liebigs Ann. Chem. **244**, 29 (1888).

184. Gerngross, O.: Über die Benzoylierung von Imidazol-Derivaten. Ber. dtsch. chem. Ges. **46**, 1908 (1913).

185. — Über den Reaktionsmechanismus bei der Aufspaltung von Imidazol-Derivaten durch Benzoylchlorid und Alkali. Ber. dtsch. chem. Ges. **46**, 1913 (1913).

186. — Über Benzoylderivate des Histidins und Histamins. Z. physiol. Chem. (Hoppe-Seyler) **108**, 50 (1920).

187. Gish, D. T. and F. H. Carpenter: *p*-Nitrobenzyloxycarbonyl Derivatives of Amino Acids. J. Amer. Chem. Soc. **75**, 950 (1953).

188. — — Preparation of Arginyl Peptides. J. Amer. Chem. Soc. **75**, 5872 (1953).

189. Glasoe, P. K., J. Kleinberg and L. F. Audrieth: Acid Catalysis in Amines. II. The Catalytic Effect of Various Butylammonium Salts on the Aminolysis of Ethyl Phenylacetate in Anhydrous *n*-Butylamine. J. Amer. Chem. Soc. **61**, 2387 (1938).

190. Goldschmidt, St.: Über die Polymerisation der 2,5-Dioxo-Oxazolidine zu höhermolekularen Polypeptiden sowie über zwei neue Peptidsynthesen. Angew. Chem. **62**, 538 (1950).

191. Goldschmidt, St. und Ch. Jutz: Über Peptid-Synthesen. III. Mitt. Eine neue Synthese des Glutathions. Chem. Ber. **86**, 1116 (1953).

192. — — Über Peptidsynthesen. IV. Mitt. Glutaminsäurepeptide. Chem. Ber. **89**, 518 (1956).

193. Goldschmidt, St. und H. L. Krauss: N-substituierte Amide der Phosphorigen und der Phosphor-Säure und ihre Verwendung zum Aufbau von Peptid-Bindungen. Angew. Chem. **67**, 471 (1955).

194. — — Über aryl-substituierte Phosphor-Stickstoff-Verbindungen. Liebigs Ann. Chem. **595**, 193 (1955).

195. Goldschmidt, St. und H. Lautenschlager: Über Peptid-Synthesen. II. Umsetzung von Phosphorazoverbindungen mit Acylaminosäuren und Acylpeptiden. Liebigs Ann. Chem. **580**, 68 (1953).

196. Goldschmidt, St. und F. Obermeier: N-Substituierte Amide der phosphorigen und Phosphorsäure und deren Reaktion mit Carbonsäuren. Liebigs Ann. Chem. **588**, 24 (1954).

197. Goldschmidt, St. und M. Wick: Über eine neue Methode zur Herstellung der Peptidbindung. Z. Naturforsch. 5b, 170 (1950).

198. — — Über Peptidsynthesen. I. Liebigs Ann. Chem. **575**, 217 (1952).

199. Grassmann, W., H. Endres und A. Steber: Esterbindungen im Prokollagen. Z. Naturforsch. 9b, 513 (1954).

200. Grassmann, W. und K. Kühn: Abbau des Kollagens und des Prokollagens mit Natriumperjodat und Phenyljodosoacetat. Z. physiol. Chem. (Hoppe-Seyler) **301**, 1 (1955).

201. Grassmann, W. und F. Schneider: Zur Spezifität der Dipeptidase. Enzymatisches Verhalten von Asparaginsäure- und Glutaminsäurepeptiden. Biochem. Z. **273**, 452 (1934).

202. Grassmann W. und E. Schulte-Uebbing: Die Synthese einiger Peptide nach dem Phthalyl-Verfahren. Chem. Ber. **83**, 244 (1950).

203. Grassmann, W. und E. Wünsch: Peptidsynthesen. I. Zur Darstellung optisch aktiver Dipeptide nach der Phosphorazomethode. Chem. Ber. (in Vorbereitung).

204. — — Peptidsynthesen. IV. Zur Darstellung von *l*-Oxyprolyl-Glycin. Chem. Ber. (in Vorbereitung).

205. GRASSMANN, W., E. WÜNSCH und P. DEUFEL: Peptidsynthesen. V. Darstellung und peptidsynthetische Verwendung von O-Benzyl-*Dl*-Serin. Chem. Ber. (in Vorbereitung).

206. GRASSMANN, W., E. WÜNSCH und G. FRIES: Peptidsynthesen. VI. Darstellung und peptidsynthetische Verwendung von O-Benzyl-*L*-Tyrosin. Chem. Ber. (in Vorbereitung).

207. GRASSMANN, W., E. WÜNSCH und G. FÜRST: Peptidsynthesen. VII. Die Synthese von Leucyl-glycyl-lysin. (Unveröffentlicht.) G. FÜRST: Dipl. Arbeit, Univ. München, 1955; E. WÜNSCH: Dissert., Univ. München, 1955.

208. GRASSMANN, W., E. WÜNSCH und A. RIEDEL: Peptidsynthesen. II. Die Darstellung optisch-aktiver höherer Peptide nach der Phosphorazomethode. Chem. Ber. (in Vorbereitung).

209. GREENSTEIN, J. P.: Studies of Multivalent Amino Acids and Peptides. IX. The Synthesis of *l*-Cystinyl-*l*-cystine. J. Biol. Chem. **121**, 9 (1937).

210. — A Synthesis of Homoarginine. J. Organ. Chem. (USA) **2**, 480 (1937).

211. — Studies of Multivalent Amino Acids and Peptides. X. Cystinyl Peptides as Substrates for Aminopolypeptidase and Dipeptidase. J. Biol. Chem. **124**, 255 (1938).

212. GREENSTEIN, J. P., S. M. BIRNBAUM and M. C. OTEY: Optical and Enzymatic Characterization of Amino Acids. J. Biol. Chem. **204**, 307 (1953).

213. GRIMMEL, H. W., A. GUENTHER and J. F. MORGAN: Phosphazo Compounds and their Use in Preparing Amides. J. Amer. Chem. Soc. **68**, 539 (1946).

214. HABERMANN, J.: Zur Kenntnis der Glutaminsäure. Liebigs Ann. Chem. **179**, 248 (1875).

215. HANBY, W. E., S. G. WALEY and J. WATSON: Synthetic Polypeptides. II. Polyglutamic Acid. J. Chem. Soc. (London). **1950**, 3239.

216. HANSON, H. und R. ILLHARDT: Über die Eignung von Phthalylpeptiden als Substrate für proteolytische Enzyme. Z. physiol. Chem. (Hoppe-Seyler) **298**, 210 (1954).

217. HARINGTON, C. R. and T. H. MEAD: Synthesis of Glutathione. Biochemic. J. **29**, 1602 (1935).

218. — — Synthesis of Peptides Containing Cystine and Glutamine, with Remarks on their Possible Bearing on the Structure of Insulin and a Note on the Amide Nitrogen of Insulin. Biochemic. J. **30**, 1598 (1936).

219. HARINGTON, C. R. and R. V. PITT RIVERS: The Synthesis of Cysteine- (Cystine-) Tyrosine Peptides and the Action Thereon of Crystalline Pepsin. Biochemic J. **38**, 417 (1944).

220. HARRIS, J. I. and J. S. FRUTON: Synthesis of Peptides of *L*-Serine and their Cleavage by Proteolytic Enzymes. J. Biol. Chem. **191**, 143 (1951).

221. HARRIS, J. I. and T. S. WORK: The Synthesis of Peptides Related to Gramicidin S and the Significance of Optical Configuration in Antibiotic Peptides. Biochemic. J. **46**, 582 (1950).

222. HARTUNG, W. H.: Palladium Catalyst. II. The Effect of Hydrogen Chloride in the Hydrogenation of Isomitroso Ketones. J. Amer. Chem. Soc. **53**, 2248 (1931).

223. HARTUNG, W. H., D. N. KRAMER and G. P. HAGER: Synthesis of Peptides *via* α-Benzyloximino Acids. J. Amer. Chem. Soc. **76**, 2261 (1954).

224. HAUPTSCHEIN, M., C. S. STOKES and E. A. NODIFF: Thiolesters of Perfluorocarboxylic Acids. J. Amer. Chem. Soc. **74**, 4405 (1952).

225. HEGEDÜS, B.: Über eine neue Synthese von Glutathion (γ-Glutaminyl-cysteinyl-glycin). Helv. Chim. Acta **31**, 737 (1948).

226. HEILBRON, SIR IAN: Concerning Amino-acids, Peptides and Purines. J. Chem. Soc. (London) **1949**, 2099.

227. Helferich, B., L. Moog und A. Jünger: Über den Ersatz reaktionsfähiger Wasserstoffatome in Zuckern, Oxy- und Aminosäuren durch den Triphenyl-methyl-rest. Ber. dtsch. chem. Ges. **58**, 872 (1925).

227a. Heyns, K. und R. Brockmann: Über die thermische Zersetzung des Leuchs-schen Körpers zu Polyglycin. Z. Naturforsch. **9**b, 21 (1954).

228. Hillmann, A. und G. Hillmann: Über leicht abspaltbare Acylreste bei der Peptidsynthese. Z. Naturforsch. **6**b, 340 (1951).

229. Hillmann-Elies, A., G. Hillmann und H. Jatzkewitz: N-Trityl-Amino-säuren und -peptide. Z. Naturforsch. **8**b, 445 (1953).

230. Hofmann, K. and M. Bergmann: The Specificity of Trypsin. II. J. Biol. Chem. **130**, 81 (1939).

230a. Hofmann, K. and A. Jöhl: Studies on Polypeptides. VI. Synthetic Confir-mation of N-Terminal Amino Acid Sequence of Corticotropin-A. J. Amer. Chem. Soc. **77**, 2914 (1955).

231. Hofmann, K., A. Lindenmann, M. Z. Magee and N. H. Khan: Studies on Polypeptides. III. Novel Routes to α-Amino Acid and Polypeptide Hydrazides. J. Amer. Chem. Soc. **74**, 470 (1952).

232. Hofmann, K., A. Rheiner and W. D. Peckham: Studies on Polypeptides. V. The Synthesis of Arginine Peptides. J. Amer. Chem. Soc. **75**, 6083 (1953).

232a. — — — Studies on Polypeptides. VII. The Synthesis of Peptides Containing Arginine. J. Amer. Chem. Soc. **78**, 238 (1956).

233. Holley, R. W. and A. D. Holley: Lactam Formation from Amino Acid Amides. Applications in Peptide Chemistry. J. Amer. Chem. Soc. **74**, 1110 (1952).

234. — — The Removal of N-*o*-Nitrophenoxyacetyl and N-Chloroacetyl Groups from Peptides. J. Amer. Chem. Soc. **74**, 3069 (1952).

235. — — A New Stepwise Degradation of Peptides. J. Amer. Chem. Soc. **74**, 5445 (1952).

236. Holley, R. W. and E. Sondheimer: The Synthesis of *L*-Histidyl Peptides. J. Amer. Chem. Soc. **76**, 1326 (1954).

237. Holly, F. W., E. W. Peel, E. L. Luz and K. Folkers: N-Phenylacetyl-*L*-cysteinyl-*D*-valine. J. Amer. Chem. Soc. **74**, 4539 (1952).

238. Hopkins, F. G.: On Glutathione: A Reinvestigation. J. Biol. Chem. **84**, 269 (1929).

239. Huber, W. F.: Mono-α-aminoacyl and Mono-α-dipeptide Triglycerides. J. Amer. Chem. Soc. **77**, 112 (1955).

240. Hugounenq et Morel: Sur la carbimide de la (*l*) leucine naturelle. C. R. hebd. Séances Acad. Sci. **140**, 505 (1905).

241. — — Soudure de la leucine naturelle à l'acide carbamique. C. R. hebd. Séances Acad. Sci. **140**, 150 (1905).

242. Hörmann, H., W. Grassmann, E. Wünsch und H. Preller: Die Veresterung von Peptiden und ihre Bedeutung für die Bestimmung carboxylendständiger Aminosauren nach der Reduktionsmethode. Chem. Ber. **89**, 933 (1956).

243. Hunt, M. and V. du Vigneaud: The Preparation of *d*-Alanyl-*l*-histidine and *l*-Alanyl-*l*-histidine and an Investigation of their Effect on the Blood Pressure in Comparison with *l*-Carnosine. J. Biol. Chem. **124**, 699 (1938).

244. — — The Synthesis of the Next Higher and Lower Homologues of *l*-Carnosine: γ-Aminobutyryl-*l*-histidine and Glycyl-*l*-histidine. J. Biol. Chem. **127**, 43 (1939).

245. Hurd, C. D. and C. M. Buess: The Formation of Polypeptides by Rearrangement of α-Carboxy Hydroxamic Acids. J. Amer. Chem. Soc. **73**, 2409 (1951).

246. Ing, H. R. and R. H. F. Manske: A Modification of the Gabriel Synthesis of Amines. J. Chem. Soc. (London) **1926**, 2348.

247. ISELIN, B., M. FEURER und R. SCHWYZER: Über aktivierte Ester. V. Verwendung der Cyanmethylester-Methode zur Herstellung von (N-Carbobenzoxy-S-benzyl-L-cysteinyl)-L-tyrosyl-L-isoleucin auf verschiedenen Wegen. Helv. Chim. Acta 38, 1508 (1955).

248. ISELIN, B. und R. SCHWYZER: Über aktivierte Ester. VI. Tetrahydro-pyranyl-Derivate von Aminosäuren und ihre Umsetzungen. Helv. Chim. Acta 39, 57 (1956).

249. JACOBS, W. A. and M. HEIDELBERGER: On Amides, Uramino Compounds, and Ureides Containing an Aromatic Nucleus. J. Amer. Chem. Soc. 39, 2418 (1917).

250. JOHN, W. D. and G. T. YOUNG: Amino-acids and Peptides. XII. α- and β-L-Aspartyl-L-valine. J. Chem. Soc. (London) 1954, 2870.

251. JUNG, S. L., J. G. MILLER and A. R. DAY: Effect of Structure on Reactivity. VIII. Aminolysis of Methyl Acetate with some β-Phenylethylamines. J. Amer. Chem. Soc. 75, 4664 (1953).

252. KAPFHAMMER, J. und H. MÜLLER: Guanidosäuren und Guanidopeptide. Z. physiol. Chem. (Hoppe-Seyler) 225, 1 (1934).

253. KARRER, P. und G. BUSSMANN: Einwirkung von Diazomethan auf Hippursäurechlorid. Helv. Chim. Acta 24, 645 (1941).

254. KATCHALSKY, A. und M. PAECHT: Phosphate Anhydrides of Amino Acids. J. Amer. Chem. Soc. 76, 6042 (1954).

255. KATCHALSKI, E.: Poly-α-Amino Acids. Adv. Protein Chem. 6, 123 (1951).

256. KATCHALSKI, E., I. GROSSFELD and M. FRANKEL: Poly-lysine. J. Amer. Chem. Soc. 69, 2564 (1947).

257. — — — Poly-condensation of α-Amino Acid Derivatives. III. Poly-lysine. J. Amer. Chem. Soc. 70, 2094 (1948).

258. KATCHALSKI, E. and A. PATCHORNIK: Poly-L-Histidine. XIV. Int. Kongr., Zürich, 1955; Ref. Bd. 191.

259. KATCHALSKI, E. and M. SELA: The Synthesis and Spectrophotometric Study of Poly-L-tyrosine and Poly-3,5-diiodotyrosine. J. Amer. Chem. Soc. 75, 5284 (1953).

260. KATCHALSKI, E. and P. SPITNIK: Poly-arginine. Nature (London) 164, 1092 (1949).

261. — — Ornithine Anhydride. J. Amer. Chem. Soc. 73 2946 (1951).

262. KATSOYANNIS, P. G. and V. DU VIGNEAUD: The Synthesis of p-Toluenesulfonyl-L-isoleucyl-L-glutaminyl-L-asparagine and Related Peptides. J. Amer. Chem. Soc. 76, 3113 (1954).

263. KENNER, G. W. and R. J. STEDMAN: Peptides. I. The Synthesis of Peptides through Anhydrides of Sulphuric Acid. J. Chem. Soc. (London) 1952, 2069.

264. KENT, L. H.: A Modified Method for the Preparation of Carbodiimides. XIV. Int. Kongr., Zürich, 1955; Ref. Bd. 192.

265. KHORANA, H. G.: The Stepwise Degradation of Peptides. Chem. and Ind. 1951, 129.

266. — The Chemistry of Carbodiimides. Chem. Rev. 53, 145 (1953).

267. KIDD, D. A. A. and F. E. KING: Preparation of Phthalyl-L-Glutamic Acid. Nature (London) 162, 776 (1948).

268. KING, F. E., J. W. CLARK-LEWIS, D. A. A. KIDD and G. R. SMITH: Syntheses from Phthalimido-acids. IV. p-Glycylaminobenzoic Acid and Derivatives. J. Chem. Soc. (London) 1954, 1039.

269. KING, F. E., J. W. CLARK-LEWIS and G. R. SMITH: Syntheses from Phthalimido-acids. V. Amides of Glycine, Dl-Alanine, and L-Glutamic Acid with Amphetamine, Benzocaine, and Procaine. J. Chem. Soc. (London) 1954, 1044.

270. King, F. E., J. W. Clark-Lewis and G. R. Smith: Syntheses from Phthalimido-acids. VI. Further Product from Phthalyl-*DL*-aspartic Anhydride, and the Preparation of Phthalylglycyl-*DL*-asparagine and -*DL*-serine. J. Chem. Soc. (London) **1954**, 1046.

271. King, F. E., B. S. Jackson and D. A. A. Kidd: Syntheses from Phthalimido-acids. II. Further Reactions of Phthalyl-glutamic Anhydride. J. Chem. Soc. (London) **1951**, 243.

272. King, F. E. and D. A. A. Kidd: A New Synthesis of Glutamine and of γ-Dipeptides of Glutamic Acid from Phthalylated Intermediates. J. Chem. Soc. (London) **1949**, 3315.

273. — — Syntheses from Phthalimido-acids. III. The Preparation of *DL*- and *L*-Asparagine from Phthalyl-*DL*- and -*L*-aspartic Anhydride. J. Chem. Soc. (London) **1951**, 2976.

274. Kögl, F. und A. M. Akkerman: Über die Synthese von Epi-G-Glutathion. Rec. trav. chim. Pays-Bas **65**, 216 (1946).

275. Kollonitsch, J., A. Hajós und F. Gábor: Neuere Synthese von Peptiden. XIV. Int. Kongr., Zurich, 1955; Ref. Bd. **193**; Nature (London) (im Druck).

276. Kossel, A. und E. L. Kennaway: Über Nitroclupein. Z. physiol. Chem. (Hoppe-Seyler) **72**, 486 (1911).

277. Kraut, K. und F. Hartmann: Über das Glycin. Liebigs Ann. Chem. **133**, 99 (1865).

278. Kuhn, R. und H. W. Ruelius: Über die Einwirkung von Diazomethan auf Peptide und Proteine. Chem. Ber. **85**, 38 (1952).

279. Kurtz, A. C.: A New Method fur Isolating *l*-(+)-Lysine. J. Biol. Chem. **140**, 705 (1941).

280. Langenbeck, W. und P. Kresse: Über eine zweckmäßige Modifikation der Peptidester-Synthese nach Bailey. J. prakt. Chem. (4) **2**, 261 (1955).

281. Lettré, H. und M. E. Fernholz: Über chemisch markierte Antigene, II. Mitt. Zur Reaktionsfähigkeit von Oxazolonen. Z. physiol. Chem. (Hoppe-Seyler) **266**, 37 (1940).

282. Leuchs, H.: Über die Glycin-carbonsaure. Ber. dtsch. chem. Ges. **39**, 857 (1906).

283. Leuchs, H. und Geiger, W.: Über die Anhydride von α-Amino-N-carbonsäuren und die von α-Aminosauren. Ber. dtsch. chem. Ges. **41**, 1721 (1908).

284. Leuchs, H. und W. Manasse: Über die Isomerie der Carbäthoxy-glycyl-glycinester. Ber. dtsch. chem. Ges. **40**, 3235 (1907).

285. Levene, P. A. and D. W. Hill: On a Dipeptide Phosphoric Acid Isolated from Casein. J. Biol. Chem. **101**, 711 (1953).

286. Levene, P. A., R. E. Steiger and R. E. Marker: Studies on Racemization. X. Action of Alkali on Ketopiperazines and Peptides. J. Biol. Chem. **93**, 605 (1931).

287. Levine, S.: Synthesis of Glycyl and Alanyl Chlorides. J. Amer. Chem. Soc. **76**, 1382 (1954).

288. Levy, A. L.: Anhydro-N-Carboxy-*DL*-β-Phenylalanine. Nature (London) **165**, 152 (1950).

289. Lindenmann, A., N. H. Khan and K. Hofmann: Studies on Polypeptides. IV. Remarks Regarding the Use of the Phenylthiocarbonyl Protecting-group in Peptide Synthesis. J. Amer. Chem. Soc. **74**, 476 (1952).

290. Lipmann, F.: Über die Bindung der Phosphorsäure in Phosphorproteinen. 1. Mitt. Isolierung einer phosphorhaltigen Aminosäure (Serinphosphorsäure) aus Casein. Biochem. Z. **262**, 3 (1933).

291. — Metabolic Generation and Utilization of Phosphate Bond Energy. Adv. Enzymology **1**, 99 (1941).

292. LIWSCHITZ, Y. and A. ZILKHA: Syntheses of Aspartyl Amides and Peptides trough N-Benzyl-*dl*-aspartic Acid. J. Amer. Chem. Soc. **76**, 3698 (1954).

293. LORING, H. S. and V. DU VIGNEAUD: The Synthesis of Crystalline Cystinyl-diglycine and Benzylcysteinylglycine and their Isolation from Glutathione. J. Biol. Chem. **111**, 385 (1935).

294. LYNEN, F. und E. REICHERT: Zur chemischen Struktur der „aktivierten Essigsäure". Angew. Chem. **63**, 47 (1951).

295. LYNEN, F., E. REICHERT und L. RUEFF: Zum biologischen Abbau der Essigsäure. VI. „Aktivierte Essigsäure", ihre Isolierung aus Hefe und ihre chemische Natur. Liebigs Ann. Chem. **574**, 1 (1951).

295a. MAGEE, M. Z. and K. HOFMANN: The Application of the Curtius Reaction to the Polymerization of Triglycine. J. Amer. Chem. Soc. **71**, 1515 (1949).

296. MAILLARD, L. C.: Synthèse de polypeptides par action de la glycérine sur le glycocolle. Ann. Chim. [9] **1**, 519 (1914).

297. McKENZIE, A. and J. K. WOOD: Observations on Some Organic Compounds of Arsenic. J. Chem. Soc. (London) **117**, 406 (1920).

298. MEGGY, A. B.: Glycine Peptides. I. The Polymerization of Piperazine-2 : 5-dione at 180°. J. Chem. Soc. (London) **1953**, 851.

299. MELVILLE, J.: Labile Glutamine Peptides, and their Bearing on the Origin of the Ammonia set Free During the Enzymic Digestion of Proteins. Biochemic. J. **29**, 179 (1935).

300. MEYER, K. H. et Y. GO: Observations roentgénographiques sur des polypeptides inférieurs et supérieurs. Helv. Chim. Acta **17**, 1488 (1934).

301. MICHAELIS, A. und G. SCHROETER: Über das Phosphorazo-benzolchlorid und dessen Derivate. Ber. dtsch. chem. Ges. **27**, 490 (1894).

302. MICHEEL, F. und CH. BERDING: Synthese eines hochpolymeren *DL*-Lysyl-*L*-tyrosyl-peptides. Chem. Ber. **88**, 1062 (1955).

303. MILLER, G. L., O. K. BEHRENS and V. DU VIGNEAUD. A Synthesis of the Aspartic Acid Analogue of Glutathione (Asparthione). J. Biol. Chem. **140**, 411 (1941).

304. MILLER, H. K. and H. WAELSCH: Benzyl Esters of Amino Acids. J. Amer. Chem. Soc. **74**, 1092 (1952).

305. — — The Synthesis of Glutamine and Asparagine Peptides and of Glutamine. Arch. Biochem. Biophys. **35**, 176 (1952).

306. MOHR, E. und TH. GEIS: Benzoylamino-isobuttersäure-lactimon. Ber. dtsch. chem. Ges. **41**, 798 (1908).

307. — — 2. Mitt. über lactonähnliche Anhydride acylierter Aminosäuren. Über das Lacton der Benzoyl-α-aminoisobuttersäure. J. prakt. Chem. **81**, 49 (1910).

308. MOHR, E. und F. STROSCHEIN: Über die Lactimone des Benzoyl-alanins und des Benzoyl-phenyl-alanins. Ber. dtsch. chem. Ges. **42**, 2521 (1909).

309. MOORE, J. A., J. R. DICE, E. D. NICOLAIDES, R. D. WESTLAND and E. L. WITTLE: Azaserine, Synthetic Studies. I. J. Amer. Chem. Soc. **76**, 2884 (1954).

309a. MOREL, A.: Soudure des acides aminés dérivés des albumines. C. R. hebd. Séances Acad. Sci. **140**, 505 (1905).

310. NAEGELI, C. und A. TYABJI: Über den Umsatz aromatischer Isocyansäure-ester mit organischen Säuren. II. Isolierung einiger Carbaminsäure-carbonsäure-anhydride. Helv. Chim. Acta **18**, 142 (1935).

311. — — Über den Umsatz aromatischer Isocyansäure-ester mit organischen Säuren. I. Theorie und Anwendung der Reaktion für die präparative Darstellung von Säureanhydriden. Helv. Chim. Acta **17**, 931 (1934).

312. NEUBERGER, A.: Stereochemistry of Amino Acids. Adv. Protein Chem. **4**, 297 (1948).

313. Neuberger, A. and F. Sanger: The Availability of the Acetyl Derivatives of Lysine for Growth. Biochemic J. **37**, 515 (1943).

314. Neuman, R. E. and E. L. Smith: Synthesis of Proline and Hydroxyproline Peptides; their Cleavage by Prolinase. J. Biol. Chem. **193**, 97 (1951).

315. Nicolaides, E. D., R. D. Westland and E. L. Wittle: Azaserine, Synthetic Studies. II. J. Amer. Chem. Soc. **76**, 2887 (1954).

316. Nicolet, B. H.: Interpretation of the Dehydration of Acetylglutamic Acid by Means of Glutamylthiohydantoin Derivates. J. Amer. Chem. Soc. **52**, 1192 (1930).

317. Nienburg, H.: Eine zweite Synthese des d-Glutamins. Ber. dtsch. chem. Ges. **68**, 2232 (1938).

318. Noguchi, J. and T. Hayakawa: A New Synthetic Method of Protein Analogs having Periodic Arrangement of Amino Acids. J. Amer. Chem. Soc. **76**, 2846 (1954).

319. O'Brien, J. L. and C. Niemann: The Acid Catalyzed Cyclization of λ-Acyl-amino Acids. J. Amer. Chem. Soc. **72**, 5348 (1950).

320. Orden, H. O. Van and E. L. Smith: The Synthesis of Arginyl Dipeptides; the Action of Papain and Trypsin. J. Biol. Chem. **208**, 751 (1954).

321. Overell, B. G. and V. Petrow: Polymers of Some Basic and Acidic α-Amino-acids. J. Chem. Soc. (London) **1955**, 232.

322. Pacsu, E. and E. J. Wilson, Jr.: Polycondensation of Certain Peptide Esters. I. Polyglycine Esters. J. Organ. Chem. (USA) **7**, 117 (1942).

323. Parham, W. E. and E. L. Anderson: The Protection of Hydroxyl Groups. J. Amer. Chem. Soc. **70**, 4187 (1948).

324. Patchornik, A., M. Sela and E. Katchalski: Polytryptophan. J. Amer. Chem. Soc. **76**, 299 (1954).

325. Pauly, H. und J. Weir: Über eine einseitige Esterbildung der Benzoyl-asparaginsäure. Ber. dtsch. chem. Ges. **43**, 661 (1910).

326. Pawlewski, B.: Neue Methode der Acetylierung von Amidoverbindungen. Ber. dtsch. chem. Ges. **31**, 661 (1898).

327. — Über die Acetylierung aromatischer Amine. Ber. dtsch. chem. Ges. **35**, 110 (1902).

328. Petersen, S.: Niedermolekulare Umsetzungsprodukte aliphatischer Diiso-cyanate. Liebigs Ann. Chem. **562**, 205 (1949).

329. Pirie, N. W.: The Cuprous Derivatives of Some Sulphydryl Compounds. Biochemic. J. **25**, 614 (1931).

330. Piutti, A. und G. Magli: Influenza della temperatura e della concentrazione sul potere rotatorio delle soluzioni acquose di alcemi aspartati monoalcolici. Gazz. chim. ital. **36**, II, 738 (1906).

331. Plapinger, R. E. and T. Wagner-Jauregg: A Nitrogen-to-Oxygen Phosphoryl Migration: Preparation of dl-Serinephosphoric and Threoninephosphoric Acid. J. Amer. Chem. Soc. **75**, 5757 (1953).

332. Plöchl, J.: Über Ortho-nitro-glycine und ihre Reduktionsprodukte. Ber. dtsch. chem. Ges. **19**, 6 (1886).

333. Popenoe, E. A. and V. du Vigneaud: The Synthesis of L-Phenyl-alanyl-L-Glutaminyl-L-asparagine. J. Amer. Chem. Soc. **76**, 6202 (1954).

334. Posternak, S.: The Phosphorus Nucleus of Caseinogen. Biochemic. J. **21**, 289 (1927).

335. Posternak, Th. et S. Grafl: Sur la préparation de quelques peptides dérivés de la phosphotyrosine et sur leur dégradation enzymatique. Helv. Chim. Acta **28**, 1258 (1945).

336. Prelog, V. und P. Wieland: l-Lysyl-glycyl-glycyl-l-glutaminsäure. Helv. Chim. Acta **29**, 1128 (1946).

337. LE QUESNE, W. J. and G. T. YOUNG: Amino-acids and Peptides. I. An Examination of the Use of Carbobenzyloxy-*L*-glutamic Anhydride in the Synthesis of Glutamylpeptides. J. Chem. Soc. (London) 1950, 1954.

338. — — Amino-acids and Peptides. II. Synthesis of α- and γ-*L*-Glutamylpeptides by the Azide Route. J. Chem. Soc. (London) 1950, 1959.

339. — — Amino-acids and Peptides. VI. Synthesis of *L*-Aspartyl Peptides from Carbobenzyloxy-*L*-aspartic Anhydride. J. Chem. Soc. (London) 1952, 24.

340. — — Amino-acids and Peptides. VII. The Autohydrolysis of Glutamylpeptides. J. Chem. Soc. (London) 1952, 594.

341. RADENHAUSEN, R.: Über Hydrazide substituierter Amidosäuren und das Hydrazid der Fumarsäure. J. prakt. Chem. 52, 446 (1895).

342. RAMSAY, H.: Neue Darstellung der Glykocyamine oder Guanidosäuren. Ber. dtsch. chem. Ges. 41, 4385 (1908).

343. RAO, K. R., S. M. BIRNBAUM, R. B. KINGSLEY and J. P. GREENSTEIN: Enzymatic Susceptibility of Corresponding Chloroacetyl- and Glycyl-*L*-amino Acids. J. Biol. Chem. 198, 507 (1952).

344. REES, P. S., D. P. TONG and G. T. YOUNG: Amino-acids and Peptides. XI. The Kinetics of the Self-condensation of Glycylglycylglycine Methyl Ester. J. Chem. Soc. (London) 1954, 662.

345. RESSLER, C. and V. DU VIGNEAUD: The Synthesis of the Tetrapeptide Amide S-Benzyl-*L*-cysteinyl-*L*-prolyl-*L*-leucylglycinamide. J. Amer. Chem. Soc. 76, 3107 (1954).

346. ROBERTS, C. W.: The Synthesis of *L*-Cysteinyl-*L*-tyrosyl-*L*-isoleucine. J. Amer. Chem. Soc. 76, 6203 (1954).

347. ROSENMUND, K. W. und F. ZETZSCHE: Über die Beeinflussung der Wirksamkeit von Katalysatoren. V. Mitt. Ber. dtsch. chem. Ges. 54, 2038 (1921).

348. ROWLANDS, D. A. and G. T. YOUNG: Amino-acids and Peptides. IX. γ-*L*-Glutamyl-*L*-alanine, -*L*-valine and -*L*-leucine. J. Chem. Soc. (London) 1952, 3937.

349. RUDINGER, J.: Aminosäure und Peptide. X, Verschiedene Derivate. und Reaktionen der 1-*p*-Toluolsulfonyl-*L*-pyrrolid-5-on-2-carbonsäure. Chem. Listy 48, 235 (1954).

350. RUDINGER, J. und H. CZURBOVÁ: Aminosäuren und Peptide. XII. Eine neue Racematspaltung der *DL*-Glutaminsäure; die Synthese von *D*-Glutamin und γ-*D*-Glutamyl-glycın. Chem. Listy 48, 254 (1954).

351. RYDON, H. N. and P. W. G. SMITH: Polypeptides. I. The Condensation-polymerisation of Some Polyglycine Esters and Azides. J. Chem. Soc. (London) 1955, 2542.

352. SACHS, H. and E. BRAND: Optical Rotation of Peptides. VII. α- and γ-Dipeptides of Glutamic Acid and Alanine. J. Amer. Chem. Soc. 75, 4608 (1953).

353. — — Benzyl Esters of Glutamic Acid. J. Amer. Chem. Soc. 75, 4610 (1953).

354. — — Optical Rotation of Peptides. VIII. Glutamic Acid Tripeptides. J. Amer. Chem. Soc. 76, 1811 (1954).

355. — — An Examination of the Use of Carbobenzyloxy-γ-*L*-glutamyl Hydrazide in the Synthesis of γ-Peptides. J. Amer. Chem. Soc. 76, 1815 (1954).

355 a. SACHS, H. and H. WAELSCH: Thioesters of Glutamic Acid. J. Amer. Chem. Soc. 77, 6600 (1955).

356. SAKAMI, W. and G. TOENNIES: The Investigation of Amino Acid Reactions by Methods of Non-aqueous Titrimetry. II. Differential Acetylation of Hydroxy Groups, and a Method for the Preparation of the O-Acetyl Derivatives of Hydroxyamino Acids. J. Biol. Chem. 144, 203 (1942).

357. SANNIÉ, C.: Méthode simple de synthèse des 2–5 dicétopiperazines (Anhydrides des acides α-aminés). Bull. soc. chim. France [5] 9, 487 (1942).

358. Schallenberg, E. E. and M. Calvin: Ethyl Thioltrifluoracetate as an Acetylating Agent with Particular Reference to Peptide Synthesis. J. Amer. Chem. Soc. **77**, 2779 (1955).

359. Schlogl, K. und H. Fabitschowitz. Konstitutionsermittelung von Peptiden. VI. Lysylpeptide. XI. Mitt. uber Peptide. Monatsh. Chem. **84**, 937 (1953).

360. Schmidt, E., F. Hitzler und E. Lahde: Zur Kenntnis aliphatischer Carbodiimide. I. Mitt. Ber. dtsch. chem. Ges. **71**, 1933 (1938).

361. Schmidt, E. und M. Seefelder: Zur Kenntnis aliphatischer Carbodiimide. IV. Mitt. Liebigs Ann. Chem. **571**, 83 (1951).

362. Schönheimer, R.: Ein Beitrag zur Bereitung von Peptiden. Z. physiol. Chem. (Hoppe-Seyler) **154**, 203 (1926).

363. Schott, H. F., J. B. Larkin, L. B. Rockland and M. S. Dunn: The Synthesis of l-(—)-Leucylglycylglycine. J. Organ. Chem. (USA) **12**, 490 (1947).

364. Schramm, G. und H. Restle: Über die Darstellung einheitlicher Polypeptide. Makromolek. Chem. **13**, 103 (1954).

365. Schramm, G. und G. Thumm: Die Darstellung höherer ringförmiger Peptide. Z. Naturforsch. 3b, 218 (1948).

366. Schulze, E. und E. Winterstein: Über die Konstitution des Arginins. Ber. dtsch. chem. Ges. **32**, 3191 (1899).

367. Schumann, I. and R. A. Boissonnas: Splitting of N-phthalyl Groups of Amino-acids with Phenylhydrazine. Nature (London) **169**, 154 (1952).

368. — — Scission, par la phénylhydrazine, du groupe "phthalyle" d'acides aminés et de peptides N-phthalylés. Helv. Chim. Acta **35**, 2235 (1952).

369. — — Synthèse de la L-valyl-L-(δ-carbobenzoxy)-ornithyl-L-leucyl-D-phenylalanyl-L-proline. Helv. Chim. Acta **35**, 2237 (1952).

370. Schütte, E.: Darstellung von Guanidosäuren und Guanidopeptiden. Z. physiol. Chem. (Hoppe-Seyler) **279**, 52 (1943).

371. Schwyzer, R.: Coenzym A. Eine einfache Synthese von S-Acylderivaten des Pantetheins. Helv. Chim. Acta **35**, 1903 (1952).

372. — Amid- und Peptid-Synthesen in verdunnter, wäßriger Lösung. Angew. Chem. **65**, 267 (1953).

373. — Coenzym A. Modellversuche zur biologischen Acylierungsreaktion. Über die Reaktionsfähigkeit von Thiolcarbonsäuren und ihren Estern. Helv. Chim. Acta **36**, 414 (1953).

374. — Darstellung von Aminosäure- und Peptid-Thioestern. Helv. Chim. Acta **37**, 647 (1954).

375. Schwyzer, R., M. Feurer und B. Iselin. Über aktivierte Ester. III. Umsetzung aktivierter Ester von Aminosäure- und Peptid-Derivaten mit Aminen und Aminosäureestern. Helv. Chim. Acta **38**, 83 (1955).

376. Schwyzer, R., M. Feurer, B. Iselin und H. Kägi: Über aktivierte Ester. II. Synthese aktivierter Ester von Aminosäure-Derivaten. Helv. Chim. Acta **38**, 80 (1955).

377. Schwyzer, R. und Ch. Hürlimann: Coenzym A. Modellreaktionen zur enzymatischen Aktivierung von Acylderivaten des Coenzyms A. Helv. Chim. Acta **37**, 155 (1954).

378. Schwyzer, R. and B. Iselin: Activated Esters as Intermediates in the Synthesis of Amide and Peptide Bonds. IV. Cyanmethylesters of Carbobenzoxy-L-leucine and of Carbobenzoxy-DL-leucine, and their Reactions with Benzylamine and Derivatives of Glycine. Ann. Acad. Scient. Fennicae Ser. A, II, **60**, 181 (1955).

379. Schwyzer, R., B. Iselin und M. Feurer: Über aktivierte Ester. I. Aktivierte Ester der Hippursäure und ihre Umsetzungen mit Benzylamin. Helv. Chim. Acta **38**, 69 (1955).

380. SELA, M. and A. BERGER: The Mechanism of Polymerization of N-Carboxy-α-amino Acid Anhydrides. J. Amer. Chem. Soc. **75**, 6350 (1953).

381. SEMB, J., J. H. BOOTHE, R. B. ANGIER, C. W. WALLER, J. H. MOWAT, B. L. HUTCHINGS and Y. SUBBAROW: Pteroic Acid Derivatives. V. Pteroyl-α-glut-amyl-α-glutamylglutamic Acid, Pteroyl-γ-glutamyl-α-glutamylglutamic Acid, Pteroyl-α-glutamyl-γ-glutamylglutamic Acid. J. Amer. Chem. Soc. **71**, 2310 (1949).

382. SHAPIRO, B. L.: The Synthesis of Peptides. Chem. and Ind. **1952** 1119.

383. SHATENSHTEIN, A. I.: A Study of Acid Catalysis in Liquid Ammonia. J. Amer. Chem. Soc. **59**, 432 (1937).

384. SHEEHAN, J. C., D. W. CHAPMAN and R. W. ROTH: The Synthesis of Stereochemically Pure Peptide Derivatives by the Phthaloyl Method. J. Amer. Chem. Soc. **74**, 3822 (1952).

385. SHEEHAN, J. C. and V. S. FRANK: A New Synthesic Route to Peptides. J. Amer. Chem. Soc. **71**, 1856 (1949).

386. — — Peptide Syntheses Using Energy-rich Phosphorylated Amino Acid Derivatives. J. Amer. Chem. Soc. **72**, 1312 (1950).

386 a. SHEEHAN, J. C., M. GOODMAN and G. P. HESS: Peptide Derivatives Containing Hydroxyamino Acids. J. Amer Chem. Soc. **78**, 1367 (1956).

387. SHEEHAN, J. C., M. GOODMAN and W. L. RICHARDSON: The Product Derived from the Cyclization of Triglycine Azide. J. Amer. Chem. Soc. **77**, 6391 (1955).

388. SHEEHAN, J. C. and G. P. HESS: A New Method of Forming Peptide Bonds. J. Amer. Chem. Soc. **77**, 1067 (1955).

389. SHEEHAN, J. C. and D. A. JOHNSON: The Synthesis and Reactions of N-Acyl Thiol Amino Acids. J. Amer. Chem. Soc. **74**, 4726 (1952).

390. SHEEHAN, J. C. and W. L. RICHARDSON: A New Method for the Synthesis of Macrocyclic Peptides. J. Amer. Chem. Soc. **76**, 6329 (1954).

391. SHEMIN, D. and R. M. HERBST: The Synthesis of Dipeptides from α-Keto Acids. J. Amer. Chem. Soc. **60**, 1951 (1938).

392. SIEFKEN, W.: Mono- und Polyisocyanate. IV. Mitt. über Polyurethane. Liebigs Ann. Chem. **562**, 75 (1949).

393. SIFFERD, R. H. and V. DU VIGNEAUD: A New Synthesis of Carnosine, with some Observations on the Splitting of the Benzyl Group from Carbobenzoxy Derivates and from Benzylthio Ethers. J. Biol. Chem. **108**, 753 (1935).

394. SIGMUND, F. und F. WESSELY: Untersuchungen über α-Amino-N-Carbonsäure-anhydride. II. Z. physiol. Chem. (Hoppe-Seyler) **157**, 91 (1926).

395. Sitz.-Berichte, Heidelberg. Akad. Wiss. **1931**, 9 Abl.

396. SMITH, E. L.: Action of Carboxypeptidase on Peptide Derivatives of *L*-Tryptophan. J. Biol. Chem. **175**, 39 (1948).

397. SMITH, E. L. and M. BERGMANN: The Peptidases of Intestinal Mucosa. J. Biol. Chem. **153**, 627 (1944).

398. SONDHEIMER, E. and R. W. HOLLEY: Imides from Asparagine and Glutamine. J. Amer. Chem. Soc. **76**, 2467 (1954).

399. — — The Synthesis of Peptides of *L*-Glutamine by the Carbobenzoxy Azide Method. J. Amer. Chem. Soc. **76**, 2816 (1954).

400. ŠORM, F. and J. RUDINGER: On Proteins and Amino-acids. III. The Synthesis of γ-*L*-Glutamyl-*L*-Tyrosine. Collect. Czech. Chem. Comm. **15**, 491 (1950).

401. STADTMAN, E. R. and F. H. WHITE, Jr.: The Enzymatic Synthesis of N-Acetylimidazole. J. Amer. Chem. Soc. **75**, 2022 (1953).

402. STAUDINGER, H.: Über aliphatische Diazoverbindungen und Ketene. Helv. Chim. Acta **5**, 87 (1922).

403. Stekol, J. A.: The Metabolism of *l*- and *dl*-α-Hydroxy-β-benzyl-thiopropionic and *dl*-α-Hydroxy-γ-benzyl-thiobutyric Acids in the Rat. J. Biol. Chem. **140,** 827 (1941).

404. Stevens, C. M. and R. Watanabe: Amino Acid Derivatives. I. Carboallyloxy Derivatives of α-Amino Acids. J. Amer. Chem. Soc. **72,** 725 (1950).

405. Stoll, A. und Th. Petrzilka: Versuche zur Synthese des Peptidteils der Mutterkornalkaloide. I. Helv. Chim. Acta **35,** 589 (1952).

406. Süss, O.: Über eine neue Peptidsynthese. Liebigs Ann. Chem. **572,** 96 (1951).

407. Swan, J. M. and V. du Vigneaud: The Synthesis of *L*-Glutaminyl-*L*-asparagine, *L*-Glutamine and *L*-Isoglutamine from *p*-Toluene-sulfonyl-*L*-glutamic Acid. J. Amer. Chem. Soc. **76,** 3110 (1954).

408. Synge, R. L. M.: Experiments on Amino-acids. III. A Method for the Isolation of Hydroxy-amino-acids from Protein Hydrolysates. Biochemic. J. **33,** 1924 (1939).

409. — The Synthesis of Some Dipeptides Related to Gramicidin S. Biochemic. J. **42,** 99 (1948).

410. Tanenbaum, St. W.: Preparation of Isoasparagine by the Phthaloyl Method. J. Amer. Chem. Soc. **75,** 1754 (1953).

411. Taurog, A., S. Abraham and I. L. Chaikoff: Synthesis of Some O-Glucuronides and O-Glucosides of Phenolic Amino Acids. J. Amer. Chem. Soc. **75,** 3473 (1953).

412. Thate, A.: Über die Einwirkung von Reduktionsmitteln auf O-Nitrophenoxyl-essigsäure. J. prakt. Chem. **29,** 145 (1884).

413. Thierfelder, H. und E. v. Cramm: Über glutaminhaltige Polypeptide und zur Frage ihres Vorkommens im Eiweiß. Z. physiol. Chem. (Hoppe-Seyler) **105,** 58 (1919).

414. Turba, F. und K. Schuster: Über eine Synthese des *d,l*-Arginins. Z. physiol. Chem. (Hoppe-Seyler) **283,** 27 (1948).

415. Vaughan, J. R., Jr.: Diethyl Chloroarsenite as a Reagent for the Preparation of Peptides. J. Amer. Chem. Soc. **73,** 1389 (1951).

416. — Acylalkylcarbonates as Acylating Agents for the Synthesis of Peptides. J. Amer. Chem. Soc. **73,** 3547 (1951).

417. — Preliminary Investigations on the Preparation of Optically Active Peptides Using Mixed Carbonic-Carboxylic Acid Anhydrides. J. Amer. Chem. Soc. **74,** 6137 (1952).

418. Vaughan, J. R., Jr. and J. A. Eichler: The Preparation of Optically-active Peptides Using Mixed Carbonic-Carboxylic Acid Anhydrides. J. Amer. Chem. Soc. **75,** 5556 (1953).

419. — — Synthesis of Model, High Molecular Weight Peptides by the Mixed Carbonic-Carboxylic Acid Anhydride Procedure. J. Amer. Chem. Soc. **76,** 2474 (1954).

420. Vaughan, J. R., Jr. and R. L. Osato: Preparation of Peptides Using Mixed Carboxylic Acid Anhydrides. J. Amer. Chem. Soc. **73,** 5553 (1951).

421. — — The Preparation of Peptides Using Mixed Carbonic-Carboxylic Acid Anhydrides. J. Amer. Chem. Soc. **74,** 676 (1952).

422. Velluz, L., G. Amiard et R. Heymès: Utilisation d'intermédiaires N-benzylés en synthèse peptidique. I. Chlorures de N-benzyl α-amino-acides. Bull. soc. chim. France [5] **21,** 1012 (1954).

422a — — — Utilisation d'intermédiaires N-benzylés en synthèse peptidique. III. Dibenzylation des α-amino-acides et dédoublement. Bull. soc. chim. France [5] **22,** 201 (1955).

422b. VELLUZ, L., J. ANATOL et G. AMIARD: Utilisation d'intermédiaires N-benzylés en synthèse peptidique. II. Passage par les anhydrides mixtes éthylcarboniques des N,N-dibenzyl α-amino-acides. Bull. soc. chim. France [5] **21**, 1449 (1954).

423. VERDIER, C. H. DE: Isolation of Phosphothreonine from Bovine Casein. Nature (London) **170**, 804 (1952).

424. VIGNEAUD, V. DU: Chem. Soc. Sympos. on Peptide Chemistry. Nature (London) **175**, 1115 (1955).

425. VIGNEAUD, V. DU, L. F. AUDRIETH and H. S. LORING: The Reduction of Cystine in Liquid Ammonia by Metallic Sodium. J. Amer. Chem. Soc. **52**, 4500 (1930).

426. VIGNEAUD, V. DU and O. K. BEHRENS: A Method for Protecting the Imidazole Ring of Histidine during Certain Reactions and its Application to the Preparation of *l*-Amino-N-methylhistidine. J. Biol. Chem. **117**, 27 (1936).

427. VIGNEAUD, V. DU, D. T. GISH and P. G. KATSOYANNIS: A Synthetic Preparation Possessing Biological Properties Associated with Arginine-Vasopressin. J. Amer. Chem. Soc. **76**, 4751 (1954).

428. VIGNEAUD, V. DU and M. HUNT: A Preliminary Study of β-*l*-Aspartyl-*l*-histidine as a Possible Biological Precursor of *l*-Carnosine. J. Biol. Chem. **125**, 269 (1938).

429. VIGNEAUD, V. DU, H. S. LORING and G. L. MILLER: The Synthesis of α-Glutamylcysteinylglycine (Isoglutathione). J. Biol. Chem. **118**, 391 (1937).

430. VIGNEAUD, V. DU and G. L. MILLER: A Synthesis of Glutathione. J. Biol. Chem. **116**, 469 (1936).

431. VIGNEAUD, V. DU, C. RESSLER, J. M. SWAN, C. W. ROBERTS and P. G. KATSOYANNIS: The Synthesis of Oxytocin. J. Amer. Chem. Soc. **76**, 3115 (1954).

432. VIGNEAUD, V. DU, C. RESSLER, J. M. SWAN, C. W. ROBERTS, P. G. KATSOYANNIS and S. GORDON: The Synthesis of an Octapeptide Amide with the Hormonal Activity of Oxytocin. J. Amer. Chem. Soc. **75**, 4879 (1953).

433. WALDEN, P.: Über die gegenseitige Umwandlung optischer Antipoden. Ber. dtsch. chem. Ges. **29**, 133 (1896).

434. WALDSCHMIDT-LEITZ, R. und K. KÜHN: Über einen Weg zur enzymatischen Synthese von Peptiden. Chem. Ber. **84**, 381 (1951).

435. WALEY, S. G.: The Use of Formyl as a Protective Group in Peptide Syntheses. Chem. and Ind. **1953**, 107.

436. WALEY, S. G. and J. WATSON: The Kinetics of the Polymerization of Sarcosine Carbonic Anhydride. Proc. Roy. Soc. A **199**, 499 (1949).

437. — — Some Peptides of Lysine. J. Chem. Soc. (London) **1953**, 475.

438. WATERS, K. L. and W. H. HARTUNG: α-Alkoximino Acids. J. Organ. Chem. (USA) **12**, 469 (1947).

439. WEAVER, W. E. and W. H. HARTUNG: Peptides *via* Amides of α-Benzyloximino Acids. J. Organ. Chem. (USA) **15**, 741 (1950).

440. WESSELY, F.: Untersuchungen über α-Amino-N-Carbonsäureanhydride. I. Z. physiol. Chem. (Hoppe-Seyler) **146**, 72 (1925).

441. WESSELY, F. und E. KEMM: Zur Kenntnis der isomeren Glycylglycin-N-carbonsäuren. Z. physiol. Chem. (Hoppe-Seyler) **174**, 306 (1928).

442. WESSELY, F., E. KEMM und J. MAYER: Zur Kenntnis der Polypeptid-N-carbonsäuren. Z. physiol. Chem. (Hoppe-Seyler) **180**, 64 (1929).

443. WESSELY, F., K. SCHLÖGL und G. KORGER: Ein Beitrag zur Konstitutionsermittlung von Peptiden. I. Monatsh. Chem. **83**, 1156 (1952).

444. WESSELY, F., K. SCHLÖGL und E. WAWERSICH: Konstitutionsermittlung von Peptiden. III. Über eine schonende Methode zur Umwandlung von Peptiden in Hydantoinpeptide. VII. Mitt. über Peptide. Monatsh. Chem. **83**, 1439 (1952).

445. Wessely, F. und F. Sigmund: Untersuchungen uber α-Amino-N-Carbonsaure-anhydride. III. (Zur Kenntnis hohermolekularer Verbindungen.) Z. physiol. Chem. (Hoppe-Seyler) **159**, 102 (1926).

446. Wetzel, F. H., J. G. Miller and A. R. Day: Effect of Structure on Reactivity. VII. The Energies of Activation and the Entropies of Activation for the Ammonolysis of Esters. J. Amer. Chem. Soc. **75**, 1150 (1953).

447. Weygand, F.: Peptidsynthesen mit N-Trifluor-acetylaminosäuren. Chem.-Ztg. **78**, 480 (1954).

448. Weygand, F. und E. Csendes: N-Trifluoracetyl-aminosäuren. Angew. Chem. **64**, 136 (1952).

448 a Weygand, F. und R. Geier: N-Trifluoracetyl-aminosáuren. IV. Mitt. N-Trifluoracetylierung von Aminosäuren in wasserfreier Trifluoressigsaure. Chem. Ber. **89**, 647 (1956).

449. Weygand, F. und E. Leising: N-Trifluoracetyl-aminosäuren. II. Mitt. Chem. Ber. **87**, 248 (1954).

450. Weygand, F. und M. Reiher: N-Trifluoracetyl-aminosäuren. III. Mitt. Zwei neue Peptidsynthesen. Chem. Ber. **88**, 26 (1955).

451. White, J.: The Synthesis of Cystinyldiglycine and Cystinyldialanine. J. Biol. Chem. **106**, 141 (1934).

452. Wieland, Th.: Peptidsynthesen. Angew. Chem. **63**, 7 (1951).

453. — Peptidsynthesen. II. Angew. Chem. **66**, 507 (1954).

454. — Auf- und Abbau von Peptiden. Angew. Chem. **65**, 40 (1953).

455. — Über S-Aminoacylverbindungen. XIV. Int. Kongr., Zürich, 1955; Ref. Bd. **204**.

456. Wieland, Th. und H. Bernhard: Über Peptidsynthesen. 3. Mitt. Die Verwendung von Anhydriden aus N-acylierten Aminosäuren und Derivaten anorganischer Säuren. Liebigs Ann. Chem. **572**, 190 (1951).

457. — — Über Peptidsynthesen. 7. Mitt. Di- und Tri-peptidverbindungen des Thiophenols. Liebigs Ann. Chem. **582**, 218 (1953).

458. Wieland, Th., E Bokelmann, L. Bauer, H. U. Lang und H. Lau. Über Peptidsynthesen. 8. Mitt. Bildung von S-haltigen Peptiden durch intramolekulare Wanderung von Aminoacylresten. Liebigs Ann. Chem. **583**, 129 (1953).

459. Wieland, Th. und K. Freter: Über Aminothiosäuren. III. Mitt. Thio-β-alanin und homologe ω-Aminothiosäuren. Chem. Ber. **87**, 1099 (1954).

460. Wieland, Th. und G. Hörlein: Synthesen einiger β-Indolacetyl-aminosäuren und -peptide. Liebigs Ann. Chem. **591**, 192 (1955).

461. Wieland, Th., W. Kern und R. Sehring: Über Anhydride von acylierten Aminosauren. Liebigs Ann. Chem. **569**, 117 (1950).

462. Wieland, Th. und H. Köppe: Synthese von S-Lactyl- und S-β-Oxybutyryl-glutathion. Liebigs Ann. Chem. **581**, 1 (1953).

463. Wieland, Th. und W. Schafer: Über Peptidsynthesen. 6. Mitt. Die Darstellung einiger Aminoacyl-thiophenole und ihre Umsetzung mit Aminen und Aminosäuren. Liebigs Ann. Chem. **576**, 104 (1952).

464. Wieland, Th., W. Schäfer und E. Bokelmann: Über Peptidsynthesen. V. Über eine bequeme Darstellungsweise von Acylthiophenolen und ihre Verwendung zu Amid- und Peptid-Synthesen. Liebigs Ann. Chem. **573**, 99 (1951).

465. Wieland, Th. und G. Schneider: N-Acyl-Imidazole als energiereiche Acylverbindungen. Liebigs Ann. Chem. **580**, 159 (1953).

466. Wieland, Th. und R. Sehring: Eine neue Peptid-Synthese. Liebigs Ann. Chem. **569**, 122 (1950).

467. Wieland, Th. und D. Sieber: Thioalanin (α-Aminothiopropionsaure) als Vertreter der Klasse der Thioaminosäuren. Naturwiss. **40**, 242 (1953).

468. WIELAND, TH., D. SIEBER und W. BARTMANN: Über Aminothiosäuren. II. Mitt. S-Analoge einiger α-Aminosäuren. Chem. Ber. **87**, 1093 (1954).

469. WIELAND, TH. und D. STIMMING: Die Aufspaltungsweise gemischter Säureanhydride mit Hydroxylamin. Liebigs Ann. Chem. **579**, 97 (1953).

470. WIELAND, TH. und H. WEIDINGER: Über Peptidsynthesen. 10. Mitt. α- und γ-Phenylthioester von N-Acylglutaminsäuren. Liebigs Ann. Chem. **597**, 111 (1955).

471. WILSON, E. J., Jr. and E. PACSU: Polycondensation of Certain Peptide Esters. II. Protein Models. Notes on the Preparation of Tripeptide Methyl Esters. J. Organ. Chem. (USA) **7**, 126 (1942).

472. WILSON, W., G. RILEY and J. H. TURNBULL: Phosphorylated Aminoacids and Peptides Related to the Phosphoproteins. XIV. Int. Kongr., Zürich, 1955; Ref. Bd. **88**.

473. WINITZ, M. and J. S. FRUTON: A Method for the Synthesis of Cyclic Polypeptides. J. Amer. Chem. Soc. **75**, 3041 (1953).

474. WOOD, J. L. and V. DU VIGNEAUD: Racemization of Benzyl-*l*-Cysteine, with a New Method of Preparing *d*-Cystine. J. Biol. Chem. **130**, 109 (1939).

475. WOODS, G. F. and D. N. KRAMER: Dihydropyrane Addition Products. J. Amer. Chem. Soc. **69**, 2246 (1947).

476. WOODWARD, R. B. and C. H. SCHRAMM: Synthesis of Protein Analogs. J. Amer. Chem. Soc. **69**, 1551 (1947).

477. WOOLLEY, D. W.: Strepogenin Activity of Derivatives of Glutamic Acid. J. Biol. Chem. **172**, 71 (1948).

478. WÜNSCH, E.: Unveröffentlichte Arbeiten (Dissert. Univ. München, 1955).

479. ZERVAS, L.: Phosphorylierung und Tritylierung von Aminosäuren. Neue Methoden für Peptidsynthesen. Vortrag XIV. Int. Kongr., Zürich, 1955.

479 a. ZERVAS, L., G. PANAYOTIS and P. G. KATSOYANNIS: N-Phosphoramino Acids and Peptides. J. Amer. Chem. Soc. **77**, 5351 (1955).

480. ZERVAS, L. and D. M. THEODOROPOULOS: N-Tritylamino Acids and Peptides. A New Method of Peptide Synthesis. J. Amer. Chem. Soc. **78**, 1359 (1956).

(Eingelaufen am 20. März 1956.)

Namenverzeichnis. Index of Names. Index des Auteurs.

Sachverzeichnis. Index of Subjects. Index des Matières.

38*

Manzsche Buchdruckerei, Wien IX.

Fortsetzung von vorhergehender Seite

Siebenter Band. Mit 12 Abbildungen. VII, 330 Seiten. Gr.-8⁰. 1950.

DM 50.40, $ 12.—, sfr. 52.—, £4. 6.0d.
Ganzleinen DM 53.70, $ 12.80, sfr. 55.50, £4.12.0d.

Inhalt: **Jeger, O.** Über die Konstitution der Triterpene. — **Heusser, H.** Konstitution, Konfiguration und Synthese digitaloider Aglykone und Glykoside. — **Niemann, C.** Thyroxine and Related Compounds. — **Cook, A. H.** Penicillin and its Place in Science. — **Stoll, A.** and **B. Becker.** Sennosides A and B, the Active Principles of Senna. — **Williams, J. W.** Some Recent Developments in the Chemistry of Antibodies.

Achter Band. Mit 47 Abbildungen. XI, 400 Seiten. Gr.-8⁰. 1951.

DM 67.—, $ 16.—, sfr. 68.80, £5.14.6d.
Ganzleinen DM 70.50, $ 16.80, sfr. 72.20, £6. 0.6d.

Inhalt: **Frey-Wyssling, A.** and **K. Mühlethaler.** The Fine Structure of Cellulose. — **Stacey, M.** and **C. R. Ricketts.** Bacterial Dextrans. — **Leloir, L. F.** Sugar Phosphates. — **Kenner, G. W.** The Chemistry of Nucleotides. — **Schinz, H.** Die Veilchenriechstoffe. — **Asahina, Y.** Neuere Entwicklungen auf dem Gebiete der Flechtenstoffe. — **Galinovsky, F.** Lupinen-Alkaloide und verwandte Verbindungen. — **Pailer, M.** Brechwurzel-Alkaloide. — **Corey, R. B.** X-Ray Diffraction Studies of Crystalline Amino Acids and Peptides. — **Zechmeister, L.** and **M. Rohdewald.** Some Aspects of Enzyme Chromatography.

Neunter Band. Mit 20 Abbildungen. XI, 535 Seiten. Gr.-8⁰. 1952.

DM 79.—, $ 18.80, sfr. 81.—, £6.14.6d.
Ganzleinen DM 82.50, $ 19.60, sfr. 84.50, £7. 0.0d.

Inhalt: **Inhoffen, H. H.** und **H. Siemer.** Synthetische Chemie der Carotinoide. — **Baxter, J. G.** Synthesis and Properties of Vitamin A and Some Related Compounds. — **Meunier, P.** Les Antivitamines. — **Stoll, A.** Recent Investigations on Ergot Alkaloids. — **Tomita, M.** Die Alkaloide der Menispermaceae-Pflanzen. — **Dean, F. M.** Naturally Occurring Coumarins. — **Borsook, H.** The Biosynthesis of Proteins and Peptides, including Isotopic Tracer Studies. — **Kalckar, H. M.** The Enzymes of Nucleoside Metabolism. — **McNutt, W. S.** Nucleosides and Nucleotides as Growth Substances for Microorganisms. — **Campbell, D. H.** and **N. Bulman.** Some Current Concepts of the Chemical Nature of Antigens and Antibodies.

Zehnter Band. Mit 19 Abbildungen. IX, 529 Seiten. Gr.-8⁰. 1953.

DM 80.—, $ 19.—, sfr. 81.70, £6.16.0d.
Ganzleinen DM 83.—, $ 19.80, sfr. 85.—, £7. 1.6d.

Inhalt: **Alder, K.** und **Marianne Schumacher.** Anwendungen der Dien-Synthese für die Erforschung von Naturstoffen. — **Mark, H.** Physical Chemistry of Rubbers. — **Asselineau, J.** et **E. Lederer.** Chimie des lipides bactériens. — **Rosenkranz, G.** and **F. Sondheimer.** Syntheses of Cortisone. — **Chatterjee, A.** Rauwolfia Alkaloids. — **Feinstein, L.** and **M. Jacobson.** Insecticides Occurring in Higher Plants.

Elfter Band. Mit 67 Abbildungen. VIII, 457 Seiten. Gr.-8⁰. 1954.

DM 71.80, $ 17.20, sfr. 74.—, £6.3.0d.
Ganzleinen DM 74.80, $ 18.—, sfr. 77.40, £6.8.6d.

Inhalt: **Peat, S.** Starch: Its Constitution, Enzymic Synthesis and Degradation. — **Freudenberg, K.** Neuere Ergebnisse auf dem Gebiete des Lignins und der Verholzung. — **Inhoffen, H. H.** und **K. Brückner.** Probleme und neuere Ergebnisse in der Vitamin D-Chemie. — **Schmid, H.** Natürlich vorkommende Chromone. — **Pauling, L.** and **R. B. Corey.** The Configuration of Polypeptide Chains in Proteins. — **Schroeder, W. A.** Column Chromatography in the Study of the Structure of Peptides and Proteins. — **Lemberg, R.** Porphyrins in Nature. — **Albert, A.** The Pteridines.

Zwölfter Band: Mit 15 Abbildungen. X, 550 Seiten. Gr.-8⁰. 1955.

DM 79.80, $ 19.—, sfr. 81.70, £ 6.16.0d.
Ganzleinen DM 82.80, $ 19.80, sfr. 85.10, £ 7. 1.6d.

Inhalt: **Haagen-Smit, A. J.** Sesquiterpenes and Diterpenes. — **Jones, E. R. H.** and **T. G. Halsall.** Tetracyclic Triterpenes. — **Tschesche, R.** Neuere Vorstellungen auf dem Gebiete der Biosynthese der Steroide und verwandter Naturstoffe. — **Haxo, F. T.** Some Biochemical Aspects of Fungal Carotenoids. — **Warren, F. L.** The Pyrrolizidine Alkaloids. — **Thompson, E. O. P.** and **A. R. Thompson.** Paper Chromatography in the Study of the Structure of Peptides and Proteins. — **Roche, J.** et **R. Michel.** Acides aminés iodés et iodoprotéines. — **Slotta, K.** Chemistry and Biochemistry of Snake Venoms. — **Beadle, G. W.** Gene Structure and Gene Action.